PHYSICS RESEARCH AND TECHNOLOGY

FOCUS ON QUANTUM MECHANICS

PHYSICS RESEARCH AND TECHNOLOGY

Additional books in this series can be found on Nova's website
under the Series tab.

Additional E-books in this series can be found on Nova's website
under the E-books tab.

PHYSICS RESEARCH AND TECHNOLOGY

FOCUS ON QUANTUM MECHANICS

DAVID E. HATHAWAY
AND
ELISABETH M. RANDOLPH
EDITORS

Nova Science Publishers, Inc.
New York

For permission to use material from this book please contact us:
Telephone 631-231-7269; Fax 631-231-8175
Web Site: http://www.novapublishers.com

Additional color graphics may be available in the e-book version of this book.

LIBRARY OF CONGRESS CATALOGING-IN-PUBLICATION DATA

Focus on quantum mechanics / editors, David E. Hathaway and Elisabeth M. Randolph.
p. cm.
Includes index.
ISBN 978-1-62100-680-0 (hardcover)
1. Quantum theory. I. Hathaway, David E. II. Randolph, Elisabeth M.
QC173.F64 2011
530.12--dc23

2011037415

Published by Nova Science Publishers, Inc. † New York

CONTENTS

PREFACE

Through the development of physical science there has never been a theory which has changed so drastically the shape of science as quantum mechanics; nor has there been a scientific theory which has had such a profound impact on human thinking. Since its inception, quantum mechanics has played a significant role in philosophical thought both as a source of metaphysical ideas and as an important example of a "scientific revolution." This book gathers the most current research in the field of quantum mechanics including quantum mechanics and realism; the quantum mechanical mechanism behind end results of the GTR; andgeometric phases and topological effects in quantum mechanics; some quantum effects in magnetic fields.

In Chapter 1, the authors generalize the results of Naber about the Fractionary Schr¨odinger Equation with the use of the theory of Tempered Ultradistributions. Several examples of the use of this theory are given. In particular they evaluate the Green's function for a free particle in the general case.

Chapter 2 presents an axiomatic modification of quantum mechanics with a *possible-worlds* semantics capable of *predicting* essential "nonquantum" features of an observable universe model - the *topology* and *dimensionality* of spacetime, the *existence*, the *signature* and a *specific form* of a metric unit, and a set of naturally preferred directions (*vistas*) in spacetime unrelated to its metric properties. It is shown that the dynamics of the modification is represented by a *hyper-Hamiltonian flow* (superposition of three dependent Hamiltonian flows).

In Chapter 3, Quantum Mechanics of the Early Universe (Planck scale) is considered as a Quantum Mechanics with a fundamental (minimal) length and, at the same time, as a deformation of the conventional Quantum Mechanics. In this context, various approaches to deformation (or different deformations) of the corresponding Heisenberg Algebra are discussed. It is shown that there is a new small dimensionless parameter, having with a reciprocal quantity an explicit physical interpretation, in terms of which the above-mentioned Heisenberg Algebra deformations are represented. It is shown that this parameter appears in thermodynamics as well. Because of this, it is demonstrated that entropy and its density play a significant role in solving the problem of the vacuum energy density (cosmological constant) of the Universe and hence the dark energy problem. It is shown that the fundamental quantities in the Holographic Dark Energy Models may be expressed in terms of a new small dimensionless parameter. It is revealed that this parameter is naturally occurring in Gravitational Thermodynamics. Using the indicated parameter within the scope of the

Generalized Uncertainty Principle, a high-energy deformation model is constructed for the particular case of Einstein's equations (Ultraviolet-limit of General Relativity). On this basis the possibility of a new approach to the problem of Quantum Gravity is considered. Besides, the results obtained on the uncertainty relation of the pair "cosmological constant - volume of space-time", where the cosmological constant is a dynamic quantity, are reconsidered and generalized up to the Generalized Uncertainty Relation.

Bosonic condensates in optical lattices have been studied intensively in recent times. In Chapter 4 the authors present an introduction to the theory of Bogoliubov quasiparticles in atomic condensates trapped in lattice potentials. The authors then give an overview of two results. The first is a stability analysis of a chain of condensates with periodic boundary conditions. Such a system can be realized by overlapping onto the lattice a standard parabolicshape potential pierced by a blue-detuned laser. Then they show that in the case of only two lattice sites, Bogoliubov theory predicts correctly the existence of a collective mode of oscillation, which turns out to be the Josephson plasmon.

The authors have to find a new concept of space that is closer to quantum physics than to relativity. It is not easy to select an appropriate title for such investigation, because the word *space* denotes something quite general. Adding further predicates, for example turning space into *deep space, primordial space* or *free space*, makes it more special. But anything similar to translocal space, should be more general than the historic space. Space-time as was understood until today, is a local, rather restricted concept. It denotes a tangent manifold, a measurable surface with a simple idea of neighbourhood. Quantum phenomena, however, involve relations between remote areas. They should 'give space' to long distance weak and strong links. That is, the authors also deal with long distance entanglement induced by the strong force. The authors shall introduce some new concepts such as the *veil* which annihilates physical properties although it is not a black hole. They factor out the neutrino. They shall relate trans-local fine structures to the forces of nature.

In Chapter 5 the authors investigate a special Peano fractal the symmetry breakage of which gives birth to time. Going into electromagnetic space graphs they show how the propagating energy carries the connection with it. The authors show that a spin-network may disclose properties of *neural nets* and *Hopfield models* with a typical nonlinear Onsager Hamiltonian. The authors go into some coarse precategory and its automorphism group. They will also show that the surface manifold they want to reconstruct by renormalization is not just the Minkowski space, but rather the Minkowski algebra. Finally the authors find out there is a *natural standard model* and derive the natural strong force fermion spinor from the constitutive graded Lie algebra of the Clifford algebra $C\ell_{3,1}$. In that image the dynamic degrees of freedom are increased, but never reduced. Therefore the authors could have chosen the title *Free Space*. But free space is ready for entanglement. It is translocal, is pregeometric and has primordial features. It is not pregiven, has no arbitrary rigid body features, is not dilatable, neither affine, nor conformal.

In Chapter 6 the authors study linear generalizations of the time-dependent Schrödinger equation in (1+1) dimensions. These generalizations include several well-known quantum-mechanical models, such as the Schrödinger equation with position-dependent mass or for an energy-dependent potential. In the first part of this work they show that generalized Schrödinger equations are related to their conventional counterparts by a particular point transformation, a fact that can be used to generate solutions of generalized Schrödinger equations. As a particular case, the authors study L^2-norm preservation and construct a point transformation that relates the stationary Schrödinger equation to its generalized counterpart. Besides providing a solution method by themselves, point transformations are used in the second part of this work, which is devoted to Darboux transformations. They show that generalized Schrödinger equations admit Darboux transformations, and describe two construction methods for these transformations. One of these methods is based on intertwining relations, while the second method uses the point transformations that were introduced previously. Several examples illustrate our considerations.

Chapter 7 presents the development of a density functional theory (DFT)-based method for accurate, reliable treatment of various resonances in atoms. Many of these are known to be notorious for their strong correlation, proximity to more than one thresholds, degeneracy with more than one minima. Therefore they pose unusual challenges to both theoreticians and experimentalists. It is well-known that DFT has been one of the most powerful and successful tools for electronic structure calculation of atoms, molecules, solids in the past two decades. While it has witnessed diverse spectacular applications for ground states, the same for excited states, unfortunately, has come rather lately and remains somehow less conspicuous relatively. Our method uses a work-function-based exchange potential in conjunction with the popular gradientcorrected Lee-Yang-Parr correlation functional. The resulting Kohn-Sham equation, in the non-relativistic framework, is numerically solved accurately and efficiently by means of a generalized pseudospectral method through a non-uniform, optimal spatial discretization. This has been applied to a variety of excited states, such as low and high states; single, double, triple as well as multiple excitations; valence and core excitations; autoionizing states; satellites; hollow and doubly-hollow states; very high-lying Rydberg resonances; etc., of atoms and ions, with remarkable success. A thorough and systematic comparison with literature data reveals that, for all these systems, the exchange-only results are practically of Hartree-Fock quality; while with inclusion of correlation, this offers excellent agreement with available experimental data as well as those obtained from other sophisticated theoretical methods. Properties such as individual state energies, excitation energies, radial densities as well as various expectation values are studied as well. This helps us in predicting many states for the first time. In summary, the authors have presented an accurate, simple, general DFT method for description of arbitrary excited states of atoms and ions.

In Chapter 8, it is argued that the usual postulates of quantum mechanics are too strong. It is conjectured that it is possible to interpret all experiments if the authors maintain the formalism of quantum theory without modification, but weaken the postulates concerning the relation between the formalism and the experiments. A set of postulates is proposed where realism is ensured. Comments on Bell's theorem are made in the light of the new postulates.

In Chapter 9 the authors present a conception that surmounts the Cartesian Cut - prevailing in science- based on a representation of the fusion of the physical 'objective' and the 'subjective' realms. The authors introduce a mathematical-physics and philosophical theory for the physical realm and its mapping to the cognitive and perceptual realms and a philosophical reflection on the bearings of this fusion in cosmology, cognitive sciences, human and natural systems and its relations with a time operator and the existence of time cycles in Nature's and human systems. This conception stems from the self-referential construction of spacetime through torsion fields and its singularities; in particular the photon's selfreferential character, basic to the embodiment of cognition ; the authors shall elaborate this in detail in perception and neurology. They discuss the relations between this embodiment, bio-photons and wave genetics, and the relation with the enactive approach in cognitive sciences due to Varela. They further discuss the relation of the present conception with Penrose's theory of consciousness related to non-computatibility -in the sense of the Goedel-Turing thesis- of quantum processes in the brain. The authors characterize quantum jumps in terms of the singularities of the torsion potential given by the differential of the complex logarithmic map (CLM) of the propagating wave satisfying the nilpotent eikonal equation of geometrical optics. They discuss the relations between the structure of these singularities, surmounting the alleged wave-particle duality in quantum physics, and the de Broglie-Vigier double solution pilot wave theory. The authors discuss the relations between torsion and semiosis, and the torsion singularities as primeval to perception and cognition. The authors introduce Matrix Logic from the torsion produced by the nonduality of the True and False Operators, and non-orientable surfaces: the Moebius band and the Klein-bottle. The authors identify the extended photon twistor representations derived from the above theory with cognitive states of the Null Operator of Matrix Logic. The CLM that generates the torsion potential is found to be embodied as the *analytical* topographic map representation (TMR) of diverse and integrated sensorial modes in the neurocortex, while the Klein-bottle appears to be the topological embodiment of self-reference in general, is the *topological* TMR. The singularities of the analytical map are the loci of the two-dimensional sections of the neurocortex in which *all* the field orientations are superposed. This is the singularity of the Klein-bottle. The authors discuss the appearance of vortical torsion structures in the striate neurocortex and its relation with Karman vortices as the 'interiorization' of the 'outer' torsion fields which describe the representations of the field of orientations of visual stimuli -as it appears in the hypercolumnar structure in the neurocortex. The Brownian motions diffusion processes produced by the torsion geometries through the CMP have correlated diffusion processes in the neurocortex that can be associated with developmental morphogenetic growth patterns. They discuss the relations between morphogenesis, neurology and development in particular of the human and mammal heart and self-reference. The authors relate torsion in Marix Logic, the resultant Logical Momentum Cognition Operator and its decomposition into the Spin Operator and the Time Operator. This relates quantum physics statements into logical statements. The Time Operator is a primeval distinction between cognitive states in this Matrix Logic as its action amounts to compute the difference between these states. As a geometric action, the Time Operator is a ninety degrees rotation in the 2-plane of all cognitive states. The authors relate the Time Operator with intention, control, will and the appearance of life, and *chronomes* (time waves and patterns in natural and human systems and phenomenae). The authors discuss cosmological and anthropological problems from this perspective. In particular, they shall confront the Myth of the Eternal Return as a self-referential process, relating it to the Time Operator, nilpotence, the enaction approach to cognition due to Varela, and the emergent consciousness proposal due to Penrose. The authors discuss the relations with the microviolations of the second law of thermodynamics in statistical thermodynamics, the torsion geometry of Brownian motions, the synthropic action of the Time Operator, the Klein-bottle non-linear topology of time, with subjectivity and angular momentum. They discuss the relation between the topology of chronomes and the so called arrow of time. The authors introduce the notion of heterarchies of Klein-bottles, or still, of quantized reentering limited domains, and discuss its relation with selfdetermination, syntropy, entropy, and the Time Operator. The authors present as fundamental examples of these heterarchies the Myth of Eternal Return and the Human Being. They shall relate the

Time Operator with the perception of depth considered in the phenomenological philosophy of Merleau-Ponty as a protodimension, and the problem of hemilateral synchronization with universal torsion chronomes (Kozyrev) in terms of the transactional interpretation of quantum mechanics, and the CMP retinotopic representation. The authors discuss *quantized* time and motion perception in relation with ATP production.

As explained in Chapter 10, a viable methodology for the exact analytical solution of the multiparticle Schrodinger and Dirac equations has long been considered a holy grail of theoretical chemistry. Since a benchmark work by Torres-Vega and Frederick in the1990's[1], the Quantum Phase Space Representation (QPSR) has been explored as an alternate method for solving various physical systems, including the harmonic oscillator[2], Morse oscillator[3], one-dimensional hydrogen atom[4], and classical Liouville dynamics under the Wigner function[5]. QPSR approaches are particularly challenging because of the complexity of phase space wave functions and the fact that the number of coordinates doubles in the phase space representation.These challenges have heretofore prevented the exact solution of the multiparticle equation in phase space. Recently, Simpao[6] has developed an exact analytical symbolic solution scheme for broad classes of differential equations utilizing the Heaviside Operational Ansatz (HOA). It is proposed to apply this novel methodology to QPSR problems to obtain exact solutions for real chemical systems and their dynamics. In his preliminary work, Simpao[6] has already applied this method to a number of simple systems, including the harmonic oscillator, with solutions in agreement to those obtained by Li [refs.2,3,4,5]. He has also demonstrated the exact solution to the radial Schrodinger Equation for an N-particle system with pairwise Coulomb interaction[7]. In addition to the Schrodinger Equation, the HOA method is capable of treating the Dirac equation[8] as well as differential systems governing both relativistic and non-relativistic particle dynamics. Applying these methods would allow us to pursue further exploration of this methodology, starting with the exact solution of multielectron atoms and moving toward complex molecules and reaction dynamics. It is believed that the coupling of HOA with QPSR represents not only a fundamental breakthrough in theoretical physical chemistry, but it is promising as a basis for exact solution algorithms that would have tremendous impact on the capabilities of computational chemistry/physics. As the theoretical foundation for spectroscopy is the Schrodinger equation, the significance of this discovery to the enhanced analysis of spectroscopic data is obvious. For example, the analysis of the Compton line in momentum spectroscopy necessitates the consideration of the momentum wavefunction for the molecular system under study. The novel methods refs. [6,7,8] allow the exact determination of the momentum[and configuration] space wavefunction from the QPSR wavefunction by way of a Fourier Transform. For example, the primary focus of [7] is the pairwise 1/rij interaction in context of the radial equation in the nonrelativistic Schrodinger case. This application of the exact solution ansatz developed above corresponds to the problem of n-particles with pairwise Coulomb interaction;scaling the parameters and variables of the problem yields the exact solution of the QPSR Schrodinger equation for the first principles general polyatomic molecular Hamiltonian. Upon a straightforward slight adaptation of this non-relativistic Schrodinger result, the QPSR Dirac equation addressed in [8] immediately yields the relativistic counterpart for the first –principles general polyatomic molecular Hamiltonian solution. These results form the cornerstone of the exact solution to the quantum dynamics of particular chemical systems, which shall appear elsewhere. Although all of the examples of dynamical systems mentioned above are at once solved as a special case of the general integral method already published in HOA, it is illuminating for applications to write out the solutions as the integrals are evaluated explicitly for the same. The HOA result is currently being used as the primary algorithm in the development of computer programs known as 'solver engines' for quantum chemistry/physics and plethora applications: to be reported elsewhere. In this note some remarks, examples and further directions, concerning HOA as a tool to solve and provide analytical insight into solutions of dynamical systems occurring in, but not limited to Mathematical Chemistry, are posited. Among these more general considerations are the developement of Novel Exact Analytical solutions of Generalized Hamiltonian/Lagrangian Dynamical systems in the Quantum Phase Space Representation and Classical Connections; also the exact analytical solutions of attendant differential/difference equations. To accomplish this, HOA[Heaviside Operational Ansatz] methods and Ehrenfest's Theorem are applied to general[i.e., Hamilton Extended Principle] quantum canonical and Lagrangian dynamical systems in QPSR [Quantum Phase Space Representation], yielding quadrature exact analytical solutions thereof. This novel result, arising at the nexus of; the Quantum Phase Space Representation [QPSR]; a non-traditional use of the Extended Hamilton Principle as manifest in the Generalised Lagrangian/Hamiltonian Formalism; the Heaviside Operational Ansatz [HOA] and Ehrenfest's Theorem, yields a prescription for exact analytical quadrature solutions to not only broad classes of closed/open quantum dynamical systems, but associated classical dynamical systems as well; what's more, to numerous and sundry attendant Functional Differential Equations [FDEs], For convenience, the authors begin with a recap of the HOA construction following from [ref. 6].

A careless analysis of the meaning of conservation laws in quantum mechanics can lead to paradoxes. In order to illustrate this assertion, in Chapter 11 the authors analyze the problem of energy conservation of a free particle. It is shown that paradoxes of this type can be solved by the adoption of a criterion referred to conservation laws which is compatible with the axiomatic of quantum mechanics and has not been previously formulated in an explicit way.

In Chapter 12, the authors review a recent development on nonperturbative analyses in imaginary-time path integral formalism. The key ingredient of our analysis resides in considering valley configurations in the phase space of path integral. It enables us in particular to treat properly negative modes around classical configurations which often break the applicability of conventional semi-classical approximations. To see how their method works, the authors apply it to quantum mechanical problems of an asymmetric double-well and a symmetric triple-well potential. The authors show that valley configurations connect perturbative and non-perturbative regions in the phase space, from which they have a manipulation to extract purely nonperturbative contributions to physical quantities. As a by-product, the authors derive a dispersion relation which relates a perturbative contribution with nonperturbative one. With their method, the authors calculate the nonperturbative corrections to the spectrum not only for the ground state but also all for the excited states beyond the dilute-gass approximation. Using the dispersion relation, the authors also derive large-order behavior of the perturbation series for the corresponding energy eigenvalues. The authors then check their results by employing perturbation theory, the WKB approximation, and consequences of N-fold supersymmetry to confirm their accuracy qualitatively and quantitatively.

As presented in Chapter 13, basing on known theoretical results (mainly the Bell and the Bell–Kochen–Specker theorems) most quantum physicists maintain that contextuality and nonlocality are unavoidable consequences of the mathematical apparatus of quantum mechanics (QM). But contextuality and nonlocality entail known paradoxes and unsolved problems in the quantum theory of measurement. A critical analysis shows, however, that the premises in the proofs of the theorems mentioned above rest on an implicit assumption (metatheoretical classical principle, or MCP) which does not fit in well with the operational philosophy of QM. IfMCP is replaced by a weaker assumption (metatheoretical generalized principle, or MGP) such proofs cannot be carried out, which opens the way to a noncontextual, hence local, interpretation of QM. This topic has been discussed in several papers, and a *semantic realism* (*SR*) interpretation of QM has been provided which is noncontextual, hence avoids the aforementioned paradoxes and the problems of the quantum theory of measurement. More recently, an *extended semantic realism* (*ESR*) model has been worked out which modifies and extends the original SR interpretation but preserves its basic features. The ESR model consists of a microscopic and a macroscopic part. The former is a noncontextual hidden variables theory for QM that provides a justification of the assumptions introduced in the macroscopic part. The latter can be considered as an autonomous theory that embodies the formalism of QM into a noncontextual (hence local) framework, reinterpreting quantum probabilities as conditional (in a nonconventional sense) rather than absolute. One can then show that the ESR model implies predictions that differ in some cases from those of QM and make it *falsifiable*. In particular the ESR model predicts that, whenever idealized measurements are performed and some additional assumptions are introduced, a *modified Bell–Clauser–Horne-Shimony–Holt* (*BCHSH*) inequality holds which is compatible with the reinterpreted quantum probabilities. Hence, the long– standing conflict between "local realism" and QM is settled in the ESR model. By referring to the macroscopic part of the ESR model one can also supply a Hilbert space representation of the generalized observables introduced by the model and a generalization of the projection postulate of elementary QM (*generalized projection postulate*). These results imply a new mathematical representation of mixtures which does not coincide with the standard quantum representation and avoids some deep interpretational problems that follow from the standard representation in QM. A further generalization of the projection postulate (*generalized Lüders postulate*) can then be provided, and the generalized projection postulate can be partially justified by introducing a nonlinear time evolution of the compound system made up of the measured object and the measuring apparatus.

In: Focus on Quantum Mechanics
Editors: David E. Hathaway et al. pp. 1-13
ISBN 978-1-62100-680-0

Chapter 1

ULTRADISTRIBUTIONS AND THE FRACTIONARY SCHRÖDINGER EQUATION*

A.L. De Paoli and M.C. Rocca
Departamento de Física, Fac. de Ciencias Exactas,
Universidad Nacional de La Plata.
C.C. 67 (1900) La Plata. Argentina.

Abstract

In this work, we generalize the results of Naber about the Fractionary Schrödinger Equation with the use of the theory of Tempered Ultradistributions. Several examples of the use of this theory are given. In particular we evaluate the Green's function for a free particle in the general case.

PACS: 03.65.-w, 03.65.Bz, 03.65.Ca, 03.65.Db.

1. Introduction

The properties of ultradistributions (ref.[6, 7]) are well adapted for their use in fractional calculus. In this respect we have shown that it is possible (ref.[2]) to define a general fractional calculus with the use of them.

Ultradistributions have the advantage of being representable by means of analytic functions. So that, in general, they are easier to work with them.

They have interesting properties.One of those properties is that Schwartz tempered distributions are canonical and continuously injected into tempered ultradistributions and as a consequence the Rigged Hilbert Space with tempered distributions is canonical and continuously included in the Rigged Hilbert Space with tempered ultradistributions.

Fractional calculus has found motivations in a growing area concerning general stochastic phenomena.These include the appearance of alternative diffusion mechanisms other

**This work was partially supported by Consejo Nacional de Investigaciones Científicas ; Argentina.*

than Brownian, as well as classical and quantum mechanics formalisms including dissipative forces, and therefore allowing an extension of the quantization schemes for non-conservative systems [3]. In particular it is interesting to study the fractional Schrödinger equation. Our aim is to extend a previous study [1] about this equation. Using an analytical definition of fractional derivative [2] we show here that it is possible to obtain a general solution for the time fractional equation, for any complex value of the derivative index. Furthermore the associated Green functions can be evaluated in a straightforward way.

This paper is organized as follow:
In section 2 we define the fractional Schrödinger equation for all ν complex with the use of the fractional derivative defined via the theory of tempered ultradistributions. In section 3 we solve this equation for the free particle and give three examples:$\nu = 1/2, \nu = 1$ and $\nu = 2$. In section 4 we realize the treatment of the potential well and we analyze the cases $\nu = 1/2, \nu = 1$ and $\nu = 2$. In section 5 we study the Green fractional functions for the free particle in three cases: the retarded Green function, the advanced Green function and the Wheeler-Green function. As an example we prove that for $\nu = 1$ these functions coincide with the Green functions of usual Quantum Mechanics.In section 6 we discuss the results obtained in the previous sections. Finally we have included three appendixes: a first appendix on distributions of exponential type, a second appendix on tempered ultradistributions and a third appendix on fractional calculus using ultradistributions.

2. The Fractional Schrödinger Equation

Our starting point in the study of the fractional Schrödinger equation is the current known Schrödinger equation:

$$i\hbar\partial_t\psi(t,x) = -\frac{\hbar^2}{2m}\partial_x^2\psi(t,x) + V(x)\psi(t,x) \tag{2.1}$$

According to ref.[1], (2.1) can be writen as:

$$iT_p\partial_t\psi(t,x) = -\frac{L_p^2 M_p}{2m}\partial_x^2\psi(t,x) + \frac{V(x)}{E_p}\psi(t,x) \tag{2.2}$$

where $L_P = \sqrt{G\hbar/c^3}$, $T_p = \sqrt{G\hbar/c^5}$, $M_p = \sqrt{\hbar c/G}$ and $E_p = M_p c^2$.
If we define $N_m = m/M_p$ and $N_v = V/E_p$ we obtain for (2.2)

$$iT_p\partial_t\psi(t,x) = -\frac{L_p^2}{2N_m}\partial_x^2\psi(t,x) + N_v\psi(t,x) \tag{2.3}$$

By analogy with ref.[1] we define the fractional Schrödinger equation for all ν complex as:

$$(iT_p)^\nu\partial_t^\nu\psi(t,x) = -\frac{L_p^2}{2N_m}\partial_x^2\psi(t,x) + N_v\psi(t,x) \tag{2.4}$$

where the temporal fractionary derivative is defined following ref.[2] (see Appendix III)

3. The Free Particle

From (2.4) for the free particle the fractionary equation is:

$$(i\partial_t)^\nu \psi(t,x) + \frac{L_p^2}{2T_p^\nu N_m} \partial_x^2 \psi(t,x) = 0 \tag{3.1}$$

By the use of the Fourier transform (complex in the temporal variable and real as usual in the spatial variable) the corresponding equation is (see Appendix II and ref.[2])

$$\left(k_0^\nu - \frac{L_p^2}{2T_p^\nu N_m} k^2 \right) \hat{\psi}(k_0,k) = b(k_0,k) \tag{3.2}$$

whose solution is:

$$\hat{\psi}(k_0,k) = \frac{b(k_0,k)}{k_0^\nu - \frac{L_p^2}{2T_p^\nu N_m} k^2} \tag{3.3}$$

and in the configuration space (anti-transforming)

$$\psi(t,x) = \oint_\Gamma \int_{-\infty}^{\infty} \frac{a(k_0,k)}{k_0^\nu - \frac{L_p^2}{2T_p^\nu N_m} k^2} e^{-i(k_0 t + kx)} dk_0 \, dk \tag{3.4}$$

where:

$$a(k_0,k) = \frac{b(k_0,k)}{4\pi^2}$$

We proceed to analyze solutions of (3.4) for some typical cases in the following section.

Examples

As a first example we consider the case $\nu = 1/2$
Let α be given by:

$$\alpha = \frac{L_p^2}{2T_p^{\frac{1}{2}} N_m} \tag{3.5}$$

From (3.4) we obtain

$$\psi(t,x) = \oint_\Gamma \int_{-\infty}^{\infty} \frac{a(k_0,k)}{k_0^{\frac{1}{2}} - \alpha k^2} e^{-i(k_0 t + kx)} dk_0 \, dk \tag{3.6}$$

or equivalently:

$$\psi(t,x) = \int_{-\infty}^{\infty} a(k) e^{-i(\alpha^2 k^4 t + kx)} dk + \int_{-\infty}^{0} \int_{-\infty}^{\infty} a(k_0,k)$$

$$\left[\frac{1}{(k_0 + i0)^{\frac{1}{2}} - \alpha k^2} - \frac{1}{(k_0 - i0)^{\frac{1}{2}} - \alpha k^2} \right] e^{-i(k_0 t + kx)} dk_0 \, dk \tag{3.7}$$

where:

$$a(k) = -4\pi i\alpha k^2 a(\alpha^2 k^4, k)$$

With some of algebraic calculus we obtain for (3.7):

$$\psi(t,x) = \int\limits_{-\infty}^{\infty} a(k)e^{-i(\omega^2 t+kx)}dk+$$

$$\int\limits_{0}^{\infty}\int\limits_{-\infty}^{\infty} \frac{a(k_0,k)}{k_0+\omega^2}e^{i(k_0 t-kx)}dk_0\, dk \tag{3.8}$$

with:

$$\omega = \alpha k^2$$

and where we have made the re-scaling:

$$-2ik_0^{\frac{1}{2}}a(-k_0,k) \to a(k_0,k)$$

The first term in (3.8) represent free particle on-shell propagation and the second term describes the contribution of off-shell modes.

As a second example we consider the case $\nu = 1$.
In this case (3.4) takes the form:

$$\psi(t,x) = \oint\limits_{\Gamma}\int\limits_{-\infty}^{\infty} \frac{a(k_0,k)}{k_0-\omega}e^{-i(k_0 t+kx)}dk_0\, dk \tag{3.9}$$

Evaluating the integral in the variable k_0 we have:

$$\psi(t,x) = \int\limits_{-\infty}^{\infty} a(k)e^{-i(\omega t+kx)}dk \tag{3.10}$$

where $a(k) = -2\pi i a(\omega, k)$. Thus we recover the usual expression for the free-particle wave function.

Finally we consider the case $\nu = 2$. For it we have

$$\psi(t,x) = \oint\limits_{\Gamma}\int\limits_{-\infty}^{\infty} \frac{a(k_0,k)}{k_0^2-\omega^2}e^{-i(k_0 t+kx)}dk_0\, dk \tag{3.11}$$

After to perform the integral in the variable k_0 we obtain from (3.11):

$$\psi(t,x) = \int\limits_{-\infty}^{\infty} a(k)e^{-i(\omega t+kx)} + b^+(k)e^{i(\omega t+kx)}dk \tag{3.12}$$

with $a(k) = -2\pi i a(\omega, k)$ and $b^+(k) = -2\pi i a(-\omega, -k)$

4. The Potential Well

We consider in this section the potential well. The fractionary equation for a particle confined to move within interval $0 \leq x \leq a$ is:

$$(iT_p)^\nu \partial_t^\nu \psi(t,x) = -\frac{L_p^2}{2N_m}\partial_x^2 \psi(t,x) \tag{4.1}$$

To solve this equation we use the method of separation of variables. Thus if we write:

$$\psi(t,x) = \psi_1(t)\psi_2(x) \tag{4.2}$$

As is usual we obtain:

$$\frac{(iT_p)^\nu \partial_t^\nu \psi_1(t)}{\psi_1(t)} = -\frac{\frac{L_p^2}{2N_m}\partial_x^2 \psi_2(x)}{\psi_2(x)} = \lambda \tag{4.3}$$

Then we conclude that $\psi_2(x)$ satisfies:

$$\partial_x^2 \psi_2(x) + \frac{2\lambda N_m}{L_p^2}\psi_2(x) = 0 \tag{4.4}$$

The solution of (4.4) is the habitual one:

$$\psi_{2n}(x) = b_n \sin\left(\frac{n\pi}{a}x\right) \tag{4.5}$$

with:

$$\lambda_n = \frac{1}{2N_m}\left(\frac{n\pi L_p}{a}\right)^2 \tag{4.6}$$

and the boundary conditions satisfied by $\psi_{2n}(x)$ are:

$$\psi_{2n}(0) = \psi_{2n}(a) = 0$$

As a consequence of (4.3),(4.5) and (4.6) the Fourier transform $\hat{\psi}_1(k_0)$ of $\psi_1(t)$ should be satisfy:

$$(k_0^\nu - \lambda_n)\hat{\psi}_{1n}(k_0) = 0 \tag{4.7}$$

whose solution is:

$$\hat{\psi}_{1n}(k_0) = \frac{c_n(k_0)}{k_0^\nu - w_n} \tag{4.8}$$

with $w_n = \lambda_n / T_p^\nu$

Therefore the final general solution for $\psi(t,x)$ is:

$$\psi(t,x) = \sum_{n=1}^{\infty} \sin\left(\frac{n\pi}{a}x\right) \oint\limits_{\Gamma} \frac{a_n(k_0)e^{-ik_0 t}}{k_0^\nu - w_n} dk_0 \tag{4.9}$$

where we have defined:

$$a_n(k_0) = \frac{b_n c_n(k_0)}{2\pi}$$

which is an entire analytic function of k_0.

Examples

As a first example we consider the case $\nu = 1/2$. For it the solution (4.9) takes the form:

$$\psi(t,x) = \sum_{n=1}^{\infty} \sin\left(\frac{n\pi}{a}x\right) \oint_{\Gamma} \frac{a_n(k_0)e^{-ik_0t}}{k_0^{\frac{1}{2}} - w_n} dk_0 \tag{4.10}$$

or equivalently:

$$\psi(t,x) = \sum_{n=1}^{\infty} a_n \sin\left(\frac{n\pi}{a}x\right) e^{-iw_n^2 t} +$$

$$\sum_{n=1}^{\infty} \sin\left(\frac{n\pi}{a}x\right) \int_{-\infty}^{0} \left[\frac{1}{(k_0+i0)^{\frac{1}{2}} - w_n} - \frac{1}{(k_0-i0)^{\frac{1}{2}} - w_n}\right] \times$$

$$a_n(k_0)e^{-ik_0t}\, dk_0 \tag{4.11}$$

After performing some algebra we have for (4.11) the expression:

$$\psi(t,x) = \sum_{n=1}^{\infty} a_n \sin\left(\frac{n\pi}{a}x\right) e^{-iw_n^2 t} +$$

$$\sum_{n=1}^{\infty} \sin\left(\frac{n\pi}{a}x\right) \int_{0}^{\infty} \frac{a_n(k_0)}{k_0 + w_n^2} e^{-ik_0t}\, dk_0 \tag{4.12}$$

Analogously as before, the second term in (4.12) represents of-shell stationary modes.

As a second example we consider $\nu = 1$. In this case:

$$\psi(t,x) = \sum_{n=1}^{\infty} \sin\left(\frac{n\pi}{a}x\right) \oint_{\Gamma} \frac{a_n(k_0)e^{-ik_0t}}{k_0 - w_n} dk_0 \tag{4.13}$$

Performing the integral in the variable k_0 we have:

$$\psi(t,x) = \sum_{n=1}^{\infty} a_n \sin\left(\frac{n\pi}{a}x\right) e^{-iw_n t} \tag{4.14}$$

Which is the familiar general solution for the infinite well.

Finally for $\nu = 2$:

$$\psi(t,x) = \sum_{n=1}^{\infty} \sin\left(\frac{n\pi}{a}x\right) \oint_{\Gamma} \frac{a_n(k_0)e^{-ik_0t}}{k_0^2 - w_n} dk_0 \tag{4.15}$$

and after to compute the integral:

$$\psi(t,x) = \sum_{n=1}^{\infty} \sin\left(\frac{n\pi}{a}x\right) \left(a_n e^{-i\sqrt{w_n}t} + b_n e^{+i\sqrt{w_n}t}\right) \tag{4.16}$$

with $a_n = a_n(\sqrt{w_n})$ and $b_n^+ = a_n(-\sqrt{w_n})$

5. The Green Function for The Free Particle

As other application that shows the generality of the fractional calculus defined with the use of ultradistributions, we give the evaluation of the Green function corresponding to the free particle. Let β be defined as:

$$\beta^2 = \frac{L_p^2}{2T_p^\nu N_m} \tag{5.1}$$

Then $G(t-t',x-x')$ should be satisfy the equation:

$$(i\partial_t)^\nu G(t-t',x-x') + \beta^2\partial_x^2 G(t-t',x-x') = \delta(t-t')\delta(x-x') \tag{5.2}$$

As G is function of $(t-t',x-x')$ it is sufficient to consider G as function of (t,x):

$$(i\partial_t)^\nu G(t,x) + \beta^2\partial_x^2 G(t,x) = \delta(t)\delta(x) \tag{5.3}$$

For the Fourier transform $\hat{G}$ of G we have:

$$(k_0^\nu - \beta^2 k^2)\hat{G}(k_0,k) = \frac{Sgn[\Im(k_0)]}{2} + a(k_0,k) \tag{5.4}$$

where $a(k_0,k)$ is as usual a rapidly decreasing analytic entire function of the variable k_0. Selecting:

$$a(k_0,k) = \frac{1}{2}$$

we obtain the equation for the retarded Green function:

$$(k_0^\nu - \beta^2 k^2)\hat{G}_{ret}(k_0,k) = H[\Im(k_0)] \tag{5.5}$$

and then:

$$G_{ret}(t,x) = \frac{1}{4\pi^2}\oint_\Gamma \int_{-\infty}^{\infty} \frac{H[\Im(k_0)]}{k_0^\nu - \beta^2 k^2} e^{-i(k_0 t + kx)}\, dk_0\, dk \tag{5.6}$$

If we take:

$$a(k_0,k) = -\frac{1}{2}$$

we obtain the advanced Green function:

$$G_{adv}(t,x) = -\frac{1}{4\pi^2}\oint_\Gamma \int_{-\infty}^{\infty} \frac{H[-\Im(k_0)]}{k_0^\nu - \beta^2 k^2} e^{-i(k_0 t + kx)}\, dk_0\, dk \tag{5.7}$$

For the Wheeler Green function (half advanced plus half retarded):

$$G_W(t,x) = \frac{1}{2}[G_{adv}(t,x) + G_{ret}(t,x)] \tag{5.8}$$

we have:

$$G_W(t,x) = \frac{1}{8\pi^2}\oint_\Gamma \int_{-\infty}^{\infty} \frac{Sgn[\Im(k_0)]}{k_0^\nu - \beta^2 k^2} e^{-i(k_0 t + kx)}\, dk_0\, dk \tag{5.9}$$

Example

When we select $\nu = 1$ we obtain the usual Green functions of Quantum Mechanics. For example for G_{ret} we have:

$$G_{ret}(t,x) = \frac{1}{4\pi^2} \oint_{\Gamma} \int_{-\infty}^{\infty} \frac{H[\Im(k_0)]}{k_0 - \beta^2 k^2} e^{-i(k_0 t + kx)} \, dk_0 \, dk \tag{5.10}$$

or equivalently:

$$G_{ret}(t,x) = \frac{1}{4\pi^2} \int_{-\infty}^{\infty} \int_{-\infty}^{\infty} \frac{1}{(k_0 + i0) - \beta^2 k^2} e^{-i(k_0 t + kx)} \, dk_0 \, dk \tag{5.11}$$

After the evaluation of the integral in the variable k_0, G_{ret} takes the form:

$$G_{ret}(t,x) = -\frac{i}{2\pi} H(t) \int_{-\infty}^{\infty} e^{-i(\beta^2 k^2 t + kx)} \, dk \tag{5.12}$$

With a square's completion (5.12) transforms into:

$$G_{ret}(t,x) = -\frac{iH(t)}{2\pi\beta\sqrt{t}} e^{\frac{ix^2}{4\beta^2 t}} \int_{-\infty}^{\infty} e^{is^2} \, ds \tag{5.13}$$

From the result of ref.[4]

$$\int_{-\infty}^{\infty} e^{is^2} \, ds = \sqrt{\pi} e^{-i\frac{\pi}{4}} \tag{5.14}$$

we have

$$G_{ret}(t,x) = -iH(t) \left(\frac{m}{2\pi i \hbar t}\right)^{\frac{1}{2}} e^{\frac{imx^2}{2\hbar t}} \tag{5.15}$$

Taking into account that for $\nu = 1$:

$$\beta^2 = \frac{\hbar}{2m}$$

we obtain the usual form of G_{ret} (see ref.[5])

$$G_{ret}(t-t^{'}, x-x^{'}) = -iH(t-t^{'}) \left(\frac{m}{2\pi i \hbar (t-t^{'})}\right)^{\frac{1}{2}} e^{\frac{im(x-x^{'})^2}{2\hbar(t-t^{'})}} \tag{5.16}$$

With a similar calculus we have for G_{adv}:

$$G_{adv}(t-t^{'}, x-x^{'}) = iH(t^{'}-t) \left(\frac{m}{2\pi i \hbar (t^{'}-t)}\right)^{\frac{1}{2}} e^{\frac{im(x-x^{'})^2}{2\hbar(t-t^{'})}} \tag{5.17}$$

and for G_W:

$$G_W(t-t^{'}, x-x^{'}) = -\frac{i}{2} Sgn(t-t^{'}) \left(\frac{m}{2\pi i \hbar |t-t^{'}|}\right)^{\frac{1}{2}} e^{\frac{im(x-x^{'})^2}{2\hbar(t-t^{'})}} \tag{5.18}$$

6. Discussion

In a earlier paper (ref.[2] we have shown the existence of a general fractional calculus defined via tempered ultradistributions. All ultradistributions provide integrands that are analytic functions along the integration path. These properties show that tempered ultradistributions provide an appropriate framework for applications to fractional calculus. With the use of this calculus we have generalized in the present work the results obtained by Naber (ref.[1]). We have defined the fractionary Schrödinger equation for all values of the complex variable ν and treated the cases of the free particle and the potential well. For $\nu = 1$ the results obtained coincide with the usual Quantum Mechanics, and the cases $\nu = 1/2$ and $\nu = 2$ have shown the appearance of extra terms, besides to those with the usual ($\nu = 1$) framework. We have obtained a general expression for the Green function of the free particle and shown that for $\nu = 1$ this Green function coincide with the obtained in ref.[5].. For the benefit of the reader we give in this paper two Appendixes with the main characteristics of n-dimensional tempered ultradistributions and their Fourier anti-transformed distributions of the exponential type, and a third Appendix about the general fractional calculus defined via the use of tempered ultradistributions.

7. Appendix I: Distributions of Exponential Type

For the sake of the reader we shall present a brief description of the principal properties of Tempered Ultradistributions.

Notations. The notations are almost textually taken from ref[7]. Let $\mathbb{R}^n$ (res. $\mathbb{C}^n$) be the real (resp. complex) n-dimensional space whose points are denoted by $x = (x_1, x_2, ..., x_n)$ (resp $z = (z_1, z_2, ..., z_n)$). We shall use the notations:

(i) $x + y = (x_1 + y_1, x_2 + y_2, ..., x_n + y_n)$; $\alpha x = (\alpha x_1, \alpha x_2, ..., \alpha x_n)$

(ii)$x \geqq 0$ means $x_1 \geqq 0, x_2 \geqq 0, ..., x_n \geqq 0$

(iii)$x \cdot y = \sum_{j=1}^{n} x_j y_j$

(iV)$| x | = \sum_{j=1}^{n} | x_j |$

Let $\mathbb{N}^n$ be the set of n-tuples of natural numbers. If $p \in \mathbb{N}^n$, then $p = (p_1, p_2, ..., p_n)$, and p_j is a natural number, $1 \leqq j \leqq n$. $p + q$ denote $(p_1 + q_1, p_2 + q_2, ..., p_n + q_n)$ and $p \geqq q$ means $p_1 \geqq q_1, p_2 \geqq q_2, ..., p_n \geqq q_n$. x^p means $x_1^{p_1} x_2^{p_2} ... x_n^{p_n}$. We shall denote by $| p | = \sum_{j=1}^{n} p_j$ and by D^p we denote the differential operator $\partial^{p_1 + p_2 + ... + p_n} / \partial x_1{}^{p_1} \partial x_2{}^{p_2} ... \partial x_n{}^{p_n}$

For any natural k we define $x^k = x_1^k x_2^k ... x_n^k$ and $\partial^k / \partial x^k = \partial^{nk} / \partial x_1^k \partial x_2^k ... \partial x_n^k$

The space $\mathcal{H}$ of test functions such that $e^{p|x|} |D^q \phi(x)|$ is bounded for any p and q is defined (ref.[7]) by means of the countably set of norms:

$$\|\hat{\phi}\|_p = \sup_{0 \leq q \leq p, x} e^{p|x|} \left| D^q \hat{\phi}(x) \right| \quad , \quad p = 0, 1, 2, ... \tag{7.1}$$

According to reference[9] $\mathcal{H}$ is a $\mathcal{K}\{M_p\}$ space with:

$$M_p(x) = e^{(p-1)|x|} \quad , \quad p = 1, 2, ... \tag{7.2}$$

$\mathcal{K}\{e^{(p-1)|x|}\}$ satisfies condition $(\mathcal{N})$ of Guelfand (ref.[8]). It is a countable Hilbert and nuclear space:

$$\mathcal{K}\{e^{(p-1)|x|}\} = \mathcal{H} = \bigcap_{p=1}^{\infty} \mathcal{H}_p \tag{7.3}$$

where $\mathcal{H}_p$ is obtained by completing $\mathcal{H}$ with the norm induced by the scalar product:

$$<\hat{\phi},\hat{\psi}>_p = \int_{-\infty}^{\infty} e^{2(p-1)|x|} \sum_{q=0}^{p} D^q\overline{\hat{\phi}}(x) D^q\hat{\psi}(x)\, dx \quad ; \quad p=1,2,... \tag{7.4}$$

where $dx = dx_1\, dx_2 ... dx_n$

If we take the usual scalar product:

$$<\hat{\phi},\hat{\psi}> = \int_{-\infty}^{\infty} \overline{\hat{\phi}}(x)\hat{\psi}(x)\, dx \tag{7.5}$$

then $\mathcal{H}$, completed with (7.5), is the Hilbert space H of square integrable functions.

The space of continuous linear functionals defined on $\mathcal{H}$ is the space Λ_∞ of the distributions of the exponential type (ref.[7]).

The "nested space"

$$\mathfrak{H} = (\mathcal{H}, H, \Lambda_\infty) \tag{7.6}$$

is a Guelfand's triplet (or a Rigged Hilbert space [8]).

In addition we have: $\mathcal{H} \subset \mathcal{S} \subset H \subset \mathcal{S}' \subset \Lambda_\infty$, where $\mathcal{S}$ is the Schwartz space of rapidly decreasing test functions (ref[10]).

Any Guelfand's triplet $\mathfrak{G} = (\Phi, H, \Phi')$ has the fundamental property that a linear and symmetric operator on Φ, admitting an extension to a self-adjoint operator in H, has a complete set of generalized eigen-functions in Φ' with real eigenvalues.

8. Appendix II: Tempered Ultradistributions

The Fourier transform of a function $\hat{\phi} \in \mathcal{H}$ is

$$\phi(z) = \frac{1}{2\pi} \int_{-\infty}^{\infty} \overline{\hat{\phi}}(x)\, e^{iz\cdot x}\, dx \tag{8.1}$$

$\phi(z)$ is entire analytic and rapidly decreasing on straight lines parallel to the real axis. We shall call $\mathfrak{H}$ the set of all such functions.

$$\mathfrak{H} = \mathcal{F}\{\mathcal{H}\} \tag{8.2}$$

It is a $\mathcal{Z}\{M_p\}$ space (ref.[9]), countably normed and complete, with:

$$M_p(z) = (1+|z|)^p \tag{8.3}$$

$\mathfrak{H}$ is also a nuclear space with norms:

$$\|\phi\|_{pn} = \sup_{z \in V_n} (1 + |z|)^p |\phi(z)| \tag{8.4}$$

where $V_k = \{z = (z_1, z_2, ..., z_n) \in \mathbb{C}^n : |Im z_j| \leqq k, 1 \leqq j \leqq n\}$

We can define the usual scalar product:

$$< \phi(z), \psi(z) >= \int_{-\infty}^{\infty} \phi(z)\psi_1(z)\, dz = \int_{-\infty}^{\infty} \overline{\hat{\phi}}(x)\hat{\psi}(x)\, dx \tag{8.5}$$

where:

$$\psi_1(z) = \int_{-\infty}^{\infty} \hat{\psi}(x)\, e^{-iz\cdot x}\, dx$$

and $dz = dz_1\, dz_2 ... dz_n$

By completing $\mathfrak{H}$ with the norm induced by (8.5) we get the Hilbert space of square integrable functions.

The dual of $\mathfrak{H}$ is the space $\mathcal{U}$ of tempered ultradistributions (ref.[7]). In other words, a tempered ultradistribution is a continuous linear functional defined on the space $\mathfrak{H}$ of entire functions rapidly decreasing on straight lines parallel to the real axis.

The set $\mathfrak{U} = (\mathfrak{H}, H, \mathcal{U})$ is also a Guelfand's triplet.

Moreover, we have: $\mathfrak{H} \subset \mathcal{S} \subset H \subset \mathcal{S}' \subset \mathcal{U}$.

$\mathcal{U}$ can also be characterized in the following way (ref.[7]): let $\mathcal{A}_\omega$ be the space of all functions $F(z)$ such that:

I- $F(z)$ is analytic for $\{z \in \mathbb{C}^n : |Im(z_1)| > p, |Im(z_2)| > p, ..., |Im(z_n)| > p\}$.

II- $F(z)/z^p$ is bounded continuous in $\{z \in \mathbb{C}^n : |Im(z_1)| \geqq p, |Im(z_2)| \geqq p, ..., |Im(z_n)| \geqq p\}$, where $p = 0, 1, 2, ...$ depends on $F(z)$.

Let Π be the set of all z-dependent pseudo-polynomials, $z \in \mathbb{C}^n$. Then $\mathcal{U}$ is the quotient space:

III- $\mathcal{U} = \mathcal{A}_\omega / \Pi$

By a pseudo-polynomial we understand a function of z of the form $\sum_s z_j^s G(z_1, ..., z_{j-1}, z_{j+1}, ..., z_n)$ with $G(z_1, ..., z_{j-1}, z_{j+1}, ..., z_n) \in \mathcal{A}_\omega$

Due to these properties it is possible to represent any ultradistribution as (ref.[7]):

$$F(\phi) =< F(z), \phi(z) >= \oint_\Gamma F(z)\phi(z)\, dz \tag{8.6}$$

$\Gamma = \Gamma_1 \cup \Gamma_2 \cup ... \Gamma_n$ where the path Γ_j runs parallel to the real axis from $-\infty$ to ∞ for $Im(z_j) > \zeta$, $\zeta > p$ and back from ∞ to $-\infty$ for $Im(z_j) < -\zeta$, $-\zeta < -p$. (Γ surrounds all the singularities of $F(z)$).

Formula (8.6) will be our fundamental representation for a tempered ultradistribution. Sometimes use will be made of "Dirac formula" for ultradistributions (ref.[6]):

$$F(z) = \frac{1}{(2\pi i)^n} \int_{-\infty}^{\infty} \frac{f(t)}{(t_1 - z_1)(t_2 - z_2) ... (t_n - z_n)}\, dt \tag{8.7}$$

where the "density" $f(t)$ is such that

$$\oint_{\Gamma} F(z)\phi(z)\,dz = \int_{-\infty}^{\infty} f(t)\phi(t)\,dt \tag{8.8}$$

While $F(z)$ is analytic on Γ, the density $f(t)$ is in general singular, so that the r.h.s. of (8.8) should be interpreted in the sense of distribution theory.

Another important property of the analytic representation is the fact that on Γ, $F(z)$ is bounded by a power of z (ref.[7]):

$$|F(z)| \leq C|z|^p \tag{8.9}$$

where C and p depend on F.

The representation (8.6) implies that the addition of a pseudo-polynomial $P(z)$ to $F(z)$ do not alter the ultradistribution:

$$\oint_{\Gamma}\{F(z)+P(z)\}\phi(z)\,dz = \oint_{\Gamma} F(z)\phi(z)\,dz + \oint_{\Gamma} P(z)\phi(z)\,dz$$

But:

$$\oint_{\Gamma} P(z)\phi(z)\,dz = 0$$

as $P(z)\phi(z)$ is entire analytic in some of the variables z_j (and rapidly decreasing),

$$\therefore \quad \oint_{\Gamma}\{F(z)+P(z)\}\phi(z)\,dz = \oint_{\Gamma} F(z)\phi(z)\,dz \tag{8.10}$$

9. Appendix III: Fractional Calculus

The purpose of this sections is to introduce definition of fractional derivation and integration given in ref. [6]. This definition unifies the notion of integral and derivative in one only operation. Let $\hat{f}(x)$ a distribution of exponential type and $F(\Omega)$ the complex Fourier transformed Tempered Ultradistribution. Then:

$$F(k) = H[\Im(k)]\int_{0}^{\infty} \hat{f}(x)e^{ikx}\,dx - H[-\Im(k)]\int_{-\infty}^{0} \hat{f}(x)e^{ikx}\,dx \tag{9.1}$$

($H(x)$ is the Heaviside step function) and

$$\hat{f}(x) = \frac{1}{2\pi}\oint_{\Gamma} F(k)e^{-ikx}\,dk \tag{9.2}$$

where the contour Γ surround all singularities of $F(k)$ and runs parallel to real axis from $-\infty$ to ∞ above the real axis and from ∞ to $-\infty$ below the real axis. According to [6] the fractional derivative of $\hat{f}(x)$ is given by

$$\frac{d^\lambda \hat{f}(x)}{dx^\lambda} = \frac{1}{2\pi}\oint_{\Gamma}(-ik)^\lambda F(k)e^{-ikx}\,dk + \oint_{\Gamma}(-ik)^\lambda a(k)e^{-ikx}\,dk \tag{9.3}$$

Where $a(k)$ is entire analytic and rapidly decreasing. If $\lambda = -1$, d^λ/dx^λ is the inverse of the derivative (an integration). In this case the second term of the right side of (9.3) gives a primitive of $\hat{f}(x)$. Using Cauchy's theorem the additional term is

$$\oint \frac{a(k)}{k} e^{-ikx} dk = 2\pi a(0) \tag{9.4}$$

Of course, an integration should give a primitive plus an arbitrary constant. Analogously when $\lambda = -2$ (a double iterated integration) we have

$$\oint \frac{a(k)}{k^2} e^{-ikx} dk = \gamma + \delta x \tag{9.5}$$

where γ and δ are arbitrary constants.

References

[1] M. Naber: *J. of Math. Phys* **45**, 3339 (2004).

[2] D. G. Barci, G. Bollini, L. E. Oxman, M. C. Rocca: *Int. J. of Theor. Phys.* **37**, 3015 (1998).

[3] F. Riewe: *Phys. Rev. E* **55**, 3581 (1997).

[4] L. S. Gradshtein and I. M. Ryzhik : "Table of Integrals, Series, and Products". Sixth edition, 3.322, 333 Academic Press (2000).

[5] L. Schiff: "*Quantum Mechanics*", **65**, McGraw-Hill Kogakusha, Ltd (1968).

[6] J. Sebastiao e Silva : *Math. Ann.* **136**, 38 (1958).

[7] M. Hasumi: Tôhoku *Math. J.* **13**, 94 (1961).

[8] I. M. Gel'fand and N. Ya. Vilenkin : "Generalized Functions" Vol. 4. Academic Press (1964).

[9] I. M. Gel'fand and G. E. Shilov : "Generalized Functions" Vol. 2. Academic Press (1968).

[10] L. Schwartz : "Théorie des distributions". Hermann, Paris (1966).

In: Focus on Quantum Mechanics
Editors: David E. Hathaway et al. pp. 15-34
ISBN 978-1-62100-680-0

Chapter 2

GEOMETRIC MODIFICATION OF QUANTUM MECHANICS

Vladimir Trifonov
American Mathematical Society

Abstract

I present an axiomatic modification of quantum mechanics with a *possible-worlds* semantics capable of *predicting* essential "nonquantum" features of an observable universe model - the *topology* and *dimensionality* of spacetime, the *existence*, the *signature* and a *specific form* of a metric on it, and a set of naturally preferred directions (*vistas*) in spacetime unrelated to its metric properties. It is shown that the dynamics of the modification is represented by a *hyper-Hamiltonian flow* (superposition of three dependent Hamiltonian flows).

1. Introduction

The technical purpose of the paper is to provide a formal definition of the notion of an *observer* and related constructs in order to make the description of quantum systems more compatible with the kinematic structure of general relativity (GR). We consider the following kinematic axioms of GR:

Axiom 1.1. Spacetime of the universe is a smooth manifold.

Axiom 1.2. The dimensionality of spacetime is four.

Axiom 1.3. Spacetime is equipped with a Lorentzian metric.

The fact that these *still* are axioms is somewhat troublesome, at least to the author, who would rather see them as corollaries of a single assertion.

We shall consider changes to the kinematic, dynamic and semantic structure of quaternionic quantum mechanics (QQM, [2]) that deal with notions pertaining to the above axioms and the following assertion:

Assertion 1.1. The logic of the observer is bivalent Boolean.

We shall show that (technically accurate versions of) the statements of the axioms are derivable from (a technically accurate version of) this assertion. We start with three simple observations.

Observation 1.1. In QQM the quaternionic Hilbert space $\mathbb{V}$ contains a natural principal bundle, $(\mathcal{V}_{\mathbb{H}}^{\circ})$, with the following components:

1. The total space is the set $\mathcal{V}_{\mathbb{H}}^{\circ} := \mathbb{V} \setminus \{\mathbf{0}\}$ of nonzero vectors of $\mathbb{V}$, with the natural manifold structure canonically generated by the linear structure of $\mathbb{V}$.

2. The base space is a quaternionic projective space $\mathcal{P}\mathcal{V}_{\mathbb{H}}^{\circ}$ whose points are quaternionic rays in $\mathbb{V}$.

3. The standard fiber is the set $\mathcal{H}^{\circ} = \mathbb{H} \setminus \{\mathbf{0}\}$ of nonzero quaternions which is a four dimensional manifold and a Lie group $\mathcal{H}^{\circ} \cong SU(2) \times \mathbb{R}^{+}$.

4. The structure group is also $\mathcal{H}^{\circ}$. It acts on $\mathcal{V}_{\mathbb{H}}^{\circ}$ from the left, and for each $\psi \in \mathcal{V}_{\mathbb{H}}^{\circ}$ its orbit (the ray through ψ) is a copy of $\mathcal{H}^{\circ}$ via the fiber diffeomorphism.

We shall refer to this bundle as the *hyperquantum bundle over* $\mathbb{V}$. The recent discovery [20] of natural relativistic structure on $\mathcal{H}^{\circ}$ turns each fiber of $(\mathcal{V}_{\mathbb{H}}^{\circ})$ into a Lorentzian manifold. This principal bundle is a generalization of a principal bundle $(\mathcal{V}_{\mathbb{C}}^{\circ})$ used in geometric quantum mechanics (GQM) [14], the *quantum bundle over* a *complex* Hilbert space $\mathbb{V}$ (see [18] and references therein). The hyperquantum bundle forms the kinematic foundation of the modification.

Observation 1.2. The total space $\mathcal{V}_{\mathbb{H}}^{\circ}$ has the structure of a hyper-Kähler manifold with the Riemannian metric and symplectic forms induced by the decomposition of values of the quaternionic Hermitian form on $\mathbb{V}$ in the canonical basis $(\mathbf{1}, \boldsymbol{i}, \boldsymbol{j}, \boldsymbol{k})$ of quaternions. Hence vector fields and flows on $\mathcal{V}_{\mathbb{H}}^{\circ}$ are subject to the hyper-Hamiltonian formalism [11], roughly a superposition of three hamiltonian evolutions on $\mathcal{V}_{\mathbb{H}}^{\circ}$, on which the dynamics of the modification is based.

Observation 1.3. The observer theory sketched in [19] supplies material for a rigorous definition of an *observer* and *perceptible* analogues of standard physical constructs such as *time*, *spacetime* and a *dynamical system*, and study the dependence of their properties on the type of logic of the observer. The *perceptible spacetime* acquires a group structure, and a *perceptible dynamical system* becomes a set with a left action of a monoid. The main result of [19] asserts that if the logic of an observer is bivalent Boolean then the perceptible spacetime is isomorphic to the Lie group of nonzero quaternions, $\mathcal{H}^{\circ}$, and a perceptible dynamical system is a set with a left action of $\mathcal{H}^{\circ}$. A modest category-theoretic generalization of this schema provides a semantic foundation of the modification.

In this paper we restate and combine these observations in a coherent way. It should be stressed that the proposed modification is not a complete theory because the treatment of certain important aspects related to quantum measurement and probability is too sketchy. Here our primary objective is *complete unambiguity*, i. e., axiomatic consistency rather than completeness. The paper is organized as follows.

Section 2. - Basic constructs and notations. The technical material is given very selectively, the purpose being an introduction of notational conventions rather than education. For example, the reader is assumed to have some familiarity with Birkhoff categories and hyper-Kähler geometry. Since this paper utilizes some conventional structures ($\mathbb{F}$-algebras) in an unconventional way, thorough knowledge of [20] is essential for constructive reading.

In particular, the notions of a *principal inner product* on an $\mathbb{F}$-algebra and a *principal metric* on a unital algebra, introduced in [20], are crucial to our treatment. $\mathbb{N}$ denotes natural numbers and zero, $\mathbb{C}$ an $\mathbb{R}$ are the fields of complex numbers and real numbers, respectively, and $\mathbb{R}$ is assumed taken with its standard linear order and euclidean topology. Small Greek indices, α, β, γ and small Latin indices p, q *always* run 0 to 3 and 1 to 3, respectively. The set of $[\,{}^{n}_{m}\,]$-tensor fields on a smooth manifold $\mathcal{M}$ is denoted $\mathcal{M}[\,{}^{n}_{m}\,]$.
Section 3. - A slightly nonstandard description of hyper-Käher manifolds and basic notions associated with quaternionic Hilbert spaces and quaternionic regular maps.
Section 4. - The semantics of the modification. We introduce the notions of an *experient* and a *reality*. The reader is warned that we shall neither discuss the philosophical issues involved, nor use the terminology and notation that usually accompany them (see for example [5], [8]). Ours is a simpler and more pragmatic task – to *formally* redefine certain technical constructs of physics in terms of *elementary experiences* of an observer, considered as *primitive entities*.
Section 5. - We study a particular species of experients called $\mathbb{F}$-*observers* and the associated notions of a *temporal reality* and a *phenomenon*.
Section 6. - A specialization of some of the above notions, namely an *observer* and a *robust reality*.
Section 7. - We define the notion of a *dynamical system*, its *evolution* and its *perceptibles*. The important notion of *propensity* is defined.
Section 8. - A special case of a robust reality, a *cosmology*. This Section contains one of the central results of the paper, namely the essential uniqueness of the cosmology of the observer. We compute several characteristics of the cosmology such as topology and dimensionality of its spacetime, as well as properties of its metric and naturally preferred directions in the spacetime.
Section 9. - We define *physical systems*, *observables* and their *measurements*. It is shown that standard quantum systems of complex quantum mechanics correspond to a *degenerate*, in a strictly defined sense, kind of physical systems, and the notions of a measurement and propensity seem to reflect certain aspects of quantum measurement and probability, respectively.
Section 10. - We conclude the paper with an informal summary of the results.

2. Technical Preliminaries

Definition 2.1. A *signature*, Σ, is an ordered pair (S, s), where S is a set of *elementary symbols* and $\mathsf{s} : \mathsf{S} \to \mathbb{N}$ is the *arity map*, assigning to each elementary symbol $s \in \mathsf{S}$ a natural number $\mathsf{s}(s)$, called the *arity* of s.

Definition 2.2. For a category $\mathcal{E}$ with products and coproducts and a signature $\Sigma = (\mathsf{S}, \mathsf{s})$, an endofunctor $\Gamma : \mathcal{E} \to \mathcal{E}$ is called a Σ-*functor* (*on* $\mathcal{E}$) if for each $\mathcal{E}$-object A,

$$\Gamma(\mathsf{A}) = \coprod_{s \in \mathsf{S}} \mathsf{A}^{\mathsf{s}(s)}, \tag{1}$$

where $\coprod$ denotes coproduct of $\mathcal{E}$-objects, and $\mathsf{A}^{\mathsf{s}(s)}$ is a product of $\mathsf{s}(s)$ copies of A.

Definition 2.3. Given an endofunctor $\Gamma : \mathcal{E} \longrightarrow \mathcal{E}$ on a category $\mathcal{E}$, an *algebra*, $\mathbb{A}$, *for* Γ is an ordered pair (A, a), where A is an $\mathcal{E}$-object, called the *carrier*, and $\mathsf{a} : \Gamma(\mathsf{A}) \longrightarrow \mathsf{A}$ is an $\mathcal{E}$-arrow, called the *structure map* of the algebra. Let $\mathbb{A} = (\mathsf{A}, \mathsf{a})$ and $\mathbb{B} = (\mathsf{B}, \mathsf{b})$ be algebras for Γ. A Γ-*morphism*, $\mathbb{A} \longrightarrow \mathbb{B}$, is a map $f : \mathsf{A} \longrightarrow \mathsf{B}$, such that the following diagram commutes:

$$\begin{array}{ccc} \Gamma(\mathsf{A}) & \xrightarrow{\Gamma(f)} & \Gamma(\mathsf{B}) \\ {\scriptstyle \mathsf{a}}\big\downarrow & & \big\downarrow{\scriptstyle \mathsf{b}} \\ \mathsf{A} & \xrightarrow{f} & \mathsf{B} \end{array} \quad . \tag{2}$$

Example 2.1. Let $\mathbb{F}$ be a field, and $\mathsf{S} = \{\mathbf{0}, +\} \cup \mathbb{F}$ and $\mathsf{s}(\mathbf{0}) = 0$, $\mathsf{s}(+) = 2$, $\mathsf{s}(r) = 1, \forall r \in \mathbb{F}$. A *vector space*, $V = (\mathsf{V}, \mathsf{v})$, *over a field* $\mathbb{F}$ is an algebra for the Σ-functor $\Gamma : \boldsymbol{Set} \longrightarrow \boldsymbol{Set}$ on the category of sets for the signature $\Sigma = (\mathsf{S}, \mathsf{s})$.

Remark 2.1. Dually, *coalgebras for an endofunctor* $\Gamma : \mathcal{E} \longrightarrow \mathcal{E}$ are defined by reversal of structure maps. Algebras and coalgebras for an endofunctor Γ and their Γ-morphisms form categories denoted $\mathcal{E}^{\Gamma}$ and $\mathcal{E}_{\Gamma}$, respectively, (see, e. g., [13]).

Example 2.2. A monoid is an example of an algebra, $M = (\mathsf{M}, \mathsf{m})$, for a Σ-functor on $\boldsymbol{Set}$ with $\mathsf{S} = \{\imath, *\}$, $\mathsf{s}(\imath) = 0$, $\mathsf{s}(*) = 2$. As with every Σ-functor, the structure map m can be split into the constituents, giving the more familiar notation $(\mathsf{M}, \imath, *)$, where $\imath$ is understood as a preferred element, the identity of M, and $*$ is a binary operation on M.

Definition 2.4. An endofunctor Γ on $\mathcal{E}$ is called a *monad* if there exist two natural transformations, $\flat : id(\mathcal{E}) \longrightarrow \Gamma$ and $\natural : \Gamma^2 \longrightarrow \Gamma$ such that the following diagrams commute:

$$\begin{array}{ccc} \Gamma^3 & \xrightarrow{\Gamma\natural} & \Gamma^2 \\ {\scriptstyle \natural\Gamma}\big\downarrow & & \big\downarrow{\scriptstyle \natural} \\ \Gamma^2 & \xrightarrow{\natural\Gamma} & \Gamma \end{array} \qquad \begin{array}{ccccc} \Gamma & \xrightarrow{\Gamma\flat} & \Gamma^2 & \xleftarrow{\flat\Gamma} & \Gamma \\ & \searrow\!\!\searrow & \big\downarrow{\scriptstyle \natural} & \swarrow\!\!\swarrow & \\ & & \Gamma & & \end{array} , \tag{3}$$

where Γ^n denotes n iterations of the functor.

Definition 2.5. The *category of algebras*, $\bar{\mathcal{E}}^{\Gamma}$, *for the monad* $\Gamma : \mathcal{E} \longrightarrow \mathcal{E}$ is a subcategory of $\mathcal{E}^{\Gamma}$ such that the following diagrams commute for each object $\mathbb{A} = (\mathsf{A}, \mathsf{a})$ of $\bar{\mathcal{E}}^{\Gamma}$:

$$\begin{array}{ccc} \Gamma^2(\mathbb{A}) & \xrightarrow{\Gamma(\mathsf{a})} & \Gamma(\mathbb{A}) \\ {\scriptstyle \natural(\mathbb{A})}\big\downarrow & & \big\downarrow{\scriptstyle \mathsf{a}} \\ \Gamma(\mathbb{A}) & \xrightarrow{\mathsf{a}} & \mathbb{A} \end{array} \qquad \begin{array}{ccc} \mathbb{A} & \xrightarrow{\flat(\mathbb{A})} & \Gamma(\mathbb{A}) \\ & {\scriptstyle id(\mathbb{A})}\searrow & \big\downarrow{\scriptstyle \mathsf{a}} \\ & & \mathbb{A} \end{array} \tag{4}$$

Example 2.3. For each monoid $M = (\mathsf{M}, \mathsf{m})$, an endofunctor Γ on $\boldsymbol{Set}$ sending a set X to the set $\mathsf{M} \times \mathsf{X}$ is a monad. An object, $\mathbb{X} = (\mathsf{X}, \mathsf{x})$, in the category of algebras, $M\boldsymbol{Set}$, for this monad is a set, X, with a left action, x, of the monoid M, also referred to as an M-*set*. For each $a \in \mathsf{M}$, we can define a map $\mathsf{x}_a : \mathsf{X} \longrightarrow \mathsf{X}$ by $\mathsf{x}_a(x) := \mathsf{x}(a, x), \forall x \in \mathsf{X}$. The arrows $(\mathsf{X}, \mathsf{x}) \longrightarrow (\mathsf{X}', \mathsf{x}')$ are *equivariant* (i. e., preserving the action) functions $f : \mathsf{X} \longrightarrow \mathsf{X}'$, making the following diagram commute:

$$\begin{array}{ccc} \mathsf{X} & \xrightarrow{f} & \mathsf{X}' \\ {\scriptstyle \mathsf{x}_a}\downarrow & & \downarrow{\scriptstyle \mathsf{x}'_a} \\ \mathsf{X} & \xrightarrow{f} & \mathsf{X}' \end{array} . \tag{5}$$

The category of M-sets is of utmost importance in our exposition.

Example 2.4. An $\mathbb{F}$-*algebra* (a linear algebra over the field $\mathbb{F}$) is an example of an algebra for a Σ-functor on $\boldsymbol{Set}$ with $\mathsf{S} = \{\mathbf{0}, +, \cdot\} \cup \mathbb{F}$ and $\mathsf{s}(\mathbf{0}) = 0, \mathsf{s}(+) = 2, \mathsf{s}(\cdot) = 2, \mathsf{s}(r) = 1, \forall r \in \mathbb{F}$. For an unconventional description of $\mathbb{F}$-algebras, as vector spaces over $\mathbb{F}$ with $[\frac{1}{2}]$-tensors on them (*structure tensors*), crucial to our technical setup, see [20].

Definition 2.6. A complete regularly co-well-powered category with regular epi-mono factorizations is called *Birkhoff* if it has enough regular projectives.

Remark 2.2. For a detailed view of Birkhoff categories see [13].

Definition 2.7. A full subcategory of a Birkhoff category is called a *Birkhoff variety* if it is closed under products, subobjects and quotients.

Definition 2.8. An endofunctor $\Gamma : \mathcal{E} \to \mathcal{E}$ is called a *varietor* if it preserves regular epis, and the forgetful functor $U : \mathcal{E}^{\Gamma} \to \mathcal{E}$ has a left adjoint.

A real vector space V induces a natural manifold structure on its carrier. This manifold which we denote $\mathcal{V}$, is referred to as the *linear manifold canonically generated by* V. Since V and $\mathcal{V}$ have the same carrier, there is a bijection $\mathfrak{J}_V : \mathcal{V} \to V$. We use the normal $(a, u, \ldots)$ and bold $(\boldsymbol{a}, \boldsymbol{u} \ldots)$ fonts, to denote the elements of $\mathcal{V}$ and V, respectively, e. g., $\mathfrak{J}_V(a) = \boldsymbol{a}$. The tangent space $T_a\mathcal{V}$ is identified with V at each point $a \in \mathcal{V}$ via an isomorphism $\mathfrak{J}_a^* : T_a\mathcal{V} \to V$ sending a tangent vector to the curve $\mu : \mathbb{R} \to \mathcal{V}, \mu(t) = a + tu$, at the point $\mu(0) = a \in \mathcal{V}$, to the vector $\boldsymbol{u} \in V$, with the "total" map $\mathfrak{J}_V^* : T\mathcal{V} \to V$. The set of nonzero vectors of V constitutes a submanifold of $\mathcal{V}$, referred to as the *punctured manifold* (*of* V), denoted $\mathcal{V}^\circ$.

Definition 2.9. For a real vector space V and a linear map $\boldsymbol{F} : V \to V$, a vector field $\boldsymbol{f} : \mathcal{V} \to T\mathcal{V}$ on $\mathcal{V}$, such that the following diagram commutes

$$\begin{array}{ccc} \mathcal{V} & \xrightarrow{\boldsymbol{f}} & T\mathcal{V} \\ {\scriptstyle \mathfrak{J}_V}\downarrow & & \downarrow{\scriptstyle \mathfrak{J}_V^*} \\ V & \xrightarrow{\boldsymbol{F}} & V \end{array} , \tag{6}$$

is called the *vector field canonically generated* by $\boldsymbol{F}$.

Remark 2.3. In particular, for a unital algebra $\mathbb{A}$, the linear manifold canonically generated by its underlying vector space A is denoted by $\mathcal{A}$ and the punctured manifold by $\mathcal{A}^\circ$.

Definition 2.10. For real vector spaces U, V, and a linear map $\boldsymbol{F} : U \to V$, the map $F : \mathcal{U} \to \mathcal{V}$, such that the following diagram commutes:

$$\begin{array}{ccc} \mathcal{U} & \xrightarrow{F} & \mathcal{V} \\ {\scriptstyle \mathfrak{d}_U}\downarrow & & \downarrow{\scriptstyle \mathfrak{d}_V} \\ U & \xrightarrow{\boldsymbol{F}} & V \end{array}, \tag{7}$$

is called the *linear induction* of $\boldsymbol{F}$.

3. Quaternionic Maps and Hyper-Hamiltonian Vector Fields

Definition 3.1. For the quaternion algebra $\mathbb{H} = (H, \mathsf{H})$ a basis on the vector space H is called *canonical* if the components of the structure tensor H are given by the entries of the following matrices

$$\mathsf{H}^0_{\alpha\beta} = \begin{pmatrix} 1 & 0 & 0 & 0 \\ 0 & -1 & 0 & 0 \\ 0 & 0 & -1 & 0 \\ 0 & 0 & 0 & -1 \end{pmatrix}, \ \mathsf{H}^1_{\alpha\beta} = \begin{pmatrix} 0 & 1 & 0 & 0 \\ 1 & 0 & 0 & 0 \\ 0 & 0 & 0 & 1 \\ 0 & 0 & -1 & 0 \end{pmatrix},$$
$$\mathsf{H}^2_{\alpha\beta} = \begin{pmatrix} 0 & 0 & 1 & 0 \\ 0 & 0 & 0 & -1 \\ 1 & 0 & 0 & 0 \\ 0 & 1 & 0 & 0 \end{pmatrix}, \ \mathsf{H}^3_{\alpha\beta} = \begin{pmatrix} 0 & 0 & 0 & 1 \\ 0 & 0 & 1 & 0 \\ 0 & -1 & 0 & 0 \\ 1 & 0 & 0 & 0 \end{pmatrix}. \tag{8}$$

Remark 3.1. The set $\{\boldsymbol{i}\}$ of canonical bases $(\boldsymbol{i}_\beta)$ is parametrized by the elements of $SO(3)$. We shall refer to a bijection $\Xi : SO(3) \to \{\boldsymbol{i}\}$ as an *array*.

Remark 3.2. For a set X, the values of a map $f : \mathsf{X} \to \mathbb{H}$ can be decomposed in a canonical basis $(\boldsymbol{i}_\beta)$, producing an ordered quadruple (f_β) of real valued maps on X which we call the *constituents* of f in the basis $(\boldsymbol{i}_\beta)$. We shall be careful to distinguish these from components of tensorial objects.

Let $\mathcal{M} = (M, \boldsymbol{g}, (\tilde{\boldsymbol{\omega}}^p))$ be a hyper-Kähler manifold. Then, for an array $\Xi : SO(3) \to \{\boldsymbol{i}\}$, we can consider the ordered quadruple $(\boldsymbol{g}, \tilde{\boldsymbol{\omega}}^1, \tilde{\boldsymbol{\omega}}^2, \tilde{\boldsymbol{\omega}}^3))$ as constituents $(\tilde{\boldsymbol{\omega}}^\beta)$, where $\tilde{\boldsymbol{\omega}}^0 := \boldsymbol{g}$, of a quaternion valued map $\tilde{\boldsymbol{\omega}} : T\mathcal{M} \times T\mathcal{M} \to \mathbb{H}$ in the standard basis $\Xi(\mathbf{1})$, assigning to each ordered pair $(\boldsymbol{u}, \boldsymbol{v})$ of tangent vectors at each point $\phi \in \mathcal{M}$ a quaternion $\tilde{\boldsymbol{\omega}}^\beta(\boldsymbol{u}, \boldsymbol{v})\boldsymbol{i}_\beta$, which allows us to define the constituents $\tilde{\boldsymbol{\omega}}^p$ in every canonical basis $(\boldsymbol{i}_\beta)$. Similarly, we can assign the ordered quadruple $(\mathcal{J}_\beta)$, where $\mathcal{J}_0$ is the identity map on $T\mathcal{M}$, and $\mathcal{J}_p$ is the complex structure corresponding to $\tilde{\boldsymbol{\omega}}^p$, to each canonical basis $(\boldsymbol{i}_\beta)$, as constituents of the hypercomplex structure $\boldsymbol{\mathfrak{J}}$.

Definition 3.2. For an array Ξ and a hyper-Kähler manifold $\mathcal{M}$, the ordered pair $\boldsymbol{\Omega} : (\tilde{\boldsymbol{\omega}}, \boldsymbol{\mathfrak{J}})$ is called a *hyper-Kähler structure* on $\mathcal{M}$ *generated by* Ξ. For a canonical basis $(\boldsymbol{i}_\beta)$ the maps $\tilde{\boldsymbol{\omega}}^p$ and $\mathcal{J}_p$ are called the *symplectic* and *complex constituents* of $\boldsymbol{\Omega}$, respectively, *in* the basis $(\boldsymbol{i}_\beta)$.

Definition 3.3. A left module V over $\mathbb{H}$ is called a *(left) quaternionic vector space*.

Remark 3.3. A *right quaternionic vector space* is defined similarly, as well as the right versions of constructs based on it.

Definition 3.4. A map $\widehat{\boldsymbol{F}} : \mathsf{V} \to \mathsf{V}$ is called a *(left) quaternion linear operator*, if

$$\widehat{\boldsymbol{F}}(\boldsymbol{a}\boldsymbol{\phi}) = \boldsymbol{a}\widehat{\boldsymbol{F}}(\boldsymbol{\phi}), \quad \forall \boldsymbol{\phi} \in \mathsf{V}, \quad \forall \boldsymbol{a} \in \mathbb{H}.$$

Definition 3.5. A *(left) quaternionic Hilbert space*, $\mathbb{V}$, is a quaternionic vector space, V, together with a map $\langle \cdot \mid \cdot \rangle : \mathsf{V} \times \mathsf{V} \to \mathbb{H}$, called a *quaternionic Hermitian inner product*, such that

$$\begin{aligned} &\langle \boldsymbol{\phi} \mid \boldsymbol{\psi} + \boldsymbol{\xi} \rangle = \langle \boldsymbol{\phi} \mid \boldsymbol{\psi} \rangle + \langle \boldsymbol{\phi} \mid \boldsymbol{\xi} \rangle, \\ &\qquad \langle \boldsymbol{\phi} \mid \boldsymbol{\psi} \rangle = \overline{\langle \boldsymbol{\psi} \mid \boldsymbol{\phi} \rangle}, \quad \langle \boldsymbol{a}\boldsymbol{\phi} \mid \boldsymbol{\psi} \rangle = \boldsymbol{a}\langle \boldsymbol{\phi} \mid \boldsymbol{\psi} \rangle, \\ &\qquad \| \boldsymbol{\phi} \|^2 := \langle \boldsymbol{\phi} \mid \boldsymbol{\phi} \rangle \in \mathbb{R}, \| \boldsymbol{\phi} \|^2 > 0, \quad \forall \boldsymbol{\phi} \neq \mathbf{0}, \\ &\qquad\qquad\qquad \forall \boldsymbol{\phi}, \boldsymbol{\psi}, \boldsymbol{\xi} \in \mathsf{V}, \forall \boldsymbol{a} \in \mathbb{H}, \end{aligned}$$

and the diagonal $\| \cdot \|$ induces a topology on V, relative to which V is separable and complete.

Definition 3.6. For a quaternionic Hilbert space $\mathbb{V}$ and a quaternion linear operator $\widehat{\boldsymbol{F}}$ on $\mathbb{V}$, its *quaternionic adjoint (with respect to* $\langle \cdot \mid \cdot \rangle$*)* is a quaternion linear operator $\widehat{\boldsymbol{F}}^{\dagger}$ on V, such that

$$\langle \boldsymbol{\phi} \mid \widehat{\boldsymbol{F}}(\boldsymbol{\psi}) \rangle = \langle \widehat{\boldsymbol{F}}^{\dagger}(\boldsymbol{\phi}) \mid \boldsymbol{\psi} \rangle, \quad \forall \boldsymbol{\phi}, \boldsymbol{\psi} \in \mathbb{V}.$$

Definition 3.7. We refer to a quaternion linear operator $\widehat{\boldsymbol{F}}$ as *quaternionic (anti-)Hermitian* if it coincides with (the negative of) its quaternionic adjoint.

Remark 3.4. Given a basis, $(\boldsymbol{e}_j)$, on an n dimensional quaternionic Hilbert space $\mathbb{V}$, and a canonical basis, $(\boldsymbol{i}_\beta)$, on H, $\mathbb{V}$ induces a real $4n$ dimensional vector space, V, and the latter canonically generates a real linear $4n$ dimensional manifold $\mathcal{V}_{\mathbb{H}}$ and the punctured manifold $\mathcal{V}^{\circ}_{\mathbb{H}}$.

Definition 3.8. For an operator $\widehat{\boldsymbol{F}} : \mathbb{V} \to \mathbb{V}$ on a quaternionic Hilbert space $\mathbb{V}$ its *expectation operator* is a map $\widehat{F} : \mathbb{V} \to \mathbb{H}$ assigning to each $\boldsymbol{\phi} \in \mathbb{V}$ a quaternion $\langle \boldsymbol{\phi} \mid \widehat{\boldsymbol{F}}(\boldsymbol{\phi}) \rangle$.

Lemma 3.1. *For each quaternionic anti-Hermitian operator,* $\widehat{\boldsymbol{F}}$*, its expectation operator* $\widehat{F}$ *has the following property:*

$$\widehat{F}(\boldsymbol{\phi}) = \mathfrak{Im}(\widehat{F}(\boldsymbol{\phi})), \quad \forall \boldsymbol{\phi} \in \mathbb{V}. \tag{9}$$

Proof.

$$\langle \boldsymbol{\phi} \mid \widehat{\boldsymbol{F}}(\boldsymbol{\phi}) \rangle = \langle -\widehat{\boldsymbol{F}}(\boldsymbol{\phi}) \mid \boldsymbol{\phi} \rangle = -\langle \widehat{\boldsymbol{F}}(\boldsymbol{\phi}) \mid \boldsymbol{\phi} \rangle = -\overline{\langle \boldsymbol{\phi} \mid \widehat{\boldsymbol{F}}(\boldsymbol{\phi}) \rangle}. \tag{10}$$

Thus, for each $\boldsymbol{\phi} \in \mathbb{V}$ the quaternion $\langle \boldsymbol{\phi} \mid \widehat{\boldsymbol{F}}(\boldsymbol{\phi}) \rangle$ coincide with the negative of its adjoint, which means that its real part is zero. □

Definition 3.9. For a quaternionic anti-Hermitian operator $\widehat{\boldsymbol{F}} : \mathbb{V} \to \mathbb{V}$ the linear induction $F : \mathcal{V}_{\mathbb{H}} \to \mathcal{H}$ of its expectation operator $\widehat{F}$ is called the *expectation* of $\widehat{\boldsymbol{F}}$.

Definition 3.10. For a quaternion linear operator $\widehat{\boldsymbol{F}} : \mathbb{V} \longrightarrow \mathbb{V}$ a vector field $\boldsymbol{f} : \mathcal{V}_{\mathbb{H}} \longrightarrow T\mathcal{V}_{\mathbb{H}}$ canonically generated by $-\widehat{\boldsymbol{F}}$ is called the *hyperfield* of $\widehat{\boldsymbol{F}}$.

Remark 3.5. The set of the integral curves of the hyperfield of $\widehat{\boldsymbol{F}}$ can be formally represented by the following differential equation

$$\dot{\psi} = -\widehat{\boldsymbol{F}}(\psi). \tag{11}$$

Hyperfields are a generalization of *Schrödinger vector fields* of geometric quantum mechanics [18]. Indeed, due to the simple relationship between eigenvalues of Hermitian and anti-Hermitian operators in the complex case we can use either kind to represent observables. If we take anti-Hermitian operators, the Schrödinger equation has the form (11), with $\widehat{\boldsymbol{F}} = \widehat{\boldsymbol{H}}$, which defines the set of integral curves of the Schrödinger vector field of the Hamiltonian operator $\widehat{\boldsymbol{H}}$.

Given an array Ξ, the quaternionic Hermitian product $\langle \cdot \mid \cdot \rangle$ induces a map $T\mathcal{V}_{\mathbb{H}} \times T\mathcal{V}_{\mathbb{H}} \longrightarrow \mathbb{H}$ whose constituents in a canonical basis $(\boldsymbol{i}_\beta)$ can be identified with the symplectic constituents in $(\boldsymbol{i}_\beta)$ of a hyper-Kähler structure $\boldsymbol{\Omega}$ on $\mathcal{V}_{\mathbb{H}}$ generated by Ξ. Similarly, the maps $(\boldsymbol{\imath}'_\beta)$ defined by

$$\boldsymbol{\imath}'_\beta(\boldsymbol{\phi}) := \boldsymbol{i}_\beta \boldsymbol{\phi}, \quad \forall \boldsymbol{\phi} \in \mathbb{V}$$

produce the complex constituents of $\boldsymbol{\Omega}$.

Example 3.1. The quaternion algebra $\mathbb{H}$ is a quaternionic vector space. Together with a natural quaternionic Hermitian inner product defined by

$$\langle \boldsymbol{a} \mid \boldsymbol{b} \rangle := \boldsymbol{a}\overline{\boldsymbol{b}}, \quad \forall \boldsymbol{a}, \boldsymbol{b} \in \mathbb{H}, \tag{12}$$

it is a quaternionic Hilbert space. Therefore $\mathcal{H}$ and $\mathcal{H}^\circ$ possess natural hyper-Kähler structures.

Definition 3.11. Let $\mathcal{M}$ and $\mathcal{N}$ be hyper-Kähler manifolds with hyper-Kähler structures $\boldsymbol{\Omega}^{\mathcal{M}}$ and $\boldsymbol{\Omega}^{\mathcal{N}}$, respectively, generated by an array Ξ. A smooth map $f : \mathcal{M} \longrightarrow \mathcal{N}$ is called *quaternionic* (*regular*) if there exists an $SO(3)$ matrix B such that

$$\sum_{p,q=1}^{3} \mathsf{B}_{pq} \mathcal{J}_q^{\mathcal{N}} \circ df \circ \mathcal{J}_p^{\mathcal{M}} = df, \tag{13}$$

where $\mathcal{J}_p^{\mathcal{M}}$ and $\mathcal{J}_p^{\mathcal{N}}$ are the complex constituents, in a canonical basis $(\boldsymbol{i}_\beta) = \Xi(\mathsf{B})$, of the hyper-Kähler structures of $\mathcal{M}$ and $\mathcal{N}$, respectively, and df is the differential of f.

Remark 3.6. Quaternionic maps generalize holomorphic functions of complex analysis (see e. g., [9]).

Definition 3.12. For a quaternionic map $f : \mathcal{M} \to \mathcal{H}$, its *hyper-Hamiltonian vector field* is a vector field $\boldsymbol{f}$ on $\mathcal{M}$, such that

$$(df_0)(\boldsymbol{u}) = \boldsymbol{g}(\boldsymbol{f}, \boldsymbol{u}), \quad \forall \boldsymbol{u} \in \mathcal{M}[\begin{smallmatrix}1\\0\end{smallmatrix}], \tag{14}$$

where $\boldsymbol{g}$ is the Riemannian metric on $\mathcal{M}$, and df_0 is the differential of f_0. The maps f and f_0 are referred to as the *generating map* and the *main generator* of $\boldsymbol{f}$, respectively.

Remark 3.7. Since f_0 is invariant under a canonical basis change, so is the definition of a hyper-Hamiltonian vector field.

Theorem 3.1. *For an array Ξ and a quaternionic Hilbert space $\mathbb{V}$ let $\boldsymbol{f}$ be the hyper-Hamiltonian vector field of a quaternionic map $f : \mathcal{V}_{\mathbb{H}} \to \mathcal{H}$, such that $\mathfrak{Im}(f)$ is the expectation of a quaternion anti-Hermitian operator $\widehat{\boldsymbol{F}}$ on $\mathbb{V}$. Then $\boldsymbol{f}$ is the hyperfield of $\widehat{\boldsymbol{F}}$, and for each canonical basis $(\boldsymbol{i}_\beta)$ there exists an ordered triple $(\boldsymbol{f}_p)$ of vector fields on $\mathcal{V}_{\mathbb{H}}$ such that*

$$\boldsymbol{f} = \boldsymbol{f}_1 + \boldsymbol{f}_2 + \boldsymbol{f}_3, \quad (df_p)(\boldsymbol{u}) = \tilde{\omega}^p(\boldsymbol{f}_p, \boldsymbol{u}), \quad \forall \boldsymbol{u} \in \mathcal{V}_{\mathbb{H}}[{}^1_0], \tag{15}$$

(no summation on p) where df_p is the differential of f_p, and $\tilde{\omega}^p$ is a symplectic constituent of the hyper-Kähler structure on $\mathcal{V}_{\mathbb{H}}$ generated by Ξ, in the basis $(\boldsymbol{i}_\beta)$.

Proof. The hyperfield of $\widehat{\boldsymbol{F}}$ is clearly a hyper-Hamiltonian vector field whose generating map is f [18]. It was shown in [11] and [17] that given a quaternionic map f, the conditions (15) are satisfied for a unique vector field $\boldsymbol{f}$, and hence the hyperfield of $\widehat{\boldsymbol{F}}$ coincides with $\boldsymbol{f}$. □

4. Semantics

The observer theory outlined in [19] describes an observer as an *experient*, as opposed to an *occupant*, of the environment. This can be presented within a framework similar to what is known as the *initial algebra* and *final coalgebra* approach to syntax and semantics of formal languages ([1], [21]). The underlying idea is that perception and comprehension are somehow dual to each other, and each *experience* can be considered as both a *percept* and a *concept*. This idea has its origin in logic and theoretical computer science where similar dualities are considered (a symbol and its meaning, syntax of a formal language and its semantics). We start with a category $\mathcal{E}$ representing the *totality of experiences*, and an endofunctor Γ on $\mathcal{E}$ interpreted as the *language of thought* of the experient (see [10] for a philosophical discussion). The categories of algebras, $\mathcal{E}^\Gamma$, and coalgebras, $\mathcal{E}_\Gamma$, for this endofunctor represent *perception* and *comprehension* of the experient, respectively. If the perception, $\mathcal{E}^\Gamma$, has an initial object, $\mathbb{I}$, the latter serves as the *syntax* of the experient's language of thought. Then an algebra, $\mathbb{A}$, for Γ is considered a *paradigm* or a *model* for the language of thought, and the unique arrow $\mathbb{I} \to \mathbb{A}$ as a *meaning function* or an *interpretation*.

Definition 4.1. An *experient*, $\boldsymbol{\mathcal{E}}$, is an ordered pair $(\mathcal{E}, \Gamma)$, where $\mathcal{E}$ is a Birkhoff category, called the *metauniverse*, and $\Gamma : \mathcal{E} \to \mathcal{E}$ is a varietor called the *construer*. The objects and arrows of $\mathcal{E}$ are called *metaphenomena* and *metalinks*, respectively. $\mathcal{E}$-elements of metaphenomena are called *reflexors*. The category of algebras $\mathcal{E}^\Gamma$ is called the *perception category*. Objects and arrows of $\mathcal{E}^\Gamma$ are referred to as *paradigms* and *shifts*, respectively. An experient is *coherent* if the metauniverse is a topos. An experient is called *Boolean* if the metauniverse is a Boolean topos.

Remark 4.1. It follows from the Definition 4.1 that $\mathcal{E}^\Gamma$ is also a Birkhoff category (see [13], p. 125).

Remark 4.2. Intuitively, metaphenomena are *complex percepts* composed out of elementary ones (reflexors). For a coherent experient $\mathcal{E} = (\mathcal{E}, \Gamma)$, we are to think of the internal logic of $\mathcal{E}$ as the *metalogic* of the experient. In particular, the metalogic of a Boolean experient is Boolean.

Definition 4.2. For a paradigm $\mathbb{A}$ of an experient $\mathcal{E}$, and a natural number n, a *reality of rank* n is an ordered pair $\mathcal{R} = (\mathbb{A}, \boldsymbol{R})$, where $\boldsymbol{R}$ is a subobject of $\mathbb{A}^n$. We refer to $\mathbb{A}$ as the *underlying paradigm*, and $\boldsymbol{R}$ is called the *ontology* of $\mathcal{R}$.

Remark 4.3. Intuitively, a reality of rank n is an n-ary relation on a collection of reflexors. From this point on we are interested exclusively in realities of rank 2, henceforth referred to simply as *realities*.

Definition 4.3. An *existence mode*, $\mathcal{E}(\mathfrak{V})$, of an experient $\mathcal{E} = (\mathcal{E}, \Gamma)$ is an ordered pair $(\mathcal{E}, \mathfrak{V})$, where $\mathfrak{V}$ is a Birkhoff variety of $\mathcal{E}^\Gamma$. A paradigm $\mathbb{A}$ such that it is (not) a $\mathfrak{V}$-object is called *(non)existent with respect to* $\mathcal{E}(\mathfrak{V})$. A shift $\mathbb{A} \to \mathbb{B}$ such that it is (not) a $\mathfrak{V}$-arrow is called *(im)possible with respect to* $\mathcal{E}(\mathfrak{V})$.

Remark 4.4. It should be noted that we thus consider every paradigm $\mathbb{A}$ of an experient also a paradigm of the experient in a certain existence mode, although $\mathbb{A}$ may be nonexistent with respect to the latter. Given an existence mode, $\mathcal{E}(\mathfrak{V})$, of an experient $\mathcal{E}$, we say that the experient *is in the mode* $\mathfrak{V}$, or that the experient *is* the *$\mathfrak{V}$-observer*.

5. $\mathbb{F}$-observers

Different existence modes possess different amounts of structure to allow for a definition of notions that we normally associate with observers. We shall focus on a class of Boolean experients whose metauniverse is $\boldsymbol{Set}$, and the construer is a Σ-functor for the following signature:

$$\Sigma = (\mathsf{S}, \mathsf{s}); \quad \mathsf{S} = \{+, \cdot, \mathbf{0}\} \cup \mathbb{F}, \quad \mathsf{s}(+) = 2, \quad \mathsf{s}(\cdot) = 2, \quad \mathsf{s}(\mathbf{0}) = 0, \quad \mathsf{s}(r) = 1, \quad \forall r \in \mathbb{F}, \tag{16}$$

where $\mathbb{F}$ is a field. They are introduced implicitly in [19], where a particular existence mode, an $\mathsf{Alg}\{\mathbb{F}\}$-observer, where $\mathsf{Alg}\{\mathbb{F}\}$ is the category of $\mathbb{F}$-algebras, is studied. $\mathsf{Alg}\{\mathbb{F}\}$-observers are sufficiently fine structured to define the fundamental notion of a *temporal reality*, and at the same time they are relatively simple: their metaphenomena and reflexors are sets and set elements, respectively. For the rest of the paper we shall deal exclusively with $\mathsf{Alg}\{\mathbb{F}\}$-observers, to whom we henceforth refer simply as *$\mathbb{F}$-observers*.

Definition 5.1. For an existent paradigm $\mathbb{A}$ of an $\mathbb{F}$-observer the underlying vector space A is called the *sensory domain* of $\mathbb{A}$, with the principal inner products of $\mathbb{A}$ referred to as *sensory forms* of the paradigm. A basis on A is called a *sensory basis* of the paradigm. The dual vector space A^* is called the *ether domain* of $\mathbb{A}$, with the elements called *ether forms*. The *motor domain* of $\mathbb{A}$ is the multiplicative subgroupoid, $M_\mathbb{A} = (\mathsf{M}_\mathbb{A}, \imath, *)$, of $\mathbb{A}$, generated by the set of nonzero reflexors of the paradigm $\mathbb{A}$. We refer to elements of $M_\mathbb{A}$ as *effectors*. An existent finite dimensional paradigm, $\mathbb{A}$, is called *rational* if its motor domain is a monoid; otherwise $\mathbb{A}$ is called *irrational*. A paradigm is called *transient* if it is neither rational nor irrational. The set $\mathcal{A}^\bullet$ of invertible reflexors of a rational paradigm $\mathbb{A}$ is called the *perception domain* of $\mathbb{A}$.

Remark 5.1. For a rational paradigm $\mathbb{A}$ of an $\mathbb{F}$-observer, the perception domain $\mathcal{A}^\bullet$ is a group, and the ontology, $\boldsymbol{R}$, of a reality $\mathcal{R} = (\mathbb{A}, \boldsymbol{R})$ is a binary relation on the carrier of $\mathbb{A}$.

Definition 5.2. A reality of an $\mathbb{F}$-observer is called *immanent* if its underlying paradigm is existent and its ontology is a partial order. Otherwise the reality is called *transcendent*.

Definition 5.3. For an $\mathbb{F}$-observer, an immanent reality $(\mathbb{F}, \preceq)$ is called a *temporal template* if $\mathbb{F}$ is an ordered field with a complete linear order $\preceq$.

Remark 5.2. Not every existence mode has a temporal template. For instance, complex numbers, $\mathbb{C}$, do not admit linear orders ([15], p. 304). Therefore the $\mathbb{C}$-observer has no temporal templates.

Definition 5.4. For a rational paradigm $\mathbb{A}$ of an $\mathbb{F}$-observer the $\mathbb{A}$-*universe* is the category of $M_{\mathbb{A}}$-sets, $M_{\mathbb{A}}\boldsymbol{Set}$ (see Example 2.3), with objects and arrows called $\mathbb{A}$-*phenomena* and $\mathbb{A}$-*links*, respectively. For an $\mathbb{A}$-phenomenon $\mathbb{X} = (\mathsf{X}, \mathsf{x})$, the metaphenomenon X is called the *propensity realm*, with elements called *propensity modes* of $\mathbb{X}$. Each metalink $\mathsf{X} \to A$, where A is the carrier of $\mathbb{A}$, is referred to as an *attribute* of $\mathbb{X}$.

Remark 5.3. We refer to the internal logic of the topos $M_{\mathbb{A}}\boldsymbol{Set}$ as the *operational logic* of the $\mathbb{F}$-observer *with respect to* $\mathbb{A}$, or as the *logic* of the $\mathbb{A}$-universe.

Definition 5.5. For each $\mathbb{A}$-phenomenon $\mathbb{X} = (\mathsf{X}, \mathsf{x})$, the $\mathcal{A}^\bullet$-set $(\mathsf{X}, \bar{\mathsf{x}})$, where $\bar{\mathsf{x}} : \mathcal{A}^\bullet \times \mathsf{X} \to \mathsf{X}$ is the restriction of x to $\mathcal{A}^\bullet$, is called the *perceptible part* of $\mathbb{X}$. For each $\phi \in \mathsf{X}$ the orbits, W_ϕ and $\overline{W}_\phi$, of ϕ, with respect to the actions of $M_{\mathbb{A}}$ and $\mathcal{A}^\bullet$, respectively, are called an *existence mode* and a *presence mode* of $\mathbb{X}$, respectively.

Remark 5.4. In other words, W_ϕ and $\overline{W}_\phi$ are the sets

$$W_\phi = \{\psi \in \mathsf{X} \quad : \quad \psi = \mathsf{x}(\xi, \phi), \quad \forall \xi \in M_{\mathbb{A}}\},$$
$$\overline{W}_\phi = \{\psi \in \mathsf{X} \quad : \quad \psi = \bar{\mathsf{x}}(\xi, \phi), \quad \forall \xi \in \mathcal{A}^\bullet\}. \quad (17)$$

Intuitively, each presence mode of an $\mathbb{A}$-phenomenon is the perceptible part of one of its existence modes.

Definition 5.6. An $\mathbb{A}$-phenomenon $\mathbb{X}$, together with a map $\sigma : \overline{W} \to \mathcal{A}^\bullet$ for each presence mode $\overline{W}$, is called a *stable phenomenon* if each σ is a bijection. We refer to σ as a *proper view* of $\overline{W}$. For a stable phenomenon, the propensity modes $\phi \in (W_\phi \backslash)\overline{W}_\phi$ are called *(im)perceptible*.

Definition 5.7. For an $\mathbb{F}$-observer, a rational paradigm $\mathbb{A}$ is *consistent* if the $\mathbb{A}$-universe is a Boolean topos; a consistent paradigm of maximal dimensionality is called a *home* paradigm of the $\mathbb{F}$-observer.

Remark 5.5. Intuitively, the consistency condition requires the logic of the $\mathbb{A}$-universe to match the logic of the metauniverse (the metalogic of the $\mathbb{F}$-observer). $\mathbb{F}$-observers without home paradigms, referred to as *Wanderers*, may or may not be of interest, but our main concern will be precisely with home paradigms, and more specifically with home paradigms of the $\mathbb{R}$-observer and their $\mathbb{A}$-universes, due to the following result.

Theorem 5.1 (Trifonov, 1995). *Every home paradigm of the $\mathbb{R}$-observer is isomorphic to the quaternion algebra $\mathbb{H}$ with a family of Minkowski sensory forms.*

Definition 5.8. For an $\mathbb{F}$-observer, a reality $\mathcal{R} = (\mathbb{A}, \boldsymbol{R})$ is called *stable* if it is immanent and $\mathbb{A}$ is a rational paradigm; otherwise the reality is called *unstable* (or *virtual*).

Remark 5.6. Given a stable realty $\mathcal{R} = (\mathbb{A}, \boldsymbol{R})$ we refer to the logic of the $\mathbb{A}$-universe also as the *logic of* $\mathcal{R}$.

Definition 5.9. For an $\mathbb{F}$-observer with a temporal template $(\mathbb{F}, \preceq)$, a reality $\mathcal{R} = (\mathbb{A}, \boldsymbol{R})$, together with an $\mathbb{F}$-valued map $\mathsf{T} : \mathbb{A} \longrightarrow \mathbb{F}$, is called a *temporal reality* if it is stable and

$$a\boldsymbol{R}b \iff (\mathsf{T}(a) \preceq \mathsf{T}(b) \wedge a \neq b) \vee (a = b), \forall a, b \in \mathbb{A}. \tag{18}$$

Otherwise the reality is called *atemporal*. The ontology of a temporal reality is called its *temporal order*, and the map T is referred to as the *global time* of the reality. The *perceptible time* of $\mathcal{R}$ is the restriction, $\mathcal{T} : \mathcal{A}^\bullet \longrightarrow \mathbb{F}$, of T to the perception domain.

Remark 5.7. It should be emphasized that a temporal reality is defined only with respect to a certain temporal template. Some $\mathbb{F}$-observers may have several temporal templates, and some may have none. For example, since the $\mathbb{C}$-observer has no temporal templates, all realities of such an observer are atemporal. It is easy to see that a temporal template and a global time uniquely determine the temporal order of $\mathcal{R}$.

Definition 5.10. For an $\mathbb{F}$-observer and a presence mode, $\overline{W}$, of his stable $\mathbb{A}$-phenomenon, the pullback, $\mathcal{T}_{\overline{W}} := \mathcal{T} \circ \sigma$, of the perceptible time $\mathcal{T}$ under the proper view σ is referred to as the *perceptible time of* $\overline{W}$.

Definition 5.11. For an $\mathbb{F}$-observer, let $\mathbb{X}$ be a stable $\mathbb{A}$-phenomenon, $\mathcal{R} = (\mathbb{A}, \boldsymbol{R})$ a temporal reality, and T a map $\mathsf{X} \longrightarrow \mathbb{F}$. The ordered triple $\mathbf{X} = (\mathbb{X}, \mathcal{R}, T)$ is called a *realization* (*of* $\mathbb{X}$ *in* $\mathcal{R}$) if the following diagram commutes for each $\overline{W}$,

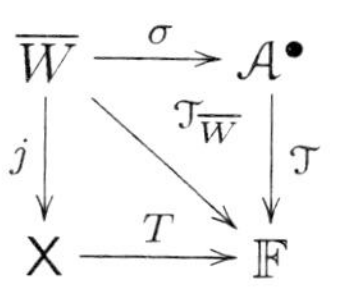

$$\begin{array}{ccc} \overline{W} & \xrightarrow{\sigma} & \mathcal{A}^\bullet \\ {\scriptstyle j}\downarrow & \searrow^{\mathcal{T}_{\overline{W}}} & \downarrow{\scriptstyle \mathcal{T}} \\ \mathsf{X} & \xrightarrow{T} & \mathbb{F} \end{array}$$

where j is the inclusion map. We refer to $\mathbb{X}$, $\mathcal{R}$ and T as the *underlying phenomenon*, the *background reality* and the *ambient time* of the realization, respectively.

6. Observers

Due to the results of [19] and [20], in the remainder of the paper we shall deal exclusively with the $\mathbb{R}$-observer, henceforth referred to simply as the *observer*. If not mentioned explicitly, it is assumed in the following that the constructs under consideration always refer to the observer.

Lemma 6.1. *Any rational paradigm of the observer is a unital algebra.*

Proof. Since nonzero elements of $\mathbb{A}$ obey associative multiplication, nonassociativity can occur only in the permutations of $(\boldsymbol{ab})\mathbf{0}$, which is impossible since $\boldsymbol{b}\mathbf{0} = \mathbf{0}, \forall \boldsymbol{b} \in \mathbb{A}$. Thus, $\mathbb{A}$ is associative, finite dimensional, and the identity of the motor domain is the identity of $\mathbb{A}$. Therefore it is unital. □

Corollary 6.1. *The perception domain $\mathcal{A}^\bullet$ of a rational paradigm $\mathbb{A}$ is a Lie group with respect to the multiplication of $\mathbb{A}$.*

Definition 6.1. For a rational paradigm $\mathbb{A}$, the linear manifold $\mathcal{A}$, canonically generated by the sensory domain A is called the *sensory manifold* of $\mathbb{A}$, and each reflexor $a \in \mathcal{A}$ is called a *viewpoint*. Viewpoints $a \in (\mathcal{A}\backslash)\mathcal{A}^\bullet$ are called *(im)proper*.

Definition 6.2. For a rational paradigm $\mathbb{A}$, a reflexor $\boldsymbol{u} \in A$, and a proper viewpoint $a \in \mathcal{A}^\bullet$, the integral curve, through a, of the left invariant vector field generated by $\boldsymbol{u}$ is called a *$(\boldsymbol{u}, a)$-vista*.

Remark 6.1. Intuitively, $(\boldsymbol{u}, a)$-vistas indicate naturally distinguished directions within the perception domain.

Lemma 6.2. *The observer has a unique temporal template.*

Proof. Indeed, there is a unique complete linear order on $\mathbb{R}$, namely the standard order $\leq$ (see, e. g., [4], p. 245). Thus, $(\mathbb{R}, \leq)$ is unique. □

Remark 6.2. The previous result makes it unnecessary to mention the temporal order explicitly, and we use the simplified notation $\mathcal{R} = (\mathbb{A}, \mathsf{T})$ for a temporal reality of the observer.

Remark 6.3. It is shown in [19] that besides the home paradigm, the observer has exactly two (up to an $\mathbb{R}$-algebra isomorphism) consistent paradigms, namely the one dimensional $\mathbb{R}$-algebra of reals, $\mathbb{R}$, and the two dimensional $\mathbb{R}$-algebra of complex numbers, $\mathbb{C}$, both subalgebras of $\mathbb{H}$.

Definition 6.3. A reality is called *robust* if it is temporal and there exists a principal metric, $\mathfrak{T}$, on the perception domain $\mathcal{A}^\bullet$ of the underlying paradigm with the perceptible time $\mathcal{T}$ as its generating function. Given a robust reality $\mathcal{R} = (\mathbb{A}, \mathsf{T})$, the structure field of $\mathcal{A}^\bullet$ is called the *structure (field)* of the reality, and we refer to $d\mathcal{T}$ and $\mathfrak{T}$ as the *ether (field)*, and the *metric* of $\mathcal{R}$, respectively. The ordered pair $\mathcal{S} = (\mathcal{A}^\bullet, \mathfrak{T})$ is called the *spacetime* of $\mathcal{R}$.

Definition 6.4. A realization $\mathbf{X} = (\mathbb{X}, \mathcal{R}, T)$ is called *robust* if the background reality is robust, and the perceptible part of the underlying phenomenon $\mathbb{X}$ is a principal $\mathcal{A}^\bullet$-bundle (X), such that the propensity realm X is its total space and each proper view $\sigma : \overline{W} \to \mathcal{A}^\bullet$

is a fiber diffeomorphism. The dimensionality of X is referred to as the *rank* of $\mathbf{X}$. For a presence mode, $\overline{W}$, of the underlying phenomenon the pullback $\mathcal{T}_{\overline{W}}$, of the metric of the background reality under the proper view σ is called the *metric* of $\overline{W}$. The ordered pair $\mathcal{W} = (\overline{W}, \mathcal{T}_{\overline{W}})$ is referred to as a *(possible) world* of $\mathbf{X}$. The bundle (X) is called the *monocosm* of the realization.

7. Dynamical Systems

Definition 7.1. A realization $\mathbf{X} = (\mathbb{X}, \mathcal{R}, T)$ is called a *dynamical system* if it is robust and its propensity realm is a Riemannian manifold $\mathcal{X} = (\mathsf{X}, \boldsymbol{g})$. The Riemannian metric $\boldsymbol{g}$ is referred to as the *propensity metric* of $\mathbf{X}$.

Remark 7.1. There is a natural connection on the monocosm $(\mathcal{X})$ of a dynamical system: the horizontal space at any point ψ is defined as the set of tangent vectors orthogonal to the world $\mathcal{W}_\psi$ with respect to the propensity metric. We refer to this connection as the *fundamental connection* of the dynamical system.

Definition 7.2. For a dynamical system $\mathbf{X} = (\mathbb{X}, \mathcal{R}, T)$ a *perceptible* is a smooth map $f : \mathcal{X} \to \mathcal{A}^\bullet$. A *temporal evolution* of a dynamical system $\mathbf{X}$ is an integral curve of a vector field $\boldsymbol{f}_T$ on $\mathcal{X}$, called the *temporal evolution vector field* of $\mathbf{X}$, such that

$$(dT)(\boldsymbol{u}) = \boldsymbol{g}(\boldsymbol{f}_T, \boldsymbol{u}), \quad \forall \boldsymbol{u} \in \mathcal{X}[{}^1_0], \tag{19}$$

The propensity realm of a dynamical system is referred to as its *state space*, and propensity modes are called *states*. A state ψ such that the vector $\boldsymbol{f}_T(\psi)$ is vertical is called the *proper* state of $\mathbf{X}$, and the possible world $\mathcal{W}_\psi$ is called an *accessible world*.

Remark 7.2. A perceptible is a smooth restriction of an attribute of the underlying phenomenon to the perception domain. Intuitively, a temporal evolution of a dynamical system is the motion of the observer's proper viewpoint across possible worlds of the system along its temporal evolution vector field. At each point of an evolution the observer encounters a possible world, a diffeomorphically perturbed copy of the spacetime of the background reality, which contains perceptible information about the system.

Definition 7.3. For a dynamical system $\mathbf{X}$, an ordered triple (f, ϕ, ψ), where f is a perceptible and ϕ, ψ are states, is called an *f-observation*. The states ϕ and ψ are called the *initial* and the *final* states, and the worlds $\mathcal{W}_\phi$ and $\mathcal{W}_\psi$ are called the *source* and the *target* worlds of the f-observation, respectively. The *propensity* $\rho(\phi, \psi) \in \mathbb{R} \cup \{\infty\}$ of an f-observation (f, ϕ, ψ) is defined by

$$\rho(\phi, \psi) := \begin{cases} \infty & : \ \mathcal{W}_\phi = \mathcal{W}_\psi \\ d^{-1} & : \ \mathcal{W}_\phi \neq \mathcal{W}_\psi \end{cases}, \tag{20}$$

where d is the length of the shortest geodesic between ϕ and ψ.

Remark 7.3. Intuitively, propensity roughly quantifies accessibility of the target world - the father away, the less accessible it is.

8. Cosmologies

Remark 8.1. A left invariant vector field $\hat{\boldsymbol{u}}$ on $\mathcal{H}^\circ$, generated by a vector $\boldsymbol{u} \in H$ with the components (u^β) in a canonical basis $(\boldsymbol{i}_\beta)$, assigns to each point $a \in \mathcal{H}^\circ$ with canonical coordinates (w, x, y, z) a vector $\hat{\mathbf{u}}(a) \in T_a\mathcal{H}^\circ$ with the components $\hat{u}^\beta(a) = (\boldsymbol{a}\boldsymbol{u})^\beta$ in the basis $(\partial_w, \partial_x, \partial_y, \partial_z)(a)$ on $T_a\mathcal{H}^\circ$:

$$\hat{u}^0(a) = wu^0 - xu^1 - yu^2 - zu^3, \quad \hat{u}^1(a) = wu^1 + xu^0 + yu^3 - zu^2,$$
$$\hat{u}^2(a) = wu^2 - xu^3 + yu^0 + zu^1, \quad \hat{u}^3(a) = wu^3 + xu^2 - yu^1 + zu^0. \quad (21)$$

Definition 8.1. A robust reality is called a *cosmology* if its underlying paradigm is isomorphic to the home paradigm of the observer.

Remark 8.2. As follows from Theorem 4.1 of [20], the choice of cosmologies of the observer is extremely limited. In fact, there is a unique, up to the functional variable T, cosmology, $(\mathbb{H}, \mathsf{T})$. Let us review its basic properties.

1. The perception domain $\mathcal{H}^\circ$ of the underlying paradigm $\mathbb{H}$ is the Lie group of nonzero quaternions with the $\mathbb{R} \times \mathbf{S}^3$ topology, the product of the real line and a three-sphere.

2. The spacetime $\mathcal{S} = (\mathcal{H}^\circ, \mathfrak{T})$ of the cosmology is a smooth four dimensional manifold with a closed FLRW metric.

3. Given a canonical sensory basis $(\boldsymbol{i}_\beta)$ and, for a differentiable function $R : \mathbb{R} \to \mathbb{R} \setminus \{0\}$, a system of natural spherical coordinates $(\eta, \chi, \theta, \varphi)$ on $\mathcal{H}^\circ$, related to the canonical coordinates by

$$w = R(\eta)\cos(\chi), \quad x = R(\eta)\sin(\chi)\sin(\theta)\cos(\varphi),$$
$$y = R(\eta)\sin(\chi)\sin(\theta)\sin(\varphi), \quad z = R(\eta)\sin(\chi)\cos(\theta),$$

the metric has the following components in the spherical frame $(\partial^R_\eta, \partial^R_\chi, \partial^R_\theta, \partial^R_\varphi)$

$$\mathcal{T}_{\alpha\beta} = diag(1, -\mathsf{a}^2, -\mathsf{a}^2\sin^2\chi, -\mathsf{a}^2\sin^2\chi\sin^2\theta), \quad \mathsf{a} := \sqrt{|\dot{\mathcal{T}}|},$$
$$\text{with} \quad R = \exp\int \frac{d\eta}{\pm\sqrt{|\dot{\mathcal{T}}|}}. \quad (22)$$

4. The perceptible time $\mathcal{T}$ is a monotonous function of η.

5. The structure field of the cosmology has the following components in the frame $(\partial^R_\eta,$

$\partial^R_\chi, \partial^R_\theta, \partial^R_\varphi$):

$$\mathcal{H}^0_{\alpha\beta} = \begin{pmatrix} \lambda & 0 & 0 & 0 \\ 0 & -\lambda^{-1} & 0 & 0 \\ 0 & 0 & -\lambda^{-1}\sin^2\chi & 0 \\ 0 & 0 & 0 & -\lambda^{-1}\sin^2\chi\sin^2\theta \end{pmatrix},$$

$$\mathcal{H}^1_{\alpha\beta} = \begin{pmatrix} 0 & \lambda & 0 & 0 \\ \lambda & 0 & 0 & 0 \\ 0 & 0 & 0 & \sin^2\chi\sin\theta \\ 0 & 0 & -\sin^2\chi\sin\theta & 0 \end{pmatrix},$$

$$\mathcal{H}^2_{\alpha\beta} = \begin{pmatrix} 0 & 0 & \lambda & 0 \\ 0 & 0 & 0 & -\sin\theta \\ \lambda & 0 & 0 & 0 \\ 0 & \sin\theta & 0 & 0 \end{pmatrix}, \quad \mathcal{H}^3_{\alpha\beta} = \begin{pmatrix} 0 & 0 & 0 & \lambda \\ 0 & 0 & 1/\sin\theta & 0 \\ 0 & -1/\sin\theta & 0 & 0 \\ \lambda & 0 & 0 & 0 \end{pmatrix},$$

where $\lambda := \dot{R}/R$.

6. The ether field of the cosmology has the following components in the frame ($\partial^R_\eta, \partial^R_\chi$ $\partial^R_\theta, \partial^R_\varphi$):

$$d\mathcal{T} = (\dot{\mathcal{T}}, 0, 0, 0).$$

7. For a canonical sensory basis $(\boldsymbol{i}_\beta)$ and the corresponding canonical coordinate system (w, x, y, z), let $\boldsymbol{u}$ be a reflexor with the components (u^β) in $(\boldsymbol{i}_\beta)$, and a - a proper viewpoint with coordinates $(\bar{w}, \bar{x}, \bar{y}, \bar{z})$. Then the $(\boldsymbol{u}, a)$-vista can be computed by solving the system of differential equations (21) with a parameter t:

$$w(t) = \exp(u^0 t)(\frac{-u^1\bar{x} - u^2\bar{y} - u^3\bar{z}}{\omega}\sin\omega t + \bar{w}\cos\omega t)$$
$$x(t) = \exp(u^0 t)(\frac{u^1\bar{w} - u^2\bar{z} + u^3\bar{y}}{\omega}\sin\omega t + \bar{x}\cos\omega t)$$
$$y(t) = \exp(u^0 t)(\frac{u^1\bar{z} + u^2\bar{w} - u^3\bar{x}}{\omega}\sin\omega t + \bar{y}\cos\omega t)$$
$$z(t) = \exp(u^0 t)(\frac{-u^1\bar{y} + u^2\bar{x} + u^3\bar{w}}{\omega}\sin\omega t + \bar{z}\cos\omega t), \quad (23)$$

where $\omega := \sqrt{(u^1)^2 + (u^2)^2 + (u^3)^2}$.

Remark 8.3. The theory of the observer we have developed so far is left invariant (utilizing left invariant vector fields on perception domains). It is easy to show that the metric of the right invariant cosmology coincides with (22), but the $(\boldsymbol{u}, a)$-vistas are different. This can be used, in principle, by the observer to determine the "chirality" of his contemporary reality.

9. Physical Systems

Definition 9.1. A dynamical system $\mathbf{X} = (\mathbb{X}, \mathcal{R}, T)$ is called a *physical system* if $\mathcal{R}$ is a cosmology and the state space is a hyper-Kähler manifold.

Definition 9.2. For a physical system $\mathbf{X} = (\mathbb{X}, \mathcal{R}, T)$, a quaternionic regular perceptible f is called an *observable*. For an observable f, we refer to its hyper-Hamiltonian vector field $\boldsymbol{f}$ as the *f-field*. A state ψ such that the vector $\boldsymbol{f}(\psi)$ is vertical is called an *f-proper* state of $\mathbf{X}$. A possible world $\mathcal{W}$ is called *f-(in)accessible* iff there is (not) an f-proper state ψ, such that $\mathcal{W} = \mathcal{W}_\psi$. For each f-proper state ψ the value $f(\psi)$ is called a *relative perceptible property* of $\mathbf{X}$.

Remark 9.1. The above definition generalizes the notions of GQM, where points at which the Schrödinger vector field of an observable becomes vertical, and the corresponding values of f parametrize the eigenvectors and eigenvalues of $\widehat{\boldsymbol{F}}$, respectively [18].

Definition 9.3. For a physical system $\mathbf{X} = (\mathbb{X}, \mathcal{R}, T)$, its *Hamiltonian* is an observable h such that the temporal evolution vector field of $\mathbf{X}$ coincides with the h-field.

Remark 9.2. A physical system can be thought of as a *sufficiently smooth fine-graining* of a cosmology. It is a natural generalization of the notion of a quantum system of GQM.

Definition 9.4. For a physical system, an f-observation (f, ϕ, ψ) is called *successful* if f is an observable, the propensity $\rho(\phi, \psi)$ exists, and the target world $\mathcal{W}_\psi$ is f-accessible, in which case we refer to $\mathcal{W}_\psi$ as the *actual world* of the f-observation. Otherwise the f-observation is called *failed* (or *unsuccessful*), and the world $\mathcal{W}_\psi$ is called *virtual*.

Remark 9.3. A successful f-observation is also referred to as a *measurement*.

Definition 9.5. For a measurement (f, ϕ, ψ) its *result* is an ordered pair $(\mathcal{W}_\psi, \mathcal{T}[\psi])$, where $\mathcal{W}_\psi$ is the actual world and $\mathcal{T}[\psi]$ is the level set of the perceptible time of $\mathcal{W}_\psi$ containing ψ. We refer to $\mathcal{T}[\psi]$ as the *hypersurface of the present*. The value $f(\psi)$ is called a *relative observable property* of the physical system.

Remark 9.4. Intuitively, the final state of an f-measurement marks the "landing spot" of the observer (or, more correctly, of his proper viewpoint) in a (target) world whose properties may differ, in a strictly defined sense, from the respective properties of the source world. Propensity carries information about the "likelihood" of the "collapse" of ϕ to ψ: the farther away ψ is from ϕ, the less likely it is that the final state is ψ. We are to think of a relative observable property as a generalization of a non-normalized eigenvalue of an observable in CQM. The main difference between the two notions is this: instead of a single real number it is an element of the quaternion algebra, which, given a canonical sensory basis, is an ordered quadruple of real numbers, the first of which can be used to represent the "landing" moment in the ambient time of the system, and the other three - to quantify various features of the system. Indeed, for a measurement to be a meaningful procedure, the time of its occurrence must be known to the observer. Even then, if a measurement of, say, a position of a particle in CQM results in a particular value of one of the coordinates, the position of the particle is still unknown unless the other two coordinates are known.

Definition 9.6. A physical system is called a *hyperquantum system* if its monocosm is a hyperquantum bundle $(\mathcal{V}^{\circ}_{\mathbb{H}})$ over a quaternionic Hilbert space $\mathbb{V}$, and the imaginary part $\mathfrak{Im}(h)$ of its Hamiltonian coincides with the restriction to $\mathcal{V}^{\circ}_{\mathbb{H}}$ of the expectation of a quaternionic anti-Hermitian operator $\widehat{\boldsymbol{H}}$ on $\mathbb{V}$.

Remark 9.5. It seems tempting to obtain quantum systems of GQM by demanding the existence of a canonical sensory basis in which the Hamiltonian has a unique nonzero constituent h_p. However, this would not be quite correct because within our framework the underlying paradigm of such systems would correspond to the two dimensional consistent (Boolean) paradigm $\mathbb{C}$, and their background reality is not a cosmology. In this sense quantum systems are a *degenerate* case of hyperquantum systems: two out of four dimensions are collapsed in each possible world.

Definition 9.7. A dynamical system is called a *quantum system* if its background reality is $(\mathbb{C}, \top)$, its monocosm is a quantum bundle $(\mathcal{V}^{\circ}_{\mathbb{C}})$ over a complex Hilbert space $\mathbb{V}$, and the complex imaginary part of its Hamiltonian coincides with the restriction to $\mathcal{V}^{\circ}_{\mathbb{C}}$ of the expectation of a complex anti-Hermitian operator $\widehat{\boldsymbol{H}}$ on $\mathbb{V}$.

Remark 9.6. It is a standard result in GQM that a Schrödinger evolution of a quantum system is also a Hamiltonian evolution with the expectation of $\widehat{\boldsymbol{H}}$ as its generating function [18], so the above definition is equivalent to the description of a quantum system in GQM. For a measurement (f, ϕ, ψ) of a quantum system the propensity $\rho(\phi, \psi)$ can be expressed in terms of probability of obtaining a particular result ([6] and references therein).

Example 9.1. The *Old World* Υ . The monocosm of Υ is the *degenerate* hyperquantum bundle with a single possible world, hence the source, target and actual worlds coincide for each measurement, and the ambient time of Υ is the perceptible time of the background cosmology. A temporal evolution of the system is orthogonal to the level sets of the perceptible time with respect to the propensity metric of Υ. Since the evolution vector field is vertical everywhere, every state of the system is proper. For each measurement of Υ the metric of its actual world is the metric of the background cosmology. In conventional terms this hyperquantum system seems to correspond to the classical coarse-grained view of the universe - an observer at rest relative to CMB in the spacetime of a spatially closed FLWR cosmology.

10. Conclusion

As we mention in the introduction, the technical purpose of the paper is to provide *formal definitions* of observer related notions which are normally considered too ambiguous for constructive discussion within mainstream physics. Below is an informal overview of some of them.

The *observer* is represented by an *existence mode* of a *Boolean experient*, and is capable of perceiving various *realities*, each based on a *paradigm*. In some realities the observer tends to distinguish *dynamical systems*, collections of experiences stable in a strictly defined sense. Dynamical systems spend most of their *ambient time* roaming their *possible worlds* according to the equation (19) which is the main dynamical equation of the modification.

An *observation* of a dynamical system perturbs its evolution resulting occasionally in a creation of an *observable property* of the system with respect to its *actual world*. For reasons that are beyond the scope of this theory, the contemporary *operational logic* of the observer seems to be bivalent Boolean *(Assertion 1)*, and hence the largest immediate environment conforming to this requirement corresponds to a *robust reality* of his *home paradigm* (a *cosmology*). Then the kinematic axioms of GR follow: the spacetime of the cosmology is a smooth manifold, because it is a Lie group, its dimensionality is indeed four, and it has a Lorentzian metric of a very special type (closed FLRW). It is curious that the requirement of Booleanity alone is sufficient, and bivalence follows (see [12], p. 121). The internal mathematics of the $\mathbb{H}$-universe, although Boolean, is nonclassical: an important version of the axiom of choice fails in the topos $\mathcal{H}^{\circ}\boldsymbol{Set}$ (see [12], p. 301), making the basic tool for obtaining, say, countable additivity of the Lebesgue measure unavailable to the observer. Since spacetime acquires a locally compact Lie group structure, the observer can use the Haar measure whose properties are less dependent on the axiom of choice and can be described constructively ([7], [3]). The nonstandard integration over spacetime may have some application to the problem of apparently non-Newtonian behavior of large structures in contemporary observational astrophysics (see [16] and references therein).

Acknowledgments. The author is grateful to the reviewers whose comments helped reduce the volume and improve clarity of the exposition.

References

[1] Aczel, P. (1997). *Lectures on Semantics*: The initial algebra and final coalgebra perspectives, in Logic of Computation, edited by H. Schwichtenberg, Springer.

[2] Adler, S. L. (1995). *Quaternionic Quantum Mechanics and Quantum Fields*, Oxford University Press, Oxford, UK.

[3] Alfsen, E. M. (1963). *Math. Scand.*, **12**, 106-116.

[4] Balcar, B., Jech, T. (2006). *Bull. of Symb. Logic*, **12** (2), 241-266.

[5] Block, N., Flanagan, O., Güzeldere, G. (1997). *The Nature of Consciousness*: Philosophical Debates, MIT Press, Cambridge, MA.

[6] Brody, D. C., Hughston L. P. (2001). *J. of Geom. and Phys.*, **38** 19-53.

[7] Cartan, H. (1940). *C. R. Acad. Sci. Paris*, **211**, 759-762.

[8] Chalmers, D. (1996). *The Conscious Mind*, Oxford University Press, New York, NY.

[9] Chen, J., Li, J. (2000). *J. of Diff. Geom.* **55**, 355-384.

[10] Fodor, J. A. (1975). *The Language of Thought*, Harvard University Press, Cambridge, MA.

[11] Gaeta, G., Morando, P. (2002). *J. of Phys.*, **A35**, 3925-3943.

[12] Goldblatt, R. (1984). *Topoi*, revised edition, in *Studies in Logic and the Foundations of Mathematics*, Vol. 98, North Holland, New York, NY.

[13] Hughes, J. (2001). *A study of categories of algebras and coalgebras*, PhD Thesis, Carnegie Mellon University.

[14] Kibble, T. W. B. (1979). *Comm. in Math. Phys.*, **65**, 189-201.

[15] Курош, А. Г. (1973). Лекции по Общей Алгебре, Второе издание, Наука, Москва.

[16] Massey, R., et al (2007). [astro-ph/0701594].

[17] Morando, P., Tarallo, M. (2003). *Mod. Phys. lett. A*, **18** (26), 1841-1847.

[18] Schilling, T. A. (1996). *Geometry of Quantum Mechanics*, PhD Thesis, Pensylvania State University.

[19] Trifonov, V. (1995). *Europhys. Lett.*, **32** (8), 621-626.

[20] Trifonov, V. (2007). *Int. J. Theor. Phys.*, **46** (2), 251-257.

[21] Turi, D. (1996). *Functorial Operational Semantics*, PhD Thesis, Free University, Amsterdam.

In: Focus on Quantum Mechanics
Editors: David E. Hathaway et al. pp. 35-61
ISBN 978-1-62100-680-0

Chapter 3

DEFORMED QUANTUM FIELD THEORY, THERMODYNAMICS AT LOW AND HIGH ENERGIES, AND DARK ENERGY PROBLEM

Alexander E. Shalyt-Margolin*
Laboratory of the Quantum Field Theory,
National Center of Particles and High Energy Physics,
Bogdanovich Str. 153, Minsk 220040, Belarus

Abstract

Quantum Mechanics of the Early Universe (Planck scale) is considered as a Quantum Mechanics with a fundamental (minimal) length and, at the same time, as a deformation of the conventional Quantum Mechanics. In this context, various approaches to deformation (or different deformations) of the corresponding Heisenberg Algebra are discussed. It is shown that there is a new small dimensionless parameter, having with a reciprocal quantity an explicit physical interpretation, in terms of which the above-mentioned Heisenberg Algebra deformations are represented. It is shown that this parameter appears in thermodynamics as well. Because of this, it is demonstrated that entropy and its density play a significant role in solving the problem of the vacuum energy density (cosmological constant) of the Universe and hence the dark energy problem. It is shown that the fundamental quantities in the Holographic Dark Energy Models may be expressed in terms of a new small dimensionless parameter. It is revealed that this parameter is naturally occurring in Gravitational Thermodynamics. Using the indicated parameter within the scope of the Generalized Uncertainty Principle, a high-energy deformation model is constructed for the particular case of Einstein's equations (Ultraviolet-limit of General Relativity). On this basis the possibility of a new approach to the problem of Quantum Gravity is considered. Besides, the results obtained on the uncertainty relation of the pair "cosmological constant - volume of space-time", where the cosmological constant is a dynamic quantity, are

*E-mail address: a.shalyt@mail.ru; alexm@hep.by. Phone (+375) 172 926034; Fax (+375) 172 926075

reconsidered and generalized up to the Generalized Uncertainty Relation.
This paper is devoted to the 80-th anniversary of Corresponding Member of Belarus National Academy of Science, Professor Lev Mitrofanovich Tomilchik.

PACS: 03.65; 05.30

Keywords: quantum field theory with UV cutoff, new small parameters,dark energy problem, gravitational thermodynamics

1. Introduction

In the last decades researchers have acquired an understanding that physical studies of the Early Universe (extremely high Plancks energies) necessitate changing of the fundamental physical theories and, in particular, quantum mechanics and quantum field theory. Inevitably, these theories should involve the notion of a fundamental length. In so doing the correspondence principle should be observed: at well-known low energies the theories with a fundamental length must represent the conventional quantum mechanics and quantum field theory with a high precision. The need for inclusion of a fundamental length into a quantum theory at Planck scales has been understood for a long time since publishing of the works devoted to a theory of strings [1]. However, this theory is still considered to be a trial theory, and its suggestions might be found inadequate. Fortunately, by now numerous publications have appeared to show, with the use of various approaches, the emergence of a fundamental length in the Early Universe [2]–[5]. Of great interest is the work [2],where on the basis of a simple gedanken experiment it has been shown that, with due regard for the gravitational interactions exhibited in the Early Universe only (Planck scales), the Heisenberg Uncertainty Principle should be extended to the Generalized Uncertainty Principle [1]–[5], in turn unavoidably bringing forth a fundamental length on the order of the Plancks length. The advent of novel theories in physics of the Early Universe is based on new parameters, i.e. on deformation of the well-known theories. The deformation is understood as an extension of a particular theory by inclusion of one or several additional parameters in such a way that the initial theory appears in the limiting transition [9]. As this takes place, the Heisenberg Algebra is, no doubt, subjected to the corresponding deformation. The basis for such a deformation may be provided by both the Generalized Uncertainty Principle [6]–[8] and the density matrix deformation [32]–[40].
In this work it is demonstrated that the new parameter α introduced into the Quantum Field Theory (QFT) with UV cutoff (fundamental length) produced by the density matrix deformation in Quantum Mechanics (QM) gives an adequate estimate of the vacuum energy density, that in turn may be directly applied to solve the dark energy problem. In this case entropy of the Universe and its dynamics play a significant role. Additionally, within the scope of a dynamic approach to Λ, its behavior associated with the Generalized Uncertainty Principle is studied for the pair "cosmological constant - volume of space-time". In what follows, there is no differentiation between the notions of the cosmological constant Λ and Vacuum Energy Density ρ_{vac}. Besides, it is demonstrated that a new small parameter occurs in Gravitational Thermodynamics. On this basis the possibility for a new approach to the problem of Quantum Gravity is discussed. Specifically, with the use of the indicated

parameter within the scope of the Generalized Uncertainty Principle, a high-energy deformation model is constructed for the specific case of Einstein's equations (Ultraviolet-limit of General Relativity).

2. Quantum Mechanics and Quantum Field Theory with UV Cutoff

As it has been repeatedly demonstrated earlier, a Quantum Mechanics of the Early Universe (Plank Scale) is a Quantum Mechanics with the Fundamental Length (QMFL)[4]–[7]. The main approach to framing of QFT with UV cutoff is that associated with the Generalized Uncertainty Principle (GUP) [1]–[8] and with the corresponding Heisenberg algebra deformation produced by this principle [6]–[8],[31]. Besides, in the works by the author [32]–[41] an approach to the construction of QMFL has been developed with the help of the deformed density matrix, the density matrix deformation in QMFL being a starting object called the density pro-matrix and deformation parameter (additional parameter) $\alpha = l_{min}^2/x^2$, where x is the measuring scale and $l_{min} \sim l_p$.
Exact definition will be as follows: [32],[33],[40]:

Definition 1. (Quantum Mechanics with the Fundamental Length [for Neumann's picture])

Any system in QMFL is described by a density pro- matrix of the form

$$\rho(\alpha) = \sum_{\mathbf{i}} \omega_{\mathbf{i}}(\alpha)|\mathbf{i}><\mathbf{i}|,$$

where

1. Vectors $|i>$ form a full orthonormal system;
2. $\omega_i(\alpha) \geq 0$ and for all i the finite limit $\lim\limits_{\alpha\to 0}\omega_i(\alpha) = \omega_i$ exists;
3. $Sp[\rho(\alpha)] = \sum_i \omega_i(\alpha) < 1$, $\sum_i \omega_i = 1.$;
4. For every operator B and any α there is a mean operator B depending on α:
$$<B>_\alpha = \sum_i \omega_i(\alpha) <i|B|i>.$$
5. The following condition should be fulfilled:
$$Sp[\rho(\alpha)] - Sp^2[\rho(\alpha)] = \alpha + a_0\alpha^2 + \ldots \approx \alpha. \tag{1}$$
Consequently, we can find the value for $Sp[\rho(\alpha)]$ satisfying the above-stated condition:
$$Sp[\rho(\alpha)] \approx \frac{1}{2} + \sqrt{\frac{1}{4} - \alpha} \tag{2}$$
and therefore

6. $0 < \alpha \leq 1/4$.

It is no use to enumerate all the evident implications and applications of **Definition 1**, better refer to [32],[33]. Nevertheless, it is clear that
for $\alpha \to 0$ the above limit covers both the Classical and Quantum Mechanics depending on $\hbar \to 0$ or not.
The explicit form of the above-mentioned deformation gives an exponential ansatz:

$$\rho^*(\alpha) = exp(-\alpha)\sum_i \omega_i |i><i|, \tag{3}$$

where all $\omega_i > 0$ are independent of α and their sum is equal to 1.
In the corresponding deformed Quantum Theory (denoted as QFT^{α}) for average values we have

$$<B>_{\alpha} = exp(-\alpha) <B>, \tag{4}$$

where $<B>$ - average in well-known QFT [37],[38]. All the variables associated with the considered α - deformed quantum field theory are hereinafter marked with the upper index α.
Note that the deformation parameter α is absolutely naturally represented as a ratio between the squared UV and IR limits

$$\alpha = (\frac{UV}{IR})^2, \tag{5}$$

where UV is fixed and IR is varying.

3. Vacuum Energy Density and New Small Parameters

The Vacuum Energy is currently a major candidate for the Dark Energy. At the same time, due to a factor of 10^{123} distinction between the experimental value ρ_{vac}^{exp} [10] and the value ρ_{vac}^{QFT} calculated using standard QFT [19],[20]: $\rho_{vac}^{QFT} \approx m_p^4$

$$\frac{\rho_{vac}^{exp}}{\rho^{QFT}} \approx 10^{-123}, \tag{6}$$

interpretation of Dark Energy as a Vacuum Energy presents great difficulties. But there are several methods enabling one to obviate the difficulties. We can name two most popular and acknowledged approaches.

3.1. Holographic Dark Energy Models

The basic relation for this model is the "energy" inequality [21]– [23]

$$E_{\overline{\Lambda}} \leq E_{BH} \rightarrow l^3\rho_{\overline{\Lambda}} \leq m_p^2 l. \tag{7}$$

Here $\rho_{\overline{\Lambda}} = \overline{\Lambda}^4$ – vacuum energy density with the UV cutoff $\overline{\Lambda}$ and l is the length scale (IR cutoff) of the system. For the equality in (7) we have the **holographic energy density**

$$\rho_{\overline{\Lambda}} \sim \frac{m_p^2}{l^2} \sim \frac{1}{(l_p l)^2}. \tag{8}$$

Also, from (7) we can get the "entropic" inequality (entropy bound)

$$S_{\overline{\Lambda}} \leq (m_p^2 A)^{3/4}, \tag{9}$$

where $A = 4\pi l^2$ is the area of this system in the spherically symmetric case.
The number of works devoted to the Holographic Dark Energy Models, beginning from the first publication [21], is ever growing [24] to relieve us from citing the whole list.

3.2. Agegraphic Dark Energy Models

Agegraphic Dark Energy Models became the subject of study only two years ago [25]. These relations were based on the result of Károlyházy for quantum fluctuations of time [26]–[28]

$$\delta t = \lambda t_{\mathrm{p}}^{2/3} t^{1/3}. \tag{10}$$

Using the uncertainty relation of "energy-time" in the flat space

$$\Delta E \sim t^{-1}, \tag{11}$$

we can obtain the **agegraphic energy density** [29], [23]

$$\rho_T \sim \frac{\Delta E}{(\delta t)^3} \sim \frac{m_{\mathrm{p}}^2}{\boldsymbol{T}^2}, \tag{12}$$

where $\boldsymbol{T}$ is an age of the Universe.
The number of publications associated with models of this type is constantly increasing too [30]. This is caused by their relative simplicity and by a sufficiently good coincidence of the agegraphic energy density ρ_T with ρ_{vac}^{exp}.

As follows from the holographic principle [42]–[45], maximum entropy that can be stored within a bounded region $\mathfrak{R}$ in 3-D space must be proportional to the value $A(\mathfrak{R})^{3/4}$, where $A(\mathfrak{R})$ is the surface area of $\mathfrak{R}$. Of course, this is associated with the case when the region $\mathfrak{R}$ is not an inner part of a particular black hole. Provided a physical system contained in $\mathfrak{R}$ is not bounded by the condition of stability to the gravitational collapse, i.e. this system is simply non-constrained gravitationally, then according to the conventional QFT $S_{\max}(\mathfrak{R}) \sim V(\mathfrak{R})$, where $V(\mathfrak{R})$ is the bulk of $\mathfrak{R}$. However in the Holographic Principle case, as it has been demonstrated originally by G. 't Hooft [42] and later by other authors (for example R. V. Buniy and S. D. H. Hsu [46]), we have

$$S_{\max}(\mathfrak{R}) \leq \frac{A(\mathfrak{R})^{3/4}}{l_p{}^2}. \tag{13}$$

In terms of the deformation parameter α introduced in Section 2, the principal values of the Vacuum Energy Problem may be simply and clearly defined. Let us begin with the Schwarzschild black holes, whose semiclassical entropy is given by

$$S = \pi R_{Sch}^2 / l_p^2 = \pi R_{Sch}^2 m_p^2 = \pi \alpha_{R_{Sch}}^{-1}, \tag{14}$$

with the assumption that in the formula for α $R_{Sch} = x$ is the measuring scale and $l_p = 1/m_p$. Here R_{Sch} is the adequate Schwarzschild radius, and $\alpha_{R_{Sch}}$ is the value of α associated with this radius. Then, as it has been pointed out in [47], in case the Fischler-Susskind cosmic holographic conjecture [48] is valid, the entropy of the Universe is limited by its "surface" measured in Planck units [47]:

$$S \leq \frac{A}{4} m_p^2, \tag{15}$$

where the surface area $A = 4\pi R^2$ is defined in terms of the apparent (Hubble) horizon

$$R = \frac{1}{\sqrt{H^2 + k/a^2}}, \tag{16}$$

with curvature k and scale a factors.
Again, interpreting R from (16) as a measuring scale, we directly obtain(15) in terms of α:

$$S \leq \pi \alpha_R^{-1}, \tag{17}$$

where $\alpha_R = l_p^2/R^2$. Therefore, the average entropy density may be found as

$$\frac{S}{V} \leq \frac{\pi \alpha_R^{-1}}{V}. \tag{18}$$

Using further the reasoning line of [47] based on the results of the holographic thermodynamics, we can relate the entropy and energy of a holographic system [49],[50]. Similarly, in terms of the α parameter one can easily estimate the upper limit for the energy density of the Universe (denoted here by ρ_{hol}) [51]:

$$\rho_{hol} \leq \frac{3}{8\pi R^2} m_p^2 = \frac{3}{8\pi} \alpha_R m_p^4, \tag{19}$$

that is drastically differing from the one obtained with well-known QFT

$$\rho^{QFT} \sim m_p^4. \tag{20}$$

Here by ρ^{QFT} we denote the energy vacuum density calculated from well-known QFT (without UV cutoff) [19]. Obviously, as α_R for R determined by (16) is very small, actually approximating zero, ρ_{hol} is by several orders of magnitude smaller than the value expected in QFT – ρ^{QFT}.
Since $m_p \sim 1/l_p$, the right-hand side of (19) is actually nothing else but the right-hand side of (8) in Holographic Dark Energy Models (subsection 2.1). Thus, in Holographic Dark Energy Models the principal quantity, **holographic energy density** $\rho_{\overline{\Lambda}}$ (8), may be estimated in terms of the deformation parameter α.
In fact, the upper limit of the right-hand side of (19) is attainable, as it has been demonstrated in [51] and indicated in [47]. The "overestimation" value of r for the energy density ρ^{QFT}, compared to ρ_{hol}, may be determined as

$$r = \frac{\rho^{QFT}}{\rho_{hol}} = \frac{8\pi}{3} \alpha_{\mathbf{R}}^{-1} = \frac{8\pi}{3} \frac{R^2}{l_p^2} = \frac{8\pi}{3} \frac{S}{S_p}, \tag{21}$$

where S_p is the entropy of the Plank mass and length for the Schwarzschild black hole. It is clear that due to smallness of α_R the value of α_R^{-1} is on the contrary too large. It may be easily calculated (e.g., see [47])

$$r = 5.44 \times 10^{122} \tag{22}$$

in a good agreement with the astrophysical data.
Naturally, on the assumption that the vacuum energy density ρ_{vac} is involved in ρ as a term

$$\rho = \rho_M + \rho_{vac}, \tag{23}$$

where ρ_M - average matter density, in case of ρ_{vac} we can arrive to the same upper limit (right-hand side of the formula (19)) as for ρ.

4. The Dynamical Approach to the Problem of Cosmological Constant and GUP

Generally speaking, Λ is referred to as a constant just because it is such in the equations, where it occurs: Einstein equations [15]. But in the last few years the dominating point of view has been that Λ is actually a dynamic quantity, now weakly dependent on time [52]–[54]. It is assumed therewith that, despite the present-day smallness of Λ or even its equality to zero, nothing points to the fact that this situation was characteristics for the early Universe as well. Some recent results [55]–[58] are rather important pointing to a potentially dynamic character of Λ. Specifically, of great interest is the Uncertainty Principle derived in these works for the pair of conjugate variables (Λ, V):

$$\Delta\Lambda\Delta V \sim \hbar, \tag{24}$$

where Λ is the vacuum energy density (cosmological constant). It is a dynamic value fluctuating around zero; V is the space-time volume. Here the volume of space-time V results from the Einstein-Hilbert action [56]:

$$S_{EH} \supset \Lambda \int d^4x\sqrt{-g} = \Lambda V. \tag{25}$$

In this case "the notion of conjugation is well-defined, but approximate, as implied by the expansion about the static Fubini–Study metric" (Section 6.1 of [55]). Unfortunately, in the proof per se (24), relying on the procedure with a non-linear and non-local Wheeler–de-Witt-like equation of the background-independent Matrix theory, some unconvincing arguments are used, making it insufficiently rigorous (Appendix 3 of [55]). But, without doubt, this proof has a significant result, though failing to clear up the situation.
Let us attempt to explain (24)(certainly at an heuristic level) using simpler and more natural terms involved with the other, more well-known, conjugate pair (E,t) - "energy - time". We use the designations of [55],[56]. In this way a four-dimensional volume will be denoted, as previously, by V.
Just from the start, the Generalized Uncertainty Principle (GUP) is used. Then a change over to the Heisenberg Uncertainty Principle at low energies will be only natural. As is known, the Uncertainty Principle of Heisenberg at Planck's scales (energies) may be

extended to the Generalized Uncertainty Principle. To illustrate, for the conjugate pair "momentum-coordinate" (p,x) this fact has been noted in many works [1]–[7]:

$$\triangle x \geq \frac{\hbar}{\triangle p} + \alpha' l_p^2 \frac{\triangle p}{\hbar}. \tag{26}$$

In [34],[40] it is demonstrated that the corresponding Generalized Uncertainty Relation for the pair "energy - time" may be easily obtained from

$$\Delta t \geq \frac{\hbar}{\Delta E} + \alpha' t_p^2 \frac{\Delta E}{\hbar}, \tag{27}$$

where l_p and t_p represent Planck length and time, respectively.
Now we assume that in the space-time volume $\int d^4x\sqrt{-g} = V$ the temporal and spatial parts may be separated (factored out) in the explicit form:

$$V(t) \approx t\overline{V}(t), \tag{28}$$

where $\overline{V}$ - spatial part V. For the expanding Universe such an assumption is quite natural. Then it is obvious that

$$\Delta V(t) = \Delta t\overline{V}(t) + t\Delta\overline{V}(t) + \Delta t\Delta\overline{V}(t). \tag{29}$$

Now we recall that for the inflation Universe the scaling factor is $a(t) \sim e^{Ht}$. Consequently, $\Delta\overline{V}(t) \sim \Delta t^3 f(H)$, where $f(H)$ is a particular function of Hubble's constant. From (27) it follows that

$$\Delta t \geq t_{min} \sim t_p. \tag{30}$$

However, it is suggested that, even though Δt is satisfying (30), its value is sufficiently small in order that ΔV be contributed significantly by the terms containing Δt to the power higher than the first. In this case the main contribution on the right-hand side of (29) is made by the first term $\Delta t\overline{V}(t)$ only. Then, multiplying the left- and right-hand sides of (27) by $\overline{V}$, we have

$$\Delta V \geq \frac{\hbar\overline{V}}{\Delta E} + \alpha' t_p^2 \frac{\Delta E\overline{V}}{\hbar} = \frac{\hbar}{\Delta\Lambda} + \alpha' t_p^2 \overline{V}^2 \frac{\Delta\Lambda}{\hbar}. \tag{31}$$

It is not surprising that a solution of the quadratic inequality (31) leads to a minimal volume of the space-time $V_{min} \sim V_p = l_p^3 t_p$ since (26) and (27) result in minimal length $l_{min} \sim l_p$ and minimal time $t_{min} \sim t_p$, respectively.
(31) is of interest from the viewpoint of two limits:
1)IR - limit: $t \longrightarrow \infty$
2)UV - limit: $t \longrightarrow t_{min}$.
In the case of IR-limit we have large volumes $\overline{V}$ and V at low $\Delta\Lambda$. Because of this, the main contribution on the right-hand side of (31) is made by the first term, as great $\overline{V}$ in the second term is damped by small t_p and $\Delta\Lambda$. Thus, we derive at

$$\lim_{t\longrightarrow\infty} \Delta V \approx \frac{\hbar}{\Delta\Lambda} \tag{32}$$

in accordance with (24) [55]. Here, similar to [55], Λ is a dynamic value fluctuating around zero.
And for the case (2) $\Delta\Lambda$ becomes significant

$$\lim_{t\longrightarrow t_{min}} \overline{V} = \overline{V}_{min} \sim \overline{V}_p = l_p^3; \lim_{t\longrightarrow t_{min}} V = V_{min} \sim V_p = l_p^3 t_p. \tag{33}$$

As a result, we have

$$\lim_{t\longrightarrow t_{min}} \Delta V = \frac{\hbar}{\Delta\Lambda} + \alpha_\Lambda V_p^2 \frac{\Delta\Lambda}{\hbar}, \tag{34}$$

where the parameter α_Λ absorbs all the above-mentioned proportionality coefficients.
For(34) $\Delta\Lambda \sim \Lambda_p \equiv \hbar/V_p = E_p/\overline{V}_p$.
It is easily seen that in this case $\Lambda \sim M_p^4$, in agreement with the value obtained using a naive (i.e. without super-symmetry and the like) quantum field theory [20],[19]. Despite the fact that Λ at Planck's scales (referred to as $\Lambda(UV)$) (34) is also a dynamic quantity, it is not directly related to well-known Λ (24),(32) (called $\Lambda(IR)$) because the latter, as opposed to the first one, is derived from Einstein's equations

$$R_{\mu\nu} - \frac{1}{2} g_{\mu\nu} R = 8\pi G_N \left(-\Lambda g_{\mu\nu} + T_{\mu\nu}\right). \tag{35}$$

However, Einstein's equations (35) are not valid at the Planck scales and hence $\Lambda(UV)$ may be considered as some high-energy generalization of the conventional cosmological constant, leading to $\Lambda(IR)$ in the low-energy limit.
In conclusion, it should be noted that the right-hand side of (26),(27) in fact is a series. Of course, a similar statement is true for (34) as well.
Then, we obtain a system of the Generalized Uncertainty Relations for the Early Universe (Plancks scales) in the symmetric form as follows:

$$\left\{ \begin{array}{lcl} \Delta x & \geq & \frac{\hbar}{\Delta p} + \alpha' \left(\frac{\Delta p}{p_{pl}}\right) \frac{\hbar}{p_{pl}} + \ldots \\ \Delta t & \geq & \frac{\hbar}{\Delta E} + \alpha' \left(\frac{\Delta E}{E_p}\right) \frac{\hbar}{E_p} + \ldots \\ \Delta V & \geq & \frac{\hbar}{\Delta\Lambda} + \alpha_\Lambda \left(\frac{\Delta\Lambda}{\Lambda_p}\right) \frac{\hbar}{\Lambda_p} + \ldots \end{array} \right. \tag{36}$$

The latter of relations (36) may be important when finding the general form for $\Lambda(UV)$, low-energy limit $\Lambda(IR)$, and also may be a step in the process of constructing future quantum-gravity equations, the low-energy limit of which is represented by Einstein's equations (35). It should be noted that a system of inequalities (36) may be complemented by the Generalized Uncertainty Relation in Thermodynamics [34],[40],[59]. Let us consider the thermodynamics uncertainty relations between the inverse temperature and interior energy of a macroscopic ensemble

$$\Delta \frac{1}{T} \geq \frac{k}{\Delta U}, \tag{37}$$

where k is the Boltzmann constant.
N.Bohr [60] and W.Heisenberg [61] have been the first to point out that such kind of uncertainty principle should take place in thermodynamics. The thermodynamic uncertainty

relations (37) were proved by many authors and in various ways [62]; their validity does not raise any doubts. Nevertheless, relation (37) was proved in view of the standard model of the infinite-capacity heat bath encompassing the ensemble. But it is obvious from the above inequalities that at very high energies the capacity of the heat bath can no longer be assumed infinite at the Planck scale. Indeed, the total energy of the pair heat bath - ensemble may be arbitrary large but finite merely as the Universe is born at a finite energy. Hence the quantity that can be interpreted as a temperature of the ensemble must have the upper limit and so does its main quadratic deviation. In other words the quantity $\Delta(1/T)$ must be bounded from below. But in this case an additional term should be introduced into (37) [34],[40],[59]:

$$\Delta\frac{1}{T} \geq \frac{k}{\Delta U} + \eta \Delta U, \tag{38}$$

where η is a coefficient. Dimension and symmetry reasons give

$$\eta \sim \frac{k}{E_p^2} .$$

As in the previous cases, inequality (38) leads to the fundamental (inverse) temperature [34],[40],[59].

$$T_{max} = \frac{\hbar}{\Delta t_{min} k} \sim \frac{\hbar}{t_p k}, \quad \beta_{min} = \frac{1}{kT_{max}} = \frac{\Delta t_{min}}{\hbar}. \tag{39}$$

In the recently published work [63]) the black hole horizon temperature has been measured with the use of the Gedanken experiment. In the process the Generalized Uncertainty Relations in Thermodynamics (38) have been derived also. Expression (38) has been considered in the monograph [64] within the scope of the mathematical physics methods.
Besides, note that one of the first studies of the cosmological constant within the scope of the Heisenberg Uncertainty Principle has been presented in several works [65] – [67] demonstrating the inference: **"vacuum fluctuation of the energy density can lead to the observed cosmological constant" [65]**. In these works, however, no consideration has been given to GUP, whereas UV-cutoff has been derived artificially.

5. Gravitational Thermodynamics in Low and High Energy and Deformed Quantum Theory

In the last decade a number of very interesting works have been published. We can primary name the works of T.Padmanbhan [66]–[77], where gravitation, at least for the spaces with horizon, is directly associated with thermodynamics and the results obtained demonstrate a holographic character of gravitation. Of the greatest significance is a pioneer work written by T.Jacobson [49]. For black holes the association has been first revealed by Bekenstein and Hawking [79],[80], who related the black-hole event horizon temperature to the surface gravitation. T.Padmanbhan, in particular in [76], has shown that this relation is not accidental and may be generalized for the spaces with horizon. As all the foregoing results have been obtained in a semiclassical approximation, i.e. for sufficiently low energies, the problem arises: how these results are modified when going to higher energies. In the context of this paper, the problem may be stated as follows: since we have some infra-red (IR) cutoff L and ultraviolet (UV) cutoff l_{min}, we naturally have a problem how the above-mentioned results on Gravitational Thermodynamics are changed for

$$L \longrightarrow l_{min}. \tag{40}$$

According to Section 3 of this paper, they should become dependent on the deformation parameter α. After all, in the already mentioned Section 3 (5) α is indicated as nothing else but

$$\alpha = \frac{l_{min}^2}{L^2}. \tag{41}$$

In fact, in several papers [81]–[87] it has been demonstrated that thermodynamics and statistical mechanics of black holes in the presence of GUP (i.e. at high energies) should be modified. To illustrate, in [86] the Hawking temperature modification has been computed in the asymptotically flat space in this case in particular. It is easily seen that in this case the deformation parameter α arises naturally. Indeed, modification of the Hawking temperature is of the following form(formula (10) in [86]):

$$T_{GUP} = (\frac{d-3}{4\pi})\frac{\hbar r_+}{2\alpha'^2 l_p^2}[1-(1-\frac{4\alpha'^2 l_p^2}{r_+^2})^{1/2}], \tag{42}$$

where d is the space-time dimension, and r_+ is the uncertainty in the emitted particle position by the Hawking effect, expressed as

$$\Delta x_i \approx r_+ \tag{43}$$

and being nothing else but a radius of the event horizon; α' – dimensionless constant from GUP. But as we have $2\alpha' l_p = l_{min}$, in terms of α (42) may be written in a natural way as follows:

$$T_{GUP} = (\frac{d-3}{4\pi})\frac{\hbar \alpha_{r_+}^{-1}}{\alpha' l_p}[1-(1-\alpha_{r_+})^{1/2}], \tag{44}$$

where α_{r_+}- parameter α associated with the IR-cutoff r_+. In such a manner T_{GUP} is only dependent on the constants including the fundamental ones and on the deformation parameter α.

The dependence of the black hole entropy on α may be derived in a similar way. For a semiclassical approximation of the Bekenstein-Hawking formula [79],[80]

$$S = \frac{1}{4}\frac{A}{l_p^2}, \tag{45}$$

where A – surface area of the event horizon, provided the horizon event has radius r_+, then $A \sim r_+^2$ and (45) is clearly of the form

$$S = \sigma \alpha_{r_+}^{-1}, \tag{46}$$

where σ is some dimensionless denumerable factor. The general formula for quantum corrections [85] given as

$$S_{GUP} = \frac{A}{4l_p^2} - \frac{\pi\alpha'^2}{4}\ln\left(\frac{A}{4l_p^2}\right) + \sum_{n=1}^{\infty} c_n \left(\frac{A}{4l_p^2}\right)^{-n} + \text{const}\,, \tag{47}$$

where the expansion coefficients $c_n \propto \alpha'^{2(n+1)}$ can always be computed to any desired order of accuracy [85], may be also written as a power series in $\alpha_{r_+}^{-1}$ (or Laurent series in α_{r_+})

$$S_{GUP} = \sigma\alpha_{r_+}^{-1} - \frac{\pi\alpha'^2}{4}\ln(\sigma\alpha_{r_+}^{-1}) + \sum_{n=1}^{\infty}(c_n\sigma^{-n})\alpha_{r_+}^{n} + \text{const} \tag{48}$$

Note that here no consideration is given to the restrictions on the IR-cutoff

$$L \leq L_{max} \tag{49}$$

and to those corresponding the extended uncertainty principle (EUP) that leads to a minimal momentum [86]. This problem will be considered separately in further publications of the author.

A black hole is a specific example of the space with horizon. It is clear that for other horizon spaces [76] a similar relationship between their thermodynamics and the deformation parameter α should be exhibited.

Quite recently, in a series of papers, and specifically in [68]–[74], it has been shown that Einstein equations may be derived from the surface term of the GR Lagrangian, in fact containing the same information as the bulk term.

And as Einstein-Hilbert's Lagrangian has the structure $L_{EH} \propto R \sim (\partial g)^2 + \partial^2 g$, in the customary approach the surface term arising from $L_{surf} \propto \partial^2 g$ has to be canceled to get Einstein equations from $L_{bulk} \propto (\partial g)^2$ [75]. But due to the relationship between L_{bulk} and L_{surf} [70]–[72],[75], we have

$$\sqrt{-g}L_{suf} = -\partial_a\left(g_{ij}\frac{\partial\sqrt{-g}L_{bulk}}{\partial(\partial_a g_{ij})}\right). \tag{50}$$

In such a manner one can suggest a holographic character of gravity in that the bulk and surface terms of the gravitational action contain identical information. However, there is a significant difference between the first case, when variation of the metric g_{ab} in L_{bulk} leads to Einstein equations, and the second case, associated with derivation of the GR field equations from the action principle using only the surface term and virtual displacements of horizons [67], whereas the metric is not treated as a dynamic variable [75].

In the case under study, it is assumed from the beginning that we consider the spaces with horizon. It should be noted that in the Fischler-Susskind cosmic holographic conjecture it is implied that the Universe represents spherically symmetric space-time, on the one hand, and has a (Hubble) horizon (16), on the other hand. But proceeding from the results of [68]– [75], an entropy boundary is actually given by the surface of horizon measured in Planck's units of area [71]:

$$S = \frac{1}{4}\frac{A_R}{l_p^2}, \tag{51}$$

where A_R is the horizon area corresponding to the Hubble horizon R (16).

To sum it up, an assumption that space-time is spherically symmetric and has a horizon is the only natural assumption held in the Fischler-Susskind cosmic holographic conjecture to support its validity. Thus the arguments in support of the Fischler-Susskind cosmic holographic conjecture are given on the basis of the results obtained lately on Gravitational Holography and Gravitational Thermodynamics.

It should be noted that Einstein's equations may be obtained from the proportionality of the entropy and horizon area together with the fundamental thermodynamic relation connecting heat, entropy, and temperature [49]. In fact [68]– [75], this approach has been extended and complemented by the demonstration of holographicity for the gravitational action (see also [76]).And in the case of Einstein-Hilbert gravity, it is possible to interpret Einstein's equations as the thermodynamic identity [77]:

$$TdS = dE + PdV. \tag{52}$$

The above-mentioned results in the last paragraph have been obtained at low energies, i.e. in a semiclassical approximation. Because of this, the problem arises how these results are changed in the case of high energies? Or more precisely, how the results of [49],[68]– [77] are generalized in the UV-cutoff? It is obvious that, as in this case all the thermodynamic characteristics become dependent on the deformation parameter α, all the corresponding results should be modified (deformed) to meet the following requirements:

(a) to be clearly dependent on the deformation parameter α at high energies;

(b) to be duplicated, with high precision, at low energies due to the suitable limiting transition;

(c) Let us clear up what is meant by the adequate high energy α-deformation of Einstein's equations (General Relativity).

The problem may be more specific.
As, according to [49],[76],[77] and some other works, gravitation is greatly determined by thermodynamics and at high energies the latter is a deformation of the classical thermodynamics, it is interesting whether gravitation at high energies (or what is the same, quantum gravity or Planck scale)is being determined by the corresponding deformed thermodynamics. The formulae (44) and (48) are elements of the high-energy α-deformation in thermodynamics, a general pattern of which still remains to be formed. Obviously, these formulae should be involved in the general pattern giving better insight into the quantum gravity, as they are applicable to black mini-holes (Planck black holes) which may be a significant element of such a pattern. But what about other elements of this pattern? How can we generalize the results [49],[76],[77]when the IR-cutoff tends to the UV-cutoff (formula (40))? What are modifications of the thermodynamic identity (52) in a high-energy deformed thermodynamics and how is it applied in high-energy (quantum) gravity? What are the aspects of using the Generalized Uncertainty Relations in Thermodynamics [34],[40],[59] (38),(38)in this respect? It is clear that these relations also form an element of high-energy thermodynamics.
By authors opinion, the methods developed to solve the problem of point (c) and elucidation of other above-mentioned problems may form the basis for a new approach to solution of the quantum gravity problem. And one of the keys to the **quantum gravity** problem is a better insight into the **high-energy thermodynamics**.

6. Possible High Energy Deformation of Einstein's Equations

Let us consider α-representation and high energy α-deformation of the Einstein's field equations for the specific cases of horizon spaces (the point (c) of Section 5). In so doing the results of the survey work ([78] p.p.41,42)are used. Then, specifically, for a static, spherically symmetric horizon in space-time described by the metric

$$ds^2 = -f(r)c^2dt^2 + f^{-1}(r)dr^2 + r^2d\Omega^2 \tag{53}$$

the horizon location will be given by simple zero of the function $f(r)$, at $r = a$.
It is known that for horizon spaces one can introduce the temperature that can be identified with an analytic continuation to imaginary time. In the case under consideration ([78], eq.(116))

$$k_B T = \frac{\hbar c f'(a)}{4\pi}. \tag{54}$$

Therewith, the condition $f(a) = 0$ and $f'(a) \neq 0$ must be fulfilled.
Then at the horizon $r = a$ Einstein's field equations

$$\frac{c^4}{G}\left[\frac{1}{2}f'(a)a - \frac{1}{2}\right] = 4\pi P a^2 \tag{55}$$

may be written as the thermodynamic identity (52)([78] formula (119))

$$\underbrace{\frac{\hbar c f'(a)}{4\pi}}_{k_B T}\underbrace{\frac{c^3}{G\hbar}d\left(\frac{1}{4}4\pi a^2\right)}_{dS}\underbrace{-\frac{1}{2}\frac{c^4 da}{G}}_{-dE} = \underbrace{Pd\left(\frac{4\pi}{3}a^3\right)}_{PdV} \tag{56}$$

where $P = T_r^r$ is the trace of the momentum-energy tensor and radial pressure. In the last equation da arises in the infinitesimal consideration of Einstein's equations when studying two horizons distinguished by this infinitesimal quantity a and $a + da$ ([78] formula (118)). Now we consider(56) in new notation expressing a in terms of the corresponding deformation parameter α. Then we have

$$a = l_{min}\alpha^{-1/2}. \tag{57}$$

Therefore,

$$f'(a) = -2l_{min}^{-1}\alpha^{3/2}f'(\alpha). \tag{58}$$

Substituting this into (55) or into (56), we obtain in the considered case of Einstein's equations in the "α–representation" the following:

$$\frac{c^4}{G}(-\alpha f'(\alpha) - \frac{1}{2}) = 4\pi P\alpha^{-1}l_{min}^2. \tag{59}$$

Multiplying the left- and right-hand sides of the last equation by α, we get

$$\frac{c^4}{G}(-\alpha^2 f'(\alpha) - \frac{1}{2}\alpha) = 4\pi P l_{min}^2. \tag{60}$$

But since usually $l_{min} \sim l_p$ (that is just the case if the Generalized Uncertainty Principle (GUP) is satisfied), we have $l_{min}^2 \sim l_p^2 = G\hbar/c^3$. When selecting a system of units, where $\hbar = c = 1$, we arrive at $l_{min} \sim l_p = \sqrt{G}$, and then (59) is of the form

$$-\alpha^2 f'(\alpha) - \frac{1}{2}\alpha = 4\pi P \vartheta^2 G^2, \tag{61}$$

where $\vartheta = l_{min}/l_p$. L.h.s. of (61) is dependent on α. Because of this, r.h.s. of (61) must be dependent on α as well, i. e. $P = P(\alpha)$.

6.1. Analysis of α-Representation of Einstein's Equations

Now let us get back to (56). In [78] the low-energy case has been considered, for which ([78] p.42 formula (120))

$$S = \frac{1}{4L_P^2}(4\pi a^2) = \frac{1}{4}\frac{A_H}{l_P^2}; \quad E = \frac{c^4}{2G}a = \frac{c^4}{G}\left(\frac{A_H}{16\pi}\right)^{1/2}, \tag{62}$$

where A_H is the horizon area. In our notation (62) may be rewritten as

$$S = \frac{1}{4}\pi\alpha^{-1}; \quad E = \frac{c^4}{2G}a = \frac{c^4}{G}\left(\frac{A_H}{16\pi}\right)^{1/2} = \frac{\vartheta}{2\sqrt{G}}\alpha^{1/2}. \tag{63}$$

We proceed to two entirely different cases: low energy (LE) case and high energy (HE) case. In our notation these are respectively given by

$$A)\alpha \to 0 \text{ (LE)}, \; B)\alpha \to 1/4 \text{ (HE)},$$
C)α complies with the familiar scales and energies.

The case of C) is of no particular importance as it may be considered within the scope of the conventional General Relativity.
Indeed, in point A)$\alpha \to 0$ is not actually an exact limit as a real scale of the Universe (Infrared (IR)-cutoff $L_{max} \approx 10^{28} cm$), and then

$$\alpha_{min} \sim l_p^2/L_{max}^2 \approx 10^{-122}.$$

In this way A) is replaced by A1)$\alpha \to \alpha_{min}$. In any case at low energies the second term in the left-hand side (61) may be neglected in the infrared limit. Consequently, at low energies (61) is written as

$$-\alpha^2 f'(\alpha) = 4\pi P(\alpha)\vartheta^2 G^2. \tag{64}$$

Solution of the corresponding Einstein equation finding of the function $f(\alpha) = f[P(\alpha)]$ satisfying(64). In this case formulae (62) are valid as at low energies a semiclassical approximation is true. But from (64)it follows that

$$f(\alpha) = -4\pi\vartheta^2 G^2 \int \frac{P(\alpha)}{\alpha^2} d\alpha. \tag{65}$$

On the contrary, knowing $f(\alpha)$, we can obtain $P(\alpha) = T_r^r$.
But it is noteworthy that, when studying the infrared modified gravity [99],[100], we have to make corrections for the considerations of point A1).

6.2. High Energy α-Deformation of General Relativity

Let us consider the high-energy case B). Here two variants are possible.

I. First variant.

In this case it is assumed that in the high-energy (Ultraviolet (UV))limit the thermodynamic identity (56)(or that is the same (52)is retained but now all the quantities involved in this identity become α-deformed. This means that they appear in the α-representation with quantum corrections and are considered at high values of the parameter α, i.e. at α close to 1/4. In particular, the temperature T from equation (56) is changed by T_{GUP} (44), the entropy S from the same equation given by semiclassical formula (62) is changed by S_{GUP} (48), and so forth:

$$E \mapsto E_{GUP}, V \mapsto V_{GUP}.$$

Then the high-energy α-deformation of equation (56) takes the form

$$k_B T_{GUP}(\alpha) dS_{GUP}(\alpha) - dE_{GUP}(\alpha) = P(\alpha) dV_{GUP}(\alpha). \tag{66}$$

Substituting into (66) the corresponding quantities $T_{GUP}(\alpha), S_{GUP}(\alpha), E_{GUP}(\alpha), V_{GUP}(\alpha), P(\alpha)$ and expanding them into a Laurent series in terms of α, close to high values of α, specifically close to $\alpha = 1/4$, we can derive a solution for the high energy α-deformation of general relativity (66) as a function of $P(\alpha)$. As this takes place, provided at high energies the generalization of (56) to (66)is possible, we can have the high-energy α-deformation of the metric. Actually, as from (56) it follows that

$$f'(a) = \frac{4\pi k_B}{\hbar c} T = 4\pi k_B T \tag{67}$$

(considering that we have assumed $\hbar = c = 1$), we get

$$f'_{GUP}(a) = 4\pi k_B T_{GUP}(\alpha). \tag{68}$$

L.h.s. of (68) is directly obtained in the α-representation. This means that, when $f' \sim T$, we have $f'_{GUP} \sim T_{GUP}$ with the same factor of proportionality. In this case the function f_{GUP} determining the high-energy α-deformation of the spherically symmetric metric may be in fact derived by the expansion of T_{GUP}, that is known from (44), into a Laurent series in terms of α close to high values of α (specifically close to $\alpha = 1/4$), and by the subsequent integration.

It might be well to remark on the following.

6.2.1. As on going to high energies we use (GUP), ϑ from equation (61)is expressed in terms of α'–dimensionless constant from GUP (26),(44):$\vartheta = 2\alpha'$.

6.2.2. Of course, in all the formulae including l_p this quantity must be changed by $G^{1/2}$ and hence l_{min} by $\vartheta G^{1/2} = 2\alpha' G^{1/2}$.

6.2.3. As noted in the end of subsection 6.1, and in this case also knowing all the high-energy deformed quantities $T_{GUP}(\alpha), S_{GUP}(\alpha), E_{GUP}(\alpha), V_{GUP}(\alpha)$, we can find $P(\alpha)$ at α close to 1/4.

6.2.4. Here it is implicitly understood that the Ultraviolet limit of Einstein's equations is independent of the starting horizon space. This assumption is quite reasonable. Because of this, we use the well-known formulae for the modification of thermodynamics and statistical mechanics of black holes in the presence of GUP [81]–[87].

6.2.5. The use of the thermodynamic identity (66) for the description of the high energy deformation in General Relativity implies that on going to the UV-limit of Einsteins equations for horizon spaces in the thermodynamic representation (consideration) we are trying to remain within the scope of **equilibrium statistical mechanics** [88] (**equilibrium thermodynamics**) [89]. However, such an assumption seems to be too strong. But some grounds to think so may be found as well. Among other things, of interest is the result from [81] that GUP may prevent black holes from their total evaporation. In this case the Plancks remnants of black holes will be stable, and when they are considered, in some approximation the **equilibrium thermodynamics** should be valid. At the same time, by authors opinion these arguments are rather weak to think that the quantum gravitational effects in this context have been described only within the scope of **equilibrium thermodynamics**[89].

II. Second variant.
According to the remark of **6.2.5.**, it is assumed that the interpretation of Einstein's equations as a thermodynamic identity (56) is not retained on going to high energies (UV–limit), i.e. at $\alpha \to 1/4$, and the situation is adequately described exclusively by **non-equilibrium thermodynamics**[89],[90]. Naturally, the question arises: which of the additional terms introduced in (56) at high energies may be leading to such a description? In the Sections 3 and 4 it has been shown that in case the cosmological term Λ is a dynamic quantity, it is small at low energies and may be sufficiently large at high energies. Then its inclusion in the low-energy case (55)(or in the α -representation (61)) has actually no effect on the thermodynamic identity (56)validity, and consideration within the scope of equilibrium thermodynamics still holds true. It is well known that this is not the case at high energies as the Λ-term may contribute significantly to make the "process" non-equilibrium in the end [89],[90].
Is this the only cause for violation of the thermodynamic identity (56) as an interpretation of the high-energy generalization of Einstein's equations? Further investigations are required to answer this question.

7. QFT with UV-Cutoff for Different Approaches and Some Comments

I. As shown by numerous authors (to start with [8]), the Quantum Mechanics with the fundamental length (UV cutoff) generated by GUP is in line with the following deformation of Heisenberg algebra

$$[\vec{x},\vec{p}] = i\hbar(1+\beta^2\vec{p}^2+\ldots) \tag{69}$$

and

$$\Delta x_{\min} \approx \hbar\sqrt{\beta} \sim l_p. \tag{70}$$

In the recent works [91] it has been demonstrated that the Holographic Principle is an outcome of this approach, actually being integrated in the approach.
We can easily show that the deformation parameter β in (69),(70) may be expressed in terms of the deformation parameter α (see Section 3 of the text) that has been introduced in the approach associated with the density matrix deformation. Indeed, from (69),(70) it follows that $\beta \sim \mathbf{1/p^2}$, and for $x_{\min} \sim l_p$, β corresponding to $x_{\min}$ is nothing else but

$$\beta \sim 1/P_{pl}^2, \tag{71}$$

where P_{pl} is Planck's momentum: $P_{pl} = \hbar/l_p$.
In this way β is changing over the following interval:

$$\lambda/P_{pl}^2 \leq \beta < \infty, \tag{72}$$

where λ is a numerical factor and the second member in (69) is accurately reproduced in momentum representation (up to the numerical factor) by $\alpha = l_{min}^2/l^2 \sim l_p^2/l^2 = p^2/P_{pl}^2$

$$[\vec{x},\vec{p}] = i\hbar(1+\beta^2\vec{p}^2+\ldots) = i\hbar(1+a_1\alpha+a_2\alpha^2+\ldots). \tag{73}$$

As indicated in the previous Section (formula (46)), parameter α has one more interesting feature:

$$\alpha_l^{-1} \sim l^2/l_p^2 \sim S_{BH}. \tag{74}$$

Here α_l is the parameter α corresponding to l, S_{BH} is the black hole entropy with the characteristic linear size l (for example, in the spherically symmetric case $l = R$ - radius of the corresponding sphere with the surface area A), and

$$A = 4\pi l^2, S_{BH} = A/4l_p^2 = \pi\alpha_l^{-1}. \tag{75}$$

This note is devoted to the demonstration of the fact that in case of the holographic principle validity in terms of the new deformation parameter α in QFT^{α}, considered above and introduced as early as 2002 [92]–[94], all the principal values associated with the Vacuum (Dark) Energy Problem may be defined simply and naturally. At the same time, there is no place for such a parameter in the well-known QFT, whereas in QFT with the fundamental length, specifically in QFT^{α}, it is quite natural [32],[33],[35], [37],[38],[40].

II. It should be noted that smallness of α_R (Section 3) leads to a very great value of r in (21),(22). Besides, from (21) it follows that there exists some minimal entropy $S_{min} \sim S_p$, and this is possible only in QFT with the fundamental length.

III.This Section is related to Section 3 in [66] as well as to Sections 3 and 6 in [67]. The constant l_{Λ} introduced in these works is such that in the case under consideration

$\Lambda \equiv l_\Lambda^{-2}$ is equivalent to R, i.e. $\alpha_R \approx \alpha_{l_\Lambda}$ with $\alpha_{l_\Lambda} = l_p^2/l_\Lambda^2$. Then expression on the right-hand side of (19) is the major term of the formula for ρ_{vac}, and its quantum corrections are nothing else but a series expansion in terms of α_{l_Λ} (or α_R)

$$\rho_{\rm vac} \sim \frac{1}{l_p^4}\left(\frac{l_p}{l_\Lambda}\right)^2 + \frac{1}{l_p^4}\left(\frac{l_p}{l_\Lambda}\right)^4 + \cdots = \alpha_{l_\Lambda} m_p^4 + \ldots \tag{76}$$

In the first variant presented in [66] and [67] the right-hand side of (76) (formulas (12),(33)) in [66] and [67], respectively)reveals an enormous additional term $m_p^4 \sim \rho_{QFT}$ for renormalization. As indicated in the previous Section, it may be, however, ignored because the gravity is described by a pure surface term. And in the case under study, owing to the Holographic Principle, we may proceed directly to (76). Moreover, in QFT^α there is no need in renormalization as from the start we are concerned with the ultraviolet-finiteness.
Moreover, a series expansion of (76) in terms of α is a complete analog of the expansion in terms of the same parameter, redetermining the measuring procedure in $QMFL^\alpha$ [33],[35],[37],[40]:

$$Sp[\rho(\alpha)] - Sp^2[\rho(\alpha)] = \alpha + a_0'\alpha^2 + \ldots \tag{77}$$

As indicated in [41], the same expansion may be used to obtain quantum corrections to the semiclassical Bekenstein-Hawking formula (51) for the black hole entropy.
IV.Besides, the Heisenberg's algebra deformations are introduced due to the involvement of minimal length in quantum mechanics. These deformations are stable in the sense of [95]. But this is not true for the unified algebra of Heisenberg and Poincare. This algebra does not carry the indicated immunity. It is suggested that the Lie algebra for the interface of the gravitational and quantum realms is in its stabilized form. Now it is clear that such a stability should be raised to the status of a physical principle. In a very interesting work of Ahluwalia - Khalilova [95] it has been demonstrated that the stabilized form of the Poincare-Heisenberg algebra [96], [97] carries three additional parameters: "a length scale pertaining to the Planck/unification scale, a second length scale associated with cosmos, and a new dimensionless constant with the immediate implication that 'point particle' ceases to be a viable physical notion. It must be replaced by objects which carry a well-defined, representation space dependent, minimal spatiotemporal extent".
Thus, within the scope of a Quantum Field Theory with the UV cutoff (fundamental length), closeness of the theoretical and experimental values for ρ_{vac} is adequately explained. In this case an important role is played by new parameters appearing in the corresponding Heisenberg Algebra deformation. Specifically, by the new small dimensionless parameter α, in terms of which one can adequately interpret both the smallness of ρ_{vac} and its modern experimental value. Besides, it is shown that the Generalized Uncertainty Principle (GUP) may be an instrument in studies of a dynamic character of the cosmological constant Λ.

8. Conclusion

In conclusion it should be noted that in a series of the authors works [32]–[41] a minimal α-deformation of QFT has been formed. By minimal it is meant that no space-time noncommutativity was required, i.e. there was no requirement for noncommutative operators associated with different spatial coordinates

$$[X_i, X_j] \neq 0, i \neq j. \tag{78}$$

However, all the well-known deformations of QFT associated with GUP (for example, [6]–[8]) contain (78) as an element of the corresponding deformed Heisenberg algebra. Because of this, it is necessary to extend (or modify) the above-mentioned minimal α-deformation of QFT $-QFT^{\alpha}$ [32]–[41] to some new deformation $\widetilde{QFT}^{\alpha}$ compatible with GUP, as it has been noted in [98].
Besides, in this paper consideration has been given to QFT with a minimal length, i.e. with the UV-cutoff. Consideration of QFT with a minimal momentum (or IR-cutoff) (49) necessitates an adequate extension of α-deformation in QFT with the introduction of new parameters significant in the IR-limit. Proceeding from point (c) of Section 5, the problem may be stated as follows:
(c) provided α-deformation of GR describes the ultraviolet (quantum-gravity) limit of GR, it is interesting to examine the deformation type describing adequately the infrared limit of GR. It seems that some indications of a nature of such deformation may be found from the works devoted to the infrared modification of gravity [99],[100].

References

[1] Veneziano,G. A Stringy Nature Needs Just Two Constants *Europhys.Lett* 1986, 2, 199–211; Amati, D.; Ciafaloni, M., and Veneziano,G. Can Space-Time Be Probed Below the String Size? *Phys.Lett.B* 1989, 216, 41–47; E.Witten, *Phys.Today* 1996, 49, 24–28.

[2] Adler,R. J.; Santiago,D. I. On gravity and the uncertainty principle. *Mod. Phys. Lett. A* 1999, 14, 1371–1378.

[3] Scardigli,F. Generalized uncertainty principle in quantum gravity from micro - black hole Gedanken experiment. *Phys. Lett. B* 1999, 452, 39–44; Bambi,C. A Revision of the Generalized Uncertainty Principle. *Class. Quant. Grav* 2008, 25, 105003.

[4] Garay,L. Quantum gravity and minimum length. *Int.J.Mod.Phys.A* 1995, 10, 145–166.

[5] Ahluwalia,D.V. Wave particle duality at the Planck scale: Freezing of neutrino oscillations. *Phys.Lett* 2000, A275, 31–35; Ahluwalia,D.V. *Mod.Phys.Lett* 2002, Interface of gravitational and quantum realms. A17, 1135–1146.

[6] Maggiore,M. A Generalized uncertainty principle in quantum gravity. *Phys.Lett* 1993, B304, 65–69.

[7] Maggiore,M. The Algebraic structure of the generalized uncertainty principle. *Phys.Lett.B* 1993, 319, 83–86.

[8] Kempf,A.; Mangano,G.; Mann,R.B. Hilbert space representation of the minimal length uncertainty relation. *Phys.Rev.D* 1995, 52, 1108–1118.

[9] Faddeev, L., Mathematical View on Evolution of Physics. *Priroda* 1989, 5, 11–18.

[10] Perlmutter, S. et al. Measurements of Omega and Lambda from 42 high redshift supernovae. *Astrophys. J* 1999, *517*, 565–586; Riess A. G. et al. Observational evidence from supernovae for an accelerating universe and a cosmological constant. *Astron. J* 1998, 116, 1009–1038; Riess A. G. et al. BV RI light curves for 22 type Ia supernovae. *Astron. J* 1999, 117, 707–724; Sahni, V.; Starobinsky, A. A. The Case for a positive cosmological Lambda term. *Int. J. Mod. Phys. D* 2000 9, 373–397; Carroll, S. M. The Cosmological constant. *Living Rev. Rel* 2001, 4, 1–50; Padmanabhan, T. Cosmological constant: The Weight of the vacuum. *Phys. Rept* 2003, 380, 235–320; Padmanabhan, T. Dark Energy: the Cosmological Challenge of the Millennium. *Current Science* 88, 1057–1071 (2005); Peebles, P. J. E.; Ratra, B. The Cosmological constant and dark energy. *Rev. Mod. Phys* 2003, 75, 559–606.

[11] Ratra,B.; Peebles, J. Cosmological Consequences of a Rolling Homogeneous Scalar Field. *Phys. Rev. D* 1988, *37*, 3406–3422; Caldwell,R. R.; Dave, R. and Steinhardt, P. J. Cosmological imprint of an energy component with general equation of state. *Phys. Rev. Lett* 1998, 80, 1582–1585.

[12] Armendariz-Picaon, C.; Damour,T.; Mukhanov, V. k - inflation. *Phys. Lett. B* 1999, 458, 209–218 ; Garriga,J.; Mukhanov,V. Perturbations in k-inflation. *Phys. Lett. B* 1999 458, 219–225 (1999).

[13] Padmanabhan, T. Accelerated expansion of the universe driven by tachyonic matter. *Phys. Rev. D* 2002 66, 021301; Bagla, J. S.; Jassal,H. K.; Padmanabhan, T. Cosmology with tachyon field as dark energy. *Phys. Rev. D* 2003, 67, 063504 ; Abramo, L. R. W.; Finelli, F. Cosmological dynamics of the tachyon with an inverse power-law potential. *Phys. Lett. B* 2003, 575, 165–171; Aguirregabiria J. M.; Lazkoz, R. Tracking solutions in tachyon cosmology. *Phys. Rev. D* 2004, 69, 123502 ; Guo, Z. K.; Zhang, Y. Z. Cosmological scaling solutions of the tachyon with multiple inverse square potentials. *JCAP* 2004 0408, 010; Copeland, E. J. et al. What is needed of a tachyon if it is to be the dark energy? *Phys. Rev. D* 2005 71, 043003.

[14] Sahni, V.; Shtanov, Y. Brane world models of dark energy. *JCAP* 2003, 0311, 014; Elizalde, E.; Nojiri, S.; and Odintsov, S. D. Late-time cosmology in (phantom) scalar-tensor theory: Dark energy and the cosmic speed-up. *Phys. Rev. D* 2004, 70, 043539.

[15] A.Einstein, *Sitzungber. Preuss. Akad. Wiss* 1917, *1*, 142–152.

[16] A.Friedmann, *Zs. Phys* 1924, 21, 326–332.

[17] A.Pais, *Subtle is the Lord... The Science and the Life of Albert Einstein*, New York: Oxford Yniversity Press, 1982.

[18] Gliner, E. B. *ZHETF* 1965 49, 542–549.

[19] Zel'dovich,Y.B. *Sov.Phys.Uspehi* 1968 11, 381–393.

[20] Weinberg, S. The Cosmological Constant Problem. *Rev. Mod. Phys* 1989 61, 1–23.

[21] Cohen,A.; Kaplan, D.; Nelson, A. Effective field theory, black holes, and the cosmological constant. *Phys. Rev. Lett* 1999, 82, 4971–4974.

[22] Myung,Y. S. Holographic principle and dark energy. *Phys. Lett. B* 2005 610, 18–22.

[23] Myung, Y. S.; Min-Gyun Seo. Origin of holographic dark energy models. *Phys. Lett. B* 2009 617, 435–439.

[24] Huang, Q.G.; Li, M. *JCAP* 2004, 0408, 013; Huang, Q.G.; Li, M. *JCAP* 2005, 0503, 001; Huang, Q.G.; Gong, Y.G. *JCAP* 2004, 0408, 006; Zhang,X.; Wu, F.Q. *Phys. Rev. D* 2005, 72, 043524; Zhang,X. *Int. J. Mod. Phys. D* 2005, 14, 1597–1606; Chang,Z.; Wu,F.Q.; and Zhang,X. *Phys. Lett. B* 2006, 633, 14–18; Wang,B.; Abdalla, E.; Su, R.K. *Phys. Lett. B* 2005, 611, 21–26; Wang,B.; Lin,C.Y.; Abdalla E. *Phys. Lett. B* 2006, 637, 357–361; Zhang,X. *Phys. Lett. B* 2007 648, 1–5; Setare, M.R.; Zhang, J.; Zhang,X. *JCAP* 2007, 0703, 007; Zhang, J.; Zhang,X.; Liu, H. *Phys. Lett. B* 2008, 659, 26–33; Chen, B.; Li, M.; Wang, Y. *Nucl. Phys. B* 2007, 774, 256–267; Zhang, J.; Zhang,X.; Liu, H. *Eur. Phys. J. C* 2007, 52, 693–699; Zhang, X.; Wu,F.Q. *Phys. Rev. D* 2007, *76*, 023502; Feng, C.J. *Phys. Lett. B* 2008, *663*, 367–371; Ma,Y.Z.; Gong,Y., *Eur. Phys. J. C* 2009, 60, 303–315; Li,M.; Lin,C.; Wang,Y. *JCAP* 2008, 0805, 023; Li,M; Li,X.D.; Lin,C.; Wang,Y. *Commun. Theor. Phys* 2009, 51, 181–186.

[25] Cai, R. G. A Dark Energy Model Characterized by the Age of the Universe. *Phys. Lett. B* 2007, 657, 228–231.

[26] Károlyházy, F. *Nuovo Cim. A* 1966, 42, 390–402.

[27] Ng,Y. J.; Van Dam, H. Limit to space-time measurement. *Mod. Phys. Lett. A* 1994, 9, 335–340.

[28] Sasakura, N. An Uncertainty relation of space-time. *Prog. Theor. Phys* 1999, 102, 169–179.

[29] Maziashvili, M. Cosmological implications of Karolyhazy uncertainty relation. *Phys. Lett. B* 2007, 652, 165–168.

[30] Wei,H.; Cai, R.G. Statefinder Diagnostic and w - w' Analysis for the Agegraphic Dark Energy Models without and with Interaction. *Phys. Lett. B* 2007, 655, 1–6; Wei,H.; Cai, R.G. Cosmological Constraints on New Agegraphic Dark Energy. *Phys. Lett. B* 2008, 663, 1–6; Neupane,I.P. Remarks on Dynamical Dark Energy Measured by the Conformal Age of the Universe. *Phys. Rev. D* 2007, 76, 123006; Maziashvili, M. Operational definition of (brane induced) space-time and constraints on the fundamental parameters. *Phys. Lett. B* 2008, 666, 364–370; Zhang,J.; Zhang,X.; Liu,H. Agegraphic dark energy as a quintessence. *Eur. Phys. J. C* 2008, 54, 303–309; Wu,J.P.; Ma,D.Z.; Ling,Y. Quintessence reconstruction of the new agegraphic dark energy model. *Phys. Lett. B* 2008, 663, 152–159; Wei,H.; Cai,R.G. Interacting Agegraphic Dark Energy. *Eur. Phys. J. C* 2008, 59, 99–105.

[31] Nouicer,K. Quantum-corrected black hole thermodynamics to all orders in the Planck length. *Phys.Lett B* 2007, *646*, 63–71.

[32] Shalyt-Margolin, A.E.; Suarez, J.G. Quantum mechanics of the early universe and its limiting transition. *gr-qc/0302119*, 16pp.

[33] Shalyt-Margolin, A.E.; Suarez, J.G. Quantum mechanics at Planck's scale and density matrix. *Intern. Journ. Mod. Phys D* 2003, 12, 1265–1278.

[34] Shalyt-Margolin, A.E.; Tregubovich, A.Ya. Deformed density matrix and generalized uncertainty relation in thermodynamics. *Mod. Phys.Lett. A* 2004, 19, 71–82.

[35] Shalyt-Margolin, A.E. Nonunitary and unitary transitions in generalized quantum mechanics, new small parameter and information problem solving. *Mod. Phys. Lett. A* 2004, 19, 391–404.

[36] Shalyt-Margolin, A.E. Pure states, mixed states and Hawking problem in generalized quantum mechanics. *Mod. Phys. Lett. A* 2004, 19, 2037–2045.

[37] Shalyt-Margolin, A.E. The Universe as a nonuniform lattice in finite volume hypercube. I. Fundamental definitions and particular features*Intern. Journ. Mod.Phys D* 2004, 13, 853– 864.

[38] Shalyt-Margolin, A.E. The Universe as a nonuniform lattice in the finite-dimensional hypercube. II. Simple cases of symmetry breakdown and restoration. *Intern.Journ.Mod.Phys.A* 2005, 20, 4951–4964.

[39] Shalyt-Margolin, A.E.; Strazhev,V.I. The Density Matrix Deformation in Quantum and Statistical Mechanics in Early Universe. In *Proc. Sixth International Symposium "Frontiers of Fundamental and Computational Physics"*,edited by B.G. Sidharth at al. Springer,2006, pp.131–134.

[40] Shalyt-Margolin, A.E. The Density matrix deformation in physics of the early universe and some of its implications. In *Quantum Cosmology Research Trends*,edited by A. Reimer, Horizons in World Physics. 246, Nova Science Publishers, Inc., Hauppauge, NY,2005, pp. 49–91.

[41] Shalyt-Margolin, A.E. Deformed density matrix and quantum entropy of the black hole. *Entropy* 2006, 8, 31–43.

[42] Hooft, G. 'T. Dimensional reduction in quantum gravity.Essay dedicated to Abdus Salam *gr-qc/9310026*, 15pp.

[43] Hooft, G. 'T. The Holographic Principle, *hep-th/0003004*,15pp.; L.Susskind, The World as a hologram. *J. Math. Phys* 1995, 36, 6377–6396.

[44] Bousso, R. The Holographic principle. *Rev. Mod. Phys* 2002, 74, 825–874.

[45] Bousso, R. A Covariant entropy conjecture. *JHEP* 1999, 07, 004.

[46] Buniy,R. V.; Hsu,S. D. H. Entanglement entropy, black holes and holography. *Phys.Lett. B* 2007, 644, 72–76.

[47] Balazs,C.; Szapudi,I. Naturalness of the vacuum energy in holographic theories. *hep-th/0603133*, 4pp.

[48] Fischler,W.; Susskind,L. Holography and cosmology. *hep-th/9806039*,7pp.

[49] Jacobson,T. Thermodynamics of space-time: The Einstein equation of state. *Phys. Rev. Lett* 1995, 75, 1260–1263.

[50] Cai,R.-G.; Kim,S.P. First law of thermodynamics and Friedmann equations of Friedmann-Robertson-Walker universe. *JHEP* 2005, 02, 050.

[51] Shalyt-Margolin,A.E.; Strazhev, V. I. Dark Energy and Deformed Quantum Theory in Physics of the Early Universe. In *Non-Eucleden Geometry in Modern Physics. Proc. 5-th Intern. Conference of Bolyai-Gauss-Lobachevsky (BGL-5)*, edited by Yu. Kurochkin and V. Red'kov,Minsk, 2007, pp. 173–178.

[52] Mukohyama, S.; Randall, L. A Dynamical approach to the cosmological constant. *Phys.Rev.Lett* 2004, 92, 211302.

[53] Cai,Rong-Gen; Hu,Bin; Zhang,Yi Holography, UV/IR Relation, Causal Entropy Bound and Dark Energy. *Commun. Theor. Phys.* 2009, 51, 954–960.

[54] Shapiro, Ilya L.; Sola, Joan. Can the cosmological "constant" run? - It may run. *arXiv:0808.0315*, 35pp.

[55] Jejjala, V.; Kavic, M.; Minic, D. Time and M-theory. *Int. J. Mod. Phys. A* 2007, 22, 3317–3405.

[56] Jejjala, V.; Kavic, M.; Minic, D. Fine structure of dark energy and new physics. *Adv. High Energy Phys.* 2007, *2007*, 21586.

[57] Jejjala, V.; Minic, D. Why there is something so close to nothing: Towards a fundamental theory of the cosmological constant. *Int.J.Mod.Phys.A* 2007, 22, 1797-1818.

[58] Jejjala, V.; Minic, D.; Tze, C-H. Toward a background independent quantum theory of gravity. *Int. J. Mod. Phys. D*, 2004, 13, 2307–2314.

[59] Shalyt-Margolin,A.E.; Tregubovich,A.Ya. Generalized uncertainty relation in thermodynamics. *gr-qc/0307018*,7pp.

[60] Bohr,N. Faraday Lectures. Chemical Society, London, 1932; pp. 349-384, 376-377.

[61] Heisenberg, W. Der Teil und Das Ganze ch 9 R.Piper, Munchen, 1969.

[62] Lindhard,J. Complementarity between energy and temperature. In: *The Lesson of Quantum Theory*; Ed. by J. de Boer, E.Dal and O.Ulfbeck North-Holland, Amsterdam 1986; Lavenda, B. Statistical Physics: a Probabilistic Approach. J.Wiley and Sons, N.Y., 1991; Mandelbrot,B. An Outline of a Purely a Phenomenological

of Statistical Thermodynamics: I.Canonical Ensembles. *IRE Trans. Inform. Theory* 1956, IT-2, 190–198; Rosenfeld L. In *Ergodic theories*; Ed. by P.Caldrirola Academic Press, N.Y., 1961; Schlogl,F. Thermodynamic Uncertainty Relation. *J. Phys. Chem. Solids* 1988, 49, 679–687; Uffink J.; Lith-van Dis,J. Thermodynamic Uncertainty Relation. *Found. of Phys.* 1999, 29, 655–679.

[63] Farmany,A. Probing the Schwarzschild horizon temperature. *Acta Phys. Pol. B* 2009, 40, 1569–1574.

[64] Carroll,R. Fluctuations, Information, Gravity and the Quantum Potential. *Fundam.Theor.Phys.148*, Springer, N.Y., 2006; 454pp.

[65] Padmanabhan,T. Vacuum fluctuations of energy density can lead to the observed cosmological constant. *Class.Quant.Grav.* 2005, 22, L107-L110.

[66] Padmanabhan,T. Darker side of the universe ... and the crying need for some bright ideas! *Proceedings of the 29th International Cosmic Ray Conference*, Pune, India,2005; pp. 47–62.

[67] Padmanabhan,T. Dark Energy: Mystery of the Millennium. *Paris 2005, Albert Einstein's century* , AIP Conference Proceedings 861, American Institute of Physics, New York, 2006; pp. 858–866.

[68] Padmanabhan,T. A New perspective on gravity and the dynamics of spacetime. *Int.Jorn.Mod.Phys* 2005, D14, 2263–2270.

[69] Padmanabhan,T. The Holography of gravity encoded in a relation between entropy, horizon area and action for gravity. *Gen.Rel.Grav* 2002, 34, 2029–2035.

[70] Padmanabhan,T. Holographic Gravity and the Surface term in the Einstein-Hilbert Action. *Braz.J.Phys* 2005, 35, 362–372.

[71] Padmanabhan,T. Gravity: A New holographic perspective. *Int.J.Mod.Phys.D* 2006, 15, 1659–1676.

[72] Mukhopadhyay,A.; Padmanabhan,T. Holography of gravitational action functionals. *Phys.Rev.D* 2006, 74, 124023.

[73] Padmanabhan,T. Dark energy and gravity. *Gen.Rel.Grav* 2008, 40, 529–564.

[74] Padmanabhan,T.; Paranjape,A. Entropy of null surfaces and dynamics of spacetime. *Phys.Rev. D* 2007, 75, 064004.

[75] Padmanabhan,T. Gravity as an emergent phenomenon: A conceptual description. *International Workshop and at on Theoretical High Energy Physics (IWTHEP 2007)*, AIP Conference Proceedings 939, American Institute of Physics, New York, 2007; pp. 114–123.

[76] Padmanabhan,T. Gravity and the thermodynamics of horizons. *Phys.Rept* 2005, 406, 49–125.

[77] Paranjape,A.; Sarkar, S.; Padmanabhan,T. Thermodynamic route to field equations in Lancos-Lovelock gravity. *Phys.Rev. D* 2006, 74, 104015.

[78] Padmanabhan,T. Thermodynamical Aspects of Gravity: New insights.arXiv:0911.5004, Invited Review to appear in Reports on Progress in Physics.

[79] Bekenstein,J.D. Black Holes and Entropy. *Phys.Rev.D* 1973, 7, 2333–2345.

[80] Hawking,S. Black Holes and Thermodynamics. *Phys.Rev.D* 1976, 13,191–204.

[81] Adler,R. J.; Chen,P.; Santiago, D. I. The generalized uncertainty principle and black hole remnants. *Gen.Rel.Grav.* 2001, 13, 2101-2108.

[82] Custodio, P. S.; Horvath, J. E. The Generalized uncertainty principle, entropy bounds and black hole (non)evaporation in a thermal bath. *Class.Quant.Grav.* 2003, 20, L197-L203.

[83] Cavaglia, M.; Das,S. How classical are TeV scale black holes? *Class.Quant.Grav.* 2004, 21, 4511–4523.

[84] Bolen, B.; Cavaglia,M. (Anti-)de Sitter black hole thermodynamics and the generalized uncertainty principle. *Gen.Rel.Grav.* 2005, 37, 1255–1263.

[85] Medved, A.J.M.; Vagenas, E.C. When conceptual worlds collide: The GUP and the BH entropy. *Phys. Rev. D* 2004, 70, 124021.

[86] Park, M.-I. The Generalized Uncertainty Principle in (A)dS Space and the Modification of Hawking Temperature from the Minimal Length. *Phys.Lett.B* 2008, 659, 698–702.

[87] Kim,Wontae.; Son,Edwin J.; Yoon, Myungseok. Thermodynamics of a black hole based on a generalized uncertainty principle. *JHEP* 2008, 08, 035.

[88] Balesku, R. Equilibruim and Nonequilibruim Statistical Mechanics,v.1,A Wiley Interscience Publications, New York-London-Sydney-Toronto, 1975.

[89] Bazarov, I.P. Thermodynamics,Moskow, Press "Higher School", 1991.

[90] Gyarmati, I. Non-Equilibruim Thermodynamics. Field Theory and Varitional Principles, Springer-Verlag, Berlin-Heidelberg-New York, 1974.

[91] Kim,Yong-Wan; Lee,Hyung Won; Myung, Yun Soo. Entropy bound of local quantum field theory with generalized uncertainty principle. *Phys.Lett.B* 2009, 673, 293-296.

[92] Shalyt-Margolin, A.E. Fundamental Length,Deformed Density Matrix and New View on the Black Hole Information Paradox. *gr-qc/0207074*, 12pp.

[93] Shalyt-Margolin, A.E.; Suarez,J.G. Density matrix and dynamical aspects of quantum mechanics with fundamental length. *gr-qc/0211083*, 14pp.

[94] Shalyt-Margolin,A.E.; Tregubovich,A.Ya. Generalized Uncertainty Relations,Fundamental Length and Density Matrix. *gr-qc/0207068*,11pp.

[95] Ahluwalia-Khalilova,D.V. Minimal spatio-temporal extent of events, neutrinos, and the cosmological constant problem. *Int.J.Mod.Phys.D* 2005, 14, 2151–2166.

[96] Vilela Mendes, R. Geometry, stochastic calculus and quantum fields in a noncommutative space-time. *J. Math. Phys* 2000, 41, 156–186.

[97] Chryssomalakos,C.; Okon, E. Generalized quantum relativistic kinematics: A Stability point of view. *Int. J. Mod. Phys D* 2004, 13, 2003–2034.

[98] Shalyt-Margolin, A.E. Entropy in the Present and Early Universe. *Symmetry* 2007, 18, 299–320.

[99] Patil,S. P. Degravitation, Inflation and the Cosmological Constant as an Afterglow. *JCAP* 2009, 0901, 017.

[100] Park, Mu-in. The Black Hole and Cosmological Solutions in IR modified Horava Gravity. *JHEP* 2009, 0909, 123.

In: Focus on Quantum Mechanics
Editors: David E. Hathaway et al. pp. 63-86
ISBN 978-1-62100-680-0

Chapter 4

THEORETICAL ASPECTS OF QUASIPARTICLE EXCITATIONS FOR BOSE-EINSTEIN CONDENSATES IN LATTICE POTENTIALS

G.S. Paraoanu
Low Temperature Laboratory, Aalto University,
P. O. Box 15100, FI-00076 AALTO, Finland.

Abstract

Bosonic condensates in optical lattices have been studied intensively in recent times. Here we present an introduction to the theory of Bogoliubov quasiparticles in atomic condensates trapped in lattice potentials. We give an overview of two results. The first is a stability analysis of a chain of condensates with periodic boundary conditions. Such a system can be realized by overlapping onto the lattice a standard parabolic-shape potential pierced by a blue-detuned laser. Then we show that in the case of only two lattice sites, Bogoliubov theory predicts correctly the existence of a collective mode of oscillation, which turns out to be the Josephson plasmon.

PACS 03.75.Fi, 74.50.+r, 05.30.Jp

1. Introduction

The achievement of Bose-Einstein condensation in trapped alkali atomic gases [1] opened the way to further extend our understanding of these macroscopic quantum phenomena in a new system: cold dilute bosonic gases. The first condensates were created in quadratic potentials realized by magnetic fields. It was however soon realized that counter-propagating laser beams would create a stationary wave which would in fact behave like a periodic potential for the atoms. Although neutral, the atoms would still feel a potential due to the ac Stark shift, which creates an increase in the internal energy of the atoms as they move away from the nodes of the interference pattern. It does not mean though that the atoms are simply confined in each of wells thus created by the laser beam. For appropriate values of the lasers' power (which is tuned experimentally using an acusto-optic modulator), the height of the potential between neighboring wells can be lowered in a controllable way, such that atoms can hop between them *via* the tunneling effect. Since there is almost dissipation and there are no impurities, these systems provide a true realization of texbook examples of many solid-state physics models [2]. An important experimental achievement related to this has been for example the demonstration of the superfluid to Mott insulator transition [3], which is based on the interplay between the hopping energy and the total interparticle interaction energy in each well.

An important situation occurs when the lattice contains only two sites (wells). In this case, we have a bosonic analog of the Josepshon effect. Although this effect refers traditionally to the case of Cooper pairs tunneling from one superconductor to another through a thin film [4], the effect is more general and occurs whenever the number of states available in the one-particle Hilbert space is restricted to two orbitals. In the case of liquid He for example, this is usually done by weakly connecting two recipients and cooling the system below the normal-superfluid transition. This generates a wealth of physical phenomena that are essentially due to collective oscillations between the two superfluids. In the case of trapped alkali gases the two superfluids can be either two condensates of the same species spatially separated by a potential barrier (external Josephson effect) [5, 6] or can correspond to spatially overlapping condensates made of atoms in two different hyperfine states (internal Josephson effect) [7]. The physics of the two-well Josephson effect with Bose-Einstein condensates has attracted significant interest from experimentalists recently. Double-well systems can be realized for example by using two pairs of Bragg beams which outcouple atoms in opposite directions [8], or using atom chip traps and dressed states created by R.F. fields [9], or by overlapping a three-dimensional harmonic potential with a periodic potential with large periodicity [10]. Some recent notable results include noise thermometry [11] and the demonstration of squeezing and entanglement in split condensates [12].

Note that all these effects are obtained essentially from using particular forms of the trapping potential. There is otherwise nothing else different from the point of view of the description of the system. The general form of the Hamiltonian for a system of interacting bosons which we will use is

$$\hat{H} = \int d\vec{r}\hat{\psi}^{\dagger}(\vec{r})\left[\frac{-\hbar^2}{2m}\nabla^2 + V(\vec{r})\right]\hat{\psi}(\vec{r}) + \frac{1}{2}\int d\vec{r}\int d\vec{r}'\hat{\psi}^{\dagger}(\vec{r})\hat{\psi}^{\dagger}(\vec{r}')U(\vec{r}-\vec{r}')\hat{\psi}(\vec{r}')\hat{\psi}(\vec{r}) \quad (1)$$

The interaction between atoms is given by scattering: if the scattering length is a, then it can be shown that the scattering potential $U(\vec{r}',\vec{r}) = g\delta(\vec{r}'-\vec{r})$, where $g = 4\pi\hbar^2/m$.

In this paper we review a number of theoretical aspects related to the description of quasiparticle excitations in systems of ultracold, Bose-Einstein condensed atoms trapped in multiple wells, either created by optical lattices or (in the case of two wells) by the technique of splitting a harmonic potential using a blue-detuned laser beam. We show that the Bogoliubov-de Gennes method offers a simple but powerful way to examine the stability of currents in such systems and allows us to calculate the fluctuations of relevant operators such as the (relative) numbers of particles and phases. In this chapter, we give a review of some of the results from [13, 14] in a self-contained presentation based on the Bogoliubov-de Gennes formalism [15]. We assume that the reader has a minimal familiarity with the field of Bose-Einstein condensation: for a more in-depth familiarization, there exists well-known review papers [16, 17] and books [18].

A similar strategy as the one presented here has been applied to the study of vortices, which is reviewed in [19]. Quantized vortices are one of the hallmarks of superfluidity, and as a result there has been a lot of experimental effort in this direction. We do expect that, naturally, systems of connected condensates such as those described here will attract a similar interest in the future. Experimentally, the first single quantized vortex was created at JILA [20] by coherent conversion between two hyperfine states of ^{87}Rb. One of the states served as a core (thus stabilizing the vortex through interatomic repulsive interactions), the other one forms the vortex around it. The existence of a phase winding of 2π, as determined by the quantization condition, was proved by a Ramsey interference technique (transferring coherently one component into the other and, at different times during the process, imaging the interference pattern). In a later series of experiments, the vortex core was removed and the precession frequency was shown to suffer only a modest change [21]. The first arrays of vortices were created by the ENS group [22] in a single-component ^{87}Rb cigar-shaped condensate by stirring it with a rotating laser beam while cooling through the transition point. The laser beam creates a small rotating attractive potential which, when combined with the trap, gives the equivalent of an anisotropic rotating "container". The angular momentum of the rotating condensate has been inferred from the measured precession of the axis of the quadrupole mode. Finally, an impressively large number of vortices (over 100) "crystallized" in an almost perfect triangular lattice - similar to the ones predicted by Abrikosov and later observed for the flux lines in type-II superconductors - has been produced shortly afterwards by the MIT group [23] in a ^{23}Na condensate stirred with a blue-detuned laser beam. Much like in the case discussed in this chapter, the properties of vortices at zero temperature can be studied by using the time-dependent Gross-Pitaevskii equation [24]. For the stability properties, one needs to go further and consider perturbations around the vortex state. For vortices created in a trap by a time-dependent and spatially-dependent perturbation the excitation spectrum was studied by a linear-response analysis in [25]. A complete numerical and analytical treatment of various instabilities of vortices formed in three-dimensional condensates in spherically and cylindrically symmetric traps can be found in [26]. The authors found that the $m = 1$ and $m = 2$ vortices are energetically (thermodynamically) unstable; for the $m = 1$ vortex they show that the instability is due to a destabilizing mode (the so-called "core" mode) predicted in [27]. Here the idea was that, once the vortex is formed, its wavefunction can be continuously changed due to particles tunneling (or thermal cloud [28]) from the vortex to the mode orthogonal to it - which is localized in the core of the vortex. This leads to a displacement of the vortex line towards the edge of the condensate. The effect of this mode can be compensated by a well-chosen rotation frequency of the trap [26]. Several studies have shown that the stability of vortices can be controlled by the strength of the interaction [29], that the thermodynamical stability can be caused by finite-temperature effects or by a pinning potential (as produced by a laser beam focused in the center) [30], and finally a phase diagram has been obtained [31]. It has been also shown that the interactions play a crucial role also in the stability of vortices against "dipole" formation [32]. Also, some unstable modes with negative energy and positive normalization of vortices in thermodynamically unstable states give rise to interesting collective oscillations, such as the precession of the vortex line around the center of the trap [33]. A careful numerical analysis of the thermodynamical instabilities has been performed in [34]. Finally, a situation in which three condensates are connected and tunneling is allowed is analyzed in [35] for arbitrary initial phases. The authors found, by numerically solving the time-dependent Gross-Pitaevskii equation, that in certain conditions a Josephson current can flow between the adjacent condensates.

2. Gross-Pitaevskii and Bogoliubov Theory: Generalities

We start by deriving the Gross-Pitaevskii equation. The easiest way is to start with the Heisenberg equations for the time-dependent field operators $\psi(\vec{r},t)$,

$$i\hbar\frac{\partial}{\partial t}\hat{\psi}(\vec{r},t) = [\hat{\psi}(\vec{r},t),\hat{H}] \tag{2}$$

$$= \left[\frac{-\hbar^2}{2m}\nabla^2 + V(\vec{r})\right]\hat{\psi}(\vec{r},t) + g\hat{\psi}^{+}(\vec{r},t)\hat{\psi}(\vec{r},t)\hat{\psi}(\vec{r},t). \tag{3}$$

The next step is simply to assume that one of the modes is highly populated (as a result of Bose-Einstein condensations); we can then expand the field operator $\hat{\psi}$ in a basis comprising that mode (call it $\Phi(\vec{r},t)$) and the rest of the modes. The mode operator $\hat{a}_{\Phi}$ is then approximated by its semiclassical value $\hat{a}_{\Phi}^{\dagger} \approx \hat{a}_{\Phi} \approx \sqrt{N}$ (since the average number of particles on this mode, in the first approximation, is equal to the total number of particles N, $\langle\hat{a}_{\Phi}^{\dagger}\hat{a}_{\Phi}\rangle \approx N$). This results in the celebrated time-dependent Gross-Pitaevskii equation:

$$i\hbar\frac{\partial}{\partial t}\Phi(\vec{r},t) = \left[\frac{-\hbar^2}{2m}\nabla^2 + V(\vec{r})\right]\Phi(\vec{r},t) + gN|\Phi(\vec{r},t)|^2\Phi(\vec{r},t). \tag{4}$$

The time-independent Gross-Pitaevskii equation is obtained from Eq. 4 by attempting a solution of the form $\Phi(\vec{r},t) = \Phi(\vec{r})\exp(-i\mu t)$, where μ is the chemical potential. We find

$$\mu\Phi(\vec{r},t) = \left[\frac{-\hbar^2}{2m}\nabla^2 + V(\vec{r})\right]\Phi(\vec{r},t) + gN|\Phi(\vec{r},t)|^2\Phi(\vec{r},t). \tag{5}$$

The time-independent Gross-Pitaevskii equation Eq. (4) can be obtained as well by the method of Lagrange multipliers. The energy functional corresponding to the Hamiltonian Eq. (1) can be obtained by replacing the field operator with the order parameter $\sqrt{N}\Phi(\vec{r})$, and after integrating by parts and using as boundary conditions a vanishing order parameter at $\pm\infty$ we obtain the functional

$$E[\Phi] = N\int d\vec{r}\left[\frac{\hbar^2}{2m}|\nabla\Phi(\vec{r})|^2 + V(\vec{r})|\Phi(\vec{r})|^2 + \frac{g}{2}|\Phi(\vec{r})|^4\right]. \tag{6}$$

We then minimize the functional Eq. (6) under the constraint that the total number of particles, as given by the normalization of the vector parameter, is fixed (equivalently, the normalization of $\Phi(\vec{r})$ is 1, as it should be for a properly defined wavefunction),

$$\frac{\delta}{\delta\Phi}E[\Phi] - \mu N\int d\vec{r}|\Phi(\vec{r})|^2 = 0. \tag{7}$$

This results in Eq. (5), with the chemical potential μ playing the role of a Lagrange multiplier. Much like the linear Schrödinger equation, the time-independent Gross-Pitaevskii equation can be solved, either numerically or analytically, if the external (trapping) potential $V(\vec{r})$ is given.

To obtain the energies of the excitations (at zero temperature) one can search for solutions of the time-dependent Gross-Pitaevskii equation in the form

$$\Phi(\vec{r},t) = e^{-i\mu t/\hbar}\left[\Phi(\vec{r}) + u(\vec{r})e^{-iEt/\hbar} + v(\vec{r})^* e^{iEt/\hbar}\right], \tag{8}$$

which is a small-amplitude complex oscillation around $\Phi(\vec{r})$. Introducing this solution into Eq. (4) and linearizing we get the Bogoliubov-deGennes equations in the general, nonuniform case,

$$Eu(\vec{r}) = \left[-\frac{\hbar^2}{2m}\nabla^2 + V(\vec{r}) - \mu + 2gN\Phi(\vec{r})^2\right]u(\vec{r}) + gN\Phi(\vec{r})^2 v(\vec{r}), \tag{9}$$

$$-Ev(\vec{r}) = \left[-\frac{\hbar^2}{2m}\nabla^2 + V(\vec{r}) - \mu + 2gN\Phi(\vec{r})^2\right]v(\vec{r}) + gN\Phi(\vec{r})^2 u(\vec{r}). \tag{10}$$

3. Gross-Pitaevskii Theory on Lattices

Consider now the case of ultracold atoms trapped in an optical lattice. We will also assume that the height of the barrier between consecutive wells is not too high - such that tunneling is possible and the system is superfluid (*i.e.* not a Mott insulator). The system we will consider is that of M identical small condensates in tunneling contact with each other and a cylindrical system of coordinates (r,θ,z). The total number of atoms is N. The centers of the condensates are positioned then at $\vec{R}_\lambda = (R,\theta_\lambda, z=0)$, where $\theta_\lambda = 2\pi\lambda/M$ and λ runs from 0 to $M-1$. As already mentioned, we assume that the phase relation between the wells has already been established, therefore it makes sense to have an order parameter for the whole system $\psi(\vec{r})$. The Hamiltonian of the system is

$$\hat{H} = \int d\vec{r}\psi^*(\vec{r})\left[-\frac{\hbar^2}{2m}\Delta + V(\vec{r}) + \frac{g}{2}|\psi(\vec{r})|^2\right]\psi(\vec{r}). \tag{11}$$

The external potential $V(\vec{r})$ is taken to be high enough around the origin, so that the atoms cannot penetrate there; it also has a number of M minima at (R,θ_λ), around which the condensed atoms tend to localize.

We also assume that each condensate has a small enough number of particles so that they are below the Thomas-Fermi regime. In this situation, the wavefunctions of each condensate depends weakly on number of atoms in the well [66], and one can apply the M-mode approximation for the field operator

$$\hat{\psi}(\vec{r}) = \sum_{\lambda=0}^{M-1}\phi(\vec{r}-\vec{R}_\lambda)\hat{a}(\lambda), \tag{12}$$

where ϕ is a solution of the Schrödinger equation for each well; the Hamiltonian (11) becomes

$$H = -t\sum_{\lambda=0}^{M-1}\left[\hat{a}^\dagger(\lambda)\hat{a}(\lambda+1) + \hat{a}^\dagger(\lambda+1)\hat{a}(\lambda)\right] + \frac{E_c}{4}\sum_{\lambda=0}^{M-1}\hat{a}^\dagger(\lambda)\hat{a}^\dagger(\lambda)\hat{a}(\lambda)\hat{a}(\lambda). \tag{13}$$

Here we have not written constant terms and the tunneling matrix element is given by

$$t = \int d\vec{r} \phi^*(\vec{r}) \left[\frac{\hbar^2}{m} \Delta + V(\vec{r}) \right] \phi(\vec{r}), \tag{14}$$

and the on-site energy is

$$E_c = \frac{8\pi\hbar^2 a}{m} \int \vec{r} |\phi(\vec{r})|^4. \tag{15}$$

The boundary conditions are periodic by construction, $\hat{a}(\lambda) = \hat{a}(\lambda + M)$. The M-mode approximation and the relation (12) can be seen also as an expansion of the field operator in the Wannier basis.

The Hamiltonian (13) can be diagonalized in the Bogoliubov formalism. We start first with identifying the solutions of the Gross-Pitaevskii equation. This can be done by seeing the $\hat{a}$'s as field operators and λ as the coordinate. Then the first term of the Hamiltonian (13) is the equivalent of the kinetic energy and the second term is the interaction energy. To find the Gross-Pitaevskii state we write

$$\hat{a}(\lambda) = \sum_{q=0}^{M-1} \chi_q(\lambda) \hat{b}_q, \tag{16}$$

where the "condensate wavefunction" $\chi_q(\lambda)$ is periodic $\chi_q(\lambda) = \chi_q(\lambda + M)$, and minimize the mean-field energy on the state $\frac{1}{\sqrt{N!}} b_q^{\dagger N} |0\rangle$ with the restriction that the total number of particles is constant

$$\frac{\delta}{\delta\chi} \left[\langle H \rangle - \mu \sum_{\lambda=0}^{M} \langle \hat{a}^\dagger(\lambda) \hat{a}(\lambda) \rangle \right] = 0. \tag{17}$$

This gives the lattice Gross-Pitaevskii equation

$$-t \left[\chi_q(\lambda+1) + \chi_q(\lambda-1) \right] + \frac{E_c}{2} N |\chi_q(\lambda)|^2 \chi_q(\lambda) = \mu \chi_q(\lambda). \tag{18}$$

The analogy with the usual continuous Gross-Pitaevskii equation is even more transparent if one notices that if $\chi(\lambda)$ does not vary much from site to site, one can define the usual lattice derivative

$$\frac{d^2}{d\lambda^2} \chi(\lambda) \simeq \chi(\lambda+1) + \chi(\lambda-1) - 2\chi(\lambda), \tag{19}$$

with the observation that the last term is in fact absent in our equations since it is effectively included in the chemical potential (in other words the mode $k = 0$ has a contribution $-2t$ to the chemical potential and not zero, as it is in the case of the continuous Hamiltonians condsidered before, where the free-particle part has a contribution $\hbar^2 k^2/2m$ which is zero if $k = 0$).

The equation (18) admits the solutions

$$\chi_q(\lambda) = \frac{1}{\sqrt{M}} e^{i \frac{2\pi}{M} q \lambda}, \tag{20}$$

$$\mu_q = -2t \cos \frac{2\pi}{M} q + \frac{E_c N}{2M}, \tag{21}$$

corresponding to uniform flow states with quasi-momentum q.

Another useful way of approaching this problem is to write the Hamiltonian (13) in the quasi-momentum basis, defined by

$$\hat{b}_k = \frac{1}{\sqrt{M}} \sum_{\lambda=0}^{M-1} e^{-i\frac{2\pi}{M}k\lambda} \hat{a}(\lambda), \tag{22}$$

$$\hat{b}_k^\dagger = \frac{1}{\sqrt{M}} \sum_{\lambda=0}^{M-1} e^{i\frac{2\pi}{M}k\lambda} \hat{a}^\dagger(\lambda), \tag{23}$$

with inverse

$$\hat{a}(\lambda) = \frac{1}{\sqrt{M}} \sum_{k=0}^{M-1} e^{i\frac{2\pi}{M}k\lambda} \hat{b}_k, \tag{24}$$

$$\hat{a}^\dagger(\lambda) = \frac{1}{\sqrt{M}} \sum_{k=0}^{M-1} e^{-i\frac{2\pi}{M}k\lambda} \hat{b}_k^\dagger. \tag{25}$$

Using the relation

$$\frac{1}{M} \sum_{\lambda=0}^{M-1} e^{i\frac{2\pi}{M}(k-k')} = \delta_{k,k'}, \tag{26}$$

we obtain

$$\hat{H} = -2t \sum_{k=0}^{M-1} \cos\left(\frac{2\pi}{M}k\right) \hat{b}_k^\dagger \hat{b}_k + \frac{E_c}{4M} \sum_{k,k',l} \hat{b}_{k+l}^\dagger \hat{b}_{k'-l}^\dagger \hat{b}_{k'} \hat{b}_k, \tag{27}$$

which looks exactly like an uniform system in which the free bosons have kinetic energy $-2t\cos(2\pi k/M)\hat{b}_k^\dagger\hat{b}_k$ instead of the usual $\hbar^2k^2/2m$.

The manybody wavefunction corresponding to formation of a condensate at the Gross-Pitaevskii level of approximation is

$$|\mathrm{GP}\rangle = \frac{1}{\sqrt{N!}}[\hat{b}_q^{\dagger N}]|0\rangle \tag{28}$$

$$= \frac{1}{\sqrt{N!}} \left[\frac{1}{\sqrt{M}} \sum_{\lambda=0}^{M-1} e^{-i\frac{2\pi}{M}q\lambda} \hat{a}(\lambda)\right]^N |0\rangle \tag{29}$$

Then the mean-field energy in the Gross-Pitaevskii state above can be, which can be readily obtained from Eq. (27) as

$$\langle \mathrm{GP}|\hat{H}|\mathrm{GP}\rangle = -2tN\cos\frac{2\pi q}{M} + \frac{E_c}{4}\frac{N(N-1)}{M}. \tag{30}$$

In contradistinction, the average energy on a fragmented state (which is analogous to the Mott insulator state in solid-state physics, see *e.g.* [36] and references therein),

$$|\Psi\rangle_\mathrm{f} = \prod_{\lambda=0}^{M-1} \frac{[\hat{a}^\dagger(\lambda)]^{N/M}}{\sqrt{(N/M)!}}|0\rangle, \tag{31}$$

is obtained as

$$ {}_f\langle\Psi|\hat{H}|\Psi\rangle_f = \frac{N(N-M)E_c}{4M}. \tag{32} $$

Note the difference between the expressions Eq. (32) and Eq. (30): the last one does not contain any term related to tunneling, as the assumption is that the state is simply a product of states localized at each well, with no effect due to intrawell tunneling. Indeed, for obvious physical reasons, this would be the case if the barrier between consecutive wells is large enough. In this case, as the number of particles in each well becomes fixed, the fluctuations in the relative phase between neighboring wells become large and the system is in an "insulating" phase.

We then find the condition

$$ 2t\cos\frac{2\pi q}{M} > \frac{E_c}{4}\left(1-\frac{1}{M}\right), \tag{33} $$

which has to be satisfied in order for the ground state of the system to be condensed. Note that in this situation we have to take seriously the possibility of fragmentation due to small number of particles involved. This condition essentially tells us that $t > E_c$ which can happen only in the Rabi and Josephson regimes (these two regimes are both characterized by $E_c \ll tN/M$).

4. Lattice Bogoliubov-deGennes Equations

We now turn to the problem of excitation spectrum. We attempt to diagonalize the Hamiltonian Eq. (27) by using the method of Bogoliubov transformations and the assumption that there exists a macroscopically occupied state with momentum q. We then approximate the mode operator corresponding to q as a classical field $\hat{b}_q \approx \hat{b}_q^\dagger \approx \sqrt{N_q}$, where N_q represents the number of atoms in the condensation mode. It is also natural to use the momentum q as a reference and index all the other momenta based on this, in other words $k\rangle q+k$. On the other hand, the total number of atoms is given by $\hat{N} = N_q + \sum_{k\neq q}\hat{b}_k^\dagger\hat{b}_k$, and therefore we can approximate

$$ \hat{b}_q \approx \hat{N} - \frac{1}{2}\sum_{k\neq 0}\hat{b}_{q+k}^\dagger\hat{b}_{q+k}. \tag{34} $$

Using these approximations, we find the following series of approximations for the Hamiltonian,

$$ \begin{aligned} \hat{H} \approx\ & -2t\cos\frac{2\pi q}{M}N_0 + \frac{E_cN_0^2}{4M} - 2t\sum_{k\neq 0}\cos\frac{2\pi(q+k)}{M}\hat{b}_{q+k}^\dagger\hat{b}_{q+k} \\ & + \frac{E_cN_0}{4M}\sum_{k\neq 0}\left(4\hat{b}_{q+k}^\dagger\hat{b}_{q+k} + \hat{b}_{q+k}^\dagger\hat{b}_{q-k}^\dagger + \hat{b}_{q+k}\hat{b}_{q-k}\right) \end{aligned} \tag{35} $$

$$ \begin{aligned} \approx\ & -2tN + \frac{E_cN^2}{4M} + \sum_{k\neq 0}\left[\left(-2t\cos\frac{2\pi(q+k)}{M} + 2t\cos\frac{2\pi q}{M}\right.\right. \\ & \left.\left. + \frac{E_cN}{2M}\right)\hat{b}_{q+k}^\dagger\hat{b}_{q+k} + \frac{E_cN}{4M}\hat{b}_{q+k}^\dagger\hat{b}_{q-k}^\dagger + \frac{E_cN}{4M}\hat{b}_{q+k}\hat{b}_{q-k}\right] \end{aligned} \tag{36} $$

Then we perform a Bogoliubov transformation in the form

$$\hat{b}_{q+k} = u_k\hat{\gamma}_{q+k} - v_k\hat{\gamma}^{\dagger}_{q-k} \tag{37}$$

$$\hat{b}^{\dagger}_{q-k} = u^*_k\hat{\gamma}^{\dagger}_{q-k} - v^*_k\hat{\gamma}_{q+k} \tag{38}$$

The operators $\hat{\gamma}_{q+k}$ are Bogoliubov operators defined around the mode q. They satisfy bosonic commutation relations, $[\hat{\gamma}_{q+k}, \hat{\gamma}_{q+k'}] = [\hat{\gamma}^{\dagger}_{q+k}, \hat{\gamma}^{\dagger}_{q+k'}] = 0$, $[\hat{\gamma}_{q+k}, \hat{\gamma}^{\dagger}_{q+k'}] = \delta_{k,k'}$. Now, the $\hat{b}$ operators are also bosonic, and as a result of this the last equation leads to the relation between u_k and v_k, $|u_k|^2 - |v_k|^2 = 1$. The purpose of this transformation is to put the Hamiltonian in the diagonal form

$$\hat{H} = \text{const.} + \sum_{k\neq 0} E_k\hat{\gamma}^{\dagger}_{q+k}\hat{\gamma}_{q+k}. \tag{39}$$

Also, when finding the ground state in the Bogoliubov-de Gennes approximation, $|\text{BdG}\rangle$, the essential property of the Bogoliubov operators is $\hat{\gamma}_k|\text{BdG}\rangle = 0$. With this condition, we obtain the celebrated Bogoliubov - de Gennes equations in the general case of excitations around a condensation mode q,

$$\left[-2t\cos\frac{2\pi}{M}(k+q) + 2t\cos\frac{2\pi}{M}q + \frac{E_cN}{2M}\right]u_k + \frac{E_cN}{2M}v_k = E_ku_k, \tag{40}$$

$$-\frac{E_cN}{2M}u_k + \left[2t\cos\frac{2\pi}{M}(k-q) - 2t\cos\frac{2\pi}{M}q - \frac{E_cN}{2M}\right]v_k = E_kv_k. \tag{41}$$

Solving these equations result in the values of u_k and v_k as well as in the dispersion relation for the quasiparticle excitations. The results are as follows:

$$E_k^{(\pm)} = 2t\sin\frac{2\pi}{M}k\sin\frac{2\pi}{M}q \pm \sqrt{\varepsilon_k\left(\varepsilon_k + \frac{E_cN}{M}\right)}, \tag{42}$$

where we have used the notation

$$\varepsilon_k = 2t\cos\frac{2\pi}{M}q\left(1 - \cos\frac{2\pi}{M}k\right). \tag{43}$$

Now, inserting the solution for E_k into (41) we obtain

$$\left[\varepsilon_k \mp \sqrt{\varepsilon_k\left(\varepsilon_k + \frac{E_cN}{M}\right)} + \frac{E_cN}{2M}\right]u_k^{(\pm)} + \frac{E_cN}{2M}v_k^{(\pm)} = 0, \tag{44}$$

$$\left[\varepsilon_k \pm \sqrt{\varepsilon_k\left(\varepsilon_k + \frac{E_cN}{M}\right)} + \frac{E_cN}{2M}\right]v_k^{(\pm)} + \frac{E_cN}{2M}u_k^{(\pm)} = 0, \tag{45}$$

with the restriction imposed by the normalization condition $|u_k|^2 - |v_k|^2 = 1$ which eventually will force us to make a choice between the two possible values indexed by $(\pm)$. An interesting observation concerns the negative-energy eigenstates; assuming that we found a positive eigenenergy $E_k^{(\pm)}$ with eigenvalues $\begin{pmatrix} u_k^{(\pm)}(\lambda) \\ v_k^{(\pm)}(\lambda) \end{pmatrix}$, the corresponding negative-energy eigenstate is $-E_k^{(\pm)}$ with eigenvalues $\begin{pmatrix} v_k^{(\pm)*}(\lambda) \\ u_k^{(\pm)*}(\lambda) \end{pmatrix}$. But $-E_k^{(\pm)} = E_{-k}^{(\mp)}$ and

$$\begin{pmatrix} v_k^{(\pm)*}(\lambda) \\ u_k^{(\pm)*}(\lambda) \end{pmatrix} = \begin{pmatrix} \frac{1}{\sqrt{M}} e^{-i\frac{2\pi}{M}(k-q)} v_k^{(\pm)*} \\ \frac{1}{\sqrt{M}} e^{-i\frac{2\pi}{M}(k+q)} u_k^{(\pm)*} \end{pmatrix} = \begin{pmatrix} u_{-k}^{(\mp)}(\lambda) \\ v_{-k}^{(\mp)}(\lambda) \end{pmatrix}, \tag{46}$$

where we have used $u_k^{(\pm)*} = v_{-k}^{(\mp)}$ and $v_k^{(\pm)*} = u_{-k}^{(\mp)}$; these last two relations can be proved readily from (40,41). They show that if say $(+)$ for a certain k is a positive-energy eigenstate, then its corresponding negative-energy eigenstate is a positive-energy eigenstate of $(-)$ for $-k$. In other words, the state $-E_k^{(-)}$ represents the quasihole (the state with minus the energy of the quasiparticle) of the state $E_{-k}^{(+)}$. Also, note that the excitations k and $-k$ propagate with different speeds around the loop, because they are excitations on top of an already existing flow in a certain direction; this phenomenon is similar to the relativistic Sagnac effect. In other words, the Sagnac effect for Bose-Einstein condensates would consist of the lifting of the degeneracy of the excitations k, $-k$ due to the rotation of the condensate.

With these observations, the values of the Bogoliubov amplitudes $u_k^{(+)}$ and $v_k^{(+)}$ are obtained from Eqs. (44-45) as

$$v_k^{(+)} = \frac{\sqrt{\varepsilon_k + \frac{E_c N}{M}} + \sqrt{\varepsilon_k}}{2\left[\varepsilon_k\left(\varepsilon_k + \frac{E_c N}{M}\right)\right]^{1/4}}, \tag{47}$$

$$u_k^{(+)} = \sqrt{1 + v_k^2} = \frac{\sqrt{\varepsilon_k + \frac{E_c N}{M}} - \sqrt{\varepsilon_k}}{2\left[\varepsilon_k\left(\varepsilon_k + \frac{E_c N}{M}\right)\right]^{1/4}}. \tag{48}$$

5. Fluctuations

The way to derive the lattice Bogoliubov - de Gennes equations is however to apply the general procedure for nonuniform potentials described in Section 2.. We start with the time-dependent Gross-Pitaevskii equation and linearize it around small fluctuations of the order parameter around the vortex. We obtain the equations

$$\begin{aligned} E_k u_k(\lambda) &= -t\left[u_k(\lambda+1) + u_k(\lambda-1)\right] + \left[-\mu_q + E_c N |\chi_q(\lambda)|^2\right] u_k(\lambda) \\ &\quad + \frac{E_c}{2} N \chi_q^2(\lambda) v_k(\lambda) \end{aligned} \tag{49}$$

$$\begin{aligned} -E_k v_k(\lambda) &= -t\left[v_k(\lambda+1) + v_k(\lambda-1)\right] + \left[-\mu_q + E_c N |\chi_q(\lambda)|^2\right] v_k(\lambda) \\ &\quad + \frac{E_c}{2} N \chi_q^{*2}(\lambda) u_k(\lambda). \end{aligned} \tag{50}$$

Compared to the form of the Bogoliubov - de Gennes equations in a nonuniform potential Eqs. (9-10) we can see that they provide a natural counterpart, with the Laplacean replaced by a lattice Laplacean as discussed also in the case of the Gross-Pitaevskii equation. The solution of this system is obtained by a Fourier transform

$$u_k(\lambda) = \frac{1}{\sqrt{M}} e^{i\frac{2\pi}{M}(k+q)} u_k \quad (51)$$

$$v_k(\lambda) = \frac{1}{\sqrt{M}} e^{i\frac{2\pi}{M}(k-q)} v_k. \quad (52)$$

This produces the same system of equations as before, namely Eq. (40-41), and the solutions are of course the same. However, the operatorial approach presented in Section 4. has the advantage of allowing us to calculate in a transparent way the fluctuations (of any order) of any operator. In general, these calculations are complicated: much simpler equations can be obtained though for the two-well problem (Josephson effect), and this will be done in a later Section.

A useful calculation though is to estimate the fluctuations of the number of particle per site (well). Since all sites are equivalent, it is enough to focus on a single site, say $\lambda = 0$. To simplify the calculation we also take $q = 0$. Then, from Eqs. (24-25) we get

$$\hat{a}(0) \approx \frac{\sqrt{N_0}}{\sqrt{M}} + \frac{1}{M}\sum_{k\neq 0} \hat{b}_k, \quad (53)$$

$$\hat{a}^\dagger(0) \approx \frac{\sqrt{N_0}}{\sqrt{M}} + \frac{1}{M}\sum_{k\neq 0} \hat{b}_k^\dagger. \quad (54)$$

Then we construct the operator $\hat{n}(0) = \hat{a}^\dagger(0)\hat{a}(0)$. The average of this operator on the ground state is in the first-order approximation

$$\langle \mathrm{BdG}|\hat{n}(0)|\mathrm{BdG}\rangle = \frac{N}{M}, \quad (55)$$

as it should be. The fluctuations can be calculated by noticing that $N_0 = \hat{N} - \sum_{k\neq 0} \hat{b}_k^\dagger \hat{b}_k$ and using the properties of the Bogoliubov operators. We get, after some algebra,

$$\langle \mathrm{BdG}|\hat{n}^2(0)|\mathrm{BdG}\rangle - \langle \mathrm{BdG}|\hat{n}(0)|\mathrm{BdG}\rangle^2 \approx \frac{N}{M^2}\sum_{k\neq 0} \frac{\varepsilon_k}{E_k}\bigg|_{q=0} \quad (56)$$

$$= \frac{N}{M^2}\sum_{k\neq 0} \frac{4t\sin^2\frac{\pi k}{M}}{\sqrt{4t\sin^2\frac{\pi k}{M}\left(4t\sin^2\frac{\pi k}{M} + E_c\frac{N}{M}\right)}} \quad (57)$$

$$= \frac{N}{2\pi M}\int_{-\pi}^{\pi} d\tilde{k} \frac{\sin^2\frac{\tilde{k}}{2}}{\sqrt{\sin^2\frac{\tilde{k}}{2}\left(\sin^2\frac{\tilde{k}}{2} + \frac{NE_c}{4tM}\right)}}, \quad (58)$$

where for the last equation we treat the modes k as a continuum, introducing $\tilde{k} = \frac{2\pi}{M}k$, of which the smallest variation is $\Delta\tilde{k} = 2\pi/M$, therefore $\sum_{\tilde{k}} \Delta\tilde{k} \to \int_{-\pi}^{\pi} d\tilde{k}$. The integral thus obtained can be solved exactly, and the result is

$$\langle \mathrm{BdG}|\hat{n}^2(0)|\mathrm{BdG}\rangle - \langle \mathrm{BdG}|\hat{n}(0)|\mathrm{BdG}\rangle^2 \approx \frac{2}{\pi}\frac{N}{M}\arctan\sqrt{\frac{4tM}{NE_c}}. \quad (59)$$

6. Stability Analysis

The complete analysis of the stability for these systems is given in [13]. Here we follow closely this reference to highlight the main results. We now perform an analysis of the stability of the persistent currents that can flow in such a system. This is can be understood simply by considering the phase parameter and varying its phase and amplitude around the value resulting from solving the Gross-Pitaevskii equation. Two types of stability can be distinguished as a result of this analysis.

6.1. Dynamical Instability

For complex-valued eigenvalues E_k the state of the system is called dynamically unstable. Indeed, an imaginary part in the E_k's would mean that any perturbation around the Gross-Pitaevskii solution is amplified exponentially in time. As such, the system is Lyapunov unstable.

The condition for dynamical stability is

$$\varepsilon_k \left(\varepsilon_k + \frac{E_c N}{M} \right) \geq 0, \tag{60}$$

which has to be satisfied for any k. We can distinguish two situations:

A] $\varepsilon_k \geq 0$ or in other words $\cos(2\pi q/M) \geq 0$; in this situation q is in the interval $[0, M/4] \bigcup [3M/4, M-1]$.

B] $\varepsilon_k < 0$ and $-\varepsilon_k \geq E_c N/M$. But now the minimum value of $-\varepsilon_k$ is obtained for $k=1$ or $k = M-1$, and it follows that

$$\cos\frac{2\pi}{M} q \leq -\frac{E_c N}{2Mt\left(1-\cos\frac{2\pi}{M}\right)}, \tag{61}$$

and therefore $E_c N/M \leq 2t\,(1-\cos 2\pi/M)$.

The system is then in a dynamically stable state if either one of the conditions A] or B] is fulfilled. Dynamically unstability means that quantum fluctuations will trigger an exponential divergent evolution away from the initial state, and therefore such states can be treated in a linearized theory only for periods of time which are logarithmic with respect to the initial state [37].

6.2. Thermodynamical Stability

Thermodynamical instability (or energetic instability) is obtained for states which are solutions of the Gross-Pitaevskii equation but are not a local minimum of the energy functional. As a result, under the action of any external interaction, the system will decay onto the true (thermodynamically stable) ground state by emitting quasiparticles. The thermodynamically unstable states have a real excitation spectrum which is characterized by the existence of eigenenergies with negative E_k (with the normalization corresponding to positive eigenstates, $|u_k|^2 - |v_k|^2 = 1$). For vortices in harmonic traps, interesting effects such as the precession of the vortex core about the condensate axis [21, 33] are predicted to occur in this regime. The mechanism is as follows: the system in a thermodynamically unstable vortex state can reduce its energy by transferring particles from the condensate to the negative-energy modes. For an un-pinned vortex for example this happens by few-particle excitations to the core mode [27].

We now consider the two cases A] and B] discussed above. The detailed analysis for each situation is given in [14]. Here we just mention the results: in the case B], we find that there is no thermodynamical stability. In case A], we obtain the condition for thermodynamical stability of a persistent current in the form

$$2t\left[1+\cos\frac{2\pi}{M}-2\cos^2\frac{2\pi}{M}q\right]\leq\frac{E_cN}{M}\cos\frac{2\pi}{M}q. \tag{62}$$

Useful particular cases can be further analyzed: for example we can see what happens in the Rabi regime, defined as $2t \gg E_cN/M$, and in the Josephson regime, where $2tM/N \ll E_c \ll 2tN/M$. These regimes are defined in the case of lattices by using the same definition as for the the external Josephson effect (see [17, 14] and also the next sections). In the Rabi regime, we search for solutions with $u_k = 1$ and $v_k = 0$. If we look at (44), it follows that we need to take the upper sign only into account. The quasiparticle energy is in this case

$$E_k = 2t\cos\frac{2\pi}{M}q - 2t\cos\frac{2\pi}{M}(k+q). \tag{63}$$

This illustrates an interesting point, namely that E_k is the difference between the "twisting" energy per particle corresponding to a quasimomentum q and the energy of an excitation around this current. Therefore, to create an elementary excitation with additional quasimomentum k relative to the flow we need to transfer an atom from the condensate to the single-particle state $k+q$. The expression Eq. (63) is positive for all k's only if $q = 0$ - meaning that all the these currents are thermodynamically unstable.

In the Josephson regime it is clear that (62) is satisfied, so we have both thermodynamical and dynamical stability for $\cos 2\pi q/M > 0$.

7. Review of the Main Results so Far

We have found the excitation spectrum around a state characterized by a persistent flow and we have discussed its stability. The spectrum thus found,

$$E_k^{(+)} = 2t\sin\frac{2\pi}{M}k\sin\frac{2\pi}{M}q + \sqrt{\varepsilon_k\left(\varepsilon_k + \frac{E_cN}{M}\right)}, \tag{64}$$

It is possible to measure this spectrum experimentally, using the techniques already demonstrated for optical lattices in [38]. The spectrum depends on the flow thus, reciprocally, it could serve as a method to detect persistent currents, similar to the case of vortices [25, 39].

8. Two-Well Lattices: External Josephson Effect

We now analyze the case of a two-site lattice. In this case, one expects oscillations of particles *via* tunnneling between the sites. If these oscillations were incoherent, nothing could be observed experimentally. However, the process of Bose-Einstein consensation ensures that all atoms have the same phase (that of the macroscopic order parameter) and therefore will move in concert. Here we will review a result [14], which demonstrates that the collective oscillations of the atoms between the wells (called Josephson plasmon in solid-state physics) can be predicted on the basis of Bogoliubov theory - in fact it is a Bogoliubov quasiparticle.

For the two-site Hamiltonian we consider the model Hamiltonian [40, 41, 5, 42, 43]

$$\hat{H} = -2t(\hat{a}^{\dagger}\hat{b} + \hat{b}^{\dagger}\hat{a}) + \frac{E_c}{4}\left[(\hat{a}^{\dagger}\hat{a})^2 + (\hat{b}^{\dagger}\hat{b})^2\right], \tag{65}$$

with no danger of confusing the site operators a and b with those used in the previous sections. More precisely, the identification is M=2 and a and b are the previous localized modes $\hat{a}(\lambda)$. Note that in order to establish the correct correspondence with Eq. (13) we have to write 2t for the tunneling Hamiltonian, due to boundary conditions which effectively doubles the term $\hat{a}^{\dagger}\hat{b} + \hat{b}^{\dagger}\hat{a}$ for two sites. It is also useful to follow the traditional notation from mesoscopic physics, and introduce a Josephson coupling energy $E_J = 2Nt$.

As before, in the case of identical wells, $\hat{a}$ and $\hat{b}$ are the annihilation operators corresponding to the condensate wavefunctions of the left and right sites. Also, as previously discussed, the parameters E_J and E_c can be evaluated, see for example [6]. Obviously, most of the previous results (with the exception of the calculation of fluctuations in Section 5., which relied on a continuum of modes) can be used now just by putting $M = 2$. An obvious result is that the state corresponding to $q = 0$ (the mode $\hat{c}$ in the notation below) is dynamically and thermodynamically stable (case A] above) when macroscopically occupied, which is expected since this is the mode on which condensation occurs. The mode $q = 1$ does not satisfy the criteria A] however, since $\cos 2\pi q/M = -1$. In principle, it could satisfy B] if $8t \geq E_c$, but in this case we have thermodynamical instability, and the system can further reduce its energy by emitting quasiparticles. This point is further illustrated in Section 10.

The Hamiltonian Eq. (65) can be also used to describe the so-called internal Josephson effect. In this case, atoms in two different hyperfine states $|1\rangle$ and $|2\rangle$ are confined in an optical or magnetic trap; to implement the Josephson coupling term a Raman transition is driven between the two levels [44]. The two modes $\hat{a}$ and $\hat{b}$ can be for example be replaced with the hyperfine states $|F{=}2, m_F{=}1\rangle$ and $|F{=}1, m_F{=}{-}1\rangle$ of ^{87}Rb for which the robustness of phase coherence has been already demonstrated [45] using a combination of microwave and R.F. fields to drive the two-photon transition, or $F = 1$, $m_F = \pm 1$ states of ^{23}Na which are miscible and can be trapped in optical traps [46]. For example, a vertical array of cold atoms trapped in the anti-nodes of an optical standing wave was used to create an analog of the ac Josephson effect [47] and a lattice of quasi-one-dimensional confining tubes formed by a pair of laser fields was employed to investigate the phase coherence between neighboring sites [48]. The physics of these experiments is based on the macroscopic coherent tunneling of atoms between the wells, and is indeed very similar to that of the Josephson effect between two superconductors connected by an insulating junction [49].

9. The Josephson Effect in the Number-phase Representation

To describe the Josephson effect it is convenient to replace the operators $\hat{a}$ and $\hat{b}$ with operators which can be related more directly to a collective oscillation between the wells.

One obvious choice is the relative number operator

$$\hat{n} = \frac{1}{2}(\hat{a}^\dagger \hat{a} - \hat{b}^\dagger \hat{b}). \tag{66}$$

The other useful operator that would be useful is the relative phase operator. To define it, we follow the approach of [50]. We define the states

$$|\varphi_p\rangle = \frac{1}{\sqrt{N+1}} \sum_{n=-N/2}^{N/2} e^{-in\varphi_p} |n\rangle, \tag{67}$$

where $|n\rangle$ is a shorthand for $|N/2+n, N/2-n\rangle$ and the phase $\varphi_p = 2\pi p/(N+1)$, $p = -N/2, \ldots, N/2$ has a discrete structure. Then we can define a phase operator by

$$e^{i\varphi} \equiv \sum_{p=-N/2}^{N/2} e^{i\varphi_p} |\varphi_p\rangle\langle\varphi_p| = \sum_{n=-N/2}^{N/2-1} |n+1\rangle\langle n| + |-N/2\rangle\langle N/2|. \tag{68}$$

It is easy to check that the commutation relation $[\hat{n}, e^{i\hat{\varphi}}] = e^{i\hat{\varphi}}$ is satisfied so one can also write $[\hat{\varphi}, \hat{n}] = i$. The eigenstates $|\varphi_p\rangle$ satisfy the orthogonality relation $\langle\varphi_p|\varphi_{p'}\rangle = \delta_{pp'}$ and also form a complete set, $\sum_{p=-N/2}^{N/2} |\varphi_p\rangle\langle\varphi_p| = 1$. We note also that in the limit of large N the phase spectrum becomes quasi-continuous and one may define a derivative $d/d\varphi$ to obtain for the number operator the phase representation $\hat{n} = -i\partial/\partial\varphi$.

Using the representation given by the states $|\varphi_p\rangle$ we can write the Hamiltonian (65) in the form of the momentum-shortened pendulum Hamiltonian,

$$\hat{H} = \frac{1}{8}E_c N^2 - E_J \sqrt{1 - \frac{4n^2}{N^2}} \cos\hat{\varphi} + \frac{1}{2}E_c \hat{n}^2. \tag{69}$$

This effective Hamiltonian becomes, in the limit of small oscillations,

$$H \approx \frac{1}{8}E_c N^2 - E_J + \frac{1}{2}E_J \varphi^2 + \frac{1}{2}\widetilde{E}_c n^2, \tag{70}$$

where the new effective energy is

$$\widetilde{E}_c \equiv E_c + \frac{4E_J}{N^2}. \tag{71}$$

Equation (70) describes a harmonic oscillator with the frequency (Josephson plasma frequency)

$$\omega_J = \sqrt{E_J \widetilde{E}_c}/\hbar = \frac{1}{\hbar N}\sqrt{E_J(4E_J + N^2 E_c)}. \tag{72}$$

We now can regard the operators $\hat{n}$ and $\hat{\varphi}$ as analogous to momenta and position; therefore one can construct the corresponding creation and annihilation operators,

$$\hat{n} = (\hat{\alpha} + \hat{\alpha}^\dagger)\left(\frac{E_J}{4\widetilde{E}_c}\right)^{1/4}, \tag{73}$$

$$\hat{\varphi} = i(\hat{\alpha} - \hat{\alpha}^\dagger)\left(\frac{\widetilde{E}_c}{4E_J}\right)^{1/4}, \tag{74}$$

where $\hat{\alpha}$ and $\hat{\alpha}^\dagger$ fulfill the bosonic commutation relation $[\alpha, \alpha^\dagger] = 1$, diagonalizing the Hamiltonian,

$$\hat{H} = \frac{1}{8}E_c N^2 - E_J + \hbar\omega_J \left(\hat{\alpha}^\dagger\hat{\alpha} + \frac{1}{2}\right). \tag{75}$$

There are three regimes that have been considered in the literature [7, 17] for the Josephson two-mode Hamiltonian according to the interaction strength E_c, namely the Rabi regime $E_c \ll E_J/N^2$, the Josephson regime $E_c/N^2 \ll E_c \ll E_J$, and the Fock regime $E_J \ll E_c$. In the Rabi regime, the phase is well-defined and the excitation is just the promotion of a single atom from the bonding (symmetric) to the anti-bonding (anti-symmetric) state. In the Josephson regime, the ground state has still a well-defined phase, but the excitation forms a collective motion, the Josephson plasmon with the plasma frequency $\sqrt{E_J E_c}/\hbar$. In the Fock regime (the equivalent of the Mott insulator), the Josephson link is dominated by the interaction energy and n is a good quantum number. Therefore the ground state has a well-defined atom number on each side, the phase is completely undefined, and the harmonic approximation (70) cannot be applied anymore.

10. Bogoliubov Quasiparticles as Josephson Oscillations

As before, in order to use the Bogolioubov approximation for the Hamiltonian (65), we first identify the condensation mode. According to the general procedure we used for lattices, this is simply the symmetric (bonding) mode, namely

$$\hat{c} = \frac{\hat{a} + \hat{b}}{\sqrt{2}}, \qquad [\hat{c}, \hat{c}^\dagger] = 1, \tag{76}$$

destroys a condensate atom. The anti-symmetric mode

$$\hat{d} = \frac{\hat{a} - \hat{b}}{\sqrt{2}} \tag{77}$$

is orthogonal to $\hat{c}$ and it will be responsible for the "depletion" caused by interaction.

Using the same general procedure Bogoliubov - de Gennes described before, we take $\hat{c}$ as a complex number and we expand up to second order in $\hat{d}$,

$$\hat{H} = \frac{1}{8}E_c N^2 - E_J + \left(\frac{1}{4}E_c N + 2\frac{E_J}{N}\right)\hat{d}^\dagger\hat{d} + \frac{1}{8}E_c N\left(\hat{d}\hat{d} + \hat{d}^\dagger\hat{d}^\dagger\right). \tag{78}$$

The Bogoliubov transformation is implemented by

$$\hat{d} = u\hat{\gamma} - v\hat{\gamma}^\dagger, \tag{79}$$
$$\hat{d}^\dagger = u^*\gamma^\dagger - v^*\hat{\gamma}, \tag{80}$$

where the parametrization $u = \cosh\chi$, $v = \sinh\chi$ ensures $|u|^2 - |v|^2 = 1$ and, thus, the canonical commutation relation $[\hat{\gamma}, \hat{\gamma}^\dagger] = 1$. The value of χ that diagonalizes the Hamiltonian is found to satisfy the equation

$$\tanh 2\chi = \frac{E_c}{E_c + 8E_J/N^2} \tag{81}$$

with the result that the Hamiltonian (78) can be written as

$$H = \frac{1}{8}E_c N(N-1) - E_J\left(1+\frac{1}{N}\right) + E_1\left(\hat{\gamma}^\dagger\hat{\gamma}+\frac{1}{2}\right). \quad (82)$$

For the quasiparticle energy we find

$$E_1 = \frac{1}{N}\sqrt{E_J(4E_J+N^2E_c} \quad (83)$$

which is exactly the energy of the Josephson plasmon calculated in Eq. (72), $E_1 = \hbar\omega_J$. To have one more consistence check, we note that by particularizing the previous results for $M = 2$ and $q = 0, k = 1$, we obtain from Eq. (43) $\varepsilon_1 = 4t$, and from Eq. (42) that $E_1 = \sqrt{4t(4t+E_cN/2}$ which, when recalling the definition $E_J = 2Nt$, gives precisely Eq. (83).

In fact, the Bogoliubov transform acts as a squeezing transformation,

$$\hat{S}(\chi) = \exp\left[\frac{1}{2}\chi(\hat{d}^2 - \hat{d}^{\dagger 2})\right], \quad (84)$$

namely

$$\hat{S}(\chi)\hat{d}\hat{S}^\dagger(\chi) = \hat{d}\cosh\chi + \hat{d}^\dagger\sinh\chi = \hat{\gamma}, \quad (85)$$

$$\hat{S}(\chi)\hat{d}^\dagger\hat{S}^\dagger(\chi) = \hat{d}^\dagger\cosh\chi + \hat{d}\sinh\chi = \hat{\gamma}^\dagger. \quad (86)$$

As a result, the ground state has then the form [14]

$$|GND\rangle = \hat{S}(\chi)\hat{D}(\sqrt{N})|0>, \quad (87)$$

where we have introduced $\hat{D}(\sqrt{N}) = \exp\left[\sqrt{N}(\hat{c}^\dagger - \hat{c})\right]$.

In fact, it can be shown [14] that the operators α and γ are identical, thus fully establishing the announced equivalence between the Josephson plasmon and the Bogoliubov excitations. Using this formalism, after some algebra one can calculate the fluctuations in the number of particles and in the relative phase in all the three regimes. These results are listed in Table 1 below.

We will now come back to the issue of stability discussed before in the beginning of Section 8. and in Eq. (63). We want to illustrate the issue of thermodynamical instability on a simple to understand situation, namely deep in the Rabi regime, where $E_c = 0$. In this case, Eq. (63) predicts a quasiparticle energy of $-4t$, identical to the one given by Eq. (72). What is the meaning of this energy? Note that in the case of zero interaction, the Bogoliubov quasiparticle operators are $\hat{c}$ and $\hat{d}$. Indeed, the Hamiltonian $-2t(\hat{a}^\dagger\hat{b}+\hat{b}^\dagger\hat{a})$ is diagonalized by the operators $\hat{c}$ and $\hat{d}$, with the result $-2t(\hat{c}^\dagger\hat{c}-\hat{d}^\dagger\hat{d})$. The minus sign between these operators makes all the difference from the point of view of the stability of the state. Imagine for example that we distribute N atoms in the mode c: then it is immediate to see that, in order to move one atom from the mode c to the mode d one needs to add an energy $4t$ (or $2E_J/N$) (the final energy is $-2t(N-1)+2t$ and the initial one is $-2tN$). Thus this manybody state is stable. Quite the opposite happens for the case when we attempt to condense on the mode d: the energy there would be $2tN$ and if we move one particle into the mode c the system will have an energy $-2t+2t(N-1)$, which is lower by $4t$ (or $2E_J/N$. Thus it is energetically favorable for the system to emit quasiparticles and get to the true ground state.

Table 1. Different tunneling regimes, as defined by the interplay between the Josephson coupling constant E_J, the on-site energy E_c, and the total number of particles N. The fluctuation in the quadrature $(\hat{d}+\hat{d}^\dagger)/\sqrt{2}$ is approximately given by $\sqrt{2/N}\Delta n$, while that of the quadrature $(\hat{d}-\hat{d}^\dagger)\sqrt{2}$ by $\sqrt{N/2}\Delta\varphi$. Even if the depletion in the Josephson regime is much smaller than N, the resulting fluctuations in the relative phase and relative number of particles are drastically different from those corresponding to the Rabi case (with many-body ground state $(1/\sqrt{2N})(\hat{a}^\dagger+\hat{b}^\dagger)^N|0\rangle$).

regime	RABI	JOSEPHSON	FOCK
defined by	$\frac{1}{N^2}E_J \gg E_c$	$\frac{1}{N^2}E_J \ll E_c \ll E_J$	$E_c \gg E_J$
excitations	$\frac{2}{N}E_J\hat{d}^\dagger\hat{d}$, $\omega_{JP}=\frac{2}{\hbar N}E_J$	$\sqrt{E_JE_c}\hat{\alpha}^\dagger\hat{\alpha}$, $\omega_{JP}=\frac{1}{\hbar}\sqrt{E_JE_c}$	$\frac{1}{2}E_c\hat{n}^2$
fluctuations	$\Delta n \simeq \frac{1}{2}\sqrt{N}$	$\sqrt{N} \gg \Delta n \simeq \frac{1}{\sqrt{2}}\left(\frac{E_J}{E_c}\right)^{1/4} \gg 1$	$\Delta n \simeq 0$
in n	$\sqrt{\frac{2}{N}}\Delta n \simeq \frac{1}{\sqrt{2}}$	$1 \gg \sqrt{\frac{2}{N}}\Delta n \simeq \frac{1}{\sqrt{N}}\left(\frac{E_J}{E_c}\right)^{1/4} \gg \frac{1}{\sqrt{N}}$	
fluctuations	$\Delta\varphi \simeq \frac{1}{\sqrt{N}}$	$\frac{1}{\sqrt{N}} \ll \Delta\varphi \simeq \frac{1}{\sqrt{2}}\left(\frac{E_c}{E_J}\right)^{1/4} \ll 1$	$\Delta\varphi \sim 2\pi$
in φ	$\frac{\sqrt{N}}{2}\Delta\varphi \simeq \frac{1}{\sqrt{2}}$	$1 \ll \sqrt{\frac{N}{2}}\Delta\varphi \simeq \frac{\sqrt{N}}{2}\left(\frac{E_c}{E_J}\right)^{1/4} \ll \sqrt{N}$	
depletion	$\langle\hat{d}^\dagger\hat{d}\rangle \simeq 0$	$1 \ll \langle\hat{d}^\dagger\hat{d}\rangle \simeq \frac{N}{8}\sqrt{\frac{E_c}{E_J}} \ll N$	—
squeezing	$\chi \simeq 0$	$\chi \simeq \frac{1}{2}\ln\frac{N}{2}\sqrt{\frac{E_c}{E_J}}$	—
type of ground state	coherent, $\hat{D}(\sqrt{N})|0\rangle$ (superfluid)	squeezed, $\hat{S}(\chi)\hat{D}(\sqrt{N})|0\rangle$ (superfluid)	fragmented, $|\frac{N}{2},\frac{N}{2}\rangle$, (Mott insulator)

11. Future Directions: Lattice Bose-Einstein Condensates as Simulators for Cosmological Effects

The physics of Bose-Einstein condensed alkali atoms in lattice potentials has moved very fast in recent times from the desktop of the theorists to the optical tables of the experimentalists. Perhaps the direction which attracts most effort nowadays is the use of such systems to achieve interesting many-body effects [51], analogous to those predicted or discovered in solid-state physics. Simulations of quantum phase transitions, as well as quantum processing of information [52], are among the expected exciting results. Besides the now classic result of Mott insulator to superfluid transition, other states of matter have been demonstrated, such as the Tonks-Girardeau gas in two-dimensional optical lattices [53] and the Dicke transition in lattices immersed in optical cavities [54].

Here we point out that an interesting alternative direction could be the study of analog cosmological effects, such as the Kibble mechanism [55] for phase transitions in the early Universe. When a system undergoes a rapid quench through a continuous phase transition, topological defects emerge as a result of the causal disconnection of the regions formed. The mechanism is the following: the correlation length of the order parameter at equilibrium is divergent through the transition point; but for a fast quench the transition is not adiabatic and thus the true correlation length (corresponding to a non-equilibrium state) is finite. The true correlation length cannot grow faster than a maximum speed, thus there will be locally causal horizons set by this speed; the field outside a local horizon cannot have a phase relation with the field inside. A defect is formed at earliest when a local horizon becomes large enough to reach other domains. This mechanism has been first proposed by Kibble in connection with the grand unified theory symmetry-breaking phase transitions in the early Universe; these transitions were expected to be responsible for the formation of long-lived topologically stable defects such as monopoles, domain walls, or strings. Later Zurek [56] has shown that the mechanism should be generic to all systems in which the order parameter suffers a symmetry breaking during the phase transition. Zurek analyzed the scenario for the case of superfluid ^{4}He and suggested that vortices can be formed by a rapid quench of in an annular geometry.

There has been remarkable experimental efforts in the last decade to test the Kibble-Zurek mechanism in different systems. In liquid crystals, the appearance of disclinations has been found through a rapid quench of a nematic liquid through the isotropic-nematic transition [57]. In ^{3}He, the appearance of vortices, as predicted by Zurek, has been observed in experiments designed to detect vortices in ^{3}He undergoing a fast superfluid phase transitions caused by exothermic neutron-induced nuclear reactions [58]; however, it has not been seen in experiments with ^{4}He crossing the λ point by fast depressurization - which was intended to create in the laboratory what supposedly happened in the early Universe in fast expansion [59], and also not in bulk high-temperature superconductors such as $YBa_2Cu_3O_{7-\delta}$ which were expected to show a spontaneous generation of flux lines during a quench through T_c [60]. Finally, a very interesting experiment was done to test the mechanism for a loop made of of 214 Josephson junctions, with a positive result [61]. The authors measured the magnetic flux generated spontaneously during cool-down to the superconducting state and found that it is not zero for all the observations, thus concluding that in these cases appropriate phase differences between consecutive states have been generated by the Kibble-Zurek mechanism. Similarly, for atomic gases there has also been recently a lot of interest in the study of coherence between condensates connected through weak links, which allow tunneling, especially in optical lattices [62, 63].

A Kibble-Zurek experiment for alkali atoms can be imagined in a variety of experimental setups. The first proposal belongs to Anglin and Zurek, who showed that during a fast quench of the atomic gas, vortex lines can be formed [64]. They also show that the phenomenological time-dependent Ginzburg-Landau theory, employed routinely to estimate the correletion length, agrees well for fast enough quenches with the results given by a kinetic theory which uses a master equation for the condensate mode similar to the one employed in the theory of multi-mode lasers. At slower quenches, the two theories are not in good agreement, and one has to supplement the GL equation with terms responsible for diffusion and dissipation (which are included in the master equation of the kinetic theory). The kinetic theory however neglects the nonlinear terms – which presumably do not play any role in the early stages of the quench, but later they stabilize the vortex. Thus, not all the vortices predicted by the kinetic theory will survive. We suggest however to create a Kibble-Zurek experiment in BEC by crafting traps which would simulate a closed array of condensates similar to the ones studied in this work, in which atoms can tunnel from one site to another, similarly to systems comprising Josephson junctions [61]. For a small number of such condensates, the experimental setup can be similar to the one of the MIT interference experiment [65], but with a spatial shape of the blue-detuned laser beam which would ensure the separation of the gas into three or more pieces. Another possibility involves using two-dimensional optical lattices in magnetic traps of the same type as those recently used to study the phase coherence between sites [62]. An approximate Kibble-Zurek situation of the type we analyze below could be created by overlapping a highly repulsive optical potential (for example generated by a blue-detuned cylindrical laser beam) in the middle of the magnetic trap. In both cases, the theoretical predictions from this paper apply which can be tested against experiments regard the issue of stability of circular currents created by connecting the condensates. We will find that currents with quasi-momentum larger than a critical value cannot be formed by the Kibble-Zurek mechanism. We will also find that certain configurations are thermodynamically unstable, thus likely to decay in a certain time.

12. Conclusions

In conclusion, we gave here an elementary introduction in the theory of quasiparticle excitation in Bose-Einstein condensates trapped in optical lattices. We discuss the stability of currents flowing in such configurations and we give a classification of the states that can be obtained in quasi-one dimensional lattices with periodic boundary conditions. We also discuss the case of two-well latices, in which case we show that the Josephson effect is recovered by an analysis of the dynamics of the quasiparticle excitations.

Acknowledgments

This work was supported by the Academy of Finland (Acad. Res. Fellowship 00857, and projects 129896, 118122, and 135135).

References

[1] M. H. Anderson, J. R. Ensher, M. R. Matthews, C. E. Wieman, and E. A. Cornell, *Science* **269** 189 (1995); K. B. Davis, M. O. Mewes, M. R. Andrews, N. J. van Druten, S. D. Drufee, D. M. Kurn, and W. Ketterle, *Phys. Rev. Lett.* **75**, 3969 (1995); C. C. Bradley, C. A. Sackett, J. J. Tollett, and R. G. Hulet, *Phys. Rev. Lett.* **75**, 1687 (1995).

[2] I. Bloch, *Nature Physics* **1**, 23 (2005).

[3] M. Greiner, O. Mandel, T. Esslinger, T. W. Hänsch, and I. Bloch, *Nature* **415**, 39.

[4] B. D. Josephson, *Phys. Lett.* **1**, 251 (1962).

[5] A. Smerzi, S. Fantoni, S. Giovanazi, and S. R. Shenoy, *Phys. Rev. Lett.* **79**, 4950 (1997).

[6] I. Zapata, F. Sols, and A. J. Leggett, *Phys. Rev. A* **57**, R28 (1998).

[7] F. Sols, in *Proceedings of the International School of Physics "Enrico Fermi"*, edited by M. Inguscio, S. Stringari, and C. E. Wieman (IOS Press, Amsterdam, 1999), p. 453.

[8] Y. Shin, G.-B. Jo, M. Saba, T. A. Pasquini, W. Ketterle, and D. E. Pritchard, *Phys. Rev. Lett.* **95**, 170402 (2005).

[9] S. Hofferberth, I. Lesanovsky , B. Fischer, J. Verdu and J. Schmiedmayer *Nat. Phys.* **2** 710 (2006).

[10] R. Gati and M. K. Oberthaler, *J. Phys. B: At. Mol. Opt. Phys.* **40**, R61 (2007).

[11] R. Gati, B. Hemmerling, J. Flling, M. Albiez, and M. K. Oberthaler *Phys. Rev. Lett.* **96**, 130404 (2006).

[12] J. Estéve, C. Gross, A. Weller, S. Giovanazzi, and M. K. Oberthaler, *Nature* **455**, 1216 (2008).

[13] Gh.-S. Paraoanu, *Phys. Rev. A* **67**, 023607 (2003).

[14] Gh.-S. Paraoanu, S. Kohler, F. Sols, and A. J. Leggett, *J. Phys. B* **34**, 4689 (2001).

[15] N. N. Bogoliubov, *J. Phys (USSR)* **11**, 23 (1947); M. D. Girardeaux and R. Arnowitt, *Phys. Rev.* **113**, 755 (1959); A. L. Fetter, *Ann. Phys.* **70**, 67 (1972).

[16] F. Dalfovo, S. Giorgini, L. P. Pitaevskii, and S. Stringari, *Rev. Mod. Phys.* **71**, 463 (1999).

[17] A. J. Leggett, *Rev. Mod. Phys.* **73**, 307 (2001).

[18] A. J. Leggett, "Quantum Liquids: Bose Condensation and Cooper Pairing in Condensed-Matter Systems", Oxford University Press, New York (2006); C. J. Pethick and H. Smith, "Bose-Einstein Condensation in Dilute Gases", University Press, Cambridge (2006).

[19] A. Fetter and A. Svidzinsky, *J. Phys.: Cond. Mat.* **13**, R 135 (2001).

[20] M. R. Matthews, B. P. Anderson, P. C. Haljan, D. S. Hall, C. E. Wieman, and E. A. Cornell, *Phys. Rev. Lett.* **83**, 2498 (1999).

[21] B. P. Anderson, P. C. Haljan, C. E. Wieman, and E. A. Cornell, *Phys. Rev. Lett.* **85** 2857 (2000).

[22] K. W. Madison, F. Chevy, W. Wohlleben, and J. Dalibard, *Phys. Rev. Lett.* **84**, 806 (2000); K. W. Madison, F. Chevy, W. Wohlleben, and J. Dalibard, *J. Mod. Optics* **47**, 2715 (2000); F. Chevy, K. W. Madison, and J. Dalibard, *Phys. Rev. Lett.* **85**, 2223 (2000).

[23] J. R. Abo-Shaeer, C. Raman, J. M. Vogels, and W. Ketterle, *Science* **292**, 476 (2001).

[24] E. Lundh, C. J. Pethick, and H. Smith, *Phys. Rev. A* **58**, 4816 (1998).

[25] R. J. Dodd, K. Burnett, M. Edwards, and C. W. Clark, *Phys. Rev. A* **56**, 587 (1997).

[26] J. J. García-Ripoll and V. M. Pérez-García, *Phys. Rev. A* **60**, 4864 (1999).

[27] D. Rokhsar, *Phys. Rev. Lett.* **79**, 2164 (1997).

[28] P. O. Fedichev and G. V. Shlyapnikov, *Phys. Rev. A* **60** R1779 (1999).

[29] H. Pu, C. K. Law, J. H. Eberly, and N. P. Bigelow, *Phys. Rev. A* **59**, 1533 (1999).

[30] T. Isoshyma and K. Machida, *Phys. Rev. A* **59**, 2203 (1999).

[31] S. Stringari, *Phys. Rev. Lett.* **82**, 4371 (1999).

[32] J. J. García-Ripoll, G. Molina-Terriza, V. M. Pérez-García, and L. Torner, *Phys. Rev. Lett.* **87**, 140403 (2001).

[33] A. A. Svidzinsky and A. L. Fetter, *Phys. Rev. Lett.* **84**, 5919 (2000).

[34] S. M. M. Virtanen, T. P. Simula, and M. M. Salomaa, *Phys. Rev. Lett.* **86**, 2704 (2001).

[35] M. Tsubota and K. Kasamatu, cond-mat/9911389.

[36] G.S.Paraoanu, *Phys. Rev. A* **77**, 041605R (2008); G. S. Paraoanu, *J. Low Temp. Phys.* **153**, 285 (2008).

[37] L. J. Garay, J. R. Anglin, J. I. Cirac, and P. Zoller, *Phys. Rev. A* **63**, 023611 (2001).

[38] C. Schori, T. Stöferle, H. Moritz, M. Köhl, and T. Esslinger, *Phys. Rev. Lett.* **93**, 240402 (2004).

[39] S. Sinha, *Phys. Rev. A* **55**, 4325 (1997); F. Zambelli and S. Stringari, *Phys. Rev. Lett.* **81**, 1754 (1998); A. A. Svidzinsky and A. L. Fetter, *Phys. Rev. A*, **58**, 3168 (1998).

[40] A. J. Leggett and F. Sols, Found. Phys. **21**, 353 (1991); F. Sols, *Physica B* **194–196**, 1389 (1994).

[41] G. J. Milburn, J. Corney, E. M. Wright, and D. F. Walls, *Phys. Rev. A* **55**, 4318 (1997); M. J. Steel and M. J. Collett, *Phys. Rev. A* **57**, 2929 (1998).

[42] S. Raghavan, A. Smerzi, S. Fantoni, and S. R. Shenoy, *Phys. Rev. A* **59**, 620 (1999).

[43] P. Capuzzi and E. S. Hernández, *Phys. Rev. A*, **59**, 3902 (1999).

[44] J. I. Cirac, M. Lewenstein, K. Mølmer, and P. Zoller, *Phys. Rev. A* **57**, 1208 (1998); M. J. Steel and M. J. Collett, *Phys. Rev. A* **57**, 2920 (1998); P. Villain and M. Lewenstein, *Phys. Rev. A* **59**, 2250 (1999).

[45] C. J. Myatt, E. A. Burt, R. W. Ghrist, E. A. Cornell, and C. E. Wieman, *Phys. Rev. Lett.* **78**, 586 (1997); D. S. Hall, M. R. Matthews, J. R. Ensher, C. E. Wieman, and E. A. Cornell, *Phys. Rev. Lett.* **81**, 1539 (1998); D. S. Hall, M. R. Matthews, C. E. Wieman, and E. A. Cornell, *Phys. Rev. Lett.* **81** 1543 (1998).

[46] J. Stenger, S. Inouye, D. M. Stamper-Kurn, H.-J. Miesner, A. P. Chikkatur, and W. Ketterle, *Nature* **396**, 345 (1998).

[47] B. P. Anderson and M. A. Kasevich, *Science* **282**, 1686 (1998).

[48] M. Greiner, I. Bloch, O. Mandel, T. W. Hänsch, and T. Esslinger, *Phys. Rev. Lett* **87**, 160405 (2001).

[49] J. Javanainen, *Phys. Rev. Lett.* **57**, 3164 (1986); M.W. Jack, M. J. Collet, and D. F. Walls, *Phys. Rev. A* **54**, R4625 (1996); F. Dalfovo, L. Pitaevskii, and S. Stringari, *Phys. Rev. A* **54**, 4213 (1996).

[50] D. T. Pegg and S. M. Barnett, *Europhys. Lett.* **6**, 438 (1988); A. Luis and L. L. Sánchez-Soto, *Phys. Rev. A* **48**, 4702 (1993).

[51] I. Bloch, J. Dalibard, and W. Zwerger, *Rev. Mod. Phys.* **80**, 885 (2008).

[52] P. Treutlein, T. Steinmetz, Y. Colombe, B. Lev, P. Hommelhoff, J. Reichel, M. Greiner, O. Mandel, A. Widera, T. Rom, I. Bloch, and T.W. Hänsch, *Fortschr. Phys.* **54**, 702-718 (2006).

[53] B. Paredes, A. Widera, V. Murg, O. Mandel, S. Fölling, I. Cirac, G. V. Shlyapnikov, T. W. Hänsch, and I. Bloch, *Nature* **429**, 277 (2004).

[54] K. Baumann, C. Guerlin, F. Brennecke, T. Esslinger, arXiv:0912.3261.

[55] T. W. B. Kibble, *J. Phys. A* **9**, 1387 (1976).

[56] W. H. Zurek, *Nature* **317**, 505 (1985); Phys. Rep. **276**, 177 (1996).

[57] I. Chuang, R. Durrer, N. Turok, and B. Yurke, *Science* **251**, 1336 (1991); M. J. Bowick, L. Chandar, E. A. Schiff, and A. M. Srivastava, *Science* **263**, 943 (1994).

[58] C. Bauerle, Y. M. Bunkov, S. N. Fisher, H. Godfrin, and G. R. Pickett, *Nature* **382** (332) (1996); V. M. H. Ruutu, V. B. Eltsov, A. J. Gill, T. W. B. Kibble, M. Krusius, Y. G. Makhlin, B. Placais, G. E. Volovik, and W. Xu, *Nature* **382**, 334 (1996).

[59] P. C. Hendry, N. S. Lawson, R. A. M. Lee, P. V. E. McClintock, and C. D. H. Williams, *Nature* **368**, 315 (1994); M. E. Dodd, P. C. Hendry, N. S. Lawson, P. V. E. McClintock, and C. D. H. Williams 81, 3703 (1998).

[60] R. Carmi and E. Polturak, *Phys. Rev. B* **60**, 7595 (1999).

[61] R. Carmi, E. Polturak, and G. Koren, *Phys. Rev. Lett.* **81**, 4966 (2000).

[62] M. Greiner, I. Bloch, O. Mandel, T. W. Hänsch, and T. Esslinger, *Appl. Phys. B* **73**, 769-772 (2001).

[63] B. P. Anderson and M. A. Kasevich, *Science* **282**, 1686 (1998); C. Orzel, A. K. Tuchman, M. L. Fenselau, M. Yasuda, and M. A. Kasevich, *Science* **291**, 2386 (2001); F. S. Cataliotti, S. Burger, C. Fort, P. Maddaloni, F. Minardi, A. Trombettoni, A. Smerzi, and M. Ignuscio, *Science* **293**, 843 (2001).

[64] J. R. Anglin and W. H. Zurek, *Phys. Rev. Lett.* **83**, 1707 (1999).

[65] M. R. Andrews, C. G. Townsend, H. -J. Miesner, D. S. Durfee, D. M. Kurn, and W. Ketterle, *Science* **275**, 637 (1997).

[66] G. J. Milburn, J. Corney, E. M. Wright, and D. F. Walls, *Phys. Rev. A* **55**, 4318 (1997).

In: Focus on Quantum Mechanics
Editors: David E. Hathaway et al. pp. 87-128
ISBN 978-1-62100-680-0

Chapter 5

TRANSLOCAL SPACE AND UNIVERSAL INTELLIGENCE

Bernd Schmeikal*
University of Vienna

Abstract

We have to find a new concept of space that is closer to quantum physics than to relativity. It is not easy to select an appropriate title for such investigation, because the word *space* denotes something quite general. Adding further predicates, for example turning space into *deep space, primordial space* or *free space*, makes it more special. But anything similar to translocal space, should be more general than the historic space. Space-time as was understood until today, is a local, rather restricted concept. It denotes a tangent manifold, a measurable surface with a simple idea of neighbourhood. Quantum phenomena, however, involve relations between remote areas. We should 'give space' to long distance weak and strong links. That is, we also deal with long distance entanglement induced by the strong force. We shall introduce some new concepts such as the *veil* which annihilates physical properties although it is not a black hole. We factor out the neutrino. We shall relate translocal fine structures to the forces of nature. We investigate a special Peano fractal the symmetry breakage of which gives birth to time. Going into electromagnetic space graphs we show how the propagating energy carries the connection with it. We show that a spin-network may disclose properties of *neural nets* and *Hopfield models* with a typical nonlinear Onsager Hamiltonian. We go into some coarse precategory and its automorphism group. We also show that the surface manifold we want to reconstruct by renormalization is not just the Minkowski space, but rather the Minkowski algebra. Finally we find out there is a *natural standard model* and derive the natural strong force fermion spinor from the constitutive graded Lie algebra of the Clifford algebra $C\ell_{3,1}$. In that image the dynamic degrees of freedom are increased, but never reduced. Therefore we could have chosen the title *Free Space*. But free space is ready for entanglement. It

*E-mail address: bettina.schmeikal@wu-wien.ac.at

is translocal, is pregeometric and has primordial features. It is not pregiven, has no arbitrary rigid body features, is not dilatable, neither affine, nor conformal.

PACS 3.65-Ud, 5.50.+q, 2.20. Sv, 75.25.-j, 2.40. Sf, 14.60 St., 03.65. Ud, 04.60.Pp, 12.39.-x, 24.85.+p,64.60.ae, 25.40.Fq, 75.10.Dg, 75.76.+j, 68.70.+w, 61.43.-j, 11.30.Hv, 11.30.-j, 11.30.Qc, 98.80.Cq, 12.38.Bx, 64.60.al, 64.60.aq

Keywords: artificial intelligence, categorical pregeometry, Clifford algebra, displacement arrow, distortion graph, entanglement, fractal, Hopfield net, inelastic scattering, Maxwell graph, particle theory, path functor, path algebra, pregeometry, natural spinor, orientation, random graph, space graph, strong entanglement, symmetry breaking, time arrow, time gate, translocal space, spin veil

AMS Subject Classification: 82C03, 51P05, 81R60, 82B20, 20B25, 05C25, 05C75, 05C12, 05C40, 18A10, 15A66, 17B45, 82C32, 82C28.

1. Prologue

No one would doubt that the moon is up there independent of the observer. NASA engineers are able to calculate trajectories connecting the missile base with a lunar landing area. We can send a spaceship from the base to the moon. If we collect all possible trajectories connecting pairs of points in space, we obtain the macroscopic continuous space-time. It takes the form of Euclidean 3-space or a Minkowski space-time having some indefinite metric, say, $\{+++-\}$. This picture is so suggestive that most of us follow the implanted local illusion without ever realizing how it is brought about. Because even the robust spacecraft is handed over from web to web in a translocal energy mesh, and the orbiting moon that is so resistant against observation is already the outcome of uncounted observations that give it stability. It is also legitimate and very important to ask what kind of thing that is - not too far away from our spacecraft - that we call 'space'. But underneath the macro-assembly of continuous structure-less point sets there are the fluctuating translocal phenomena of the small quantum world. What kind of void is it that allows for scale-free, long-distance weak links in the apparent surface manifold of space-time? There have been constructed various routes towards a theory of quantum gravity. Some of those are spin networks (Markopoulou and Smolin 2000), others use statistical geometry (Bombelli 2001), random graphs (Antonsen 1994), random lattices (Lee 1986), quantum topology (Isham 1989), quantum gravity in the form of dissipative deterministic systems (t'Hooft 1999), cellular networks (Requardt 1998), fuzzy lumps (Requardt 2000, 2004), random metric spaces, loop quantum gravity (Rovelli 2006, Nicolai, Peeters, Zamaklar 2005) and general translocal networks, that is, continuous spaces with translocal shortcuts (Requardt 2004).

While the above analyses discuss random graph statistics, topology, scaling, coarse graining and renormalization, the present paper asks about the phenomenology of long-distance connections and force fields. Is there a fundamental reason for the emergence of long-distance links? Is there a basic mechanism that gives rise to primordial erratic structures of space and time and standard phenomena of high energy physics? My answer is that the functioning of such a mechanism is clear and easy to understand. That is, we have more than, say, mere topological statistics. Rather, there is an evident process, that lets things appear as they do. But a translocal primordial net has some salient phenomenological features that cannot be gained by compiling stochastic properties alone.

The translocal scale-free dynamic assembly has some unexpected physical appearance. Its long-range connections and the local Hamiltonian energy contributions can be calibrated and balanced in a natural way. In such an equilibrated field the space-time vanishes. It enters so to say a state of being less than void manifold. It becomes a no-thing without metric, orientation and duration. We shall call this no-thing a *veil* . The zero-state of a primordial net should annihilate physical qualities. Thus, the extreme vacuum state is much less and at the same time much more than space-time. This appearance implies so to speak the annihilation of a fluctuating pregeometric net by its own zero-point activity. It becomes "invisible". But beyond that, the translocal net resembles a dynamic Hopfield net, and a hydrodynamic or Maxwell graph with the well known nonlinear Hamiltonian function capable of critical transitions. These phenomena occur when the net is incited by a local symmetry breakage. In this picture, both microspace and macrospace appear to contain holes of no-thing-ness. Those holes are void of space-time and manifold. They have no orientation, no measure, no metric, no mass, no spin, no charge and so on. Yet, continuum physics is not ruled out by that approach, but it can be reconstructed from the primordial dynamic system by a geometric renormalisation process as Requardt (2004) has shown. All this sounds like magic which to some extent it actually is.

1.1. Arbitraryness of Methods

There do exist two quite different approaches. The first relies strongly on the analysis of the surface manifold and can be looked at as the more traditional top down approach. The second is the new bottom up oriented method. It is often said that the first method imposes quantum theory on a quasi God-given schema of space-time. Even topological dynamic graphs may appear as embedded into preexisting smooth manifolds as for example in loop quantum gravity (Requardt 2004, p. 2). In contrast to that we should look at dynamic graphs and strongly fluctuating irregular dynamic systems that should lead to course grained scale independent networks and reconstructed continua. But that model of a primordial discrete substratum is no less God-given. Therefore, we should prefer the word "arbitrary" to "God-given". Both bottom up and top down methods may be more or less arbitrary. They may or may not be aided by experiments. They may or may not have majority support by scientists. Anyway, I would not use the predication of God-given for the Minkowski space.

2. Random Graph and Peano Fractal

To see how it works it is helpful to consider some figures. We have no preexisting manifold, but interacting constituents, the points which may have an internal structure. The most simple random graph for a coarse grained primordial substratum is a graph G with n points but no edges

It shall turn out to be an important question, if such a random graph is of any interest, since there is no relation between the points. Is it observable at all? As Connes (2000) has said we should *"allow them to speak to each other"*. We should give importance to relations

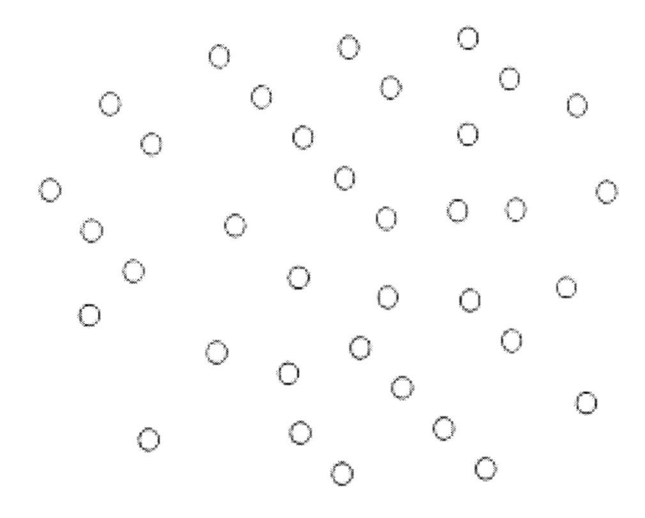

Figure 1. Points with no edges.

between points. Therefore, we consider dynamic random graphs G_m with n points, m actual edges and a maximal number of $N = n(n-1)/2$ possible edges.

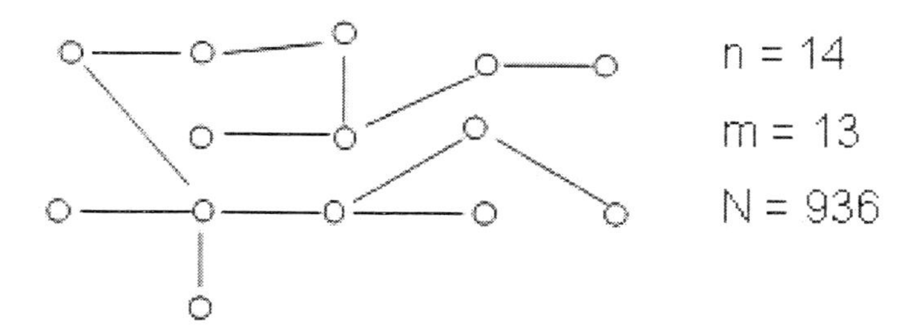

Figure 2. Random graph.

Those graphs look somewhat featureless and show no apparent relation to physical events. The number of graphs with m edges has a binomial distribution $B(m/p, N)$ where p is the independent edge probability. A single random graph G_m as in figure 2. occurs with probability $p(G_m) = p^m(1-p)^{N-m}$. This is the same formula as in Requardt (2004, ch. 4.2, p. 13).

Next we shall consider the picture of a directed random graph in order to account for a fundamental scale independent property of point relations. We use to denote that property as spin. We shall allow for spin at all possible scales. Accordingly we can draw knots with different size in order to indicate that the points have interior structure like grains or lumps. Consider some rather symmetric, recurrent structure as depicted in figure (2.) below.

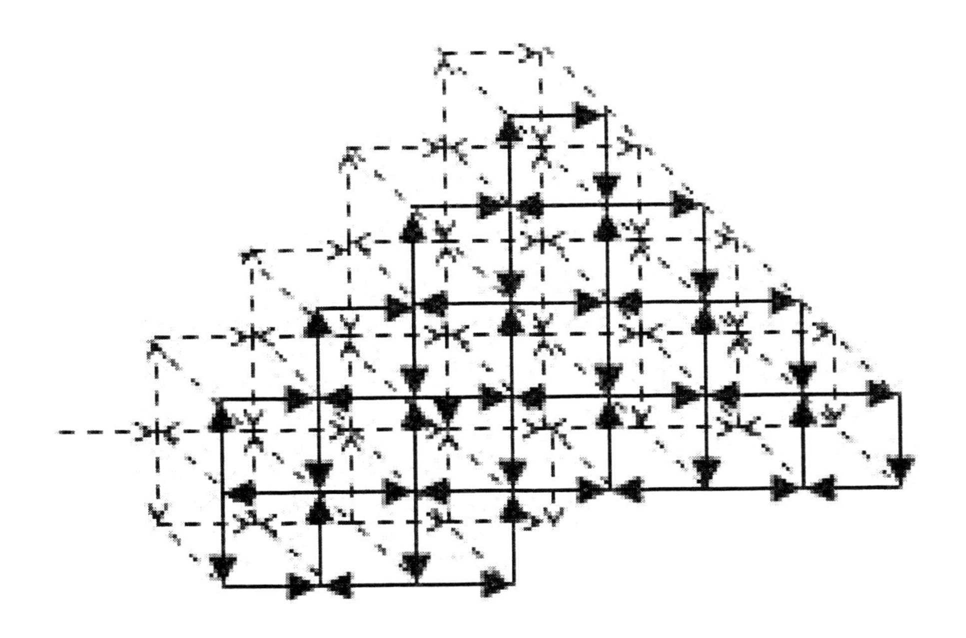

Figure 3. Plane zero spin network.

As I have used this graph in many papers, but not published any one, I should first tell you how it came that I introduced this special pre-category of primordial space. It is obvious that it contains some special dynamic elements which are important for both mathematics and physics research. The planar 4-cycle turned out to be an element of a special Peano-fractal. In 1995 Nottale using some results by Feynman derived a spatial fractal dimension 2 for quantum trajectories. He wrote *"The discovery that the typical quantum mechanical paths are continuous but nondifferentiable and may be characterized by a fractal dimension 2 may be attributed to Feynman (1965, Schweber 1986). Though Feynman evidently did not use the word 'fractal' that was coined in 1975 by Mandelbrot, his description of quantum mechanical paths fully corresponds to this concept.* We must be aware the problem of dimension of a particle trajectory has not yet been answered. I believe that the pregeometric investigation of field graphs can help us to understand what goes on. Such a graph as in figure (2.) is called a diagram scheme or a precategory. I have drawn it in such a way that it looks similar to a cubic grid. Nevertheless, it is a mere graph. The graph is fundamental as it allows to introduce the concepts of self-similarity and scale-freedom. Figure (2.) consists of two layers each of which is based on a *generative Peano curve*. We can think of an elementary arrow as a spatial event having Planck's length l_P of $1,61624 \times 10^{-35}m$ with a very small indefinite thickness so that it constitutes an open set in a topological space. But that is not a must. The generative Peano curve looks as below.

With that generative line-element we can built up quasi planar Peano curves of step n.

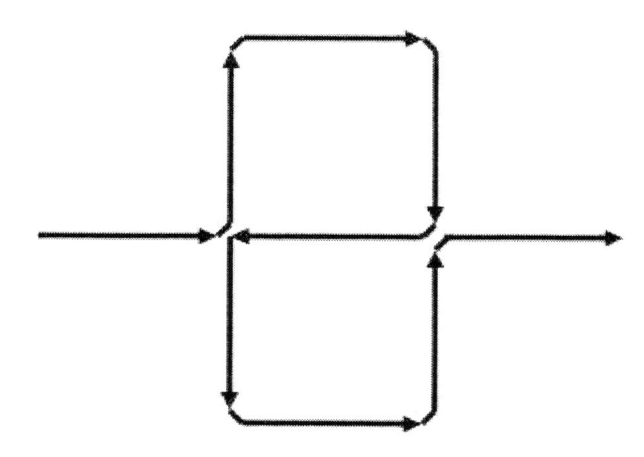

Figure 4. Step 1 - generative Peano curve.

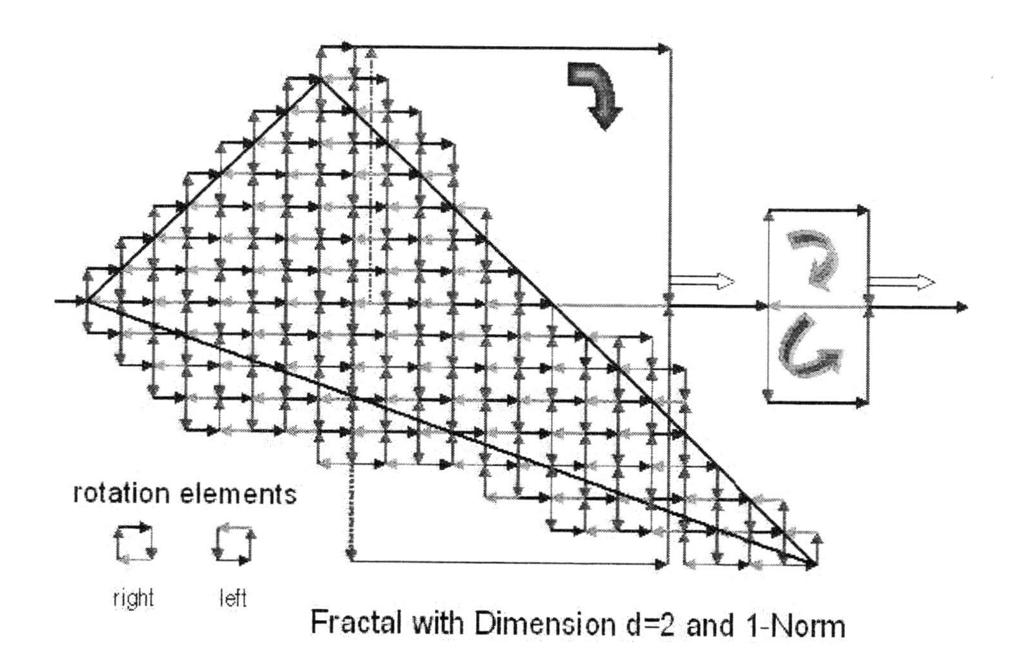

Figure 5. Step 3 - Peano curve.

3. Space-time Annihilating Veil

3.1. Translocal Fine Structure as Related to Force

In recent research on transpositions (Schmeikal 2004) we have described the pregeometric conditions that are imposed on the space-time by high energy physics events. These are given by inner scale, inhomogenous, involutive automorphisms or *grade transpositions* which commute microscopic vector fields with small scale space-time areas and -volumes. These microscopic events can be made responsible for a peculiar Lie structure of the Clifford algebra of space-time resulting in the typical standard model of HEPhy. In other words, summing up the microevents of the strong force, we obtain the standard model of HEPhy as a space-time group.

We would like to investigate the translocal fine structure of microscopic space-time as related to force. This fine structure should be held responsible for the appearance of quantum gravity and quantum fields as its *coarse grained consequence*. We are repeating here the words said by Manfred Requardt who followed the difficult *bottom up oriented avenue* (Requardt 2004, p. 2) to establish the existence of a microscopic wormhole structure compatible with *a certain Machian flavor* of such finding. With Requardt and others we share the working hypothesis that collective interactive behaviour is resulting from some basic principles. This holds for biology and social science just as well as for spin networks. We have to understand the fundamental priciples of interaction in the quantum vacuum. Our Ansatz is compatible with that of Requardt insofar as we also assume the effectiveness of strongly fluctuating dynamic graphs. A Continous manifold is supposed to appear as coarse grained limit, the existence of which is bound to geometrically critical states an the associated scale-free networks.

3.2. Coarsening by Factoring Out

In every nonlinear theory capable of self-sustaining structural organization there are interacting elements. In the case of physical space those elements are called points. An explicit notion of interacting points does not exist in classical differential geometry. But there is the widespread concept of quotienting out the collection of all equivalence classes induced by an epimorphism $\phi : X \rightarrow Y$. This can be applied to metric spaces and to their Clifford algebras. Given a space or geometric algebra X with an equivalence relation $R \subset X \times X$, also briefly written $\sim$, the equivalence class of an element $x \in X$ is the subset $\{x\} = \{x' \in X/x' \sim x\}$. The collection of all equivalence classes is denoted as $X/\sim$ and is called the quotient space (or algebra). If X is an algebra it is expected that the algebraic operations (Clifford multiplication, idempotency, orthogonality etc.) are well definded as mappings between the cosets. An epimorphism $\phi : X \rightarrow Y$ induced an equivalence relation of X via relation $x \sim x'$ if and only if $\phi(x) = \phi(x')$. To each element $y \in Y$ there corresponds a unique equivalence class $\phi^{-1}(y) \subset X$. If ϕ is an epimorphism, then $X/\sim$ is a quotient space (or algebra), where $\sim$ is the above equivalence. The map ϕ is an isomorphism between $X/\sim$ and Y. Conversely, if $\sim$ is an equivalence relation on X such that $X/\sim$ is a quotient space (algebra), then the natural injection $\phi : X \rightarrow X/\sim$ definded as $\phi(x) = \{x\}$ is an epimorphism. This is indeed a procedure of coarsening and provides orbits, leaves, stranded braids etc. as equivalence classes. Let us consider the surface manifold, that is, the Clifford algebra, in order to obtain some information about the dynamics of graphs in relation to the HEPhy phenomena.

3.3. The Surface Manifold of Spacetime Algebra

We assume there exists a reasonable method of coarsening, a renormalization process (Requardt 2003, 2004), to reconstruct a macroscopic *and* mesoscopic fixed point that we perceive as a continuous spacetime manifold. Nevertheless, it remains an open question what that manifold should look like. It is almost certain that this manifold is not simply the Minkowski space. Because quantum phenomena, even only on the surface, move in both space and isospin space. Therefore we must respect as surface the whole Clifford algebra. In any theory of strong interaction we have to describe movements in the isospin spaces which, at the same time, classify the elementary particles. As we have shown, it is the constitutive Lie algebra $l^{(2)} \subset C\ell_{3,1}$ which provides the root spaces A_2 of the special unitary group $SU(3)$. The Lie algebra $l^{(2)}$ gives rise to both color-rotations and flavor rotations as well as Lorentz transformations. Force transformations and motion in spacetime occur in the Minkowski algebra, not in Minkowsi space. This algebra represents the surface, and it is this algebra that we are going to reconstruct by renormalization. In section 13 we have pinned the $C\ell_{3,1}$-generators of the $l^{(2)}$. It has a subgroup $\mathrm{SU_{Cl}}(2)$ which carries out trigonal rotations and thereby executes the Pauli principle in strong force fields and it has a $\mathrm{SL_{Cl}}(2)$-component isomorphic with the spin-group $\mathbf{Spin}(2,1)$ responsible for the unfolding of Lorentz boosts.

3.4. Factoring Out the Massless Neutrino

In accordance with the three previous subsections we consider quotient spaces derived from the action of the group $L^{(2)}$ given by the exponential map $exp(l^{(2)})$. We have the isomorphism $L^{(2)} \simeq \mathrm{SL_{Cl}}(3)$. The index 'Cl' tells us that the group $L^{(2)}$ may be either $\mathrm{Sl}(3,\mathbb{R})$ or $\mathrm{Sl}(3,\mathbb{C})$ depending on the definition of the Clifford algebra, that is, on the field over which the algebra is defined. The equivalence classes are the orbits and leaves derived by conjugation $x' = \{g \circ x \circ g^{-1} / g \in L^{(2)}\}$ (where $\circ$ denotes Clifford multiplication). So the $L^{(2)}$ acts like a spin group. However, it properly transforms the elements of the whole Minkowski algebra, that is, its isospin subspaces including the generating Minkowski space. We can calibrate the Lie algebra such that the inner spaces of the low energy quarks u, d, s can be represented in standard notation by the following three mutually annihilating (orthogonal) primitive idempotents

$$f_u = \frac{1}{2}(1 - e_1)\frac{1}{2}(1 + e_{24}), \qquad f_d = \frac{1}{2}(1 - e_1)\frac{1}{2}(1 - e_{24}),$$
$$f_s = \frac{1}{2}(1 + e_1)\frac{1}{2}(1 - e_{24}), \qquad \text{represent u, d, and s-quarks.} \tag{1}$$

These can be collected into one equivalence class of low energy fermions having baryon number $\frac{1}{3}$ by a trigonal flavor rotation (Schmeikal 2010, equation 37) from the subgroup $\mathrm{SU_{Cl}}(2)$. The algebra contains a partition into three leaves of orthogonal quark-triples. We can factor out equivalence classes with equal baryon numbers. The differentiable manifold of such a leave on a stranded braid is constituted by orbits on

$$\mathcal{SB} = \left\{\frac{1}{4}\right\} \oplus \frac{\sqrt{3}}{4} S^8 \subset C\ell_{3,1} \tag{2}$$

where S^8 denotes the unit 8-sphere. The idea to construct the Lie algebra $l^{(2)}$ was initiated by Roy Chisholm who organized the first world conference on Clifford algebras in 1985. He demanded that *"there is no distinction between spacetime and the internal interaction space"* (1993, p. 371). This was a great challenge. We had to show in which way the inner symmetries and dynamics of strong interaction were connected with Lorentz transformations in subnuclear dimensions. So it became a sophisticated question whether and how the stability of baryons could be connected with the stability of spacetime. We followed an old idea by Greider and Weiderman(1988), Chisholm and Farewell (1992) and Schmeikal (1996) to construct a tetrahedral symmetry of the four standard primitive idempotents of $C\ell_{3,1}$ in order to represent trigonal rotations of quantumchromodynamics. We assume that the tetrahedral rotation stems from a hyperoctahedral reorientation symmetry of the idempotents. Thus the multiplets of $\mathrm{SU}(3)$ could turn out as real geometric phenomena in flat spacetime. The charge and baryon numbers are conserved. Since these numbers are conceived as geometric quantum numbers of algebraic eigenstates, they shall be invariant not only under the usual $\mathrm{SU}(3)$ transformations, but also under reorientation of spacetime algebra. We no longer held the view that at subnuclear level algebraic, topological and metric properties of spacetime were pregiven and defined from the outset. Instead we spoke of local quantum mechanical states of spacetime which were purified by strong interactions. For the concept of a *"dynamical purification of states"* read Narnhofer and Thirring 1996. In this model the fermion states of the standard model of HEPhy are essentially determined by the six isomorphic Cartan subalgebras and respectively the 'chromatic spaces' of the Clifford algebra $C\ell_{3,1}$. One of the central features of the idempotent-model becomes apparent, as soon as we attempt to investigate orbits of the neutrino state f_1 to which the $l^{(2)}$ is calibrated. It has to turn out that f_1 annihilates $l^{(2)}$. We have

$$f_1 \circ l^{(2)} = \{0\} \tag{3}$$

and respectively for the group

$$f_1 \circ L^{(2)} = \{f_1\}. \tag{4}$$

This refers to a very peculiar situation. As Connes and Requardt have emphasized there do exist lots of interesting quotient spaces with highly irregular orbits. We have pinned some such subspaces in $C\ell_{4,1} \simeq \mathrm{SL}(4,\mathbb{C})$ (Schmeikal 2010, chap. 48). Resultingly the induced topology might be trivial. The only open or closed sets are then the total and the empty space. That refers to the indiscrete or *coarse* topology. Exactly that is happening as soon as we factor out the 'neutrino' f_1 and consider

$$\{f_1\} = \{g \circ f_1 \circ g^{-1} / g \in L^{(2)}\}. \tag{5}$$

The equivalence class induced by all possible $l^{(2)}$ transformations of the primitive idempotent f_1 contains only the fix point f_1. Thus $L^{(2)}$ is a rank 2 stabilizer group of the field. So our subtle attempt to derive the standard model from a spacetime group in the Minkowski algebra led to the surprising result that there must be an inner space that takes the form of a primitive idempotent which annihilates the whole tangent space of $l^{(2)}$ at the origin. It is this property described by equation (4) which we kiddingly denoted as the omnipotency of the field f_1 (Schmeikal 2006, p. 80). Though this may not tell us much more than that the f_1 is a center of some rather complicated motion in space-time and isospin space, it is striking how that 'neutrino' - provided it is one - seems to refuse to partake in motion and strong interaction.

The 'f_1-neutrino' does not participate in any flavor rotation, and its impulse does not seem to partake in any Lorentz boost as is provided by the generators of $l^{(2)}$. However, it is capable of $\frac{1}{2}$-fermion spin. Suppose it has (almost) zero mass. The 'f_1-neutrinos' obey the Pauli principle. They will move extremely fast, may be much faster than light, and under a peculiar condition their spin vectors will flip with extreme velocity. We are tempted to construct a primordial space graph representing the f_1-neutrino field with compensated spin and annihilated energy. It looks as follows We draw a rectangular array

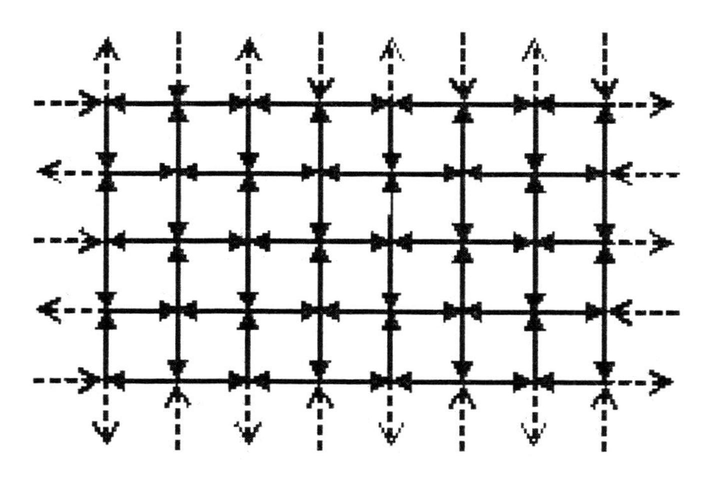

Figure 6. Zero energy grid.

with $2m$ horizontal and $2n$ vertical arrows. Those may be interpreted in various different ways. If the grid represents impulse, we obtain a total zero impulse grid. If it represents energy we obtain a zero energy grid as in figure 2.. If the arrows are supposed to represent spin, the spin vectors sum up to zero. They satisfy the Pauli principle as each spin $1/2$ has a neighbour with spin $-1/2$.

A zero energy grid can be understood as an unstable configuration of interacting spins like in the Ising-model. It has a nonlinear Hamiltonian energy as in (6) and is capable of phase transitions. In the case of the planar grid those transitions have been calculated by Lars Onsager (1944).

$$\mathbf{H} = -J_{ij}s_i s_j - H\sum_{i=1}^{N} s_i \tag{6}$$

with magnetic field H and interaction strength or exchange coupling function J_{ij}. We shall not consider the grid as pregiven, but we shall allow the arrows to appear, disappear and relocate. Therefore we shall denote the function J_{ij} as the 'relocation tensor'. As we shall generalize the model to obtain 3- and 4-dimensional grids we must expect that the Ising model does not provide analytic solutions for critical transitions. The balanced planar zero energy grid provides Maxwell solutions and light without shadow as some of the photons may not interact with the matter field.

3.5. The Veil

The veil is nothing simple. It has no mass, no energy and spin zero. In its unperturbed ground state it is entirely invisible to all forces. It provides no measure, no basis, no frame and not any of the known quantum numbers. The zero spin network veils space and time. It neglects the features of space, time and matter. It seems to be a no-thing. Yet, though it has a zero sum of spin it can flip between two states while resting in the zero total spin ground state. There are many features that we can realize in the veil.

The veil perfectly fits the Pauli principle. By its regular up-down arrangement it can transfer small energy fluctuations in zero time. Thus it provides the basis for entanglement phenomena. If we try to single out one spin and try to flip it, immediately all neighbouring spins are reacting. Thus by one single flip we can release a total net flip. We sketched such a net flip in figure (3.5.). As the network may have any possible size - it does not matter even if the edges can be counted or not - every zero spin net is capable of a flip. So it also has two spin eigenstates though, in a way, it does not exist. Such a graph or net which veils the expected observables we shall denote as 'insisting' rather than 'existing'. The energy impuls hidden in a veil may be large, but it is insisting as long as it is veiled. Since any edge and location is immediately reached by any impuls triggering flip of spin, every knot is within reach of every other. In this sense it is legitimate to perform a transition from the zero energy graph to a complete network figure (3.5.) which can provide the substratum for a Hopfield net.

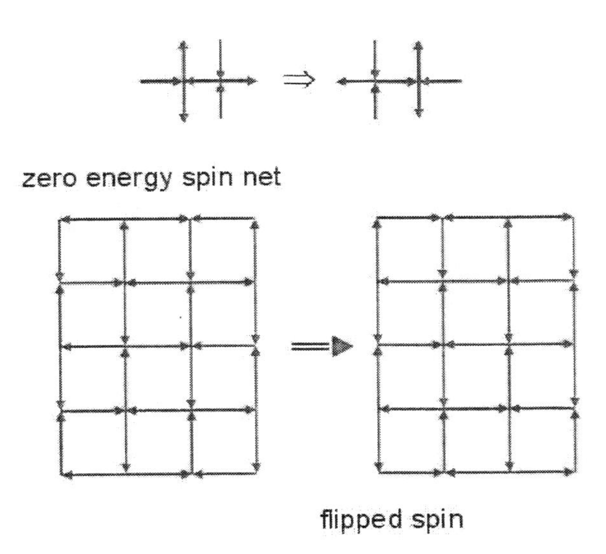

Figure 7. Network spin flip.

As we have shown elsewhere (2010, chapters 24, 27), the Cartan approach to strong interaction brings to light certain features of rigid body displacement. The veil has such a property. In one way it is an optimal fluid as it reacts instantenously like a fluid without viscosity. In another way it acts like a rigid body. We can visualize it as follows: pick out

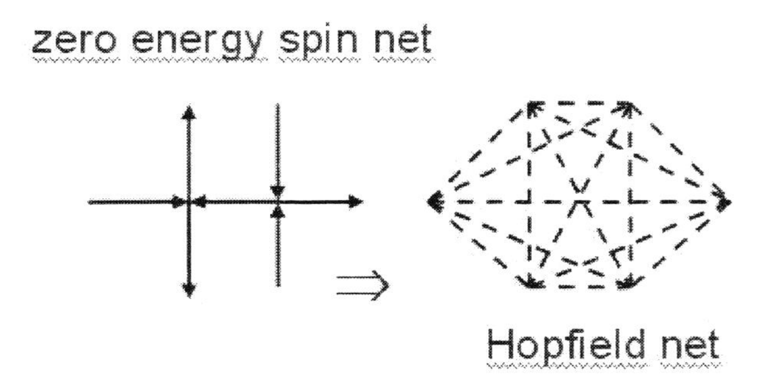

Figure 8. Transition to complete graph.

one spin arrow and turn it. Immediately the neighbouring arrows will react by rotation like wheels in a clockwork. Thus the veil emphasizes a peculiar dichotomy: something maximally rigid can nevertheless react maximally fluid. By the way, that comes close to the tetrahedral structure of water polymers. One can even build models (of the type Buckminster Fuller) to demonstrate these effects. The veil has the additional property of zero mass, zero energy, zero spin, zero measure and indifinite orientation.Though it is insisting rather than existing in one way, it will react if we excite it. It will show certain rigid body features, extreme fluidity and extreme fastness, transmit fluctuations immediately, enforce entanglements. As it induces universal connectedness, we can turn over from a zero net image to a complete graph such like a Hopfield net figure (3.5.). A large Hopfield net (wormhole) is the carrier of translocal spatial interaction. Its energy account is actually determined by a Hamiltonian function as in equation (6).

4. Creating Time

The raison d'être of the veil is given by a global neutrino equivalence. It veils the advanced and retarded motion of a spin wave as we show in figure (4.). The dynamic element figure (2.) does allow for a circular macro net flow. This figure clarifies why directional movement is promoted by a symmetry breakage. We obtain a circular oriented sequence of arrows when we separate the advanced propagation from the retarded . Alongside the trajectory the spin is preserved. You can easily see that the discontinuously approached ellipse is a zero energy spacetime net. It is only by breaking the symmetry between advanced and retarded wave (a/r-symmetry breakage) that we obtain elliptic spatial propagation in forward time. But that suggests that there exist two trajectories on which spin energies propagate with reversed time relative to each other. The directed arrow of time can only result from a strong suppression of the retarded spin wave. Such a symmetry breakage stands for the exclusion of one spin direction. This is in perfect accordance with the handedness of weak currents.

Fortunately, our model does not eliminate Majorana-neutrinos in favour of Dirac-neutrinos, but it explains their relation. As long as a/r-symmetry is only slightly broken - the arrow of time is in appearance but not yet definite enough - there do exist four chiral

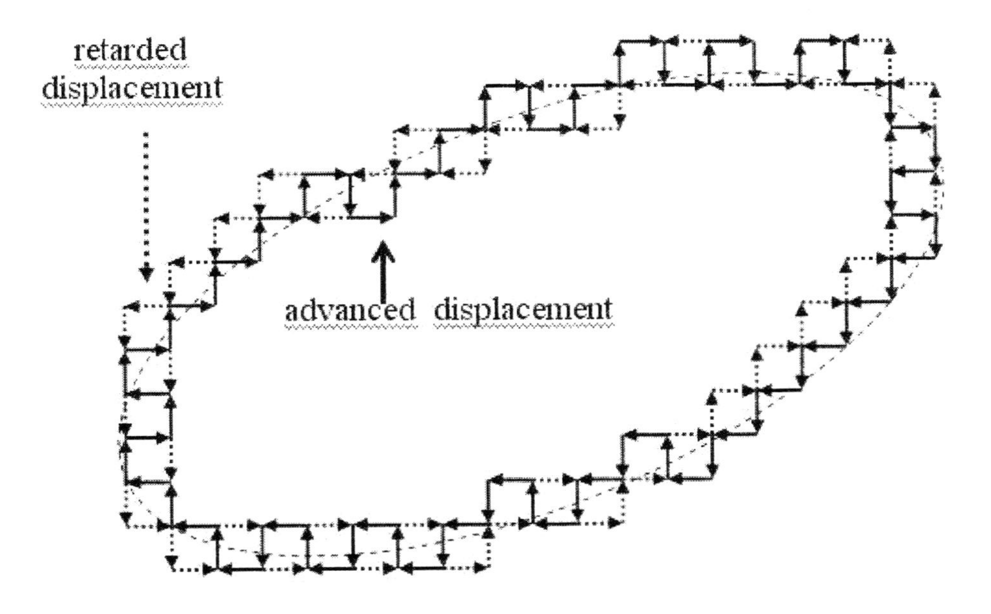

Figure 9. Fractal elliptic trajectory.

states $\nu^{\uparrow}$, $\bar{\nu}^{\uparrow}$, $\nu^{\downarrow}$, $\bar{\nu}^{\downarrow}$. However, as the neutrinos participate only in weak interaction, we should observe in zero mass $m_{\nu} = 0$ only two states with opposite helicity $\nu^{\uparrow}$ and $\bar{\nu}^{\downarrow}$. The opposite spin states $\nu^{\downarrow}$ and $\bar{\nu}^{\uparrow}$ should not be realised in nature. This is true, but in our model it is the outcome of an extensive symmetry breakage. The veil hides such asymmetry and so mystifies the true Dirac nature of the neutrinos. With the emergence of time the neutrinos change from Dirac- to Majorana-neutrinos. In reality we should be able to identify six leptons and three neutrinos in the form of structural peculiarities of the translocal network. We are able to identify those states as geometric traits in the surface Clifford algebra. It should be possible to find mathematical equivalents within the more erratic structure if we think about the possible complexity of intermediate graphs that carry their own complexity, orientation, metric structure and dimension. Here we can use the Internal Scaling Dimension (Requardt 2004, p. 18) to see there is no deficiency of complexity.

5. The Hidden ELM Structure

The appearance of a zero energy grid in correlation with a global validity of the Pauli principle can be considered as an intelligent balance of energy. Because, suppose the orientation of arrows could change. The fractal structure would be there, but the grid would have non-zero energy. There is actually hidden an ELM structure as a self annihilating potentially dynamic structure in a 3-dimensional zero energy net which incorporates what we may call a flow-distortion. This is represented in figure (5.). There are various elementary structures that can be correlated with the dynamics of force fields. Consider figure (5.). It is constituted by the generative element of figure (2.) operating in two perpendicular planes. We can regard the red colored lines (broken lines) as a magnetic rotation and the green ones (unbroken) as electric. Similar such graphs are well known within the Computational Electromagnetics community. They apply the *Finite Difference Time-Domain method* (FDTD). The Micro-Engineers use the *geometry of Time-Stepping* with its *primary and secondary grid* (Matiussi 2001). The green and red energy flows are counter rotating. They are blocked or 'inhibited'. We can unmake the inhibition. We can turn it into an exhibition of the electric field. We can unblock the magnetic circulation or both. There are non-zero electric field strengths in figure (12). We just have to reverse a few arrows. Thus a zero energy state is diminished.

Hidden ELM structure and distortion

in the plane areas parallel to e_{12}

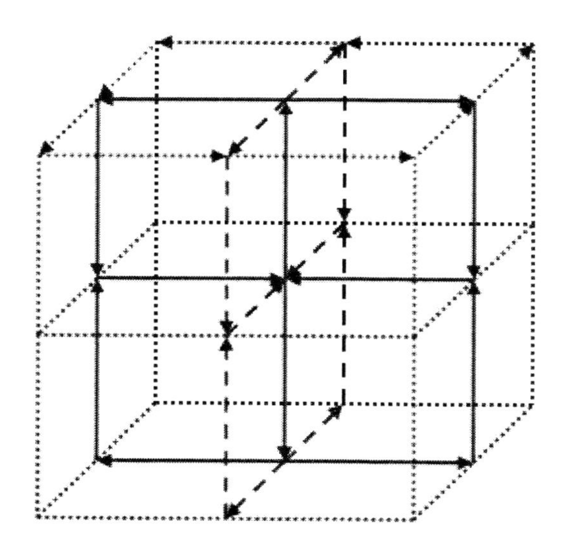

Figure 10. ELM structure and distortion plane.

We should explain what we could mean by a flow-distortion. If we consider figure (2.) we can enumerate the arrows as we proceed through the figure. Though the direction - which for now we would interpret as a spatially directed flow of energy - may change, the order of oriented arrows is preserved. We obtain arrows directed forward labelled from 1 to 9. When a sequence closes we obtain a circle. Then we would speak of an undisturbed flow in the graph. So, if we look at the top of figure 10, we can see that the arrows do not allow for definite orientation in the plane areas, but their orientation would be distorted. Though we could have circular flows in the two plane areas e_{13}, e_{23}, we could not uphold a circular flow in the third plane e_{12}. They same dynamic element does, however allow for a circular macro net flow as is depicted in figure (4.). Especially that figure makes it clear that

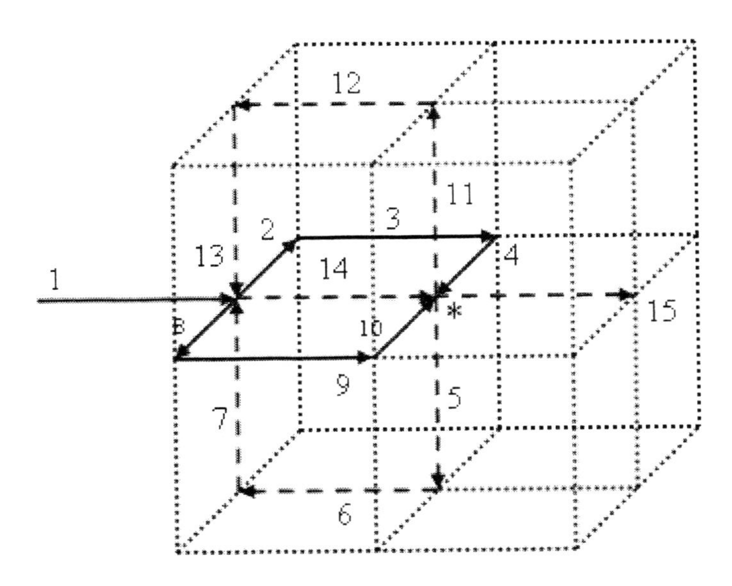

Figure 11. Transversal dynamic ELM element inhibited.

directional movement is promoted by a symmetry breakage. We obtain a circular oriented sequence of arrows when we separate the advanced propagation from the retarded. You can easily see that the discontinuously approached ellipse is a zero energy spacetime net. It is only by breaking the symmetry between advanced and retarded wave that we obtain elliptic propagation in time.

There is a second rather universal denotation - of comparable importance like symmetry breakage - namely that of inhibition and exhibition of dynamic elements. This can remind us of neural networks and especially the directed graphs built up by McCulloch-Pits neurons. We are indeed referring to these neuroinformatic models as we are speaking about universal intelligence. Actually the knots in spacetime are very peculiar nano-electronic components with a special sensibility for stable structures and some unfamiliar degree of freedom to decide on the nature of an energy-connection. It may be interpreted as either incoming or outgoing or, in the words of neurology as a dendrite or an axon.

When we first thought about those fractal features we were not convinced that they would support new insights into phenomena such as scale independent spin, annihilated space and so on. Since then the situation has changed. It was in October 2006 that I sent a first draft of the concept in categorical notation to Zbigniew Oziewicz. I solved several elementary graph theoretic problems of representation and translated some of the old Clifford algebra problem of reorientation into the categorical language. It became obvious that the Ising-model for self organizing spin interaction would be important. Only lately this year I found out to my own surprise that an artificial intelligence approach would be appropriate to understand the appearance of stable energetic patterns such as

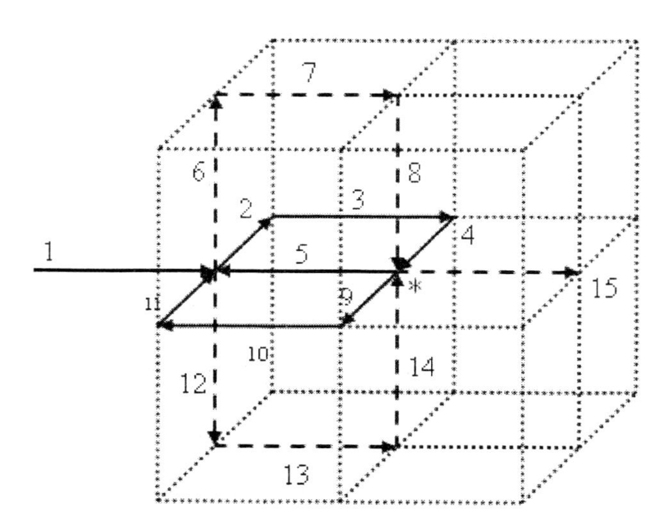

Figure 12. Electric dynamic element exhibited.

electromagnetic fields, neutrinos, fermions and so on. I knew how the standard model could be derived from geometric features of the Minkowski algebra. But now there appeared an additional aspect that required something astonishingly new.

Namely, first, space and time exist only there where they actually appear. They are brought forth by symmetry breakage of zero energy nets, and where we can observe them their appearance is indeed an act of intelligence. In other words neither matter, nor space and time must be separated from the universal intelligence. Briefly before I wrote that down, exactly three years after the discovery, I found out the artificial intelligence people, more precisely John Hopfield, had made known his feedback network with symmetric 'synaptic weighs' in 1982. These symmetric weighs are important because in the meta-stable homogeneous zero energy net we would expect every small region having essentially the same dynamic properties as any other. A Hopfield net has the same energy function as the Ising model. Just the coupling function J_{ij} is to be replaced by the weigh-matrix of the net. This weigh matrix allows for exhibitory and inhibitory connections between vertices.

Note in both cases, figures 10 and 11, we have a unit energy output, though in one case both electric and magnetic rotations are blocked, whereas in the other the electric exhibited. There are at least two ways we can represent the vertices which we have marked by a star. The first and most obvious representation of a *-vertex as in Figure 10 by a McCulloch-Pitts neuron might look like in figure 13.

We obtain $in_j = x_1 w_{1j} + ... + x_5 w_{5j} = +1 - 1 + 1 - 1 + 1 = 1$ and therefore $out_j = \phi(1 - \frac{1}{2}) = \phi(\frac{1}{2}) = 1$. This is a unit impulse at the outgoing spin arrow (the axon).

Consider the McCulloch-Pitts *"space-time neuron"* (with vertex *) in figure 12. Here we have unblocked the electric field, and to our surprise the design of the neuron does not

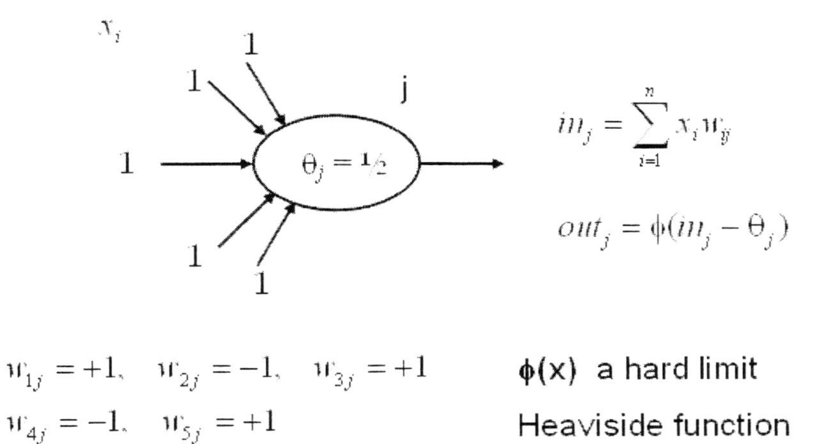

Figure 13. Space vertex as McCulloch-Pitts neuron.

change. It is essentially given by 3 exhibitory channels and 2 inhibitory ones. Notice, this structure can be enlarged in order to unblock an electric wave front. See the figure 14.

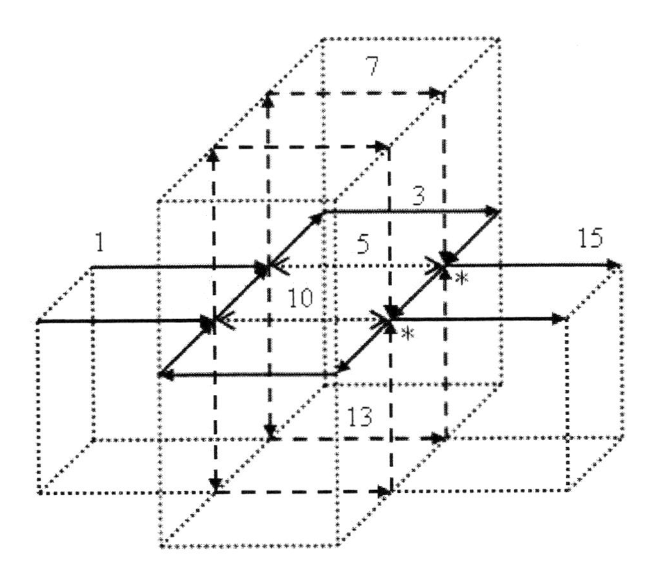

Figure 14. Electric field front.

6. Arrow Controlling Symmetry Breakage

To achieve a symmetry breakage as in figure 9 we can use the *distorted* edges in directions parallel to e_2. Those act inhibitory on the energy transport in the plane parallel to e_{13} and cause a unidirectional flow. We treat the zero energy vertex like an electronic component. In figure 15 you can see how the control 'electrode' acts on the four elementary flows.

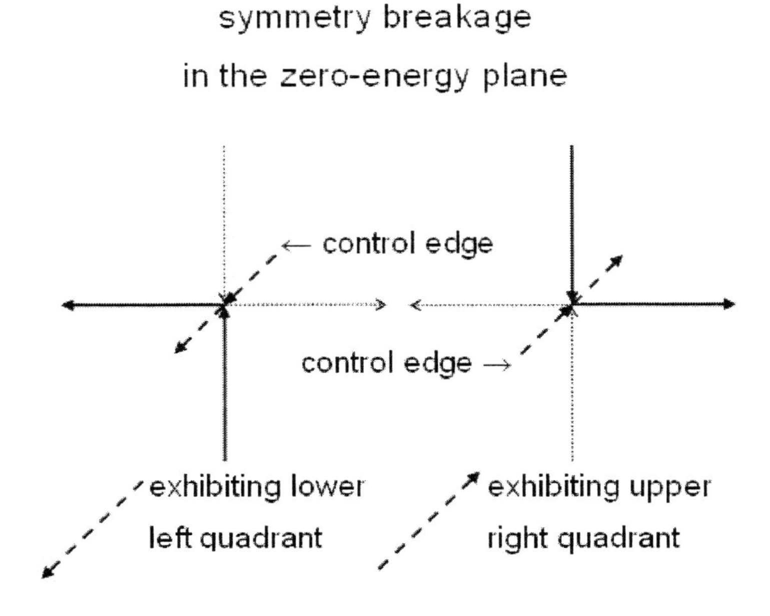

Figure 15. Control of symmetry breakage by arrow.

In figure 16 we simulate the symmetry breaking by two electronic components in an integrated circuit. We use two fieldeffecttransistors, one of them a p-Fet the other an n-Fet, to control the emergent flow geometry.

We can assume that a small Planck region when activated actually discloses the structure of a double field effect layer sensitive for zero point fluctuations of gate voltage. In this way, the fundamental symmetry breakage necessary to promote directional motion is brought forth by lateral inhibition in an artificial neural network.

7. The Timegate

The zero energy net is essentially composed by the elementary cross of two incoming and two outgoing arrows. The vertical lines now represent virtual pairs e^- e^+ in a vacuum fluctuation. The two horizontal lines act like the grid in a vacuum tube or the gate in a FET. They may oscillate about a zero point or built up a stable potential which controls a current emanating from the vertex of vacuum polarization. The disturbance coming in on the diagonal dashed line polarizes the vertex so that in accordance with the Heisenberg

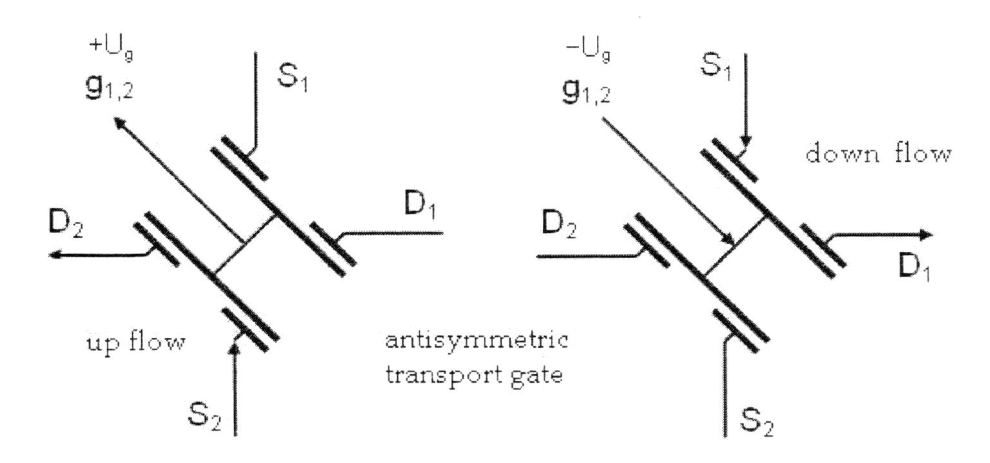

Figure 16. Control of symmetry breakage by two FETs.

uncertainty principle the flux of particles e^- can overshoot the virtual e^+. The vertex acts so to say like a cold cathode in a triode. During a small period of imbalance that is caused by the disturbance a comparatively weak potential occurring at the gate can control a strong acceleration of the low energy electrons. At any case a small zero point fluctuation at the gate will cause a large fluctuation of current caused by the outer field whether electric (by U^+) or magnetic. This polarization diminishes the zero energy gross yield and involves some nonlinear effects. The time gate acts as a vacuum polarizer and amplifier of created currents. It also operates as a component in an intelligent network which is capable to 'internalize' peculiar patterns of energy. As it annihilates the zero balance of the arrows it also wipes out the compensation of advanced and retarded currents and thus delivers time. It is therefore that we shall denote that small structure as a timegate.

8. Displacement and Torsion

In noncommutative geometry we are aware the relations between identifyable points (x, y) are more important than the initial space X which is supposed to contain those points. Intuitively, in agreement with the philosophy around figure 1, or for convenience as Requardt said, we assume that in the relation $R \subset X \times X$ each x has at least one outgoing edge. If the base set X is (made) countable, we can represent relations in terms of non-symmetric adjacency matrices. In the next sections we shall discover some of those basic adjacency matrices which seem to be constitutive for the emergence of differential geometry.

In the unstable, not metric and non-oriented diagram scheme there may emerge displacement of energy, like a displacement current. A force quantum can be displaced from one location to another. But the predicates of location in terms of differential geometry are not well defined. *Displacement* is a pre-categorical feature of forces. It is an arrow in the category of graphs. We shall denote it as *pre-categorical.* Recall that a

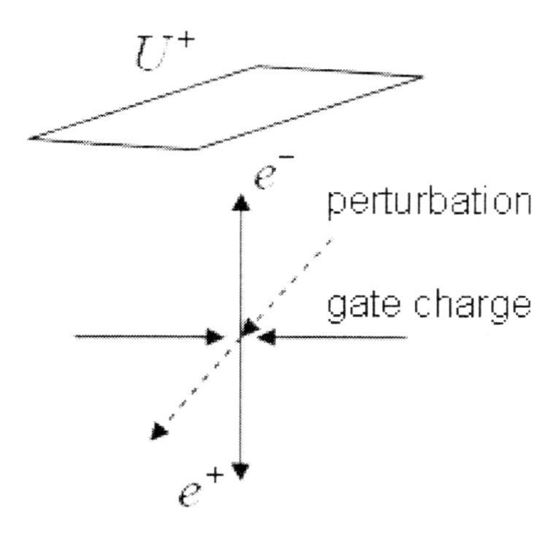

Figure 17. Timegate.

category is conceived as a graph with some additional functions. A directed graph G is a set of vertices and a set of arrows (edges). The morphisms of graphs are the arrows of the category **Grph**. Graphs are often pictured by diagrams of vertices and edges. Therefore a graph is called a diagram scheme or a *precategory* [Mac Lane 2000, p. 48].

Definition 1. ***Displacements***

are arrows f, g, h... in the precategory of graphs: $. \rightarrow .$

Definition 2. ***Torsion***

is a directed cycle concatenation of arrows, Triangle T or square S.

Definition 3. ***Distortion***

is a counter-directed pair of arrows, or two half cycles working in opposite direction or counter-rotating cycles.

We do not know what a location is. Euclidean space is not pre-given and the Lorentz metric is not either. It is not evident that any particle may go from a location, say, a, to another, say, b, at any time. The physical space is not that which physicists had in their minds as synchronous complete concepts. Euclidean 3-space and Minkowski space are such ready concepts with rather restricted validity. Physical space is not given by a basis, but rather it is processed and recreated. Therefore

Definition 4. ***Locations a, b, c ...***

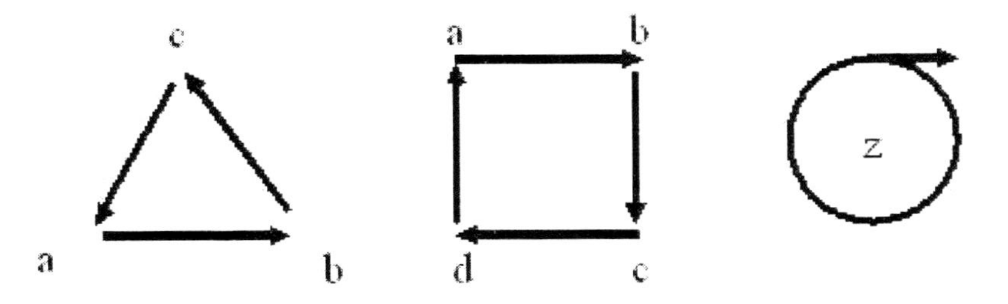

Figure 18. Displacement by torsion in **grph.**

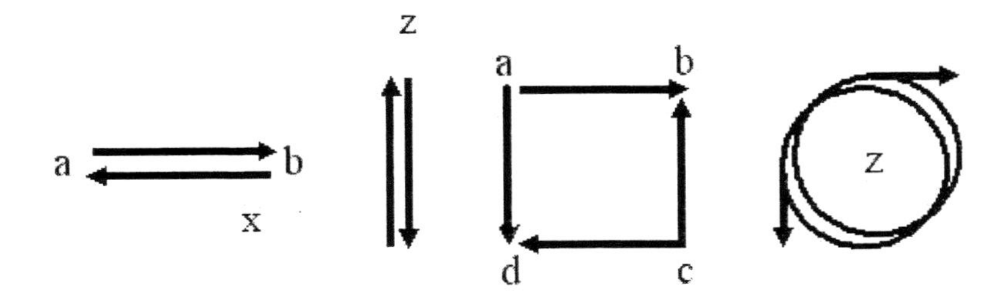

Figure 19. Displacement by counter-dislocation in **grph.**

are vertices in a directed graph. They are represented by operations of Domain and Codomain which assign to the arrow f in the graph an element $a = \mathrm{dom}\, f$ *and an element* $b = \mathrm{cod}\, f$.

Definition 5. ***Spaces between different locations***

are arrows f, g, h in the precategory of graphs.

Mathematically, an arrow is an object of a precategory, category or metacategory. But physically it represents a qualitative and quantitative difference, that is, a relation between different locations in the field. It also has a dynamic interpretation as a translation or propagation of a field quantity from one location to another. As an *interval* it may possess different possible scales. The interval might be 1.) qualitative, binary or nominal. In that case we only differ between two or more locations. But we do not know more about the difference. Locations may further be measured 2.) on an ordinal scale, in which case we observe some order relation. Or the difference may have 3.) an interval- or absolute metric scale. So we can move from a discrete topology without orientation and metric to a vector space or even geometric algebra with given metric. But in pregeometry we make use of the discrete concept of graph. From this we gain insight into emerging features of connectivity, orientation, dimension, symmetry, continuity, metric and other predicates brought forth by the field. Before we go deeper into this, for philosophical completeness, let us take a brief view into a concept of pre-pregeometry and eventual states beyond energy and space.

9. Precategories of Space

Let us step out intuitively by creating primary statements for pregeometry.

Concept 1. ***regions***

Represent ***regions*** *in pregeometric physical space by finite or infinite directed graphs in the precategory.*

Concept 2. ***scale freedom***

Infinite graphs having an infinite number of vertices shall involve ***self-similar graphs****.* [1]

Demand 1. ***algebra***

Represent a pregeometric concept of space by a ***graph algebra*** *from which the underlying space graph can be recovered.* (Shallon 1979)[2]

Demand 2. ***dualizability***

The pregeometric concept of a physical space requires that its ***graph algebra be dualizable****.*

9.1. Graph Algebra

Given a directed graph G with a set of vertices, usually $V \in \mathbb{Z}$, the graph algebra $A(G)$ is the algebra with universe $V \cup \{0\}$, where $0 \notin V$, and its multiplication is derived from the edges.

Definition 6. ***binary operation in*** *$A(G)$:*

$$a \circ b = \begin{cases} a & \text{if } a, b \in V \text{ and an edge is directed from } a \text{ to } b \\ 0 & \text{otherwise} \end{cases}$$

The graph algebra has just this operation. In order to watch the path of a field displacement we need some further algebra which I shall denote as

9.2. Path Algebra

Let $\mid V \mid = n$ be the cardinality of a vertex set. A path algebra is a subalgebra of the matrix algebra $\mathrm{Mat}(n, \mathbb{Z})$ of $n \times n$ matrices over the integers. The path algebra $P(G) \subset \mathrm{Mat}(n, \mathbb{Z})$ of a graph G is generated by the adjacency matrix α. It is a matrix whose rows and columns are indexed by the vertices in natural address order and whose i-j th entry is the number of edges leading from vertex i to vertex j.

---

[1] *Self-similarity of fundamental solutions of field equations promote the study of self-similar graphs. Turbulence and other dynamic phenomena in fundamental solutions of discrete field equations such as discrete Navier-Stokes equations support this idea.*

[2] *In her thesis C. R. Shallon introduced Graph Algebras to construct algebras (Davey 1999) with unexpected properties such as finite algebras which are not finitely based .*

9.3. Field Graph

The field graph is a generative element which covers information about the kinematic displacements induced by the force transacted in that field. In abstract terms it is a pregeometric graph as described in the previous sections. It will be denoted by symbols F or G. As a directed graph G it typically consists of a set V of locations (vertices) and a set E of arrows f (force displacements) and a pair of functions $E \rightrightarrows V$

$$\partial_0 : E \to V \qquad \partial_1 : E \to V$$

$$\partial_0 f = \mathrm{dom}\, f \qquad \partial_1 f = \mathrm{cod}\, f$$

A morphism of field graphs L: $G \to G'$ is a pair of functions L_V: $V \to V'$ and L_E: $E \to E'$ such that

$$L_V \partial_0 f = \partial_0 L_E f \qquad L_V \partial_1 f = \partial_1 L_E f$$

for every arrow $f \in E$. These morphisms are the arrows of the *category of field graphs*. The graph G represents a diagram scheme for the field dynamics.

9.4. Composition

The following procedures are basic procedures of category theory. Consider field graphs or equivalently V-graphs to a fixed set V of regions. We are interested in automorphisms $V \to V$. If E and H are the sets of force displacements of two V-graphs, the product over V is defined as usual

$$E \times_V H = \{\langle g, f\rangle | \partial_0 g = \partial_1 f, g \in E, f \in H\}$$

as a set of composable pairs of arrows as force displacements f, g: $.\to.\to.$ with defining relations

$$\partial_0 \langle g, f\rangle = \partial_0 f \quad and \quad \partial_1 \langle g, f\rangle = \partial_1 g$$

that make the product $E \times_V H$ a V-graph. Any V-graph G can be used to generate the free category $C(G)$ of field graphs on the same set V of regions. The arrows of the free field category $C(G)$ are the *strings* of composable arrows of G. Thus any arrow from location a to location b represents a path from b to a, consisting of successive displacements (directed edges) of G. This construction is in direct correlation with section II.7 in Mac Lane. It is obvious that we obtain a functor from the free category of graphs to the path-algebra $\mathrm{Mat}(n, \mathbb{Z})$ which can be seen as a category **Matr**$\mathbb{Z}$ of square matrices with entries in $\mathbb{Z}$.

9.5. Path Functor of Free Category of Field Graphs

In the identity morphism of V-graphs $V \rightrightarrows V$ both functions domain and range are given by the identity map $V \to V$. There is a special V-graph V for which there is an isomorphism $E \cong E \times_V V$, given by $f \mapsto \langle f, \partial_0 f\rangle$. We also have $E \cong V \times_V E$. A category with locations V is represented by a V-graph E equipped with the morphisms of composition and identity of V-graphs:

$$\text{com:} \quad E \times_V E \to E \quad \text{and} \quad \text{id:} \quad V \to E$$

Together with the composition of matrix multiplication in $\mathrm{Mat}(n, \mathbb{Z})$ and the identity matrix we thus obtain the functorial relation

$$PF : C(G) \to \mathbf{Matr}\mathbb{Z}$$

where the fundamental object- and arrow-functions are injective from $C(G)$ into $\mathrm{Matr}\mathbb{Z}$.

10. Dualizable Graph Algebra of Torsion Displacement

The first step towards a pregeometric understanding of *QED* must consist in the construction of an adequate precategory of the *curl graph* which is characteristic for the Maxwell theory of electrodynamic displacement. We can represent the elementary pregeometric curl by a T or S-graph as in figure 18. But we also have to respect the second demand. Murskii (1996), in 1965, proved the existence in three valued logic of a closed class of terms which cannot be generated by a finite complete system of identities. Following that rigor, Shallon (1979) observed, the goupoid constructed by Murskii can be represented by a graph algebra. She proved that there are many other finite graph algebras which are not finitely based. Davey, Idziak, Lampe and McNulty (1999) showed the equivalence of the following statements:

Demand 3. *: equivalence of properties*

(i) graph algebra is dualizable,
(ii) each connected component of G is either complete, bipartite complete, or a loose vertex,
(iii) G is 4-transitive or
(iv) $A(G)$ is finitely based.

Starting from the assumption that pregeometric physical space is correlated with the field we assume, that the category of space can be finitely generated by some small graph. Though it may be the case that globally the space category is discontinous and fractal, still it can be generated by some elementary precategorical structure. From this there follows that the smallest torsion graph is a 4-transitive cycle as is constitutive for figure (2.). The graph algebra $A(S)$ of the 4-cycle has the following multiplication table.

$$A(S) = \begin{pmatrix} 0 & 0 & 0 & 0 & 0 \\ 0 & 0 & 1 & 0 & 0 \\ 0 & 0 & 0 & 2 & 0 \\ 0 & 0 & 0 & 0 & 3 \\ 0 & 4 & 0 & 1 & 0 \end{pmatrix}.$$

The adjacency matrix α is the integer matrix

$$\alpha = \begin{pmatrix} 0 & 1 & 0 & 0 \\ 0 & 0 & 1 & 0 \\ 0 & 0 & 0 & 1 \\ 1 & 0 & 0 & 0 \end{pmatrix} \quad \text{with} \quad \alpha^4 = \begin{pmatrix} 1 & 0 & 0 & 0 \\ 0 & 1 & 0 & 0 \\ 0 & 0 & 1 & 0 \\ 0 & 0 & 0 & 1 \end{pmatrix}, \tag{7}$$

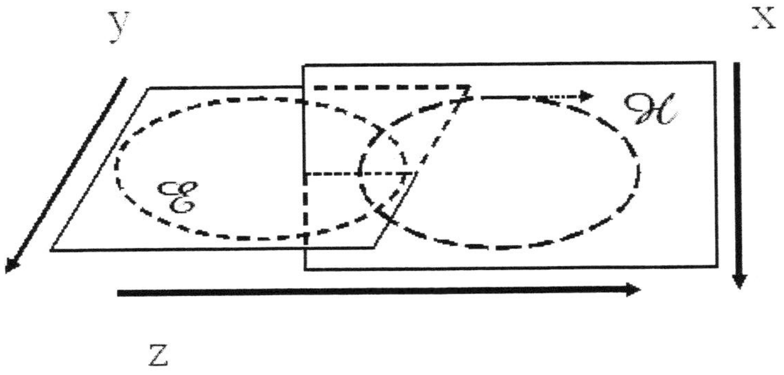

Figure 20. Idealistic image of curls E and H in QED.

The matrix α generates the group $\mathbb{Z}_4 \subset \mathrm{Mat}(4, \mathbb{Z})$. Now, most of us are familiar with an image like figure (20). We see two interlocked electromagnetic curls propagating in direction $\vec{z}$. That is, they have a common displacement alongside $\vec{z}$. The local field graph will consist of two different 4-cycles which share one edge. It is an extension of the 4-cycle, that is, two 4-cycles with a common edge. They can be thought to correlate with planes $\{x, y\}$, $\{y, z\}$ and roughly pinned as below

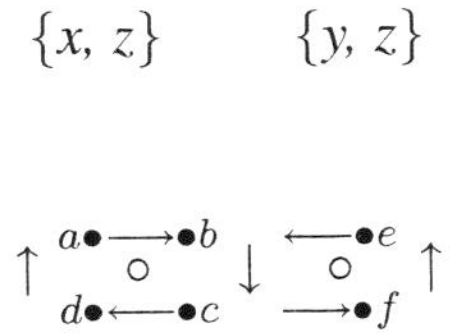

$$\{x, z\} \qquad \{y, z\}$$

$$\uparrow \begin{matrix} a\bullet \longrightarrow \bullet b \\ \circ \\ d\bullet \longleftarrow \bullet c \end{matrix} \downarrow \begin{matrix} \longleftarrow \bullet e \\ \circ \\ \longrightarrow \bullet f \end{matrix} \uparrow$$

The extended 4-cycle reminds us of two orthogonal faces of a 3-cube. Adding two edges we obtain a graph which provides the path element of a Peano-fractal. If we triple each edge and substitute it by this generating local graph and procede recursively, we obtain a fractal sequence. Tripling edges gives a sequence of 3, 3^2, 3^3 a.s.o. edges. Carrying out the recursion, we obtain a planar grid of counter rotating torsions. In the metric plane this would be a recursive bisecting of the grid. The Peano fractal line turns out to have fractal dimension 2. So, provided we consider the directed edges as scaled lineelements, the counter rotating planar graph would indeed represent a fractal decomposition of dynamic displacement with zero sum.

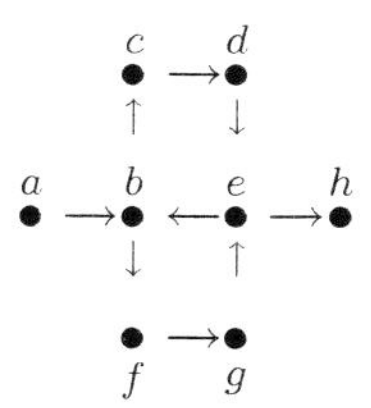

So the planar 4-cycle turns out to be an element of that special Peano-fractal P of figure (2.). We must be aware the problem of dimension of a particle trajectory has not been fully answered. I believe that the pregeometric investigation of field graphs can help us to understand the phenomena that bring about dimension once we begin to understand the role of the distorion field. We can think of a discontinuous plane as a decomposition of $\mathbb{R}^2$ into continuously many counter-rotating 4-cycles. Assigning an elementary energy unit to each infinitesimal directed edge, summing it all up, we obtain zero energy. Now suppose, this zero energy arrangement turns out unstable, in that case a photon or fermion may find a straight path to travel trough the unstable array.

11. Space Graphs

11.1. Discrete Structure and Freedom from Surface Symmetry

Pregeometric graphs are not just any graphs, but they comply with physical demands. To some extent their properties correlate with geometric and physical features of the surface manifold. For instance we know that the Euclidean surface manifold fulfills certain geometric constraints. The surface dynamics accords with symmetry elements. Historically this advised the construction of Lie manifolds and their symmetry groups. On the other hand microscopic quantum phenomena disclose further degrees of freedom. The long distance weak links seem to violate the surface symmetries. At least this is our first impression. But it may also be the case that the global dynamics based on discrete structures has just further degrees of freedom. Mycroscopic motion must violate surface principles implied by a surface Lie algebra, yet it may involve a larger symmetry group which incorporates surface algebra gained by renormalization. How that goes we can study by the aid of a peculiar discrete struture, the electromagnetic space graph S. The automorphism group of this graph is generated by a few transpositions of coarse grained regions. By transposition we do not mean linear affine parallel transport or rotation, but rather an exchange of vertices. Some analogous procedure - transposition as exchange of base units having unequal grades - had caused a further degree of freedom which is responsible for the emergence of the standard model within the Clifford algebra of Minkowski space. That is to say, degrees of freedom are increased by introducing transpositions in both the surface and the pregeometric structure (as for transpositions in Clifford algebra, read Schmeikal 2004).

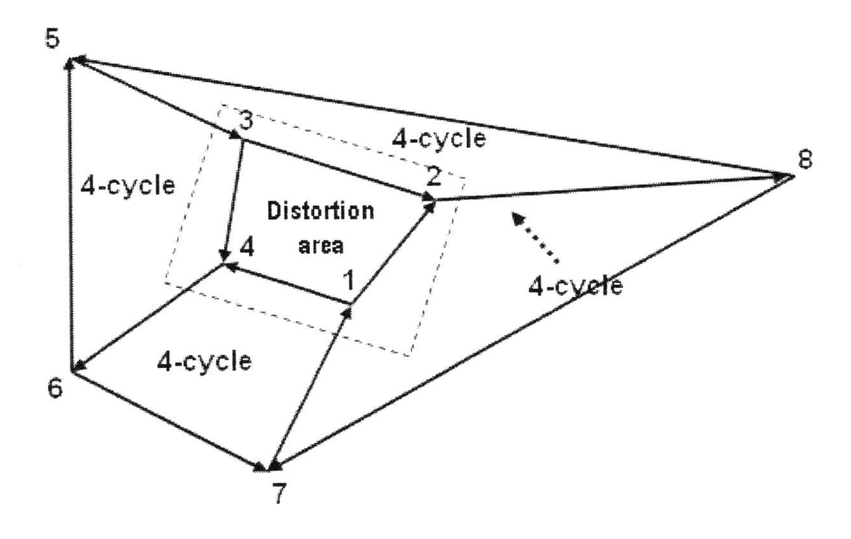

Figure 21. Planar space graph.

11.2. The Electromagnetic Space Graph

The space graph S has to be an electrodynamic graph with a 4-vertex distortion subgraph. It involves four counter directed planar 4-cycles in the vertical faces and two distortion graphs on the horizontal faces (figure 11.2.). We are using the words *vertical* and *horizontal* in an intuitive way to challenge the imagination. It should be noted, that those discrete structures we are using are not structures embedded in a preexisting smooth manifold. [3] It is however not unimportant to realize that the electrodynamic space graph is so to say a relative within an edge-model of polyhedra, that is, some cube with directed edges which turns out to be a planar graph.

This means, a certain important type of spatial net is obtained from the extension of graphs that obey Kuratowski's planarity theorem. The space graph demonstrates that the harmony of the planar counter rotating decomposition cannot be generalized to 3-space. You cannot construct a tetrahedron with four harmonious 3-cycles. At some vertex there must appear a distortion, that is, counter directed arrows. But there is a planar graph that consolidates the 4 cycles with the distortion area as pinned down in figure (11.2.). I looked into the historic literature. I found out, it is essentially the Whittaker form of the electrodynamic equations of motion which makes clear that in Euclidean space a planar distortion graph is necessary to link the counter rotating vertical planes. Those distortions are represented by vectors $\boldsymbol{f}$ and $\boldsymbol{g}$ of the Whittaker equations of motion (1904). Consider for

[3]This correlates with what Requardt (2004) wrote on page 2: Discrete structures like e.g. (topological) graphs occur of course elsewhere in (quantum) gravity research but usually as structures embedded in a preexisting smooth manifold, sharing sometimes even some of the metric properties of the ambient space or derive from certain triangulations, that is, being typically very regular.

example, up to automorphism, the adjacency matrix of S

$$\alpha(S) = \begin{pmatrix} 0 & 1 & 0 & 1 & 0 & 0 & 0 & 0 \\ 0 & 0 & 0 & 0 & 0 & 0 & 0 & 1 \\ 0 & 1 & 0 & 1 & 0 & 0 & 0 & 0 \\ 0 & 0 & 0 & 0 & 0 & 1 & 0 & 0 \\ 0 & 0 & 1 & 0 & 0 & 0 & 0 & 0 \\ 0 & 0 & 0 & 0 & 1 & 0 & 1 & 0 \\ 1 & 0 & 0 & 0 & 0 & 0 & 0 & 0 \\ 0 & 0 & 0 & 0 & 1 & 0 & 1 & 0 \end{pmatrix}.$$

Path matrices of order m are given by terms $\alpha(S)^m$. Matrices thus generated are denoted as $\alpha_0, \alpha_1, ..., \alpha_n$ with

$$\alpha_n = \alpha_0^{n+1} = 2^k \alpha_{mod_4(n)} \quad \text{for even } n = 2k \quad \text{or odd } n = 2k+1, \tag{8}$$

For example we have $\alpha_{11} = \alpha_0^{12} = 2^5\alpha_3 = 16\alpha_3$. There reappears the cycle $\{\alpha_1, \alpha_2, \alpha_3, \alpha_4\}$. Let $\mathbb{M}_4 \rightleftharpoons \{\alpha_1, \alpha_2, \alpha_3, \alpha_4\}$. We have an epimorphism

$$\phi : \{\alpha\} \longrightarrow \mathbb{M}_4 \quad \text{respectively } \mathrm{Mat}(8, \mathbb{Z}) \longrightarrow \mathbb{M}_4, , \tag{9}$$

Consider the example of paths with lenght 10:

$$\alpha_{10} = \alpha_0^{11} = \begin{pmatrix} 0 & 0 & 0 & 0 & 32 & 0 & 32 & 0 \\ 16 & 0 & 16 & 0 & 0 & 0 & 0 & 0 \\ 0 & 0 & 0 & 0 & 32 & 0 & 32 & 0 \\ 16 & 0 & 16 & 0 & 0 & 0 & 0 & 0 \\ 0 & 0 & 0 & 0 & 0 & 16 & 0 & 16 \\ 0 & 32 & 0 & 32 & 0 & 0 & 0 & 0 \\ 0 & 0 & 0 & 0 & 0 & 16 & 0 & 16 \\ 0 & 32 & 0 & 32 & 1 & 0 & 1 & 0 \end{pmatrix}.$$

These matrices allow us to calculate distances and internal dimension of networks.

11.3. Reorientation and Space Graph Automorphisms

The reorientation group of Euclidean 3-space is the octahedral group $\mathbf{O}$. A Dreibein in Euclidean space has octahedral automorphisms. The 24 elements of $\mathbf{O}$ provide the space congruences of a 3-cube. Clearly, these symmetries preserve the *'rigid body'* property. They represent rotations of the whole space and therefore keep the region (cube) intact. Giving up the rigid body property, going beyond the local coherence of space we pave the way for quantum phenomena. Note the following geometric phenomenon: Like the cube, so also the 2-plane has a reorientation group, namely the dihedral group D_{2d} or D_4 of the space congruences of a square. The dihedral group is an 8 element non-abelian group. It is also preserving the connection - the rigid body feature - neighbouring quadrants are not torn apart by the action of any symmetry element of D_4. But, recall, the four quadrants of a plane coordinate system allow for $4! = 24$ permutations. This is the same number of elements as the $\mathbf{O}$ is providing. Actually both groups are isomorphic: $\mathbf{O} \simeq \mathbf{S}_4$ However, the octahedral group, acting as permutations of quadrants, tears the plane square apart while the dihedral group preserves the connectivity of quadrants. Seen from the restricted dihedral symmetry, the coarse spatial elements of a plane square, the quadrants, have gained a dynamic degree of freedom which is somehow measured by the index of the dihedral group ($= 3$) within the octahedral. Now look at figure (11.2.). The space graph is essentially built up by four 4-cycles and one 4-distortion involving 8 vertices of two distinct types: the first kind of vertices has one outgoing and two incoming directed edges, the second kind of vertices has two outgoing and one incoming edge. Enumerating the eight vertices and denoting group elements as transpositions of vertices (2-cycles) we have eight generating elements (Coxeter reflections) which read

$$\begin{aligned} g_1 &= ((1,3)), \quad g_2= ((1,6)), \quad g_3 = ((2,4)), \quad g_4= ((2,7)), \\ g_5 &= ((3,8)), \quad g_6= ((4,5)), \quad g_7 = ((5,7)), \quad g_8 = ((6,8)) \end{aligned} \tag{10}$$

These reflections generate the automorphism group of the space graph. This is some hyperoctahedral group $\mathbf{O}(S)$ having order n = 216. The octahedral subgroup $\mathbf{O} \subset \mathbf{O}(S)$ has the index 9 relative to the cardinality of the larger group $\mathbf{O}(S)$. This index shows the importance of trigonal permutations in the space graph. The index of the subgroup $\mathbb{Z}_3$ is 8 in $\mathbf{O}$, but 72 in the automorphism group $\mathbf{O}(S)$. A cyclic element of $\mathbb{Z}_4 \subset \mathbf{O}(S)$ has the order 54 relative to $|\mathbf{O}(S)|$. The ratio of indices of $\mathbb{Z}_3$ over $\mathbb{Z}_4$ in the space graph automorphism group is equal to $\frac{72}{54} = \frac{4}{3}$, the same as in the Euclidan octahedral group. Notice, we have $|D_4| : |\mathbf{O}| : |\mathbf{O}(S)| = 1 : 3 : 3^2$. In contrast to the euclidean octahedral group the graph automorphism group $\mathbf{O}(S)$ contains the subgroup Dic_3, the dicyclic group of order 12, the semidirect product of $\mathbb{Z}_3$ and $\mathbb{Z}_4$. Thus, 4-cycles and 3-cycles play a dominant role in translocal electromagnetic phenomena. We have seen in the previous sections that the 4-cycles correlate with electromagnetic phenomena. But what is the meaning of the dominant 3-cycles? The group $\mathbb{Z}_3$ indicates the presence of some kind of flavor in the electromagnetic primordial domain. From the viewpoint of dynamic systems the $\mathbb{Z}_3$ and $\mathbb{Z}_4$ represent three types of energy and two modes of linear propagation of energy. But while the first fixes the existence of a transversal electromagnetic photon as in figure (5.), the other propagator signifies a longitudinal photon (AIAS 2000). Let us consider the figure (11.3.).

It is a snapshop. What shall we see on the photograph? Above all we should see a finite graph. Second we should see it is a planar graph. It can be drawn in a plane having coordinates $\{x, y\}$ such that no two edges traverse each other. Next we should see that the plane has four quadrants each of which embeds a smaller planar graph. The smaller graph is an extension of the space graph from figure (11.2.). It looks as if the space graph extends towards the inner. Like graph (11.2.) it is a relative within the edge-model of polyhedra, namely it can be thought to represent two cubes, one behind the other in the vertical z-line, resting on the $\{x, y\}$-plane. Now we must be careful with our interpretation. Is it possible that the whole graph is representing nothing else than a cubic grid? Is it eight cubes or octants constituting a cube with larger (double) size resting on the $\{x, y\}$-plane? No! This is not possible, since even only a two cube grid cannot generally be flattened. The cubes must be erected vertically on the plane. Arbitrary many cubes may form a linear chain of dices. Otherwise they cannot form a planar graph. That cubic string can be posed

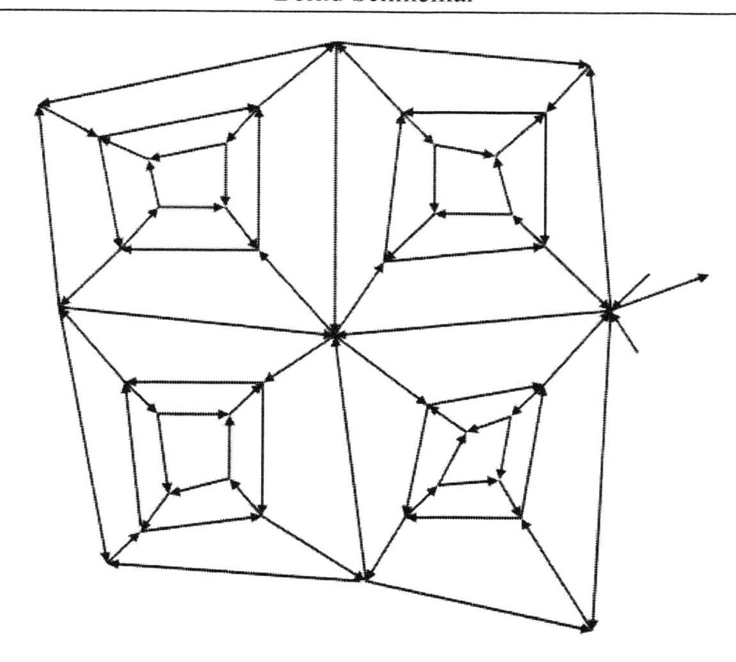

Figure 22. Deep planar space graph.

orthogonally on the plane so that it rests on the face of the last cube. We call that array a cubic string. A cubic string can be flattened into the plane on which it is posed. Four cubic strings may also be flattened. But their faces must be separate. Imagine their edges touching in the plane on which they are fixed as in figure (11.3.). Suppose this denotes a dynamic process which it should anyway. The connection of strings in the plane may be there in the moment we shoot the photograph, but in the next it vanishes. The whole 3-dimensional arrangement may be rather unstable, a temporary illusion! The strings, moving in z-direction, may so to say press ahead with an illusion of 3-dimensional local connectedness, while in reality the space tends to decay almost everywhere. Such is the nature of the electromagnetic space graph.

From this we obtain a new image for the propagation of electromagnetic energy in the surface manifold. While the energy, transversal photon, is propagating in direction of e_3, electric curls unfolding in the plane area parallel to e_{13} and the magnetic in e_{23}, the space connection of the field is actually manufactured at the wavefront where the photon is localized. In the 'past' region, the propagation does not contribute to connectivity. So the 3-dimensional conflation decays. We can speak of a photonic space concatenation or an electromagnetic space-interface. The propagator carries the connection with it.

12. Strong Force Entanglement

We would like to represent the strong force in the Clifford algebra of the Minkowski space in indefinite signature, that is, $C\ell_{3,1}$. For that purpose we use the primitive idempotents f_s, f_d, f_u from equations (1) which, we know, obey the desired flavor- and color symmetries. However, in order to be able to work in the Dirac picture of quantum states in Hilbert space, we assume that the primitive idempotents represent the density matrices of the states. This is natural since the real Clifford algebra can be represented by 4×4 matrices with real entries, the so called Majorana algebra $Mat(4, \mathbb{R})$. Therefore, let

$$|u\rangle = \frac{1}{c}\begin{pmatrix}0\\0\\1\\0\end{pmatrix} \quad \text{and} \quad |d\rangle = \frac{1}{c}\begin{pmatrix}0\\0\\0\\1\end{pmatrix}, \quad c \text{ a complex number with } \quad |c| = 1 \qquad (11)$$

Note, we leave the first entry for a lepton and the second for the strange quark. We obtain for the density matrices

$$\rho_u = |u\rangle\langle u| = f_u = \begin{pmatrix}0&0&0&0\\0&0&0&0\\0&0&1&0\\0&0&0&0\end{pmatrix} \quad \text{and} \quad \rho_d = |d\rangle\langle d| = f_d = \begin{pmatrix}0&0&0&0\\0&0&0&0\\0&0&0&0\\0&0&0&1\end{pmatrix}, \qquad (12)$$

In a strong force entanglement the isospin states of fermions disclose correlations in comparatively small subnuclear regions. Although we have been unable to explain these phenomena they do not seem different from the familiar spin entanglement and can, in principle, be approached in a similar way, namely on the basis of non-metric directed graphs. What makes these correlations so impressive is that they signify a communication, that is, they exist in superpositions just like the quantum states themselves. When the superposition is destroyed by the interaction - in any HEPhy experiment - the correlation is communicated between the carriers of isospin. That is what is meant by "points speaking to each other". There is no reason why we should presume a metric space, in that special case, rather we have a veil in which fermions appear and disappear while they propagate the connection. This is especially important in strong interaction since it provides the connection at very small scales, even though the phenomenon is probably scale independent. Recall, the basic scale freedom can be derived from a potential for spin flips at any size of the veil. In agreement with Feynman's approach and the Heisenberg uncertainty we have every reason to consider spin-veils for all the forces we know at present. In their ground states they may have no measurable properties, but in action they disclose all the astonishing phenomena we are observing.

Consider a mixed state

$$|\psi\rangle = \frac{1}{\sqrt{2}}|d\,,d\rangle + \frac{1}{\sqrt{2}}|u\,,d\rangle = \frac{1}{\sqrt{2}}\begin{pmatrix}0\\0\\0\\0\\0\\0\\0\\1\\1\end{pmatrix} \qquad (13)$$

having density matrix

$$\rho_\psi = \frac{1}{2}\begin{pmatrix} 0&0&0&0&0&0&0&0\\ 0&0&0&0&0&0&0&0\\ 0&0&0&0&0&0&0&0\\ 0&0&0&0&0&0&0&0\\ 0&0&0&0&0&0&0&0\\ 0&0&0&0&0&0&0&0\\ 0&0&0&0& &0&1&1\\ 0&0&0&0&0&0&1&1 \end{pmatrix} \quad \text{and another state} \tag{14}$$

$$|\xi\rangle = \frac{1}{\sqrt{2}}|d\,,d\rangle + \frac{1}{\sqrt{2}}|u\,,u\rangle = \frac{1}{\sqrt{2}}\begin{pmatrix}0\\0\\0\\0\\0\\1\\0\\0\\1\end{pmatrix} \tag{15}$$

having density matrix

$$\rho_\xi = \frac{1}{2}\begin{pmatrix} 0&0&0&0&0&0&0&0\\ 0&0&0&0&0&0&0&0\\ 0&0&0&0&0&0&0&0\\ 0&0&0&0&0&0&0&0\\ 0&0&0&0&1&0&0&1\\ 0&0&0&0&0&0&0&0\\ 0&0&0&0&0&0&0&0\\ 0&0&0&0&1&0&0&1 \end{pmatrix} \tag{16}$$

The density matrix ρ_ψ can be factorized as tensor product

$$\rho_\psi = \frac{1}{2}f_d \otimes \begin{pmatrix} 0&0&0&0\\ 0&0&0&0\\ 0&0&1&1\\ 0&0&1&1 \end{pmatrix}, \tag{17}$$

but ρ_ξ cannot. It is obvious that fermions that communicate the same isospin are entangled most. Those states cannot be factorized.

12.1. Grade Transposition and Entanglement

We want to show that the entanglement involves a new degree of freedom, namely the transposition or exchange of spatial domains with a different grade. Bear in mind that translocal dynamics involves the Clifford algebra of the Minkowski space. A transposition, formally, is derived from the primitive idempotents which span the color spaces ch_1 to ch_6. Each color space is a linear space spanned by three commuting base units of the six isomorphic Cartan subalgebras of $C\ell_{3,1}$ including the unit scalar (Schmeikal 2004, p. 357f., 2010, ch. 22). Select the following primitive idempotents from Cartan subalgebras 2 and 6

$$f_{23} = \frac{1}{4}(1 - e_1)(1 + e_{34}) \qquad f_{62} = \frac{1}{4}(1 + e_3)(1 - e_{24}) \tag{18}$$

and form 'reflections' $s_{23} = 1 - 2f_{23}$ and $s_{62} = 1 - 2f_{62}$. You obtain the Clifford number

$$T = s_{23}s_{62} = \frac{1}{2}(Id + e_1 - e_{34} + e_{134})\frac{1}{2}(Id - e_3 + e_{24} - e_{234}) \tag{19}$$

Using standard representation

$$e_1 = \begin{pmatrix} 1 & 0 & 0 & 0 \\ 0 & -1 & 0 & 0 \\ 0 & 0 & -1 & 0 \\ 0 & 0 & 0 & 1 \end{pmatrix} \qquad e_2 = \begin{pmatrix} 0 & 1 & 0 & 0 \\ 1 & 0 & 0 & 0 \\ 0 & 0 & 0 & 1 \\ 0 & 0 & 1 & 0 \end{pmatrix},$$

$$e_3 = \begin{pmatrix} 0 & 0 & 1 & 0 \\ 0 & 0 & 0 & -1 \\ 1 & 0 & 0 & 0 \\ 0 & -1 & 0 & 0 \end{pmatrix} \quad \text{and} \quad e_4 = \begin{pmatrix} 0 & -1 & 0 & 0 \\ 1 & 0 & 0 & 0 \\ 0 & 0 & 0 & -1 \\ 0 & 0 & 1 & 0 \end{pmatrix}, \tag{20}$$

The flavor rotation $T \in Mat(4, \mathbb{R})$ takes the form

$$T = \begin{pmatrix} 1 & 0 & 0 & 0 \\ 0 & 0 & 1 & 0 \\ 0 & 0 & 0 & 1 \\ 0 & 1 & 0 & 0 \end{pmatrix}, \tag{21}$$

Verify $T|d\rangle = |u\rangle$ and the entangled fermion state is

$$|\xi\rangle = \frac{1}{\sqrt{2}}(|d, d\rangle + |T\,d,\, T\,d\rangle) \tag{22}$$

12.2. The New Degree of Freedom

We suggest that long distance weak links and translocal interaction of spatial domains involve some new degrees of freedom of motion. What that freedom is in the case of the strong force entanglement, can be seen from the action of the trigonal flavor rotation T and T^{-1} on the elements of the Cartan subalgebras and the color spaces respectively. Using the equation (19) we verify the following flavor transformations

Tringonal T acts on idempotents in terms of conjugations (23)

$$
\begin{aligned}
T^{-1} f_{14} T &\longrightarrow f_{13} \quad \text{implying spinor mapping} \quad \hat{T}|d\rangle \longrightarrow |u\rangle \\
T^{-1} f_{13} T &\longrightarrow f_{12} \quad \text{implying} \quad \hat{T}|u\rangle \longrightarrow |s\rangle \\
T^{-1} f_{12} T &\longrightarrow f_{14} \quad \text{implying} \quad \hat{T}|s\rangle \longrightarrow |d\rangle
\end{aligned}
\tag{24}
$$

where $\hat{T}$ denotes the matrix. There is a significant difference to the elder approaches to spin gauge theory which has given legitimity to the claim that our standard model is a space time group. This is in the fact that transformations of flavor and color not only transform the isospin spaces, but they do involve the spacetime manifold. The trigonal operator acts on the space-time domain as a 'grade-unfolder', namely we have

$$
T e_1 T^{-1} = e_{24} \qquad T e_{24} T^{-1} = e_{124} \qquad T e_{124} T^{-1} = e_1 \tag{25}
$$

It is this mapping in the surface manifold which gives space to both small distance strong entanglements, long distance weak links and scale free discontinuous mappings from trajectory lines to spacetime areas and spacetime volumes. We keep in mind

$$
T: \quad e_1 \longrightarrow e_{24} \longrightarrow e_{124} \longrightarrow e_1 \tag{26}
$$

This grade transformation is characteristic for strong interacting elastic and inelastic scattering of fermions in the selected Cartan subalgebra. Each Cartan subalgebra corresponds with one color space and one of the six leptons. For example e_1, e_{24}, e_{124} denote the first Cartan subalgebra and signify the electron. Each lepton plays a special part. The stable electron annihilates the constitutive Lie algebra $l^{(2)}$ and thus represents a fixed point for the constitutive group $L^{(2)}$ isomorphic with either $\mathrm{SL}(3, \mathbb{R})$ or $\mathrm{SU}(3, \mathbb{C})$. We have

$$
\forall_{g \in L^{(2)}}\, g\, f_{11}\, g^{-1} = f_{11} \quad \text{and so} \quad T\, f_{11} T^{-1} = f_{11} \tag{27}
$$

Under such transformations Heisenberg's uncertainty relation

$$
\triangle x \triangle p_x \geq \frac{\hbar}{2} \quad \text{turns into} \quad \triangle V_{2,1} \triangle \left(\frac{E}{c^2 t^2} \right) \geq \hbar \tag{28}
$$

where $V_{2,1}$ is a spacetime volume associated with the trivector e_{124} measured by a quantity $xy(ct)$ and having physical SI dimension $[\mathrm{m}^3]$ and HEPhy dimension eV^{-3}. Now the stable lepton, in this calibration an e^-, experiences a different situation than the quarks which interact weakly and strongly.[4] Yet we encounter similar situations in the weakly decaying neutron and in the *infinite momentum frame* of an incoming proton. We must

[4] SI denotes the International system of units. The HEPhy dimensions are mediated by rescaling $c = 1$ and $\frac{h}{2\pi} = 1$. Then lenght and time obtain the dimension of reciprocal energy $197,33 MeV = 1\, fm^{-1}$

differ between electrons in a Stern-Gerlach experiment, weakly interacting neutrons, elastic electron-nucleon scattering and ineleastic electron- quark scattering. The following examples may purvey a brief guesstimate.

1.) Double-slit: the lifetime of the electrons is infinite. The $\triangle(E/c^2t^2)$ goes to zero. The uncertainty of spacetime volume $\triangle V_{2,1}$ covers long distance entanglement. The same argument applies to the photon. It tends to develop space.

2.) Decay of a free neutron: n $\rightarrow$ p $+e^- + \bar{\nu}_e$ is mediated by a weak interaction which turns $|u\,d\,d\,\rangle$ into $|u\,u\,d\,\rangle$. The picture is well known. The incoming neutron transmutes into a proton by emiting a W^- Boson at the left vertex of the Feynman graph. At the right vertex the W^- field quantum decays into the outgoing e^- and $\bar{\nu}_e$. This requires a weakly induced flavor rotation from $|d\rangle$ into $|u\rangle$ and respectively a developing grade $e_1 \rightarrow e_{124}$. We are not yet entirely sure about the nature of the neutron decay. The lifetime of the $|d\rangle$ quark seems to be infinite. That would imply that the spacetime volumes $V_{2,1}$ of $|d\rangle$ quarks in neutron rays reach far beyond the estimated diameter of neutrons (about $1,5 \times 10^{-15}$m) and thus support long range entanglement.

3.) Inelastic electron-nucleon scattering: Feynman originated the idea - a high energy proton (or neutron) can be looked at as a beam of collinear free quarks. The energy-difference between a proton and a number of quarks, whose total impulse is determined by the proton impulse, amounts

$$\triangle E = \sqrt{P^2 + M^2} - \sum_i \sqrt{k_{L,i}^2 + k_{T,i}^2 + m_i^2} \tag{29}$$

where the first term on the right denotes the on shell energy of the proton, the $k_{L,i}$ the longitudinal fermion impulse parallel to the beam, $k_{T,i}$ their transversal impulses and m_i their constituent masses. In large hadron colliders the absolute value of $|P|$ becomes very large. With it correlate the $k_{L,i}$ and the squareroots can be series expanded into

$$\triangle E = \frac{M^2}{2|P|} - \sum_i \frac{k_{T,i}^2 + m_i^2}{2\,k_{L,i}} \tag{30}$$

Equation (30) implies that in a reference system having high proton impulses $\triangle E$ becomes zero and is not measurable. In such an *infinite momentum frame* we observe vanishing violation of energy conservation, and the lifetime of the *intermediary collinear quarks* tends towards infinity. The proton approaches a stable state with inner transversal zero impulses. It is only when one of the quarks obtained a big transversal impulse in some inelastic scattering that the quarks drive apart and the fragmentation sets in. Because of the large lifetime of collinear quarks the spacetime volume of entanglement excedes the volume of the proton by far. This induces a coherence of collinear fermions in the ray.

4.) 1964 weak decay of an Ω^- in Brookhaven National Laboratory: The Ω^- has strangeness $S = 3$ since it consists of three $|s\,\rangle$-quarks. At any weak decay the strangeness has to change by one unit. Therefore the particle decays in a cascade. The production of

Ω^- happens in a collision $K^- + p \to \Omega^- + K^+ + K^0$ and creates a cascade

$$\begin{aligned} \Omega^- &\to \Xi^0 + \pi^- \\ \Xi^0 &\to \Lambda + \pi^0 \\ \Lambda &\to \pi^- + p \end{aligned} \tag{31}$$

The first event in cascade (31) correlates with the discovery of the Ω^- while scanning 100000 bubble chamber images. Its invariant mass had been $1686 \pm 12\ MeV$ as Gell-Mann had predicted. Its lifetime amounted $7 \times 10^{-11}\ s$. Looking at the trajactory as a cylinder, we would obtain a $\triangle V_{2,1} \simeq 2,454 \times 10^{-18}[\mathrm{m}^2\mathrm{s}]$. Even that is much larger than the volume limited by a femptometer diameter times the lenght of the event.

5.) Events in strong interaction involve resonances and particles with extremely short lifetimes. Volumina $\triangle V_{2,1}$ then tend to retrack to the smaller scales of the subnuclear range. It may be that subnuclear small scale fermion entanglement is characteristic for strong interaction.

12.3. Long Distance Strong Links

Spin phenomena are not restricted to microscopic regions. Spinors govern the system dynamics of matter at all possible scales. Eli Cartan had become aware of the importance of spinors fifteen years before Paul Dirac introduced them into quantum mechanics. He defined the pure spinor in terms of *polarized isotropic multivectors*. Strictly, Cartan's theory introduces the concept by constructing an isotropic 3-vector (x_1, x_2, x_3) the components of which give us a zero lenght $x_1^2 + x_2^2 + x_3^2 = 0$. This can be satisfied in a nontrivial way if the components are given, for example, by two complex quantities $x_1 = \xi_0^2 - \xi_1^2,\ x_2 = i(\xi_0^2 + \xi_1^2),\ x_3 = -2\xi_0\xi_1$. The pair (ξ_0, ξ_1) constitutes a spinor. How the mathematics develops from that point onwards and why it represents a scale-free concept, can be found in Schmeikal (2010, ch. 24ff.). Like the spin, so the isospin, and even spin of cosmic matter is capable of double values and translocal entanglement. To give a full discussion of the macroscopic nature of a flavor-spinor would go beyond the scope of this chapter. Let me, therefore, just give you a small outline.

Consider a spacecraft with a proton accelerator or neutrino gun capable of a stable flavor-oscillation. We fire a beam to a 10^7m distant target where locations are measured. While the inner state is rotated, the location may turn from a 10^7m vector in the reference system of the spacecraft to a non-local area with spacetime volume $V_{2,1} = 10^7\mathrm{m}^3$. During one oscillation the location may turn from almost point-like to unfurled and spread out. This is a consequence of the SM-nature of spacetime algebra.

13. The Natural Standard Model

The Clifford algebra of the spacetime has its own standard model. Most of us have not yet realized this. The inner fermion states u, d, s are essentially given by the equations (1) and the lepton $f_1 = \frac{1}{4}(1 + e_1)(1 + e_{24})$. Isomorphic images exist in the other Cartan subalgebras (chromatic spaces). The primitive idempotents represent density matrices. We shall later pose the question, if they possess spinors in the traditional sense. Now we ask what is the constitutive algebra $l^{(2)}$? In order to visualize and comprehend the meaning of the special unitary symmetry $\mathrm{SU}(3)$ in connection with the Clifford algebra it is useful to understand the switching between a compact form of the Lie group $\mathrm{SU}(3,\mathbb{C})$ and a non-compact real form of $\mathrm{SL}(3,\mathbb{R})$ (Magnea 2002, p. 17). The generators of the algebra $\mathrm{sl}(3)$ can be written as a Clifform, that is, as a multivectorial form of Clifford numbers independent of a definite representation over some determined number field:

$$\begin{array}{lll}
\lambda_1 = \frac{1}{4}(e_{34} - e_{134}) & \lambda_2 = \frac{1}{4}(-e_{23} + e_{123}) & \underline{\lambda_3 = \frac{1}{4}(e_{24} - e_{124})} \\
\lambda_4 = \frac{1}{4}(e_2 + e_{14}) & \lambda_5 = -\frac{1}{4}(e_4 + e_{12}) & \lambda_6 = \frac{1}{4}(e_3 - e_{234}) \\
\lambda_7 = \frac{1}{4}(e_{13} + e_j) & \underline{\lambda_8 = -\frac{1}{4}(e_1 + e_{24})} & \text{all in } C\ell_{3,1}
\end{array} \tag{32}$$

This is an abstract representation of a 'Clifform' $\mathrm{su}_{\mathrm{Cl}}(3)$ in the standard basis of the Clifford algebra $C\ell_{3,1}$. As we know that the real algebra $C\ell_{3,1}$ is isomorphic with the Majorana algebra of real 4×4 matrices, we can be sure this is a non-compact form of $\mathrm{sl}(3,\mathbb{R})$ standing for the compact form $\mathrm{su}(3,\mathbb{C})$. In the real representation $Mat(4,\mathbb{R})$ the underlined elements are the diagonal matrices denoting isospin and hypercharge.

To compare with the standard matrix representations use the Majorana matrices (20). The λ_1 to λ_8 are Clifford algebra forms of the familiar algebra $\mathrm{sl}(3,\mathbb{R})$. As the algebra is usually described by a list of non-vanishing root commutation relations, we prefer the following linear combinations:

$$\begin{array}{llll}
T_{0,1} = -2\lambda_3 & T_{0,2} = 2\lambda_8 & \text{diagonal matrices} & \\
T_{+,1} = -\lambda_1 - \lambda_2 & T_{+,2} = \lambda_4 + \lambda_5 & T_{+,3} = -\lambda_6 - \lambda_7 & \text{shift elements} \\
T_{-,1} = \lambda_2 - \lambda_1 & T_{-,2} = \lambda_4 - \lambda_5 & T_{-,3} = \lambda_7 - \lambda_6 &
\end{array} \tag{33}$$

Using the above basis we obtain the standard representation of the $\mathrm{SL}(3,\mathbb{R}) \subset \mathrm{Mat}(4,\mathbb{R})$ by positive definite shift operators and their diagonal operators.

$$T_{0,1} = \begin{pmatrix} 0 & 0 & 0 & 0 \\ 0 & 1 & 0 & 0 \\ 0 & 0 & -1 & 0 \\ 0 & 0 & 0 & 0 \end{pmatrix} \qquad T_{0,2} = \begin{pmatrix} 0 & 0 & 0 & 0 \\ 0 & 0 & 0 & 0 \\ 0 & 0 & 1 & 0 \\ 0 & 0 & 0 & -1 \end{pmatrix}$$

$$T_{+,1} = \begin{pmatrix} 0 & 0 & 0 & 0 \\ 0 & 0 & 1 & 0 \\ 0 & 0 & 0 & 0 \\ 0 & 0 & 0 & 0 \end{pmatrix} \qquad T_{-,1} = \begin{pmatrix} 0 & 0 & 0 & 0 \\ 0 & 0 & 0 & 0 \\ 0 & 1 & 0 & 0 \\ 0 & 0 & 0 & 0 \end{pmatrix}$$

$$T_{+,2} = \begin{pmatrix} 0 & 0 & 0 & 0 \\ 0 & 0 & 0 & 0 \\ 0 & 0 & 0 & 1 \\ 0 & 0 & 0 & 0 \end{pmatrix} \qquad T_{-,2} = \begin{pmatrix} 0 & 0 & 0 & 0 \\ 0 & 0 & 0 & 0 \\ 0 & 0 & 0 & 0 \\ 0 & 0 & 1 & 0 \end{pmatrix}$$

$$T_{+,3} = \begin{pmatrix} 0 & 0 & 0 & 0 \\ 0 & 0 & 0 & 1 \\ 0 & 0 & 0 & 0 \\ 0 & 0 & 0 & 0 \end{pmatrix} \qquad T_{-,3} = \begin{pmatrix} 0 & 0 & 0 & 0 \\ 0 & 0 & 0 & 0 \\ 0 & 0 & 0 & 0 \\ 0 & 1 & 0 & 0 \end{pmatrix}$$

$$\begin{array}{lll} [T_{+,j}, T_{-,j} = T_{0,j} & [T_{\pm,j}, T_{0,j} = \mp 2T_{\pm,j} & \text{j= 1, 2} \\ [T_{+,1}, T_{0,2}] = \pm T_{\pm,1} & [T_{+,2}, T_{0,1}] = \pm T_{\pm,2} & \\ [T_{\pm,1}, [T_{\pm,1}, T_{\pm,2}]] = [T_{\pm,2}, [T_{\pm,2}, T_{\pm,1}]] = 0 & & \end{array} \tag{34}$$

To obtain a compact form of $\mathrm{su}(3,\mathbb{C})$ in $\mathbb{C} \otimes C\ell_{3,1}$ we just have to construct a linear combination of the diagonal generators and instead use the hypercharge generator.

$$\lambda_8 = \frac{1}{2\sqrt{3}}(-2e_1 + e_{24} + e_{124}) \tag{35}$$

In agreement with the Weyl trick where non-compact real forms are derived from compact complex forms, we invert the procedure and multiply the second, fifth and eight matrix by the unit imaginary. Thus we obtain up to a factor 2 a standard representation within the complex Clifford algebra. The various forms of $\mathrm{sl}(3,\mathbb{C})$, $\mathrm{sl}(3,\mathbb{R})$ and $\mathrm{su}(3,\mathbb{C})$ are equivalent with respect to classification of hadrons since they possess the same root space, namely A_2. The Clifform we use is, however, independent from the chosen field of representation. In that sense it is indeterminate where the predicate of compactness is concerned. If we do not know the field over which the Clifford algebra is constructed, we cannot decide on the special group algebra that shall be constitutive for the physical phenomena. Anyway, the largest constitutive Lie algebra in $\mathbb{C} \otimes C\ell_{3,1}$ is $\mathrm{sl}(4,\mathbb{C})$ and respectively $\mathrm{su}(4,\mathbb{C})$. The reduction of $\mathrm{su}(4,\mathbb{C})$ to $\mathrm{su}(3,\mathbb{C})$ is brought on by the fixing of one primitive idempotent as a neutrino state that does not interact with the strong force. In the matrix algebra this means that one row and column, say, row 1, column 1, have zero entries. The exponential map produces a unit at the same entry.

We obtain a Clifform of $\mathrm{sl}(3,\mathbb{R})$ which I prefer to denote as $\mathrm{sl}_{\mathcal{C}\ell}(3)$ - together with a representation in the Majorana algebra. The $\mathrm{sl}_{\mathcal{C}\ell}(3)$ is a closure of the Lie-product of form $\mathrm{sl}_{\mathcal{C}\ell}(2) \times \mathrm{so}_{\mathcal{C}\ell}(3)$. This Clifform is indeed determined by the inhomogeneous graded spin- and isospin subalgebras. It is generated by the Lie product of $\mathrm{sl}_{\mathcal{C}\ell}(2)$ and $\mathrm{so}_{\mathcal{C}\ell}(3)$, independently of the chosen matrix representation and field. So the emergence of the $\mathrm{SU}(3)$ root-lattices is entirely traced back to the Lie product of two well known algebras of orthogonal space groups where the first factor generates the spin group of the 3-dimensional Lorentz group $Spin_+(2,1)$ and the second the rotation group $\mathrm{SO}(3)$.

Consider the lepton state $f_1 \in C\ell_{3,1}$ and the corresponding minimal ideal $S_1 = f_1\, C\ell_{3,1}$. Therein we find the space of isotropic spinors which satisfy $\xi \circ \xi = 0$. We define

Definition 7. ***isotropic spinors***

$$\textit{Let } S_1 = f_1\, C\ell_{3,1} \quad \textit{and} \quad H_1 = \{\xi \,/\, \xi \in S_1 = f_1\, C\ell_{3,1}\} \textit{ and } \xi \circ \xi = 0$$

and ask for the general solution of the equation

$$\phi \circ \hat{\phi} = f_1 \qquad \text{there follows} \quad \phi \in H_1 \quad \text{where} \quad \hat{\phi} \text{ is the gradeinverse of } \phi. \tag{36}$$

This is formulated in Clifford algebra, but is quite analogous to equations (12). Spinor spaces of form H_1 and others are discussed in my book on primordial space (2010) in chapter 25. The spinors themselves are discussed in a forthcoming work on fermions.

13.1. Derivation of the Natural Strong Force Fermion Spinor

Note, that all the color states of the type $f_{\chi k}$ ($\chi = 1, \ldots 6; k = 1, \ldots, 4$), that is, all fermion densities actually signify the *inner states* which are strictly separate from their outer spinor spaces. The inner spaces are, for example spanned by $ch_1 = \{e_1, e_{24}, e_{124}\}$ while the spinor space associated with ch_1 is in $span\{e_3, e_{13}, e_{234}, e_{1234}\}$. This has a deep meaning. While the fermion performs two movements, - one called a fractal extension, another an angular shift dynamics, its density is bound to the first Cartan subalgebra e_1, e_{24}, e_{124}. This is a rule in free space: motion in outer space is othogonal to the inner. The orthogonality, however, is somewhat complex and has to be defined in terms of annihilation in the graded algebra. The problem (36) has a definite solution for all chromatic spaces. We solved the equations with MAPLE Clifford. We shall write down the solution for the first chromatic space:

$$\begin{aligned}
|\nu_e\rangle &= y_1 = \pm\frac{i}{\sqrt{2}}\,\varphi\,\iota_+ \\
|s\rangle &= y_2 = \pm\frac{i}{\sqrt{2}}\,\hat{\varphi}\,\iota_+ \\
|u\rangle &= y_3 = \pm\frac{i}{\sqrt{2}}\,\hat{\varphi}\,\hat{\iota}_+ \\
|d\rangle &= y_4 = \pm\frac{i}{\sqrt{2}}\,\varphi\,\hat{\iota}_+
\end{aligned} \tag{37}$$

where $\varphi = \frac{1}{2}(e_1 + e_{24})$ is an *area extender* of direction e_1 and $\iota_+ = \frac{1}{\sqrt{2}}(e_3 + e_{13})$ a *positive angular shift operator* associated with direction e_1. We have

$$[\varphi, \iota_+] = \iota_+ \qquad [\varphi, \iota_-] = -\iota_- \text{ with } \iota_- = \frac{1}{\sqrt{2}}(e_3 - e_{13})$$

It can easily be verified that the y_i are isotropic direction fields. We have:

$$y_i \circ y_i = 0 \qquad \text{and} \quad y_i \circ \hat{y}_i = f_{1i} \qquad i = 1, \ldots 4 \tag{38}$$

The 'hat' $\hat{y}$ denotes the grade involuted y. The first index of f_{1i} equal 1 denotes the first Cartan subalgebra and chromatic space. Evaluating the standard spinor by Clifford multiplying the factors in (38) we obtain of course Clifford numbers that are not easily seen through, for example

$$y_3 = \mp \frac{i}{\sqrt{2}}(e_3 - e_{13} - e_{234} + j) \tag{39}$$

It is the decomposition into area extender and angular shift that makes the term understandable. Note that it represents the standard form belonging to f_{13} which turns the spinor y_3 into a manifold by some parallel transposition or any group action of $L^{(2)}$. Pay attention to the matrix representation of components and their peculiar occupation of entries

$$e_3 = \begin{pmatrix} 0 & 0 & 1 & 0 \\ 0 & 0 & 0 & -1 \\ 1 & 0 & 0 & 0 \\ 0 & -1 & 0 & 0 \end{pmatrix} \qquad e_{13} = \begin{pmatrix} 0 & 0 & 1 & 0 \\ 0 & 0 & 0 & 1 \\ -1 & 0 & 0 & 0 \\ 0 & -1 & 0 & 0 \end{pmatrix}$$

$$e_{234} = \begin{pmatrix} 0 & 0 & -1 & 0 \\ 0 & 0 & 0 & -1 \\ -1 & 0 & 0 & 0 \\ 0 & -1 & 0 & 0 \end{pmatrix} \qquad j = \begin{pmatrix} 0 & 0 & -1 & 0 \\ 0 & 0 & 0 & 1 \\ 1 & 0 & 0 & 0 \\ 0 & -1 & 0 & 0 \end{pmatrix}$$

Two of these components are symmetric, two are antisymmetric and therefore isotropic direction fields themselves. All four superimpose such that we obtain

$$\nu_e = \pm \begin{pmatrix} 0 & 0 & i & 0 \\ 0 & 0 & 0 & 0 \\ 0 & 0 & 0 & 0 \\ 0 & 0 & 0 & 0 \end{pmatrix} \qquad |u\rangle = \pm \begin{pmatrix} 0 & 0 & 0 & 0 \\ 0 & 0 & 0 & 0 \\ i & 0 & 0 & 0 \\ 0 & 0 & 0 & 0 \end{pmatrix}$$

$$d = \pm \begin{pmatrix} 0 & 0 & 0 & 0 \\ 0 & 0 & 0 & i \\ 0 & 0 & 0 & 0 \\ 0 & 0 & 0 & 0 \end{pmatrix} \qquad |s\rangle = \pm \begin{pmatrix} 0 & 0 & 0 & 0 \\ 0 & 0 & 0 & 0 \\ 0 & 0 & 0 & 0 \\ 0 & i & 0 & 0 \end{pmatrix}$$

This is the standard representation of natural fermion spinors. They satisfy equations (38).

References

[1] AIAS, Authors.; Representation of the Vacuum Electrodynamic Field in Terms of Longitudinal and Time-like Potentials: Canonical Qantization; *Journal of New Energy*, Vol. 4, no 2, 2000

[2] Antonsen, F. Random Graphs as a Model for Pregeometry; *Int J Theor Phys* **33**, 1994, 1189.

[3] Bombelli, L. *Statistical Geometry of Random Weave States*; gr-qc/0101080

[4] Chisholm, J.S.R.; Farwell, R.S. Tetrahedral structure of idempotents of the Clifford algebra $C\ell_{3,1}$; In *Clifford Algebras and their Applications in Mathematical Physics*; A. Micali; et al.; Eds.; Kluwer, Dordrecht, 1992, 27–32.

[5] Chisholm, J.S.R., Farwell, R.S.; Spin Gauge Theories: Principles and Predictions; In: *Clifford Algebras and their Application in Mathematical Physics*; F. Brackx; et al.; Eds.; Dortrecht, 1993, 367–374.

[6] Connes, A. *Noncommutative Geometry*; Acad Pr, New York, 1994.

[7] Connes, A. *Noncommutative Geometry Year 2000*; QA/0011193

[8] Davey, B. A.; Idziak, P. M.; Lampe, W. A.; McNulty, G. F. *Dualizability and Graph Algebra*; int., School of Mathematics, La Trobe University, Bundoora, Victoria 3083, Australia, available: B.Davey@latrobe.edu.au, March 1999.

[9] Greider, K.; Weiderman, T. *Generalised Clifford Algebras as Special Cases of Standard Clifford Algebras*; l'UCD Preprint 16, 1988. (cited after Chisholm 1992), 32.

[10] t' Hooft, G. Quantum Gravity and Dissipative Deterministic Systems; *Class Quant Grav* **16**, 1999, 3263, gr-qc/9903084

[11] Isham, C. J. *An Introduction To General Topology and Quantum Topology*; Summer Inst. on Physics, Geometry and Topology, Banff, August 1989, Imperial/TP/88-89/30

[12] Lee, T. D. *Random Lattices to Gravity*; In G. Feinberg: Selected Papers , vol 3, Birkhäuser, Boston, 1986.

[13] Markopoulou, F.; Smolin, L. Causal Evolution of Spin Networks; *Nucl. Phys. B* **508**, 1997, 409, gr-qc/9702025

[14] Matiussi, C. The Geometry of Time-Stepping; Progress In Electromagnetics Research, **PIER 32**, 2001, 123-149.

[15] Murskii, V.L. The existence in three-valued logic of a closed class without a finite complete system of identities; *Doklady Akad Nauk SSSR* **163**, 1996, 815-818.

[16] Nicolai, H.; Peeters, K.; Zamaklar,M. Loop quantum gravity: an outside view; *Class Quant Grav* **22**, 2005, R193, hep-th/0501114

[17] Nottale, L. *Scale relativity, fractal space-time and quantum mechanics*; In: Quantum Mechanics, Diffusion and Chaotic Fractals; M.S. El Naschi, O.E. Rossler, I. Prigogine et al.; Eds.; Oxford, 1995, pp. 51-78.

[18] Requardt, M. Cellular Networks as Models for Planck Scale Physics; *J Phys A: Mat Gen* **31**, 1998, 7997, hep-th/9806135

[19] Requardt, M.; Roy, S. Quantum Space-Time as a Statistical Geometry of Lumps in Random Networks; *Class Quant Grav* **17**, 2000, 2029, gr-qc/9912059

[20] Requardt, M. A Geometric Renormalization Group and Fixed Point Behavior in Discrete Quantum Space-Time; *JMP* **44**(2003)5588, gr-qc/0110077

[21] Requardt, M. *Wormhole Spaces, Connes' "Points, Speaking to Each Other", and the Translocal Structure of Quantum Theory*; arXiv:hep-th/0205168v3, 2004.

[22] Rovelli, C. *What is Time? What is Space?*; Di Renzo Editore, Roma, 2006.

[23] Schmeikal, B.; *The generative process of space-time and strong interaction - quantum numbers of orientation*; In: Clifford Algebras with Numeric and Symbolic Computations. R. Abłamowicz, P. Lounesto, J.M. Parra; Eds.; Birkhäuser, Boston, 1996, 83-100.

[24] Schmeikal, B. *Transposition in Clifford Algebra*; In: Clifford Algebras - Applications to Mathematics, Physics and Engineering, R. Ablamowicz; Eds.; Birkhäuser, Boston-Basel-Berlin, 2004, 351-372.

[25] Schmeikal, B. *Adv Appl Cliff Alg*. 2006, 16, 69-83.

[26] Schmeikal, B. *Adv Appl Cliff Alg*, 2007, 17, 107-135.

[27] Schmeikal, B. *Primordial Space*; Nova Science Publishers: New York, NY, USA, 2010 (forthcoming).

[28] Shallon, C.R. *Nonfinitely Based Binary Algebras Derived from Lattices*; Ph.D. thesis, University of California at Los Angeles, 1979.

[29] Schweber, S.S. *Feynman and the visualization of space-time processes*; *Rev Mod Phys* **58**, 449-508 (1986)

[30] Whittaker, E. T.; On an Expression of the Electromagnetic Field due to Electrons by means of two Scalar Potential Functions; *Proceedings of the London Mathematical Society*, Vol. 1, 1904, p. 367-372.

In: Focus on Quantum Mechanics
Editors: David E. Hathaway et al. pp. 129-156
ISBN 978-1-62100-680-0

Chapter 6

LINEAR GENERALIZATIONS OF THE TIME-DEPENDENT SCHRÖDINGER EQUATION: POINT TRANSFORMATIONS, DARBOUX TRANSFORMATIONS, AND THEIR PROPERTIES

Axel Schulze-Halberg*
Department of Mathematics and Actuarial Science
Indiana University Northwest
3400 Broadway
Gary, IN 46408
United States of America

Abstract

We study linear generalizations of the time-dependent Schrödinger equation in (1+1) dimensions. These generalizations include several well-known quantum-mechanical models, such as the Schrödinger equation with position-dependent mass or for an energy-dependent potential. In the first part of this work we show that generalized Schrödinger equations are related to their conventional counterparts by a particular point transformation, a fact that can be used to generate solutions of generalized Schrödinger equations. As a particular case, we study L^2-norm preservation and construct a point transformation that relates the stationary Schrödinger equation to its generalized counterpart. Besides providing a solution method by themselves, point transformations are used in the second part of this work, which is devoted to Darboux transformations. We show that generalized Schrödinger equations admit Darboux transformations, and describe two construction methods for these transformations. One of these methods is based on intertwining relations, while the second method uses the point transformations that were introduced previously. Several examples illustrate our considerations.

PACS 03.65.Ge.

Keywords: time-dependent Schrödinger equation, point transformation, Darboux transformation.

*E-mail address: xbataxel@gmail.com

1. Introduction

Several fundamental equations of Theoretical Physics are linear generalizations of the quantum mechanical time-dependent Schrödinger equation. A famous example for such a generalization is the Schrodinger equation with position-dependent mass [22] [23], which has become an important model for describing transport phenomena in semiconductors [10]. Another popular example is the Schrödinger equation for energy-dependent potentials [6]. Such potentials have proved to resemble the behaviour of interactions in heavy quark systems with very high accuracy [4] [11]. These two models, although completely unrelated, share two key properties regarding their solvability. The first property is the fact that the underlying equations of both models are related to the Schrödinger equation by means of a point transformation. The second, less known property is the existence of a Darboux transformation [3] for both models, which we comment on below. As is well-known, a point transformation in general is simply a change of the dependent and the independent variables in the underlying differential equation. In applications regarding Schrödinger equations, one often considers point transformations that leave the form of the equation invariant, but affect only the potential. Clearly, this provides a method for generating solutions, if one applies the point transformation to a solvable Schrödinger equation. There is a vast amount of literature on point transformations applied to Schrödinger equations, the reader may consult [21] for an overview regarding the stationary Schrödinger equation. The time-dependent case and more general equations are considered in [9], while point transformations for the Schrödinger equation for position-dependent mass are studied in [19]. While point transformations are based on coordinate changes, this is not true for the Darboux transformation, which consists in the application of a linear differential operator that maps solutions of a differential equation onto solutions of another differential equation. Typically, these equations are of similar form, but differ in a nonconstant parameter only, such as in the case of two Schrödinger equations that are the same up to their potential. Since originally [3] the Darboux transformation was constructed for ordinary differential equations only, in the context of Quantum Mechanics it was found applicable to the stationary Schrödinger equation [12]. Besides, the Darboux transformation arises naturally within the theory of supersymmetry [2], and is therefore often referred to as supersymmetrical transformation. There is very much literature on Darboux transformations, especially when applied to stationary Schrödinger equations, the reader may refer to the monograph [7] and references therein. Nowadays, the Darboux transformation has become one of the most powerful methods for designing spectra and generating solutions of stationary and time-dependent Schrödinger equations. While only some differential equations admit Darboux transformations, it is interesting to note that the two above mentioned equations of Schrödinger type (the position-dependent mass case and the equation for energy-dependent potentials) do so. This is no coincidence, as will be demonstrated in this work. In section 3 we consider a linear generalization of the time-dependent Schrödinger equation that comprises the two aforehand mentioned models and many more. In section 4 we derive point transformations for our generalized Schrödinger equations and discuss their properties such as L^2-norm preservations and special cases. Section 5 is devoted to showing that Darboux transformations can be constructed for our generalized Schrödinger equation. In particular, we first use the well-known intertwining relation for extraction of the first-order Darboux transformation. Since especially in the higher-order case the analysis of intertwining relations is a tedious procedure, we show that if point transformations are put to work correctly, the construction of arbitrary-order Darboux transformations is turned into a straightforward process.

2. Linear Generalizations of the Schrödinger Equation

The purpose of this section is to introduce the equation that we will study in the present article. Besides commenting on equivalent forms that our equation can take, we identify special cases that correspond to well-known Schrödinger-type models.

2.1. The Generalized Schrödinger Equation

Let us consider the following equation in (1+1) dimensions, that is, in one spatial and one time variable:

$$i\,\Psi_t + f\,\Psi_{xx} + g\,\Psi_x - V\,\Psi \;=\; 0, \tag{1}$$

where $\Psi = \Psi(x,t)$ stands for the solution and the indices denote partial differentiation. Furthermore, the coefficients $f = f(x,t)$, $g = g(x,t)$ and $V = V(x,t)$ are assumed to be arbitrary functions. We will call equation (1) generalized, time-dependent Schrödinger equation, as it generalizes several known cases of Schrödinger-type equations. Since in these equations the function V plays the role of a physical interaction, we will refer to V as potential. In general, our equation (1) itself does not have an immediate physical interpretation, while its special cases do, as we will see below. Note that for now we do not impose constraints on the coefficients f, g and V in our equation (1), but it is understood that generally f and g will have to be sufficiently smooth, while V must be at least continuous. The generalized equation (1) can be given in different, but equivalent forms, the first of which reads as follows:

$$i\,h\,\Psi_t + f\,\Psi_{xx} + g\,\Psi_x - V\,\Psi \;=\; 0, \tag{2}$$

introducing a function $h = h(x,t)$. Clearly, the form (1) of our equation can be converted into (2) through a multiplication by h and afterwards replacing $fh \rightarrow f, gh \rightarrow g, Vh \rightarrow V$. Conversely, we recover the initial form (1) from (2) through a division by h and an obvious renaming of coefficients. In the stationary - that is, time-independent - case, our equation in the form (2) can be identified with a Schrödinger equation for an energy-dependent potential, as will be discussed in the next section. Note that although the form (2) of our equation has one arbitrary coefficient more that its initial counterpart (1), the number of independent coefficients remains three. Now, besides (2), there is another form that equation (1) can take:

$$i\,h\,\Psi_t + f\,\Psi_{xx} + f_x\,\Psi_x - V\,\Psi \;=\; 0, \tag{3}$$

This form is obtained by simply letting $g = f_x$ in (2). Despite the coefficients of Ψ_{xx} and Ψ_x now being related to each other, the number of independent coefficients remains the same as in the forms (1) and (2) of our equation. In the stationary case, (3) has the self-adjoint form that is commonly used for stating equations within the context of Sturm-Liouville problems.

2.2. Special Cases

Let us now see important special cases that emerge from our generalized Schrödinger equation.

- Clearly, the conventional time-dependent Schrödinger equation is a special case of (1). In fact, we obtain

$$i\,\Psi_t + \Psi_{xx} - V\,\Psi \;=\; 0,$$

by employing the settings $f = 1$ and $g = 0$ in (1).

- Another particular case of (1) is the Schrödinger equation for position-dependent mass, that appears in the context of transport phenomena in crystals [10]. This equation reads

$$i\,\Psi_t + \frac{1}{2\,m}\,\Psi_{xx} - \frac{m_x}{2\,m^2}\,\Psi_x - V\,\Psi \;=\; 0, \tag{4}$$

where the mass $m = m(x,t)$ is usually taken to be a smooth, positive function. We obtain the position-dependent mass case from (1) by setting $f = 1/(2m)$ and $g = f_x$.

- The next special case of our generalized Schrödinger equation that we consider here will be derived from the equation's form (2). We set $f = 1$, $g = 0$ and assume the equation to be stationary and to have undergone variable separation with the stationary solution being called $\psi = \psi(x)$. If we further redefine $V = V_2(x)$ and $h = 1 - V_1(x)$, then our equation (2) takes the following form:

$$\psi''(x) + (E - E\,V_1(x) - V_2(x))\,\psi(x) \;=\; 0,$$

where the prime denotes the derivative and V_1, V_2 can be interpreted as components of a linearly energy-dependent potential. Such potentials are used for modeling the interaction of heavy quark systems [4] [11].

- Finally, a special case of our generalized Schrödinger equation is given by the stationary Fokker-Planck equation [13], which reads

$$\frac{d^2}{dx^2}\,(V_1\,\psi) - \frac{d}{dx}\,(V_2\,\psi) + c\,\psi \;=\; 0.$$

Here $\psi = \psi(x)$ denotes the solution, obtained from the stationary version of (1) after separation of variables. The functions $V_1 = V_1(x)$ and $V_2 = V_2(x)$ stand for the diffusion coefficient and the drift potential, respectively, while c is a constant. The Fokker-Planck equation can be brought into the form (1) after multiplication by a suitable factor.

These examples shall be sufficient for demonstrating the relevance of our generalized Schrödinger equation.

3. Point Transformations

A point transformation is a change of the dependent and independent variables in a differential equation. The goal of such a change of variables can be to simplify the underlying equation, such as in the presence of Lie symmetries [14]. Another reason for applying a point transformation is to establish an interrelation between the solution of an equation and solutions of a different equation, which is precisely what we will do in this section. Our two equations are of the same type, namely (1), while the coefficients f, g and the potential V differ in both equations. This does not only allow for generating solvable equations of the form (1), but will also be an important tool for the construction of the Darboux transformation in section 4.

3.1. Point Transformations for Generalized Schrödinger Equations

We will now construct a point transformation that converts our initial equation (1) to its counterpart

$$i\,\Phi_v + \hat{f}\,\Phi_{uu} + \hat{g}\,\Phi_u - \hat{V}\,\Phi \;=\; 0, \tag{5}$$

where u and v stand for the new variables. Furthermore, the functions $\hat{f}$, $\hat{g}$, $\hat{V}$ depend on u, v, and differ from the corresponding coefficients f, g, V in the initial equation (1). We start the transformation process by applying the following change of the dependent variable to our generalized Schrödinger equation (1):

$$\Psi \;=\; \exp(F)\;\Phi, \tag{6}$$

where $F = F(x,t)$ is a function that is to be determined. It is well-known [9] that the transformation (6) is the most general change of the dependent variable such that equation (1) maintains its linearity. Substitution of (6) into (1) renders the latter equation in the following form:

$$i\,\Phi_t + f\,\Phi_{xx} + (g + 2\,f\,F_x)\;\Phi_x + \left(i\,F_t + g\,F_x + f\,F_x^2 + f\,F_{xx} - V\right)\;\Phi \;=\; 0. \tag{7}$$

Before we impose constraints on our transformation function F, we perform a change of the independent variables, that is, we switch from the coordinates x and t to new coordinates $u = u(x,t)$ and $v = v(t)$. This is necessary, as the transformation (6) does not give a sufficient number of parameters that would allow to change all the coefficients in our generalized Schrödinger equation, and so convert it into a different equation of the same kind. Recall that we need to change three coefficients independently, which are f, g and V. One of these coefficients can be changed by means of our transformation function F and another coefficient can be handled via the potential V, after which we run out of free parameters and so cannot change the last remaining coefficient in our equation. But, as we will see now, a change of coordinates will provide the missing parameter. Note that the change of the time variable is merely a scaling, since the new coordinate v does not depend on x. We have to enforce this type of dependency in order to avoid second-order derivatives with respect to v in our equation, which would lose its Schrödinger character.

Taking into account the following relations between the derivative operators in the old and new coordinates, which are immediate consequences of the chain rule:

$$\begin{aligned} \partial_t &= u_t\,\partial_u + v_t\,\partial_v \\ \partial_x &= u_x\,\partial_u \\ \partial_{xx} &= u_x^2\,\partial_{uu} + u_{xx}\,\partial_u. \end{aligned}$$

Hence, after substitution of these derivatives, our equation (7) takes the following form:

$$i\,\Phi_v + \frac{f\,u_x^2}{v'}\,\Phi_{uu} + \frac{1}{v'}\,(i\,u_t + g\,u_x + 2\,f\,F_x\,u_x + f\,u_{xx})\,\Phi_u + \\ + \frac{1}{v'}\left(i\,F_t + g\,F_x + f\,F_x^2 + f\,F_{xx} - V\right)\,\Phi = 0. \tag{8}$$

Note that although we express the solution Φ through the new variables u and v, for the sake of simplicity we maintain its name. We must now redefine the coefficients in (8), such that it takes the form (5). Comparison of the latter two equation immediately brings up the following conditions:

$$\hat{f} = \frac{f\,u_x^2}{v'} \tag{9}$$

$$\hat{g} = \frac{1}{v'}\,(i\,u_t + g\,u_x + 2\,f\,F_x\,u_x + f\,u_{xx}) \tag{10}$$

$$\hat{V} = \frac{1}{v'}\left(i\,F_t + g\,F_x + f\,F_x^2 + f\,F_{xx} - V\right). \tag{11}$$

We will now choose our free parameter functions u and F in order to fulfill the conditions (9) and (10). The first condition is solved as follows:

$$u = \sqrt{v'}\int\sqrt{\frac{\hat{f}}{f}}\,dx. \tag{12}$$

The second condition (10) can be resolved by means of our transformation function F, as it was introduced in (6):

$$F = \int\left(-\frac{g}{2\,f} + \frac{\hat{g}\,v'}{2\,f\,u_x} - \frac{u_{xx}}{2\,u_x} - i\,\frac{u_t}{2\,f\,u_x}\right)\,dx \tag{13}$$

We could state F in a more explicit form by taking into account that u is already known explicitly, see (12). However, for the sake of simplicity we omit to substitute u and its derivatives into (13). It remains to consider the last condition (11), which is already fulfilled, as it relates the transformed potental $\hat{V}$ to its initial counterpart V. In summary, the solutions Ψ and Φ of our generalized Schrödinger equations (1) and (5) are interrelated by means of (6), together with switching to new coordinates u and v, given by (12) and an arbitrary function $v = v(t)$, respectively:

$$\Psi = \exp\left(\int\left(-\frac{g}{2\,f} + \frac{\hat{g}\,v'}{2\,f\,u_x} - \frac{u_{xx}}{2\,u_x} - i\,\frac{u_t}{2\,f\,u_x}\right)\,dx\right)\,\Phi, \tag{14}$$

where the function u is given by (12), and the potential $\hat{v}$ in the transformed equation has the form (11). Let us now demonstrate how our point transformation works in a concrete case.

3.2. Example: Schrödinger Equation for Position-dependent Mass

We will now show how our point transformation interrelates two different special cases of our generalized Schrödinger equation (1). In particular, we will convert a conventional Schrödinger equation into a Schrödinger equation for position-dependent mass.

The Schrödinger equations. Our starting point the the equation

$$i\,\Psi_t + \frac{1}{2\,m}\,\Psi_{xx} - a\,x^2\,\Psi = 0, \tag{15}$$

where m and a denote positive constants. Clearly, the conventional Schrödinger equation (15) is a special case of our generalized Schrödinger equation (1), if we employ the following settings:

$$f = \frac{1}{2\,m} \qquad g = 0 \qquad V = a\,x^2. \tag{16}$$

Our goal will be to map equation (15) onto a position-dependent mass Schrödinger equation of the form

$$i\,\Phi_v + \frac{1}{2\,p\,u^2}\,\Phi_{uu} - \frac{1}{p\,u^3}\,\Phi_u - \hat{V}\,\Phi = 0, \tag{17}$$

where u, v stand for the new variables, $p = p(v)$ is an arbitrary, positive, time-dependent function and $\hat{V}$ represents the transformed potential, which is to be determined. Equation (17) emerges from the generalized Schrödinger equation (5) provided the following settings are used:

$$\hat{f} = \frac{1}{2\,M} \qquad \hat{g} = -\frac{M_u}{2\,M^2} \qquad M = p\,u^2. \tag{18}$$

Thus, (17) can be identified with a Schrödinger equation for position-dependent mass M. Before we start constructing our point transformation, let us mention a set of explicit solutions to our initial equation (15). These solutions, called Ψ_n, are parametrized by a non-negative integer $n \geq 0$ and read

$$\Psi_n = \exp\left(-\sqrt{\frac{a\,m}{2}}\,x^2 - i\,\sqrt{\frac{a}{2\,m}}\,(1+2\,n)\,t\right)\,H_n\left((2\,a\,m)^{\frac{1}{4}}\,x\right). \tag{19}$$

Here, H_n denotes the Hermite polynomial of order n [1].

Construction of the point transformation. For the sake of simplicity, let us define the new time coordinate to be $v(t) = t$. This is a minor restriction, but will make subsequent calculations much more transparent. Next, we find the new spatial coordinate u, as defined in (12). At first sight, the latter formula seems to involve not more than an indefinite integration. Unfortunately, the situation is a bit more complicated, as can be recognized as follows: first observe that the right hand side of (12) contains the function $\hat{f}$ under the integral. This function - as displayed in (18) - is already given in terms of the new coordinate u, which we are just trying to find. So, formula (12) is not just a simple integral, but a nonlinear integral equation with respect to the function $u = u(x, t)$: on inserting the settings (16) and (18) for f and $\hat{f}$, respectively, into (12), we get (recall $v = t$):

$$u = \int \sqrt{\frac{m}{p\,u^2}}\,dx. \tag{20}$$

We see that the sought function u appears on both sides of the equation, where on the right hand side it forms part of an integrand. Fortunately, the integral equation (20) can be solved. To this end, differentiate both sides with respect to x in order to obtain a differential equation, which can then be found to have the following solution:

$$u = \left(\frac{4\,m}{p}\right)^{\frac{1}{4}} \sqrt{x}. \tag{21}$$

Note that an irrelevant constant of integration was set to zero. We have now determined the new coordinates u and v that will be used in our point transformation. It remains to compute the transformation function F, that appears in (6), and the explicit form of which is given in (13). We will need the inverse of the function (21), which reads

$$x = \sqrt{\frac{p}{4\,m}}\,u^2. \tag{22}$$

Now, we substitute the functions given in (16), (18) and (12) into (13). Note that before integration, all coordinates u must be replaced by (21), since the integration variable is x. We get after simplification

$$\begin{aligned} F &= \int -\left(\frac{1}{4\,x} + i\,\frac{m\,p'}{2\,p}\,x\right)\,dx \\ &= -\frac{1}{4}\log(x) + i\,\frac{m\,p'}{4\,p}\,x^2 + c \end{aligned} \tag{23}$$

where $c = c(t)$ denotes a constant of integration.

Transformed solution and corresponding potential. Now that we have determined our point transformation completely, let us compute the transformed Schrödinger equation (17) in explicit form. In particular, we are interested in its potential $\hat{V}$ and in the transformed solution that will be generated from (19). We start by constructing the potential, which can be obtained after substitution of $v = t$, f, g, F and V, as given in (16) and (23), respectively, into (11). We will first express the transformed potential through the initial coordinates x and t:

$$\hat{V} = \left(-\frac{5}{32\,m}\right)\frac{1}{x^2} + \left(a - \frac{m\,(p')^2}{8\,p^2} + \frac{m\,p''}{4\,p}\right)\,x^2 - i\,\left(\frac{p'}{8\,p} + c'\right).$$

Now we switch to the new coordinate u by substituting (22) into the latter expression:

$$\hat{V} = \left(-\frac{5}{8\,p}\right)\frac{1}{u^4} + \left(\frac{a\,p}{4\,m} - \frac{(p')^2}{32\,p} + \frac{p''}{16}\right)\,u^4 - i\,\left(\frac{p'}{8\,p} + c'\right). \qquad (24)$$

It remains to find solutions of our transformed equation (17). To that end, we make use of our point transformation (14). Since we have a solution Ψ - given in (19) - and we want to find the transformed solution Φ, let us solve (14) for Φ, which after renaming it to Φ_n reads

$$\Phi_n(u,t) = \exp\left(-F(x,t)\right)\,\Psi_n(x,t)\,\Big|_{x=x(u,t)},$$

where the function $x = x(u,t)$ is given in (22). Thus, we express the exponential on the right hand side through the new coordinate u:

$$F = -\frac{1}{4}\,\log\left(\sqrt{\frac{p}{4\,m}}\,u^2\right) + i\,\frac{p'}{16}\,u^4 + c, \qquad (25)$$

and we do the same with the functions Ψ_n, as given in (19):

$$\Psi_n = \exp\left(-p\,\sqrt{\frac{a}{32\,m}}\,u^4 - i\,\sqrt{\frac{a}{2\,m}}\,(1+2\,n)\,t\right)\,H_n\left(\sqrt{p}\,\left(\frac{a}{8\,m}\right)^{\frac{1}{4}}\,u^2\right). \qquad (26)$$

Now, on combining (25) and (26), we obtain a set of exact solutions of our transformed equation (17):

$$\Phi_n = \left(\frac{4\,m}{p}\right)^{\frac{1}{8}}\sqrt{\frac{1}{u}}\,\exp\left(-p\,\sqrt{\frac{a}{32\,m}}\,u^4 + i\,\left(\frac{p'}{16}\,u^4 - \sqrt{\frac{a}{2\,m}}\,(1+2\,n)\,t\right) + c\right) \times$$
$$\times\; H_n\left(\sqrt{p}\,\left(\frac{a}{8\,m}\right)^{\frac{1}{4}}\,u^2\right). \qquad (27)$$

In summary, we have transformed the conventional Schrödinger equation (15) into the position-dependent mass Schrödinger equation (17), the potential and solution of which are displayed in (24) and (27), respectively.

3.3. Transformation to the Stationary Case

As we have seen in the previous example, by means of our point transformation (6) we were able to construct solutions of a generalized Schrödinger equation from known solutions of a conventional Schrödinger equation. The latter equation has been thoroughly studied regarding its integrability and many explicit solutions have been found. This is especially true for the stationary Schrödinger equation, that is, if the potential does not depend on the time variable. It is therefore useful to know which time-dependent potentials can be generated out of stationary potentials.

Conventional Schrödinger equations. The starting point for our considerations is the following result [5] that establishes a relation between the stationary Schrödinger equation and its time-dependent counterpart. In particular, consider the time-dependent Schrödinger equation

$$i\,\Psi_t + \Psi_{xx} - V\,\Psi = 0, \tag{28}$$

and assume that the potential V of this equation is given by the expression

$$V = F_2\,x^2 + F_1\,x + F_0 + \exp\left(8\int A\,dt\right)\hat{V}, \tag{29}$$

where F_2, F_1, F_0 are real-valued and given

$$F_2 = A' - 4\,A^2 \qquad F_1 = B' - 4\,A\,B \qquad F_0 = C' - B^2. \tag{30}$$

Furthermore, in (30), the arbitrary functions $A = A(t)$, $B = B(t)$ and $C = C(t)$ are real-valued, and $\hat{V}$ is an arbitrary function that depends on its argument as follows:

$$\hat{V} = \hat{V}\left(\exp\left(4\int A\,dt\right)x + 2\int \exp\left(4\int A\,dt\right)B\,dt\right). \tag{31}$$

Then the time-dependent Schrödinger equation (28) can be converted into another time-dependent Schrödinger equation with a time-independent potential, using our point transformation (6), (13), (12). Since we apply this transformation to conventional Schrödinger equations, our settings in (12) and (13) are

$$f = \hat{f} = 1 \qquad g = \hat{g} = 0.$$

Plugging these settings into (12) and (13), we can determine the coordinate change u and from there the transformation function F, respectively:

$$u = \sqrt{v'}\,x + d \tag{32}$$

$$\begin{aligned} F &= -\frac{i}{2}\int\left(\frac{d'}{\sqrt{v'}} + \frac{v''}{2\,v'}\,x\right)dx \\ &= -i\,\frac{v''}{8\,v'}\,x^2 - i\,\frac{d'}{2\,\sqrt{v'}}\,x + c, \end{aligned} \tag{33}$$

where $c = c(t)$ and $d = d(t)$ are constants of integration. Now, we choose the arbitrary function v and constants of integration $d = d(t)$ and $c = c(t)$ that arise from (12) and (13), respectively, as follows

$$\begin{aligned} v &= \int \exp\left(8\int A\,dt\right)dt \\ c &= -i\,C + 2\int A\,dt \\ d &= 2\exp\left(-4\int A\,dt\right)\int \exp\left(4\int A\,dt\right)B\,dt. \end{aligned} \tag{34}$$

Using these settings and applying our point transformation (6), (32), (33) to the time-dependent Schrödinger equation, we obtain the equation

$$i\,\Phi_v + \Phi_{uu} - \hat{V}\,\Phi = 0,$$

where the potential $\hat{V} = \hat{V}(u)$ is stationary. If we now separate off the time variable by means of

$$\Phi = \exp(-i\,E\,v)\,\phi,$$

where $\phi = \phi(u)$ and E is a constant, then we arrive at

$$\phi'' + (E - \hat{V})\,\phi = 0, \tag{35}$$

which is the stationary Schrödinger equation for the potential $\hat{V}$ and energy E. Thus, once the solution of this equation is known, then the solution Ψ of our time-dependent Schrödinger equation (28) can be constructed by combining the above findings:

$$\begin{aligned}\Psi &= \exp\left(-i\,\frac{v''}{8\,v'}\,x^2 - i\,\frac{d'}{2\,\sqrt{v'}}\,x - i\,C + 2\int A\,dt - I\,E\int \exp\left(8\int A\,dt\right)\,dt\right)\times\\ &\times\ \phi\left(\sqrt{v'}\,x + 2\,\exp\left(-4\int A\,dt\right)\int \exp\left(4\int A\,dt\right)\,B\,dt\right).\end{aligned} \tag{36}$$

In summary, if the potential V of a conventional, time-dependent Schrödinger equation is of the form (29), then our point transformation (6) with the above settings yields a stationary Schrödinger equation. We will now extend this concept to our generalized Schrödinger equation (1)

Generalized Schrödinger equations. The principal idea of mapping generalized, time-dependent Schrödinger equations onto stationary Schrödinger equations consists in the composition of two point transformations P_1 and P_2. The first transformation P_1 takes the generalized Schrödinger equation into a conventional Schrödinger equation, which is then mapped onto its stationary counterpart by means of the second point transformation P_2. Thus, the complete transformation P is given by $P = P_2 \circ P_1$, as shown by the following commutative diagram in figure 1. Since we have the two point transformations

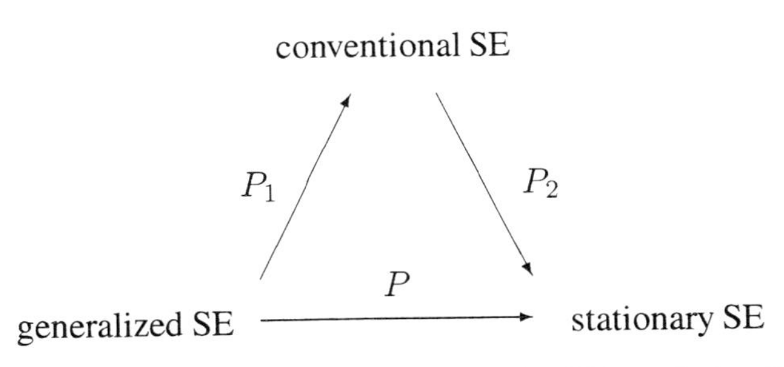

Figure 1. Definition of the point transformation P (SE stands for Schrödinger equation).

P_1 and P_2 in explicit form, the remaining task is to construct its composition P and the potential of the generalized Schrödinger equation that allows for reduction to a stationary potential. Regarding the potentials, let us recall that the point transformation P_1, given by the formulas (6), (12), (13) with $\hat{f} = 1, \hat{g} = 0$, yields a conventional Schrödinger equation for the transformed potential (11), which we rename to $\hat{V}_1$:

$$\hat{V}_1 = \frac{1}{v'}\left(i\,F_t + g\,F_x + f\,F_x^2 + f\,F_{xx} - V\right), \tag{37}$$

where F is given by (13). From the previous paragraph we know that this potential can only be reduced to a stationary one, if it has the form (29). After renaming to V_1 and rewriting in coordinates of the generalized Schrödinger equation (1), the potential (29) takes the form

$$V_1 = F_2\left(\sqrt{v'}\int\sqrt{\frac{1}{f}}\,dx\right)^2 + F_1\left(\sqrt{v'}\int\sqrt{\frac{1}{f}}\,dx\right) + F_0 + \exp\left(8\int A\,v'\,dt\right)\hat{V}. \tag{38}$$

As mentioned before, the integrals that suddenly appear in this expression, stem from the fact that the coordinates x and t in (29) had to be replaced by (12) for $\hat{f} = 1$ and v, respectively. Now, if we want to convert (37) into a stationary potential, it must be the same as (38), which gives a condition on V, the potential of the initial generalized Schrödinger equation. We get the equation $\hat{V}_1 = V_1$, which reads explicitly

$$\frac{1}{v'}\left(i\,F_t + g\,F_x + f\,F_x^2 + f\,F_{xx} - V\right) = \\ = F_2\left(\sqrt{v'}\int\sqrt{\frac{1}{f}}\,dx\right)^2 + F_1\left(\sqrt{v'}\int\sqrt{\frac{1}{f}}\,dx\right) + F_0 + \exp\left(8\int A\,v'\,dt\right)\hat{V}$$

The solution for V is then given by

$$V = i\,F_t + g\,F_x + f\,F_x^2 + f\,F_{xx} - v'\,F_2\left(\sqrt{v'}\int\sqrt{\frac{1}{f}}\,dx\right)^2 - \\ - v'\,F_1\left(\sqrt{v'}\int\sqrt{\frac{1}{f}}\,dx\right) - v'\,F_0 - v'\exp\left(8\int A\,v'\,dt\right)\hat{V}. \tag{39}$$

This is the form that the potential V of the generalized Schrödinger equation (1) must have, such that it allows reduction to the stationary potential $\hat{V}$ by means of our point transformation P. In the last step we will now determine this point transformation P. To this end, we need to build the composition of (14) and (36), where the latter transformation must be expressed through the coordinates x and t, given by (12) for $\hat{f} = 1$ and v, respectively. After combining (14), (36) and taking into account the explicit form of u, given by (12) for

$\hat{f} = 1$, we get

$$\Psi = \exp\left(\int \left(-\frac{g}{2f} - \frac{f_x}{4f} - i\,\frac{v''}{4\sqrt{f}\,v'} \int \sqrt{\frac{1}{f}}\,dx - i\,\frac{1}{2\sqrt{f}} \int \left(\sqrt{\frac{1}{f}}\right)_t dx \right) dx \right. -$$
$$- i\,\frac{\hat{v}''\,v'}{8\,\hat{v}'} \left(\int \sqrt{\frac{1}{f}}\,dx \right)^2 - i\,\frac{d'\,\sqrt{v'}}{2\,\sqrt{\hat{v}'}} \int \sqrt{\frac{1}{f}}\,dx - i\,C + 2 \int A\,v'\,dt -$$
$$\left. - \;i\,E \int \exp\left(8 \int A\,v'\,dt \right) v'\,dt \right) \times$$
$$\times\;\phi\left(\sqrt{\hat{v}'\,v'} \int \sqrt{\frac{1}{f}}\,dx + 2\,\exp\left(-4 \int A\,v'\,dt \right) \int \exp\left(4 \int A\,v'\,dt \right) B\,v'\,dt \right), \tag{40}$$

where the function $\hat{v}$ is given by (34). Note that we had to change its name, as otherwise it is easy to confuse the arbitrary function v and the already defined function $\hat{v}$. In summary, if the generalized Schrödinger equation (1) has a potential of the form (39), then its solution is given by the function (40), where ϕ solves the conventional, stationary Schrödinger equation for a potential $\hat{V}$.

Example: the stationary harmonic oscillator potential. We will now apply the above developed formalism for generating generalized, time-dependent Schrödinger equations that admit exact solutions. To this end, let us take the coventional, stationary Schrödinger equation (35) for the harmonic oscillator potential:

$$\hat{V} = u^2. \tag{41}$$

The stationary Schrödinger equation (35) has the following set of solutions ϕ that we parametrize through a nonnegative integer n:

$$\phi_n = \exp\left(-\frac{1}{2}\,u^2 \right)\; H_n(u)\,. \tag{42}$$

Here H_n denotes the Hermite polynomial of order n [1]. The corresponding energies $E = E_n$ read

$$E_n = 2\,n + 1. \tag{43}$$

We will now use our point transformation (40) to construct solutions for a position-dependent mass Schrödinger equation (4), which constitutes a special case of our generalized Schrödinger equation. To this end, we first need to define the position-dependent mass, that is, the coefficient functions f and g in (1). Clearly, the position dependent mass $m = m(x,t)$ enters in the coefficients of (1) as follows:

$$f \;=\; \frac{1}{2\,m} \qquad g \;=\; -\,\frac{m_x}{2\,m^2}. \tag{44}$$

Let us choose the following position-dependent mass:

$$m \;=\; \alpha\,\exp(x), \tag{45}$$

where $\alpha = \alpha(t)$ is an arbitrary, real and positive function. According to (44), this gives the following coefficients f and g:

$$f = \frac{\exp(-x)}{2\,\alpha} \qquad g = -\frac{\exp(-x)}{2\,\alpha}. \tag{46}$$

Before we substitute these settings into formula (40), we simplify subsequent calculations by assigning the following values to our arbitrary functions A, B and C:

$$A = -\frac{\alpha'}{8\,\alpha} \qquad B = 0 \qquad C = 0. \tag{47}$$

We now insert the latter settings and (46) into (40). Note that due to the number of expressions involved, we do not state details of the substitution and evaluation process, but instead give the result. For details the reader may refer to [17]. We arrive at the following solution Ψ, parametrized through a nonnegative integer n:

$$\Psi_n = \exp\left(-4\,\exp(x) + \frac{1}{4}\,x - i\,(2\,n+1)\int \frac{1}{\alpha}\,dt\right)\,H_n\left(\sqrt{8}\,\exp\left(\frac{1}{2}\,x\right)\right) \tag{48}$$

This time-dependent function solves the position-dependent mass Schrödinger equation (4) for the mass (45) and the potential V, which can be extracted from (39) after insertion of (47) and (46):

$$V = \frac{8}{\alpha}\,\exp(x) - \frac{3}{32\,\alpha}\,\exp(-x). \tag{49}$$

Note that this potential depends on the function α, such that we have generated a time-dependent equation together with its solutions from a stationary Schrödinger equation for the harmonic oscillator potential.

3.4. Norm Preservation

It is well-known that point transformations between conventional Schrödinger equations and position-dependent mass Schrödinger equations preserve the L^2-norm [18]. In this section we will show that this is no longer true for generalized Schrödinger equations, but only if certain restrictions apply. Let us consider the point transformation in its general form (6), (12), (13). The norm of the transformed solution Φ on a domain D is then given by

$$\begin{aligned}
||\Phi|| &= \int_D |\,\Psi\,|^2\,du \\
&= \int_D |\,\exp(-F)\,\Psi\,|^2\,du \\
&= \int_D \left|\exp\left(-\int\left(-\frac{g}{2\,f} + \frac{\hat{g}\,v'}{2\,f\,u_x} - \frac{u_{xx}}{2\,u_x} - i\,\frac{u_t}{2\,f\,u_x}\right)dx\right)\Psi\right|^2 du.
\end{aligned}$$

If u is an exponential, that is, $u = \exp(kx)$ for a constant k, then the third term under the inner integral becomes constant and will not affect the normalizability of Φ. We can therefore assume that this term is not constant and integrate it:

$$\|\Phi\| = \int_D \left| \sqrt{u_x} \exp\left(-\int \left(-\frac{g}{2\,f} + \frac{\hat{g}\, v'}{2\,f\,u_x} - i\,\frac{u_t}{2\,f\,u_x} \right) dx \right) \Psi \right|^2 du.$$

Next, we assume that the functions u and f are real-valued. In this case, the last term under the inner integral is purely imaginary and cancels out when taking the absolute value:

$$\|\Phi\| = \int_D u_x \left| \exp\left(\int \left(\frac{g}{2\,f} - \frac{\hat{g}\, v'}{2\,f\,u_x} \right) dx \right) \Psi \right|^2 du.$$

Now we switch variables from u to x in the outer integral and denote by $x(D)$ the transformed domain:

$$\|\Phi\| = \int_{x(D)} u_x^2 \left| \exp\left(\int \left(\frac{g}{2\,f} - \frac{\hat{g}\, v'}{2\,f\,u_x} \right) dx \right) \right|^2 |\Psi|^2 \, dx. \tag{50}$$

The expression on the right hand side cannot be simplified any further, as long as we are not given more details on the coefficient functions of our generalized Schrödinger equations and on the norm of Ψ. Only in special cases as the following we can say more about (50):

- If the solution Ψ of our initial generalized Schrödinger equation is L^2-normalizable, then it is sufficient for the norm of Φ to exist if the exponential term times $u_x^2 \sim \hat{f}/f$ stays bounded in $x(D)$. This is so, since we have

$$\|\Psi\| = \int_{x(D)} |\Psi|^2 \, dx. \tag{51}$$

  Thus, if the exponential times u_x^2 in (50) can be estimated to yield a constant, $\|\Phi\|$ will obviously exist.

- Let us assume that the transformed equation is a conventional Schrödinger equation, that is, $\hat{f} = 1$ and $\hat{g} = 0$. Furthermore, assume that $g = f_x$, as e.g. in the position-dependent mass Schrödinger equation. Then, after insertion of u as given in (12), we obtain from (50)

$$\begin{aligned} \|\Phi\| &= v' \int_{x(D)} \frac{1}{f} \left| \exp\left(\int \frac{f_x}{2\,f} \, dx \right) \right|^2 |\Psi|^2 \, dx \\ &= v' \int_{x(D)} |\Psi|^2 \, dx. \end{aligned} \tag{52}$$

  Note that the integral in the exponential gives either a logarithm or a constant, depending on whether f is an exponential or not. In both cases, the first and second factor under the integral cancel. If we now assume that Ψ is normalizable, so will be Φ due to (51). In other words, point transformations between conventional Schrödinger equations and generalized Schrödinger equations with $g = f_x$ always preserve the L^2-normalizability.

- Even if Ψ is not normalizable, this does not imply the same for its transformed counterpart Φ. Looking at (50), we see that the coefficients in (5) can be chosen such that the integrals still exist.

Hence, the use of point transformations for mapping generalized Schrödinger equations onto each other does in general not preserve L^2-normalizability. In fact, the point transformation might break normalizability, generate or preserve it.

4. Generalized Darboux Transformations

In this section we show how Darboux transformations for generalized Schrödinger equations can be constructed. We will refer to such Darboux transformations as generalized Darboux transformations. The principal tool we will be using is the point transformation that we derived and discussed in the previous section. Before we begin constructing generalized Darboux transformations, let us review the well-known, conventional case.

4.1. The Conventional Darboux Transformation

Consider the time-dependent Schrödinger equation

$$i\,\Psi_t + \Psi_{xx} - V\,\Psi = 0, \tag{53}$$

where $V = V(x,t)$ denotes the potential. The $n-$th order Darboux transformation of a solution Ψ to (53) is defined as

$$D_{n,(u_j)}(\Psi) = L\,\frac{W_{n,(u_j),\Psi}}{W_{n,(u_j)}}, \tag{54}$$

where $L = L(t)$ is arbitrary, the family (u_j) of n solutions to (53) are such that $(u_1, u_2, ..., u_n, \Psi)$ is linearly independent, and $W_{n,(u_j)}$, $W_{n,(u_j),\Psi}$ denote the Wronskians of the families (u_j) and of (u_j, Ψ), respectively. Then, the function $\Phi = D_{n,(u_j)}(\Psi)$ solves the time-dependent Schrödinger equation

$$i\,\Phi_t + \Phi_{xx} - U\,\Phi = 0, \tag{55}$$

where the potential U reads

$$U_n = V + i\,\frac{L'}{L} - 2\,\left[\log\left(W_{n,(u_j)}\right)\right]_{xx}. \tag{56}$$

Thus, the Darboux transformation establishes a relation between the TDSEs (53) and (55), and as such allows for the generation of solvable Schrödinger equations. For performing a Darboux transformation, one needs the solutions u_j, $j = 1, ..., n$, which are therefore called auxiliary solutions. Finally, let us mention the explicit form of first-order Darboux transformations, that is, (54) for $n = 1$. We have

$$D_{1,(u_1)}(\Psi) = L\,\frac{W_{1,(u_1),\Psi}}{W_{1,(u_1)}} = L\left(-\frac{(u_1)_x}{u_1}\,\Psi + \Psi_x\right). \tag{57}$$

Note that the Wronskian $W_{1,(u_1)}$ is just the function u_1 itself. The corresponding transformed potential U_1 can be extracted from the general case (56), setting $n = 1$:

$$U_1 = V + i\,\frac{L'}{L} - 2\,[\log(u_1)]_{xx}\,. \tag{58}$$

This relation is well-known in the SUSY formalism of Quantum Mechanics, where the potentials U_1 and V are referred to as supersymmetric partners.

4.2. The Generalized Darboux Transformation

We will now extend the concept of Darboux transformations to our generalized Schrödinger equation. In the first part of this section, we construct first-order Darboux transformations by means of intertwining relations, while in the second part we show an alternative method of derivation. Throughout this section, we will consider our generalized Schrödinger equation to be in its form (3), more precisely, we have

$$i\,\Psi_t + \frac{f}{h}\,\Psi_{xx} + \frac{f_x}{h}\,\Psi_x - \frac{V_1}{h}\,\Psi = 0. \tag{59}$$

Recall that writing our equation in the form (59) is not a restriction, as we have seen in section 3.1.

4.2.1. Construction via Intertwining Relations

Our goal is to obtain generalized Darboux transformations of first order that relate equation (59) to its counterpart

$$i\,\Phi_t + \frac{f}{h}\,\Phi_{xx} + \frac{f_x}{h}\,\Phi_x - \frac{U_1}{h}\,\Phi = 0, \tag{60}$$

by means of intertwining relations. In order to simplify the notation in subsequent calculations, let us define the following operators:

$$H_1 = -\frac{f}{h}\,\partial_{xx} - \frac{f_x}{h}\,\partial_x + \frac{V_1}{h} \tag{61}$$

$$H_2 = -\frac{f}{h}\,\partial_{xx} - \frac{f_x}{h}\,\partial_x + \frac{U_1}{h}. \tag{62}$$

We will refer to these operators as (generalized) Hamiltonians. Clearly, these operators render our equations (59) and (60) in the form

$$(i\,\partial_t - H_1)\,\Psi = 0 \qquad (i\,\partial_t - H_2)\,\Phi = 0.$$

We are looking for a first-order Darboux transformation between these generalized two Schrödinger equations, which can be written in the form

$$\Phi = \mathcal{L}\,\Psi \qquad \mathcal{L} = L_0 + L_1\,\partial_x, \tag{63}$$

where the functions $L_0 = L_0(x,t)$ and $L_1 = L_1(x,t)$ are to be determined suitably by means of the intertwining relation.

The intertwining relation. In order to determine L, we require it to convert solutions of the first TDSE (59) into solutions of the second TDSE (60). The intertwining relation involving the operator L and the Hamiltonians H_1 and H_2 defined in (61) and (62), respectively, is given by

$$(i\,\partial_t - H_2)\,\mathcal{L} = \mathcal{L}\,(i\,\partial_t - H_1)\,. \tag{64}$$

If we apply this operator equation to a solution of the Schrödinger equation (59), the right hand side vanishes. Consequently, the left hand side must also vanish, which implies that $\mathcal{L}\Psi$ is a solution of equation (60). However, before this can happen, the operator $\mathcal{L}$ must be determined correctly. To this end, we expand the intertwining relation in order to get conditions for the sought operator $\mathcal{L}$. Let us assume that $\mathcal{L}$ is given in the form (63), substitution of which in combination with (61) and (62) renders (64) in the form

$$\begin{aligned}\left(i\,\partial_t + \frac{f}{h}\,\partial_{xx} + \frac{f_x}{h}\,\partial_x - \frac{U_1}{h}\right)(L_0 + L_1\,\partial_x) &= \\ &= (L_0 + L_1\,\partial_x)\left(i\,\partial_t + \frac{f}{h}\,\partial_{xx} + \frac{f_x}{h}\,\partial_x - \frac{V_1}{h}\right).\end{aligned} \tag{65}$$

We will now expand both sides of this intertwining relation and find the coefficients of the derivative operators. The intertwining relation can only be fulfilled if the coefficients of a derivative operator are the same on both sides, which gives conditions on the coefficients. Let us first evaluate the left hand side of the latter intertwining relation:

$$\begin{aligned}&\left(i\,\partial_t + \frac{f}{h}\,\partial_{xx} + \frac{f_x}{h}\,\partial_x - \frac{U_1}{h}\right)(L_0 + L_1\,\partial_x) = \\ =\; & \frac{f}{h}\,L_1\,\partial_{xxx} + \left[\frac{f}{h}\,L_0 + 2\,\frac{f}{h}\,(L_1)_x + \frac{f_x}{h}\,L_1\right]\partial_{xx} + i\,L_1\,\partial_{xt} + \\ +\; & \left[2\,\frac{f}{h}\,(L_0)_x + \frac{f_x}{h}\,L_0 + i\,(L_1)_t + \frac{f}{h}\,(L_1)_{xx} + \frac{f_x}{h}\,(L_1)_x - \frac{U_1}{h}\,L_1\right]\partial_x + \\ +\; & i\,(L_0)_t\,\partial_t + \left[i\,(L_0)_t + \frac{f}{h}\,(L_0)_{xx} + \frac{f_x}{h}\,(L_0)_x - \frac{U_1}{h}\,L_0\right].\end{aligned} \tag{66}$$

Next, we process the right hand side of (65) in the same way:

$$
\begin{aligned}
& (L_0 + L_1\,\partial_x)\left(i\,\partial_t + \frac{f}{h}\,\partial_{xx} + \frac{f_x}{h}\,\partial_x - \frac{V_1}{h}\right) = \\
= \;& \frac{f}{h}\,L_1\,\partial_{xxx} + \left[\frac{f}{h}\,L_0 + \frac{f_x}{h}\,L_1 - \frac{f\,h_x}{h^2}\,L_1 + \frac{f_x}{h}\,L_1\right]\partial_{xx} + i\,L_1\,\partial_{xt} + \\
+ \;& \left[\frac{f_x}{h}\,L_0 + \frac{f_x}{h}\,L_1 - \frac{f\,h_x}{h^2}\,L_1 - \frac{V_1}{h}\,L_1\right]\partial_x + \\
+ \;& i\,L_0\,\partial_t + \left[-\frac{V_1}{h}\,L_0 - \frac{(V_1)_x}{h}\,L_1 + \frac{V_1\,h_x}{h^2}\,L_1\right].
\end{aligned}
\tag{67}
$$

Again, the intertwining relation (65) can only hold if its two sides (66) and (67) are the same. It is easy to see that the terms associated with the derivatives ∂_{xxx}, ∂_{xt} and ∂_t are already equal on both sides and therefore cancel in the intertwining relation. Since there are more terms in the coefficients that cancel in the same way, let us now recombine (66) and (67) after simplification, that is, without equal terms that appear on both sides.

$$
\begin{aligned}
& 2\,\frac{f}{h}\,(L_1)_x\,\partial_{xx} + \left[i\,(L_1)_t + 2\,\frac{f}{h}\,(L_0)_x + \frac{f}{h}\,(L_1)_{xx} + \frac{f_x}{h}\,(L_1)_x - \frac{U_1}{h}\,L_1\right]\partial_x + \\
+ \;& \left[i\,(L_0)_t + \frac{f}{h}\,(L_0)_{xx} + \frac{f_x}{h}\,(L_0)_x - \frac{U_1}{h}\,L_0\right] \\
= \;& \left[\frac{f_x}{h}\,L_1 - \frac{f\,h_x}{h^2}\,L_1\right]\partial_{xx} + \left[\frac{f_x}{h}\,L_1 - \frac{f\,h_x}{h^2}\,L_1 - \frac{V_1}{h}\,L_1\right]\partial_x + \\
+ \;& \left[-\frac{V_1}{h}\,L_0 - \frac{(V_1)_x}{h}\,L_1 + \frac{V_1\,h_x}{h^2}\,L_1\right]
\end{aligned}
\tag{68}
$$

As mentioned before, we now collect the coefficients of each derivative operator on both sides of the latter intertwining relation and require the coefficients to be the same.

Resolution of the constraints. Since there are only three different derivative operators left in our intertwining relation (68), namely, ∂_{xx}, ∂_x and the multiplication (derivative of order zero), we obtain three equations. These equations have the following form:

$$
2\,\frac{f}{h}\,(L_1)_x = \frac{f_x}{h}\,L_1 - \frac{f\,h_x}{h^2}\,L_1 \tag{69}
$$

$$
i\,(L_1)_t + 2\,\frac{f}{h}\,(L_0)_x + \frac{f}{h}\,(L_1)_{xx} + \frac{f_x}{h}\,(L_1)_x - \frac{U_1}{h}\,L_1 = \frac{f_x}{h}\,L_1 - \frac{f\,h_x}{h^2}\,L_1 - \frac{V_1}{h}\,L_1 \tag{70}
$$

$$
i\,(L_0)_t + \frac{f}{h}\,(L_0)_{xx} + \frac{f_x}{h}\,(L_0)_x - \frac{U_1}{h}\,L_0 = -\frac{V_1}{h}\,L_0 - \frac{(V_1)_x}{h}\,L_1 + \frac{V_1\,h_x}{h^2}\,L_1. \tag{71}
$$

We will now solve this system of equations with respect to the coefficients L_0 and L_1 in our operator $\mathcal{L}$, recall its form as given in (63). Since we are dealing with three equations, we will need a third function as a variable, which we take to be the potential U_1 of the TDSE

(60). In order to solve the above three equations, we start with (69) and determine L_1:

$$\begin{aligned} \frac{2\,(L_1)_x}{L_1} &= \left(\frac{f}{h}\right)_x \frac{h}{f} \\ L_1 &= L\sqrt{\frac{f}{h}}, \end{aligned} \tag{72}$$

where $L = L(t)$ is an arbitrary constant of integration. It remains to solve equations (70) and (71) by determining L_0 and the potential U_1, which will be done by elimination of the potential difference. In order to do so, we first need to write equations (70) and (71) in a slightly different form:

$$\begin{aligned} i\,(L_1)_t + 2\,\frac{f}{h}\,(L_0)_x + \frac{f}{h}\,(L_1)_{xx} + \frac{f_x}{h}\,(L_1)_x + \frac{f_x}{h}\,L_1 - \frac{f\,h_x}{h^2}\,L_1 &= \frac{U_1 - V_1}{h}\,L_1 \\ i\,(L_0)_t + \frac{f}{h}\,(L_0)_{xx} + \frac{f_x}{h}\,(L_0)_x + \frac{(V_1)_x}{h}\,L_1 - \frac{V_1\,h_x}{h^2}\,L_1 &= \frac{U_1 - V_1}{h}\,L_0. \end{aligned}$$

Now we multiply the first and the second of these equations by L_0 and L_1, respectively, such that the right hand sides of these equations become the same. Consequently, the left hand sides must also be the same, and we can equate them to each other. This results in the following equation:

$$\begin{aligned} & i\,(L_1)_t\,L_0 + 2\,\frac{f}{h}\,(L_0)_x\,L_0 + \frac{f}{h}\,(L_1)_{xx}\,L_0 + \frac{f_x}{h}\,(L_1)_x\,L_0 + \frac{f_x}{h}\,L_0\,L_1 - \frac{f\,h_x}{h^2}\,L_0\,L_1 = \\ &= i\,(L_0)_t\,L_1 + \frac{f}{h}\,(L_0)_{xx}\,L_1 + \frac{f_x}{h}\,(L_0)_x\,L_1 + \frac{(V_1)_x}{h}\,L_1^2 - \frac{V_1\,h_x}{h^2}\,L_1^2. \end{aligned} \tag{73}$$

We will now solve this equation with respect to L_0. To this end, we will introduce a new function K defined by $K = L_0/L_1$. Before we substitute this function into (73), we first evaluate the following expressions, which we will need in the substitution:

$$\frac{2\,(L_1)_x}{L_1} = \frac{h}{f}\left(\frac{f}{h}\right)_x, \quad \frac{(L_1)_{xx}}{L_1} = -\left(\frac{h}{f}\right)_{xx}\frac{f}{2\,h} + \frac{3}{4}\left[\left(\frac{h}{f}\right)_x \frac{f}{h}\right]^2, \tag{74}$$

where we have used the explicit form (72) of L_1. We use this explicit form and the derivatives (74) in order to rewrite equation (73), where we substitute L_0 by $L_0 = KL_1$. After simplification we arrive at the following equation:

$$i\,K_t = \left(-\frac{f}{h}\,K_x + \frac{f}{h}\,|K|^2 - \frac{f_x}{h}\,K - \frac{V_1}{h}\right)_x. \tag{75}$$

We see that our former equation (73), which depends on L_0 and L_1, has been converted to an equation that depends on K only. Unfortunately, we cannot solve (75), as it is an equation of Riccati type for K, which is not integrable in a general case like ours [8]. Still, for practical reasons it makes sense to linearize (75) by means of the following setting:

$$K = -\frac{(u_1)_x}{u_1}, \tag{76}$$

introducing a new function $u_1 = u_1(x,t)$. Assuming that u_1 is twice continuously differentiable, implying $(u_1)_{xt} = (u_1)_{tx}$, we substitute (76) in (75) and get after simplification the following equation for the function u:

$$\left[i\,\frac{(u_1)_t}{u_1} + \frac{f}{h}\,\frac{(u_1)_{xx}}{u_1} + \frac{f_x}{h}\,\frac{(u_1)_x}{u_1} - \frac{V_1}{h} \right]_x = 0\,. \tag{77}$$

Clearly, this equation holds if the expression in square brackets does not depend on x. We integrate on both sides and multiply with u_1:

$$i\,(u_1)_t + \frac{f}{h}\,(u_1)_{xx} + \frac{f_x}{h}\,(u_1)_x - \frac{V_1}{h}\,u_1 = C\,u_1, \tag{78}$$

where $C = C(t)$ is a purely time-dependent constant of integration. Equation (78) is identical to the initial equation (59) for $C = 0$. However, setting C to zero is not a restriction, since solutions to (78) with $C \neq 0$ and $C = 0$ differ from each other only by a purely time-dependent factor, which cancels out in (76). Thus, our equation (78) can be taken in the form

$$i\,(u_1)_t + \frac{f}{h}\,(u_1)_{xx} + \frac{f_x}{h}\,(u_1)_x - \frac{V_1}{h}\,u_1 = 0. \tag{79}$$

The function u_1 is called auxiliary solution of the TDSE (59), as it is needed for determining the function L_0 that appears as a coefficient in the operator L. Once a solution u_1 of (79) is known, then the function K can be found from (76), which in turn determines the sought coefficient L_0 by means of $L_0 = KL_1$. Taking into account the explicit form (72) of L_1, we obtain

$$L_0 \;=\; -L\,\sqrt{\frac{f}{h}}\,\log\,(u_1)_x\,. \tag{80}$$

Thus, with the coefficients L_0 and L_1 we have determined the sought operator $\mathcal{L}$, as given in (63), completely.

Potential difference and the operator $\mathcal{L}$. Before we state the operator L in its explicit form, let us find the potential U_1 by solving (70):

$$U_1 \;=\; V_1 + \frac{(L_1)_{xx}}{m\,L_1} + \frac{2\,f\,(L_0)_x}{L_1} + f_x\,\frac{(L_1)_x}{L_1} - h\left(\frac{f_x}{h}\right)_x + i\,h\,\frac{(L_1)_t}{L_1}. \tag{81}$$

This can be specified more by inserting the explicit form of L_1 and L_0, as given in (72) and (80), respectively. We obtain

$$U_1 = V_1 + i\,\frac{L'}{L} - i\,\frac{h}{2}\,\log\left(\frac{h}{f}\right)_t - 2\,\sqrt{f\,h}\,\left(\sqrt{\frac{f}{h}}\,\frac{(u_1)_x}{u_1}\right)_x - \sqrt{f\,h}\left[\frac{1}{h}\left(\sqrt{f\,h}\right)_x\right]_x. \tag{82}$$

The operator $\mathcal{L}$ is given explicitly by

$$\mathcal{L} \;=\; L\,\sqrt{\frac{f}{h}}\left[\partial_x - \log(u_1)_x\right]. \tag{83}$$

Finally, the solution $\Phi = L(\Psi)$ of the second TDSE (60) can now be given using the latter form of L:

$$\Phi = \mathcal{L}(\Psi) = L\sqrt{\frac{f}{h}}\left[\Psi_x - \log(u_1)_x\,\Psi\right]. \tag{84}$$

Hence, if Ψ is a solution of the first TDSE (59), then $\Phi = \mathcal{L}(\Psi)$ is a solution of the second TDSE (60), provided the potential U_1 is related to the potential V_1 via (82). Thus, we have succeeded in constructing a first-order Darboux transformation for our generalized Schrödinger equation (59). After verifying that the above findings are compatible with the conventional Schrödinger equation and presenting a simple application of our Darboux transformation, we will pass on to another construction method, which will yield Darboux transformations of higher order.

Reduction to the conventional case. Let us briefly verify that our expressions for the operator $\mathcal{L}$ and the potential U_1 reduce correctly to their conventional counterparts that are given in (57) and (58), respectively. To this end, we observe that in the conventional case we have $f = h = 1$. On substituting this setting into (84), we recover immediately the correct expression (57). As for the potential U_1, plugging $f = h = 1$ into its explicit form (82), we obtain

$$\begin{aligned} U_1 &= V_1 + i\,\frac{L'}{L} - 2\left(\frac{(u_1)_x}{u_1}\right)_x \\ &= V_1 + i\,\frac{L'}{L} - 2\,\log(u_1)_{xx}, \end{aligned}$$

which coincides with the desired expression (58).

Application. Let us now give an example of how our formalism can be applied. We consider a generalized Schrödinger equation given by

$$i\,\frac{p}{\lambda\,x}\,\psi_t + \psi_{xx} - \left(\frac{p}{x} - q^2\right)\psi = 0, \tag{85}$$

where $q,\ p > 0$ and $\lambda = \lambda(t)$ is an arbitrary function that is purely time-dependent. This is a special case of equation (59) for the following coefficient functions:

$$\begin{aligned} f &= 1 \\ h &= \frac{p}{\lambda\,x} \qquad (86) \\ V_1 &= \frac{p}{x} - q^2, \qquad (87) \end{aligned}$$

note that the equation (85) was multiplied by the function h. We want to apply our generalized Darboux transformation to (85), for which we need two of its solutions Ψ and u_1 that are linearly independent. Once we have these solutions, we compute the new solution Φ

and potential U_1 by means of (83) and (82), respectively. Now, the two solutions Ψ and u_1 are chosen as follows:

$$\Psi = \sin(q\,x)\,\exp\left(-i\int \lambda\,dt\right) \tag{88}$$

$$u_1 = \cos(q\,x)\,\exp\left(-i\int \lambda\,dt\right). \tag{89}$$

Clearly, the two solutions (88) and (89) are linearly independent. After substitution of our settings (86) and (87) into our operator $\mathcal{L}$ and the potential difference, as given in (63) and (82), respectively, we get

$$\Phi = L\sqrt{\frac{\lambda}{p}\,x}\,\frac{q}{\cos(q\,x)}\,\exp\left(-i\int \lambda\,dt\right) \tag{90}$$

$$\begin{aligned} U_1 &= \frac{p}{x} - q^2 + \frac{i\,p}{\lambda\,x}\left(\frac{L'}{L} - \frac{h_t}{2\,h}\right) + 2\sqrt{h}\left(\frac{K}{\sqrt{h}}\right)_x - \sqrt{h}\left[\frac{\left(\sqrt{h}\right)_x}{h}\right]_x \\ &= \frac{p}{x} - q^2 - \frac{1}{4\,x^2} + \frac{i\,p}{\lambda\,x}\left(\frac{L'}{L} + \frac{\lambda'}{2\,\lambda}\right) + 2\,\frac{q^2}{\cos^2(q\,x)} + \frac{q}{x}\,\tan(q\,x). \end{aligned}$$

If we take $L = 1/\sqrt{\lambda}$ then imaginary part of the potential U_1 becomes zero:

$$U_1 = \frac{p}{x} - q^2 - \frac{1}{4\,x^2} + 2\,\frac{q^2}{\cos^2(q\,x)} + \frac{q}{x}\,\tan(q\,x). \tag{91}$$

This potential is stationary, as its initial counterpart V_1 in (87). If we wanted the potential U_1 to be time-dependent, then we would have to satisfy $L \neq 1/\sqrt{\lambda}$, since the time-dependent terms are in the potential's imaginary part. Finally, if we wanted the potential U_1, as given in (91), to be time-dependent and real-valued, then we would have to write the arbitrary function L as follows:

$$L = \exp\left[i\int\left(M - \frac{\lambda'}{2\,\lambda}\right)\,dt\right],$$

introducing a function $M = M(t)$, which renders the potential (91) in the form

$$U_1 = \frac{p}{x} - q^2 - \frac{1}{4\,x^2} + 2\,\frac{q^2}{\cos^2(q\,x)} + \frac{q}{x}\,\tan(q\,x) - \frac{p\,M}{\lambda\,x}.$$

The fact that L has now turned complex does not matter, since it appears only in the solution Φ, as displayed in (90).

4.2.2. Construction via Point Transformations

In this section we show that our generalized Darboux transformation (63) can be constructed without evaluating intertwining relations. While for the first-order case this might not seem useful, the method we will describe below allows for the straightforward derivation of higher-order Darboux transformations. Such transformations can also be obtained by means of intertwining relations, but their evaluation would be very complicated due to the arbitrarily high number of terms involved. Before we start to calculate, let us describe the construction scheme for our generalized Darboux transformations.

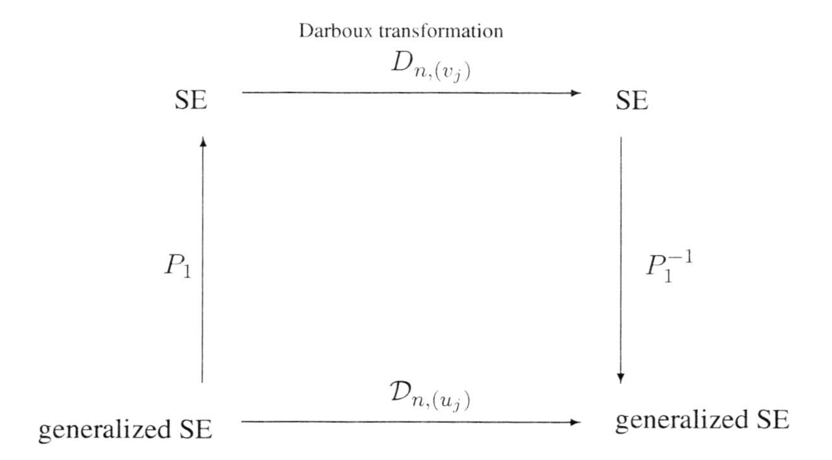

Figure 2. Definition of the point transformation P (SE stands for Schrödinger equation).

Outline of our construction. The way we obtain our Darboux transformation can be understood by means of the commutative diagram in figure 2. The abbreviation SE stands for Schrödinger equation, while P_1 denotes a special case of our point transformation (6), (13), (12) for $\hat{f} = 1$, $\hat{g} = 0$, that takes the generalized Schrödinger equation into the conventional case. Note that P_1 was used for reduction to the stationary case, see section 3.3. for details. We then construct our generalized Darboux transformation $\mathcal{D}_{n,(u_j)}$ by first converting the generalized Schrödinger equation into its conventional form, then applying the conventional Darboux transformation $D_{n,(v_j)}$, and finally reinstalling the generalized form by means of applying the inverse point transformation P_1^{-1}. In summary, we have the following relation between the conventional and the generalized Darboux transformation:

$$\mathcal{D}_{n,(u_j)} = P^{-1} \circ D_{n,(v_j)} \circ P. \tag{92}$$

Clearly, the first order transformation is $\mathcal{D}_{1,(u_j)} = \mathcal{L}$, as defined in (83).

The point transformation. We can evaluate (92) explicitly, since we know the point transformation P_1 that turns (1) into the conventional Schrödinger equation. As we know from section 3.3., application of this transformation yields the conventional Schrödinger equation

$$i\,\psi_t + \psi_{xx} - \hat{V}_1\,\psi = 0, \tag{93}$$

where the potential $\hat{V}_1$ is given by (37), that is,

$$\hat{V}_1 \;=\; \frac{1}{v'}\left(i\,F_t + g\,F_x + f\,F_x^2 + f\,F_{xx} - V_1\right).$$

Note that the solution in (93) is called ψ instead of Φ, as we reserve the latter name for the solution of our final, generalized Schrödinger equation after the Darboux transformation.

The conventional Darboux transformation. Now the conventional Darboux transformation (54) is applicable to equation (93), in particular, to its solution ψ:

$$D_{n,(v_j)}(\psi) = l \frac{W_{n,(v_j),\psi}}{W_{n,(v_j)}}, \tag{94}$$

where $l = l(v)$ is an arbitrary function, (v_j) is a family of n auxiliary solutions of (28), and $W_{n,(v_j),\psi}$, $W_{n,(v_j)}$ are the n-th order Wronskians of the auxiliary solution family (v_j) and of the solution ψ of equation (28). The function $\phi = D_{n,(v_j)}(\psi)$ solves the Schrödinger equation

$$i\,\phi_v + \phi_{uu} + \frac{1}{v'\,h}\left(i\,F_t\,h + F_x^2\,f + F_{xx} + F_x\,g - \hat{U}_1\right)\phi = 0, \tag{95}$$

where the transformed potential function $\hat{U}_1$ reads

$$\hat{U}_1 = V_1 + i\,v'\,h\,\frac{l'}{l} - 2\,v'\,h\,\left[\log\left(W_{n,(v_j)}\right)\right]_{uu}. \tag{96}$$

Note that the unusual factor $v'h$ cancels with the same factor in the coefficient of ϕ in (95). Clearly, here V is understood to be expressed in the new variables u and v. The task is now to rewrite the Darboux transformation (94) and the transformed potential (96) in the variables x and t.

The generalized Darboux transformation. Starting with the Darboux transformation, we need to know how the Wronskians transform under the inverse of the point transformation (36). Let (u_j) be a family of n auxiliary solutions of the generalized Schrödinger equation (59) that is related to the family (v_j) via the point transformation P_1. We then have [20]

$$\begin{aligned} W_{n,(v_j),\psi} &= \exp\left(-(n+1)\,F\right)\left(\frac{1}{u_x}\right)^{\frac{1}{4}n(n+1)} W_{n,(u_j),\Psi} \\ W_{n,(v_j)} &= \exp\left(-n\,F\right)\left(\frac{1}{u_x}\right)^{\frac{1}{4}n(n-1)} W_{n,(u_j)}. \end{aligned} \tag{97}$$

We employ these results in the Darboux transformation (94), which takes the form

$$D_{n,(v_j)}(\Psi) = l\,\exp\left(-F\right)\left(\frac{1}{u_x}\right)^n \frac{W_{n,(u_j),\Psi}}{W_{n,(u_j)}}. \tag{98}$$

Since the last function is still a solution of (28), we have to multiply it by $\exp(F)$, as to invert the mutiplicative part of the point transformation (36). After doing so and inserting the explicit form (12) of u we obtain the final result

$$\begin{aligned} \mathcal{D}_{n,(u_j)}(\Psi) &= \exp\left(F\right)\,D_{n,(v_j)}(\phi) \\ &= L\left(\frac{f}{h}\right)^{\frac{n}{2}} \frac{W_{n,(u_j),\Psi}}{W_{n,(u_j)}}, \end{aligned} \tag{99}$$

where $L = l/\sqrt{v'}$. This is our generalized Darboux transformation of arbitrary order.

The transformed potential. Next, we determine the transformed potential (96), making use of (97):

$$
\begin{aligned}
U_1 &= V_1 + i\,v'\,h\,\frac{L'}{L} - 2\,v'\,h\left[\log\left(\exp(-n\,F)\left(\frac{1}{u_x}\right)^{\frac{1}{2}n(n-1)} W_{n,(u_j)}\right)\right]_{uu} \\
&= V_1 + i\,v'\,h\,\frac{L'}{L} - 2\,v'\,h\left[-n\,F + \frac{1}{2}\,n(n-1)\,\log\left(\frac{f}{h\,v'}\right) + \log\left(W_{n,(u_j)}\right)\right]_{uu} \\
&= V_1 + i\,v'\,h\,\frac{L'}{L} - 2\,f\left[-n\,F + \frac{1}{2}\,n(n-1)\,\log\left(\frac{f}{h\,v'}\right) + \log\left(W_{n,(u_j)}\right)\right]_{xx} + \\
&+\; 2\,h\left(\frac{f}{h}\right)_x\left[-n\,F + \frac{1}{2}\,n(n-1)\,\log\left(\frac{f}{h\,v'}\right) + \log\left(W_{n,(u_j)}\right)\right]_x. \qquad (100)
\end{aligned}
$$

We can cast the potential (100) in a slightly different form:

$$
\begin{aligned}
U_1 &= V_1 + i\,v'\,h\,\frac{L'}{L} - 2\,f\left[\log\left(W_{n,(u_j)}\right)\right]_{xx} + \qquad (101) \\
&+\; 2\,h\left(\frac{f}{h}\right)_x\left[\log\left(W_{n,(u_j)}\right)\right]_x - \frac{f}{h\,v'}\left[-2\,n\,F + n(n-1)\,\log\left(\frac{f}{h\,v'}\right)\right]_{xx} + \\
&+\; h\left(\frac{f}{h}\right)_x\left[-2\,n\,F + n(n-1)\,\log\left(\frac{f}{h\,v'}\right)\right]_x.
\end{aligned}
$$

The transformed potential (100) can also be written in the following compact form that we will use in subsequent considerations:

$$
\begin{aligned}
U_1 &= V_1 + i\,v'\,h\,\frac{L'}{L} - 2\,\sqrt{f\,h}\left[\sqrt{\frac{f}{h}}\left[\log\left(\left(\frac{f}{h\,v'}\right)^{\frac{\kappa}{2}} W_{n,(u_j)}\right)\right]_x\right]_x + \\
&+\; 2\,n\,f\left(F_{xx} + \frac{F_x}{2}\left[\log\left(\frac{f}{h}\right)\right]_t\right), \qquad (102)
\end{aligned}
$$

where the constant κ is defined as $\kappa = n(n-1)/2$. In summary, the generalized Darboux transformation (99) maps the Schrödinger equation (59) onto its counterpart (60) for the potential (100) or one of its equivalent representations.

Reduction to the first-order case. Finally, let us briefly verify that our Darboux transformation (99) and its transformed potential (100) reduce correctly to the first-order expressions (83) and (82), respectively, if we set $n = 1$. Starting with (99), we get

$$
\begin{aligned}
\mathcal{D}_{1,(u_1)}(\Psi) &= L\left(\frac{f}{h}\right)^{\frac{1}{2}} \frac{W_{1,(u_1),\Psi}}{W_{1,(u_1)}} \\
&= L\,\sqrt{\frac{f}{h}}\left(-\frac{(u_1)_x}{u_1}\,\Psi + \Psi_x\right) \\
&= \mathcal{L}\,\Psi.
\end{aligned}
$$

Hence, the Darboux transformation reduces to the correct expression. Regarding the potential (100), we obtain for $n = 1$

$$U_1 = V_1 + i\,\frac{L'}{L} - i\,\frac{h}{2}\,\log\left(\frac{h}{f}\right)_t - 2\,\sqrt{f\,h}\,\left(\sqrt{\frac{f}{h}}\,\frac{(u_1)_x}{u_1}\right)_x - \sqrt{f\,h}\,\left[\frac{1}{h}\left(\sqrt{f\,h}\right)_x\right]_x,$$

which coincides with (82) if we choose the arbitrary function v as $v(t) = t$.

5. Conclusion

The purpose of the present review is to explain in detail how generalized and conventional Schrödinger equations are related to each other, and how this relation can be used to generate their solutions. Loosely speaking, one could say that generalized Schrödinger equations are just conventional ones, written in different coordinates. While this fact already provides one of the most popular solution methods, namely, the point transformation, it can be used to modify another popular solution method - the conventional Darboux transformation - such that it becomes applicable to generalized Schrödinger equations. Although the Darboux transformation can be obtained solely by analyzing intertwining relations, this tedious process can be turned into a straightforward calculation by means of the point transformation. Finally, from the previous argument it becomes clear that Darboux transformations can be defined for any equation that is related to the conventional Schrödinger equation by means of a point transformation, such as the Fokker-Planck equation [16] or the nonhomogeneous Burgers equation [15].

References

[1] Abramowitz, M.; Stegun, I.A. *Handbook of mathematical functions with formulas, graphs, and mathematical tables;* Dover Publications, New York, USA, 1964.

[2] Bagrov V.G.; Samsonov, B.F. *Phys. Lett. A* 1996, 210, 60-64.

[3] Darboux, M.G. *Comptes Rendus Acad. Sci. Paris* 1882, 94, 1456-1459.

[4] de Sanctis, M.; Quintero, P. *Eur. Phys. J. A* 2009, 39, 1434-6001.

[5] Finkel, F.; Gonzalez-Lopez, A.; Kamran, N.; Rodriguez, M.A. *J. Math. Phys.* 1999, 40, 3268- 3274.

[6] Formanek, J.; Lombard,R.J.; Mares, J. *Czech. J. Phys.* 2004, 54, 289-315.

[7] Junker, G. *Supersymmetric methods in quantum and statistical physics;* Springer, Berlin, Germany, 1995.

[8] Kamke, E. *Differentialgleichungen - Lösungsmethoden und Lösungen;* B.G. Teubner, Stuttgart, Germany, 1983.

[9] Kingston, J.G.; Sophocleous, C. *J. Phys. A* 1998, 31, 1597-1619.

[10] Landsberg, G.T. *Solid state theory: methods and applications*; Wiley Interscience, London, Great Britain, 1969.

[11] Lombard, R.J.; Mares, J. *Phys. Lett A* 2009, 373, 426-429.

[12] Natanzon, G.A. *Vestnik Leningrad. Univ.* No. 16 Fiz. Him. Vyp. 1977, 3, 33-39.

[13] Risken, H. *The Fokker-Planck equation: methods of solution and applications*; Springer, Berlin, Germany, 1996.

[14] H. Stephani, *Differential equations - their solutions using symmetries,* Cambridge University Press, 1989.

[15] Schulze-Halberg, A.; Carballo Jimenez, *J.M. Phys. Scripta* 2009, 80, 065014 (8pp).

[16] Schulze-Halberg, A.; Garcia-Ravelo, J.; Morales, J.; Pena, J.J.; Roy, P. *Phys. Lett. A* 2009, 373, 1610-1615.

[17] Schulze-Halberg, A. (2008). Reduction of time-dependent Schrödinger equations with effective mass to stationary Schrödinger equations. *Research Letters in Physics,* 2008 (article ID 589269, 4 pp.), DOI:10.1155/2008/589269.

[18] Schulze-Halberg, A. Internat. *J. Modern Phys. A* 2007, 22, 1735-1769.

[19] Schulze-Halberg, A. *Cent. Eur. J. Phys.* 2005, 3, 591-609.

[20] Schulze-Halberg, A. *J. Phys. A* 2005, 38, 5831-5836.

[21] Ushveridze, A.G. *Exactly-solvable models in Quantum Mechanics*; Taylor and Francis, New York, USA, 1994.

[22] von Roos, O.; Mavromatis, H. *Phys. Rev. B* 1985, 31, 2294-2298.

[23] von Roos, O. *Phys. Rev. B* 1983, 27, 7547-7552.

In: Focus on Quantum Mechanics
Editors: David E. Hathaway et al. pp. 157-195
ISBN 978-1-62100-680-0

Chapter 7

A DENSITY FUNCTIONAL METHOD FOR GENERAL EXCITED STATES IN ATOMS

Amlan K. Roy *
Division of Chemical Sciences,
Indian Institute of Science Education and
Research (IISER), Block FC, Sector III,
Salt Lake, Kolkata-700106, India.

Abstract

This chapter presents the development of a density functional theory (DFT)-based method for accurate, reliable treatment of various resonances in atoms. Many of these are known to be notorious for their strong correlation, proximity to more than one thresholds, degeneracy with more than one minima. Therefore they pose unusual challenges to both theoreticians and experimentalists. It is well-known that DFT has been one of the most powerful and successful tools for electronic structure calculation of atoms, molecules, solids in the past two decades. While it has witnessed diverse spectacular applications for ground states, the same for excited states, unfortunately, has come rather lately and remains somehow less conspicuous relatively. Our method uses a work-function-based exchange potential in conjunction with the popular gradient-corrected Lee-Yang-Parr correlation functional. The resulting Kohn-Sham equation, in the non-relativistic framework, is numerically solved accurately and efficiently by means of a generalized pseudospectral method through a non-uniform, optimal spatial discretization. This has been applied to a variety of excited states, such as low and high states; single, double, triple as well as multiple excitations; valence and core excitations; autoionizing states; satellites; hollow and doubly-hollow states; very high-lying Rydberg resonances; etc., of atoms and ions, with remarkable success. A thorough and systematic comparison with literature data reveals that, for all these systems, the exchange-only results are practically of Hartree-Fock quality; while with inclusion of correlation, this offers excellent agreement with available experimental data as well as those obtained from other sophisticated theoretical methods. Properties such as individual state energies, excitation energies, radial densities as well as various expectation values are studied as well. This helps us in predicting many states for the first time. In summary, we have presented an accurate, simple, general DFT method for description of arbitrary excited states of atoms and ions.

*E-mail address: akroy@iiserkol.ac.in, akroy@chem.ucla.edu

1. Introduction

A central objective in modern quantum chemistry, much of materials science, condensed matter physics and various other branches in science today, is, stated simply, to understand structure, dynamics, energetics through the application of rigorous principles of quantum mechanics. While these results usually complement the information obtained from physical, chemical and biological experiments, in many cases, this could also be used to predict and unveil many hitherto unobserved phenomena. Some of the most commonly used properties are: calculating the potential energy surface (energy changes as a function of structural parameters), interaction energies (absolute as well as relative), electronic charge distributions, dipole and higher multipole moments, spectroscopic quantities such as vibrational frequencies, NMR chemical shifts, ESR g tensors, hyperfine coupling constants, chemical reactivity, mechanistic routes of a reaction, cross sections for collision with other particles, behavior of atoms/molecules under an external field, such as a strong laser field, etc. The starting point for most such quantum mechanical studies is the non-relativistic Schrödinger equation:

$$H\Psi(\mathbf{r},t) = E\Psi(\mathbf{r},t) \tag{1}$$

Here H is the Hamiltonian operator including various energy components. The solution of this equation yields total energy E, as well as the many-particle wave function, which contains all relevant informations about the system under investigation. The ultimate goal of achieving *exact* solution of this equation is essentially beyond our reach except for a small number of highly simplified model cases. Our systems of interest contain many atoms and electrons, where the solution easily becomes unmanageable. The main difficulty arises due to the presence of electron-electron interaction terms, and approximation methods must be invoked. This lead to the development of multitude of *ab initio* methods in today's electronic structure theory; one of the very first and successful being the Hartree-Fock (HF) method.

Every conceivable property of a many-electron interacting system can be obtained as a *functional* of the *ground-state electron density*, $\rho(\mathbf{r})$, thus replacing the complicated, intractable many-particle wave function. Stated otherwise, in principle, this scalar function of position, determines all informations embedded in the many-body wave function of ground and all excited states. Existence of such functionals lies at the heart of density functional theory (DFT) and as such, the electron density is defined as,

$$\rho(\mathbf{r}_1) = N\int\cdots\int|\Psi(\mathbf{x_1},\mathbf{x_2},\cdots,\mathbf{x_N})|^2\, ds_1 d\mathbf{x_2}\cdots d\mathbf{x_N} \tag{2}$$

The multiple integration involves integral over spin coordinates of all electrons (spin integration implies summation over the possible spin states) and all but one of the spatial variables. Clearly $\rho(\mathbf{r})$ is a real, non-negative, visualizable function of only three spatial coordinates (in contrast to the 4N dimensional many-electron wave function for an N electron system) with direct physical significance (can be measured experimentally unlike the complex-valued, wave function), vanishes at infinity and integrates to the total number of electrons:

$$\begin{aligned} \rho(\mathbf{r} \to \infty) &= 0 \\ \int \rho(\mathbf{r}) d\mathbf{r} &= N \end{aligned} \quad (3)$$

Because of its transparency in dealing with the problematic inter-electronic repulsion in a rigorous quantitative way plus favorable computational cost, DFT has been the most popular and beloved of quantum mechanical methods for atoms, molecules and solids for more than three decades or so.

The first attempt to use electron density as basic variable in the context of atoms, molecules rather than wave function is almost as old as quantum mechanics itself. In the original quantum statistical model of Thomas and Fermi [1, 2], kinetic energy of electrons is approximated as an explicit functional of density, whereas nuclear-electron attraction and electron-electron repulsion contributions are treated in a classical manner. The following simple expression of kinetic energy is derived by assuming electrons to be in the background of an idealized, non-interacting homogeneous electron gas, i.e., a fictitious model of system of constant electron density:

$$T_{TF}[\rho(\mathbf{r})] = \frac{3}{10}(3\pi^2)^{2/3} \int \rho(\mathbf{r})^{5/3} \, d\mathbf{r} \quad (4)$$

Combining this with the classical terms, one can obtain the so-called celebrated Thomas-Fermi energy functional for electrons in an external potential $v_{ext}(\mathbf{r})$ as follows,

$$E_{TF}[\rho(\mathbf{r})] = T_{TF}[\rho(\mathbf{r})] + \int v_{ext}(\mathbf{r})\rho(\mathbf{r}) \, d\mathbf{r} + \frac{1}{2} \int\int \frac{\rho(\mathbf{r})\rho(\mathbf{r}')}{|\mathbf{r}-\mathbf{r}'|} \, d\mathbf{r}d\mathbf{r}' \quad (5)$$

Here the last term represents classical electrostatic repulsion. Now, minimization of this functional $E[\rho(\mathbf{r})]$ for all possible $\rho(\mathbf{r})$ subject to the constraint on total number of electrons,

$$\int \rho(\mathbf{r}) \, d\mathbf{r} = N \quad (6)$$

leads to the ground-state density and energy. Since $T_{TF}[\rho]$ is only a very coarse approximation to true kinetic energy, as well as exchange and correlation effects are completely ignored, results obtained using this approach are rather crude. It misses the essential physics and chemistry, such as shell structure of atoms and molecular binding. However, this illustrates the important fact that energy of an interacting system can be written *completely* in terms of single-particle density *alone*. This was further extended by Dirac [3] to incorporate exchange effects, leading to the so-called local density approximation (LDA). This gives significant improvements over the original TF method and very much is in use, still today,

$$E_{TFD}[\rho(\mathbf{r})] = T_{TF}[\rho(\mathbf{r})] + \int v_{ext}(\mathbf{r})\rho(\mathbf{r}) \, d\mathbf{r} - \frac{3}{4}\left(\frac{3}{\pi}\right)^{1/3} \int \rho(\mathbf{r})^{4/3} d\mathbf{r} + \frac{1}{2} \int\int \frac{\rho(\mathbf{r})\rho(\mathbf{r}')}{|\mathbf{r}-\mathbf{r}'|} \, d\mathbf{r}d\mathbf{r}'. \quad (7)$$

Soon, the provocative simplicity of above density approach, compared to traditional wave-function-based methods influenced a considerably large number of calculations.

However, due to the lack of a rigorous foundation (e.g., no variational principle was established as yet) combined with the fact that fairly large errors were encountered in solid-state and molecular calculations, the theory somehow lost its appeal and charm; realistic electronic structure calculation within such a deceptively simple route seemed a far cry, leading to very little practical impact on chemistry. However, the situation was about to change after the landmark paper by Hohenberg and Kohn [4], where this was put on a firm theoretical footing. This earns it the status of an *exact* theory of many-body system and eventually, laid the groundwork of all of modern DFT. The first HK theorem simply states that the external potential $v_{ext}(\mathbf{r})$ in a many-electron interacting system is *uniquely* determined, to within a constant, by ground-state density $\rho(\mathbf{r})$. Now, since H is fully determined (except to a constant), it easily follows that many-body wave functions for *all* states (ground and excited) are also determined. Thus all properties of the system are completely determined by $\rho(\mathbf{r})$ only. The proof is based on *reductio ad absurdum* and skipped here for brevity.

Note that this is only an existence theorem and as such, completely unhelpful in providing any indication of how to predict the density of a system. The answer lies in the second theorem which states that, for any external potential $v_{ext}(\mathbf{r})$, one can define a functional, $E[\rho]$ in terms of $\rho(\mathbf{r})$, as follows,

$$E[\rho] = E_{ne}[\rho] + (T[\rho] + E_{ee}[\rho]) = \int v_{ext}(\mathbf{r})\rho(\mathbf{r})\, d\mathbf{r} + F_{HK}[\rho] \tag{8}$$

Here, the energy has been separated into two parts: one that depending on the actual system, i.e., the potential energy because of nuclear-electron attraction, and a universal term (in a sense that the form is independent of N, R_A, Z_A, or in other words, *same for all electrons*), consisting of the kinetic and electron repulsion energy components. For a given $v_{ext}(\mathbf{r})$, global minimum of this functional provides ground-state energy while, the density minimizing this functional corresponds to exact ground-state density as,

$$E_0 = \min_{\rho \to N} \left(F[\rho] + \int \rho(\mathbf{r}) v_{ext}(\mathbf{r})\, d\mathbf{r} \right); \quad F[\rho] = \min_{\Psi \to \rho} \langle \Psi | T + E_{ee} | \Psi \rangle. \tag{9}$$

It is worth noting that, this apparently simple-looking *universal* or *Hohenberg-Kohn* functional, $F_{HK}[\rho]$ is the holy grail of DFT. It remains absolutely silent about the explicit forms of functionals for both $T[\rho]$ and $E_{ee}[\rho]$. Design of their accurate forms remains one of the major challenges and lies at the forefront of modern development works in DFT. Further partitioning of energy is possible through the following realization,

$$E_{ee}[\rho] = \frac{1}{2} \int \int \frac{\rho(\mathbf{r})\rho(\mathbf{r}')}{|\mathbf{r} - \mathbf{r}'|}\, d\mathbf{r} d\mathbf{r}' + E_{nc}[\rho] = J[\rho] + E_{nc}[\rho]. \tag{10}$$

Here $J[\rho]$ signifies classical Coulomb repulsion whereas the last term is associated with the *non-classical* contribution to electron repulsion, containing all the effects of exchange, correlation as well as self-interaction correction.

Despite the charm and simplicity of HK theorem, very little progress could be made in terms of practical applications to realistic atoms/molecules. *All* this tells us is that, in principle, a unique mapping between ground-state density and energy exists. However, there is no guideline whatsoever, about the construction of this functional which delivers the ground-state energy. In so far as theoretical prediction of molecular properties are concerned through computational DFT, no conspicuous changes could be observed with the culmination of these theorems. Because one is still left with the difficult problem of solving many-body system in presence of $v_{ext}(\mathbf{r})$; calculations are as hard as before the HK theory. The variational principle in second theorem also calls for caution. In any real calculation, in absence of the exact functional, one is invariably left with no choice, but to use some approximate forms. The variational theorem, however, applies only to the case of *exact* functionals, implying that in DFT world, energy delivered by a trial functional has absolutely no meaning. This is in sharp contrast to the conventional wave function-based, variational methods such as HF or CI, where the lower an energy E, better a trial function approximates the true wave function.

A year later, in a ground breaking work, Kohn and Sham [5] proposed a route to approach the hitherto unknown universal functional. The central part of their ingenious idea stems from the realization that the original many-body problem of an interacting system could be replaced by an auxiliary, fictitious non-interacting system of particles. This ansatz, then, in principle, holds the promise for exact calculations of realistic systems using only an independent-particle picture of non-interacting fermions, which are *exactly soluble* (in practice by numerical methods). The non-interacting reference system is constructed from a set of one-electron orbitals, facilitating the major portion of kinetic energy to be computed to a good accuracy (exact wave functions of non-interacting fermions are Slater determinants). The residual part of kinetic energy, which is usually fairly small, is then merged with the non-classical component of electron-electron repulsion, which is also unknown,

$$\begin{aligned} F[\rho] &= T_s[\rho] + J[\rho] + E_{xc}[\rho] \\ E_{xc}[\rho] &= (T[\rho] - T_s[\rho]) + (E_{ee}[\rho] - J[\rho]) = T_c[\rho] + E_{nc}[\rho]. \end{aligned} \tag{11}$$

Here $T_s[\rho]$ corresponds to the exact kinetic energy of a hypothetical non-interacting system having the same electron density as that of our real interacting system. The exchange-correlation (XC) functional $E_{xc}[\rho]$ now contains everything that is unknown. Thus it includes not only the non-classical electrostatic effects of electron-electron repulsion, but also the difference of true kinetic energy $T_c[\rho]$ and $T_s[\rho]$. Now we are in a position to write down the celebrated Kohn-Sham (KS) equation in its standard form,

$$\left[-\frac{1}{2}\nabla^2 + v_{eff}(\mathbf{r})\right] \psi_i(\mathbf{r}) = \varepsilon_i \psi_i(\mathbf{r}) \tag{12}$$

where the "effective" potential $v_{eff}(\mathbf{r})$ includes the following terms,

$$v_{eff}(\mathbf{r}) = v_{ext}(\mathbf{r}) + \int \frac{\rho(\mathbf{r}')}{|\mathbf{r} - \mathbf{r}'|} \, d\mathbf{r}' + v_{xc}(\mathbf{r}). \tag{13}$$

Here $v_{eff}(\mathbf{r})$ and $v_{ext}(\mathbf{r})$ signify the effective and external potentials respectively. Literature in DFT is very vast; many excellent books and review articles are available. Here we refer the readers to [6, 7, 8, 9, 10, 11, 12, 13, 14, 15, 16, 17, 18, 19, 20] for a more detailed and thorough exposition on the subject.

So far, we have restricted our focus to the ground states. Now let us shift our attention to the calculation of excited states. It is well known that DFT has been one of the most powerful and successful tools in predicting numerous ground-state properties of many-electron systems such as atoms, molecules, solids, over the past four decades. However, mainly due to its inherent weaknesses, extension to excited-state problems has been quite difficult and rather less straightforward, so much so that DFT is often dubbed as a *ground-state theory*. Results on excited states remained very scarce until very recently. Although considerable progress has been made by employing several different strategies to address the problem lately, there are many crucial unresolved issues as yet, which require further attention.

At this point, however, it may be appropriate to discuss the major difficulties in dealing with excited states within the realm of Hohenberg-Kohn-Sham (HKS) DFT. As already mentioned in the beginning, very foundation of DFT relies on a presupposition that the electron density alone is sufficient to describe *all states* (both ground and excited) of a desired system. Indeed, $\rho(\mathbf{r})$ contains all relevant informations on excited states as well besides the ground state, but the problem is that no practical way to extract this information has been found out as yet. Moreover, there is no HK theorem for excited state parallel to ground state. This is presumably due to the fact that, for a general excited state, the wave function (in general, a complex quantity) can not be bypassed through the pure state density (a real quantity). Because from a hydrodynamical point of view, *phase* part of the hydrodynamic function is *constant* for ground and some excited states (the *static* stationary states), but not so for a general excited state. Working completely in terms of single-particle density in contrast to the state function may be advantageous for ground states, but it is disadvantageous for excited states, because an individual excited state can not be characterized solely in terms of density. Besides, approximate functional forms of $T[\rho]$ and $E_{xc}[\rho]$, valid for both and ground states are unknown; certainly there are insurmountable difficulties in constructing *exact* functionals for these. Note that, there is no reason that this functional will have same general form for both ground and an arbitrary excited state. Finally one also has to deal with the nagging and tedious problem of ensuring both Hamiltonian and wave function orthogonalities, as in any standard variational calculation. Due to these reasons, excited states within DFT remains a very challenging and important area of research.

The purpose of this chapter is to present a detailed account on an excited state density functional method, which has been shown to be very promising for arbitrary excited states of atoms. This relies on a *time-independent* SCF procedure in contrast to the so-called *time-dependent* DFT (TDDFT). Its success and usefulness has been well documented in a series of papers for diverse atomic excited states [21, 22, 23, 24, 25, 26, 27, 28, 29, 30, 31, 32, 33]. Section II gives a brief review of some of the most prominent DFT methods for excited states currently available in the literature. Section III presents the methodology and computational implementation used in our present work. Results from our calculation are discussed in Section IV, with reference to other theoretical methods as well as with experiments, wherever possible. Finally we conclude with a few remarks on the past, present and future of this method.

2. A Brief Review of Excited-state DFT

In this section we will briefly mention some of the excited state DFT methods, with emphasis on those having realistic practical applications. However, before that, a very obvious question arises: Can the original KS equation (12) be excited? To answer this, we rewrite the expression for "effective" potential in an atom in the following equivalent form,

$$v_{eff}(\mathbf{r}) = -\frac{Z}{r} + \int \frac{\rho(\mathbf{r}')}{|\mathbf{r}-\mathbf{r}'|}\, d\mathbf{r}' + \frac{\delta E_{xc}[\rho]}{\delta \rho}. \tag{14}$$

Here Z denotes nuclear charge on the atom. There is an inherent degeneracy in terms of electron spin. Moreover the other angular momentum quantum number (l,m) information required to characterize an individual excited state is clearly missing. So it is not possible to select an excited state of a given space-spin symmetry corresponding to a particular electronic configuration. Next immediate question comes: can one possibly calculate an average of a set of degenerate excited states? That depends on the form of XC functional used. As mentioned before, exact forms of XC functionals are unknown as yet and almost all existing functionals are designed for ground states only. The validity of these functionals for arbitrary excited states is unknown and in fact, common sense dictates that ground and excited state functional may not have same same; in all possibility they would be different.

Introduction of spin density, $\rho^S(\mathbf{r}) = \rho^{\alpha}(\mathbf{r}) - \rho^{\beta}(\mathbf{r})$, made it possible for the development of more flexible potentials. Here ρ^{α}, ρ^{β} signify electron densities corresponding to the up, down spins respectively. It is easy to show that for different spin densities, $v_{eff}^{\alpha}(\mathbf{r})$, $v_{eff}^{\beta}(\mathbf{r})$ will be different; self-consistent solution of the spin-polarized KS equation can be obtained. A thermodynamical version of the spin-density formalism has been presented [34], which proves that standard DFT methods could be used for calculation of lowest lying state of each spatial or spin irreducible representation of a given system, since, in a sense, these represent the "ground state" in that particular symmetry. However there are inherent problems where the non-interacting case does not reduce to a single determinant. Many excited states can not be described by a single determinant and intrinsically need a multi-configurational description. For example, consider the open-shell configuration p^2, as in the ground state of carbon atom, giving rise to three multiplets $^3P, ^1D, ^1S$. In accordance with the above discussion, one can obtain the energy of ^3P and ^1D terms, but not ^1S. This can be understood from a consideration of the fact that, in absence of spin-orbit coupling, all valid states of an atom must be simultaneous eigenfunctions of not only the Hamiltonian operator H, but also angular momentum operators L^2, L_z, S^2, S_z, as well as parity operator Π, such that any given state is characterized by the following quantum numbers L, M_L, S, M_S, π, associated with these operators. In traditional wave functional methods, these informations are carried by wave functions. But in DFT, the basic variable is electron density; so, in principle, it should be $E_{xc}[\rho]$, which should contain this dependence on these above quantum numbers. On the other hand, most of the functionals in DFT depend solely in terms of charge or spin densities; therefore clearly lacking any of these mentioned quantities needed for a complete description of the state of an atom. Stated otherwise, we are facing a very important conceptual problem here: how can one describe the state of a many-electron system, which are eigenfunctions of all these operators working entirely in terms of density with no access to the correct N-electron wave function and its associated symmetry characteristics?

This problem is partially resolved by making reference to Slater's transition state theory [35, 36]. Following this prescription, an *ad-hoc* approach to solve the multiplet problem was proposed, the so-called *sum method* [37, 38]. Here the working equations are exactly like the KS equations, but density is assumed to correspond to a fictitious transition state where one or more orbitals are fractionally occupied. These authors argue that energy of a term, not representable by a single determinant, but requires a linear combination of determinants, can not be computed by spin densities of the corresponding state functions, but can be written as a weighted sum of determinantal energies as follows:

$$E(M_j) = \sum_i A_{ji} E(D_i) \tag{15}$$

Here $E(M_i)$, $E(D_i)$ refer to the energies of different multiplets and determinants respectively. Note that the sum method has no firm theoretical justification; rather an empirical extension of the HF to the $X\alpha$ method. By using some elegant group theoretical method, a semi-automatic protocol has been developed to obtain the weights of various determinants [39, 40].

The above intuitive ideas was utilized on a rigorous foundation to formulate a scheme in which the density is a sum of M lowest-energy eigenstate densities with equal weightage [41]. Now we discuss at some length the *ensemble density* or *fractional occupation approach* to excited states [42, 43, 44]. This suggests to work with an ensemble of densities rather than a pure-state density and the subspace formulation [41] can be considered a special case of it.

The generalized eigenvalue problem for a time-independent Hamiltonian H with M eigenvalues $E_1 \leq E_2 \leq \cdots \leq E_M$ for its M low-lying states is:

$$H\Psi_k = E_k \Psi_k \quad (k = 1, 2, \cdots, M) \tag{16}$$

Applying the Rayleigh-Ritz variational principle, one can write the ensemble energy as:

$$\mathcal{E} = \sum_{k=1}^{M} w_k E_k, \quad 0 \leq w \leq 1/M, \quad 1 \leq g \leq M-1 \tag{17}$$

where $w_1 \geq w_2 \geq \cdots \geq w_M \geq 0$ are the weighting factors chosen such that: $w_1 = w_2 = \cdots = w_{M-g} = \frac{1-w}{M-g}$, $w_{M-g+1} = w_{M-g+2} = \cdots = w_M = w$. The limit $w = 0$ corresponds to the eigenensemble of $M-g$ states ($w_1 = w_2 = \cdots = w_{M-g} = \frac{1}{M-g}$ and $w_{M-g+1} = w_{M-g+2} = \cdots = w_M = 0$), whereas $w = 1/M$ leads to the eigenensemble of M states ($w_1 = w_2 = \cdots = w_M = 1/M$).

The generalized HK theorem can be established; as well as the KS equation for ensembles following the variational principle in standard manner,

$$\left[-\frac{1}{2}\nabla^2 + v_{KS}\right] u_i(\mathbf{r}) = \varepsilon_i u_i(\mathbf{r}) \tag{18}$$

where the ensemble KS potential, given below,

$$v_{KS}(\mathbf{r};\rho_w) = v_{ext}(\mathbf{r}) + \int \frac{\rho_w(\mathbf{r})}{|\mathbf{r}-\mathbf{r}'|}\, d\mathbf{r} + v_{xc}(\mathbf{r}; w, \rho_w) \tag{19}$$

could be defined as a functional of the ensemble density as follows,

$$\rho_w^I(\mathbf{r}) = \frac{1-wg_I}{M_{I-1}} \sum_{m=1}^{M_I-g_I} \sum_j \lambda_{mj}\, |u_j(\mathbf{r})|^2 + w \sum_{m=M_I-g_I+1}^{M_I} \sum_j \lambda_{mj}\, |u_j(\mathbf{r})|^2. \tag{20}$$

Here g_I denotes degeneracy of the Ith multiplet, $M_I = \sum_{i=1}^{I} g_i$ defines multiplicity of the ensemble, λ_{mj} are occupation numbers, with $0 \leq w \leq 1/M_I$. Density matrix is defined as,

$$P^{M,g}(w) = \sum_{m=1}^{M} w_m |\Psi_m\rangle\langle\Psi_m|. \tag{21}$$

The XC potential v_{xc} is the functional derivative of ensemble XC energy functional E_{xc},

$$v_{xc}(\mathbf{r}; w, \rho) = \frac{\delta E_{xc}[\rho, w]}{\delta\rho(\mathbf{r})} \tag{22}$$

One can then express excitation energies in terms of one-electron energies ε_j,

$$\overline{E}^I = \frac{1}{g_I} \frac{d\mathcal{E}^I(w)}{dw}\bigg|_{w=w_I} + \sum_{i=2}^{I-1} \frac{1}{M_I} \frac{d\mathcal{E}^I(w)}{dw}\bigg|_{w=w_i} \tag{23}$$

where

$$\frac{d\mathcal{E}^I(w)}{dw} = \sum_{j=N+M_{I-1}}^{N-1+M_I} \varepsilon_j - \frac{g_I}{M_{I-1}} \sum_{j=N}^{N-1+M_{I-1}} \varepsilon_j + \frac{\partial E_{xc}^I}{\partial w}\bigg|_{\rho_w} \tag{24}$$

Clearly, excitation energies cannot be calculated as a difference of the one-electron energies; there is an extra quantity (the last term) that needs to be determined.

The two-particle density matrix of the ensemble is obtained as an weighted sum of two-particle density matrices of ground and excited states as follows,

$$\Gamma^{M,g,w}(\mathbf{r}_1, \mathbf{r}_2; \mathbf{r}'_1, \mathbf{r}'_2) = \sum_{m=1}^{M} w_m \Gamma^m(\mathbf{r}_1, \mathbf{r}_2; \mathbf{r}'_1, \mathbf{r}'_2) \tag{25}$$

The total ensemble density then takes the following form,

$$\mathcal{E}_w^{M,g} = \mathrm{Tr}\{P^{M,g}(w)H\} = \mathrm{Tr}\{P^{M,g}(T + E_{ee})\} + \mathrm{Tr}\{P^{M,g}(w)V\} = F^{M,g}(w) + \int \rho(\mathbf{r}) v_{ext}(\mathbf{r}) d\mathbf{r} \tag{26}$$

where $\rho(\mathbf{r})$ is the ensemble density; $V = \sum_{i=1}^{N} v_{ext}(\mathbf{r}_i)$. The ensemble XC energy is given by,

$$E_{xc}^{M,g}[w, \rho] = F^{M,g}[w, \rho] - T_s^{M,g}[w, \rho] - J[\rho]. \tag{27}$$

Here the last two terms denote ensemble non-interacting kinetic and Coulomb energies.

Solution of the ensemble KS equation (18) requires knowledge of the ensemble XC potential, exact form of which remains unknown and several approximations have been proposed. In [42, 43, 44], excitation energies of He were studied using the quasi-local density approximation [45]. First excitation energies of several atoms [46] as well higher excitation energies [47] have been reported using the parameter-free exchange potential of Gáspár [48], which depends explicitly on spin orbitals. Several ground-state LDA functionals have been employed for this purpose: *viz.*, Gunnarsson-Lundqvist-Wilkins [49], von Barth-Hedin [50], Ceperley-Alder [51], local density approximations parametrized by Perdew and Zunger [52] and Vosko *et al.* [53]. It is found that, in general, spin-polarized calculations provide better results compared to the non-spin-polarized ones; however, in most cases, estimated excitation energies are highly overestimated. Generally, these functionals provide results which are in close agreement with each other. These references use minimum (0) and maximum values of the weighting factor w. Any w value satisfying the inequality in (18) is appropriate, provided that one uses the *exact* XC energy. However since the latter is unknown, one has to take recourse to approximate functionals; thus different excitation energies are obtained with different w. This variation is studied in detail in [47]; in some occasions the change is small, while for others considerably large variation is observed. Simple local *ensemble potential* has been proposed [54] for this purpose as well,

$$v_x(\rho_w, w) = -3\alpha(w)\left(\frac{3\rho_w}{8\pi}\right)^{1/3}; \quad E_x[\rho_w, w] = -\frac{9}{4}\left(\frac{3}{8\pi}\right)^{1/3}\alpha(w)\int \rho_w^{4/3}\mathbf{dr} \tag{28}$$

However, calculated excitation energies are still very far from the actual values. This leads us to the conclusion that like the ground-state DFT, search for accurate XC functional again remains one of the major bottlenecks in the success of ensemble or fractional occupation approach to excited-state energies and densities.

In another development [55, 56], KS equations were obtained by partitioning the wave function into following two components,

$$\psi(\mathbf{r}_1, \mathbf{r}_2, \cdots, \mathbf{r}_N) = \phi(\mathbf{r}_1, \mathbf{r}_2, \cdots, \mathbf{r}_N) + \tilde{\psi}(\mathbf{r}_1, \mathbf{r}_2, \cdots, \mathbf{r}_N) \tag{29}$$

such that the two-particle density matrix becomes,

$$\rho_2(\mathbf{r}', \mathbf{r}) = \rho_2^0(\mathbf{r}', \mathbf{r}) + \tilde{\rho}_2(\mathbf{r}', \mathbf{r}) \tag{30}$$

where

$$\begin{aligned} \rho_2^0(\mathbf{r}', \mathbf{r}) &= N(N-1)\int |\phi(\mathbf{r}', \mathbf{r}, \mathbf{r}_3, \cdots, \mathbf{r}_N|^2 \, d^4\mathbf{r}_3 \, d^4\mathbf{r}_4 \cdots d^4\mathbf{r}_N \\ \tilde{\rho}_2(\mathbf{r}', \mathbf{r}) &= N(N-1)\int [\phi^*\tilde{\psi} + \phi\tilde{\psi}^* + \tilde{\psi}\tilde{\psi}^*] \, d^4\mathbf{r}_3 \, d^4\mathbf{r}_4 \cdots d^4\mathbf{r}_N \end{aligned} \tag{31}$$

A factor of 2 is included in ρ_2^0 and $\tilde{\rho}_2$. The symbol $\cdots \int d^4\mathbf{r}_j$ stands for real-space integration and spin summation for the jth particle. The spin-independent one-particle density $\rho_s(\mathbf{r})$ is,

$$\rho_s(\mathbf{r}) = \frac{1}{N-1}\sum_{s'}\int \rho_2^0(\mathbf{r}', \mathbf{r}) \, \mathbf{dr}' \tag{32}$$

As a result,

$$\sum_{s'}\int \tilde{\rho}_2(\mathbf{r}', \mathbf{r}) \, \mathbf{dr}' = 0 \tag{33}$$

The above properties uniquely define the two components $\phi(\mathbf{r}_1, \mathbf{r}_2, \cdots, \mathbf{r}_N)$ and $\tilde{\psi}(\mathbf{r}_1, \mathbf{r}_2, \cdots, \mathbf{r}_N)$ of any eigenstate $\psi(\mathbf{r}_1, \mathbf{r}_2, \cdots, \mathbf{r}_N)$. Now, the variational optimization involving the N-particle Hamiltonian yields the following KS type equation,

$$\left[\frac{1}{2}\nabla^2 + v_{ext}(\mathbf{r}) + v_H(\mathbf{r}) + v_{xc}^s(\mathbf{r})\right] \psi_{is}(\mathbf{r}) = \varepsilon_{is}\psi_{is}(\mathbf{r}) \tag{34}$$

where the three v terms denote external, Hartree and XC potentials respectively,

$$v_{xc}^{s}(\mathbf{r}) = -\sum_{s'} \int \frac{\rho_{s'}(\mathbf{r}')}{|\mathbf{r}-\mathbf{r}'|} \left[f_{s',s}(\mathbf{r}',\mathbf{r}) + \frac{1}{2} \frac{\sum_{s''} \delta f_{s',s''}(\mathbf{r}',\mathbf{r}) \rho_{s''}(\mathbf{r})}{\delta \rho_{s}(\mathbf{r})} \right] d\mathbf{r}'. \tag{35}$$

Here $f_{s's}(\mathbf{r}',\mathbf{r}) = 1 - g_{s's}(\mathbf{r}',\mathbf{r})$ and $g_{s's}(\mathbf{r}',\mathbf{r})$ are the pair correlation functions, defined as $\rho_2^{s's}(\mathbf{r}',\mathbf{r}) = \rho_{s'}(\mathbf{r}')\rho_s(\mathbf{r})g_{s's}(\mathbf{r}',\mathbf{r})$. The method has produced reasonably good agreements with experimental as well as other density functional methods for total ground-state energies of free atoms, ionization and affinity energies, etc. [56]. However, there are significant difficulties as far as practical computations are concerned for general excited states.

A configuration-interaction scheme restricted to single excitations (CIS) has been used in the realm of DFT for electronic excitations [57]. HF orbital energies in the matrix elements of CIS Hamiltonian are then replaced by the corresponding eigenvalues obtained from gradient-corrected KS calculations. Additionally it requires three empirical parameters determined from a reference set to scale the Coulomb integrals and shifting the diagonal CIS matrix elements. Even though this also suffers from a lack of a solid theoretical foundation, resultant excitation energies of molecules obtained by this method show fairly good agreement. This has also been extended to multi-reference CI schemes [58].

Apart from the methods, several other attempts have been made to calculate individual excited states. Some important ones are mentioned below. A time-independent quantal density functional theory (Q-DFT) of singly or multiply excited bound non-degenerate states has been proposed [59]. Existence of a variational KS DFT, with a minimum principle, for the self-consistent determination of an individual excited state energy and density has been established [60]. A perturbative approach is also suggested [61, 62], where the non-interacting KS Hamiltonian serves as the zeroth-order Hamiltonian. Two variants of perturbation theory (PT) were used: (a) the so-called *standard DF PT*, where the zeroth-order Hamiltonian takes the form $H_0 = T + V_{ext} + V_H + V_{xc}$, and the perturbation is given by $H_1 = V_{ee} - V_H - V_{xc}$, (b) *the coupling-constant PT* is based on the adiabatic connection of the Hamiltonian where a link is made between KS Hamiltonian and fully interacting Hamiltonian keeping ground-state density constant, independent of α, such that, $H^{\alpha} = T + V_{ext} + V_H + V_{xc} + \alpha(V_{ee} - V_H - V_x) - V_c^{\alpha}$. Here V_c^{α} is second order in α and equals the correlation potential when $\alpha = 1$. Zeroth-order Hamiltonian is again the KS Hamiltonian; the perturbing Hamiltonian contains a term $H_1^{(1)} = \alpha(V_{ee} - V_H - V_x)$, which is linear in perturbation parameter α and a component $-v_c^{\alpha}$, which contains second and higher order contributions. In another development, accurate calculation of correlation energies of excited states has been proposed via a suitable multi-reference DFT method (such as MC-SCF including the complete active space). They successfully describe the *non-dynamical* correlation; the fraction of dynamic correlation can be taken into account by DFT [63]. Applicability of subspace DFT for atomic excited states has been studied [64]. Theories for individual excited states have been proposed elsewhere as well [65]. A localized HF-based DFT has been put forth for excitation energies of atoms, molecules [66]. This is based on separating the electron-electron interaction energy of KS wave function of a given excited state as Coulomb and exchange energy as follows [67]. The former is given as,

$$U = \sum_{\Gamma,a,\Lambda,b} f_a^\Gamma f_b^\Lambda \times \sum_{\gamma,\lambda} \int d\mathbf{r}' d\mathbf{r}''\, 2\, \frac{\phi_a^{\Gamma,\gamma}(\mathbf{r}')\phi_a^{\Gamma,\gamma}(\mathbf{r}')\phi_b^{\Lambda,\lambda}(\mathbf{r}'')\phi_b^{\Lambda,\lambda}(\mathbf{r}'')}{|\mathbf{r}'-\mathbf{r}''|} = \frac{1}{2}\int d\mathbf{r}' d\mathbf{r}''\, \frac{\bar{\rho}(\mathbf{r}')\bar{\rho}(\mathbf{r}'')}{|\mathbf{r}'-\mathbf{r}''|}. \tag{36}$$

The totally symmetric part $\bar{\rho}(\mathbf{r})$ of electron density being given by,

$$\bar{\rho}(\mathbf{r}) = 2 \sum_{\Gamma,a,\gamma} f_a^\Gamma \phi_a^{\Gamma,\gamma}(\mathbf{r})\phi_a^{\Gamma,\gamma}(\mathbf{r}) \tag{37}$$

Here $\phi_a^{\Gamma,\gamma}(\mathbf{r})$ denotes the spatial part of orbital which belongs to the energy level a of irreducible representation Γ and which transforms under symmetry operations according to the symmetry partner γ of Γ. Occupation number of energy level a of Γ is denoted by f_a^Γ. Summation indices a, b run over all at least partially occupied, i.e., not completely unoccupied levels of Γ. The exchange energy is then the reminder of electron-electron interaction energy of the KS wave function. The corresponding open-shell localized HF exchange potential is then expressed as $v_x^{OSLHF}(\mathbf{r}) = v_x^s(\mathbf{r}) + v_x^c(\mathbf{r})$, where the two terms represent a generalized Slater potential and a correction term respectively.

Recently an optimized effective potential approach and its exchange-only implementation for excited states has been reported [68]. This uses a bifunctional DFT for excited states [60, 69] that employs a simple method of taking orthogonality constraints into account (TOCIA) [70, 71] for solving eigenvalue problems with restrictions. A ΔSCF approach, or ΔKS approach [72], wherein the excitation energy is simply the difference in energy between ground- and excited-state HF or KS calculation, has been employed as well, and found to be especially successful for core-excited states.

So far all the methods we have discussed lie within the purview of *time-independent* DFT. Now we move on to the formalisms for excitation energies within the TDDFT framework. Many excellent articles and reviews are available on the subject [73, 74, 75, 76, 77, 78, 79, 80, 81, 82, 83, 84]; here we mention only the essential details. Consider the unperturbed, ground-state of a many-electron system characterized by an external potential $v_0(\mathbf{r})$, subject to a TD perturbation $v_1(\mathbf{r},t)$, such that at a later time, the external potential (a functional of the TD density), is given by $v_{ext}(\mathbf{r},t) = v_0(\mathbf{r}) + v_1(\mathbf{r},t)$. The density-density response function takes the form,

$$\chi(\mathbf{r},t;\mathbf{r}',t') = \left.\frac{\delta\rho[v_{ext}](\mathbf{r},t)}{\delta v_{ext}(\mathbf{r}',t')}\right|_{v[\rho_0]} \tag{38}$$

where the functional derivative needs to be evaluated at the external potential corresponding to an unperturbed ground-state density ρ_0. The first-order, linear density response to the perturbation $v_1(\mathbf{r},t)$ is then given by,

$$\rho_1(\mathbf{r},t) = \int dt' \int d\mathbf{r}' \chi(\mathbf{r},t;\mathbf{r}',t')\, v_1(\mathbf{r}',t'). \tag{39}$$

Now, realizing that the Runge-Gross theorem also holds for non-interacting particles moving in an external potential $v_s(\mathbf{r},t)$, one can write the KS response function of a non-interacting, unperturbed many-electron density ρ_0 as,

$$\chi_s(\mathbf{r},t;\mathbf{r}',t') = \left.\frac{\delta\rho[v_s](\mathbf{r},t)}{\delta v_s(\mathbf{r}',t')}\right|_{v_s[\rho_0]}. \tag{40}$$

With this definition, $\chi_s(\mathbf{r},t;\mathbf{r}',t')$ is expressed in terms of static KS orbitals $\{\phi_k\}$,

$$\chi_s(\mathbf{r},\mathbf{r}';\omega) = \sum_{j,k}(f_k - f_j)\frac{\phi_k^*(\mathbf{r})\phi_j(\mathbf{r})\phi_j^*(\mathbf{r}')\phi_k(\mathbf{r}')}{\omega-(\varepsilon_j-\varepsilon_k)+i\delta}. \tag{41}$$

Here f_k, f_j denote occupation numbers of KS orbitals; $\varepsilon_j, \varepsilon_k$ signify KS orbital energies; and ω is the frequency obtained after applying a Fourier transform with respect to time. Summation includes both occupied and unoccupied orbitals, plus the continuum states. Now, the first-order density change $\rho_1(\mathbf{r},t)$ in terms of linear response of the non-interacting system to the effective perturbation $v_{s,1}(\mathbf{r},t)$ can be written in terms of frequency ω as,

$$\begin{aligned}\rho_1(\mathbf{r},\omega) &= \int \chi_s(\mathbf{r},\mathbf{r}';\omega)\, v_1(\mathbf{r}',\omega)\, d\mathbf{r}' + \\ &\int\!\!\int \chi_s(\mathbf{r},\mathbf{r}';\omega) \times \left(\frac{1}{|\mathbf{r}'-\mathbf{r}''|} + f_{xc}[\rho_0](\mathbf{r}',\mathbf{r}'';\omega)\right)\rho_1(\mathbf{r}'',\omega)\, d\mathbf{r}'\, d\mathbf{r}''.\end{aligned} \tag{42}$$

It is established that the frequency-dependent linear response of a finite interacting system has discrete poles at the true excitation energies $\Omega_m = E_m - E_0$ of an unperturbed system. So the idea is to calculate the shifts in KS orbital energy differences $\omega_{jk} = \varepsilon_j - \varepsilon_k$, which are poles of the KS response function. True excitation energies (Ω) are generally *not* identical with the KS excitation energies ω_{jk}. The exact density response ρ_1, however, has poles at true excitation energies $\omega = \Omega$. True excitation energies can then be described by those frequencies where the eigenvalues $\lambda(\omega)$ of the following equation,

$$\int d\mathbf{r} \int d\mathbf{r}' \chi_s(\mathbf{r}'',\mathbf{r};\omega)\left(\frac{1}{|\mathbf{r}-\mathbf{r}'|} + f_{xc}[\rho_0](\mathbf{r},\mathbf{r}';\omega)\right)\xi(\mathbf{r}',\omega) = \lambda(\omega)\,\xi(\mathbf{r}',\omega) \tag{43}$$

satisfy $\lambda(\Omega) = 1$. For practical purposes, one needs to expand Ω about one particular KS energy difference $\omega_\nu = \omega_{jk}$:

$$\begin{aligned}\chi_s(\mathbf{r}'',\mathbf{r};\omega) &= 2\alpha_\nu \frac{\Phi_\nu(\mathbf{r}'')\Phi_\nu^*(\mathbf{r})}{\omega-\omega_\nu} + 2\sum_{k\neq\nu}\alpha_k \frac{\Phi_k(\mathbf{r}'')\Phi_k^*(\mathbf{r})}{\omega_\nu-\omega_k+i\delta} + \cdots \\ f_{xc}[\rho_0](\mathbf{r},\mathbf{r}';\omega) &= f_{xc}[\rho_0](\mathbf{r},\mathbf{r}';\omega_\nu)\ \left.\frac{df_{xc}[\rho_0](\mathbf{r},\mathbf{r}';\omega)}{d\omega}\right|_{\omega_\nu}(\omega-\omega_\nu) + \cdots \\ \xi(\mathbf{r}'',\omega) &= \xi(\mathbf{r}'',\omega_\nu) + \left.\frac{d\xi(\mathbf{r}'',\omega)}{d\omega}\right|_{\omega_\nu}(\omega-\omega_\nu) + \cdots \\ \lambda(\omega) &= \frac{A(\omega_\nu)}{\omega-\omega_\nu} + B(\omega_\nu) + \cdots\end{aligned} \tag{44}$$

The index $\nu = (j,k)$ denotes a contraction implying a single-particle transition ($k \rightarrow j$), i.e., $\Phi_\nu(\mathbf{r}) = \Phi_k^*(\mathbf{r})\Phi_j(\mathbf{r})$ and $\alpha_\nu = n_k - n_j$. Assuming that the true excitation energy is not too far away from ω_ν and inserting Laurent expansions for $\chi_s, f_{xc}, \xi, \lambda$ into the above expressions, one finds that,

$$\begin{aligned}A(\omega_\nu) &= M_{\nu\nu}(\omega_\nu) \\ B(\omega_\nu) &= \left.\frac{dM_{\nu\nu}}{d\omega}\right|_{\omega_\nu} + \frac{1}{M_{\nu\nu}(\omega_\nu)}\sum_{k\neq\nu}\frac{M_{\nu k}(\omega_\nu)M_{k\nu}(\omega_\nu)}{\omega_\nu-\omega_k+i\delta}.\end{aligned} \tag{45}$$

where the matrix elements are given by,

$$M_{k\nu}(\omega_\nu) = 2\alpha_\nu \int\int \phi_k^*(\mathbf{r}) \left(\frac{1}{|\mathbf{r}-\mathbf{r}'|} + f_{xc}(\mathbf{r},\mathbf{r}';\omega) \right) \phi_\nu(\mathbf{r}')d\mathbf{r}d\mathbf{r}'. \quad (46)$$

So, the condition $\lambda(\Omega) = 1$ and its complex conjugate then, leads to, in lowest order,

$$\Omega = \omega_\nu + \Re M_{\nu\nu} \quad (47)$$

Just like the time-independent case, now one has to approximate the TD XC potential. The simplest construction is the adiabatic approximation, which makes use of ground-state XC potential, but replaces ground-state density $\rho_0(\mathbf{r})$ with the instantaneous TD density $\rho(\mathbf{r},t)$,

$$v_{xc}^{ad}([\rho];\mathbf{r},t) = \left. \frac{\delta E_{xc}[\rho_0(\mathbf{r})]}{\delta \rho_0(\mathbf{r})} \right|_{\rho_0(\mathbf{r})=\rho(\mathbf{r},t)} \quad (48)$$

Within the adiabatic approximation, the XC kernel can be calculated from,

$$f_{xc}^{ad}(\mathbf{r},t;\mathbf{r}',t') \equiv \frac{\delta v_{xc}([\rho_0];\mathbf{r})}{\delta \rho_0(\mathbf{r}')} \delta(t-t') \quad (49)$$

The kernel above is *local* in time, but *not necessarily local* in space. Clearly, this approximation completely neglects the frequency dependence arising from the XC vector potential; consequently the retardation and dissipation effects are completely ignored in this picture. This has been widely used for single-particle excitation energies (see, for example, [85, 86, 87, 88], for some recent work) with good success, although it performs rather poorly for multiple excitations and charge-transfer states. For explicit functionals of density, it is straightforward to calculate XC kernel. However, for orbital-dependent, such as meta-generalized gradient approximated (GGA) or hybrid functionals, it is not so and may be evaluated with the help of optimized effective potential or other simple, accurate approach.

In practice, modern TDDFT excitation energies ω and corresponding response functions $\mathbf{X}$, $\mathbf{Y}$ are generally obtained by solving a non-Hermitian eigenvalue equation,

$$\begin{pmatrix} \mathbf{A} & \mathbf{B} \\ \mathbf{B} & \mathbf{A} \end{pmatrix} \begin{pmatrix} \mathbf{X} \\ \mathbf{Y} \end{pmatrix} = \omega \begin{pmatrix} \mathbf{1} & \mathbf{0} \\ \mathbf{0} & \mathbf{-1} \end{pmatrix} \begin{pmatrix} \mathbf{X} \\ \mathbf{Y} \end{pmatrix} \quad (50)$$

Here **X,Y** are the excitation vectors representing excitation, deexcitation components of electronic density change, whereas the elements of **A,B** are given by,

$$A_{ai\sigma,bj\sigma'} = \delta_{ab}\delta_{ij}\delta_{\sigma\sigma'}(\varepsilon_{a\sigma} - \varepsilon_{i\sigma'}) + K_{ai\sigma,bj\sigma'}, \qquad B_{ai\sigma,bj\sigma'} = K_{ai\sigma,jb\sigma'} \quad (51)$$

where σ, σ' denote spin indices, $\varepsilon_{p\sigma}$ is the pth KS molecular orbital energy. Indices $i, j, \cdots$ and $a, b, \cdots$ correspond to occupied, virtual orbitals. Matrix element $K_{ai\sigma,bj\sigma'}$ is given by,

$$K_{pq\sigma,rs\sigma'} = (pq\sigma|rs\sigma') - c_x\delta_{\sigma\sigma'}(pr\sigma|qs\sigma') + f_{pq\sigma rs\sigma'}^{xc}. \quad (52)$$

Here $p, q, \cdots$ indicate general MOs and $(pq\sigma|rs\sigma')$ identifies a two-electron repulsion integral in the Mulliken notation, whereas c_x is a mixing parameter of HF exchange integral in case of hybrid functionals. The last term, $f^{xc}_{pq\sigma rs\sigma'}$ represents a Hessian matrix element of the XC energy functional E_{xc} in terms of density, in the adiabatic approximation,

$$f^{xc}_{\sigma\sigma'} = \frac{\delta^2 E_{xc}}{\delta\rho_\sigma(\mathbf{r}_1)\delta\rho_{\sigma'}(\mathbf{r}_2)}. \tag{53}$$

Finally note that, if the orbitals are assumed real, following matrices can be defined,

$$\mathbf{F} = (\mathbf{A}-\mathbf{B})^{-1/2}(\mathbf{X}+\mathbf{Y}), \quad \Omega = (\mathbf{A}-\mathbf{B})^{1/2}(\mathbf{A}+\mathbf{B})(\mathbf{A}-\mathbf{B})^{1/2}, \tag{54}$$

to express the problem in compact form,

$$\Omega\mathbf{F} = \omega^2\mathbf{F}. \tag{55}$$

Although, in general, the non-adiabatic correction is needed for even in low-frequency limit, it has been demonstrated that, at least for smaller systems, the largest source of error for accurate excitation energies, arises from the approximation to static XC potential. This justifies validity and wellness of adiabatic approximation for low-lying excitations in atoms, molecules. Applicability and performance of other functionals such as adiabatic, non-empirical meta-GGA as well other adiabatic hybrid functionals have been reported lately [89]. XC functionals with varying fractions of HF exchange [90, 91], self-interaction correction [92] or other combinations [93], "short-range" corrected functionals [94], methods such as CIS(D) which use exact exchange [95] have been suggested for better representation.

Numerous other variations of TDDFT for excited states have been proposed. In a resolution of identity (RI-J) approach to analytical TDDFT excited-state gradients [96], classical Coulomb energy and its derivatives are computed in an accelerated manner by expanding the density in an auxiliary basis. The Lagrangian of the excitation energy is derived, which is stationary with respect to all electronic degrees of freedom. Now the excited-state first-order properties are conveniently obtained because the Hellmann-Feynmann theorem holds. A state-specific scheme for TDDFT based on Davidson algorithm has been developed [97] to reduce the rank of response matrix and efficient memory use, without loss of accuracy. In another work [98], two-body fragment MO method (FMO2) was combined with TDDFT by dividing the system into fragments and electron density of each of these latter being determined self-consistently. In another work [99], the excitation spectrum was calculated by means of Tamm-Dancoff approximation and the spin-flip formalism [100, 101]. A double-hybrid DFT for excited states [102] is also available, where a mixing of GGA XC with HF exchange and a perturbative second-order correlation part (obtained from KS GGA orbitals and eigenvalues) is advocated. TDDFT within the Tamm-Dancoff approximation is also implemented using a pseudospectral method to evaluate the two-electron repulsion integrals [103]. On a separate work, a subspace formulation of TDDFT within the frozen-density embedding framework has been presented [104]. This allows to incorporate the couplings between electronic transitions on different subsystems which becomes very important in aggregates composed of several similar chromophores, e.g., in biological or biomimetic light-harvesting systems. An occupation number averaging scheme [105] for TDDFT response theory in frequency domain has been prescribed lately, where an average of excitation energies over the occupation number is adopted; this leads to equations of non-symmetric matrix form. Another work [106] combines generalized orbital expansion operators (designed to generate excited states having well-defined multiplicities) and the non-collinear formulation of DFT, for treatment of excited states.

3. The Work-Function Route to Excited States

In this section, we present a simple DFT method for ground and arbitrary excited states of an atom. This has been tremendously successful for many excited states of many-electron atoms. The approach is simple, computationally efficient, and has been overwhelmingly successful for an enormous number of atomic states, such as singly, doubly, triply excited states; low-, moderately high- and high-lying Rydberg states; valence as well as core excitation; autoionizing resonances and satellite states; negative atoms as well. This is within a time-independent framework and results have been presented in the references [21, 22, 23, 24, 25, 26, 27, 28, 29, 30, 31, 32, 33].

In this approach, a physical understanding of KS theory via quantum mechanical interpretation of electron-electron interaction energy functional, $E_{ee}^{KS}[\rho]$, and its functional derivative (potential), $v_{ee}^{KS}(\mathbf{r}) = \delta E_{ee}^{KS}[\rho]/\delta\rho(\mathbf{r})$, is established in terms of fields arising from source charge distributions (quantum mechanical expectations of Hermitian operators). Further, clear provisions are made to distinguish Pauli-Coulomb correlation (due to Pauli exclusion principle and Coulomb repulsion) and kinetic energy correlation components of the energy functional and potential; each components arises from a separate field and source charge distribution [107, 108, 109, 110]. It may be recalled that $E_{ee}^{KS}[\rho]$ in KS theory represents Pauli and Coulomb correlations as well as correlation contributions to the kinetic energy. The corresponding local potential $v_{ee}^{KS}(\mathbf{r})$ (obtained as a functional derivative), consists of two separate contributions: (i) a *purely* quantum mechanical (Pauli and Coulomb) electron-electron correlation component $W_{ee}(\mathbf{r})$, and (ii) a correlation kinetic energy component $W_{t_c}(\mathbf{r})$.

The interaction potential, $v_{ee}^{KS}(\mathbf{r})$, is defined as the work done to bring an electron from infinity to its position ar $\mathbf{r}$ against a field $\mathcal{F}(\mathbf{r})$,

$$v_{ee}^{KS}(\mathbf{r}) = \frac{\delta E_{ee}^{KS}[\rho]}{\delta\rho(\mathbf{r})} = -\int_{\infty}^{\mathbf{r}} \mathcal{F}(\mathbf{r}') \cdot d\mathbf{l}'. \tag{56}$$

The field is a sum of two fields: $\mathcal{F}(\mathbf{r}) = \mathcal{E}_{ee}(\mathbf{r}) + Z_{t_c}(\mathbf{r})$. The first term originates from Pauli and Coulomb correlations as its quantum mechanical source-charge distribution is the pair-correlation density $g(\mathbf{r},\mathbf{r}')$, while the second terms arises from a kinetic energy-density tensor $t_{\alpha\beta}(\mathbf{r})$. The latter accounts for the difference of the fields derived from tensor for the interacting and KS non-interacting system. The electron-electron interaction potential, $v_{ee}^{KS}(\mathbf{r})$, is expressed as a sum: $v_{ee}^{KS}(\mathbf{r}) = W_{ee}(\mathbf{r}) + W_{t_c}(\mathbf{r})$, where,

$$W_{ee}(\mathbf{r}) = -\int_{\infty}^{\mathbf{r}} \mathcal{E}_{ee}(\mathbf{r}') \cdot d\mathbf{l}', \qquad W_{t_c}(\mathbf{r}) = -\int_{\infty}^{\mathbf{r}} Z_{t_c}(\mathbf{r}') \cdot d\mathbf{l}'. \tag{57}$$

The functional derivative in Eq. (56) can be identified as the work done due to the fact that $\nabla v_{ee}^{KS}(\mathbf{r}) = -\mathcal{F}(\mathbf{r})$, so that the sum of two works $W_{ee}(\mathbf{r})$ and $W_{t_c}(\mathbf{r})$ is path-independent. The latter is rigorously valid provided the field $\mathcal{F}(\mathbf{r})$ is smooth, i.e., it is continuous, differentiable and has continuous first derivative. It is also implicit that curl of the field vanishes, i.e., $\nabla \times \mathcal{F}(\mathbf{r}) = 0$. Furthermore, for certain systems, such as closed-shell atoms, jellium metal clusters, jellium metal surfaces, open-shell atoms in central-field approximation, etc., the work done $W_{ee}(\mathbf{r})$ and $W_{t_c}(\mathbf{r})$ are separately path independent, i.e., $\nabla \times \mathcal{E}_{ee}(\mathbf{r}) = \nabla \times Z_{t_c}(\mathbf{r}) = 0$.

Now, it is known that the pair-correlation density $g(\mathbf{r},\mathbf{r}')$ is not a static, but rather a *dynamic* charge distribution, whose structure varies as a function of electron position. This dynamic nature is incorporated into the definition of local potential (through the force field $\mathcal{E}_{ee}(\mathbf{r})$) in which electrons move, via Coulomb's law as,

$$\mathcal{E}_{ee}(\mathbf{r}) = \int \frac{g(\mathbf{r},\mathbf{r}')(\mathbf{r}-\mathbf{r}')}{|\mathbf{r}-\mathbf{r}'|^3}\, d\mathbf{r}'. \tag{58}$$

So, one can define the component $W_{ee}(\mathbf{r})$ as work done to bring an electron from infinity to its position at $\mathbf{r}$ against this force field, as given in Eq. (57). However, this can be further simplified by recognizing that pair-correlation density, $g(\mathbf{r},\mathbf{r}')$ can be expressed as a sum of density $\rho(\mathbf{r}')$ and Fermi-Coulomb hole charge density $\rho_{xc}(\mathbf{r},\mathbf{r}')$: $g(\mathbf{r},\mathbf{r}') = \rho(\mathbf{r}') + \rho_{xc}(\mathbf{r},\mathbf{r}')$. The field $\mathcal{E}_{ee}(\mathbf{r})$ is then constituted of two fields, namely, the Hartree ($\mathcal{E}_H(\mathbf{r})$) and XC ($\mathcal{E}_{xc}(\mathbf{r})$) fields as: $\mathcal{E}_{ee}(\mathbf{r}) = \mathcal{E}_H(\mathbf{r}) + \mathcal{E}_{xc}(\mathbf{r})$. These fields are defined again as:

$$\mathcal{E}_H(\mathbf{r}) = \int \frac{\rho(\mathbf{r}')(\mathbf{r}-\mathbf{r}')}{|\mathbf{r}-\mathbf{r}'|^3}\, d\mathbf{r}', \qquad \mathcal{E}_{xc}(\mathbf{r}) = \int \frac{\rho_{xc}(\mathbf{r},\mathbf{r}')(\mathbf{r}-\mathbf{r}')}{|\mathbf{r}-\mathbf{r}'|^3}\, d\mathbf{r}'. \tag{59}$$

The component $W_{ee}(\mathbf{r})$ is a sum of works $W_H(\mathbf{r})$ and $W_{xc}(\mathbf{r})$, done to move an electron in the corresponding Hartree and XC fields as $W_{ee}(\mathbf{r}) = W_H(\mathbf{r}) + W_{xc}(\mathbf{r})$, with,

$$W_H(\mathbf{r}) = -\int_{\infty}^{\mathbf{r}} \mathcal{E}_H(\mathbf{r}')\cdot d\mathbf{l}', \qquad W_{xc}(\mathbf{r}) = -\int_{\infty}^{\mathbf{r}} \mathcal{E}_{xc}(\mathbf{r}')\cdot d\mathbf{l}'. \tag{60}$$

The work $W_H(\mathbf{r})$ is path-independent, $\nabla \times \mathcal{E}_H(\mathbf{r}) = 0$, and also it is recognized as the Hartree potential $v_H(\mathbf{r})$ in DFT. The functional derivative of Coulomb self-energy functional $E_H[\rho]$ can be physically interpreted as work done in the field of electron density. The component $W_{ee}(\mathbf{r})$ is then given as a sum of Hartree potential and the work done to move an electron in the field of quantum mechanical Fermi-hole charge distribution: $W_{ee}(\mathbf{r}) = v_H(\mathbf{r}) + W_{xc}(\mathbf{r})$. The latter is path independent for symmetrical density systems as mentioned previously, since $\nabla \times \mathcal{E}_{xc}(\mathbf{r}) = 0$ in all those cases. However, note that $\rho_{xc}(\mathbf{r},\mathbf{r}')$ that gives rise to the field $\mathcal{E}_{xc}(\mathbf{r})$ need not possess the same symmetry for arbitrary electron position.

Finally, the KS electron-interaction energy $E_{ee}^{KS}[\rho]$ can also be expressed in terms of above fields (and hence source charge distribution) as follows. The quantum mechanical electron-interaction energy is,

$$E_{ee}[\rho] = \int d\mathbf{r}\, \rho(\mathbf{r})\mathbf{r}\cdot \mathcal{E}_{ee}(\mathbf{r}), \tag{61}$$

which can be further reduced to its Coulomb self-energy and XC components,

$$E_H[\rho] = \int d\mathbf{r}\, \rho(\mathbf{r})\mathbf{r}\cdot \mathcal{E}_H(\mathbf{r}), \qquad E_{xc}[\rho] = \int d\mathbf{r}\, \rho(\mathbf{r})\mathbf{r}\cdot \mathcal{E}_{xc}(\mathbf{r}), \tag{62}$$

and the correlation-kinetic energy component is,

$$T_c[\rho] = \frac{1}{2}\int d\mathbf{r}\ \rho(\mathbf{r})\mathbf{r}\cdot Z_{t_c}(\mathbf{r}). \tag{63}$$

Such a description for XC potential in terms of the source charge distribution, gives a hope of writing KS equation of an interacting many-electron system which could, in principle, be applicable for both ground and excited states. Because, this procedure leads to a *universal* prescription, independent of any state, as it does not have a definite functional form; it is completely and *uniquely* determined by the electronic configuration of a particular state in question. Hence the applicability for ground as well as excited states; the same equation gives it all. Although the present method falls within the spirit of exchange energy as defined in Slater's theory via Fermi-hole charge distribution, it is expected to offer improvement over the Hartree-Fock-Slater (equivalent to LDA method in DFT) theory, because current scheme accounts for the dynamic nature of charge distribution. This definition gives the expected falling off $(1/r)$ of exchange potential at large r. Since at $r \rightarrow \infty$, the Coulomb-hole contributions to the interaction in Eq. (62) is already zero, this implies that current method should give almost *exact* results in the asymptotic region.

Now, we proceed to some details of the actual numerical implementation. Note that the work $v_x(\mathbf{r})$ against the force field due to a Fermi-hole charge can be determined exactly since the latter is known explicitly in terms of orbitals as,

$$\rho_x(\mathbf{r},\mathbf{r}') = -\frac{|\gamma(\mathbf{r},\mathbf{r}')|^2}{2\rho(\mathbf{r})}, \qquad \gamma(\mathbf{r},\mathbf{r}') = \sum_i \phi_i^*(\mathbf{r})\phi_i(\mathbf{r}'). \tag{64}$$

Here the terms have following meaning. $\gamma(\mathbf{r},\mathbf{r}')$ refers to the single-particle density matrix spherically averaged over electronic coordinates for a given orbital angular quantum number, $\phi_i(\mathbf{r}) = R_{nl}(r)\ Y_{lm}(\Omega)$ signifies the single-particle orbital, and $\rho(\mathbf{r})$ is the total electron density expressed in terms of occupied orbitals, $\rho(\mathbf{r}) = \sum_i |\phi_i(\mathbf{r})|^2$. For spherically symmetric systems, exchange part in Eq. (59) can be simplified as,

$$\mathcal{E}_{x,r}(r) = -\frac{1}{4\pi}\int \rho_x(\mathbf{r},\mathbf{r}')\ \frac{\partial}{\partial r}\ \frac{1}{|\mathbf{r}-\mathbf{r}'|}\ d\mathbf{r}' d\Omega_r. \tag{65}$$

Now one can use the well-known expansion,

$$\frac{1}{|\mathbf{r}-\mathbf{r}'|} = 4\pi \sum_{l'',m''} \frac{1}{2l''+1} Y^*_{l''m''}(\Omega)\ Y_{l''m''}(\Omega')\ \frac{r_<^{l''}}{r_>^{l''+1}}, \tag{66}$$

to obtain

$$\begin{aligned}\mathcal{E}_{x,r}(r) &= \frac{1}{2\pi\rho(r)}\int \sum_{n,l,m,n',l',m',l''} R_{nl}(r)R_{nl}(r')R_{n'l'}(r)R_{n'l'}(r')\left[\frac{\partial}{\partial r}\frac{r_<^{l''}}{r_>^{l''+1}}\right] \\ & r'^2 dr' \frac{(2l+1)}{(2l'+1)} \times C^2(ll''l';m,m'-m,m')C^2(ll''l';000),\end{aligned} \tag{67}$$

where $R_{nl}(r)$ denotes radial part of the single-particle orbitals and C's are the Clebsch-Gordan coefficients [111]. Now the exchange integral in Eq. (60) can be written as an integral over radial coordinates only,

$$v_x(r) = -\int_{\infty}^{r} \mathcal{E}_{x,r}(r')\,dr'. \tag{68}$$

While the exchange potential $v_x(\mathbf{r})$ can be accurately calculated through the procedure as delineated above, the correlation potential $v_c(\mathbf{r})$ is unknown and must be approximated for practical calculations. In the present work, two correlation functionals are used, (i) a simple, local, parametrized Wigner-type potential [112] (ii) a slightly more complicated, generalized gradient-corrected correlation energy functional of Lee-Yang-Parr (LYP) [113].

With this choice of $v_x(\mathbf{r})$ and $v_c(\mathbf{r})$, the following KS-type differential equation is solved self-consistently to produce a self-consistent set of orbitals, from which $\rho(\mathbf{r})$ is constructed,

$$\left[-\frac{1}{2}\nabla^2 + v_{es}(\mathbf{r}) + v_{xc}(\mathbf{r})\right]\phi_i(\mathbf{r}) = \varepsilon_i\phi_i(\mathbf{r}), \tag{69}$$

where $v_{es}(\mathbf{r})$ represents the usual Hartree electrostatic potential including electron-nuclear attraction and inter-electronic Coulomb repulsion, whereas $v_{xc}(\mathbf{r}) = v_x(\mathbf{r}) + v_c(\mathbf{r})$. Total energy is then obtained as a sum of following terms in the usual manner,

$$T = -\frac{1}{2}\sum_i \int \phi_i^*(\mathbf{r})\nabla^2\phi_i(\mathbf{r})\,d\mathbf{r}, \qquad E_{es} = -Z\int \frac{\rho(\mathbf{r})}{r}\,d\mathbf{r} + \frac{1}{2}\int\int \frac{\rho(\mathbf{r})\rho(\mathbf{r}')}{|\mathbf{r}-\mathbf{r}'|}\,d\mathbf{r}\,d\mathbf{r}'. \tag{70}$$

Two-electron Hartree and exchange energies can be simplified further,

$$\begin{aligned}
E_H &= \frac{1}{2}\sum \int\int R_{nl}^2(r)R_{n',l'}^2(r')\,\frac{r_<^{l''}}{r_>^{l''+1}}\,r^2 r'^2\,dr\,dr' \\
&\quad \times C(ll''l;mom)\,C(ll''l;ooo)\,C(ll''l';m'om')\,C(l'l''l'';ooo) \\
E_x &= \sum(\text{pairs with parallel spin})\int\int R_{nl}(r)R_{n',l'}(r)R_{nl}(r')R_{n',l'}(r')\,\frac{r_<^{l''}}{r_>^{l''+1}}\,r^2 r'^2\,dr\,dr' \\
&\quad \times C^2(ll''l';m,m-m',m')\,C^2(ll''l';ooo)\times\left(\frac{2l+1}{2l'+1}\right)
\end{aligned} \tag{71}$$

Now a few words should be mentioned regarding numerical solution of the KS equation for orbitals. In earlier stages of the development of this method [21, 22, 23, 24, 25, 26, 27, 28, 29], a Numerov-type finite difference (FD) scheme was adopted for the discretization of spatial coordinates. It is, however, well-known that, due to existence of Coulomb singularity at the origin and presence of long-range nature of the Coulomb potential, FD methods require a large number of grid points to achieve decent accuracy even for ground-state calculations. Certainly excited states (especially those higher-lying Rydberg ones) would need much more grid points to properly describe their long tail. Here, we describe the extension of a generalized pseudospectral (GPS) method for *nonuniform* and optimal spatial discretization and solution of KS equation, Eq. (69). This procedure has been demonstrated to be capable of providing high precision solution of eigenvalues and wave functions for a variety of *singular* as well as non-singular potentials, like Hulthen, Yukawa, Spiked harmonic oscillators, logarithmic, Hellmann potentials; very accurate results have also been obtained for static and dynamic calculation in Coulomb singular systems (like atoms, molecules) such as electronic structure, multi-photon processes in strong fields, Rydberg atom spectroscopy and dynamics, etc. [30, 114, 115, 116, 117, 31, 118, 119, 32, 120, 33, 121, 122, 123]. In addition, the GPS method is computationally orders of magnitude faster than the equal-spacing FD methods. In what follows, we briefly outline the GPS procedure appropriate for our present DFT study. General discussion on the approach could be found in additional references [124, 125].

The most important feature of this method is to approximate an *exact* function $f(x)$ defined on the interval $[-1,1]$ by an Nth-order polynomial $f_N(x)$,

$$f(x) \cong f_N(x) = \sum_{j=0}^{N} f(x_j)\; g_j(x), \tag{72}$$

such that the approximation be *exact* at *collocation points* x_j,

$$f_N(x_j) = f(x_j). \tag{73}$$

We chose to employ the Legendre pseudospectral method where $x_0 = -1$, $x_N = 1$, and $x_j (j = 1,\ldots,N-1)$ are determined by roots of first derivative of the Legendre polynomial $P_N(x)$ with respect to x, i.e., $P'_N(x_j) = 0$. In Eq. (72), $g_j(x)$ are the cardinal functions satisfying a unique property $g_j(x_{j'}) = \delta_{j'j}$, and defined by,

$$g_j(x) = -\frac{1}{N(N+1)P_N(x_j)}\;\frac{(1-x^2)\;P'_N(x)}{x-x_j}, \tag{74}$$

The general eigenvalue problem for our radial KS-type equation can now be written as,

$$\hat{H}(r)\psi(r) = E\psi(r), \tag{75}$$

with

$$\hat{H}(r) = -\frac{1}{2}\;\frac{d^2}{dr^2} + V(r), \tag{76}$$

For structure and dynamics calculations this involves Coulomb potential, which typically has a singularity problem at $r = 0$, as well as the long-range $-1/r$ behavior. This usually requires a large number of grid points in the *equal-spacing* finite-difference methods, which are not feasible to extend to Rydberg state calculations. This can be overcome by first mapping the semi-infinite domain $r \in [0,\infty]$ into a finite domain $x \in [-1,1]$ by a mapping transformation $r = r(x)$ and then using the Legendre pseudospectral discretization technique. At this stage, following algebraic nonlinear mapping [126, 127] is used,

$$r = r(x) = L\;\frac{1+x}{1-x+\alpha}, \tag{77}$$

where L and $\alpha = 2L/r_{max}$ are the mapping parameters. Further, introducing,

$$\psi(r(x)) = \sqrt{r'(x)} f(x) \tag{78}$$

and following a symmetrization procedure, a transformed Hamiltonian is obtained as,

$$\hat{H}(x) = -\frac{1}{2}\frac{1}{r'(x)}\frac{d^2}{dx^2}\frac{1}{r'(x)} + V(r(x)) + V_m(x), \tag{79}$$

where

$$V_m(x) = \frac{3(r'')^2 - 2r'''r'}{8(r')^4}. \tag{80}$$

The advantage of this mapping scheme is that this leads to a *symmetric* matrix eigenvalue problem. Note that for the mapping used here, $V_m(x) = 0$. Therefore, discretizing our Hamiltonian by GPS method leads to the following set of coupled equations,

$$\sum_{j=0}^{N}\left[-\frac{1}{2}D^{(2)}_{j'j} + \delta_{j'j}\,V(r(x_j)) + \delta_{j'j}\,V_m(r(x_j))\right]A_j = EA_{j'}, \quad j = 1,\ldots,N-1, \tag{81}$$

$$A_j = r'(x_j)\ f(x_j)\ [P_N(x_j)]^{-1} = \left[r'(x_j)\right]^{1/2}\psi(r(x_j))\ [P_N(x_j)]^{-1}. \tag{82}$$

Here $D^{(2)}_{j'j}$ represents symmetrized second derivative of the cardinal function in respect to r,

$$D^{(2)}_{j'j} = \left[r'(x_{j'})\right]^{-1} d^{(2)}_{j'j}\left[r'(x_j)\right]^{-1}, \tag{83}$$

and

$$\begin{aligned} d^{(2)}_{j',j} &= \frac{1}{r'(x)}\frac{(N+1)(N+2)}{6(1-x_j)^2}\frac{1}{r'(x)}, \quad j = j', \\ &= \frac{1}{r'(x_{j'})}\frac{1}{(x_j - x_{j'})^2}\frac{1}{r'(x_j)}, \quad j \neq j'. \end{aligned} \tag{84}$$

The orbitals $\{\phi_i(\mathbf{r})\}$ obtained from self-consistent solution of KS equation (69) are used to construct various determinants for a given electronic configuration of an atom, which, in turn, could be employed to calculate the associated multiplets related to this configuration. Here we use Slater's diagonal sum rule for the multiplet energies [128]. Similar strategy for multiplets has been adopted earlier [37, 129, 130, 131, 40, 132].

4. Results and Discussion

At first, we give some sample results for singly excited $1s^2ns$ ^2S, $1s^2np$ ^2P states of Li, as well $1s^22sns$ ^3S Be, $1s^22snp$ ^3P states of Be, in Table 1. Note that in this and all other following tables, we present only non-relativistic results; state energies are in atomic units, while excitation energies in eV. For all these calculations, a convergence criteria of 10^{-5} and 10^{-6}, as well a radial grid of 500 points have been used. In the literature generally excitation energies are reported, while individual state energies are given seldom. However, in a DFT study of low-lying singly excited states of some open-shell atoms (B, C, O, F, Na, Mg, Al, Si, P, Cl) [27], excitation energies from X-only and numerical HF methods were found to be in good agreement with each other. Surprisingly, though, the two correlation

Table 1. Comparison of singly excited-state energies of Li and Be (in a.u.) with literature data. Numbers in parentheses denote absolute percentage deviations with respect to reference values. Taken from ref. [30].

State	$-E$(X)	$-E$(XC)	$-E$(Literature)	State	$-E$(X)	$-E$(XC)	$-E$(Literature)
Li				Be			
$1s^23s\ ^2S$	7.30966 7.31021[1]	7.35773 (0.05)	7.35394[2]	$1s^22s3s\ ^3S$	14.37798	14.42917 (0.20)	14.42629[5]
$1s^25s\ ^2S$	7.25996	7.30466 (0.01)	7.30339[2]	$1s^22s5s\ ^3S$	14.30562	14.34996	
$1s^22p\ ^2P$	7.36486 7.36507[1]	7.41204 (0.03)	7.41016[3]	$1s^22s2p\ ^3P$	14.51068 14.51150[4]	14.56660 (0.03)	14.56223[5]
$1s^24p\ ^2P$	7.26859	7.31262 (0.01)	7.31190[3]	$1s^22s4p\ ^3P$	14.31462 14.31464[4]	14.35910	

[1] HF result, ref. [133]. [2] Ref. [134]. [3] Ref. [135]. [4] HF result, ref. [136]. [5] Ref. [137].

energy functionals (Wigner and LYP) did not show any considerable improvements in excitation energies although excited-state energies were dramatically improved. Therefore, in this work, we consider both the state energies and excitation energies. In this table, two sets of calculations are performed; solution of Eq. (69) with (i) $v_{xc} = v_x$ (exchange-only or E(X)) and (ii) $v_x + v_c$ (exchange plus correlation or E(XC)). These states as well the other ones in proceeding tables are of great significance in atomic physics; thus have been studied by both experimentalists and theoreticians by employing a multitude of techniques and formalisms. Some prominent reference values are quoted for comparison, wherever possible. The X-only results are fairly close to the HF values [133, 136], errors ranging from 0.0057% to as low as 0.0001% for Be $1s^22s4p\ ^3P$, indicating the accuracy in our calculation. The doublet S states of Li are compared with the full-core-plus-correlation method with multi-configuration interaction wave functions [134], while doublet P states with a combined configuration-interaction-Hylleraas method [135]. For Be, literature is quite scanty and present density functional results match very closely with the multi-configuration calculations [137]. One finds some overestimation in total energy caused by the LYP correlation functional employed here; errors ranging from 0.052%–0.003%. For a more detailed discussion, see [30].

Next, in Table 2, some even-parity doubly excited states ($ns^2\ ^1S$, $np^2\ ^1D$, n=2–17) of He are presented. Many of these have been identified to be autoionizing in nature, e.g., $ns^2\ ^1S$. It is seen that the calculated energy values have never fallen below the quoted results. In the former case, DFT results are comparable to literature data for smaller n and tends to increase gradually with an increasing n, as evident from the absolute per cent deviations given in the parentheses. This could occur either because of the inadequate description of long-range nature of correlation potential employed or some deficiencies in the work-function formalism itself. Finally we see that while accuracy of doubly excited state calculation is not as good as that of singly excited state, error in the former still remains well within 3.6%. More details on these could be found in ref. [30].

Now, single and double excitation energies of some selected states of He, Be, are displayed in Table 3, along with the reference values. These are estimated with respect to our calculated, non-relativistic ground-state energies of He, Be, i.e., -2.90384 and -14.66749 a.u. (as obtained from the same KS equation (69)). No experimental results could be found

Table 2. Calculated doubly excited-state ($ns^2\ ^1S$, $np^2\ ^1D$) energies of He (in a.u.) along with literature data for comparison. Numbers in parentheses denote absolute percentage errors with respect to literature data. Adopted from ref. [30].

State	$-E$(XC)	$-E$(Literature)	State	$-E$(XC)	$-E$(Literature)
$2s^2\ ^1S$	0.76637 (1.48)	0.77787[1]	$2p^2\ ^1D$	0.69272 (1.31)	0.70195[3]
$3s^2\ ^1S$	0.34578 (2.19)	0.35354[1]	$3p^2\ ^1D$	0.31540 (0.04)	0.31554[3]
$4s^2\ ^1S$	0.19659 (2.19)	0.20099[1]	$4p^2\ ^1D$	0.18095	
$5s^2\ ^1S$	0.12754 (2.12)	0.13030[2]	$5p^2\ ^1D$	0.11610	
$6s^2\ ^1S$	0.08808 (3.05)	0.09085[2]	$6p^2\ ^1D$	0.08115	
$7s^2\ ^1S$	0.06524 (3.35)	0.0675[2]	$7p^2\ ^1D$	0.05980	
$9s^2\ ^1S$	0.03889		$9p^2\ ^1D$	0.03604	
$11s^2\ ^1S$	0.02503		$11p^2\ ^1D$	0.02414	
$13s^2\ ^1S$	0.01811		$13p^2\ ^1D$	0.01728	
$15s^2\ ^1S$	0.01348		$15p^2\ ^1D$	0.01297	
$17s^2\ ^1S$	0.01132		$17p^2\ ^1D$	0.01010	

[1] Ref. [138]. [2] Ref. [137]. [3] Ref. [139].

Table 3. Single and double excitation energies of He and Be (in a.u.) compared with literature data. Numbers in parentheses denote absolute percentage errors with respect to the best theoretical data available. Adopted from ref. [30].

State	Present Work	$\Delta\varepsilon_{KS}$	Other theory	Experiment
Single excitation of He and Be				
He 1s2s ^3S	0.72839 (0.02)	0.7460[1]	0.72850[2]	0.72833[3]
He 1s2s ^1S	0.75759 (0.02)		0.75775[2]	0.75759[3]
He 1s2p ^3P	0.77041 (0.02)	0.7772[1]	0.77056[2]	0.77039[3]
He 1s2p ^1P	0.77971 (0.02)		0.77988[2]	0.77972[3]
He 1s3s ^3S	0.83494 (0.01)	0.8392[1]	0.83504[2]	0.83486[3]
He 1s3s ^1S	0.84231 (0.02)		0.84245[2]	0.84228[3]
He 1s3p ^3P	0.84548 (0.02)	0.8476[1]	0.84564[2]	0.84547[3]
He 1s3p ^1P	0.84841 (0.02)		0.84858[2]	0.84841[3]
Be $1s^22s2p\ ^3P$	0.10089	0.1327[1]		0.100153[3]
Be $1s^22s3s\ ^3S$	0.23832(0.63)	0.2444[1]	0.236823[4]	0.237304[3]
Be $1s^22s4p\ ^3P$	0.30839	0.3046[1]		0.300487[3]
Be $1s^22s5s\ ^3S$	0.31753	0.3153[1]		0.314429[3]
Double excitation of He				
He $2s^2\ ^1S$	2.13747(0.54)		2.1259[5],2.1285[6]	
He $3s^2\ ^1S$	2.55806(0.61)		2.5425[7],2.5496[6]	
He $4s^2\ ^1S$	2.70725(0.48)		2.6942[7],2.7017[6]	
He $5s^2\ ^1S$	2.77630(0.10)		2.7735[7]	
He $6s^2\ ^1S$	2.81576(0.10)		2.8129[7]	
He $7s^2\ ^1S$	2.83860(0.09)		2.8362[8]	
He $2p^2\ ^1D$	2.21120(0.68)		2.1961[8],2.2082[6]	
He $3p^2\ ^1D$	2.58844(1.60)		2.5477[8],2.5595[6]	
He $4p^2\ ^1D$	2.72289(1.08)		2.6938[8]	

[1] Ref. [140]. [2] Ref. [141, 142]. [3] Ref. [143]. [4] Ref. [144]. [5] Ref. [145]. [6] Ref. [146]. [7] Ref. [137]. [8] Ref. [147].

for doubly excited resonances. In some occasions, our calculated excitation energies have fallen below the experimental results. This is not surprising keeping in mind that the present methodology is *non-variational*. As a consequence, the variational restriction on a particular excited state being the lowest of a given space-spin symmetry does not hold good. Here we also report the single-particle KS energies (obtained from the difference of KS eigenvalues) for single excitations in He, Be [140], which, of course, do not show the multiplet separation. The excitation energies from true KS potential for He, Be are clearly quite good. However, those from some other commonly used approximate exchange energy functionals (such as LDA) produce large errors in excitation energy [140]. Also, it may be mentioned here that, for excitation energies in Ne satellites [26], both LDA and one of the most commonly used gradient-corrected exchange functional of Becke [148] (enormously successful for numerous DFT applications) have been found to be absolutely unsuitable for such studies, producing very large errors. Besides, present results are not corrected for relativistic effects, which is included in the experimental values. Single excitations in He show reasonably good agreement with both theory and experiment, while for Be the corresponding discrepancy is somehow larger. Nevertheless, the overall agreement between current results and literature data is quite satisfactory. Apart from the errors in XC potentials as discussed earlier, another possible source could be rooted in the *single-determinantal* nature of our method. Assumption of spherical symmetry in dealing with the exchange potential could also bring some inaccuracies. Stated differently, the solenoidal component of the electric field $\mathcal{E}_x(\mathbf{r})$ may not be negligible compared to the irrotational component for these states, although this usually holds quite well for atoms [110]. As a further check, some representative radial expectation values for singly and doubly excited states of He as well as singly excited states of Li, Be have also been studied [30]. These match quite well with the HF values [149], once again reassuring the accuracy in our calculation.

Now we move on to the triple excitations. For this purpose, we compare our DFT excitation energies for all the eight $2l2l'2l''$ ($n=2$ intrashell) triply excited states, *viz.*, $2s^22p$ $^2P^o$; $2s2p^2$ $^2D^e$, $^4P^e$, $^2P^e$, $^2S^e$; and $2p^3$ $^2D^o$, $^2P^o$, $^4S^o$ of selected members of Li-isoelectronic series, i.e., B^{2+}, N^{4+} and F^{6+} in Table 4. At this stage, it may be appropriate to illustrate the details of a multiplet calculation from individual determinants by an example. For this, we consider all the four multiplets ^2D, ^4P, ^2P, ^2S associated with a $2s2p^2$ configuration. This gives rise to 30 determinants which satisfy following relations (left and right correspond to the multiplet and determinantal energies),

$$\begin{aligned}
^2D &= (0^+1^+1^-) \\
^4P &= (0^+1^+0^+) \\
^2D+^4P+^2P &= (0^+1^+0^-)+(0^+1^-0^+)+(0^-1^+0^+) \\
^2D+^4P+^2P+^2S &= (0^+1^+-1^-)+(0^+1^--1^+)+(0^-1^+-1^+)+(0^+0^+0^-) \quad (85)
\end{aligned}$$

where the numbers denote m_l values while $(+,-)$ m_s values. For more details, see [24].

In this case, we no more report individual state energies as these are very difficult to compare directly; instead only excitation energies are reported. To put our results in proper perspective, all triple excitation energies in this table are computed with respect to the accurate non-relativistic ground state of [150]. No experimental results have been reported as yet in the literature and appropriate theoretical results are quoted here. All

Table 4. Comparison of calculated excitation energies (in eV) of the n=2 intrashell triply excited states of B^{2+}, N^{4+} and F^{6+} relative to the non-relativistic ground states of [150]. GPS signifies present work. See ref. [32] for details.

State	B^{2+}		N^{4+}		F^{6+}	
	GPS	Ref.	GPS	Ref.	GPS	Ref.
$2s^22p\ ^2P^o$	436.588	436.07[1],436.59[2]	894.541	893.93[1],894.12[3]	1514.229	1515.67[1],1514.90[3]
$2s2p^2\ ^4P^e$	436.917	436.69[1],436.89[2]	894.876	894.54[1],894.51[3]	1514.474	1516.33[1],1515.16[3]
$2s2p^2\ ^2D^e$	441.893	441.34[1],442.00[2]	902.655	901.93[1],902.32[3]	1524.898	1526.43[1],1525.86[3]
$2p^3\ ^4S^o$	443.852	443.86[1],444.15[2], 443.63[4]	905.329	905.15[1],905.15[3], 904.43[4]	1528.187	1530.42[1],1529.35[3], 1528.51[4]
$2s2p^2\ ^2S^e$	445.387	445.11[1],445.75[2]	907.930	907.41[1],907.87[3]	1531.822	1533.61[1],1533.18[3]
$2s2p^2\ ^2P^e$	445.814	445.35[1],446.21[2]	908.455	907.99[1],908.59[3]	1532.717	1534.55[1],1534.17[3]
$2p^3\ ^2D^o$	446.173	446.02[1],446.58[2]	909.089	909.02[1],909.37[3]	1533.816	1536.01[1],1535.35[3]
$2p^3\ ^2P^o$	450.088	450.04[1],450.65[2]	915.023	915.00[1],915.47[3]	1541.946	1543.95[1],1543.46[3]

[1] Ref. [151]. [2] Ref. [152]. [3] Ref. [153]. [4] Ref. [154].

these states are autoionizing except the $2p^3\ ^4S^o$, which is bound, metastable against autoionization by conservation of parity and angular momentum. These are studied through a multi-configuration-interaction type formalism within a Rayleigh-Ritz variational principle [154]. Recently a perturbation theory method (1/Z expansion) [151] as well a truncated diagonalization method [152, 153] have been employed to determine the position of all these states. It is gratifying that our current positions for all these 8 states for these 3 ions follow the same orderings as in [151, 152, 153], which clearly demonstrates the reliability in our calculation. All these excitation energies show excellent agreement with the literature data, with a maximum discrepancy of 0.125%; for the three ions the deviation ranges are 0.0005–0.125%, 0.007–0.049% and 0.044–0.099% respectively. Both underestimation and over-estimation is observed in excitation energies.

Table 5. $2s^2ns\ ^2S^e$ and $2s^2np\ ^2P^o$ resonances of Li. State energies and excitation energies, relative to the ground state of [155]. GPS signifies present work. See [31] for details.

n	$\langle A,ns\rangle\ ^2S^e$				$\langle A,np\rangle\ ^2P^o$			
	−E(a.u.)		Excitation energy(eV)		−E(a.u.)		Excitation energy(eV)	
	GPS	Ref.	GPS	Ref.	GPS	Ref.	GPS	Ref.
2					2.2448	2.2503[1],2.247[2], 2.2428[3]	142.385	142.255[1],142.439[3],142.12[4], 142.310[5],142.35[6],142.33[7]
3	1.9871	2.0048[3,8], 2.0102[12]	149.396	148.632[9],148.822[5], 148.788[12],148.914[3]	1.9740	1.9935[10],1.991[2], 1.9879[3]	149.753	149.241[10],149.07[11], 149.222[5],149.374[3]
6	1.9094	1.9165[3]	151.510	150.855[9],151.317[3], 151.025[5]	1.9075	1.9214[10],1.9145[3]	151.562	151.203[10],150.88[11], 151.371[3],151.057[5]
8	1.9004	1.9072[3]	151.755	151.092[9],151.570[3], 151.263[5]	1.8996	1.9068[3]	151.777	151.11[11],151.581[3], 151.285[5]
10	1.8965	1.9037[3]	151.861	151.190[9],151.665[3]	1.8961	1.9034[3]	151.872	151.20[11],151.673[3]
12	1.8944	1.9018[3]	151.918	151.247[9],151.717[3]	1.8943		151.921	
16	1.8925		151.970	151.296[9]	1.8925		151.970	
20	1.8917		151.992	151.318[9]	1.8918		151.989	
22	1.8914		152.000	151.325[9]	1.8916		151.994	

[1] Ref. [156]. [2] Ref. [157]. [3] Ref. [152]. [4] Ref. [158]. [5] Ref. [159]. [6] Ref. [160].
[7] Ref. [161]. [8] Ref. [162]. [9] Ref. [163]. [10] Ref. [164]. [11] Ref. [165]. [12] Ref. [166].

Table 6. Selected $3l3l'nl''$ term energies (in a.u.) and excitation energies (in eV) of Li, relative to the ground state of [155]. Adopted from ref. [31].

State	−E	Exc. energy	State	−E	Exc. energy	State	−E	Exc. energy
$3s^24s\ ^2S^e$	0.90054	178.959	$3p^3\ ^4S^o$	1.00055	176.238	$3p^24s\ ^4P^e$	0.89860	179.012
$3s^26s\ ^2S^e$	0.85729	180.136	$3s3p4s\ ^4P^o$	0.93313	178.072	$3p^26s\ ^4P^e$	0.85744	180.132
$3s^23p\ ^2P^o$	1.01210[1]	175.924[2,3]	$3s3p5s\ ^4P^o$	0.90193	178.921	$3p^24s\ ^2D^e$	0.87282	179.713
$3s^26p\ ^2P^o$	0.85558	180.182	$3s3p5p\ ^4D^e$	0.89924	178.994	$3p^24p\ ^4D^o$	0.89642	179.071
$3s^23d\ ^2D^e$	0.97108	177.040	$3s3p6p\ ^4D^e$	0.88767	179.309	$3p^25p\ ^4D^o$	0.86846	179.832
$3s^26d\ ^2D^e$	0.85352	180.238	$3s3p5d\ ^4F^o$	0.89400	179.137	$3p^25p\ ^2F^o$	0.84343	180.513
$3s3p^2\ ^4P^e$	1.02288[4]	175.630	$3s3p6d\ ^4F^o$	0.88517	179.377	$3p^26p\ ^2F^o$	0.83225	180.817

Now we present results for triply excited hollow resonances, $2l2l'nl''(n \geq 2)$ in Li. In [31], 12 such resonance series, *viz.*, $2s^2ns\ ^2S^e$, $2s^2np\ ^2P^o$, $2s^2nd\ ^2D^e$, $2p^2ns\ ^2D^e, ^4P^e$, $2s2pns\ ^4P^o$, $2s2pnp\ ^4D^e$, $2p^2np\ ^2F^o, ^4D^o$, $2p^2nd\ ^2G^e$, $^4F^e$ and $2s2pnd\ ^4F^o$, covering a total of about 270 low-, moderately high- and high-lying states (with n as high as 25) were studied in some detail. These represent the model case of a highly correlated, multi-excited three electron system in presence of a nucleus, and hence a four-body Coulombic problem. These are often termed as *hollow* states, as all three electrons reside in higher shells leaving the K shell empty. They have many fascinating properties, as well are very difficult for both theory and experiments. For example, these are difficult to produce from ground state by single photon absorption or electron impact excitation; also they have proximity to more than one thresholds; moreover there are infinite open channels associated with these resonance. Table V gives some representative state energies and excitation energies of even-parity $2s^2ns\ ^2S^e$ and odd-parity $2s^2np\ ^2P^o$ resonances in Li (n=2–22). The latter is calculated relative to the accurate ground state of Li, using full core plus correlation within a multi-configuration interaction wave function [155]. The reference energy value -7.47805953 a.u. is to be compared with our present value of -7.4782839 a.u. In literature, these states are conveniently classified using our independent particle model classification [170, 152] where the six core Li^+ n=2 intrashell doubly excited states, *viz.*, $2s^2\ ^1S^e$, $2s2p\ ^3P^o$, $2p^2\ ^3P^e$, $2p^2\ ^1D^e$, $2s2p\ ^1P^o$ and $2p^2\ ^1S^e$ are denoted by A, B, C, D, E and F respectively. For the former, no experimental results are available as yet. Lower members of the former series have been studied in considerable detail by a variety of techniques, such as a hyper-spherical coordinate method [162], a combination of saddle point and complex coordinate rotation [166]. Of late, an eigenphase derivative technique in conjunction with a quantum defect theory [163] reported the low and high resonances up to $n = 22$, whereas the same up to $n = 12$ were done by a truncated diagonalization method [152]. Our DFT results are in good agreement with these references; state energies lie about 0.36–0.88% above [152], whereas the excitation energies are higher by 0.41–0.51% from those of [163]. The $\langle A, ns \rangle\ ^2P^o$ resonances are the most widely studied series in Li, both theoretically and experimentally. Position of the lowest state in this series has been experimentally measured at 142.33 eV [161], 142.35 eV [160]. These are generally supported by theoretical calculations, e.g., combined saddle-point and complex coordinate rotation approach [156], a complex scaling method having correlated basis functions constructed from B-splines [157], an R-matrix theory [159], etc. Present excitation energy is only 0.02% above the experimental results. Other members of the series with $n = 3 - 7$ are in reasonably good agreement with complex coordinate rotation calculations [164]. Our state energies are underestimated by 0.24–0.98% with respect to those of [156, 164], leading to higher excitation energies (deviations with respect to [159, 165] being only 0.05–0.32% and 0.43–0.46% for $n = 2-9$ and $n = 3-10$ respectively). For further discussion on these and other hollow resonances see [31].

Table 6 extends the method for some higher lying triply excited hollow resonances of Li having both K and L shells empty, the so-called *doubly hollow* states, *viz.*, $3l3l'nl''$(3≤n≤6) ($^2S^e$, $^2P^o$, $^2D^e$, $^2F^o$, $^4S^o$, $^4P^{e,o}$, $^4D^{e,o}$, $^4F^o$). For $2l2l'nl''$ resonances, several accurate, reliable experimental and theoretical results are available; however, the same for $3l3l'nl''$ resonances, are very limited mainly because of the greater challenges encountered. These have very distinctive features: (a) they are weak (by about an order of magnitude compared to the lower hollow states), broad and having much larger widths [169]. The principal difficulties with these higher hollow states at larger photon energies are mainly due to a very rapid increase in the density of possible triply excited and other lower states of same symmetry, as well as of the number of open channels available, giving rise to very strong and quite complicated electron correlation effects. Nevertheless some attempts have been made, which are quoted here. The energies and decay rates of N^{4+} and N^{2+} $3l3l'3l''$ have been studied using a CI approach [171]. Positions and widths of N^{4+} (3,3,3) $^2S^{e,o}$ states are investigated by a space partition as well as a stabilization procedure both of which use the L^2 discretization [172]. A large scale state specific theory calculations for 11 n=3 resonances of He^- has been suggested [173]. Critical issues in the theory and computation of the lowest three n=3 intrashell states, *viz.*, $3s^23p$ $^2P^o$, $3s3p^2$ $^4P^e$ and $3s3p^2$ $^2D^e$ of Z=2–7 in the light of state specific theory, has been published [168]. Energies, widths and Auger branching ratios for eight He^- $3l3l'3l''$ states are calculated by complex rotation method [168]. A semi-quantitative analysis of the angular correlation of 64 n=3 intrashell states of a model three-electron atom confined on the surface of a sphere were presented recently [174]. The only result available for such triply photo-excited (3,3,3) KL hollow state for Li are the $3s^23p$ $^2P^o$ and $3s3p^2$ $^4P^e$, both theoretically, whereas only the former experimentally. The former's position has been measured at 175.25 and 175.165 eV by synchrotron radiation measurement [169] and photo-ion spectroscopy [167]. In a saddle-point calculation with R-matrix approximation [167], a 570-term 25 angular component wave function gives an energy of -1.043414 a.u., and position at 174.11 eV. This is in reasonable agreement with the complex rotation calculation of -1.043 a.u., [157], and the state specific result [168] of -1.0409856 a.u., as well as, with the multi-configuration Dirac-Fock [169] excitation energy of 174.14 eV. DFT energy value of -1.01210 a.u., gives its position at 175.940 eV, (about 0.67 eV above the experimental value of [169]) and matches well with the state specific result of 175.15 eV [168]. Calculated $3s3p^2$ $^4P^e$ state energy of -1.02288 a.u., matches closely with the state specific result of -1.0393859 a.u. [168]. Leaving aside a few of those as mentioned above, most of these can not be compared directly due to lack of any reference values and we expect that these results may be useful in future studies of these resonances. Note that our result gives $3s3p^2$ $^4P^e$ as the lowest n=3 resonance rather than the $3s^23p$ $^2P^o$, the former lying 0.0108 a.u., below the latter which coincides with the ordering found in other calculation such as complex rotation for He^- [175] and CI calculation for N^{4+} [171]. However this disagrees with the state specific calculation of [168], where the ordering is reversed and separation for Li being about 0.0016 a.u. Clearly, more accurate calculation with better correlation functionals would be required to achieve such

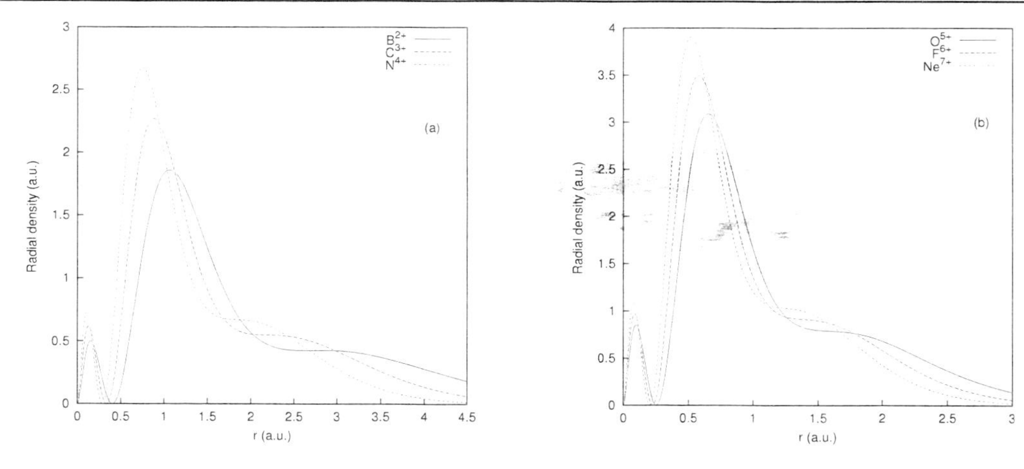

Figure 1. The radial densities (a.u.) of $2s^23s\ ^2S^e$ states for (a) B^{2+}, C^{3+}, N^{4+} and (b) O^{5+}, F^{6+}, Ne^{7+} respectively. Taken from ref. [32].

smaller separations (of the order of 1×10^{-3} a.u.) within this DFT formalism to reach a more authentic conclusion. Now Fig. 1 depicts the radial densities for some representative (a) $2l2l'nl''$ and (b) $3l3l'nl''$ hollow states; as expected, they show the characteristic shell structures (superpositions of orbital radial densities).

Now Table 7 reports DFT energies for ground and excited states of some negative atoms, namely, Li^- and Be^-. For the former, 4 states are considered, *viz.*, [He]$2s^2\ ^1S^e$, $1s2s2p^2\ ^5P^e$, [He]$2p^3\ ^5S^o$, $1s2s2p3p\ ^5P^e$; while for the latter 3 states, i.e, [He]$2s2p^2\ ^4P^e$, [He]$2p^3\ ^4S^o$, $1s2s2p^3\ ^6S^o$. Two sets of energies are reported *viz.*, X-only (non-correlated) and XC (correlated, with LYP functional). Excepting the core-excited even-parity $1s2s2p3p\ ^5P^e$ of Li^- (reported lately), the rest 4 states have been investigated quite extensively. Comparisons with literature data are made, wherever possible. Per cent deviations are given in parentheses; for X-only case these are relative to the lone literature results in column 4; for XC case, these are with respect to the recent variational Monte Carlo (VMC) [176] values except for the 4th state of Li^-, where such a result is unavailable; and this is given in reference to the saddle-point calculation of [177]. Our X-only ground state of Li^- is higher from the accurate HF calculation of [178] by a marginal 0.0004 a.u. The XC energy shows fairly good agreement (slightly above) with the accurate correlated MCHF-*n* expansion considering all expansions [178], as well as VMC method [176]. These seem to be in considerable disagreement with the earlier result of [179]. X-only results for the core-excited high-spin even-parity $^5P^e$ and odd-parity $^5S^o$ states of Li^- also show excellent agreement with the HF energy [176], while XC energies match well with literature values, such as VMC [176], CI [180], variational multi-configuration calculation [181], saddle-point [177], MCHF [182], etc. Note that XC energies for these two states are lower than all of these reference results by 0.171 and 0.137%, respectively giving maximum deviations in our calculation. As already verified, the X-only results are practically of HF quality; hence this overestimation is probably caused by the approximate correlation potential used. Note that while ours is a single determinantal method, some of these correlated calculations are highly elaborate and extensive; for example, in [181], a 45 angular component 1004-term wave function was used, [180] used a 320-term CI and [177] used 7-50 angular spin components with 541–

Table 7. Calculated ground and excited states of Li^-, Be^- along with literature data. Numbers in the parentheses denote absolute per cent deviations. Adopted from [33].

Ion	State	−E(a.u.)			
		X-only		XC	
		This work	Ref.	This work	Ref.
Li^-	$[He]2s^2\ ^1S^e$	7.4278(0.005)	7.4282[1]	7.4984(0.009)	7.4553[2],7.5008[1], 7.4991[3]
	$1s2s2p^2\ ^5P^e$	5.3640(0.006)	5.3643[3]	5.3925(0.171)	5.3866[4,5], 5.3833[3],5.3865[6], 5.3863[11]
	$1s2p^3\ ^5S^o$	5.2223(0.004)	5.2225[3]	5.2608(0.137)	5.2561[4,5], 5.2536[3],5.2560[6], 5.2558[11]
	$1s2s2p3p\ ^5P^e$	5.3289		5.3683(0.007)	5.3679[5]
Be^-	$[He]2s2p^2\ ^4P^e$	14.5078(0.008)	14.5090[7]	14.5806(0.062)	14.5779[8],14.5716[3], 14.5708[7],14.5769[10]
	$[He]2p^3\ ^4S^o$	14.3272(0.002)	14.3275[7]	14.4081(0.049)	14.4063[8],14.4010[3], 14.4002[7]
	$1s2s2p^3\ ^6S^o$	10.4279(0.009)	10.4288[7]	10.4758(0.092)	10.4662[3], 10.4615[7], 10.4711[9]

[1]Ref. [178]. [2]Ref. [179]. [3]Ref. [176]. [4]Ref. [181]. [5]Ref. [177]. [6]Ref. [180].
[7]Ref. [183]. [8]Ref. [184]. [9]Ref. [185]. [10]Ref. [186]. [11]Ref. [182].

1298 linear parameters for the former. Nevertheless, as clearly seen, our results exhibit a rather small discrepancy from these references. Even-parity $^5P^e$ state has been considered rather lately [177]. Current energy value shows very good matching with the reference, where the authors used a saddle-point restricted variational method with accurate multi-configurational wave functions built from STO basis sets. Be^- does not have a $[He]2s^22p$ ^2P ground state; three metastable bound states found in the discrete spectrum are given here. $[He]2s2p^2\ ^4P^e$, $[He]2p^3\ ^4S^o$ and $1s2s2p^3\ ^6S^o$ states lie below the Be $1s^22s2p\ ^3P$, $1s^22p^2\ ^3P$ and $1s2s2p^2\ ^5P$ excited states. X-only energies are again in excellent agreement with the HF results [183]. A decent number of sophisticated accurate theoretical results are found in literature for the correlated case. Notable amongst them are the CI calculation of [183] including single, double, triple, quadrupole sub-shell excitations. First two states have also been studied through a method of full core plus correlation and restricted variation approach [184]. XC energies, in this case, show better agreements with literature results than those for Li^-; however, as in Li^- again falling below the reference values for all three states. Also a combined Rayleigh-Ritz and a method of restricted variation exists for the $^6S^o$ state [185]. The transition wave lengths of Li^- $1s2s2p^2\ ^5P^e \rightarrow 1s2p^3\ ^5S^o$ and Be^- $[He]2s2p^2\ ^4P^e \rightarrow [He]\ 2p^3\ ^4S^o$ have also been found to be in good agreement with literature results [33].

Before passing, a few remarks should be made on the present approach. Although DFT has enjoyed remarkable success for studying properties of atoms, molecules, solids, clusters in ground states, the same for excited states has been less conspicuous, partly because of complete abandoning of the state-function concept. Other major problems are cumbersome wave function and Hamiltonian orthogonality requirements between a given excited state and other lower states of same space-spin symmetry, as well as the unavailability of universal XC energy density functional. Despite all these difficulties, numerous attractive and elegant attempts have been made over the years to tackle these issues. However, until only very recently, most of the proposed methods have been either found to be computationally rather difficult to implement or producing large errors in excitation energies, except TDDFT where decent excitation energies are obtained. Moreover, it is not a straightforward task to extract the radial density. Also, most of these methods have dealt with lower and singly excited states; multiple and higher excitations, especially the Rydberg series as studied here, have not been reported so far by any other DFT approach except the current method. Also note that while the recently popular TDDFT route provides accurate excitation energies efficiently, extraction of individual state energies as well as densities, expectation values are not easy. The present scheme however, offers a simple, attractive way to produce energies, excitation energies as well as densities and expectation values with very good accuracy. In the current approach, all the well-known problems of DFT have essentially been bypassed by bringing the traditional wave function concept within DFT, so that the atomic orbital and electronic configuration pictures are retained. Fermi-hole charge distribution and hence local exchange potential is known precisely in terms of orbitals, which will be different for ground and excited states. Not being explicitly dependent on any functional form, the exchange potential is *universal*; it is fixed by the given electronic configuration of a particular state. Thus the same KS equation is now valid for both ground and excited states, obviating the necessity for obtaining the exchange potential as a functional derivative of exchange energy density functional. Instead it is now directly obtained from Fermi hole charge distribution. Due to the locality of XC potential, SCF solution of KS equation is computationally much easier than the HF equations, which involve a nonlocal integral operator. Yet, as demonstrated above, our X-only results are practically of HF quality and with correlation included, results go much beyond the HF level. Unlike many other sophisticated quantum mechanical methods, the present methodology does not involve basis set dependence, continuum mixing or explicit r_{12} dependence; it works essentially in the single determinantal framework.

In the HKS DFT, all many-body effects are incorporated into a local multiplicative potential, obtained as a functional derivative $\delta E_{xc}[\rho]/\delta\rho$ within the variational principle. Although the exact form of this $E_{xc}[\rho]$ is unknown, good approximations exist; however, with these approximations, bounds of total energy are no longer rigorous. Therefore the work-function prescription is not derived from the variational principle for energy, in the sense that it is not expressible as $\delta E_{xc}[\rho]/\delta\rho$, but is based on a *physical* interpretation for the local many body potential that an electron moves in an electrostatic potential arising from the Fermi-hole charge distribution. So even though a KS-type equation is solved with the work-function potential, this procedure is not subject to a variational bound. Thus the variational restriction on the excited state being the *lowest* state of a particular space-spin symmetry is not applicable here. Furthermore, although the existence of a local effective potential is guaranteed in KS DFT, no mathematical proof for the existence of such a potential for excited states has been known. Therefore, a key assumption is that excited states can also be described by a local potential. This is based on the fact that the physical argument used for the construction of ground state potential, can also be equally applied for excited states. Further discussion on the method and its application could be found in [21, 22, 23, 24, 25, 26, 27, 28, 29, 30, 31, 32, 33].

5. Conclusion

A simple DFT methodology has been presented for accurate, reliable, efficient calculation of ground and excited states of neutral, positive, negative atoms. Nonrelativistic energies, excitation energies, radial densities, radial expectation values, transition wave lengths are reported and compared with the best theoretical and experimental results available till date. The work-function exchange in conjunction with a GPS scheme for the solution of resulting KS equation makes it a simple and computationally efficient route for these important challenging systems. The accuracy achieved within this single determinantal framework is quite comparable to those from more elaborate and extensive calculations available in the literature. Success and usefulness of the method has been clearly demonstrated for a wide variety of excitations from single to multiple and low to very high Rydberg resonances as well as the satellite states, hollow, doubly hollow states etc. Computed quantities show excellent agreement with literature results. Almost all of these systems are highly correlated. Since the exchange potential is treated quite accurately (almost as good as HF), a major source of error in the present work is certainly due to the inefficiency of LYP potential in incorporating the delicate and intricate correlation effects, which could be further improved or replaced by more accurate energy density functionals for better accuracy. The assumption of spherical symmetry in calculating the exchange potential could also account for partial errors as well. In other words, the rotational component of electric field may not have insignificant contribution compared to the irrotational component for these states, in general, although this usually holds true. To summarize, this work presented a current account of a simple general and efficient DFT-based method for accurate and faithful description of multiply excited atomic systems.

Acknowledgements

I thank Prof. F. Columbus, for the kind invitation to present some of my recent works in this exciting book. I gratefully acknowledge the support of Prof. B. M. Deb in numerous ways. It is in his laboratory that I got introduced to this fascinating area of DFT. Earlier developments of this work took place there, when I was working as a graduate student. I thank Prof. P. Panigrahi for critical reading of the manuscript. It is a pleasure to thank my current colleagues at IISER-Kolkata, for their kind support.

References

[1] L. H. Thomas. *Proc. Camb Phil. Soc.*, **23**:542, 1927.

[2] E. Fermi. *Rend. Accad. Lincei*, **6**:602, 1927.

[3] P. A. M. Dirac. *Proc. Camb. Phil. Soc.*, **26**:376, 1930.

[4] P. Hohenberg and W. Kohn. *Phys. Rev.*, **136**:B864, 1964.

[5] W. Kohn and L. J. Sham. *Phys. Rev.*, **140**:A1133, 1965.

[6] R. G. Parr and W. Yang. *Density Functional Theory of Atoms and Molecules*. Oxford University Press, New York, 1989.

[7] R. O. Jones and O. Gunnarsson. *Rev. Mod. Phys.*, **61**:689, 1989.

[8] R. M. Dreizler and E. K. U. Gross (Eds). *Density Functional Theory: An Approach to the Quantum Many-Body Problem*. Springer-Verlag, Berlin, 1990.

[9] D. P. Chong (Eds). *Recent Advances in Density Functional Methods*. World Scientific, Singapore, 1995.

[10] J. M. Seminario (Eds). *Recent Developments and Applications of Modern DFT*. Elsevier, Amsterdam, 1996.

[11] D. Joulbert (Eds). *Density Functionals: Theory and Applications*. Springer, Berlin, 1998.

[12] J. F. Dobson and G. Vignale and M. P. Das (Eds). *Density Functional Theory: Recent Progress and New Directions*. Plenum, New York, 1998.

[13] Á. Nagy. *Phys. Rep.*, **1**:298, 1998.

[14] W. Kohn. *Rev. Mod. Phys.*, **71**:1253, 1999.

[15] W. Koch and M. C. Holthausen. *A Chemist's guide to Density Functional Theory*. John Wiley, New York, 2001.

[16] R. G. Parr and K. D. Sen (Eds). *Reviews of Modern Quantum Chemistry: A Celebration of the Contributions of Robert G. Parr*. World Scientific, Singapore, 2002.

[17] C. Fiolhais and F. Nogueira and M. Marques (Eds). *A Primer in Density Functional Theory*. Springer, Berlin, 2003.

[18] N. I. Gidopoulos and S. Wilson. *The Fundamentals of Electron Density, Density Matrix and Density Functional Theory in Atoms, Molecules and the Solid State*. Springer, Berlin, 2003.

[19] R. M. Martin. *Electronic Structure: Basic Theory and Practical Methods*. Cambridge University Press, Cambridge, 2004.

[20] D. S. Sholl and J. A. Steckel. *Density functional Theory: A Practical Introduction*. John-Wiley, Hoboken, NJ, 2009.

[21] R. Singh and B. M. Deb. *J. Mol. Struct. (Theochem)*, **361**:1321, 1996.

[22] R. Singh and B. M. Deb. *J. Chem. Phys.*, **104**:5892, 1996.

[23] A. K. Roy, R. Singh, and B. M. Deb. *J. Phys. B*, **30**:4763, 1997.

[24] A. K. Roy, R. Singh, and B. M. Deb. *Int. J. Quant. Chem.*, **65**:317, 1997.

[25] A. K. Roy and B. M. Deb. *Phys. Lett. A*, **234**:465, 1997.

[26] A. K. Roy and B. M. Deb. *Chem. Phys. Lett.*, **292**:461, 1998.

[27] R. Singh, A. K. Roy, and B. M. Deb. *Chem. Phys. Lett.*, **296**:530, 1998.

[28] R. Singh and B. M. Deb. *Phys. Rep.*, **311**:47, 1999.

[29] Vikas, A. K. Roy, and B. M. Deb. *Ind. J. Chem. Sec. A (Special Issue)*, **39**:32, 2000.

[30] A. K. Roy and S. I. Chu. *Phys. Rev. A*, **65**:052508, 2002.

[31] A. K. Roy. *J. Phys. B*, **37**:4369, 2004.

[32] A. K. Roy. *J. Phys. B.*, **38**:1591, 2005.

[33] A. K. Roy and A. F. Jalbout. *Chem. Phys. Lett.*, **445**:355, 2007.

[34] O. Gunnarsson and B. I. Lundqvist. *Phys. Rev. B*, **13**:4274, 1976.

[35] J. C. Slater. *Adv. Quant. Chem.*, **6**:1, 1972.

[36] J. C. Slater. *The Self-Consistent Field for Molecules and Solids, Vol. IV.* McGraw-Hill, New York, 1974.

[37] T. Ziegler, A. Rauk, and E. J. Baerends. *Theor. Chim. Acta*, **43**:261, 1977.

[38] U. von Barth. *Phys. Rev. A*, **20**:1693, 1979.

[39] C. Daul. *Int. J. Quant. Chem.*, **52**:867, 1994.

[40] A. C. Stückl, C. A. Daul, and H. U. Güdel. *Int. J. Quant. Chem.*, **61**:579, 1997.

[41] A. K. Theophilou. *J. Phys. C*, **12**:5419, 1979.

[42] E. K. U. Gross, L. N. Oliveira, and W. Kohn. *Phys. Rev. A*, **37**:2805, 1988.

[43] E. K. U. Gross, L. N. Oliveira, and W. Kohn. *Phys. Rev. A*, **37**:2809, 1988.

[44] L. N. Oliveira, E. K. U. Gross, and W. Kohn. *Phys. Rev. A*, **37**:2821, 1988.

[45] W. Kohn. *Phys. Rev. A*, **34**:737, 1986.

[46] Á. Nagy. *Phys. Rev. A*, **42**:4388, 1990.

[47] Á. Nagy. *J. Phys. B*, **24**:4691, 1991.

[48] R. Gáspár. *Acta Phys. Hung.*, **35**:213, 1974.

[49] O. Gunnarsson, B. I. Lundqvist, and J. W. Wilkins. *Phys. Rev. B*, **10**:1319, 1974.

[50] U. von Barth and L. Hedin. *J. Phys. C*, **5**:1629, 1972.

[51] D. M. Ceperley and B. J. Alder. *Phys. Rev. Lett.*, **45**:566, 1980.

[52] J. P. Perdew and A. Zunger. *Phys. Rev. B*, **23**:5048, 1981.

[53] S. H. Vosko, L. Wilk, and M. Nusair. *Can. J. Phys.*, **58**:1200, 1980.

[54] Á. Nagy. *J. Phys. B*, **29**:389, 1996.

[55] L. Fritsche. *Phys. Rev. B*, **33**:3976, 1986.

[56] J. Cordes and L. Fritsche. *Z. Phys. D*, **13**:345, 1989.

[57] S. Grimme. *Chem. Phys. Lett.*, **259**:128, 1996.

[58] S. Grimme and M. Waletzke. *J. Chem. Phys.*, **111**:5645, 1999.

[59] V. Sahni, L. Massa, R. Singh, and M. Slamet. *Phys. Rev. Lett.*, **87**:113002, 2001.

[60] M. Levy and Á. Nagy. *Phys. Rev. Lett.*, **83**:4361, 1999.

[61] A. Görling. *Phys. Rev. A*, **54**:3912, 1996.

[62] C. Filippi, C. J. Umrigar, and X. Gonze. *J. Chem. Phys.*, **107**:9994, 1997.

[63] E. San-Fabián and L. Pastor-Abia. *Int. J. Quant. Chem.*, **91**:451, 2003.

[64] F.Tasnádi and Á. Nagy. *Int. J. Quant. Chem.*, **92**:234, 2003.

[65] A. Görling. *Phys. Rev. A*, **59**:3359, 1999.

[66] V. Vitale, F. Della Sala, and A. Görling. *J. Chem. Phys.*, **122**:244102, 2005.

[67] F. Della Sala and A. Görling. *J. Chem. Phys.*, **118**:10439, 2002.

[68] V. N. Glushkov and M. Levy. *J. Chem. Phys.*, **126**:174106, 2005.

[69] Á. Nagy and M. Levy. *Phys. Rev. A*, **63**:052502, 2001.

[70] V. N. Glushkov. *J. Math. Chem.*, **31**:91, 2002.

[71] V. N. Glushkov. *Opt. Spectrosc.*, **93**:11, 2002.

[72] N. A. Besley, A. T. B. Gilbert, and P. M. W. Gill. *J. Chem. Phys.*, **130**:124308, 2009.

[73] M. E. Casida. In D. P. Chong, editor, *Recent Advances in Density Functional Methods*, page 155, Singapore, 1995. World Scientific.

[74] R. Bauernschmitt and R. Ahlrichs. *Chem. Phys. Lett.*, **256**:454, 1996.

[75] M. Petersilka, U. J. Gossmann, and E. K. U. Gross. *Phys. Rev. Lett.*, **76**:1212, 1996.

[76] D. J. Tozer and N. C. Handy. *J. Chem. Phys.*, **109**:10180, 1998.

[77] S. J. A. van Gisbergen, F. Kootstra, P. R. T. Schipper, O. V. Gritsenko, J. G. Snijders, and E. J. Baerends. *Phys. Rev. A*, **57**:2556, 1998.

[78] R. E. Stratmann, G. E. Scuseria, and M. J. Frisch. *J. Chem. Phys.*, **109**:8218, 1998.

[79] A. Görling, H. H. Heinze, S. P. Ruzankin, M. Staufer, and N. Rösch. *J. Chem. Phys.*, **110**:2785, 1999.

[80] K. Yabana and G. F. Bertsch. *Int. J. Quant. Chem.*, **75**:55, 1999.

[81] H. H. Heinze, A. Görling, and N. Rösch. *J. Chem. Phys.*, **113**:2088, 2000.

[82] F. Furche. *J. Chem. Phys.*, **114**:5982, 2001.

[83] S. Grimme. In K. B. Lipkowitz and D. B. Boyd, editors, *Reviews in Computational Chemistry, Vol. 20*, page 153, New York, 2004. Wiley-VCH.

[84] K. Burke, J. Werschnik, and E. K. U. Gross. *J. Chem. Phys.*, **123**:062206, 2005.

[85] A. Dreuw, J. L. Weisman, and M. Head-Gordan. *J. Chem. Phys.*, **119**:2943, 2003.

[86] D. J. Tozer. *J. Chem. Phys.*, **119**:12697, 2003.

[87] D. Jacquemin, M. Bouhy, and E. A. Perpete. *J. Chem. Phys.*, **124**:204321, 2006.

[88] R. D. Adams, B. Captain, M. B. Hall, E. Trufan, and X. Z. Yang. *J. Am. Chem. Soc*, **129**:12328, 2007.

[89] J. Tao, S. Tretiak, and J-X. Zhu. *J. Chem. Phys.*, **128**:084110, 2008.

[90] A. Nakata, Y. Imamura, and H. Nakai. *J. Chem. Phys.*, **125**:064109, 2006.

[91] A. Nakata, Y. Imamura, and H. Nakai. *J. Chem. Thoery Comput.*, **3**:1295, 2007.

[92] Y. Imamura and H. Nakai. *Int. J. Quant. Chem.*, **107**:23, 2007.

[93] Y. Imamura and H. Nakai. *Chem. Phys. Lett.*, **419**:297, 2006.

[94] J.-W. Song, M. A. Watson, A. Nakata, and K. Hirao. *J. Chem. Phys.*, **129**:184113, 2008.

[95] F. A. Asmuruf and N. A. Besley. *Chem. Phys. Lett.*, **463**:267, 2008.

[96] D. Rappoport and F. Furche. *J. Chem. Phys.*, **122**:064105, 2005.

[97] M. Chiba, T. Tsuneda, and K. Hirao. *Chem. Phys. Lett.*, **420**:391, 2006.

[98] M. Chiba, D. G. Fedorov, and K. Kitaura. *J. Chem. Phys.*, **127**:104108, 2007.

[99] J. Guan, F. Wang, T. Ziegler, and H. Cox. *J. Chem. Phys.*, **125**:044314, 2006.

[100] F. Wang and T. Ziegler. *J. Chem. Phys.*, **121**:12191, 2004.

[101] F. Wang and T. Ziegler. *J. Chem. Phys.*, **122**:204103, 2005.

[102] S. Grimme and F. Neese. *J. Chem. Phys.*, **127**:154116, 2007.

[103] C. Ko, D. K. Malick, D. A. Braden, R. A. Friesner, and T. J. Martínez. *J. Chem. Phys.*, **128**:104103, 2008.

[104] J. Neugebauer. *J. Chem. Phys.*, **126**:134116, 2007.

[105] C. Hu and O. Sugino. *J. Chem. Phys.*, **126**:074112, 2007.

[106] O. Vahtras and Z. Rinkevicius. *J. Chem. Phys.*, **126**:114101, 2007.

[107] M. Harbola and V. Sahni. *Phys. Rev. Lett.*, **62**:489, 1989.

[108] V. Sahni and M. Harbola. *Int. J. Quant. Chem. Symp.*, **24**:569, 1990.

[109] A. Holas and N. H. March. *Phys. Rev. A*, **51**:2040, 1995.

[110] V. Sahni. *Phys. Rev. A*, **55**:1846, 1997.

[111] M. E. Rose. *Elementary Theory of Angular Momentum*. Wiley, New York, 1957.

[112] G. Brual and S. M. Rothstein. *J. Chem. Phys.*, **69**:1177, 1978.

[113] C. Lee, Y. Wang, and R. G. Parr. *Phys. Rev. B*, **37**:785, 1988.

[114] A. K. Roy and S. I. Chu. *Phys. Rev. A*, **65**:043402, 2002.

[115] A. K. Roy and S. I. Chu. *J. Phys. B*, **35**:2075, 2002.

[116] A. K. Roy. *Phys. Lett. A*, **321**:231, 2004.

[117] A. K. Roy. *J. Phys. G*, **30**:269, 2004.

[118] A. K. Roy. *Pramana-J. Phys.*, **38**:2189, 2005.

[119] A. K. Roy. *Int. J. Quant. Chem.*, **104**:861, 2005.

[120] K. D. Sen and A. K. Roy. *Phys. Lett. A.*, **357**:112, 2006.

[121] A. K. Roy, A. F. Jalbout, and E. I. Proynov. *J. Math. Chem.*, **44**:260, 2008.

[122] A. K. Roy, A. F. Jalbout, and E. I. Proynov. *Int. J. Quant. Chem.*, **108**:827, 2008.

[123] A. K. Roy and A. F. Jalbout. *J. Mol. Struct: Theochem*, **853**:27, 2008.

[124] D. Gottlieb, M. Yousuff, and S. A. Orszag. In R. G. Voigt, D. Gottlieb, and M. Y. Hussaini, editors, *Spectral Methods for Partial Differential Equations*, Philadelphia, 1984. SIAM.

[125] C. Canuto, M. Y. Hussaini, A. Quarteroni, and T. A. Zang. *Spectral Methods in Fluid Dynamics*. Springer, Berlin, 1988.

[126] G. Yao and S. I. Chu. *Chem. Phys. Lett.*, **204**:381, 1993.

[127] J. Wang, S. I. Chu, and C. Laughlin. *Phys. Rev. A*, **50**:3028, 1994.

[128] J. C. Slater. *Quantum Theory of Atomic Structure, Vol.II*. McGraw-Hill, New York, 1960.

[129] J. H. Wood. *J. Phys. B*, **13**:1, 1980.

[130] M. Lannoo, G. A. Baraff, and M. Schlüter. *Phys. Rev. B*, **24**:943, 1981.

[131] R. M. Dickson and T. Ziegler. *Int. J. Quant. Chem.*, **58**:681, 1996.

[132] C. Pollak, A. Rosa, and E. J. Baerends. *J. Am. Chem. Soc.*, **119**:7324, 1997.

[133] W. A. Goddard III. *Phys. Rev.*, **176**:106, 1968.

[134] Z.-W. Wang, X.-W. Zh, and K. T. Chung. *Phys. Rev. A*, **46**:6914, 1992.

[135] J. S. Sims and S. A. Hagstrom. *Phys. Rev. A*, **11**:418, 1975.

[136] A. W. Weiss. *Phys. Rev. A*, **6**:1261, 1972.

[137] N. Koyama, H. Fukuda, T. Motoyama, and M. J. Matsuzawa. *J. Phys. B*, **19**:L331, 1986.

[138] A. Bürgers, D. Wintgen, and J-M. Rost. *J. Phys. B*, **28**:3163, 1995.

[139] E. Lindroth. *Phys. Rev. A*, **49**:4473, 1994.

[140] A. Savin, C. J. Umrigar, and X. Gonze. *Chem. Phys. Lett.*, **288**:391, 1998.

[141] G. W. F. Drake. In F. S. Levin and D. A. Micha, editors, *Casimir Forces: Theory and Esperiments on Atomic Systems*, New York, 1993. Plenum.

[142] G.W. F. Drake and Z. C. Yan. *Chem. Phys. Lett.*, **229**:486, 1994.

[143] S. Bashkin and J. O. Stoner Jr. *Atomic Energy Levels and Grotrian Diagrams, Vol. I.* North-Holland, Amsterdam, 1975.

[144] D. Begue, M. Merawa, and C. Pouchan. *Phys. Rev. A*, **57**:2470, 1998.

[145] Y. K. Ho. *Phys. Rev. A*, **23**:2137, 1981.

[146] D. R. Herrick and O. Sinanoglu. *Phys. Rev. A*, **11**:97, 1975.

[147] H. Fukuda, N. Koyama, and M. Matsuzawa. *J. Phys. B*, **20**:2959, 1987.

[148] A. D. Becke. *Phys. Rev. A*, **38**:3098, 1988.

[149] C. F. Fischer. *The Hartree-Fock Method for Atoms*. John Wiley, New York, 1977.

[150] Z.-C. Yan, M. Tambasco, and G. W. F. Drake. *Phys. Rev. A*, **57**:1652, 1998.

[151] U. I. Safronova and R. Bruch. *Phys. Scr.*, **57**:519, 1998.

[152] M. J. Conneeely and L. Lipsky. *At. Data Nucl. Data Tables*, **82**:115, 2002.

[153] M. J. Conneeely and L. Lipsky. *At. Data Nucl. Data Tables*, **86**:35, 2004.

[154] B. F. Davis and K. T. Chung. *Phys. Rev. A*, **42**:5121, 1990.

[155] K. T. Chung. *Phys. Rev. A*, **44**:5421, 1991.

[156] K. T. Chung and B. C. Gou. *Phys. Rev. A*, **52**:3669, 1995.

[157] L. B. Madsen, P. Schlagheck, and P. Lambropoulos. *Phys. Rev. A*, **62**:062719, 2000.

[158] S. Diehl, D. Cubaynes, J.-M. Bizau, L. Journel, B. Rouvellou, S. Al Moussalami, F. J. Wuilleumier, E. T. Kennedy, N. Berrah, C. Blancard, T. J. Morgan, J. Bozek, A. S. Schlachter, L. Vo Ky, P. Faucher, and A. Hibbert. *Phys. Rev. Lett.*, **76**:3915, 1996.

[159] K. Berrington and S. Nakazaki. *J. Phys. B*, **31**:313, 1998.

[160] Y. Azuma, H. Hasegawa, F. Koike, G. Kutluk, T. Nagata, E. Shigemasa, A. Yagishta, and I. A. Sellin. *Phys. Rev. Lett.*, **74**:3768, 1995.

[161] L. M. Kiernan, M.-K. Lee, B. F. Sonntag, P. Sladeczek, P. Zimmermann, E. T. Kennedy, J.-P. Mosnier, and J. T. Costello. *J. Phys. B*, **28**:L161, 1995.

[162] T. Morishita and C. D. Lin. *Phys. Rev. A*, **67**:022511, 2003.

[163] H. L. Zhou, S. T. Manson, P. Faucher, and L. Vo Ky. *Phys. Rev. A*, **62**:012707, 2000.

[164] K. T. Chung and B. C. Gou. *Phys. Rev. A*, **53**:2189, 1996.

[165] L. Vo Ky, P. Faucher, H. L. Zhou, A. Hibbert, Y.-Z. Qu, J.-M. Li, and F. Bely-Dubau. *Phys. Rev. A*, **58**:3688, 1998.

[166] Y. Zhang and K. T. Chung. *Phys. Rev. A*, **58**:1098, 1998.

[167] S. Diehl, D. Cubaynes, K. T. Chung, F. J. Wuilleumier, E. T. Kennedy, J.-M. Bizau, L. Journel, C. Blancard, L. Vo Ky, P. Faucher, A. Hibbert, N. Berrah, T. J. Morgan, J. Bozek, and A. S. Schlachter. *Phys. Rev. A*, **56**:R1071, 1997.

[168] N. A. Piangos and C. A. Nicolaides. *Phys. Rev. A*, **67**:052501, 2003.

[169] Y. Azuma, F. Koike, J. W. Cooper, T. Nagata, G. Kutluk, E. Shigemasa, R. Wehlitz, and I. A. Sellin. *Phys. Rev. Lett.*, **79**:2419, 1997.

[170] M. J. Conneely and L. Lipsky. *Phys. Rev. A*, **61**:032506, 2000.

[171] N. Vacek and J. E. Hansen. *J. Phys. B*, **25**:883, 1992.

[172] H. Bachau. *J. Phys. B*, **29**:4365, 1996.

[173] C. A. Nicolaides and N. A. Piangos. *J. Phys. B*, **34**:99, 2001.

[174] T. Morishita and C. D. Lin. *Phys. Rev. A*, **64**:052502, 2001.

[175] K. T. Chung. *Phys. Rev. A*, **64**:052503, 2001.

[176] F. J. Gálvez and E. Buendía and A. Sarsa. *Eur. Phys. J. D*, **40**:161, 2006.

[177] Z. B. Wang, B. C. Gou, and F. Chen. *Eur. Phys. J. D*, **37**:345, 2006.

[178] C. F. Fischer. *J. Phys. B*, **26**:855, 1993.

[179] A. W. Weiss. *Phys. Rev.*, **166**:70, 1968.

[180] C. F. Bunge. *Phys. Rev. A*, **22**:1, 1980.

[181] H. Y. Yang and K. T. Chung. *Phys. Rev. A*, **51**:3621, 1995.

[182] C. F. Fischer. *Phys. Rev. A*, **41**:3481, 1990.

[183] D. R. Beck, C. A. Nicolaides, and G. Aspromallis. *Phys. Rev. A*, **24**:3252, 1981.

[184] J.-J. Hsu and K. T. Chung. *Phys. Rev. A*, **52**:R898, 1995.

[185] J.-J. Hsu and K. T. Chung. *J. Phys. B*, **28**:L649, 1995.

[186] A. V. Bunge. *Phys. Rev. A*, **33**:82, 1986.

In: Focus on Quantum Mechanics
Editors: David E. Hathaway et al. pp. 197-208
ISBN 978-1-62100-680-0

Chapter 8

REALISTIC INTERPRETATION OF QUANTUM MECHANICS

Emilio Santos
Departamento de Física. Universidad de Cantabria.
Santander. Spain

Abstract

It is argued that the usual postulates of quantum mechanics are too strong. It is conjectured that it is possible to interpret all experiments if we maintain the formalism of quantum theory without modification, but weaken the postulates concerning the relation between the formalism and the experiments. A set of postulates is proposed where realism is ensured. Comments on Bell´s theorem are made in the light of the new postulates.

1. Introduction

After the discovery of quantum mechanics a warm debate took place about its interpretation, but since 1927 the Copenhagen interpretation (CI), due to Bohr, dominated in the scientific community and the debate almost ceased, although a few critical voices remained like Einstein and Schrödinger (see [1] for reprints of the relevant papers of that period). But it is interesting that the CI was not understood in the same way by different people. In particular Bohr did not attempt to apply quantum mechanics to the macroscopic domain, whilst von Neumann did it [2], and even made a model of measurement on this basis. The debate reappeared with Bohm's work in 1952 and, with more strength, after Bell's paper in 1964 [3] and it lasts until today. In the last few decades the CI has been progresively abandoned and the so-called many worlds interpretation (MWI) is taken its place, specially amongst cosmologists on the one hand and workers in quantum information on the other. I include in MWI the interpretations in terms of decoherence[4] or consistennt histories[5]. In my opinion they are just (important) elaborations within MWI. Still, some people claim that no interpretation is really needed [6]. The reason for the variety of interpretations of quantum mechanics is that many predictions of the theory have a paradoxical character, and people have attemped to solve these paradoxes by different means, without complete success till now in my opinion.

In my view the first step towards a satisfactory solution of the paradoxes is to investigate what is the minimal set of postulates of quantum mechanics which are really indispensable for the interpretation of observations and experiments. In some sense this approach is what the CI attempted and it is close to the "no-interpretation" above mentioned [6]. However, at a difference with the merely instrumentalist (pragmatic, sometimes named positivistic) character of the CI, I propose including the requirement that quantum mechanics is universal and realistic. By universal I mean that quantum mechanics applies to both the microscopic and the macroscopic domain (although maybe not to the universe as a whole, see below). This contrasts with the CI (at least Bohr´s) view that the referent of the theory is always the union of a microscopic system plus a macroscopic context (including the measuring apparatus), but the macroscopic systems should be treated according to classical theories.

Realistic means that we assume that *physics* (or science in general) *makes assertions about the world and not only about the results of observations or experiments*. In particular this implies that we shall give an *ignorance interpretation* to the probabilities predicted by the theory, at least the probabilities about properties of macroscopic bodies.

In order to expose my proposal I shall divide this paper in five parts with the following aims:

1) Pointing out that a part of the postulates of quantum theory are unnecessarily strong,

2) Analyzing some experiments in order to discover what postulates are really indispensable,

3) Proposing new postulates leading to a minimal realistic interpretation,

4) Studying the relation with other interpretations: CI, many worlds and hidden variables,

5) Showing, with the example of Bell´s theorem, how the conceptual problems are alleviated.

2. The Standard Postulates of Quantum Mechanics

Any theory of physics contains a (mathematical) formalism plus postulates giving the connection with observations or experiments (semantical rules). In quantum mechanics the formalism is the theory of Hilbert spaces combined with relativistic (Lorentz) invariance plus some particular postulates (e.g. masses and coupling constants of elementary particles). (Following a common practice I shall speak about "quantum mechanics" in the rest of this article, but I really mean "quantum theory". Also it is known that quantum fields cannot be formulated in Hilbert spaces, but require the more general framwork of C* algebras, but I shall ignore this point of mathematical rigour).

Most textbooks do not attempt to make precise the connection of the formalism with the observations or experiments, but we might try to divide the traditional semantical rules in two classes:

1) *The correspondences operators-observables and vectors-states.* They establish that we must associate a vector of an appropriate Hilbert space to every state of a physical system, and a self-adjoint operator to every observable. This statement alone gives very little information about the connection of the formalism with the experiments because no mention is made of the values of the physical quantities. It should be necessarily complemented with postulates about the measurement.

2) *The theory of measurement*, which establish that "when we measure the observable associated to the operator A in the state with vector (wavefunction) ψ we obtain one of the eigenvalues of A, say a_j, with probability $|\langle\psi\,|\,\psi_j\rangle|^2$ ".

It is increasingly obvious that *we should not postulate* a theory of mesurement. The measurement is just an interaction between some physical system and a macroscopic apparatus and therefore the study of the measurement should be *derived* from the remaining postulates if quantum mechanics is to be regarded as a fundamental theory of nature. The problem of including the measurement within the postulates of a theory is that makes it subjective and ambiguous. Because, what is really a measurement?, at what time is it made exactly?, a bad experiment, would not give results in disagreement with those postulated ?. These, and other arguments, have been brillantly exposed by John Bell, who proposed even the removal of the word "measurement" from physical theories [3]. Nevertheless it is a fact that the relation between operators and observables in quantum mechanics is always stated with reference to the possible values which may be obtained in measurements. This fact contrast with the situation in classical physics, for instance classical statistical mechanics. In that theory we also associate states of physical systems to some elements of the theory, namely probability distributions in phase space. Also we associate observables with other elements, namely functions of the phase space variables. The semantical rules are complete when we assume that all variables have values simultaneously and give an ignorance interpretation to the probabilies. But a similar procedure cannot be used in quantum mechanics as explained in the following.

According to the quantum-mechanical formalism, combined with the standard correspondence between measurable values of the dinamical variables and the eigenvalues of the corresponding operators, the variables cannot have a value when they are not measured. Indeed this is the essential containt of the Kochen-Specker theorem (see e.g. [7] or [8]). Consequently we cannot make statements about observables outside the context of a measurement, which contradicts the desire of removing the theory of measurement. I think that the existence of several, quite different, interpretations of quantum mechanics derives from that difficulty. CI (Bohr´s) assumes that the connection formalism-experiments should always involve macroscopic apparatuses which must be treated according to classical physics. This implies that macroscopic variables *do possess values* independently of measurement, and the postulates refer to these *objective* values. John von Neumann´s interpretation (included in CI by some people) is actually different. It *does not* attribute values to macroscopic variables from the start, but assumes that there is a "collapse of the wavefunction" which *objectifies* the values (maybe by the action of the mind or consciousness of a human observer). MWI solves the problem with an appeal to many branches of the "wavefunction of the universe", *all relevant variables having values* in each individual branch. But for me all these interpretations are unsatisfactory. CI, both Bohr´s and von Neumann´s, because it establish an "infamous boundary" between the macro and the microscopic domain (or between matter and mind). MWI because it contains assumptions which cannot be tested empirically and, in addition, look rather bizarre (a copy of each observer lives in every branch of the universe´s wavefunction). This is why I am proposing an alternative having elements of both, CI and MWI, but trying to remove their difficulties.

At this moment a comment of mathematical character is in order. As is well known, in the standard approach the *states* of physical systems are associated to *vectors* of the

Hilbert space and the *observables* to *self-adjoing operators*. Actually a generalization is possible if we associate observables to normalized positive operator valued *(POV) measures* [9]. On the other hand, all vectors which are obtained by multiplication of a given vector times complex numbers are assumed to represent the same state. Therefore it is appropriate to speak about *rays* of the Hilbert space rather than about vectors. Nevertheless, I shall retain the more common, although mathematically less correct, use of the words vector and operator because here I am putting the emphasis in the conceptual, philosophical, questions rather than in the formal, mathematical, ones.

Let us look more closely to the correspondence *states-vectors* and *observables-selfadjoing operators*. They are usually assumed (explicitly or implicitly) to be *one to one*, except for the superselection rules. The one-to-one character of the states-vectors correspondence is sometimes named *superposition principle* and it is enuntiated "if two vectors correspond to states, any vector which is a linear combination of the former also corresponds to a state, except for the superselection rules". Actually if the correspondence rays-states is one to one, also the correspondence *observables-selfadjoing operators* is one to one and viceversa. (This result is related to Gleason´s theorem [7]). In my opinion such postulates are unnecessarily strong and even absurd, because if we assume that every self-adjoint operator represents an observable, we are making postulates about "what may be measured *in principle*, that is what *could* be achieved in the laboratory in a more or less distant future". I think that the correspondence can be only in one direction, and so is stated in carefull textbooks. That is, we might assume that for every possible state which may be found in nature or manufactured in the laboratory there is an associated vector, and that for every observable which can actually be measured there is a self-adjoint operator, but not viceversa. This may be represented as follows

states $\mapsto$ *vectors, observables* $\mapsto$*self-adjoint operators.*

However, this correspondence is still too strong in my opinion. We should just admit

states $\mapsto$ *density matrices, dynamical variables* $\mapsto$ *self-adjoint operators*

That is, states should be associated to density operators and only rarely the density operators would correspond to vectors in Hilbert space, the so-called pure states . Also we should speak about dynamical variables rather than observables, because observability is a *practical* question which should not be postulated for all dinamical variables. A more precise statement of the postulates which I propose will be made in section 4, but in order to motivate them I shall mention a few typical experiments.

3. The Interpretation of Experiments

In the following I consider some experiments in order to see that rather weak postulates are indispensable:

1. Static properties of atoms, nuclei, molecules and solids.

Probably the most dramatic qualitative and (modulo some unavoidable approximations due to the complexity of the calculations) quantitive success of quantum theory is the interpretation of the physics of atoms, nuclei, molecules and solids, in particular their static properties. For instance, the prediction of the form, size and binding energy of molecules or crystals (hence their stability), the electric and magnetic properties (if external fields are included), etc., is the basis for most of theoretical chemistry and solid state physics.

In order to get these predictions from quantum mechanics, it is enough to take into account the evolution of the Schrödinger (or Schrödinger-Pauli) equation for electrons and nuclei, coupled to quantized Maxwell equations for the electromagnetic field. If we impose appropriate boundary conditions (e.g. no radiation coming from infinity), we may *derive* the existence of an unique stationary state ("the ground state"). Therefore, there is no need to *postulate* that the ground state is the eigenstate of the Schrödinger equation with the lowest eigenvalue, this fact following from the formalism. After that, the solution of the stationary Schrödinger equation gives all the required information, provided we assume that the expectation value of the energy of the system is given by the standard rule

$$E = \langle \Psi | H | \Psi \rangle ,$$

where H is the Hamiltonian operator and Ψ the vector state. In particular it is not necessary to assume that all self-adjoint operators represent observables, or that all eigenvectors of the Hamiltonian represent physical states. Similar arguments apply to the static properties of atoms or nuclei.

4. Collisions

For the study of (elastic o inelastic) scattering it is enough to consider the evolution of two or more systems (atoms, molecules, nuclei,...), both in the ground state, which initially are at a macroscopic distance and approach each other. Aside from the quantum evolution equations plus postulates of classical physics for the interpretation of the final results, we only need rules for preparation of the initial state and the interpretation of the final state.

The initial state usually consists of a beam whose preparation may be described in terms of classical physics (e. g. a macroscopic accelerator). Hence we should derive the density matrix corresponding to the (usually microscopic) quantum systems in the incoming beam and the detector. Nevertheless I do not think that it is possible to propose any general postulate saying how to do that. We should use the method or trial and error with the only general rule that that the appropriate density matrix is the one havint the greatest von Neumann entropy compatible with our information. After that we must use the quantum formalism in order to compute the evolution of the initial density matrix until the detector. Then we should apply to the detection process the quantum formalism, which most times could be approximated by classical equations. At the end we arrive at a density matrix for the final state of the measuring apparatus. Decoherence theory[4] shows that the apparatus density matrix is, to a very good approximation, diagonal in the coordinates representation. That density matrix is interpreted as a probability distribution (with the ignorance interpretation.) I think that this is the way how physicists in labs actually interpret the collision experiments.

This interpretation is very good *for all practical purposes* (FAPP), but presents a fundamental difficulty (Bell pointed out the important difference between FAPP and fundamental assertions[3]). In fact the final density matrix of the measuring apparatus is only approximately, but not exactly, diagonal. Therefore either we renounce to the interpretation of its diagonal elements as true probabilities (this should be the position of MWI) or we break quantum mechanics at the macroscopic level (this would be the position of CI). My position departs from both MWI and CI by assuming that quantum mechanics itself is an approximation to a more fundamental theory not yet known. An extremely good approximation, indeed. I eleborate more on that below.

5. Spectroscopy

This technique gives rise to the most spectacular agreement between quantum predictions and experimental results, the precision being sometimes better than 1 part in 10^9. This happens, e.g. in atomic spectra with visible light (electronic transitions) or microwaves (hyperfine transitions). The experiments of atomic spectroscopy are frequently interpreted as measurements of the energy levels of atoms. However, that interpretation is not necessary (although some people argüe that it is suggested by the formalism). We may simply assume that we are dealing with the evolution, governed by the quantum equations, of a beam of incoming radiation interacting with a material system (atom, nucleus, etc.). After all spectroscopy is a particular example of scattering experiment where a light beam is substituted for the incoming beam of particles. Both the incoming light and the outgoing light may be usually treated as classical.

6. Interference

These experiments are currently taken as the most dramatic examples of non-classical (quantum) behaviour. Nevertheless we need rather weak postulates for their interpretation. Again, it is enough to know the initial state, the evolution (including the interaction of microobjects with macro or mesoscopic devices like a grating or a detector) and the interpretation of the final results as in collision experiments (e.g. interpretation of what we see in a photographic plate as blackening of grains by the action of the incoming particles). We do not need to speak about whether a *particle* goes through one or both slits.

In all these examples we see that the standard postulates about the existence of discrete energy states, and their correspondence with vectors in Hilbert space, or about the association of observables with operators, are not really necessary. This leads us to conjecture that the (mathematical) *formalism* of quantum mechanics, the standard *postulates of macroscopic physics* for the connection formalism-experiments, plus some *particular hypotheses* (like the assumption that an atom consists of a nucleus plus electrons) are sufficient for the interpretation of all experiments.

Of course, for the applications it is more economical to use some "practical recipes", like Feynman's rule of summing probability amplitudes of indistinguishable paths in experiments of interference, but summing probabilities if the paths are distinguishable. The problem is that conceptual difficulties arise when the practical rules are taken as postulates. For example, the wave-particle duality appears as highly counterintuitive in experiments showing, alternatively, recombination and anticorrelation. In a typical experiment [10] one sends "individual photons" to a beam-splitter and verifies anticorrelated detection, that is either the detector in the transmitted beam or the detector in the reflected beam fires, but not both. This apparently proves the corpuscular nature of the photon. However, recombination of the two beams gives rise to interference, which apparently shows that the photon has gone by both paths at the same time. The experiment appears as mind boggling because it cannot be explained either assuming that something travels along both paths or assuming that there is propagation only along one path. However, there are alternative interpretations where something *real* (an electromagnetic field actually) propagates by the two paths, which explains interference, but some mechanism prevents the firing of both detectors at the same time, which explains anticorrelated detection [11].

Another crucial point of our proposed approach is that it is not necessary to *postulate* the existence of discrete energy states of quantum systems. Such states may be just mathematical intermediates in the calculation of the evolution. This is the case, for instance, with Fock states of the radiation field (e.g. single photon states). The states appear as mathematical constructs, not necessarily representing anything real, and are similar to the Fourier components in the standard solution of linear partial differential equations. A typical example of these is the diffusion equation, where the Fourier components of the series solution may not be positive definite and this does not imply that probabilities are negative, because only the sum of the series represents a probability. (There is a difference with the Schrödinger equation, however, in the fact that no theorem of positivity exists here, similar to the theorem stating that the fundamental solution of a diffusion equation is always positive.)

The moral of our analysis is that the standard postulates of quantum mechanics (in particular the *universal* correspondence between vectors and states) *constrains* the possible interpretations of quantum mechanics. My conjecture is that weaker postulates may allow for alternative interpretations free from conceptual difficulties.

7. Proposed Postulates

Firstly I admit without any change the usual formalism of quantum theory (Hilbert space, equations of Dirac, Klein-Gordon, Maxwell, etc.)

The postulates of connection with experiments are reduced to:

1) *To every physical system we associate a Hilbert space in the standard form.*

2) *To every "preparation" we associate a density operator.*

A preparation is a set of well specified laboratory manipulations. But I claim that it is necessary to specify the actual operations and the full macroscopic context. It is not enough to say, for instance, "I take a pair of photons of such and such properties". We should say something like "I take a crystal of specified kind, cut in such or such form, at which we direct a laser with specified properties", etc. That is we should specify the full macroscopic context. States are defined by the preparation like chemical species are defined by the recipe for obtaining them in the laboratory (either extracting them from natural products or by synthesis). Thus I propose that

2') *The density operator corresponding to a preparation is the one having maximum von Neumann entropy compatible with our knowledge about the system.*

I do not make any distinction between "pure states" and "mixtures", but claim that we shall treat them on the same footing. I do *not* assume that there is a physically realizable state corresponding to every vector in Hilbert space. In this sense I reject the standard form of the *superposition principle.* However the break which I consider for that principle is not of the "superselection rule" type, but deeper. My conjecture is that, in most cases, *physical states will have a positive (nonzero) von Neumann entropy*, pure states being just mathematical constructs. But I do not propose this condition as a postulate, I want just to remove the postulate that the opposite is true. In any case I do not assume that "pure states" provide a complete information about a single system, but about an ensemble of systems (i. e. those corresponding to a given preparation procedure). In this sense my proposal may be classified within the so-called *statistical interpretation* of quantum mechanics[12]

I do not postulate anything about observables (e.g. that the possible values obtained in a measurement are the eigenvalues of some selfadjoint operator). But I do *not* claim that such statements cannot be a part of the theory, I only claim that all statements of that kind, if true, *should be derived* from the remaining postulates. Observable is anything that can be actually observed (measured). Therefore the observables should be defined by a specific method of measurement. And measurement is a physical interaction which should be studied using the remaining postulates of quantum theory.

We should look at the process of measurement as follows. We have a system prepared in a specified form (that is, in some "state" represented by a density operator) and an experimental (macroscopic) context, also represented by a density operator. Both the system and the context evolve in interaction until they arrive at a final state. At the end of the measurement we observe a *macroscopic* system and this observation does not require special postulates (in addition to those of classical physics). Any macroscopic apparatus is in contact with the environment (it is always an open systems) and it is possible to prove [13] that the evolution gives, after a long enough time, a reduced density matrix (resulting from taking the partial trace with respect to the environment) which is diagonal in the coordinates representation for the macroscopic variables (e.g. the center of mass of macroscopic bodies).

Thus the final postulate is:

3) *The center of mass of any macroscopic body, or any macroscopic part of it, has a definite position in space at every time. The probability distribution is given by Born´s rule,* that is it equals the modulus squared of the wave function in the position representation.

This is the objectification postulate needed for a realist interpretation of quantum mechanics. Certainly it is open to criticism. Firstly the word macroscopic has not a sharp meaning. A solution (admittedly not very good) is to define as macroscopic any system with mass greater than, say, one microgram. Another possible criticism is that the postulate refers directly to macroobjects. Now, postulates about macroscopic objects may be derived from postulates about microobjects, but the inverse process is not trivial. Therefore finding the consequences of our three postulates for the microscopic domain (the one most properly quantal) may be difficult or impossible. I am aware of these problems, but I might argüe that they are less dramatic than the difficulties associated to CI or MWI, as commented above. Also I suppose that the difficulties of my approach are related to the fact that quantum mechanics is an approximation of a more fundamental theory. This is suggested by the fact that the conceptual difficulties are alleviated when we pass from (elementary) quantum mechanics to quantum field theory. For instance the stationary states with sharp energy which appear in the solution of Schrödinger equation are rather bizarre, but it is known that the states are neither estationary nor sharp in energy when the interaction with the quantized radiation field is taken into account. My conjecture is that a fundamental theory free from difficulties will be obtained only when the unification of quantum field theory and general relativity is achieved.

Our third postulate implies that the reduced quantum density matrix of macroscopic variables should be interpreted as a probability distribution. That is we propose an *ignorance interpretation* of the density matrix, which cannot be derived either in the CI or in the MWI. In fact, in both approaches a fundamental value is attributed to "pure states" of physical systems. They correspond to vectors (more correctly rays) of the Hilbert space, whilst only non-idempotent density matrices are seen as associated to lack of information. As is well known this leads to the impossibility of objectification as correctly stressed by P. Mittelstaedt [9].

In my proposal all density matrices (including those corresponding to rays) have an operational meaning: They are associated to physically realizable preparations, and they take account of the actual *information* that we have about the system. In this sense we leave open the possibility that the information is incomplete, even about a system represented by a "quantum pure state", that is we leave open the possibility of hidden variables. Furthermore, our objectification postulate requires that the initial information about the system to be measured is already incomplete, although I shall not elaborate further on this point in this paper. Another consequence of the objectification postulate is to view decoherence[4] as a loss of information closely similar to the "coarse graining" which happens in classical physics when we average over the degrees of freedom which are out of our control.

In my approach the concept of "state" has an epistemological, rather than ontological, character. States are rather similar to the probability distributions (ensembles) used in classical statistical mechanics. They refer to our knowledge about the system. However, that knowledge has a fundamentally objective character because it rests upon an objectively defined preparation procedure. The lack of ontological commitement derives from my rejection of statements of principle. That is I consider meaningless expressions like "the maximum information which may be obtained *in principle*". The actual information is what matters.

8. Relation with Copenhagen, Many Worlds and Hidden Variables Interpretations

The approach here presented has some similarities with the CI and also with the MWI but, at a difference with these, it allows (almost requests) for hidden variables. Let us analyze in some detail these points.

With the Copenhagen interpretation I share: 1) the emphasis on the need of speaking always about the macroscopic context, and 2) an operational approach to state (preparation) and observable (measurement). Indeed, I propose to remove all postulates relating directly the elements of the theory (electrons, photons, etc.) with actual physical objects. In this sense the postulates about the connection formalism-experiments refer always to a (possibly microscopic) *system plus* its *macroscopic context* as in the CI. In some sense the proposed postulates go farther than Heisenberg, for whom it is nonsense to speak about *trajectories* of electrons. The proposed intepretation avoids even assuming from the start that the electron itself (or the photon, etc.) is a *real* object, although I do not assume the opposite either. In the minimal realist interpretation the microscopic entities are assumed to be "theoretical (human) constructs" useful for the description of nature at the microscopic level, although they are related to some objective reality.

However there are three sharp differences between this approach and the CI: 1) I assume that *the formalism* of quantum mechanics should be applied both to micro and to macroscopic bodies, in contrast with CI (at least with Bohr´s view), 2) I do not exclude the existence of a subquantum level (hidden variables) which in the future might be accesible to our knowledge, and 3) related to this is the fact that the proposed interpretation is not considered the final word, it is just a provisional one to be used until we have a more fundamental theory.

With the MWI (or relative state interpretation) I share the assumption of the full validity of the quantum *formalism* even for the macroscopic world. However, the objectification postulate implies that the macroscopic variables always possess values (all macroscopic measurements may be reduced to position measurements). But the objectification postulate applies to a reduced density operator, obtained by taking the partial trace with respect to the environment. Consequently it does not apply to the whole universe, which has no environment. Also, as mentioned above, I do not postulate any fundamental relevance for the "quantum pure states", which seems to be a (maybe implicit) assumption of the MWI.

9. Quantum Mechanics and Local Realism: Bell's Theorem

It seems obvious, at least to me, that the best interpretation of quantum mechanics would be in terms of local hidden variables, if this were possible. (More properly, it would be desirable to have a local realist substitute for quantum mechanics, in a similar way that general reltivity is a local substitute for Newtonian gravitation. I believe that this was Einstein´s desideratum). Consequently Bell´s theorem is the biggest problem for a satisfactory interpretation of quantum mechanics. But I claim that local hidden variables have not yet been excluded by performed experiments.

The proof of Bell's theorem requires to assume that there are states such that: 1) they are physically realizable in the laboratory, and 2) they violate a Bell inequality . As I do not admit as a postulate of quantum mechanics the realizability of any particular state, the derivation of the theorem would involve proving that such state may be produced. That is I demand a detailed prosposal of an experiment where such state may be manufactured before a rigorous claim may be made about the incompatibility of local realism with the empirical predictions of quantum mechanics. On the other hand, a detailed experimental proposal is proved to be truly reliable only when the experiment is actually made. Consequently no claim of incompatibility may be made until such experiment is performed.

In the meantime Bell´s is a purely mathematical theorem (purely means without direct implications for the real world) that shows the incompatibility between two formalisms: 1) the Hilbert space formalism *plus the postulate that all vectors may be physically realized* (except for the superselection rules), and 2) the Bell formalism for local hidden variables theories. This does not mean that Bell's theorem is irrelevant. On the contrary, I think that *it is one of the most important discoveries in the physics of the 20th century*. But its relevance consists of being a guide for possible experiments able to test quantum theory versus local realism. As is well known (or rather, it *should* be well known) *no (loophole-free) experiment has been performed able to refute local realism up to now*. It is remarkable that this happens more than forty years after Bell´s work, which shows that the empirical disrproof of local realism is not a trivial matter. Actually the optical photon experiments are unable to truly test the Bell inequalities due to the detection loophole[14]. More suitable seem to be experiments with atomic qubits. One such experiment has already been performed[15], but it did not close the locality loophole and presents other difficulties[16].

My conjecture is that no experiment will ever refute quantum mechanics. But I also guess that decoherence and other sources of noise (e.g. quantum zeropoint fluctuations) might prevent the violation of local realism. That is, I still believe that quantum mechanics and local hidden variables are compatible, provided we define quantum mechanics with a set of postulates far weaker than is made usually, in the line shown in this paper.

References

[1] J. A. Wheeler and W. Zurek (eds.), *Quantum theory and measurement* (Princeton University Press, Princeton, NJ, 1983).

[2] J. von Neumann, *Mathematische Grundlagen der Quantenmechanik* (Springer, Berlin, 1932). (English translation, Princeton University Press, Princeton, NJ, 1955)

[3] J. S. Bell, *Speakable and unspeakable in quantum mechanics* (Cambridge University Press, Cambridge, 1987).

[4] W. H. Zurek, *Rev. Mod. Phys.* **75**, 715 (2003)

[5] M. Gell-Mann and J. B. Hartle, in *Physical origins of Time Asymmetry* (J. J. Halliwell, J. Pérez-Mercader and W. H. Zurek, eds.. Cambridge U. P. 1994).

[6] C. A. Fuchs and A. Peres, *Physics Today* **53** n° 3, page 70 (2000).

[7] M. Redhead, *Incompleteness, nonlocality and realism* (Clarendon Press, Oxford, 1987).

[8] N. D. Mermin, *Rev. Mod. Phys.* **65**, 803 (1993).

[9] P. Busch, P. J. Lahti and P. Mittelstaedt, *The quantum theory of measurement*, Springer-Verlag, Berlin, 1991.

[10] P. Grangier, G. Roger and A. Aspect, *Europhys. Lett.* **1**, 173 (1986).

[11] T. W. Marshall and E. Santos, *Found. Phys.* **18**, 185 (1988).

[12] L. E. Ballentine, *Quantum mechanics* (Prentice-Hall, Englewood Cliffs, NJ, 1990).

[13] R. Omnès, *Interpretation of quantum mechanics* (Princeton University Press, Princeton, 1994).

[14] E. Santos, *Found. Phys.* **34**, 1643 (2004).

[15] D. N. Matsukevich et al., *Phys. Rev. Lett.* **100**, 150404 (2008).

[16] E. Santos, *Phys. Rev.* **100**, 044104 (2009).

In: Focus on Quantum Mechanics
Editors: David E. Hathaway et al. pp. 209-294
ISBN 978-1-62100-680-0

Chapter 9

SURMOUNTING THE CARTESIAN CUT FURTHER: TORSION FIELDS, THE EXTENDED PHOTON, QUANTUM JUMPS, THE KLEIN-BOTTLE, MULTIVALUED LOGIC, THE TIME OPERATOR CHRONOMES, PERCEPTION, SEMIOSIS, NEUROLOGY AND COGNITION

Diego L. Rapoport*
DC&T, Univ. Nacional de Quilmes, Bernal,
Buenos Aires, Argentina

Abstract

We present a conception that surmounts the Cartesian Cut -prevailing in science- based on a representation of the fusion of the physical 'objective' and the 'subjective' realms. We introduce a mathematical-physics and philosophical theory for the physical realm and its mapping to the cognitive and perceptual realms and a philosophical reflection on the bearings of this fusion in cosmology, cognitive sciences, human and natural systems and its relations with a time operator and the existence of time cycles in Nature's and human systems. This conception stems from the self-referential construction of spacetime through torsion fields and its singularities; in particular the photon's self-referential character, basic to the embodiment of cognition ; we shall elaborate this in detail in perception and neurology. We discuss the relations between this embodiment, bio-photons and wave genetics, and the relation with the enactive approach in cognitive sciences due to Varela. We further discuss the relation of the present conception with Penrose's theory of consciousness related to non-computatibility -in the sense of the Goedel-Turing thesis- of quantum processes in the brain. We characterize quantum jumps in terms of the singularities of the torsion potential given by the differential of the complex logarithmic map (CLM) of the propagating wave satisfying the nilpotent eikonal equation of geometrical optics. We discuss the relations between the structure of these singularities, surmounting the alleged wave-particle duality in quantum physics, and the de Broglie-Vigier double solution pilot wave theory.

*E-mail address: diego.rapoport@gmail.com. Senior Fellow, Telesio Galilei Academy of Sciences, London

We discuss the relations between torsion and semiosis, and the torsion singularities as primeval to perception and cognition. We introduce Matrix Logic from the torsion produced by the nonduality of the True and False Operators, and non-orientable surfaces: the Moebius band and the Klein-bottle. We identify the extended photon twistor representations derived from the above theory with cognitive states of the Null Operator of Matrix Logic. The CLM that generates the torsion potential is found to be embodied as the *analytical* topographic map representation (TMR) of diverse and integrated sensorial modes in the neurocortex, while the Klein-bottle appears to be the topological embodiment of self-reference in general, is the *topological* TMR. The singularities of the analytical map are the loci of the two-dimensional sections of the neurocortex in which *all* the field orientations are superposed. This is the singularity of the Klein-bottle. We discuss the appearance of vortical torsion structures in the striate neurocortex and its relation with Karman vortices as the 'interiorization' of the 'outer' torsion fields which describe the representations of the field of orientations of visual stimuli -as it appears in the hypercolumnar structure in the neurocortex. The Brownian motions diffusion processes produced by the torsion geometries through the CMP have correlated diffusion processes in the neurocortex that can be associated with developmental morphogenetic growth patterns. We discuss the relations between morphogenesis, neurology and development in particular of the human and mammal heart and self-reference. We relate torsion in Marix Logic, the resultant Logical Momentum Cognition Operator and its decomposition into the Spin Operator and the Time Operator. This relates quantum physics statements into logical statements. The Time Operator is a primeval distinction between cognitive states in this Matrix Logic as its action amounts to compute the difference between these states. As a geometric action, the Time Operator is a ninety degrees rotation in the 2-plane of all cognitive states. We relate the Time Operator with intention, control, will and the appearance of life, and *chronomes* (time waves and patterns in natural and human systems and phenomenae). We discuss cosmological and anthropological problems from this perspective. In particular, we shall confront the Myth of the Eternal Return as a self-referential process, relating it to the Time Operator, nilpotence, the enaction approach to cognition due to Varela, and the emergent consciousness proposal due to Penrose. We discuss the relations with the microviolations of the second law of thermodynamics in statistical thermodynamics, the torsion geometry of Brownian motions, the synthropic action of the Time Operator, the Klein-bottle non-linear topology of time, with subjectivity and angular momentum. We discuss the relation between the topology of chronomes and the so called arrow of time. We introduce the notion of heterarchies of Klein-bottles, or still, of quantized reentering limited domains, and discuss its relation with self-determination, syntropy, entropy, and the Time Operator. We present as fundamental examples of these heterarchies the Myth of Eternal Return and the Human Being. We shall relate the Time Operator with the perception of depth considered in the phenomenological philosophy of Merleau-Ponty as a protodimension, and the problem of hemilateral synchronization with universal torsion chronomes (Kozyrev) in terms of the transactional interpretation of quantum mechanics, and the CMP retinotopic representation. We discuss *quantized* time and motion perception in relation with ATP production.

1. Introduction

In this chapter we shall present a theory in which the 'exterior' [1] world of physics, particularly the constitution of spacetime through the phenomenon of quantum jumps and extended photon structures, is fused with the 'interior' world of perception, cognition and subjectivity at large. This will surmount the Cartesian Cut conception which separates the world into an objective theatre on which consciousness plays a passive role as a bearer of information of the 'exterior' world [2]

Our conception is radically different; it is based on self-reference: The subject cognizes the world and simultaneously establishes himself as a singularity (an irreductible form which is also a process) through cognition and perception stemming from distinctions, differences that make a difference in the sense of Bateson [8]. These are distinctions which on being perceived, generated, cognized, abstracted or interpreted, generate or unfold higher-order differences, which amount to the universe of all manifestations, either virtual, processual, operational, algorithmic, formal, conceptual or real; for further developments of a differential epistemology for science that departs from this notion due to Bateson, see Johansen [58]. Without distinctions in its manifold manifestations, the world would be homogeneous and imperceptible [109], and definitively, there would be no thing or process to cognize nor Cartesian subject to bear cognition, nor consciousness.

Returning to our discussion on the prevailing conception, we wrote as customary 'information' to indicate the Cartesian take on cognition, which erases the ideative aspect of knowledge of a lifeworld (lived world) of all traces of subjectivity. Instead, through the semiotic action of breaking 'information' by introducing the linking sign '-', we indicate the presence of intention as a generative field, whose consequence is the emergence of a function derived from cognition; thus 'in-formation' is not about the data contained by the subject as a mere receptacle which is no more than the subject qua object of the Cartesian conception. [3] The latter introduction is an example of the fact that signs encode energy as discussed by Pattee [91] and Taborsky [138]. Most physicists working towards understanding cognition, apply physical models in their Cartesian mindset. There are several works that propose the origin of consciousness in the *brain's* electromagnetic field; again,

[1] To surmount the Cartesian Cut we take initially the trend of expliciting its most notorious dualisms, postponing to a section below and to Rapoport [109] and the epistemology by Johansen [58] in which they are unnecessary from the very beginning; they both depart from the primeval notions of distinction and boundary.

[2] Or at best a physical world that arises from information (Wheeler's it from *bit*), yet devoided of subjectivity and in particular of intentionality, and thus is still inside of the Cartesian Cut epistemology, as is the case of the observer, in the Copenhagen interpretation of quantum physics, a participant of the quantum measurement that 'collapses' the probability wave function.

[3] The linking sign, -, appears as a conveyor of adjacency of two terms A and B which have equal qualities. The adjacency relation is written as $A - B$, and the linking sign together with deduction, induction and abduction, give rise to an associative *synduction* (Greek, syn -, with) logic. In fact, synduction is a creative logical operator. Synduction logic appears in the context of the *perplex* numbers system due to Chandler [15]. This system was constructed by abstracting logical diagrams from the electrical properties of the atomic numbers. The diagrammatic logic of discrete perplex numbers is analogous to the diagrammatic logic of category theory for continuous variables. In the present conception, the linkage sign is a conveyor of *in*tention, or still, of will; the will of providing meaning through the suffix *in* linked to *formation*. As we shall see in the course of this work, it is not an abstract adjacency sign, but the semiotic embodiment of a Time Operator which stands for willful action.

the brain, not the body [94, 148, 77], neglecting, among other aspects, insights claimed by Traditional Chinese Medicine [84], based in a vortex in-formation interconnected meridian system that regulates *all* physiological processes and embodies emotions organically . [4] McFadden's scientifical, poetical and humane reflections on will are a far cry from the usual academic parlance. Yet, it is framed in the Cartesian mindset, linking will and intention to non-computability of quantum processes on neurons, following Penrose [92], not to the action of a time operator as we shall elaborate in the present chapter. The bottom-line for this idea of non-computability is the Hypothesis of the Continuum of Mathematics which heavily relies in the Antidiagonal Number Construction of Cantor's Theorem; see page 22 in [47]. We quote N. Hellerstein's comments on this number: "The number thus constructed leads to consider an infinity of infinites; so surely it must, within itself, contain an infinite amount of *information* about all these infinities. Otherwise, the silly thing is just bluffing us !". This *infinite* amount of information of the continuum is what is at stake in Penrose's proposal of consciousness as an *emergent* physical-algorithmic phenomena. [5] Hellerstein dilligently follows the query to conclude in few lines of page 106 that the bluff is evident. Indeed, the Antidiagonal Number has a paradoxical bit at a certain place N, until this place is a mere *finite* dyadic and henceforth it has an infinite paradoxical section made of an identical paradoxical time wave [61] which appears in the 4-valued logic that follows

[4]As the expression "venting ones' spleen" means, consonantly with this system, to vent ones' ire.

[5]Of course, this infinite information idea runs counter with the conception of a quantum space and time, which is the core of torsion, yet remarkably this has escaped the attention of the researchers working in quantum formulations of consciousness: A contradiction in the very cornerstone of a conception which we understand to be ill-formulated; for a discussion in terms of phenomenological philosophy and the Klein-bottle see [117]. This contradiction is carried further in the attempts to blend quantum physics with Einstein's GR which has null torsion and thus refers to a continuum, alike the one we are presently examining; so GR has to be *quantized*, a futile enterprise till today. In 'Shadows of the Mind', Penrose presents his theory based in the Goedel theorem and still Turing's theory of computation. Penrose addresses the former from its alleged protoform, the paradoxical diagonal construction known as Cantor's Theorem, reaching the proposition that a computation does *and* does not terminate; see page 75 [92]. Would we assume as Penrose *implicitly* did -*without* remarking that it was tantamount to a particular *choice* of a logic, namely the Aristotle-Boole logic- the principle of non-contradiction of this logic (i.e., for any proposition, p, p *and* not p is false) we would conclude with him that this is not possible and indeed the computation does not terminate. Thus, we *do* obtain a protoform of the Goedel Theorem; for the details see page 75 [92]. Now, in a self-referential construction as the one produced by Penrose albeit unacknowledged as such, and as further elaborated by Hellerstein and the present author, the corresponding logic is not Boolean but multivalued; it is the Klein-bottle logic [109]. In fact there is a number whose computation does terminate (it has a *finite* initial dyadic expansion) *and* also does *not* terminate; indeed, its finite dyadic expansion is followed by an *identical* infinite *paradoxical* expansion and thus is *redundant* with exception of its first term; yet, for the effect of better approximation we could still repeat it as far as we wished. (Thus, the principle of non-contradiction chosen by Penrose clearly misses the whole point of his arguments.) This paradoxical numbers is precisely the Antidiagonal Number produced by Cantor's Theorem. Hellerstein proves further that in this multivalued logic framework, which essentially coincides with the logic constructed by the present author from the protologic that stems from a primeval distinction and the self-referential extension to the Klein-bottle, there are an *infinite* quantity of real numbers which have this same property and there only exist a *finite* quantity of reals that are actually infinite. For a discussion on self-reference, the Klein-bottle and the need of an ontology which is lacking in Goedel's Theorem we refer to Johansen (2006) [58]. Our criticism to Penrose's approach is essentially the same as Johansen's: The anchoring to an ontology in which self-reference plays a generative role is lacking and thus the rejection of multivalued logic and self-reference is persued as if they would not be present in the developments, while these developments are essentially self-referential. Our point of departure to reach the understanding of the role of the Klein-bottle presented in the present conception was from this work due to Johansen.

from raising the calculus of distinctions of Spencer-Brown originated from a primeval distinction to encompass a self-referential equation which topologically is the Klein-bottle and from which we derived Matrix Logic [109]. Thus,"Cantor's Theorem is hereby exposed as not only superfluous, but actually ridiculous. The continuum is countable; Cantor's Paradox detects bit-flip at a dyadic. Therefore I propose a down-to-earth alternative to Cantor's tottering cardinal tower; a single countable infinity with paradoxical logic". This he calls, most appropiately, *Mathematics for Mortals*;" see page 108 in [47] [6] Hence generalizing to a multivalued logic with paradox, "we get a much more simple theory; a sign of elegance". Thus, the hypothesis of Penrose further raised by many researchers, that non-computability is a source for consciousness and free will [7] appears to be ill-conceived as we have just discussed, though it contains, in our understanding, some truth, in the fact that will and paradoxical time waves, which appear in the Antidiagonal Number Construction and in most (actually infinite) real numbers, are indeed related; see page 107 [47]. Hence, we feel obliged to note that we have just unveiled that time as a wave (in fact, a reentrance operator of a form on itself, the Klein-bottle) appears in Mathematics conceived as a system through the Continuum Hypothesis. Thus, a *time operator* is present in Mathematics as a system, inside the Continuum Hypothesis. Time, as an operator, reappears in the rotational recursive structure of the natural numbers (and in particular in the self-referential primes),

[6]Following our adherence to Merleau-Ponty's and S. Rosen's philosophies in which cognition is embodied, this grounding of abstraction implies an embodiment of Mathematics in the human lifeworld. While Computer Science by definition works with discrete finite numbers which first appeared in Physics through *finite* big numbers, the usual corpus of Mathematics ignores this embodiment which in fact is the core of a Platonian lifeworld, not its denial as usually considered. In fact, it dissociates itself from Physics in doing so. An alternative approach is to work with appropiate (say, Mersenne primes) p-adic number fields as developed in the excellent work by Pitkanen [93].

[7]The problem of the lack of constraints of an emergent physically-algorithmic will is still a very difficult if not impossible to resolve issue in a Cartesian mindset. This mindset ignores the joint constitution of the world and the subject. How can the subject in its finiteness determine the lack of constraints? If it is a matter of asserting its undetermination vis-à-vis a pledged independent reality of his/her subjectivity, which is believed to emerge from its undeterminateness, we find that in taking a conceptual route that ignores self-reference, then the wished freedom is left undetermined as well, and we are led to a paradoxical situation in which the subject and his/her lack of constraints become tied self-referentially, thus belying the initial conception. No theoretical exercise whatever its mathematical acuteness and complexity as in Penrose's extraordinary corageous and thought provoking works can surmount this problem if it ignores self-reference in its outset. (In fact, Penrose takes a whole chapter III to justify his conclusions and use of Cantor's Theorem, notably using self-referential constructions -and very complex argumentations which he uses to claim that these constructions are not grounded on self-reference! Thus, would we apply to Penrose's theory his own choice of logic for constructing his theory, i.e. the principle of non-contradiction, it is clear that it would be rendered absurd.) Which is the absolute Undetermined Source from which the subject can claim its lack of constraints and how is it that the subject can assert the lack of constraints of the Source? This leads to an infinite antiregress of emergence in which the world and the subject are pushed away one from the other -perpetuating the alienation of the Cartesian mindset- in each attempt to assert their claimed independence. This is the disolution of the grounding of an emergent consciouness approach which is very much ingrained in the works by physicists studying consciousness, and of course, is no matter of concern for the working Cartesian scientist. He/she is pleased enough in finding their efficacy in *controlling* the world, without even able to imagine nor grasp that control and will are undisolubly fused to a time operator which is grounded in the Klein-bottle meta-algorithmic fusion of object-with-subject. So their free will in practicing this control is related to a phantasy that transpires hubris which makes of the world a token for predation in which the scientist reinforces its own alienation. We shall reencounter this phantasy in a cosmological (!) setting in our study below of the Myth of Eternal Return as a self-referential system.

in the remarkable work by Johansen [60]. In examining the natural numbers by following Johansen's approach we shall find ourselves very far from the trivial sequential notion of linear time parameter, and in particular in its embodiment through the Peano construction of the natural numbers. In distinction to Hilbert's proposal of constructing an axiomatisation for Physics and the working mathematician's daily practices, axiomatics can only provide for the algorithmic rules. These rules embody the legislation for the formalities [8] of the game in which Mathematics is daily played on the background of an absent self-referential epistemology. Yet, the axiomatical approach will never provide for the meta-algorithmic lifeworld. This is another example of the Cartesian Cut emergence antiregress we commented before.

Our proposal in first approximation only requires to ponder what the meaning of intention is, to establish the link between will, intention and the time operator. [9] In the Cartesian mindset, language is considered to be informational, not in-formational. Interpretation is ontologically inexistent. Subjects qua objects, mere containers of data, i.e. information. Furthermore, hermeneutics, the inquisitive intention of the subject in search for interpretation for achieving understanding to sustain the logic of her/his own being together with the logic of the Universe with whom the subject is enacting themselves, is neglected in the Cartesian mindset. Thus, the lifeworld of Being is shattered to broken bits which henceforth will force cognitive scientists to frame consciousness in terms of some kind of emergence. These shattered bits produced by the act of irreflection will never be recomposed because the Cartesian mindset has no operation-operand to provide for the glueing, nor the Klein-bottle which stands not for a mere reparation of the Cartesian Cut, but for the lifeworld of Being. So physicalism and emergence -which we have already discussed- or some form of subjectivism will be the core of its paradigmatical framework, or we may encounter a so called 'third way', such as the emergence of the 'Great Doubt' in which the Buddhist conception will want us to disolve [115]. Thus, the subject is turned into an object though a thinking one as in Descartes, yet a subject for whom thinking is a process untraceable to its origins, as in the physicalist emergence proposal, or the subject disappears as in the 'Great Doubt', Varela's et al proposal in cognitive sciences to surmount the Cartesian Cut. We have already presented a conception which surmounts the Cartesian Cut based on geometrical-logo-physical self-referential fields (torsion fields), phenomenological philosophy, second-order cybernetics (the cybernetics of controlling systems, i.e. which include the controller), multivalued logics, non-orientable surfaces such as the Moebius and Klein-bottle surfaces, and its relations to neurology [109].

In [111] we showed that extended photons are codified as cognitive states in a multi-valued Matrix Logic (originally due to A. Stern [135]), which has as particular cases quantum, fuzzy and Boolean logics. This codification establishes a relation between quanta and thought. We introduced Matrix Logic in two ways. In the first way as the topology of

[8]Musès' criticism of the non-standard analysis (A. Robinson, Edward Nelson nowadays) approach to infinitesimals and infinities as purely legislative in distinction of his treatment through hypernumbers [82] is an excellent showoff of how the management of this axiomatisation approach becomes a clear example of the infinite antiregress we mentioned before, leading to a vicious circle of unending complexity in ignorance of the fact that the existance of time operators is essential to Mathematics as we encountered above.

[9]In this chapter we shall propose in these terms, an explanation of an ancient myth, the Myth of Eternal Return, as an illustration of an all encompassing cosmology that includes natural, and particularly, human systems.

paradox in a protologic (related to the work of Spencer-Brown [133]) which follows from a primeval cleavage-distinction [10] (which thereby acts as a semiotic -i.e. through a sign-codification of the torsion quantum field that embodies the fusion of object and subject and the joint genesis of both) in a two-dimensional plane of all potentialities, giving rise to the appearance of a 4-state valued logic in which in addition to the Boolean states we have two primeval time waves enacting self-reference as new paradoxical states. In this first approach, the topology of torsion fields is given by the Klein-bottle (of Eternal Return fame). In the second way, we introduced Matrix Logic as the logic generated by two times two matrices representing all possible logical operators (not mere scalar connectives), from the torsion that surges from the commutator of the True and False *Operators* that extend the Boolean values; these operators are non-dual, in distinction to the Aristotelian-Boolean true and false logical values. This torsion can be also seen appearing in the guise of the non-orientability of the Moebius and Klein-bottle surfaces, which thus provides Matrix Logic with *superposition* [11] topo-logical cognitive states for its foundations. In particular, quantum field operators take a logical representation as nilpotent operators (i.e. their self-multiplication is null, i.e. equal to 0, which is not to be confused with *nil*, the value for nothingness.). We established a two-way transformation between the eigenstates of Null Operator of Matrix Logic and the twistors representation of the extended photon arising from the quaternionic wave propagation and eikonal equations and viceversa [111]. Thus, the primeval distinction gives rise to the most fundamental joint constitution of object with subject: the transformation of the quantum photon states into the eigenstates of the Null Operator, and vice versa. We showed that a Logical Momentum Operator, also called the Cognition Operator (since it can be alternatively represented as a variation of cognitive value, basically its derivative), appears from the non-duality of True and False, or still, from the above mentioned non-orientability. We found that the Cognition Operator decomposes as a Spin Operator plus a Time Operator. This allows to express quantum physics statements into logical statements and vice versa. Thus, the Cognition Operator introduces a fundamental Time Operator which represents the most primeval distinction between cognitive states in this Matrix Logic, since its action on two cognitive states amounts to compute the difference between them. This Time Operator admits, from its matrix representation, an interpretation as a primeval ninety degrees rotation in the 2-plane of all cognitive states. In this article we shall relate Time to intention-will-purpose, to the origins of life, to the Myth of the Eternal Return as a logo-physical process, and time-waves in natural and in particular human systems (chronomes), especially to the problem of the constitution of stereoscopic

[10]This notion of a primeval distinction-cleavage or still of boundary in the sense of the protologic due to Spencer-Brown, was further extended to a 3-valued logic by introducing the semiotic codification of the reentrance of a form on itself (essentially, the Klein-bottle which for the ancient traditions was the Eternal Return Ouroboros, or the Phrygian cap (of French Revolution fame) from the homonimous civilization of Anatolia, or still the Pelican Christus [98]), as a *third* logical value. Indeed, this value appears in addition to two other values; these are given by the distinguished (by the primeval distinction) state on a plane devoided of other signs (the plane of all potential states or forms, the Plenum), and the unndistinguished state (the 'empty state', to Spencer-Brow) in which the plane is taken without the semiotic codification by the cleavage-distinction (thus, an untagged state); in the Boolean interpretation we intepret them as the usual false and true logical values [146]. It was still extended to a 4-valued logic by considering the reentrance of a form on itself with a delay [61] from which we derived Matrix Logic [109].

[11]A more fundamental topo-logical rendering of superposition states than the usual approach in quantum physics.

vision and the perception of depth. In the phenomenological philosophy of Merleau-Ponty depth is a protodimension, and its appearance in the perception of the Necker cube shows that depth is related to time and the appearance of multivalued logic (since there exist two alternative interpretations-perceptions of the Necker cube, both being true), surpassing thus Aristotelian dualism. Thus, the Necker Cube is rendered as a surface of paradox, alike the Moebius band and the Klein-bottle.

Yet, in these previous works, while the role of the Klein-bottle in the neurocortex structure was pointed out, the relation between the theory of the physical world -in terms of torsion- and the world of subjectivity in terms of perception, was only summarily introduced to argue the gestaltic identity between the physical theatre, logical and cognitive realms and visual perception embodied in the Klein-bottle. In the present chapter we shall extend our works [109, 111], establishing a relation between the self-referential geometry of spacetime constituted by photon fields (as basic example of torsion fields, essentially vortical fields), and the somatotopic, visual and integrated sensorial modes representation in the neurocortex. We shall further relate this to the complex logarithm map which will appear in the characterization of quantum jumps in terms of the singularities of a gradient logarithm of photon waves satisfying the eikonal equations for light rays, and will reappear as the in-formational topographic map in the neurocortex. We shall see that this mapping from the 'outer' (skin surface, limbs, retina, clochea, integrated body sensorium) to the 'inner' world through a topographic representation on the neurocortex, is embodied in two superposed integrated maps. One of these topographical maps of the sensorium on the neurocortex is analytical (we shall qualify this soon), the other one is topological; these two maps are closely related between themselves and to the establishment of selfhood, as shall follow from our discussions in the final section of this chapter. The topological map is associated to the non-orientability of the Moebius and Klein-bottle self-referential surfaces. The analytical map is provided by the complex logarithmic function which, as in the physical realm, has singularities (in the physical case the loci for quantum jumps), which in the neurological case have for correlates vortical structures in the neurocortex (more specifically, the so called hypercolumnar structure); in the latter case, these singularities stand for the points in the neurocortex in which the stimuli orientation mapping given by the logarithmic map becomes multivalued. Yet, this map is the analytical representation of the $3D$ outer body (and its surface) to a unique $2D$ plane in the neurocortex in which each point codifies a whole hypercolumn vortex which is not anatomically distinguishable [54]. In the present conception, it will transpire that in distinction to other approaches to consciousness, which claim additional noetic higher-dimensions for consciousness following the Einstein tradition of treating complexity through the introduction of additional dimensions [3, 21], the dimension 2 common to the Klein-bottle, the space of all cognitive states in Matrix Logic, the undistinguished plane from which Matrix Logic arises from the primeval distinction, the dimension of the phase space associated to will-self-reference-control, the dimension for holography in general [124] and in particular in the neurocortex [73, 137, 141], and the complex plane (or still, the Riemann sphere) for the representation of the complex logarithm, will be singled out.

To resume: The multivalued singular structures of spacetime vortical torsion structures which are the locus for quantum jumps and as such are the most primitive distinctions that make possible the constitution of a spacetime, have in the perceptual mapping of the sensorium, given by the complex logarithm mapping, a representation in the neurocortex. This representation is provided by the same complex logarithm in which the quantum jumps are turned into multivalued points of stimuli orientation representation of the neurocortex (or still, their hypercolumn vortical structures). Hence, this conception establishes the embodiment of the Universe jointly with subjectivity, the subject, perception, cognition and thought.

Our presentation of this conception and its unfolding in the present article will be completely different to the one presented in [109]. In that work, the unfolding of the conception started with the philosophical aspects, mainly departing from second-order cybernetics to further link with phenomenological philosophy and its relation with (paradoxical) logic, depth perception (departing from the work of M. Merleau-Ponty [79]), time and subjectivity (departing from the work of Heidegger, and more relevant to the present conception, to the works of Hegel [46] and G. Gunther [39]), to finally connect the present conception with the lifeworld embodied in the Klein-bottle, following the conception of *radical recursion* in the work by S. Rosen [116]. This previous presentation naturally led through the protodimension given by depth as elaborated by Merleau-Ponty and Rosen to visual perception and the Klein-bottle [109]. From those preliminaries we introduced the notion of primitive distinction and its identification as the semiotic codification of a torsion field that generates space, and furthermore time by considering the paradoxical equation in the calculus of distinctions that arises from this primeval distinction. We further introduced the time oscillations that arise from the solutions of this paradoxical equation (the Klein-bottle) and the multivalued logic that arises from them, the *Klein-bottle logic*. We then introduced Matrix Logic due to Stern [135] to show that it is associated with the torsion in cognitive space introduced by the non-duality of the True and False Operators. We further studied the transformation of cognitive operations into quantum spin operations and viceversa.

The present chapter will take for point of departure, instead of second-order cybernetics, phenomenological philosophy and visual perception as we did in [109], rather with the most basic embodiment of the fusion of object and subject that embodies (quantum) action and perception as the result of an integrative process of the 'exterior objective' and 'subjective' worlds: the photon. Indeed, the photon is not seen but it is the seeing [156], so at the fundamental level of constitution of reality, action and perception are unseparable which thus appears to be the foundation of coorigination of the world and the subject. [12] So quantum physics will be our departure point though as we have just disgressed, physical reality cannot be separated from perception [70, 44] nor from thinking!

Our conception is somewhat related to the concept of *enaction* proposed by Varela and associates [115] for surmounting the Cartesian Cut by conceiving embodied cognition, following the phenomenological philosophy due to M. Merleau-Ponty [79]. We coincide with these authors that '...the self becomes an objectified subject and a subjectified object'; page 242 [115]). Yet, how this transformation is produced is left unexplained; in particular the

[12]Thus in the photon we find a fusion of Kant's nouminal (the 'objective external' world) and phenomenal (the 'internal' perceptual) realms. Due to the relation between the eigenstates of the Null Logical Operator (with all matrix elements being 0) of Matrix Logic with the twistor representations of the extended photon, we have claimed that the photon is seeing-thinking. This is because of the above mentioned relation with the cognitive states of the Null Operator of Matrix Logic, which we named the *Mind Apeiron* since it embodies all the potential cognitive logical states [111].

relation of enaction and the Klein-bottle is altogether ignored, as well as the essential role of the latter in perception. Furthermore, it is quite remarkable that in the conception of Varela [13] no relation between the photon as the embodiment of the self-referential fusion of action and perception is proposed. We shall present this relation in this chapter, in fact it will be our starting approach as we shall explain soon. By failing to notice this primeval *joint* constitution of the Self with the quantum photon (we shall find this later in the Myth of Eternal Return), they take the philosophy of Buddhism to elaborate an extreme form of nihilism, despite of their pledges on the contrary sense. Varela and associates base their proposal of enaction for surmounting the Cartesian Cut on the 'Great Doubt', which is the disolution of the Self and thus of the claimed fusion of object-with-subject. Hence, in their disgressions self-hood is found to be inexistent but a mysterious illusion that, in contradiction with their profession of fusion of object-with-subject in their embodied cognition enaction approach, is left unembodied, with no explanation for the origin of its reification being delivered. In this regard, it is remarkable still that in this proposal not even the body's cells embody cognition, due to the fact that they are short lived and thus -in their own words- are similar to the wooden planks of a ship that on decomposition due to the inclemence of the environment and hard use, are condemned to ephimery and to be replaced until the ship itself is no longer. Thus, with this conceptual background they are lead to question if the continuously replaced ship in its components is the same one than the previous one to substantiate the permanence of Self. We would like to comment on the Cartesian mindset still implicit to the proposal of Varela and associates for an embodiment of cognition which has no body for Being (we shall later find this proposal in the Myth of Eternal Return as an ontological mistake). In their take of enaction, the Self if biologically grounded should be *inert*: the cells are mere mechanical pieces, and thus time and light, as the primal organization fields of fusion of subjectivity and objectivity, are disconnected to Self; thus, implicit to this conception, is the resignation to physicalism. This neglect of the role of light in the constitution of the body, not to mention embodiment, runs counter with the knowledge on bio-photonics by Popp et al [95] at the time of the writing of Varela and associates, that originated in the discoveries of ultraweak photon emission from living systems by Alexander Gurwitsch circa 1923 in the USSR [40], and especially the fact that DNA emits electromagnetic waves [34]. Furthermore, as unveiled by Gariaev and associates, DNA has in the 98 percent strands unparticipating in the bio-chemical functions (the so-called 'junk DNA') [14], the structure of a language. These findings were confirmed in the work of Mantegna et al [72]. Thus, cells have down to DNA a cognition-like structure which is based on the photon's fusion of object-with-subject and in a certain linguistic structuring. Thus, interpretation and meaning, as well as intention [15] are biologically grounded in the bio-photonic

[13]Which is superposed with an apologetics of a particular religion -Buddhism- as a conceptual basis for cognition and sciences at large, and still the path to the experience of wholeness which should lead us to the creation of a better world, as well as a philosophy of pragmatics which thus includes ethics.

[14]An expression of the hubris implicit to the interpretation of the standard dogma in genetics, yet not uncommon to the practices of scientists of all professions.

[15]In the conception presented in this chapter, language is not a mere conveyor of information, but an informed and in-forming field by its self-referential essential character. It semiotically embodies in-formation as well as produces it, embodying cognition and the will to in-form. Language has an indicative -and in some instances imperative- character, and thus embodies purpose-will-intention. We shall later see the connection of this with time operators and life.

structure-process of DNA. Furthermore, communication between distant cells is produced by electromagnetic signals [95, 26]. For further crucial studies on the DNA electromagnetic structure-process and its holographic behaviour we refer to Marcer and Schempp [74, 124]; a remarkable non-Newtonian systemic rendering of bio-physics can be found in Simeonov [128]. Thus we are lead to suggest that the physical-biological basis for the embodiment of Self is already present in the bio-photon. [16] We shall find this again when discussing the embodiment of perception through neurocortex topographical maps. Returning to the philosophical aspects of Varela's proposal, the achievement of the experience of the 'Great Doubt' is claimed to be the core of Buddhism by its authorities and practitioners, and in particular of these authors' approach to cognitive sciences, leaving their 'enaction' absolutely ungrounded. Remarkably, the experience of 'Great Doubt' is claimed to be achieved through the so-called 'illumination', which makes the interpretation and theoretization of their conception, which led to the 'Great Doubt', a contradiction. In our understanding, the 'Great Doubt' is the specular nihilist image of the positivism of the dualistic approach, in which by taking the principle of no contradiction as its conceptual basis, all the universe of discourse is placed on the positive affirmative truth value which encompasses *all* the universe of discourse, as explained by Gunther[39]. Thus, in the dualistic approach, subjectivity and its relation with time and multivalued logics (in particular the self-complex which Varela and generally Buddhism want to leave groundless) is rendered without an ontological locus [39, 109]. This follows a quite puzzling -to this author- tradition of Tibetan Buddhism, in which though light and its experience is discussed quite extensively [38], yet its fundamental self-referential character is not mentioned at all (see especially pages 52 and 83 of Guenther's philosophical treatise), to the of our best knowledge, although we admit to be far from being scholarly versed in the subject. Anyway, it is remarkable that the great scientist Varela, which was very close to the maximum exponents of Tibetan Buddhism, would not be aware of this fundamental character of light to omit mentioning it all in his work [115]. [17]

[16] We shall later relate the appearance of life to the willful action of timing, a time operator, that produce higher complexity structures-processes, countering thus the second law of thermodynamics, and in fact production syntropic processes of self-determination and self-differentiation of ever higher-degrees orders. We shall return to discuss this issue on introducing the notion of chronotopology and discuss its non-linear being. So, returning to the linking operator in synduction logic,-, which we anticipated to be related to the willful action of a time operator, the bio-photon expresses through light emmisions of cells and more complex biological structures, higher order self-regulations through light.

[17] We shall later reencounter this attempt to ommit the self-referential being of light, in the Myth of Eternal Return. All religions, in spite of their cognitive richness, have a fundamental problem: the difficulty of grounding them socially, where the social character already finds its grounds in languages and the intentions embodied by them, and of course, in social organizations as well. Thus the former invariance of the lack of grounding of belief systems and religions, might perhaps be the hiatus between cognitive states, which are not the True state, and volition; both cognition and volition are actions, having both a perceptual and ideo-logical framework. This hiatus is related to the impossibility of disolving self-reference into the undistinguished -under a primeval cleavage- state, the void to the Buddhist tradition, which is the plenum of all potential states, the Unmanifest, Apeiron. We shall return to this issue further below in relation with chronomes, universal times waves in natural and human systems. As with religions, any conception needs to be integrally grounded, and the economic and social realms have to be embodied into non-alienating self-organizations. For the studies of second-order economics and the *surmounting* critique of Marx's theory of capital, proposing an alternative grounding for the science of economics and its social implementation, we refer to the work by Johansen [60]; for the electromagnetic potential based geometry of written and spoken language we refer to the work of Doucet

Yet, in the present setting related to the fundamental inhomogeneities given by photons as the fundamental case of torsion, and wavefront propagation of singularities which give a non-trivial spacetime which is not based on the notion of metrics but in the primeval notion of distinction and difference producing differences (as argued by Bateson [8]), which also is the starting point for the differential epistemology due to Johansen [58]. The unfolding of the present chapter following this Introduction, will be on mathematical terms, rather than philosophical and perceptual, to which we shall arrive as a byproduct of the present departure with the photon, as the primeval gestalt of a participative universe, in which subjectivity is primeval. Then, the unfolding of the present conception in terms of light, will be the natural means for establishing the connection between the physical notion of quantum jumps as a primeval quantum distinction associated to torsion, with visual and somatosensory perception which stem from these primeval differences.

Thus this chapter will be separated into three main parts: Firstly we shall present the theory for the constitution of spacetime in terms of photon torsion fields and the relation with quantum jumps, comprising Sections II to IV, which we shall later discuss. Secondly we shall present Matrix Logic in the framework of the torsion in cognitive space that arises from the non-duality of the True and False Operators, and the codification of extended photons as eigenstates of the Null Operator of Matrix Logic. [18] Let us recall that Matrix Logic was in the already described two possible approaches for its presentation, both related to the Klein-bottle, as we already discussed above. In one of these two possible approaches, we departed from the torsion that surges from the commutator of the True and False *Operators* that extend the Boolean values and are non-dual (in distinction with Aristotelian-Boolean logic), that stem from the non-orientability of the Moebius and Klein-bottle surfaces, which thus has superposition topo-logical cognitive states for its foundations. This leads to the definition of a Logical Momentum Operator which is this commutator of the True and False Operators, which expresses their non-duality. In turn, this Logical Momentum Operator decomposes into the sum of the Spin Operator and the Time Operator. The former Spin operator allows to express quantum physics statements into logical statements and vice versa. The Time Operator represents a distinction between cognitive states in this Matrix Logic as its action on two cognitive states amounts to compute the difference between them. We find again, the differences that produce differences. [19] The Time Operator through its matrix representation has a dynamical interpretation in terms of rotations, since it turns to be a primeval ninety degrees rotation in the 2-plane of all cognitive states. Thus, also angular momentum is related to it. This will have a crucial role in the third series of topics that make this article. We shall start by relating this subjective Time Operator to intention-control, to the Myth of Eternal Return as a self-referential process, and to the existence of chronomes, time waves and patterns in natural and human systems and phenomenae. We shall relate the Time Operator with the perception of depth considered in the phenomenological philosophy of Merleau-Ponty as the primeval dimension, and further relate it with a possible

[23] and, most emphatically, to Musès [83].

[18] The philosophically inclined reader unincumbed with mathematical intrincacies might skip these sections and proceed to the more conceptual sections that follow.

[19] The introduction of a time operator in Quantum Mechanics was a very controversial issue following Pauli's initial attempts. It reappeared in the work of the Brussels school leadered by Prigogine [5], and more recently in the work by Schommers through the time-energy uncertainty relation [125].

solution to the problem of hemilateral eyes synchronization through universal chronomes (time structures and processes) such as the Kozyrev resonance entanglement due to torsion fields [66, 107, 134]. We shall discuss the relation of time and space perception in terms of ATP's metabolism to relate them to the energetics of neurons. We shall extend this to the quantization of time and the appearance of the perception of motion. We shall characterize cognition as the projection on the plane of all cognitive states in Matrix Logic of a vortex torsion structure. We shall discuss the relation between the complex logarithmic map that gives rise to torsion potential vortex and its singularities as the fundamental quanta, as the *analytical* map that allows to represent the visual (in the foveal area), somatotopic, motor, auditory and integrated perceptions in a plane in the neurocortex. The singularities of this map will appear to be the loci of the two-dimensional sections of the neurocortex in which *all* the field orientations are superposed, yielding another example of the *plenumpotence* of 0, which already appeared in the light eigenstates of the Null Logical Operator, **0**, the Mind Apeiron. This is the singularity of the Klein-bottle, the hole for which the whole unfolds to return to the singularity. We shall see that the Klein-bottle yields the *topological* map representation of the sensorium, a topographic map superposed to the analytical complex logarithm to which is related, as we shall discuss below. Further, we shall discuss the appearance of vortical torsion structures in the striate neurocortex and its relation with Karman vortices in viscous fluid dynamics as the 'interiorization' of the 'outer' torsion fields, which describe the representations of the field of orientations of visual stimuli -as it appears in the hypercolumnar structure in the striate neurocortex proposed by Hubel and Wiesel [54]. We shall relate these discontinuities with the natural appearance in the striate neurocortex of the Klein-bottle to the enaction of continuity of the representations of orientations. We shall further see that the complex logarithm map lead to diffusion processes in the neurocortex that can be associated with developmental morphogenetic growth patterns. Also we shall discuss the relation between the complex logarithmic map retinotopic representation and its function to integrate elementary eye distinctions to yield the binocular depth perception, thus returning to the integration of torsion vortex fields and stereoscopic vision, and the appearance of time discussed before, but from a neurological point of view that reaffirms its previous quantum explanation as entanglement through a torsion resonance effect. Finally, we shall discuss the relations between morphogenesis and anatomy-physiology and non-orientability in the Moebius and Klein-bottle surfaces, in particular of the human and mammal heart.

In all these developments the Klein-bottle plays an essential generative role, as the embodiment of self-reference and torsion [62], and of thought as much as crucial to perception and to neurology [137, 109]. Already in Rosen's work we find that the idea that the origins of his thought on self-reference and paradox can be established in terms of a topological phenomenology that is traced back to Merleau-Ponty and Heidegger, which Rosen establishes in terms of the Klein-bottle. In contrast with Aristotelian dualism this (genus 0 Riemann) surface is both open *and* closed, continuous *and* discontinuous, inside and outside are fused since it has a single side, and fundamentally non-orientable. This surface can be seen -in a first approach- as having an uncontained part (the subject), a contained part (the object) and a containing part (space) [62, 116]. Yet, there is an in-formation flow from the uncontained to the contained part which we could inquire if it can be reduced to the Cartesian cut view of space as a container. The Moebius band which is another surface of paradox can be actually realized in three-dimensional space by taking a band, twisting it and glueing its opposite extremes and thus we have a surface of paradox which is non-orientable as the Klein-bottle, but it can still be thought as satisfying the classical figure of object on space independent of the subject. Yet, in distinction with the Moebius band, the Klein-bottle is not a Cartesian object. [20] Those objects are space occupying and space is thus a mere container. The Cartesian conception counters the experimental and theoretical studies in visual perception that go back to the master painters of Renaissance and more contemporarily in the work of Luneburg [70], Heelan (a former student of Heisenberg) [44] and Indow [56]. The construction of the Klein-bottle can be algorithmized as an initial identification of two sides of a rectangle (or more generally by a closed cleavage by a distinction of plane free of singularities which is thus topologically deformable to a rectangle) by glueing two opposite sides with the same orientation and identifying the other sides having *opposite* orientations which makes this figure impossible to be actually constructed in $3D$ in distinction with the Moebius band which can be seen as contained in space in a Cartesian sense, though it is non-orientable. [21] This creates a topologically imperfect model in $3D$ since a hole has to be produced so that its construction already introduces singularities which then through the in-formation flow produces the whole structure, so that the whole structure is produced from a hole, and this returns to the singularity to complete the flow. Remarkably, this will manifest the torsion embodied in the Klein-bottle jointly with the singularity. This is further related to D. Bohm's holomovement [13, 99] and the integrality of the paradoxical structure to his implicate order, while the singularity is related to the explicate structure, in a first approximation. Indeed, in this wholeness surging from a singularity and back to it, what is at stake is the integral structure from which the implicate and explicate orders are instances and interchangeable through the flow [13, 99]. In this sense, our repetitious expression of fusion of object and subject should not be a conceived as a mere reparation of the Cartesian cut, but rather an indication of the integral structure of which both are instances in a process in which they are unseparable. Indeed, this is an holographic structure, and as such is constructed by the neurocortex as we already discussed before. The Cartesian mindset attitude to this would be to view the Klein-bottle as embedded in $4D$ where the hole is no longer necessary and in doing this, the concrete real figure is cast into an idealization which cannot be manifested by the subject (the Cartesian minded mathematician who thus keeps detached from this abstract ideal in-formation now ideally contained in the Cartesian view). Rosen's stance, to which we adhere, is instead to keep the hole -so that singularities are unavoidable as in quantum physics or already in the geometrical model of the photon we briefly presented above- as the starting point for questioning the Cartesian stance. In distinction with a Moebius band, a torus or any other object in $3D$, the loss of continuity of the Klein-bottle is necessary showing that $3D$ space is unable to contain the surface in the Cartesian stance for ordinary objects. So instead of abstracting by incorporating a fourth additional dimension (as is the proposal of Special and General Relativity), we keep the hole that produces the wholeness and instead of an additional dimension, we think of the depth dimension as the primeval dimension which becomes the source for the Cartesian dimension.[22] To re-

[20] In spite of the repeated claims of J. Lacan's acolytes in contradiction of their guru's writings.

[21] Actually, the Klein-bottle can be weaved uni-dimensionally, a feat accomplished by Sheilah Morgan; our gratitude to Dr. Melanie Purcell for pointing this out to the author.

[22] This is already apparent in the multivalued perception and logic elicited by the Necker Cube [85, 116].

sume, the Klein-bottle instead of being contained in space it contains itself and the flow of in-formation (of action, to start with) that is associated to this topology, is the manifestation of this self-containment, this paradoxical situation which becomes real through the production of a singularity which produces the whole structure. By doing this, it supersedes the Cartesian cut and the Aristotelian dualism, by superseding the dicothomy of container and contained, and in semiotic terms, of interpreter and interpreted. We can additionally discern from the previous discussion, that the in-formation process of self-reference, i.e. of consciousness which through the laws of thought which are not longer those of Aristotelian dualism, transforms the 'outside' into the 'inside' world (this transformation is the fourth ontological locus that Gunther proposed for surmounting Aristotelian dualism: thought as a process [39]; for further discussion see [109]), and this transformation produces a relation between thought, logic and the physical world (and thus is essentially logo-physical), and topo-logical superposition states have a genetic role, and is further related with the actual process of transformation of the 'outside' and 'inside' realms. Since discontinuity can be seen as the source for wholeness, one can enquire on the role of quantization associated to the topology of the Klein-bottle and the in-formation process that is associated to this singularity and the self-referential topology, and furthermore the role of quantization with regards to the multivalued logic that is associated to this in-formation structure and its paradoxical character. We shall deal with these questions below. Already Rosen established a link between the Klein-bottle and quantization and still with Musès hypernumbers which incorporate not only non-trivial square roots of -1 but also of $+1$, the latter being associated to spinors, and more concretely, with the Pauli matrices of quantum mechanics [116] (2008)[82]), further applied to a cosmology placed in terms of the hypernumbers which are positive square roots of $+1$.

Returning to the issue of the organization of the presentation of the conception in this chapter, the first topic (comprising Sections II-IV) will deal with a geometrical theory for the characterization of quantum jumps in terms of spacetime singularities produced by a torsion field given by a closed yet not exact differential one-form given by the logarithmic differential of a wave function propagating on spacetime as a lightlike singularity described by a nilpotence condition: the eikonal equation of geometrical optics for light rays.[23] We shall show that quantum jumps are produced precisely when the complex logarithm of a wave function that acts as the source of this torsion singularity (a spacetime dislocation) becomes singular due to the nodes of this wave. These geometrical structures with trace-torsion field including the Hertz potential which has subluminal and superluminal so-

As we stated already, in principle, instead of considering additional dimensions in disregard of the actual link between perception, self-reference, time and paradoxical logic, we follow the phenomenological approach that threads them to a single lifeworld, and thus dimension 2 appears to be the appropiate dimension for consciousness, until ontological and perceptual considerations indicate otherwise. Rather than following the path of considering more complex geometries, we suggest to follow Musès, in considering more general number systems, hypernumbers, and the non-associative algebras related to them [82]. The suggestion of Musès is founded in the observation that the complexity of thinking intuitively follows the increasing complexity of algebraic operations, in which we relinquish commutativity, as we have done by passing from Boolean to Matrix Logic, and further to relinquish associativity, and to consider algebras rich with idempotents and nilpotent elements, which is already the case of Matrix Logic. This can be seen already in the introduction of the algebraic structures that sustain the geometrical methods in theoretical physics, such as Clifford algebras.

[23]The reader unincumbent with mathematical complexities may give it a try skipping the first topic and proceed to the second one in Section V and VI.

lutions of the Maxwell equations, yield a theory of unification of spacetime geometries, non-relativistic and relativistic quantum mechanics [107], Brownian motions and fluid and magnetofluid-dynamics [103] and the application to obtain representations for the solutions of the Navier-Stokes equation [103, 105], non-equilibrium and equilibrium statistical thermodynamics [104], a torsion based theory of the electroweak interactions [113] and still the strong interactions as characterized by Hadronic Mechanics [108]. [24] The relations between torsion and the Coriolis force have been studied in [104, 42], and those of torsion and spin in classical mechanics in [112].

Let us examine in the final paragraphs of this Introduction, the background for the first part of this article. In his theory of gravitation that stemmed from his criticism of General Relativity (GR), V. Fock showed that light rays described by the eikonal equations of geometrical optics, were at the basis for the possibility of introducing 'objective' [25] spacetime coordinates and furthermore for the construction of a theory of gravitation based on characteristic hypersurfaces of the Einstein equations of GR [31]. These equations being hyperbolic partial differential equations have propagating wavefronts that arise as singularities of spacetime which are identical to the wavefronts singular solutions of Maxwell's covariant equations of electromagnetism: they are all characterized by the solutions of the eikonal equation. These singular propagating fields stand for the inhomogenities of the otherwise uniform spacetime that the geometry of GR based on metrics lead to; this is also a common feature with a theory of spacetime conceived in terms of Cartan geometries with torsion (which is more fundamental, as the Bianchi equations show [32]) rather than the curvature produced by a metric. Without inhomogenities it is impossible to give sense to a geometrical locus as argued by Fock and separately, from a perspective based on torsion in [109]. In fact, Fock further proved that the Lorentz transformations of special relativity arise together with the Moebius (conformal) transformations as the unique solutions of the problem of establishing a relativity principle for observers described by inertial fields. As showed by Fock, it is *not* the Lorentz invariance of the Maxwell's equation what makes Lorentz invariance so important in special relativity paving the way to a diffeomorphism invariant theory of gravitation which Einstein insisted in relating to special relativity, but rather the fact that the *singular* solutions of the Maxwell equations are invariant by the Lorentz transformations and still, by the full conformal group. We must recall, that already in 1910, Bateman discovered the invariance of Maxwell's equations by this fifteen dimensional Lie group. The equivalence class of reference systems transformable by Lorentz

[24]To resume, torsion appears with a Janus face as the logo-physical-geometrical field which is incorporated into the codification by signs of the non-integrable equations of constraint that bridge the Cartesian (epistemic) Cut (and thereby allow for memory, measurement, and control in systems theory) [91], as the primitive distinction in the calculus of distinctions in the protologic of Spencer-Brown. Topologically embodied as the Klein-bottle, they generate Matrix Logic derived from a non-dualistic approach through paradox leading to multivalued logics and quantum superposition in cognition [109]. In this work, quantum field operators described by nilpotent hypernumbers, are associated to logical operators, establishing thus a connection between quantum field theory, nilpotents and multivalued logics. Nilpotence, which in our view rather should be called as plenumpotence as much as the vacuum should be called the Plenum, has a crucial role in the nilpotent universal rewrite system [119].

[25]Fock's takes an approach based in dialectical materialism. In the phenomenological philosophical and dialectical approach by the present author for surmounting the epistemic cut, the photon is not an 'objective' particle, but the very signature of the fusion of object with subject, the latter being absent in the geometry of GR and unacknowledged by Fock due to his mantainance of the epistemic cut .

transformations preserve the singular solutions which further have the essential property of being the invariance of the fusion of subject-with-object singularities propagating at a finite constant invariant speed equal to c [31]. The velocity of light waves is no longer constant for observers transformable under conformal transformations, but can be infinite [2]. For all observers related by a Lorentz transformation, if anyone would identify a propagating discontinuity with velocity c, all of them would likewise identify the phenomena. Thus, while the Maxwell equations are well defined with respect to *all* diffeomorphic observers, the singular solutions with speed c are well defined for all *Lorentz* group related observers. Most importantly, the singular sets $N(\phi) = \{x \in M : \phi(x) = 0\}$ were introduced by Fock in terms of scalar fields which are solutions ϕ of the eikonal equation

$$(\frac{\partial \phi}{\partial x})^2 + (\frac{\partial \phi}{\partial y})^2 + (\frac{\partial \phi}{\partial z})^2 - (\frac{\partial \phi}{\partial t})^2 = 0, \tag{1}$$

which in the more general case of a space-time manifold provided with an arbitrary Lorentzian metric, say g, can be written as $g(d\phi, d\phi) = 0$, from which in the case of g being the Minkowski metric lead to the light-cone differential equation $(dt)^2 - (dx^1)^2 - (dx^2)^2 - (dx^3)^2 = 0$. Notice that eq. (1) is a nilpotence condition on the field $d\phi$ with respect to the Lorentzian metric g. But while the Maxwell equations are invariant by these two groups (Lorentz and Moebius-conformal) transformations, one could look for propagating waves that remain solutions of the propagation equation determined by the metric-Laplace-Beltrami operator, $\triangle_g$, which we shall describe below- under arbitrary perturbations. Instead of considering solutions of the wave equation $\triangle_g \phi = 0$, which form a linear space, we want to investigate the class of solutions which are further invariant under action given by composition of arbitrary (with certain additional qualifications) perturbations f (real or complex valued) acting on the ϕ's by composition, $f(\phi)$ that verify the same propagation equation: $\triangle_g f(\phi) = 0$. Notice that in these considerations we are concerned with singularities propagating on spacetime which is seemingly torsionless; we shall prove in the course of this work that this is not the case: The composite functions $f(\phi)$ and the ϕ themselves will be shown to generate torsion under certain conditions to be established below. We start introducing the geometrical-analytical setting with torsion.

2. Riemann-Cartan-Weyl Geometries with Torsion Fields and Their Laplacians

In this section that follows [110] M denotes a smooth compact orientable n-dimensional manifold (without boundary) provided with a linear connection described by a covariant derivative operator $\tilde{\nabla}$ which we assume to be compatible with a given metric g on M, i.e. $\tilde{\nabla} g = 0$. Given a coordinate chart (x^α) $(\alpha = 1, \ldots, n)$ of M, a system of functions on M (the Christoffel symbols of $\tilde{\nabla}$) are defined by $\tilde{\nabla}_{\frac{\partial}{\partial x^\beta}} \frac{\partial}{\partial x^\gamma} = \Gamma(x)^\alpha_{\beta\gamma} \frac{\partial}{\partial x^\alpha}$. The Christoffel coefficients of $\tilde{\nabla}$ can be decomposed as $\Gamma^\alpha_{\beta\gamma} = \{^{\ \alpha}_{\beta\gamma}\} + \frac{1}{2} K^\alpha_{\beta\gamma}$ [7,9,10]. The first term in this decomposition stands for the metric Christoffel coefficients of the Levi-Civita connection ∇^g associated to g, i.e. $\{^{\ \alpha}_{\beta\gamma}\} = \frac{1}{2}(\frac{\partial}{\partial x^\beta} g_{\nu\gamma} + \frac{\partial}{\partial x^\gamma} g_{\beta\nu} - \frac{\partial}{\partial x^\nu} g_{\beta\gamma}) g^{\alpha\nu}$, and $K^\alpha_{\beta\gamma} = T^\alpha_{\beta\gamma} + S^\alpha_{\beta\gamma} + S^\alpha_{\gamma\beta}$, is the cotorsion tensor, with $S^\alpha_{\beta\gamma} = g^{\alpha\nu} g_{\beta\kappa} T^\kappa_{\nu\gamma}$, and $T^\alpha_{\beta\gamma} = (\Gamma^\alpha_{\beta\gamma} - \Gamma^\alpha_{\gamma\beta})$ the skew-symmetric torsion tensor. We are interested in (one-half) the Laplacian operator associated to $\tilde{\nabla}$, i.e. the operator acting on smooth functions, ϕ, defined on M by (see [103, 105])

$$H(\tilde{\nabla})\phi := 1/2\tilde{\nabla}^2\phi = 1/2g^{\alpha\beta}\tilde{\nabla}_\alpha\tilde{\nabla}_\beta\phi. \tag{2}$$

A straightforward computation shows that $H(\tilde{\nabla})$ only depends in the trace of the torsion tensor and g, so that we shall write them as $H(g, Q)$, with

$$H(g, Q)\phi = \frac{1}{2}\triangle_g\phi + \hat{Q}(\phi) \equiv \frac{1}{2}\triangle_g + Q.\nabla\phi, \tag{3}$$

where $Q := Q_\beta dx^\beta = T^\nu_{\nu\beta}dx^\beta$ the trace-torsion one-form, and $\hat{Q}$ is the vector field associated to Q via g: $\hat{Q}(\phi) = g(Q, d\phi) = Q.\nabla\phi$, (the dot standing for the metric inner product) for any smooth function ϕ defined on M; in local coordinates, $\hat{Q}(\phi) = g^{\alpha\beta}Q_\alpha\frac{\partial\phi}{\partial x^\beta}$. Finally, $\triangle_g$ is the Laplace-Beltrami operator of g: $\triangle_g\phi = \mathrm{div}_g\nabla\phi$, $\phi \in C^\infty(M)$, with div_g and ∇ the Riemannian divergence and gradient operators ($\nabla\phi = g^{\alpha\beta}\partial_\alpha\phi\partial_\beta$), respectively; of course, on application on scalar fields, $\tilde{\nabla}$, ∇^g are identical: it is in taking the second derivative that the torsion term appears in the former case. Thus for any smooth function, we have $\triangle_g\phi = (1/|det(g)|)^{\frac{1}{2}}g^{\alpha\beta}\frac{\partial}{\partial x^\beta}(|det(g)|^{\frac{1}{2}}\frac{\partial\phi}{\partial x^\alpha})$. Thus $H(g, 0) = \frac{1}{2}\triangle_g$, is the Laplace-Beltrami operator, or still, $H(\nabla^g)$, the Laplacian of Levi-Civita connection ∇^g given by the first term in eq. (3). The connections $\tilde{\nabla}$ defined by a metric g and a purely trace-torsion Q are called RCW (after Riemann-Cartan-Weyl) connections with Cartan-Weyl trace-torsion one-form, hereafter denoted by Q [103, 105].

3. The Quantum Jumps Functional and Torsion

In the following we shall take g to be a Lorentzian metric on a smooth time-oriented space-time four-dimensional manifold M which we assume compact and boundaryless; we have the associated volume n-form given by $\mathrm{vol}_g = |\det(g)|^{\frac{1}{2}}dx^1\wedge\wedge dx^2\wedge dx^3\wedge dx^4$, where (x^1, x^2, x^3, x^4) is a local coordinate system. The solutions of the wave equation constitute a linear space. Furthermore, the germ of solutions of the wave equation in a neighborhood of a point form a linear space. Thus, the algebra generated by a single solution of the wave equation

$$\triangle_g\phi = 0, \tag{4}$$

consists of solutions of this equation if and only if ϕ satisfies in addition the eikonal equation of geometrical optics

$$(\nabla\phi)^2 := g(\nabla\phi, \nabla\phi) = 0. \tag{5}$$

Indeed, if f is of class C^2 (twice differentiable) and ϕ is real-valued, or still, if f is analytic and ϕ is complex valued, then the following identity is valid

$$\triangle_g(f(\phi)) = f'\triangle_g\phi + f''(\nabla\phi)^2. \tag{6}$$

The solutions of the system of equations

$$\begin{aligned} \triangle_g \phi &= 0 & (7)\\ (\nabla\phi)^2 &= 0, & (8) \end{aligned}$$

are called monochromatic waves. They represent pure light waves; we discussed already above their relevance. A set of monochromatic waves having the structure of an algebra, will be called a monochromatic algebra. In Fock's approach, they are called *electromagnetic signals* [31]. Notice that the eikonal equation is a nilpotence condition for $d\phi$, the differential of ϕ, or equivalently its gradient, $\nabla\phi$, under the square multiplication defined by the metric. From the identity $e^{-i\phi}\triangle_g e^{i\phi} = i\triangle_g\phi - (\nabla\phi)^2$, we obtain , if $\triangle_g\phi = 0$, $(\nabla\phi)^2 = -e^{-i\phi}\triangle_g e^{i\phi}$, Let us consider the mapping $\phi \to e^{i\phi} = \psi$ which transforms the linear space of solutions of the wave equation into a multiplicative $U(1)$-group, in which the kinetic energy integrand in the Lagrangian functional $(\nabla\phi)^2$ is transformed into $-\frac{\triangle_g\psi}{\psi}$, which has the familiar form of the quantum potential of Bohm, yet in a relativistic domain [13, 106]. If the ϕ are real valued, then the ψ are bounded and we can embed the above group in the Banach algebra under the supremum norm that it generates under pointwise operations and further completion [6]. To distinguish between them we call the original linear space the functional phase space **S** and the Banach algebra defined above as the algebra of wave states **A**, or simply the functional algebra of states. It is simple to see that the critical points of the functional

$$J(\psi) = \int \frac{\triangle_g\psi}{\psi}\text{vol}_g \tag{9}$$

are those ψ which satisfy

$$\triangle_g \ln\psi = 0, \tag{10}$$

i.e., those whose phase function satisfy the wave equation. Those intrinsic states will be called *elementary states*. The new representation has two advantages over the original one. It is richer in structure and in elements, as **S** is mapped into a subset of the set of invertible elements Ω of **A**, and so, by taking logarithm pointwise, on the elements of Ω, we obtain an enlargement of **S** by possibly multivalued functions. The second advantage, that actually justifies the whole construction, is that the integrand of the Lagrangian $-\frac{\triangle_g\psi}{\psi}$, when integrated, exhibits jumps across the boundary $\partial\Omega$ of Ω. These jumps do correspond to kinetic energy changes, but in the interpretation of the integrand as a quantum potential, these changes represent a change due to the holographic information of the system present in the whole Universe, in D. Bohm's conception [13]. Let **A** be a Banach algebra of continuous complex-valued functions defined on a four-dimensional Lorentzian manifold (M, g), containing the constant functions, closed under complex-conjugation, with the algebraic operations defined pointwise and the supremum norm and containing a dense subset $\mathbf{A_2}$ of twice differentiable functions which are mapped by the Laplace-Beltrami operator $\triangle_g$ into **A**. Assume further $f \in \mathbf{A}$ is invertible with inverse $f^{-1} \in \mathbf{A}$ if and only if $\inf_{\mathrm{M}}|f(x)| > 0$. The set of invertible elements is denoted by Ω. Furthermore, assume a positive linear functional, denoted by λ such that $\lambda : \mathbf{A_2} \cap \Omega \to C$ (the complex numbers) defined by

$$\lambda(\phi) = \int \frac{\triangle_g\phi}{\phi}\text{vol}_{\text{g}} \tag{11}$$

The critical elements of λ are those u such that

$$\text{div}(\frac{\text{grad}u}{u}) = 0, \quad i.e. \ \frac{\triangle_g u}{u} - (\frac{\text{grad}u}{u})^2 = 0. \tag{12}$$

If the linear functional is strictly positive, i.e. $\lambda(\phi) = 0$ if and only if $\phi \equiv 0$, these two identities are to hold in **A**, otherwise in the sense of the inner product defined by λ on **A**. By eq. (12) the set **C** of critical points of λ is clearly a subgroup of Ω. The monochromatic functions of **A** are as before, those $w \in \mathbf{A}_2$ satisfying the system of eqs. (7, 8) and their set is denoted by **M**. From eq. (10) the composition function given by $f(w)$ belongs to **M** again if f is an analytic function on a neighborhood of the set of values taken by w on M. Since by eq. (12) $\mathbf{M} \cap \Omega \subset \mathbf{C}$, we have that $uf(w) \in \mathbf{C}$ if $w \in \mathbf{M}$ and $f(w) \in \Omega$. The spectrum $\sigma(v)$ for any $v \in \mathbf{A}$, is defined by $\sigma(v) = \{z \in C/|v - ze| \notin \Omega\}$ and therefore, by a previously assumed property, is the closure of the set of values $v(x)$ taken by v on M. It is obviously a compact non-void subset of **C**. Ω has either one or else infinitely many maximal connected components, of which Ω_0 is the one containing the identity, e, defined by $e(x) \equiv 1$. Two elements f, h belong to the same component of Ω, if and only if $fh^{-1} \in \Omega_0$. Further, $f \in \Omega - \Omega_0$ if and only if its spectrum $\sigma(f)$ separates 0 and ∞. The logarithm function, as a mapping from **A** into **A** is defined only on Ω_0. With these preliminaries, we can now show that the quantum jumps arise as a generalized form of the standard argument principle. [26]

Theorem. Let $u \in \mathbf{C}, w \in \mathbf{M} \cap \Omega$, i.e, it is an invertible monochromatic function. Denote by $H_1, H_2, \ldots$, the maximal connected components of the complement of $\sigma(w)$. Then there exists fixed numbers $q_i, i = 1, \ldots$, depending on u and w only, such that for any function $f(z)$ analytic in a neighborhood of $\sigma(w)$ and with no zeros in $\sigma(w)$, we have

$$\lambda(uf(w)) = \lambda(u) + \sum_i (N_i - P_i)q_i, \tag{13}$$

where N_i, P_i are the number of zeros and poles, respectively, of f in H_i, $i = 1, 2, \ldots$. In particular choosing $\alpha_i \in H_i$, the q_i are given by

$$q_i = 2\int g(\frac{\nabla u}{u}, \frac{\nabla w}{w - \alpha_i})\text{vol}_g, i = 1, 2, \ldots. \tag{14}$$

Proof.[27] Let $f = f(w) \in \mathbf{M}$ with $f(z)$ as in the hypothesis. A computation yields

$$\frac{\triangle_g(uf)}{uf} - \frac{\triangle_g u}{u} = 2g(\frac{\nabla u}{u}, \frac{\nabla f}{f}), \tag{15}$$

which we note that it is another way of writing

$$\frac{\triangle_g(uf)}{uf} = \frac{1}{u}H(g, \frac{df}{f})(u), \tag{16}$$

[26] The following result [110] is a simpler geometrical version of a theorem proved by Nowosad in the more intricate setting of non-compact manifolds and functionals on generalized curves in L.C. Young's calculus of variations for curves with velocities having a probability distribution (Young measures) [86]. In our approach that surmounts the epistemic cut, we are interested in a particular Riemann surface, the Klein-bottle.

[27] An example. Take a compact submanifold of Minkowski space and plane waves with adequate boundary periodicity conditions. Take $u = e^{ik.x}, w = e^{ik_0.x}, k_0^2 = 0, k_0.k \neq 0$ and the spectrum $\sigma(w) = S^1$, where S^1 is the unit circle; then $\lambda(u) = -k^2$ (minus the mass squared) and eq. (13) becomes $-\lambda(e^{ik.x}f(e^{ik_0.x})) = k^2 + 2(k_0.k)(N - P)$, where N and P are the number of zeros and poles of f inside of the unit circle.

where we have introduced in the r.h.s. of eq. (15) the laplacian defined in eq. (3) by a RCW connection defined by the metric g and the trace-torsion Q = $\frac{df}{f}$. Integrating eq. (15) yields,

$$\lambda(uf) - \lambda(u) = 2\int g(\frac{\nabla u}{u}, \frac{\nabla f}{f})\mathrm{vol_g}. \tag{17}$$

In particular this shows that q_i in eq. (14) are well defined. From (17) one gets directly

$$\begin{aligned} \lambda(ufh) - \lambda(u) &= [\lambda(uf) - \lambda(u)] + [\lambda(uh) - \lambda(u)] && (18)\\ \lambda(uf^{-1}) - \lambda(u) &= -[\lambda(uf) - \lambda(u)], && (19) \end{aligned}$$

where $h = h(w)$ as well as we recall $f = f(w)$, are the composition functions, from now onwards. Now if $f \in \Omega_0$, then $\ln f \in \mathbf{A}$ and $\nabla \ln f = \frac{\nabla f}{f}$, which substituted in eq. (17) gives, upon integration,

$$\lambda(uf) - \lambda(u) = 2\int g(\frac{\nabla u}{u}, \nabla \ln f)\mathrm{vol_g} = -2\int f\mathrm{div_g}(\frac{\nabla u}{u})\mathrm{vol_g} = 0, \tag{20}$$

by eq. (12). Hence

$$\lambda(uf) = \lambda(u), \text{ if } f \in \Omega_0. \tag{21}$$

If now f, h belong to the same component of Ω we can write $uh = (uf)(hf^{-1})$, and since $hf^{-1} \in \Omega_0$, the previous result yields

$$\lambda(uh) = \lambda(uf). \tag{22}$$

This shows that $\lambda(uf(w))$ is locally constant in Ω as f varies in the set of analytic functions. Let now $f(z) = z - \nu$ with $\nu \in H_i$. Then $z - \nu$ can be changed analytically into $z - \alpha_i$ without ν leaving H_i, which means that $w - \nu e$ and $w - \alpha_i e$ are the same connected component of Ω, with $e \equiv 1$. Therefore from eqs. (22, 17) and eq. (14) follows that

$$\lambda(u(w - \nu e)) - \lambda(u) = q_i, \tag{23}$$

and by eq. (22)

$$\lambda(u(w - \nu e)^{-1})) - \lambda(u) = -q_i. \tag{24}$$

On the other hand, if ν belongs in the unbounded component of the complement of $\sigma(w)$, we may let $\nu \to \infty$ without crossing $\sigma(w)$ so that

$$\begin{aligned} \lambda(u(w - \nu e)) - \lambda(u) &= 2\int g(\frac{\nabla u}{u}, \frac{\nabla w}{w - \nu e})\mathrm{vol_g} \\ &= \lim_{\nu\to\infty} 2\int g(\frac{\nabla u}{u}, \frac{\nabla w}{w - \nu e})\mathrm{vol_g} = 0. && (25) \end{aligned}$$

Therefore, if $f(z) = c_0\Pi_{i=1}^N(z - a_i).\Pi_{j=1}^p\frac{1}{z-b_j}, c_0 \neq 0, a_i, b_j \notin \sigma(w)$, then eq. (13) follows from eqs. (18, 24, 25) In the general case , if $f(z)$ is an holomorphic function in a neighbourhood of $\sigma(w)$,without zeros there, we can find a rational function $r(z)$ such that

$$|f(z) - r(z)| < \min_{\sigma(w)}|f(z)| \text{ in } \sigma(w), \tag{26}$$

by Runge's theorem in complex analysis. Then, $r(z)$ has no zeros in $\sigma(w)$ too, and $r(w)$ and $f(w)$ are in the same component of Ω, so that eq. (13) holds for $f(w)$ too. The proof is complete.

Observations. The quantization formula (13) tells us how the basic functional changes when we perturb the elementary state u into $uf(w)$ with f analytic near and on $\sigma(w)$. Changes occur only when zeros or poles of $f(z)$ reach and eventually cross the boundary of $\sigma(w)$, and these changes are integer multiples of fixed quanta q_i, each one attached to the hole H_i whose boundary is reached and crossed, while u, v remain fixed. Two more aspects are important. The first one being that the actual jump is measured modulo the product of the q_i by a classical difference (where by 'classical' we stress that we mean that it is the substraction, in distinction to the quantum difference given by the commutator of operators) of poles and zeros; at the level of second quantization quantum jumps appear in terms of the difference of the creation and annihilation operators which defines a time operator in a logic in which the commutator of the true and false logical operators coincide with their classical difference, establishing thus a non-null torsion in cognitive space [109]. The second aspect is the actual form of the q_i which are given by integrating the internal product of the trace-torsion one-form $Q = \frac{du}{u}$ defined by the critical state u, with another almost logarithmic differential of the form $dw/(w - \alpha_i)$.

3.1. The Appearance of Torsion

Let C_u denote the linear operator $h \to uh, h \in \mathbf{A}, u \in \Omega$. The very simple analysis above hinges on the fact that $C_u^{-1} \circ \triangle_g \circ C_u - C_{\frac{\triangle_g u}{u}}$ is a derivation on the germ $\mathbf{F}(w)$ of functions of w (see eq. (11) and also eq. (16) to see how it is related to the torsion geometry), which are analytic in a neighbourhood of $\sigma(w)$, and it could have been performed abstractly without further mention to the special case under consideration. The general abstract theory of variational calculus extending the functional λ for quantum jumps when specialized to second order differential operators, say $\triangle_g$ or still $H(g, Q)$, shows that the condition $w \in \mathbf{M}$ in not only sufficient but also necessary in order to the quantum behaviour of λ occur [86].

The set of linear mappings $C_{f^{-1}}$ of $\mathbf{A}$ defined by $h \to f^{-1}h, h \in \Omega$, f defined on M, is a group which maps each connected component of Ω onto another one. In terms of functions defined on M it changes locally the scale of the functions, i.e. the ratio of any function at two distinct points is changed in a given proportion, and it therefore a gauge transformation of the first kind. Under this transformation we have that

$$\triangle_g \to C_{f^{-1}} \triangle_g C_f = \triangle_g + 2\frac{\nabla f}{f}.\nabla + \frac{\triangle_g f}{f} = 2H(g, \frac{df}{f}) + 2V_f, \tag{27}$$

where $H(g, \frac{df}{f})$ is the RCW laplacian operator of eq. (4) with trace-torsion 1-form $Q = \frac{df}{f}$ and $V_f = \frac{\triangle_g f}{f}$ is the relativistic quantum potential defined by f^2 [106]. Now noting that for vector fields $A = A^i \partial_i, B = B^i \partial_i$, with $A^i, B^i, i = 1, \ldots, 4$ complex valued functions on M, with the hermitean pairing defined by the metric g on M, i.e. $\int g(\bar{A}, B)\text{vol}_g = \int g(B, \bar{A})\text{vol}_g$ so that $A^\dagger = \bar{A} = \bar{A}^i \partial_i$. Therefore for the gauge transformation $d \to d + \frac{df}{f}$, since $\triangle_g = -d^\dagger d$ (see [9,10,15]), we further have the transformation

$$-d^\dagger d \rightarrow -(d + \frac{df}{f})^\dagger (d + \frac{df}{f}) = -(d^\dagger + \overline{(\frac{df}{f})}.)(d + \frac{df}{f}). \quad (28)$$

where $d^\dagger$ is the adjoint operator,the codifferential, of d with respect to this hermitean product so that $d^\dagger = -\text{div}_g$ on vectorfields [32]. If we assume that $\overline{(\frac{df}{f})} = -\frac{df}{f}$, so that $|f(x)| \equiv 1$ and thus f is a phase factor, $f(x) = e^{i\phi(x)}$, i.e. a section of the $U(1)$-bundle over M then the r.h.s. of eq. (28) can be written as

$$-(d^\dagger - \frac{\nabla f}{f}.)(d + \frac{df}{f}) = \triangle_g + 2\frac{\nabla f}{f}. + (\frac{df}{f})^2 + \frac{\triangle_g f}{f} - (\frac{df}{f})^2$$
$$= (\triangle_g + 2\frac{\nabla f}{f}.) + \frac{\triangle_g f}{f} = 2H(g, \frac{df}{f}) + \frac{\triangle_g f}{f} = C_{f^{-1}} \circ \triangle_g \circ C_f. \quad (29)$$

Consequently, if f is a phase factor on M, then under the gauge transformation of the first kind $h \rightarrow f^{-1}h$, the change of $\triangle_g$ into $C_{f^{-1}} \circ \triangle_g \circ C_f$ can be completely determined by the transformation $d \rightarrow d + \frac{df}{f}$ which is nothing else than the gauge-transformation of second type, from the topological (metric and connection independent) operator d to the covariant derivative operator $d + \frac{df}{f}$, of a RCW connection whose trace-torsion is $\frac{df}{f}$, equivalent to the gauge transformation $d \rightarrow d + A$ in electromagnetism [32].

In summary, when f is a phase factor, the gauge transformations of the first and second type are equivalent, and gives rise to the exact Cartan-Weyl trace-torsion 1-form. Whenever the metric g is Minkowski or positive-definite, these gauge transformation produce a transformation of a Brownian motion with zero drift to another Brownian motion with drift given by $\frac{\nabla f}{f}$. The node set of f, which coincides with the locus of quantum jumps, becomes an impenetrable barrier for this Brownian motion [7]. If we further impose on f the condition similar to the one placed for the electromagnetic potential 1-form, A, to satisfy the Lorenz gauge $\delta A = 0$, i.e. $\delta(\frac{\text{grad} f}{f}) = 0$, we find that this is nothing else than the condition on f to be an elementary state i.e. a critical point of the the functional $\lambda(f)$ given by (11). Therefore, when f is a phase factor, both the first and second kind of gauge transformations are equivalent and they give rise to a Cartan-Weyl one-form $Q = \frac{df}{f}$.

When $\frac{df}{f}$ cannot be written globally as $d\ln f$, f is said to be a non-integrable phase factor. When f belongs to the algebra $\mathbf{A}$, this is equivalent to saying that f does not have a logarithm in $\mathbf{A}$, which means that $f \in \Omega - \Omega_0$. In any case, the 2-form of intensity $F = d(\frac{df}{f})$ is always identically 0 because $\frac{df}{f}$ can be *locally* written as $d\text{Log}f$, where Log is a pointwise locally defined logarithm determination.[28]

Consider now all the connected components Ω_α of Ω. Any such component can be transformed into Ω_0 by a gauge-transformation of the first kind: it suffices to take $f \in \Omega_\alpha$ and consider $h \rightarrow f^{-1}h$, which is indeed a diffeomorphism of Ω. This choice of the component, is a choice of gauge, and of course, there is no preferred gauge. That is, the topological operator d of one observer becomes the covariant derivative operator $d + \frac{df}{f}$ of a RCW connection for the other observer. We can interpret the difference of gauges as being

[28]The relation between Cartan torsion, singularities and dislocations in condensed matter physics is well known [63].

equivalent to the presence of the trace-torsion 1-form $\frac{df}{f}$ in the second observer's referential. However as the electromagnetic 2-form $F \equiv 0$, this is an instance of the Aharonov-Bohm phenomena: non-null effects associated with identically zero electromagnetic fields. As we said before, this difference has a Brownian motion correlate in which null drift for the former is transformed into the drift $\frac{\nabla f}{f}$ which at the level of random dynamics is a non-trivial transformation. In fluid-dynamics as described by the Navier-Stokes equations for a velocity vector field u, there is a similar transformation from a drift independent purely noise Brownian process, in which the velocity term is subsumed in a Laplacian with no interaction non-linear term as a purely diffusive process, into the Navier-Stokes Brownian process with drift given by u; see [103] (2002).

That there appear non-null effects is checked by our previous analysis of the functional $\lambda(uf(w))$, where u is any elementary state and f, besides being a phase factor, is also monochromatic. In this case λ, which is locally constant depends on which $\Omega_\alpha f$ belongs to, that is to say, on the choice of the gauge.

Finally, according to the two ways of interpreting a linear operator (as a mapping on the vector space or as a change of referential frames) we have two possibilities. Indeed let $w \in \mathbf{M}$ and let $f_t(w), t \in [0, 1]$ with $f_t(z)$ analytic in a neighbourhood of $\sigma(w)$, be a continuous curve on $\mathbf{A}$. For any $u \in \mathbf{C}$ we consider the curve of elementary states $uf_t(w)$; we described in eq. (13) the behaviour of $\lambda(uf_t(w))$ along this curve. In particular we considered $uf_t(w)$ as a perturbation, or excitation, of u evolving in time (here time may not be the time coordinate of a Lorentzian manifold but the universal evolution parameter introduced first in quantum field theory by Stuckelberg, and further elaborated by Horwitz and Piron [51].) We can also regard $u \rightarrow C_{f_t(w)}u$ as a continuous curve of gauge transformations of first kind acting on a fixed elementary state u, which, when f_t crosses $\partial\Omega$, determines a change of gauge. When that happens, it is obvious that f_t cannot be made a phase factor for all t, so that no electromagnetic interpretation can be given all along the evolution in t. However if, say, the initial states f_0 and f_1 are phase factors (i.e. $|f_i(x)| \equiv 1, i = 0, 1$), this change of gauge is equivalent to the appearance of a non-trivial trace-torsion one-form, which we can interpret as an electromagnetic potential, between the initial and final states. In any of these interpretations, a non-null effect is detected by a jump in λ as given by eq. (13); this quantum transition is interpreted in the first case as an excitation of the state u, and in the second state as a change of gauge of u, materialized by the appearance of the corresponding Cartan-Weyl one-form as an electromagnetic Aharonov-Bohm potential with zero intensity and non-null effects [1]. Thus, in this interpretation, quantum jumps are the signature of a non-trivial geometrical structure, the appearance of torsion.

3.2. Singular Sets

Finally we examine the dimensions of singular sets $N(f)$ of monochromatic functions. Recall that a C^2 real or complex-valued function f defined on (M, g) is a monochromatic wave, $f \in \mathbf{M}$, if it satisfies the system given by eqs. $(7, 8)$. In the real-valued case, all C^2 functions of f, and in the complex case, all analytic or anti-analytic functions of f belong to $\mathbf{M}$ again, by eq. (4) (we changed here our notation there, pointing precisely to $f = f(w)$ for $w \in \mathbf{M}$, as above) . If f is real, smooth and $df \neq 0$, then $N(f)$ is locally three-dimensional. If it is complex and $\mathrm{Re}(f)$ and $\mathrm{Im}(f)$ are functionally independent $N(f)$ is two-dimensional. Yet the Newtonian picture of a photon as an isolated point-like singularity moving with the speed of light in the vacuum, requires a one-dimensional singular set $N(f)$. Can we achieve this by going to hypercomplex, say quaternionic functions, or still Musès' hypernumbers which are rich in divisors of 0? The answer to the former question is negative; in the quaternionic framework, the photon is a propagating three-dimensional singularity with lower dimensional singularities, but still undivisely extended [111]; we shall present these issues in the following section.

3.3. Partial Conclusions to These Sections

This initial part of the present work started by considering the fundamental role of spacetime singularities and particularly light rays in establishing the physical world jointly with the act of (self) perception of the subject: Indeed, the photon is not seen, but is about seeing [156]. We argued that this allows for a start the introduction of spacetime itself, in terms of the fundamental role that differences and more generally inhomogeneities play. In our presentation we argued in terms of what originally appeared to be two distinct conceptions for producing this joint constitution of reality and the subject. The first conception relied on the self-referential role of the photon as a singularity which embodies the fusion of reality, cognition and perception, while the second one relied on torsion, which is also linked to self-reference [109]. Light rays are described by the eikonal equations which appear as wavefront singularities in the Maxwell and Einstein's equations of electromagnetism and General Relativity, respectively. Quantum jumps play a fundamental role in this joint constitution of physics and consciousness, as they appear as fundamental differences from which spacetime is constructed as a physical reality, while they aso have a fundamental role in visual perception. Without these quantum jumps, cognition as the difference produced from a primeval difference [109] would not occur [8]; a differential epistemology for science, departing from Bateson's concept of information as difference that produces difference, was elaborated by Johansen [58]. While propagating waves play a primeval role, the notion of closure by composition by perturbations led us to consider the action of analytical or alternatively C^2 functions on them, which produced that the propagation equation was supplemented by the eikonal equations for light rays, and thus we were lead to the class of monochromatic functions.[29] Thus we were able to give a formula that characterizes quantum jumps and further discovered that these compositions produce a torsion potential. These jumps occur whenever the perturbation ceases to be non-zero and thus become the locus for the singularities of the logarithmic map, establishing thus the generative role of 0. We shall reencounter in Sections VI and VII below, this generative role of 0, as a quantum-logical operator with all zero entries in Matrix Logic which is further associated with quantum fields. It will also appear in the visual complex logarithmic map representation in the neurocortex, in which the singularities appear as the loci where the orientation field provided by the map can take all possible values, thus becoming multivalued

[29] It is crucial for the later connection of the developments presented in the previous sections, with perceptual codification in the neurocortex, to remark that what is accessible to the visual and still peripheric sensory systems, such as the skin, are precisely the family of all perturbations waves that we have discussed in the previous sections. This will be the starting point to connect the physical and the physiological realms. With respect to the auditory system, it is important also to remark that the complex logarithmic function which acts on monochromatic waves, is also the topographic representation of this sensory mode [126, 127, 139].

in those singularities. In this setting, the lightlike singularities are non-pointlike extended structures which we shall later represent by twistors; in turn, these twistors appear to be the eigenstates of the null quantum-logical operator, which can still be represented by cognitive states and vice versa [111]. In the course of this work we discovered that propagating lightlike singularities which are fundamental to the establishment of special relativity, electromagnetism and still General Relativity, produce a torsion potential, and thus the two approaches which the present work took as its unfolding rationale, became unified. Furthermore, this potential which becomes singular in its node sets, acts as the drift (i.e., the average veocity) of Brownian motions which cannot penetrate the infinite barrier posed by these singularities. Finally, we established that this torsion potential naturally produces an Aharonov-Bohm effect, which is further associated with the quantum jumps so that non-null effects, fundamental differences in the sense discussed above, are manifested despite that the electromagnetic fields produced by the torsion potential is null.

Remarkably, the complex logarithm map plays a fundamental role in a topographical geometrical model of visual perception [126] -and as we shall discuss in this chapter, for other sensory modes and their integration- in which the singularities of the topographical orientation visual mapping -an order parameter as in superconductivity-, is associated to the singularities of vortex dislocations [127], i.e. trace-torsion singularities which we shall discuss below). These singularities led to a topological solution of the topographical visual -as well as other sensorial modes- map in terms of the Klein-bottle [137] [141] which is the basis for multivalued logics and its relation with quantum mechanics and quantum field operators [109]. Furthermore, it led to Gabor wavelet holographical representations on the visual cortex [73], substantiating the holographic quantum paradigm, due to Pribram, in neurophysiology [96]. The complex logarithm has been carried to quantum holography in terms of the coadjoint orbits representations of the Heisenberg group in quantum mechanics, and has led to important technological developments, in the work due to Schempp [124]. [30]

In considering the semiclassical theory of gravitation, quantum jumps produce discontinuities in the energy-momentum tensor leading to the existence of a cosmological time associated with a quantum-jumps time, in a *global* canonical decomposition of spacetime [76]. This approach due to Mashkevich departs from the incompleteness of the Cauchy

[30]There is an essential point that has already appeared in our previous considerations, in the treatment of quantum jumps through the complex plane, or still, the Riemann sphere, in the Klein-bottle, and finally, in the holographic codification in the visual cortex, and through the Heisenberg group representations that sustain holography: The common, basic dimension to all of them, is 2. We shall retrieve this all along the present chapter, in which this fact will reappear in all instances: 1) In the cognitive plane for Matrix Logic that allows logical operators to be treated as Musès hypernumbers and further quantum field operators ; 2) in the characterization of cognition as the projection to this plane of a vortex structure; 3) in the codification of sensorial modes in the neurocortex; 4) in the synchronization of hemilateral vision leading to stereoscopic visual representation which requires a depth variable associated to time and paradox yet can be subsumed into a single plane; 5) in the developmental growth of the body and, particularly, the anatomy-physiology of the heart. So rather than partaking in the usual practice of following the meme already installed by Special Relativity, of incorporating additional dimensions to account for complexity and more recently the particular case of consciousness, what will emerge in relief from the development of the present chapter, is the ever present dimension 2 associated to self-reference and the imaginary numbers. This will not result from any hypothesis, but from the unfolding of the mathematical structures that will be applied to the several issues. This appears to be another manifestation of the possible ruling on Nature, of the famous Occam's Razor Principle. For a different approach through the incorporation of additional dimensions to characterize a noetic space, we refer to the important work by Amoroso [3]

problem for the Einstein equations of GR: they provide only six equations for the ten components of the metric. For curved spacetime, Mashkevich proved that the diffeomorphism invariance of the solutions of the Einstein equation is not valid; only in the case of Ricci flat spacetime this is assured. This underdetermination is resolved by the canonical complementary conditions, which in the semiclassical approach are provided by nonlocal quantum jumps whilst in Fock's theory they are provided by four equations as eq. (4), the so-called harmonic coordinates. Thus, according to Mashkevich, quantum jump nonlocality is essential for GR, they occur in nonempty spacetime where the underdetermination problem arises and actually solves this problem. Furthermore, quantum jumps lead to a universe with complete retrodiction in which only partial prediction is possible [76]. This establishes a remarkable relation, which requires further ellucidation, with the theory of anticipatory systems introduced and developed by Dubois and others [24].

For closing remarks to this section, we note that quantum jumps were obtained here in terms of the quantum potential which stands for a holographic in-formation of the whole universe.

4. Monochromatic Hypercomplex Functions

In this Section we shall discuss the problem of the impossibility of localizing the photon as a point-like structure and determine its geometrical-analytical characterization; detailed proofs can be found in [111]. The starting basic issue is that for f defined on M as above, it can be proved that $N(f)$, the node set of f reduces to a single set. Let us introduce the quaternionic units $\vec{i_1}, \vec{i_2}, \vec{i_3}$ given by the multiplication rules

$$\begin{aligned} \vec{i_1}\vec{i_2} &= \vec{i_3}, \vec{i_2}\vec{i_3} = \vec{i_1}, \vec{i_3}\vec{i_1} = \vec{i_2} \\ \vec{i_j}\vec{i_k} &= -\vec{i_k}\vec{i_j}, k \neq j, \vec{i_k}^2 = -1, j, k = 1, 2, 3. \end{aligned} \tag{30}$$

Notice here that we could chose here the logical quaternions introduced before, and thus the structures we shall produce below, can be conceived as spacetime structures which are both 'inner' and 'outer' representations of the self-referential character of photons (though the neutrino is also considered below). We shall introduce the notation $(\phi, \psi) \in \mathbf{M}$ to mean that $\phi, \psi \in \mathbf{M}$ (i.e. they satisfy eqs. (7, 8)) *and* furthermore

$$g(\nabla\phi, \nabla\psi) = 0, \tag{31}$$

which is the requirement that any algebraic combination of ϕ, ψ belong in $\mathbf{M}$ as well. It will also be assumed that ϕ and ψ are functionally independent, to rule out the trivial cases. We then have the following theorem.

Theorem 1. Any monochromatic quaternion valued function F defined on (M, g) is determined by a triple of real valued functions (ϕ, f, ρ) such that

$$(\phi, f + i\rho) \in \mathbf{M}, \text{ i.e. } g(\nabla\phi, \nabla f + i\nabla\rho) = 0, \tag{32}$$

and each of ϕ, f, ρ satisfy the system

$$\triangle_g \kappa = 0, \ (\nabla\kappa)^2 = 0, \tag{33}$$

has the form

$$F = f + \rho[\vec{i_1}G(\phi) + \vec{i_2}H(\phi) + \vec{i_3}P(\phi)] \tag{34}$$

where G, H, P are real valued functions satisfying

$$P^2 + H^2 + G^2 = 1. \tag{35}$$

Thus, F is a section of a $R \times R \times S^2$-bundle over (M, g), where S^2 denotes the two-dimensional sphere. [31]

4.1. Maximal Monochromatic Algebras

A monochromatic algebra is called maximal monochromatic if it is not a proper subalgebra of a monochromatic algebra. In our context, the importance of maximal monochromatic algebras is obvious, in particular with respect to the question of singular sets. The main result in this respect is the following; the proof can be found in [111].

Theorem 2. The maximal C^2 algebras in (M, g) are precisely those generated by a single pair (see eq. (31))

$$(\phi, f + i\rho) \in \mathbf{M}, \tag{36}$$

with ϕ, f, ρ real, and are constituted by C^2-functions of the form

$$\xi(f, \phi, \rho) + \eta(f, \phi, \rho)[\vec{i_1}K(\phi) + \vec{i_2}H(\phi) + \vec{i_3}P(\phi)] \tag{37}$$

in the quaternionic case, and

$$\xi(f, \rho, \phi) + i\eta(f, \rho, \phi), \tag{38}$$

in the complex case, where for each fixed ϕ, $\xi + i\eta$ is an intrinsic analytic (or antianalytic) function of $f + i\rho$, the C^2-dependence on ϕ is arbitrary and $K^2 + H^2 + P^2 = 1$,with K, H, P of class C^2, but otherwise arbitrary. Thus, in the quaternionic case, it is given by a C^2-section of a $R \times R \times S^2$-bundle over M. In the complex case, non intrinsic functions are allowed.

[31] We have constructed the quaternions in terms of logical operators in Matrix Logic [109]. So, we can represent this result as either an 'objective' space representation of the objective-subjective photon, or as a 'subjective' representation of it in terms of a quaternionic structure which stems from the laws of thought; [109]. Remarkably, from the quaternions in any of the two representation, the logical quaternions in Matrix Logic [109] or the usual quaternions built by Hamilton as unrelated to logic, we can obtain some cosmological solutions. Indeed, the natural metric in the Lie group of the invertible quaternions, can be parametrized as the closed Friedmann-Lemaitre-Robertson-Walker metrics [144] which constitute one of the most important classes of solutions of Einsteins equations and furthermore, as the Carmeli metric of Rotational Relativity. We recall that the latter was introduced to explain spiral galaxies rotation curves and 'dark matter' [17]. We stress that these derivations do *not* require solving the Einstein's equations of GR but are intrinsic to the quaternions, or if wished, to Matrix Logic. This raises the question on what are we actually representing: Is it an 'outer' world, or a Klein-bottle cosmological fusion of the physical and the noetic realms, as this chapter claims to be the case? We shall return to this issue further below, on introducing the Bohr-like quantization of the Logical Momentum Operator in Matrix Logic.

4.2. General form of Singular Sets and Their Physical Interpretations

We are now in conditions for completing the objective of this section, namely, the characterization of the node set of complex and quaternionic monochromatic function. According to the above results the most general form for singular sets N of monochromatic complex or quaternionic functions is given by the conditions

$$\begin{aligned} \xi(f,\rho,\phi) &= 0, \\ \eta(f,\rho,\phi) &= 0, \ \ (\phi, f+i\rho) \in \mathbf{M} \end{aligned} \tag{39}$$

Although N is locally at least two-dimensional we have now the possibility of locating a higher-order zero on a bicharacteristic line.

For instance, the singular set of $\phi.(f+i\rho)$ is the union of the 2-dimensional set defined by $f=\rho=0$, and the 3-dimensional set $\phi=0$, and since $(\nabla\phi)^2=0$, their intersection $f=\rho=\phi=0$ is a bicharacteristic line carrying an isolated zero of higher order. The corresponding phase function has a higher order singularity located at a single point in three-space, moving with the speed of light along the singular line $f=\rho=0$, accompanied by the wave-front singularity $\phi=0$.

Observations This result is remarkable in many ways. Firstly, in the present analytic approach, it is apparent that the photon cannot exist per-se as a point-like singularity, since in fact the most general maximal monochromatic algebra is three-dimensional when we go to the quaternionic case and then we can describe it as built in the larger singularity. Thus, we have in a four-dimensional Lorentzian manifold, three real dimensions to describe the eikonal wave discontinuities. We have associated them with spinors and twistors, which themselves are related to conjugate minimal surfaces [32]

4.2.1. Photon, Nodal Lines, Monopoles

Until know we have described the singular sets of quaternionic and complex solutions of the eikonal and wave propagation equations. We have discussed already the fact that the case of the photon is not that of an 'objective' particle but the actual fusion of object-with-subject. The extended character of the photon was understood in terms of this fusion and the extended character of the subject. It is the prerogative of the subject to establish from this a coordinate system, i.e. to use light-singularities as a way to establish a primeval coordinate system, and this is in fact what sentient beings do. [33] A typical case is to establish a

[32] This is most remarkable since it points out to the existence of a Platonian world, with generic geometrical surfaces associated to the fusion of object-with-subject which the absorbed photon is. We recall that a minimal surface is one for which the mean curvature vanishes. Common to light rays, quantum jumps and minimal surfaces, is the fact that they all appear as solutions of variational problems and thus are established in terms of *extremal* solutions. On the opposite conceptual framework we find a geometrical approach to classical mechanics in which these extremal principles are absent yet the equations of motion even of null-mass particles are derivable [112]. The possible derivation of quantum jumps in terms of this more fundamental approach is an open problem.

[33] This was one of the main points of Fock's critique of Einstein's General Relativity [31], that it is a theory of uniform space, and thus singularities are needed to establish a real spacetime. We have discussed already that the primeval distinction that encodes the torsion field is such a singularity. Thus, in analyzing the hyperbolic nature of the Einstein's partial differential equations of GR and the Maxwell equations, Fock was lead to propose as a starting point the eikonal and propagation wave equations, whose wavefronts correspond precisely

Cartesian coordinate system $(x, y, z, t) \in R^{1,3}$ (in Minkowski space) given by taking

$$f + i\rho = y + iz, \tag{40}$$

and

$$\phi = f(r) - t, \text{with } r = (x^2 + y^2 + z^2)^{\frac{1}{2}}, \tag{41}$$

and f is a monotonic function of the radius r. In this case, for the function

$$(y + iz)(f(r) - t) \tag{42}$$

the singular set consists of a spherical wave front in 3-space moving with the speed of light and cutting the singular x-axis $y = z = 0$ at a single point in the positive semi-axis $0 \leq x$, where therefore lies a higher-order singularity.

This higher-order singular point, piloting a lower order singular spherical wave, along a lower order singular line is now liable to represent the photon, conceived as a moving point singularity carrying energy, in agreement with the experimentally observed corpuscular behaviour of the photon at a metallic plate, (obtaining pictures of its trajectories in cloud chambers). On the other hand the weaker singularity carried by the spherical wave front $f(r) - t = 0$ is responsible for the diffraction patterns in the typical slit experiments, according to Huygens's law of propagation of singularities (eikonal equation), and so accounts for the experimentally observed wave nature of the photon. In this way the purely analytical characterization of the maximal monochromatic algebras leads us unequivocally, to the correct conclusions as regards to the physical nature of the photon and express its *purportedly* dual wave-corpuscular nature as a simple mathematical fact. It is essentially a *wave*, the particle being a *factor* of it, but *not* dual in any intrinsic sense. Remarkably, this stands in contrast with the de Broglie-Vigier double solution theory [153]. In that theory, the wavelike pattern is associated to a linear propagation (alike eq. (7)) while the particle was treated as a propagating singularity ascribed to a non-linear equation, which in the present theory is eq. (8). Thus the present theory fleshes out in a completely geometric setting the double solution theory, which appears in the torsion geometry of the linear and nonlinear Dirac-Hestenes equation [106], yet it *relinquishes* duality. Hence, the present conception is radically different to the Copenhagen interpretation of quantum mechanics.

The line $y = z = 0$ in 3-space carries a singularity too, but this is a standing one, independent on time, and therefore, its presence is detected through different effects. Actually this line is called a *nodal* line of the wave function $\phi \equiv y + iz$ ([12]) or a *dislocation* line of the planes of constant phase of ψ. Around this line occur vortices of the flux of the trace torsion one-form $d\ln\phi$ of the phase function (when the circulation of this flux around a

to propagating singularities, that we have further associated with torsion. Though we cannot because of lack of space show that this is the case, i.e. that from the form of maximal monochromatic quaternionic functions we can indeed establish a coordinate system, in which the functions that produce this form act as a coordinate system! This can be established without recourse to an ad-hoc non-geometric energy-momentum tensor as Einstein's inception of it in GR, but using the energy-momentum tensor of the electromagnetic field, giving thus a self-referential construction of the metric by solving the Einstein's equations with light as a source for matter described by curvature derived from the metric in the Levi-Civita connection, according to Einstein. Thus, metrics are *not* primitive to the constitution of spacetime, as we have already argued. Self-referentially, it is the joint constitution of spacetime and the subject through torsion as a logo-physical sign and field.

nodal line is non-zero), described in detail by Hirschfelder [49]. Alternatively, Dirac found these nodal lines when considering singularities of wave functions, upon imposing the only requirement that the complex-valued functions ψ (in our example equal to $y+iz$) be single-valued and smooth, but not necessarily with single-valued argument, and then quantized them in terms of the winding number of the vector-field $(\nabla \ln Re(\psi), \nabla \ln Im(\psi))$ along a closed curve around the line. Dirac found that one could remove the non-zero circulation by means of a gauge transformation of the second kind, and that the electromagnetic vector potential associated with this transformation was precisely the same electromagnetic potential produced by a magnetic monopole at the initial point. Dirac then equated the effect of the circulation around the nodal line in the original gauge to the effect of a monopole in the new gauge [22]. His quantization by the winding number is actually just a special case of the general quantization theorem above, and his gauge interpretation is thus a concrete exemplification of the meaning of the analysis given there. Remarkably enough, a relation between the Kozyrev phenomenae -which is related to torsion [107, 134]- and Dirac monopoles has been established by Shakhparonov and has extremely interesting technological applications [123]. So, in the present setting, the torsion-trace one-form plays the role of the electromagnetic potential leading to quantization and the appearance of magnetic monopoles [113].

The variety of types of singular sets defined by the representation given in eq. (38) is very great, as exemplified in the pioneering work by Nye and Berry [12]. Besides the singular sets that we previously identified with the photon (spherical wave front plus a nodal line) there is also a remarkable singular set of the monochromatic wave constructed out of $\phi = f(r) - t$ and $f + i\rho = y + iz \in R^{1,3}$, by the following sum

$$\epsilon e^{i\omega[f(r)-t]} - (y + iz), \epsilon > 0. \tag{43}$$

Its singular set is given by

$$y = \epsilon \cos\omega[f(r) - t], z = \epsilon \mathrm{sen}\omega[f(r) - t]. \tag{44}$$

This represents an helicoidal line lying on the cylinder $y^2 + z^2 = \epsilon^2$, and moving with (variable) speed of light along its tangent direction at each of its points. (For simplicity, we can assume that $f(r) = t$ in order to get a better visualization: the speed is then constant and the helicoid has then a constant step.) Taking $y - iz$ instead, we get a screw motion with opposite handedness. The singular set is thus a moving screw in 3-space that can be right or left handed, and may carry the energy associated with a quantum jump, as shown above. It seems therefore that a monochromatic wave line like this can represent appropiately a right or left handed neutrino, concretely identified with its singular set. It has then quite distinct properties from those associated with a photon. For it is given by an infinitely long moving right of left handed helicoidal line in 3-space (which by the way, it is a minimal surface; additional discussions on dislocations, minimal surfaces and turbulence, shall be presented elsewhere) while the photon is given by a point piloting a spherical wave. In particular if the singular screw line of above is associated with an elementary state u and carries energy E, it also carries the angular momentum $\epsilon^2 E\omega/c^2$ directed along a x-axis, in the given referential. Hence the neutrino carries angular momentum while the photon does not. On the other hand, according to this description, the neutrino should not have (primary) diffraction patterns as the photon does, which should explain why it is so difficult of detect.

4.2.2. Distinction of Maximal Monochromatic Algebras and Their Twistor Representations

Lemma 2. The maximal monochromatic algebras $\mathbf{M}_1$ and $\mathbf{M}_2$ with generators $(f, \phi + i\psi)$ and $(\tilde{f}, \tilde{\phi} + i\tilde{\psi})$ respectively, are distinct if and only if

$$g(\nabla f, \nabla \tilde{f}) = 0. \tag{45}$$

Consider the generators $(\phi, f + i\rho) \in \mathbf{M}$ of a maximal monochromatic algebras. The vector fields $\nabla\phi$ and $\nabla(f + i\rho)$ are, respectively, real and complex isotropic fields, mutually orthogonal on (M, g); here M is a generic spin-manifold provided with a Lorentzian metric g. By the previous analysis $\nabla\phi$ is given by a spin vector field ω^A in the spinor form

$$\nabla\phi = \omega^A \bar{\omega}^{A'} \ \ (\text{or} - \omega^A \bar{\omega}^{A'}). \tag{46}$$

and since $\nabla(f + i\rho)$ is isotropic and orthogonal to $\nabla\phi$ then we have

$$\nabla(f + i\rho) = \omega^A \bar{\pi}^{A'}, \tag{47}$$

where π^A is another spin vector field. Consequently the pair $(\nabla\phi, \nabla(f + i\rho))$ of vector fields is completely determined by the ordered pair of spin vector fields

$$(\omega^A, \pi^{A'}), \tag{48}$$

but we have a fourfold map here since we have already altogether four different ways of building the vector fields according to eq. $(46, 47)$ out of the ordered pair given by (48). The correspondence (46), extended to complex vectors shows that the second choice in (46) reverses $\nabla\phi$ from, say , a future-pointing isotropic vector to a past-pointing isotropic vector while in (46) it chooses the complex-conjugate $\nabla(f - i\rho)$ instead of $\nabla(f + i\rho)$, reversing the roles of analytic and anti-analytic functions, which means inversion of handedness. Choosing locally a given time orientation and a given handedness, corresponds to a particular choice of the assignements in eqs. (46,47). The ordered pair (48) of spin vector fields at a point in (M, g) is called a local twistor and the corresponding field a local twistor field. From the twistor field we determine the real and complex vector fields by eqs. $(46, 47)$, which, upon integration yield an equivalent pair $(\phi, f + i\rho)$. This means that we can characterize completely a maximal monochromatic algebra (and consequently the light quanta it represents) in terms of a twistor field with divergence freee associated vector fields. This can be of course produced, by working from the start with coclosed differential forms. The new representation is even richer as it has built in an extra degree of freedom, namely, polarization, due to the factor $e^{i\phi}$mentioned before. This result, showing that we have identified as light quanta are indeed given by twistor fields, substantiates the belief of Penrose that twistors are the appropiate tool to describe zero rest-mass particles and to effect the connection between gravitation and quantum mechanics. Further below we shall see that this connection extends to the laws of thought.

5. Self-reference, the Klein-Bottle, Torsion, Matrix Logic, the Laws of Thought and the Twistor Representation of the Cognitive Plenum

In the last sections we have elaborated a theory of torsion and photons which, would we have ommited our discussion on the self-referential being of light, a Cartesian minded physicist would have read it as customary to that mindset: As a theory of an 'objective' realm, implicitly, in which subjectivity does not participate, or still does not exist at all in the universe of discourse. Yet, we have shown that both torsion and the photon are very closely related to self-reference, and thus to consciousness [109]. Furthermore, the semiotic codification of torsion as a distinction sign produces through the incorporation of paradox, a multivalued logic which is associated with the Klein-bottle and time waves [109]. From this logic, it was proved that the most general matrix-tensor logic that has as particular cases quantum, fuzzy, modal and Boolean logics [135] stem from these time waves. In this theory which stems from abandoning the scalar logic theory of Aristotle and Boole, promoting it to logical operators, we find that the Klein-bottle plays a fundamental role as an in-formation operator, which coincides with the Hadamard gate of quantum computation. The role of this gate is to transform the vector Boolean states to superposed states, the latter being associated with the torsion of cognitive space and the non-orientability of this space due to its constitution in terms of the self-referential non-orientable Moebius and Klein-bottle surfaces. Furthermore, the logical Cognition Operator which leads to quantization of cognition is generated by the torsion produced from the commutator of the TRUE and FALSE logical operators (at the start of this chapter, we wrote them as True and False, respectively), which self-referentially gives the difference between these two operators, as we shall see below. The picture that stems is that Matrix Logic can be seen as the self-referential logical code which stands at the foundation of quantum physics to which is indisolubly related. We have elaborated the relations between matrix logic, self-reference, non-orientability and the Klein-bottle, nilpotent hypernumber representations of quantum fields that represent some logical operators. Thus, in this theory, matter quantum field theories are logical operators, and vice versa, and a transformation between quantum and cognitive logical observables has been established; this theory has produced a new fundamental approach to the so-called mind-matter problem , establishing its non-separateness, and the primacy of consciousness which thus cannot be purported to be an epiphenomenon of physical and more complex fields [109]. By promoting the 'truth tables' of usual Boolean logic to matrix representations, the founder of Matrix Logic A. Stern was able to produce an operator logic theory in which logical operators may admit inverses, and the operations of commutation and anticommutation are natural [135]. Furthermore, logical operators can interact by multiplication or addition and, in some cases, being invertible [34], they yield thus a more complex representation of the laws of thought that the one provided by the usual Boolean theory of logical connectives. This representation was named the *Intelligence Code* by

[34] The fact that there are logical operators that are represented by determinant zero matrices posits a separation between the invertible and non-invertible operators which is associated with an irreversibility of thinking which is implicit to this separation. Thus we shall have a logic and an antilogic which we can perhaps relate to the isoduality transformation in Hadronic Mechanics due to Santilli [122], due to the association between nilpotent logical operators and quantum field operators. We shall discuss this very important issue elsewhere.

Stern [135]. The latter can be related to Quantum Mechanics for two-state sytems as we shall describe below. Matrix Logic is naturally quantized, since its eigenvalues take discrete values which are $\pm 1, 0, 2, \pm\phi$, with ϕ the Golden number [135]. In this setting, the null quantum-cognitive observable is the 2×2 matrix, $\mathbf{0}$, with identical entries given by 0; the latter is the *Mind Apeiron* as we discussed in [111] which we shall retake below. The relation with quantum field operators and this Mind Apeiron observable , is their role in polarizing this cognitive-quantum Apeiron through non-null square roots which can be represented by nilpotents (we rather prefer to call them plenumpotents), i.e. hypernumbers whose square is $\mathbf{0}$. [35] In distinction with the other cognitive-quantum observables, is that the eigenstates of $\mathbf{0}$ are no longer quantized, but rather give an orthogonal complex two-dimensional nullvector space. In this way the Plenum is no longer represented by a single point, $\mathbf{0}$, but rather becomes an extended object or zero-brane. This will allow to map the twistor representations of the extended photon presented in (48) with its representation in a cognitive state and vice versa!

Let us present this briefly in this Section. If we consider a space of all possible cognitive states (which in this context replace the logical variables) represented in this plenum as the set of all Dirac bras $< q| = (\bar{q} \;\; q)$, where $\bar{q} + q = 1$ [36], is a continuous cognitive logical

[35] We shall later encounter the plenumpotence of the Mind Apeiron when discussing the Myth of Eternal Return as a self-referential process. The present theory shares with the *nilpotent universal rewrite system* (NURS) due to Rowlands and further extended to biology as the *Code of Life* by Hill [119] [48] , an understanding of the generative role of 0. The theory of these authors (which has no time operator) is constructed in terms of recursion, which in their context is the *reduction* of self-reference to the 'objective' side of the Cartesian Cut. This recursion is not the full embodiment of self-reference with physics, semiotics, multivalued logic, perception, cognition and neurology. Remarkably, the reduction of self-reference to recursion in the NURS is produced -somewhat as a natural tradeoff, in recalling Pattee and Taborsky's conceptions- through an *empowerment* of semiosis, providing thus very enrichening features to this system. More specifically, mathematical symbols -conceived independently of the embodied energetics and nonintegrability claimed by Pattee [91] and Taborsky [138] which we associated to torsion and multivalued logics [109]-, are presented as a self-writing system of the objective world. The point which though unobserved by their authors is crucial to their claim of being the scripture of the 'objective' realm, is that the NURS requires an *interpreter* to become a map of this realm, alike any text, and thus the *seemingly* absence of the subject and its lifeworld is in fact embodied in the actual interpretation of the symbology by which the system is indeed of rewrite. Thus, this theory if conceived free of subjectivity as in Aristotelian dualism, which is already implicit in some degree in the reduction of self-reference to recursion, may actually produce the illusion of this independence of subjectivity, of the pledged to be a scripture of the objective realm. This is the usual conception of the Platonian realm in which the most fertile and original mathematical, artistic and creative works claim to find/receive their origin. In this chapter we are presenting a very different conception which does not deny this realm. It rather integrates it into the full fusion of the objective and subjective realms and its embodiment as a lifeworld, in which life and world are richer than abstractions, but rather a supraobjectivity and suprasubjectivity: Embodiment at *all* scales. The relations between this remarkable theory and the present conception will hopefully be presented elsewhere.

[36] Notice that a difference with the definition of qubits in quantum computation, is that for them we have the normalization condition for complex numbers of quantum mechanics. In this case, the values of q are arbitrary real numbers, which leads to the concept of non-convex probabilities. While this may sound absurd in the usual frequentist interpretation, when observing probabilities in non-orientable surfaces, say, Moebius surfaces, then if we start by associating to both sides of an orientable surface -from which we construct the Moebius surface by the usual procedure of twisting and glueing with both sides identified- the notion of say Schroedinger's cat being dead or alive in each side, then for each surface the probability of being in either state equals to 1 and on passing to the non-orientable case, the sum of these probabilities is 2. While this is meaningless in an orientable topology, in the non-orientable case which actually exist in the macroscopic world, this value is a consequence of the topology. In this case, the superposed state ' being alive and being dead' or 'true plus false' which is excluded in Aristotelian dualism, is here the case very naturally supported by the fact that we have a non-trivial

value not restricted to the false and true scalar values, represented by the numbers 0 and 1 respectively. In fact, q can take arbitrary values as we shall elaborate further below. Still, the standard logical connectives admit a 2×2 matrix representation of the their 'truth tables' and now we have that for such an operator, L, we have the action of L on a ket $|q>=\begin{pmatrix} \bar{q} \\ q \end{pmatrix}$ is denoted by $L|q>$ alike the formalism in quantum mechanics, and still we have a scalar truth value given by $<p|L|q>$, where $<p|$ denotes another logical vector. We can further extend the usual logical calculus by considering the TRUE and FALSE operators, defined by the eigenvalue equations $\mathrm{TRUE}|q>=|1>$ and $\mathrm{FALSE}|q>=|0>$, where $|1>=\begin{pmatrix} 0 \\ 1 \end{pmatrix}$ and $|0>=\begin{pmatrix} 1 \\ 0 \end{pmatrix}$ are the true and false vectors. It is easy to verify that the eigenvalues of these operators are the scalar truth values of Boolean logic. We can represent these operators by the matrices

$$\mathrm{TRUE}=\begin{pmatrix} 0 & 0 \\ 1 & 1 \end{pmatrix}, \mathrm{FALSE}=\begin{pmatrix} 1 & 1 \\ 0 & 0 \end{pmatrix} \tag{49}$$

These operators are non-self adjoint; in distinction with usual (hermitean) quantum observables, logical operators are generally non-hermitean although they may have representations as quantum field operators and viceversa. A consequence of this non-hermiticity is that in contrast with the trivial duality of the true and false scalars of Boolean logic, 1 and 0 respectively (which is represented by the relations $\bar{1}=0$ and $\bar{0}=1$), by defining the *complement* $\bar{L}$ of a logical operator L, by $I-L$, where I is the identity operator, hence $\overline{\mathrm{TRUE}} \neq \mathrm{FALSE}$ and $\overline{\mathrm{FALSE}} \neq \mathrm{TRUE}$. This notion of complementarity when restricted to scalar fields coincides with the dual operation of Boolean logic transforming conjuction into disjunction and viceversa which affirms the principle of non-contradiction (i.e. a proposition *and* its negation are false) of Aristotelian-Boolean logic, and thus the previous result proves the non-duality of TRUE and FALSE.

We note that the spaces of bras and kets do not satisfy the additivity property of vector spaces -while keeping the property that one is the dual of the other- due to the fact that normalization is not preserved under addition. A superposition principle is necessary. If $|p<$ and $|r>$ are two normalized states, then the superposition defined as follows

$$|q>=c|p>+\bar{c}|r>, \quad \text{where } \bar{c}+c=1, \tag{50}$$

also defines a normalized logical state. We can interpret these coefficients as components of a logical state $|c>$ or still a probability vector, termed denktor, a German-English hybrid for a thinking vector. The normalization condition is found as follows: Multiply the states $|p>$ and $|r>$ by $\bar{c}$ and c, respectively. By definition, the normalization condition on the sum $|q>$ with coefficients $\bar{c}, c$ leads to

$$\begin{pmatrix} \bar{q} \\ q \end{pmatrix}=c\begin{pmatrix} \bar{p} \\ p \end{pmatrix}+\bar{c}\begin{pmatrix} \bar{r} \\ r \end{pmatrix}=\begin{pmatrix} c\bar{p}+\bar{c}\bar{r} \\ cp+\bar{c}r \end{pmatrix}, \tag{51}$$

non-orientable topology. As for the case of negative probabilities, we see in the previous example that -1 is the probability value complement of the value 2.

yet, since $\bar{q}+q=c\bar{p}+\bar{c}\bar{r}+cp+\bar{c}r=c(\bar{p}+p)+\bar{c}(\bar{r}+r)=c.1+\bar{c}.1$ and thus $c+\bar{c}=1$ since $|q>$ is a normalized state by assumption. So, through this superposition principle we can give a vector space structure to normalized cognitive states. We now can identify under these prescriptions, the tangent space to the space of bras (alternatively, kets) with the space itself. [37]

5.1. Torsion in Cognitive Space, the Klein-Bottle, Quantum Mechanics and Quantum Field Theory, Logic and Hypernumbers, the Cognition, Time and Spin Operators

Returning to the vector space structure provided by the superposition principle, and thus the identification of its tangent space with the vector space itself, it follows that a vector field as a section of the tangent space can be seen as transforming a bra (ket) vector into a bra (ket) vector through a 2×2 matrix, so we can identify the tangent space which with the space of logical operators. We have as usual the commutator of any such matrices $[A, B] = AB - BA$ and the anticommutator $\{A, B\} = AB + BA$. In particular we take the case of $A = \text{FALSE}$, $B = \text{TRUE}$ and we compute to obtain

$$[\text{FALSE}, \text{TRUE}] = \text{FALSE} - \text{TRUE}, \tag{52}$$
$$\{\text{FALSE}, \text{TRUE}\} = \text{FALSE} + \text{TRUE}. \tag{53}$$

Thus in the subspace spanned by TRUE and FALSE we find that the commutator that here coincides with the Lie-bracket of vectorfields defines a torsion vector given by the vector $(1\ \ -1)$, and that this subspace is integrable in the sense of Frobenius: Indeed, we find that $[[\text{FALSE}, \text{TRUE}], \text{TRUE}] = [\text{FALSE}, \text{TRUE}]$ and $[[\text{FALSE}, \text{TRUE}], \text{FALSE}] = [\text{TRUE}, \text{FALSE}]$. Furthermore, on account that $\text{TRUE}^2 = \text{TRUE}$ and $\text{FALSE}^2 = \text{FALSE}$, i.e. both operators are idempotent, then the anticommutators also leaves this subspace invariant.

The remarkable aspect here is that the quantum distinction produced by the commutator, exactly coincides with the classical distinction produced by the difference (eq. (52)), while the same is valid for the anticommutator with a classical distinction which is represented by addition (eq, (53)). We notice that in distinction of quantum observables, these logical operators are not hermitean and furthermore they are noninvertible. Furthermore, we shall see below how this torsion is linked with the creation of cognitive superposed states, very much like the coherent superposed states that appear in quantum mechanics.

Now, if we denote by M the commutator [FALSE, TRUE] so that from eq. (52) we get

$$M = \begin{pmatrix} 1 & 1 \\ -1 & -1 \end{pmatrix}, \tag{54}$$

we then note that it is nilpotent, (in fact a nilpotent hypernumber, since $M = \epsilon_2 + i_1 = \sigma_z + i_1$). Indeed,

$$M^2 = \begin{pmatrix} 0 & 0 \\ 0 & 0 \end{pmatrix} \equiv \mathbf{0}, \tag{55}$$

[37] Here it is simple to see that if $|q>, |q'>$ are two superpositions, then for any operator L, $L|(q+q')> = L|q> + L|q'>$.

thus yielding the identically zero matrix, representing the universe of all possible cognitive states created by a non-null divisor of zero, which thus creates a polarization of the Plenum, precisely through the fact that the torsion is a superposed state which cannot be fit into the scheme of Boolean logic, but can be obtained independently by the loss of orientability of a surface which thus allows for paradox. Since M coincides with the classical difference between TRUE and FALSE, which are not hermitean, then we can think of this non-invertible operator as a *Cognition Operator* related to the variation of truth value of the cognitive state, as we shall prove further below that $M = -\frac{d}{dq}$.

We would like to note that this polarization of the Plenum $\mathbf{0}$ is not unique, there are many divisors of $\mathbf{0}$, the Plenum, for instance the operator

$$\text{ON} = \begin{pmatrix} 0 & 0 \\ 1 & 0 \end{pmatrix} := a^\dagger, \tag{56}$$

and

$$\text{OFF} = \begin{pmatrix} 0 & 1 \\ 0 & 0 \end{pmatrix} := a \tag{57}$$

satisfy

$$a^2 = \mathbf{0}, (a^\dagger)^2 = \mathbf{0}, \tag{58}$$

and furthermore, $\{a, a^\dagger\} = I$, so they can be considered to be matrix representations of creation and annihilation operators, $a^\dagger$ and a as in quantum field theory. In fact, if we consider the wave operators given by the exponentials of $a, a^\dagger$ we have

$$e^a = I + a = \begin{pmatrix} 1 & 1 \\ 0 & 1 \end{pmatrix} = \text{IMPLY}, e^{a^\dagger} = I + a^\dagger = \begin{pmatrix} 1 & 0 \\ 1 & 1 \end{pmatrix} = \text{IF}, \tag{59}$$

where IMPLY $=\rightarrow$ is the implication, and IF $=\leftarrow$ is the converse implication: $x \leftarrow y = \bar{x} \rightarrow \bar{y}$. Thus the implication and the converse implication logical operators are both wave-like logical operators given by the exponentials of divisors of $\mathbf{0}$, and in fact they are derived from quantum field operators of creation and annhilation in second-quantization theory, $a^\dagger$ and a, respectively, which in fact can be represented by nilpotent hypernumbers. Indeed, $a = \frac{1}{2}(\epsilon_3 - i_1) = \frac{1}{2}(\sigma_x - i_1)$ and $a^\dagger = \frac{1}{2}(\epsilon_3 + i_1) = \frac{1}{2}(\sigma_x + i_1)$; see [109].

Now we wish to prove that the interpretation of M as the Logical Momentum Operator is natural since $M = -\frac{d}{dq}$. Indeed,

$$-\frac{d}{dq}|q> = -\frac{d}{dq}\begin{pmatrix} 1-q \\ q \end{pmatrix} = \begin{pmatrix} 1 \\ -1 \end{pmatrix} = \begin{pmatrix} 1 & 1 \\ -1 & -1 \end{pmatrix}\begin{pmatrix} \bar{q} \\ q \end{pmatrix} = M|q> \tag{60}$$

so that for any normalized cognitive state $|q>$ we have the identity

$$M = -\frac{d}{dq}, \tag{61}$$

which allows to interpret the cognitive operator as a kind of logical momentum. Thus, in this setting which is more general but less primitive than the calculus of distinctions, it is the non-duality of TRUE and FALSE that produces cognition as variation of the continuous cognitive state. We certainly are facing a situation that is far from the one contemplated by Aristotle with his conception of a trivial duality of (scalar) true and false, and which produced the elimination (and consequent trivialization) of time and of subjectivity, as argued in [109].

Now consider a surface given by a closed oriented band projecting on the xy plane. Thus to each side of the surface we can associate its normal unit vectors, $(1 \ \ 0)$ and $(0 \ \ 1)$. Suppose that we now cut this surface and introduce a twist on the band and we glue it thus to get a Moebius surface. Now the surface has lost its orientability and we can identify one side with the other so that we can generate the superpositions

$$< 0| + < 1| =< (1 \ \ 1)| =< S_+|, \quad < 0| - < 1| =< (1 \ \ -1)| =< S_-|. \tag{62}$$

which we note that the latter corresponds to the torsion produced by the commutator of TRUE and FALSE operators. Theses states are related by a change of phase by rotation of 90 degrees. What the twisting and loss of orientability produced, can be equivalently produced by the fact that TRUE and FALSE are no longer as in Boolean logic and the Aristotelian frame, they are no longer dual and what matters is their difference, which in the case of scalar truth values does not exist. The second state in eq. (62), also can be interpreted as a state that represents the fact that the states as represented by vectors, have components standing for truth and falsity values which are independent, so that the Aristotelian link that makes one the trivial reflexive value of the other one is no longer present: they each have a value of their own. In that case then $(0 \ \ 0)$ is another state, 'neither false nor true' (which is the case of the Liar paradox as well as Schroedinger's cat), which together with $(1 \ \ 1)$, 'false and true' state together with $(0 \ \ 1)$, true, and $(1 \ \ 0)$, false states, we have a 4-state logic in which the logical connectives have been promoted to operators.

Now consider for an arbitrary normalized cognitive state q the expression

$$\begin{aligned} [q, M]|q> &= [q, -\frac{d}{dq}]|q> = -q\frac{d}{dq}|q> + \frac{d}{dq}q|q> = -q\frac{d}{dq}\begin{pmatrix} 1-q \\ q \end{pmatrix} \\ &+ \frac{d}{dq}\begin{pmatrix} q-q^2 \\ q^2 \end{pmatrix} = \begin{pmatrix} q \\ -q \end{pmatrix} + \begin{pmatrix} 1-2q \\ 2q \end{pmatrix} = |q>, \end{aligned} \tag{63}$$

for any normalized cognitive state q so that we have the quantization rule

$$[q, M] = I, \tag{64}$$

where $I = \begin{pmatrix} 1 & 0 \\ 0 & 1 \end{pmatrix}$, the identity operator. Instead of the commutation relations of quantum mechanics $[q, p] = i\hbar$ for $p = -i\frac{\partial}{\partial q}$ and those of diffusion processes associated to the Schroedinger equation, $[q, p] = \sigma$ where $p = \sigma\frac{\partial}{\partial x}$ with σ asvthe diffusion tensor given by the square-root of the metric g on the manifold with coordinates x on which the diffusion takes place so that $\sigma \times \sigma^\dagger = g$ [102], we have that the commutation of a normalized cognitive state where the Cognitive (Momentum) Operator is always the identity, thus yielding a fixed point. Indeed, consider the function $F_M(q) = [q, M]$, then $F_M(F_M(F_M(FM(\ldots))))(|q>) = |q>$, for any normalized cognitive state $|q>$. Thus,

$F_M(q)$ defines what is called in system's theory an eigenform, albeit one which does not require infinite recursion but achieves a fix point already in the first step of the process, by the formation of the commutator $[q, M]$. This is the structure of Self, which whatever operation it may suffer by the action of logical operators, retains its invariance by the quantization of logic as expressed above by eq. (64).

Now we want to return to the superposed states, S_+ and S_-, the latter being the torsion produced by the commutator of the TRUE and FALSE operators, to see how they actually construct the cognitive operator. First a slight detour to introduce the usual tensor products of two cognitive states, $|p><q|$ which as the tensor product of a vector space and its dual is isomorphic to the space of linear transformations between them, we can think as an operator L acting by left multiplication on kets and by right multiplication on bras. So that if $L = |p><q|$ then $<y|L|x> = <y, p><q|x>$, for any $<y| = \bar{y}<0| + y<1|$ and $|x> = \bar{x}|0> + x|1>$, where $<x|y> = \delta_{xy}$ is equal to 1 for $x = y$, and equal to 0 for $x \neq y$ and $\sum_i |x_i><x_i| = I$. Then,

$$M = |S_+><S_-|, \qquad (65)$$

which shows that the cognitive operator that arises from the quantum-classical difference between the TRUE and FALSE operators can be expressed in terms of the tensor products of the superposition states, being the sum of the true and false states and the torsion produced in the quantum commutator of the TRUE and FALSE operators.

Starting with the Logical Momentum or still the Cognition Operator, M, that satisfies $[q, M] = I$ for any cognitive variable q, we can link the quantization rule in cognitive space to the quantization rule of Bohr-Sommerfeld. We must stress at this point that this rule is not restricted to the quantization of the microphysical scales but also valid for astrophysical structures; for a thorough review of this we refer to [16, 120, 88] and the series of articles in [18] and references therein, notably Pitkanen's contribution [93]. As noted by Stern [135], the logical potential carrying the logical energy could be linked to the Bohr energy of atomic (to which, we add astrophysical structures) structures in the following way: $\infty(k) = \oint M dq = 2\pi(n + 1/2) = k\pi$, where q is a logical variable (if it is zero than the contour integral runs a full great circle on the Riemann sphere of zeros), n is the winding number specifying the numbers of times the closed curve runs round in an anticlockwise sense; n runs the bosonic numbers $0, 1, 2\ldots$ and $(n + 1/2)$ the fermionic numbers, $\frac{1}{2}, \frac{3}{2}, \frac{5}{2}, \ldots$. The topo-logical potential is an odd multiple $(2n + 1)\pi$ of the elemental (topo)logical phase π and is $\hbar^{-1}$ times the Bohr energy of the quantum oscillator: $\oint p dx = 2\pi\hbar(n + 1/2)$, where the position and momentum operator satisfy the standard quantum commutation relation: $[x, p] = i\hbar$. As we see, the to-pological potential, multiplied by the factor $\hbar$, gives the Bohr quantum energy opening up the possibility to treat atomic and astrophysical structures (the latter arising from the multivalued character of the Planck constant [87][93]) as a dynamical logic in a fundamental sense, where quantization stems from the closed topology or self-observation feature at this fundamental level of reality. This dynamical logic is Matrix Logic which we recall that can be derived from the protologic of Spencer-Brown (with the primeval distinction being the semiotic codification of torsion) in considering the paradoxical Klein-bottle renntrance of a form on itself [109]. Another interesting conjecture -proposed by Dienes [21]- which follows is, since matter, as energy, ($E = mc^2$) is a topo-logically transformed logical energy, the mass of an object is basically the in-formation contained in the holomatrix which projects the object and its mass out from the ground state. This is a more fundamental ontology for mass, with in-formational topo-logical roots, rather than the usual physical one. It stems from the differentiation of the Plenum placed by the semiotic codification of torsion and the associated non-orientable self-referential topology: the Klein-bottle, or still by the action of the latter on the polarization of the Mind Apeiron provided by its twistor eigenstates which themselves are torsion fields embodying the fusion of the physical and subjective realms.

Let us now introduce the operator defined by

$$\triangle = a - a^{\dagger} \tag{66}$$

so that is follows that its matrix representation is

$$\triangle = \begin{pmatrix} 0 & 1 \\ -1 & 0 \end{pmatrix}. \tag{67}$$

and furthermore

$$\triangle = \rightarrow - \leftarrow . \tag{68}$$

We shall call $\triangle$ the TIME Operator. Note that *alternatively* we can define TIME by changing the sign in the above definition. [38] We notice that it is unitary and antisymmetric:

$$\text{TIME}^{\dagger} = \text{TIME}^{-1} = -\text{TIME}. \tag{69}$$

As an hypernumber TIME $= -i_1$, minus the unique 2×2 matrix representing a 90 degrees rotation, the old commutative square root of -1 from which complex numbers appeared. The reason for considering this operator given by the difference of nilpotents is because it plays the role of a *comparison* operator. Indeed, we have

$$< p|\text{TIME}|q > \;=\; \bar{p}q - \bar{q}p = (1-p)q - (1-q)p = q - p = \bar{p} - \bar{q}. \tag{70}$$

TIME appears to be unchanged for unaltered states of consciousness:

$$< q|\text{TIME}|q >= 0, \tag{71}$$

and if we have different cognitive states p, q, then $< p|\text{TIME}|q > \neq 0$. So this operator does represent the appearance of a primitive difference on cognitive states: further, it is antisymmetric and unitary. It is furthermore linked with a difference between annihilation and creation operators and thus stands for what we argued already as a most basic difference that leads to cognition and perception: *the appearance of quantum jumps*. Without them, no inhomogeneities or events are accessible to consciousness. The very nature of self-reference as consciousness of consciousness requires such an operator for the joint constitution of the subject and the world. Thus its name, TIME operator; it stands clearly in the subject side of the construction of a conception that overcomes the Cartesian cut, yet a subject that has superposed paradoxical states.

[38] Remarkably, $-2i$TIME is the hamiltonian operator of the damped quantum oscillator in the quantum theory of open systems; see N. Gisin and I. Percival, arXiv:quant-ph/9701024v1. In this theory based on the stochastic Schroedinger equation the role of torsion is central [107].

Let us consider next the eigenvalues of TIME, i.e. the numbers λ such that $\text{TIME}|q>=\lambda|q>$; they are obtained by solving the characteristic equation $\det|\text{TIME}-\lambda I|=\lambda^2+1=0$, so that they are $\lambda=\pm i$ with complex eigenstates

$$\begin{pmatrix} 1 \\ i \end{pmatrix}, \begin{pmatrix} i \\ 1 \end{pmatrix}. \tag{72}$$

They are not orthogonal, but self-orthogonal; thus, they are spinors, and the complex space generated by them generates a two-dimensional null space. We diagonalize TIME by taking

$$\begin{pmatrix} 1 & i \\ i & 1 \end{pmatrix} \text{TIME} \begin{pmatrix} 1 & i \\ i & 1 \end{pmatrix}^{-1} = \begin{pmatrix} i & 0 \\ 0 & -i \end{pmatrix} \tag{73}$$

so that

$$\text{TIME}_{\text{diag}} = \begin{pmatrix} i & 0 \\ 0 & -i \end{pmatrix}, \tag{74}$$

which as a hypernumber implies that $\text{TIME}_{\text{diag}} = i_2$, so that $\text{TIME}^2_{\text{diag}} = -I$. We want finally to comment that TIME is not a traditional clock, yet it allows to distinguish between after and before ($\rightarrow - \leftarrow$), forward and backwards, in *logical causation*. There is no absolute logical time, nor a privileged direction of it. To have a particular direction it must be asymetrically balanced towards creation or annihilation. This can be computed as the complement of the operator phase[39]

$$\overline{cos(2a^\dagger) + sen(2a^\dagger)} = a^\dagger - a, \tag{75}$$

from which it follows that $\text{TIME} = \overline{\leftarrow^2} = \rightarrow - \leftarrow$, as we stated before.

Let us now retake the Cognition Operator M and decompose it as

$$M = \text{TIME} + \sigma, \text{ or still} \tag{76}$$

$$\begin{pmatrix} 1 & 1 \\ -1 & 1 \end{pmatrix} = \begin{pmatrix} 0 & 1 \\ -1 & 0 \end{pmatrix} + \begin{pmatrix} 1 & 0 \\ 0 & -1 \end{pmatrix}. \tag{77}$$

Then we have that

$$< q|M|q > = < q|\sigma|q > . \tag{78}$$

Indeed, since $< q|\text{TIME}|q >= 0$, so that the proof of eq. (78) follows. Furthermore we note that

$$< q|\sigma|q > = \bar{q}^2 - q^2 = (\bar{q} - q)(\bar{q} + q) = \bar{q} - q. \tag{79}$$

from the normalization condition. Note here that the identity given by eq. (79) is a kind of quadratic metric in cognitive space which due to the normalization condition looses its quadratic character to become the difference in the cognitive values: $\bar{q} - q = 1 - 2q$ which becomes trivial in the undecided state in which $\bar{q} = q = \frac{1}{2}$.

[39]The complement of a logical operator L, is defined by $\bar{L} = I - L$.

The role of σ is that of a SPIN operator, as we shall name it henceforth, which coincides with the hypernumber ϵ_2 (or as a Pauli matrix is σ_z), so that $\sigma^2 = I$, i.e. the non-trivial square root of hypernumber $I = \epsilon_0$, which is the usual Pauli matrix σ_z in the decomposition of a Pauli spinor in the form $\sigma_x e_x + \sigma_y e_y + \sigma_z e_z$, for e_x, e_y, e_z the standard unit vectors in R^3 and we write their representations as hypernumbers

$$\sigma_x = \begin{pmatrix} 0 & 1 \\ 1 & 0 \end{pmatrix} = \epsilon_3, \sigma_y = \begin{pmatrix} 0 & -i \\ i & 0 \end{pmatrix} = \epsilon_1, \sigma_z = \begin{pmatrix} 1 & 0 \\ 0 & -1 \end{pmatrix} = \epsilon_2. \quad (80)$$

We can rewrite this average equation $< q|M|q >=< q|\sigma|q >$, as an average equation where the l.h.s. takes place in cognitive space of normalized states $|q>$ and the r.h.s. takes place in a Hilbert space of a two-state quantum system, say, spin-up $\psi(\uparrow)$, spin-down $\psi(\downarrow)$, so that the generic element is of the form

$$\psi = \psi(\uparrow)|0> + \psi(\downarrow)|1> . \quad (81)$$

Indeed, if we write

$$|q>= \overline{\psi(\uparrow)}\psi(\uparrow)|0> + \overline{\psi(\downarrow)}\psi(\downarrow)|1>, \quad (82)$$

then the r.h.s. of eq. (79) is $\bar{q}^2 - q^2$, with $\bar{q} = \overline{\psi(\uparrow)}\psi(\downarrow)$, and $q = \overline{\psi(\downarrow)}\psi(\uparrow)$, so that eqs. (76, 78) can be written as

$$< q|M|q >=< \psi|\sigma|\psi > \quad (83)$$

where the average of M is taken in cognitive states while that of the SPIN operator is taken in the two-state Hilbert space. [40]

We revise the previous derivation for which the clue is the relation between cognitive states $|q>$ and elements of two-state of Hilbert state $|\psi>$ is that the former are derived from the latter by taking the complex square root of the latter, so that probability$(|0>) = \bar{q} = \overline{\psi(\uparrow)}\psi(\uparrow)$ and probability$(|1>) = q = \overline{\psi(\downarrow)}\psi(\downarrow)$, so that $< \psi|\sigma|\psi >= \bar{q} - q = (\bar{q} - q)(\bar{q} + q) = \bar{q}^2 - q^2$. Therefore, by using the transformation between real cognitive states q defined by the complex square root of ψ, i.e. $q = \bar{\psi}\psi$, we have a transformation of the average of the Cognitive Operator M on cognitive states, on the average of the SPIN Operator on two-states quantum elements in Hilbert state, i.e. eq. (76). This is a very important relation, established by an average of the Cognition Operator (which transforms an orientable plane into a non-orientable Moebius surface due to the torsion introduced by M, as represented by eq. (62), and the spin operator on the Hilbert space of two-state quantum mechanics. It is an identity between the action of the cognizing self-referential mind and the quantum action of spin. Thus the cognitive logical processes of the subject become related with the physical field of spin on the quantum states. This is in sharp contrast with the Cartesian cut, and we remark again that this is due to the self-referential classical-quantum character of M as evidenced by eq. (62) which produces a torsion on the

[40] According to Conte, "quantum mechanics is a theoretical formulation that necessarily includes also the cognitive function as it is shown by using a Clifford algebraic formulation of this theory. Therefore, quantum mechanics is also a physical theory of cognitive processes and of the profound existing link between cognitive dynamics and physical reality per se" [19].

orientable cognitive plane of coordinates (true, false) to a plane which is torsioned to yield a superposed state, S_-. The relation given by eqs. (72, 76) establishes a link between the operations of cognition and the quantum mechanical spin. This link is an interface between the in-formational and quantum realms, in which topology, torsion, logic and the quantum world operate jointly. Yet, due to fact that for the Klein-bottle there is no inside nor outside, the exchange can go in both ways, i.e. the quantum realm can be incorporated into the classical cognitive dynamics, while the logical elements can take part in the quantum evolution. Indeed, if we have a matrix-logical string which contains the momentum product, say, $\ldots < x|A|y >< q|M|q >< z|B|s > \ldots = \ldots < x|A|y >< \psi|\sigma|\psi >< z|B|s > \ldots$. Thus, the factor $< \psi|\sigma|\psi >$ entangles with the rest of the classical logical string creating a Schrodinger cat superposed state, since we have a string of valid propositions where one may be the negation of the other.

There is still another very remarkable role of these superposed states in producing a topological representation of a higher order form of self-reference, produced from oppositely twisted Moebius surfaces. So we shall consider the Cartesian modulo 2 sum of the superposed states

$$\mathcal{H} := |S_+ > \oplus |S_- >= \begin{pmatrix} 1 & 1 \\ 1 & -1 \end{pmatrix}, \tag{84}$$

which we call the topological in-formation operator which is an hypernumber; indeed, $\mathcal{H} = \sigma_x + \sigma_z = \epsilon_3 + \epsilon_2$. We could have chosen the opposite direct sum or still place the minus sign on the first row in any of the columns and obtain a similar theory, but for non-hermitean operators unless the minus sign is on the first matrix element. Notice that it is a hermitean operator, which essentially represents the topological (or still, logo-topological) in-formation of a Klein-bottle formed by two oppositely twisted Moebius surfaces. [41] The in-formation matrix satisfies $\mathcal{H}\mathcal{H}^\dagger = \mathcal{H}\mathcal{H}^{-1} = 2I$. We recognize in taking $\frac{1}{\sqrt{2}}\mathcal{H}$ the Hadamard gate in Quantum Computation [4], which due to the introduction of the $\frac{1}{\sqrt{2}}$ factor is hermitean and unitary. Now we have two orthogonal basis given by the sets $\{|0>, |1>\}$ and $\{|S_- >, |S_+ >\}$ of classical and superposed states respectively, the latter un-normalized for which a factor $\frac{1}{2}$ has to be introduced but still does not give a unitary system as in quantum theory. An important role of the Klein-bottle is precisely to transform these orthogonal basis, from classical states to superposed states which are nor classical nor quantum, but become quantized by appropiate normalization with the $\frac{1}{\sqrt{2}}$ factor. Indeed,

$$\mathcal{H}|0 >= |S_+ >, \mathcal{H}|1 >= |S_- >, \tag{86}$$

and

$$\frac{1}{2}\mathcal{H}|S_+ >= |0 >, \frac{1}{2}\mathcal{H}|S_- >= |1 > . \tag{87}$$

For the logical space coordinates $(true, false)$ we have rotated the state $|0>$ clockwise by 45 degrees through the action of $\mathcal{H}$ and multiplied it its norm by 2, and for the state

[41] Alternatively, instead of $\mathcal{H}$ we can introduce another in-formation matrix for the Klein-bottle, namely

$$\mathcal{H} := |S_+ > \oplus |S_- >= \begin{pmatrix} 1 & 1 \\ -1 & 1 \end{pmatrix}, \tag{85}$$

which is non-hermitean.

$|1>$ we have rotated it likewise after being flipped. In reverse, the superposed states are transformed into the classical states by halving the in-formation matrix of the Klein-bottle, producing 45 degrees counterclockwise rotations, one with a flip. Now classical and quantum states are functionally complete sets of eigenstates spanning each other. The classical states $|0>$ and $|1>$ can be easily determined to be the eigenstates of AND, and and the superposed states $|S_->$, $|S_+>$ are the eigenstates of NOT. It is known that the logical basis of operators $\{\text{AND}, \text{OR}\}$ is functionally complete, generating all operators. Hence our system of classical and superposed (or still, quantum by appropiate normalization by $\frac{1}{\sqrt{2}}$) eigenstates constitute together a functionally complete system: all operators of matrix logic can be obtained from them. This system is self-referential. Furthermore, there are operators which produce the rotation of one orthogonal system on the other orthogonal system. The logical Cognition Operator M defined by the commutator [FALSE, TRUE], or still eq. (61), transforms classical states $|x>=\bar{x}|0>+x|1>$ into $|S_-$ and still the anti-commutator $\{\text{FALSE}, \text{TRUE}\}$ which coincides with the matrix $\mathbf{1} = \begin{pmatrix} 1 & 1 \\ 1 & 1 \end{pmatrix}$ transforms $|x>$ into $|S_+>$, i.e.

$$M|x>=|S_->, \mathbf{1}|x>=|S_+> . \tag{88}$$

This can be rephrased by saying that M evidences on its action on a classical state the torsion in the quantum commutator of FALSE and TRUE while the ONE operator $\mathbf{1}$ transforms $|1>$ into $\begin{pmatrix} 1 \\ 1 \end{pmatrix} = |S_->$. Since both M and $\mathbf{1}$ are non-invertible, we shall use instead the fact that $\mathcal{H}^{-1} = \frac{1}{2}\mathcal{H}$, so that in addition of the transformation by the Klein-bottle of the classical basis in eq. (86), the reversed transformation from the superposed to the classical states is achieved by

$$\frac{1}{2}\mathcal{H}|S_+>=|0>, \frac{1}{2}\mathcal{H}|S_->=|1> . \tag{89}$$

Yet, we stress again that these transformations are not unitary. This is easily resolved by the $\frac{1}{\sqrt{2}}$ factor and then we have a transformation of classical into quantum states and viceversa. In the latter case, the renormalized Klein-bottle acts like a quantum operator producing coherent quantum states, a topological Schroedinger cat state which does not decohere.

6. The Eigenstates of the Null Operator, Cognitive Apeiron, the Extended Photon and Their Twistor Representations

In this section we shall retake the ellaboration in [111] following a modification of a sketchy yet ground-breaking idea by Dienes [21], elaborated in [111]. We have discussed the eigenstates of some logical operators, and now we shall discuss the eigenstates of the Mind Apeiron, namely the 2×2 identicaly zero matrix, which we denoted as $\mathbf{0}$. In distinction with the other logical operators the eigenstates of $\mathbf{0}$ (as a linear transformation from C on C, so that $\mathbf{0}$ becomes a point of C^2, its origin) are no longer quantized, but rather give an orthogonal complex two-dimensional nullvector space. [42] In this way the Plenum is no longer represented by a single point, **0**, but rather becomes an *extended* object or zero-brane. This phenomenon is well known in complex Clifford bundles. To create the correct complex vector bundle we modify C^2 to make different fibres (unit 2-spheres) disjoint, with no common origin. To achieve this we replace the origin of C^2 by a copy of the entire Riemann sphere, so that instead of having just one zero, we have a whole Riemann spheres worth of zeros, one for each fiber, giving the zero section of the bundle. This procedure is known as blowing up the origin of C^2, which amounts to take a projective subspace of $C^2 \times S$, with S the Riemann sphere (i.e. the complex projective space $CP(1)$), defined as the set of all elements of the form $\{(z_1, z_2, \xi) \in C^2 \times S/z_1 = z_2\xi\}$. Since zero is now allowed on the fibres, continuous cross-sections of the bundle exist [21].[43] These cross-sections represent the spinor (and twistor) fields on $S \simeq S^2$, the two-dimensional sphere, giving a 2- complex-dimensional vector space, which can be mapped to the 2-dimensional logic space of matrix logic by stereographic projection. We shall apply now this to the twistor representation of the extended photon through the maximal monochromatic algebra as described by (46, 47) which has an equivalent representation as a pair of divergenceless orthogonal spinor vectors $(\omega^A, \pi^{A'})$, $A, A' = 1, 2$ by (48). By stereographic projection of this twistor representation of the extended photon, we obtain a basis of cognitive space, or vice versa: We can use the Boolean basis $< 0|$ and $< 1|$, or still the superposed states $< S_+|$ and $< S_-|$, to represent the maximal monochromatic algebra by taking the inverse of the stereographic projection; in any case via the normalized Klein-bottle Hadamard in-formation matrix, we generate all the operators of Matrix Logic. In this we see how the extended photon, which we purported to be a subjective-objective fused structure, is represented as a basis for cognitive space, and conversely, from cognitive space we are able to codify the maximal monochromatic algebra representation of the extended photon. This establishes the full self-referential construction of a world which is perceived through quantum jumps, i.e. differences that produce differences, or still, in terms of cognitive states of the Mind; yet, we recall that the (absorbed) photon is already tantamount to embodiment. Furthermore, we have seen above that the role of the Planck constant $\hbar$ is precisely to connect the transformation of the quantum world (whatever the scale microphysical or astrophysical is) into the world of the mind, bridging thus the material and mind domains. The Riemann sphere is not only instrumental to codify this joint constitution by codifying the extended photon as a cognitive state. It is also the manifold in which the complex logarithmic function takes multivalued values to quantize the quantum jumps in terms of the different branches of the logarithm, allowing thus to

[42] We have already seen this an identical situation in the eigenstates of TIME. Thus the eigenstates of the Mind Apeiron are given by a nilpotence condition alike the eikonal equation; we shall see that this likeness is in fact an identity

[43] The blowing up of the origin, transforming its point-like structure to yield a manifold has profound consequences. For example, the blowing up of the origin in R^2 is the Moebius surface, which as we already saw is basic to the Intelligence Code. Two oppositely twisted Moebius bands generate the Klein-bottle by glueing them. Thus the blowup of the origin gives rise to the higher order surface of paradox whose matrix representation, up to a normalizing constant, is the Hadamard gate of Quantum Computation, which together with the phase conjugator, allows to generate all quantum gates [4]. In taking in account that DNA performs quantum computation [44] as the works of Peter Gariaev and associates shows [34] and its relation to holography [74, 78] which is already performed by sensory processing cells of the human neurocortex, evidences the relevance of the importance of this gate and the multivalued logic we presented above which is derived from the Klein-bottle; see [109] and references therein.

codify the 'outer' and 'inner' worlds. Further elaborations in relation to the transactional interpretation of Quantum Mechanics, cosmological Kozyrev torsion fields and entanglement, and brain synchronization in binocular vision, and the visual and somatosensory representations on the neurocortex, will be elaborated below. [45]

7. The Surmountal of the Cartesian Cut and Its Manifold Manifestations

7.1. Torsion, the Primeval Distinction, the Klein-Bottle, Cognition, Control, Will and Time Operators

While current science has been built in terms of the Cartesian Cut in its manifold expressions, we have unveiled a lifeworld that surmounts the Cartesian Cut which stems from incorporating into the very foundations of the constitution of space, time, thought, cognition and perception, the essential phenomenon which is the basis for consciousness: self-reference.

In second-order cybernetics self-reference transcends the cut between observer and controlled system by the semiotic codification through a primeval distinction (in the sense of the calculus of distintions of Spencer-Brown) of the torsion geometry associated to the anholonomic variables (controlling variables that cannot be separated from the system that they control) variables which Pattee considered precisely -we say, paradoxically- to encode the Cartesian Cut [91, 109]. This fuses subject-with-object into an implicate *and* explicate (in the sense of Bohm) meta-algorithmic process-form which is the Klein-bottle [109, 13]. Departing from this primeval semiotic codification of torsion through a distinction on a plane, this introduces two states: the 'empty' undistinguished state which is given by the plane itself of all potential undistinguished states (the Plenum, rather than the void in spite of being empty to perception), and the 'distinguished' state codified as the distinction sign. The cleavage of the state of all potentialities under the distinction sign generates a process through the sign as a boundary that sets an 'outer' and 'inner' domains and thus the distinction becomes a logo-physical generator. Thus is established a non-dual process of content and context mutual transformation, in which the distinction is both operand and operator, and form and function become fused. By functioning as a boundary, this transversion of the sign-boundary is an abstraction of osmosis, which is related to torsion [105]. In terms of logic, in first examination this leads to Aristotelian-Boolean logic [133]; on raising this distinction through self-referential paradox we obtain the Klein-bottle [103] and this leads to the generation through time-waves of a 4-state logic from which was derived Matrix Logic [109]. This reverses completely the historical Western tradition of disregarding paradox as nonsensical, and thus paradox was proved to be the basis of logic. Again, this is possible because content and context conform an holonomic process-structure, rather than distinct related instances.

[45] For a different conception in terms of Endophysics we refer to [118], and to the theory of fractal time due to S.Vrobel [149]. Remarkably, inasmuch matrix logic has a projective structure as well as the eigenstates of the mind apeiron, Saniga has proposed a theory of 'altered' mind states in terms of Cremona transformations which arise as well as blowups [14]. For a geometric formulation of noetic space we refer to Amoroso [3].

This participatory constitution of the geometry of space and time stands in sharp distinction with the Cartesian Cut in which space is exterior to the subject and a mere container of objects, while time is also considered to be external, though in General Relativity it is constrained by relations with the space variables. In Aristotelian thought, the cradle of Western scientific thought, we have as its backbone a dualist conception which expresses itself in the two-valued logic and the principle of non-contradiction, which is no longer valid in Matrix Logic nor in the calculus of distinctions with paradox. The dualist conception eliminates subjectivity from the universe [109]. Gunther's stance -departing from the Hegelian conception that time is related to logic- is that the elimination of time is related to the principle of non-contradiction [39], consistently with the findings in our work that its violation leads to time-waves. Further, in Matrix Logic time appears as an operator, TIME which we introduced in eq. (66), which is the matrix representation of the commutative square root of minus one which is the basis of imaginary numbers, an anticlockwise $\frac{\pi}{2}$ rotation in the cognitive plane of all cognitive (vector) states which we encountered above. We have seen already that the quantization of any cognitive value and Self as a fixed point of any cognitive state cognition arises from the Logical Momentum (Cognition) Operator, M, defined by the commutator of the False and True Operators -not (dual) scalars as in Boolean logic- which decomposes as $M = \text{TIME} + \text{SPIN}$. A torsion vector appears in two guises: as the coefficients of the self-referential structure produced by these operators, and as a superposed state of the difference between the normal unit vectors to a Moebius surface, i.e. it is given by $S_- = (1 \ -1)$, while the anticommutator of these operators yield their sum with coefficients hence given by the other superposition state given by the sum of the normal unit vectors [46]. Now, we recall that $M = |S_+ >< S_-|$, the tensor product, and the direct sum modulo 2 yields the Klein-bottle in-formation matrix, $\mathcal{H} = |S_+ > \oplus |S_- >$, which -up to a normalizing factor- is non-other than the Hadamard gate of quantum computation; both played a major role in the theory. From the association of SPIN with space torsion and since TIME is a rotation on the cognitive plane generated by the true and false states, it is clear that cognition is represented by a vortex structure-process projection on two-dimensional cognitive space! We shall see later on that this is the case of somatosensory and visual codification in the neurocortex, in which each point of the planar neurocortex codifies through a whole columnar arrangement on which a Karman vortex structure is occuring.

7.2. Control, Will, Self-reference, Life and the Time Operator of Matrix Logic

We want to explore the relation between will and TIME as introduced in Matrix Logic which is a self-referential operator (either by distinguishing two arbitrary cognitive states by computing their difference or by geometrically being a simple ninety degrees rotation, which thus if iterated by another three rotations leads to the identity) in a self-referentially constructed logic that has as its roots the Klein-bottle. This will be of paramount importance in examining later the role of time and self-reference in cosmological and human systems, and basically with regards to the existence of time waves -which we already discussed in [109] that are appearing in very diverse natural (and particularly, human) systems (including Mathematics), and cosmology as well.

[46] $[\text{False}, \text{True}] = \text{False} - \text{True} \neq 0; \{\text{False}, \text{True}\} = \text{False} + \text{True}$

We start with a disgression on the construction of number systems from the notion of primeval distinction. Indeed, with this concept of distinction and its interpretation as a boundary, we can construct numerical systems in the so-called boundary mathematics [57]. To construct the complex numbers as spatial forms, only three distinctions are necessary to generate them. Hence, this conception under this extra provision of three distinctions allows to construct the number system which allows to treat time and subjectivity : the imaginary numbers [135]. Furthermore, the time oscillations which in the present theory based on a single distinction and the Klein-bottle generate the 4-state imaginary logic of the calculus of distinctions which itself generates Matrix Logic, appear without recurring to the paradoxical equation that leads to the Klein-bottle, as an oscillation of the empty state and the *phase form* (which appears to be its self-inverse) created by the juxtaposition of the three distinctions [57]. If we think of each spatial distinction and its meaning as a boundary initiating a process which has a time interpretation as is the calculus of distinctions, then we have three different time distinctions associated to these three spatial boundaries. The first distinction creates a primeval cleavage, space, and the process of going through it, velocity; the second distinction creates a cleavage on the velocity, i.e. acceleration as a process through it, and finally the third distinction creates a cleavage on the accelerations which as a process introduces the time derivative of acceleration, thus a third order time derivative on the original undifferentiated (which is empty, to Spencer-Brown [133]) plenum state is associated to the three cleavages which allow us to construct Matrix Logic, and hence relate it to quantum field theory and the Klein-bottle. But a third order time derivative introduces a derivative of force which is not associated with determinism but rather with *control*, which is absent in Newtonian physics due to the fact that it cannot be used to predict. Yet, being related to control, it thus has an element extraneous to the Newtonian conception. Indeed, it is a subjective element linked to will, intention, purpose, which are essential characteristics of life, by which monads (i.e. entities which come to be through the semiosis of having a primeval distinction constituting their identity) can become autopoietic [109, 146]. Without will, intention, purpose, action would be purely automatic and assimilable to determinism. Thus, it would not be related to self-organization and self-reference, and the quantum world would be inexistent. As we discussed already, this would be tantamount to the disolution of Being.

Remarkably, A.Young -following a lead by Eddington- introduced a $2D$ flat space (a plane which we can trivially identify point to point with the space of cognitive states in Matrix Logic) of phase representation (basically, a circle or better still a *cycle* in the complex plane), the cycle of action, which stands in place of *processual time* [47] as related to *action*, by rotations in the range of 0 to 2π (where we recall that the 2π *uncertainty* factor appears in quantum physics [48] in the commutator relation $[p, q] = 2\pi h$); Young with unique

[47]Young points out that processual time is about *timing*, or, in our terms, through a willful operation by and through time which in our conception is embodied in the joint constitution of space, time and subjectivity. There is a conception of a time operator with different roots than the present one, proposed by the Austin-Brussels school led by Prigogine [97, 5]; we shall discuss it further below.

[48]In Eddington's theory of processes, it is the value of the curvature of space in the hypersphere with the added phase dimension, thus unifying quantum physics and gravitation. Eddington's approach departs from considering that both waves (as in the present theory) and curvature (as in General Relativity, GR) can be used to represent energy and momentum distributions, and they should not be confused. We have shown in the first part of this chapter and in several works [103, 105, 106, 107, 104] that spacetime geometry is asso-

insight associates this phase with a *choice* of timing. We observe here that Young takes the direction of action to be anticlockwise and we can as well introduce the time operator TIME in Matrix Logic, with the opposite definition as well, and thus coincide with Young's determination of direction of rotation and thus of processual time. In the elaboration by Young of the setting due to Eddington, the null rotation establishes space (with dimensions of length, L), the first ninety degrees rotation establishes velocity (dimensions, $\frac{L}{T}$, L length, T universal time [51, 52, 53]), the additional iteration of another second ninety degrees rotation establishes acceleration (dimension $\frac{L}{T^2}$), and the third iteration which amounts to a $\frac{3}{4}$ turn of the full cycle establishes control, which requires a third-order derivative in T, since the velocity and acceleration require first and second-order derivatives in T, respectively. Thus, a three time iteration from the initial position that stands for space distinction, by three ninety degrees turn, leads to control and thus to will, and by applying a final turn, we return to the initial distinction, which is space, and the rebirth of action. (This fourth turn is the Eternal Return of self-reference by the action of self-control, and thus it rep-

ciated to waves and it is fundamentally quantum, in distinction with GR, and yet in consonance with Fock's interpretation of energy-momentum densities associated to propagating singularities. These torsion geometries can be introduced by simple scale transformations on reference frames of *flat* spacetime by the action of wave functions [102, 113]. So, in distinction with GR, spacetime can be metrically flat and yet have a non-trivial geometrical structure, which we have seen is fundamentally quantum and related to the fusion of an objective and subjective realms, which of course, is absolutely absent in GR. An essential manifestation of these geometries is the appearance of Brownian motions as spacetime constituting quantum fluctuations, fusing geometry and randomness [102], the latter itself is both 'objective' and 'subjective' [11]. Thus, scale uncertainties play a basic role since they will yield uncertainties on the distribution of energy-momentum; in this conception, these uncertainties are unseparable to the joint constitution of spacetime and subjectivity through torsion . Since probability distributions come in conjugate -in the sense of complex numbers- pairs, their product gives the probability distribution of energy-momentum, and the variable conjugate to the scale is the *phase*, such as appears in the Schroedinger equation and its connection with torsion generated Brownian motions which have both a future-oriented and past-oriented waves [107]. Hence, time is already incorporated in these probability distributions [102]. Since there is to be considered the case of a reduction of scale to particular eigenvalues, the scale is related to momentum and to the phase as a coordinate. This phase coordinate (basically the imaginary exponent S -so that the uncertainty factor is related to *oscillations*- of the relativistic Schroedinger wave function [104]) is represented as a fifth-coordinate to (a now flat) spacetime. Metrics have no role whatsoever on this generation of spacetime; as the Bianchi structure equations show, torsion is more primitive than curvature [33]. Thus, scale and phase are invariant by Lorentz transformations, and more generally, by its double covering group, $\mathrm{Sl}(2, C)$, which is the group of rotations for Special Relativity and relativistic quantum mechanics, embodied in the Dirac equation and spinors. The scale uncertainty is a fluctuation of the standard, actually creating spacetime starting out of the undistinguished Plenum. These fluctuations are reflected in the measured characteristics of the system. If we take as the original standard to be provided by angular momentum, thus scale momentum is an angular momentum and the corresponding phase coordinate is an angle. (In fact, in the infinite dimensional spinor representations of $\mathrm{Sl}(2, C)$ [36], in terms of which the Dirac equation of relativistic quantum mechanics comes to the fore [113], these phase coordinates are a hidden invariant degree of freedom for spinors. Thus, the phase coordinate is indeed related to (quantized) angular momentum, spin, and gives rise to a universal spin-phase-bundle on spacetime [113], as an additional phase degree of freedom for each point of spacetime, exactly as originally conceived by Eddington.) So *fluctuations* of the phase assume a value between 0 and 2π in which the fluctuations take a uniform distribution over the whole range of 2π. This range takes the role of providing a fundamental scale upon which it is possible to pass from the quantum discrete domain to one in which we can assume continuity. We note here that fluctuations are described in terms of probabilities. Yet, as stressed by Costa de Beauregard, the concept of probability is jointly objective and subjective, " providing the hinge around which matter and mind interact"; [11]. So there is a fusion of objective and subjective realms in the introduction of the space of phases, which is reinforced through its connection to spinors and the Dirac equation, as explained before.

resents the jump from first-order to second-order cybernetics [146] where now action and self-reference are identified bringing forth the joint creation of spacetime and subjectivity [109].) This can be seen in taking in account that since $T^4 = 1$ (the fourth derivative is the identity, yielding position since the action of control is purposefully applied in view of a spatial coordinate) in the space of phases; the control of force is parametrized by the multiplication of their units, i.e. by computing the multiplication of control with that of force: $\frac{L}{T^3}\frac{ML}{T^3} = \frac{ML^2}{T^5} = \frac{ML^2}{T}$, the units of action. In our conception, in regard to the previous introduction of TIME in Matrix Logic (which is a $\frac{\pi}{2}$ anticlockwise rotation in the cognitive plane of bras and kets), this is the second-order cybernetic self-referential action by which we return to the initial point of the cycle, by applying a final ninety degrees rotation, the action of TIME applied on TIME^3, to yield $\mathrm{TIME}^4 = I$, the identity. [49] This coincides with the action of processual time to return to Self: Self-reference, returning to space -the initial distinction-, closing thus a 4-cycle in phase-space; see page 51 in Young [156].

These ideas were crucial to the development by Young of the control system for the helicopter. As Young remarks with outstanding insight, control is the essential property of *Life*; its being is related to purpose-intention and volition, in other words, to will, and as such it is anticipative yet uncomplete, requiring another $\frac{\pi}{2}$ rotation in phase-space to reach back to the self, i.e. to become self-referential. Thus we return to our previous disquisitions, in which we associated purpose-intention to a time operator, as the meanings of these nouns clear suggest. Young still associates this to the fourth level of descent of processes on which degrees of freedom are progressively lost, starting with the self-referential photon as the exponent of the highest potentiality. Thus is established an evolution from the homogeneity (of the undistinguished Plenum), the photon being the source of all inhomogeneities, as we elaborated already) to inhomogeneity (say, complex molecules, DNA, etc.) as the fundamental units from which Life emerges, and which Young with great insight associates with will, as an essential precondition. So Young argues in regards to the cycle in the space of phases (processual time) that, on reaching self-referentially to the identity, we find a final degree in which the monads (i.e. the entities formed by the primeval distinction) start to build up their own structure countering the action of increasing entropy, as is the case of complex molecules, such as the case of viruses, *through timing*. In our conception, as we have seen above, this is due to the action of TIME on the control phase. Returning to Young's view, which we state again with a slightly different wording: In this evolution from the photon -which has the maximal degree of freedom-, from which a process of individuation starts into the more complex structures, as those that we associate with Life, such as complex molecules which have lesser degree with respect to the photon which is the building process from which all more complex structures are formed of, there occurs a return by willful action (i.e. self-control) by which complex molecules chose their processual timing to counter the environment in which the monads are in exchange relation through their primeval boundaries, exercising thus their sole freedom left to reverse the condemning determinism initiated by the first two derivatives in universal time, through control and self-control, which amounts to the self-referential action of TIME. we encountered in

[49] That time might operate in the quantum domain producing interference (and thus indeed is associated with control), is an experimentally verified fact; its theoretical framing is due to Horwitz; see [53], and references therein.

Matrix Logic. [50] This is essentially linked to will and self-determination. Hence, the "free will" so much purported in the history of philosophy and diverse theologies, is the will of self-determination by the action of TIME. This is an original conception; we shall retake it in examining the Myth of Eternal Return, departing from this conception. [51] We shall also see the relation of Life as process-form, in the explicitation of the Myth of the Eternal Return, which we shall elaborate further below. We must recall here that the initial process that lead to TIME as appeared in Matrix Logic and our previous considerations, which we have identified with the processual time of any system defined by a boundary creating thus the multivalued logic that arises from paradox, is the Klein-bottle. Therefore, the *topology* of the time operator embodied in all systems and particles of any scale in their phase dimension which leads to control and self-control, is represented by the Klein-bottle. We shall reencounter it when dealing with the physiology and growth process of the neurocortex, sensorial codification and self-referential growth of bodily organs. [52]

7.3. Chronotopology, the Myth of the Eternal Return, Self-reference, Time Operators and the Klein-Bottle

Chronotopology is a far reaching concept which was first proposed by Musès in his outstanding work 'Destiny and Control in Human Systems' [83] in which he anticipates a theoretical framework for the understanding of time patterns and structures (chronomes),

[50] We are exercising a very different conception than the emergent non-computability conception due to Penrose for which Life will be another unexplained emergent factor requiring other factors, and thus ad nauseum. Our conception also greatly differs with Prigogine's which, while attempting to introduce a time-operator in a general systemic approach based in dynamical systems theory and chaos, choses the second law of thermodynamics as a *metalaw* with causative precedence to any other law of physics and Nature at large [97]. We shall return to the implications in terms of ontology of this choice.

[51] Thus, the so-called *arrow* of time which has been extensively discussed in physics and biology, is *subservient* to this timing process, or otherwise we would have reintroduced a dualism which we have already shown that it is not the case. Related to this issue we shall discuss in next Section the notion of *chronotopology*, the topology of processual time. Since this topology is non-linear, basically the Klein-bottle, the arrow of time is determined by the linear sections of the chronotopology of structures-processes which by self-referential timing come to be. It is clear that third-order derivatives notoriously absent in physics and biology point out to processes which may violate the second law of thermodynamics. While it is claimed that the first and second-order derivatives in time equations of physics are time-reversal, the Brownian motion quantization of the Schroedinger equation in which the drift is described by the trace-torsion, may *not* be time-reversal. (Remarkably, it is through the random fluctuations (say structured as Brownian motions) of the internal paramenters of a system, even in the minimalistic case of absence of heat exchange through its outer boundary, that a system can violate the second law of thermodynamics, through fluctuations of the random entropy; we shall return to this issue below). Indeed, when the trace-torsion has, in addition to the logarithmic gradient term encountered already for describing quantum jumps and the Aharonov-Bohm effect associated to it, other electromagnetic potential terms, then the Brownian motion is not time-reversal [106, 104]. Furthermore, the actual construction of the solutions of the Schroedinger equation require initial and final conditions on a time interval. Note added in proof: On completion of the revision of this article, we found in [50] very interesting discussions on processual time, Bergson's duration and the physics of life, which seem to be quite in agreement with some of the conclusions presented here.

[52] One can ponder on the possibility of Mathematics (as a system) having a time operator acting through rotations which is further related to self-reference. We already discussed in the Introduction that this is already the case of the Continuum Hypothesis [47] and also the generative structure of the natural numbers and in particular of *all* primes, from a planar structure of an original set of natural numbers, which by rotations produce the complete system simply by recursion [60] .

and their relation with psychological archetypes. The initial discussions in this pioneering work focuse on natural systems which are found to have non-linear forms-functions to later proceed to analyze human systems. Musès was very much concerned that -to his understanding- Mathematics was not the adequate language for human systems, and in this unique work he proceeds to analyze a very ancient myth. Musès shared with Young the idea that myths are accounts of processes and as such they should be studied. Myths have played a fundamental role in framing human history and till today they continue to do so. Polymath Raju proceeds one step ahead of Musès in his fundamental work, 'The Eleven Pictures of Time', to show that the action of myths is all pervasive, up to the point that contemporary theoretical physics is very much framed in terms of them. In particular, the reigning conception that time is a mere parameter without any functional role, and furthermore it is *linear*, instead of having a non-trivial topology, is one such conception that has its roots in the political management of myths. Raju traces this prevailing conception to political-theological-religious issues that were decided in the Council of Nicea, at Byzantium, circa 340 C.E.; this linearity has far reaching implications even in framing our socioeconomical and cultural mindset and organizations [101]. This present chapter is intended to be a contribution to correct these apriori's which are not of the Kantian type, but rather forcefully implicit to our lives due to decisions that acted out an all encompassing anthropological engineering operation, which hundreds of years later is still efficaciously weaved well enough into the present mindset [53] to the point of have gone unnoticed for our current generation at least, until Raju's work, to the author's best knowledge. [54] Returning to Musès studies

[53]Who has not encountered the proverbial fool that passionately utters "Bing-Bang" to triumphally hastely mumble next "God"?

[54]The history and current developments in science, particularly in physics, are rife with these cultural engineering operations. They became especially notorious with the advent of postmodernism. We have nowadays thousands of physicists working in a purported "Theory of Everything", the so called string theories, which in fact there are an infinite number of them. Remarkably, none of them can be subjected to any kind of experimental validation or attempt to refutation. So these theories came to be qualified as 'not even wrong' by its critiques. Thus, this postmodern deconstruction of physics, which in this setting ceases to be related to experience, is a totalitarian posture claiming to be a theory of everything that there is to be known (about physics, or is it about 'everything'? ...), a claim that does not stand the most candid consideration. (Of course, would we fall into the megalomany of claiming that a 'Theory of Everything' is at the disposition of science, then a first test of consistency would be to check its ability for explaining how the 'theory' was created in the first place! Of course, Goedel's theorem will be invoked to dodge the test, but then the second test will be to check if the 'theory' is able to explain how the Goedel theorem came to be. Then, as explained in the Introduction, arguments like those expounded by Penrose will be invoked, falling thus into the infinite antiregress that we alluded in discussing these arguments for an emergent consciousness, or we may consider the logical shortcomings of his approach related to his choice of Aristotelian-Boolean logic's dualism. Consequently, it follows that the purported 'Theory of Everything' is an unending theory of emergence, or simply logically untenable, and then totality is proved to be unattainable, contrary to its claims.) Certainly, such a theory of 'physics' has no qualms whatsoever in being unrelated to cognition in any sense, with exception of the Cartesian mode, which finds in this intellectual exercise its epitome of total disocciation with reality (the res extensa of Being). If the question is on its any relation with psyche, rather than physics, the answer is positive, yet it is the unfolding of a mind which though embodied in the physicist's body, has resigned to all criteria of validation with exception of its self-*satisfaction* on its intellectual power, a form of narcissism. It is an alienated self, that claims complete dominion, through the theoretical framework, on the universe. Nothing else than the nineteenth century' positivism, extended from determinism to the purported quantum domain. Another similar operation that appeared concurrently to strings theories, followed the proposal by Prigogine of a theory for everything, based on dynamical systems, chaos and the imposition of the second law of thermodynamics as a metalaw for Nature. Similar to the flockings of physicists to the *appeal* of strings (indeed, fetishism is the case), postmodern aca-

in interpreting myths as processes, he chose one myth which has a cosmological setting, but embraces all natural systems including human systems, of course. Musès claims that his rendering of the myth in question goes back to civilizations dated thousands of years ago, and was the result of his arduous -and non-conformist with respect to other scholars take- investigation in the Tayyibite Shi'a tradition of Islam, preserved in Medieval Yemen traditions and manuscripts dispersed through European libraries [37]; Musès in addition of being an outstanding mathematician (alas, not recognized much as such besides few) was a man of extraordinary qualifications and insight. It is none other than the Myth of Eternal Return, in a particular setting. This is an awesome read, and proceeding from the conception presented in this chapter, we shall unfold the issue in terms of self-reference. Musès account is very different to the more popular and somewhat naive account due to Eliade [29] in which there is no systemic approach. Musès processual approach was focused in trying to clarify a most difficult problem: To elicit and understand the origin of Evil (and consequencely also of Redemption) which he links to the will of a Third-Intelligence-Entity produced from differentiation of the Unmanifest, to self-referentially fuse with Apeiron and control it. The Unmanifest (which in Greek philosophy corresponds to Apeiron), the field of all potentialities, in our conception is the undifferentiated state (the *Plenum*) which upon introducing a primeval cleavage (torsion) will embody the fusion of object with subject and the cocreation of world and subjectivity, similarly in the latter issue to Spencer-Brown's protologic. The introduction of this primeval cleavage in this myth finds its expression in the First-Intelligence-Entity ('Aql, in Arabic), which is associated to Mind (with its entire gamut, including understanding, insight and spirit. It is also called Sabiq, the one who goes before time or space; according to Musès, in German it corresponds to *Verstand* rather than *Vernunft*), placing the first differentiation in subjectivity, as a transform of the Unmanifest, Apeiron. Our interpretation, of the First-Intelligence-Entity, which stems from our conception, is as follows: It is the primeval distinction that we have encountered as the sign in the Plenum, signifying the torsion field and the appearance of the Klein-bottle logic, in considering the paradoxical reentrance of the First-Intelligence-Entity in itSelf. It appears before time or space because it embodies the will of the fusion of subject with object in an integral process that finds its reification as both space, time, and mind, now as the psyche of Apeiron, in the account by Musès. The Second-Intelligence-Entity is a further stage of differentiation, produced by an *emanation* of the First-Intelligence-Entity, and thus light is brought into the world as *will* (which we have already related to the action of a time operator in all systems following Young's musings and to torsion in both cognitive space (TIME Operator, see eq. 68 and the explanation after eq. (71)- and spacetime, both pro-

demics -especially, in the human sciences- were happy enough to find in those disgressions on chaos, order and the action of a time operator, a fertile territory for poesis; Prigogine, in his theory of processes [97], claimed to have established a physical territory for its reification [130]. These kind of operations were traced back to obscure aspects -and also to the genuine ethical quest to reach for an embodied conception by its founders!- of the Copenhagen interpretation of quantum mechanics- Beller, "....The opponents of the postmodernist cultural studies of science conclude confidently from the Sokal affair that "the emperors have no clothes...But who, exactly, are all those naked emperors? At whom should we be laughing? " [10]. (Rather, we understand that the issue is: At what we should be laughing?, granted that the natural attitude with regards to our folly, natural and engineered ignorance, and our graced quest for integrity, is spontaneous self-referential laughter, rather than crying.) We shall later return to Prigogine's proposal. For an attempt of saving strings theories of the inconsistency of its basic postulates (albeit not from its lack of grounding, an enterprise of no concern to strings' theorists), rooted in the topological phenomenology further developed in the present chapter, we refer to [117].

ducing quantum jumps as differences producing differences) of trascendence through its reification implicit in Mind. The Third-Intelligence-Entity, according to Musès, is a further emanation of the Unmanifest, " is called *Adam Ruhani* (in Hebrew), or the Spiritual Prototype of humanity, in whose image we were formed. Thus the Third-Intelligence is the Divinity of our universe which has not yet come into Being". Its attainment is Redemption. "It corresponds to the great Anthropos Megos of Valentinian Gnosis". We continue, Musès: "Now there was no problem even conceivably arising with the first two Intelligences who realized that the source of their being was in the Unmanifest". Further, "...the main line of the dénouement of the origin of Evil in respect to the nature of Time, the Third Divine Entity sought to encompass its own origin and to plumb the very depths of being- a route that perforce have to lead into the Unmanifest, the Mistery of Mysteries that by its very nature cannot be unveiled with impunity to the one so seeking; the *veiling is inherent* and necessary for the eternal provision of immortal being. The notion echoes in the words of the great Goddess inscribed on the portals of the now lost Temple of Sais, preserved to us by the records of ancient travelers: "None can lift my veil and live". To seek to manifest the source of life out of the Unmanifest, could end only in cutting himself off from the circuit and flow of life in so trying, even though unwittingly, to preempt it. No manifest being can contain the Unmanifested". We already observed [109] that in the present conception of this chapter, the undistinguished protological state (the Plenum) cannot be merged with the distinguished state under the primeval distinction, unless both states would logically collapse into nothingness together with the 'objective' and 'subjective' realms. Musès: "The Third Divine Entity dreamed such a dream of finding that source explicitly and *controlling* [55] it to be within himself (as he mistakenly thought was the case with the Second and First-Intelligences)". [56] Thus the Third Divine Entity willed to disolve into the Unmanifest while keeping the will to self-referentially complete its own cycle. "That dream and wish momen-

---

[55]Our italics, which we introduce to remark that we already related control to a distinctive logo-physical action which is related to subjectivity and the operation of TIME in Matrix Logic.

[56]The mistake attributed to the Second-Intelligence is the 'Great Doubt', the belief of the disolution of light-Self to nothingness, proposed by Varela and associates in their take on enaction. Thus, it is the mistake of nihilism. The mistake attributed to the First Intelligence is the belief that the primeval distinction can self-obliterate itself, which in first examination, *formally*, it is no mistake at all. It is the cancellation rule in the calculus of distinctions of Spencer Brown [133]. in which the self-operation of the distinguished state on itself gives the undistinguished state, Apeiron; it is a primeval rule of nilpotence [109]. Yet, this return to the Unmanifest through this operation does not erase the ideative process which is used to produce the primeval distinction and its recursive application; it only erases the *sign* of the primeval distinction. Therefore, if there is a legitimate reason to qualify the self-obliteration of the First-Intelligence as a mistake, it is founded in the belief that *all* phenomenology is exhausted in the sign. A sign requires a signifier, they are fused, as is the case of the primeval distinction. Thus, the mistake attributed to the First-Intelligence is the belief that its Self, i.e. Self-reference, is not trascendental. (Furthermore, it is the mistake of separating immanence from trascendence, in essence, the Cartesian Cut.) For example, it is the mistake produced by Penrose in his defence of his approach to consciousness as emergent by *using* self-reference, while *denying its significance and Being*, as manifested in Chapter III of his 'Shadows of the Mind', as we explained already. It is also the mistake of physicalism and, no less, of fetishism. From our previous discussion, it appears that nilpotence, which we chose to call plenumpotence, for reasons that have been manifested along this chapter, is a trascendental property, and that nilpotent operators are trascendental; thus, the great importance of algebras rich in nilpotent elements: They represent higher-orders of Self [82]. Such is the case of the Cognition Operator, M. The present findings give retroactive support to the approach that stemmed from the eikonal equation of light rays, our departing point in the present chapter, and to the work of Rowlands and Hill [119]. Thus, the present conception can explain its self-generation as the trascendental action of self-reference.

tary, had on the level of *power and perception* dire consequences, the first of which was *retarding* of the consciousness of the Third Intelligence by reason of this thus introduced blockage or fallacy that by nature could not advance, but only hold back.....The reason for such grave consequences of a released desire on the part of the Third-Archangelic-Power is bound with the fact embedded in ancient traditions preserved in Homeric Greece of the mere wish or propensity of a god being equivalent to the determined or implementing focussed will of man". (Here we find an issue that we will reencounter later -when discussing visual hemispheric synchronization to attain stereoscopic vision-, namely the idea of hierarchies of Klein-bottles (which we shall call it quantized reentering limited domains, that we shall link with the syntropic TIME Operator) which we recall require a certain Planck action 'constant', since it is multivalued [87, 88, 93] to establish themselves as primeval distinctions reentering in different stages, a fusion of macrocosmos (a god) with microcosmos (man-woman), for better illustration). Musès: "But the implications go deeper, since the reason that it is so depends on the facts that the gods are not in our kind of time". (Again, the issue of a hierarchy of Kleinbottles that we have just claimed). "Duration of things [57], yes, and changes, too- but all without waiting time, which is the chief characteristic of what humans call time. We must wait for any idea or plan to be enacted and then mature to fruition or full manifestation". We remark here that in Musès' interpretation, the wish to *control* the source, the Unmanifest, was the source for Evil, and that *timing* (here again the time operator, which we identified with TIME of Matrix Logic, introduced in the discussion of the Eddington-Young phase space) is crucial for the manifestation of the Divinity implicit in the Third Intelligence Entity. Thus timing is essential to the very process of evolution as the unfolding of the primeval distinctions into the manifest forms-functions that make the universe, and is the will of complex structures to reverse determinism by seeking to self-referentially determine themselves through self-organization by countering the increase of entropy, as proposed by Young.[58] Recall that in the previous section we

[57]We here reencounter time as duration as in Bergson's critique to the time parameter of Einstein [90, 11]. Yet, according to Costa de Beauregard, Bergson's duration is still associated with a linear topology in his belief that the future can in no sense pre-exist and in his insistence on the irreversibility of physical causation [11], which contradicts the Myth of Eternal Return. We shall later reencounter a quantization of time perception following Bergson's intuitions. For a ellaboration on a hierarchy of causation which appears to be the one corresponding to chronotopology and includes physical causation as a particular case, we refer to the differential epistemology by Johansen [58]. We shall not, due to lack of space, enter into considerations of this fundamental problem, though we shall touch upon it along our presentation of our conception.

[58]Thus, since by timing Life be-comes, the second law of thermodynamics cannot have the 'upper hand' in the universe, as an entropy increase, as usually claimed. (As we said above, our conception is, in this sense, opposite to the one advocated by Prigogine, which pledged a metalaw character of the second law of thermodynamics [97].) Though not widely known, the mathematical theory of statistical thermodynamics already provides a proof of this: The *random* entropy corresponding to *internal* random parameters of a system (which may have its origins in the fluctuations of Apeiron) of a system which can even exchange no heat with the environment, can fluctuate and *decrease*, so that "...microviolations of the second law of thermodynamics are possible"; see pages 44 and 38 in [136]. We all know too well that Life makes of these 'violations' its *rule*, rather than the exception that the Cartesian minded physicist is eager to assert as the uncontestable truth. In fact, as kindly pointed out to the author by Prof. Jeremy Dunning-Davies, a notable scholar and contributor to the foundations of thermodynamics: "It's also important to realise that the claim that entropy never decreases is NOT a statement of the Second Law; it is merely a result that may be deduced from that law in certain circumstances"; we are deeply grateful to Prof. Dunning- Davies. We repeat, these microviolations (the random entropy will decrease) can occur even in the case of a system which does *not* exchange heat with the environment [136] for which the second law is valid so that the entropy of the system does not decrease; of course, the

have linked intention-control to the action of the time operator (TIME) acting in the third state (aceleration) to produce and thus control variations of acceleration : The first dis-

---

latter description corresponds to the *outer* boundary of exchange of a system with its environment, the usual classical mechanics conception that neglects the exchanges of a system with its *internal* frontier, which is the Plenum. These internal exchanges are at the basis of claims of over-unity energy devices based on the zero-point fluctuations of Apeiron. (We recall here that randomness is both subjective and objective [11], and that when randomness is structured as Brownian motion, which is the one of the main cases considered in statistical thermodynamics (as processes verifying the so-called Markovian condition)[136], it is associated to geometries (either on spacetime or the manifold of thermodynamical variables) with trace-torsion [104, 102, 103, 105, 106]; these are the zero-point fluctuations of Apeiron, and in the case of the electron, they appear as the jittery Zitterbewegung [107]. So, the microviolation of the second law of thermodynamics installing the primacy of Life and syntropy, can be traced back to the torsion fields which constitute the primary joint constitution of the 'objective' and 'subjective' realms.) Furthermore, the microscopic character of the internal parameters of a system, claimed to be the case by statistical thermodynamics, strictly speaking can be taken to be the internal parameters in an arbitrary pair of Klein-bottle, one contained in the other, which can be established as tagged with different scales. Brownian motion need not be microscopical but is rather universal as shown in the works of the present author in which is related to torsion geometries at any scales, which as we already explained, fuse the 'objective' and 'subjective' domains. Thus, returning to the pair of Klein-bottle in a heterarchy of quantized reentering limited domains (heterarchy rather than hierarchy, because they can have different qualities, say, one corresponds to a human being, another to the society in which he is part of), in which one is contained in the other one. the smaller one has in its own scale, internal fluctuating parameters which will produce with respect to the exterior Klein-bottle of the pair, a syntropic self-organizing process, which will also violate the increase of entropy in his own inner scale, while perhaps increasing the entropy on the scale of the outer domain. We know by watching through the Hubble Telescope, that Brownian motion in astronomical and cosmological scales coexist with syntropic organization. Would we chose the microscope instead of the telescope, the observation would be the similar. The scale-dependance of the increase of entropy valid in a linear time scale does not entail its promotion to a fundamental law of Nature nor Being. The usual approach to thermodynamics - that stemmed from the mechanicistic ideas of the 18th century, the Carnot cycle- was extended by the statistical approach which fuses the objective and subjective realms, and thus surmounts the dualistic ontology, through self-reference. Returning to the ontological problems posed by the Myth of Eternal Return, we have a cognitive decision problem related to the choice of either the true or false Boolean states, which we already know are result of the Hadamard-Klein-bottle operator on the superposed non-orientable states; see eqs. (86, 87): Indeed, in the terms of the Myth of Eternal Return, we either chose to acknowledge the fact that the First, Second and Third-Intelligence-Entities exist and thus Will-Life-TIME is the case and consequently the entropy increase is the case for *linear* sections of the non-linear chronotopology of living systems -rather than a universal all encompassing law-, or we dont. If we dont, then Will-Life-TIME have no ontological locus; Will and Life lose their ontological loci and thus nihilism will be established setting the course for the proposals of both through emergency; furthermore, if this choice is made, there are no time-operators -in the sense introduced in this chapter, hence chronotopologies can only be time-linear (which, by the way, neglects the existence of waves verifying non-linear evolutions equations [83], notoriously manifest and devastating as tsunamics are). Instead, the usual physicalist approach to Life in terms of *negentropy*, which is further rooted in the conception of a mechanical universe, is installed inside the 'objective' side of the Cartesian Cut's discourse. (Note added in completion of this chapter: A notable exception, there might be more, unknown to the author, is the work by Mae, in which processual time is discussed. Prigogine also discusses a processual time, from the mechanical pledge of the second law of thermodynamics as having the upper hand on Nature [97].) This second choice is the one taken by positivism and nihilism, and already incorporated into the Myth of Eternal Return. As trees are known by its fruits, the result of the second choice is the current predominant, but not for long, predatory mode. In the other hand, the present ontological considerations point out to the predominance of *syntropic* processes, rather than entropic processes. These are the product of the TIME Operator as the action of will-intension to reach self-reference. In terms of the phase-space of Young-Eddington, we already remarked that can relate it to the internal invariant phase hidden in the spinor infinite dimensional representations of $Sl(2, C)$, the (double covering) group of Lorentz transformations. We remark that these transformations are the linear complex (i.e. the matrix elements are complex numbers) transformations with determinant equal to 1 on the complex plane, which of course, we can chose to be the cognitive plane in Matrix Logic ! So, again, we find a representation tinction created space from Apeiron, the second one created velocity, the third one created acceleration and the fourth created control which was the final distinction since further iteration returned to the original distinction, since the fourth time derivative gives the identity. We repeat, the rotational action on the fourth distinction (control, derivative of acceleration, the third time derivative of space, will, Self) with the final recursion by TIME returns this fourth cleavage, control, to the identity through self-control, which is embodied in the formula $\mathrm{TIME}^4 = I$). Coincidently, Musès will find in the Third-Intellingence-Deity's attempt to return and control the Infinite undifferentiated Plenum the origin of the generic time-operator. The impossibility in achieving this goal, resides in that the TIME Operator returns the controlling Self to itSelf which, as we have just showed, it returns to Identity, being thus self-referential: $\mathrm{TIME}^4 = I$. Hence, its achieved goal by this final recursive action, retakes the already cleaved state that fuses the objective and subjective realms in cocreation,. Yet, *cannot return* to the undistinguished state. The Null Operator in Matrix Logic (the Mind Apeiron), whose eigenstates are identified with the twistor representations of the extended photon that is the basis for the joint constitution of objectivity and subjectivity, plays its role. The Unmanifest is thus unreachable through the direct action of TIME and thus timing comes to Being and Becoming of all systems. [59] To Musès, this impossible fusion produced the reification of evil and the downfall (we prefer to call it an *unfolding*, as in the conception due to Bohm) of the Third-Intelligence-Entity in a process of indi-

---

which fuses the physical (the double covering group of the transformation group of Special Relativity, which is the group of rotations operating in the physical realm) with the subjective (will, control, through the action of the TIME Operator) realms. Related ideas on the predominance of syntropic processes have been proposed by Johansen, expressed in a private communication.

[59]Then, as it is impossible to return through the action of TIME to the origin, the Mind Apeiron $\mathbf{0}$, the only way of returning to it is through the manifold polarizations of 0 and $\mathbf{0}$, the Mind Apeiron. Particularly, through the nilpotent (self-referential) polarizations. We have presented very important nilpotent polarizations already: the Cognition Operator, M, produced by the torsion of cognitive space, which contains TIME and SPIN in its decomposition, being the most important one. So, embedded in the Cognitive Operator resides the timing that allows systems to self-organize countering the increase of entropy; notice that this is a topo-logo-physical process. Thus, we can return to the Mind Apeiron by considering in addition to TIME also SPIN, and thus now their sum gives the Cognition Operator, M, which being nilpotent, $M^2 = \mathbf{0}$, the return to $\mathbf{0}$ is achieved! Hence, SPIN plays a fundamental role in the evolutionary individuation process that appears in the Myth of the Eternal Return. Returning to Matrix Logic, SPIN and M in their action on either cognitive and quantum states, exchange quantum and cognitive computations through averages; see eq. (83) and the following paragraph. We recall that M has together with the identity $\mathbf{1}$ the fundamental role of projecting the Boolean states to the superposition states as characterized by eq. (88) whose addition modulo 2 allows to retrieve the Hadamard quantum gate representation of the Klein-bottle ; see eq. (84) and thus the self-referential process of creation is manifested. Other very important polarizations of 0: Also, the eigenstates of $\mathbf{0}$, which we associated to the twistor representations of the photon that arose from the propagation wave and nilpotent equation of geometrical optics with which we started the developments of this chapter. Furthermore, the polarizations that have been presented in the universal nilpotent rewrite system due to Rowlands and Hill, which we have briefly discussed already. Thus, Self can return to ItSelf, but not to the Origin (the Mind Apeiron), through the action of TIME, and torsion as the embodiment of Self-reference establishes TIME included in a polarization of the Origin which establishes Cognition. The multiple polarizations and in particular the interchangeability of M and SPIN in logo-quantum computations, yield the basis for the several (seven in Young's rendering of natural systems [156] and in the Myth of the Eternal Return presented by Musès) differentiation stages that accompany and produce the return of the Third-Intelligence Entity to ItSelf. Therefore, the Unmanifest, First, Second and Third Intelligence Entities are the primeval examples of the Klein-bottle heterarchy we mentioned before. As we can see, the full reentrance of the present conception into itself and the emanation of it from the origin, are thus produced.

viduation which in seven distinct logo-physical-psyche-archetypical phases will regain full self-reference, completing thus the Eternal Return in which the Tenth-Intelligence-Entity returns to the Third-Intelligence-Entity which is not the same state as the initial departure; in completing its full process of individuation beings become Being. Thus Redemption is achieved in this awesome cosmology, that we have presently unfolded in terms of the fundamental principles we have proposed in this work, differentiation, time-operator, self-reference, paradox and the Klein-bottle. [60] In sharp distinction with the usual dualistic platitudes that place Evil as dual to God-ness, justifying thus the former and countering by this take the natural action of Time, for Musès -and the present conception as well-there is a fundamental asymmetry, which is the parasitic character of Evil vis- à-vis Deity. Indeed, without that asymmetry, the recursion would be a repetition of sameness (in fact, there would be no recursion at all!), the trivial homogeneity which we claimed to be imaginary and the Return would be unachievable because the process of individuation, under the condition of homogeneity would not come to be because it requires an initial cleavage to start with. Thus, following Musès conception of chronotopolgy and its manifestations in natural and human systems, TIME and the Klein-bottle are the operators and operands of the Eternal Return and the individuation of God and Nature. [61]

7.4. In-formation, Torsion Fields, Semiotics, Interpretation, Subjectivity, Designed and Human Systems, Time Waves and Chronomes

In the present self-referential conception no device or designed object is psycheless. We promptly add that this psychological character resides on the homo-sapiens-faber side of the Cartesian Cut constituting the device and further relating to it, and not in its atoms, molecules, parts, nor in the material aggregate of them and their constitutive relations that appears as a distinct object. [62] Indeed, etymologically, ‘design’ means to signify, an in-

[60] Musès further relates these ten Intelligence-Entities with the ten spheres of Kabbalah.

[61] The content of this myth is somewhat common to many religions though with different cosmologies, and incorporated into some ideologies, as is the case of the the communist society for Marx and his acolytes. It appears as the hope of reaching a state in which Wisdom, Justice, Beauty and Goodness reign on Earth yearned by all kind hearted people. Even in some inspiringly beautiful literature works: The return of the self-ostracized-king-to-be Aragorn in reaching self-acceptation of his royal being, in Tolkien’s trilogy. It appears in philosophy, as is the case of S. Rosen’s reflections on the Apeiron’s self-forgetfullness that achieves self-redemption, through the process of individuation to return to full self-awareness and, of course, in Nietzsche. It is further more remarkable that the conception of Musès on the origin of evil, in principle finds a reflection in the seven different divinities’ (in the Babylonean pantheon) archetypes, their taxonomy and operation through time waves conforming historical-cultural patterns, that was discovered through the work of mathematician-sophiologist Páles (together in joint critical examinations and reconfirmation of his findings, by his critics at the Slovakian Academy of Sciences) [89]. The existence of cultural time patterns was discovered by anthropologist Kroeber [67] and independently by sociologist Sorokin [132], they are present as well in the economics of the capitalist system (the Kondriatev cycles), and more recently in the most diverse natural systems [41, 129]. This is the contemporary scientific field of *chronomics*. These findings point out to the universality of anticipative, incursive and hyperincursive phenomenae in the sense of Dubois [24] which are examples of the chronotopological differentiations. We shall return to chronomics in relation with in-formation, as well as to designed and natural systems, such as visual perception.

[62] Though following our previous discussion, *all* systems have a time operator which is essential to their constitution and regulation so there is a will of all entities. While we also add that the self-referential character Klein-bottle gestalt goes down -scalewise- in its applicability to the neutron which in the conception of Young, may have its proper timing as any other particle or system. Indeed, the Klein-bottle is the topology in the

tention which has a semiotic materialization which is the codification of energy, as Pattee argumented [91]; we have argued that this primeval sign is the primitive logo-physical-perceptual torsion field. For a deep study on the dynamics of semiosis, its social interpretation and perception, its logical valorization and its architecture as a gestalt we refer to the work by Taborsky [138].

To avoid misunderstandings of our statement of the psychological character of deviced objects, we shall provide an example to illustrate that it has to do with in-formation, cognitive states and interpretation and thus, we reiterate, belongs to the subjectivity realm. The example is provided by the testimonies of one of the bleakest pages in human history; we believe it clarifies the previous statement. According to the Spanish narratives, upon the arrival of the Spaniard conquerors to the site of the Inca emperor of Perú, Atahualpa, following their initial criminal incursions, they introduced themselves in a rather multivalued paradoxical -in some levels- way. According to the narratives, a priest accompanying the invaders uttered to the emperor -through the services of a translator- a demand that he should acknowledge the over-all rule of their Love-based creed and their king, and proceeded to deliver the Bible to him. (As we know all too well, religious creeds no less than ideologies, have been used and abused to provide the legitimization of unconfessable drives.) Atahualpa grabbed what for *him* was a priori a mere object, examined it meticulously, and dropped it with contempt. What followed was a carnage of the Inca population by the invading forces and Atahualpa's own demise. To Atahualpa in-formed and handy with the intrincacies of knots and its mathematics (as suggested by archeological findings) yet un-informed on the Spaniards' perversely manipulated creed and its valorative, political and power implications in place (though surely he did understand what was meant by the act of delivering to him the object, a demand of surrender and more), there was no message conveyed by the invaders as the Bible was no message to *him*. For the Spaniards, the message was delivered *completely and irrefutably*. For Atahualpa, his examination of the object had not rendered him the meaning that the interpreter had demanded to comply with, and thus was unable to interpret it (yet, he was in-formed previously to find improbable the truth value of the ethical aspects of the message demanding submission). [63] Notably, the shocked Spaniard narrators explicited the fact that both Spaniards and Incas were aware and spoke out in similar instances that this was the case. As we know from daily life, action,

Rutherford-Animalu-Santilli compressed electron-proton model of the neutron in Hadronic Mechanics, the theory of the strong interactions due to Santilli [122]. This theory -according to its founder- succeeded by superseding Quantum Mechanics deemed to be inapplicable- to give account of the strong interactions. For its relation to torsion and Brownian motions see [103]. There is a saying by anthropologists, *refering* to self-reference: It is turtles all the way down (and up, we further comment). The pun by the author with "refering" to "self-reference", was not accidental here but fully intensional. We are indicating the self-referential being of all languages, spoken, written, gestural, etc., which somehow we fail to identify.

[63] An utterance becomes a message only if a meaning is ascribed to it and shared at least partially by the would-be participants in the communication; an act of interpretation is essential to this transformation. An interpretation requires an intention, to cognize, and an action, the actual cognition. This is not the take of the modern theory of communication based on Shannon's work, for which messages are subject-independent objects, i.e. Cartesian objects, then admit a binary coding, as in Boolean logic. With respect to the logical valorization we pointed out to it by mentioning the irrefutability issue: Atahualpa by dropping the object questioned the truth value of the invaders' claims of Love, and their true intentions with regards to his people and himself. Here appears another element absent in the dualist mindset: A cognitive state is assigned to the utterance in becoming a message and it carries a truth value in it as we described above. Hermeneutics may precede the assignation of truth value.

interpretation, truth value, cognition and perception, even if incomplete, are unseparable.

To resume: this psychic in-formational character of designed objects embodies a semiotic action jointly with intention, interpretation, a cognitive state, actual cognition and perception. As we have argued, it is a codification of energy through a torsion field from which multivalued logic, its transformation into quantum physics, and time waves appear.

Returning to our explanation, any artifact, device or even a social organization embodies purpose-intention through design (demanding thus to the beholder, user or social actor an interpretation), a fact rarely acknowledged -but crucial to an integral approach to life-, which is the intent of control and thus of anticipation. The *identification* of a subject with the reification of this psychic in-formational character is a regressed form - fetishism- of animism, all too common and a major driver of technological inventions, as well as of all kinds of undesirable events (we have seen the purport of fetishism in the previous tragic history). To wit, this reification (not the identification of the subject with it) materializes the intention of the in-formation, and since intention is associated to the action of the time operator, this process of manifestation of the in-formation is the result of the action of TIME in Matrix Logic, and is further related to perception as we shall elaborate below. The identification of the subject with this reification of the in-formation, consistently with Hegel's conception that subjectivity is related to time [46] while for Gunther this relation with time is established in connection with the abandonment of the principle of non-contradiction [39, 109], is tantamount to the assimilation of subjectivity and time to the Aristotelian universe of discourse, as transpires from direct examination of this process of fetishism. [64]

Returning to our discussion of examples of in-formation of designed objects, at a larger scale of humankind and in the antipode of largely low degrees of freedom (until nanotechnologies appeared) embodied in the transformation of raw materials to machines, or further artifacts by purpose-intention, we consider events which as A. Young rightly remarks are associated with light [156]. These are events that may appear from the infinite potentialities of differentiation and complexization that may arise from it. It is the logo-physical-photon-primeval-distinction-torsion-field that creates events as the unfoldment of a Klein-bottle, acting as a major social and historic field, as argued by M. Purcell [98], creating patterns and processes, thus embodying forms and functions, physis and psyche. [65] Since control is related to purpose-intention, we have a big-scale coherent anticipative field which conforms a Zeitgeist (spirit of the epoch), actions over resources and a self-action of humankind: the Klein-bottle in social and historical scales. Yet, purpose-intention is man-

[64]A particular notoriously active form of this fetishism is provided by the fact that usually criminals -as the previous tragic story shows- purport to be acting for good or some superior source, thus collapsing logical cognitive states to one value which is false, while pathologically claimed to be true. The reification of their identification, a second-order reification (the first one being the reification of their intentions yet with certain degrees of freedom which is followed by the exhaustion of all degrees of freedom), is what in different theologies, and in the human experience, has the lifeworld called Hell .

[65]In the topological phenomenological *radical recursion* philosophy of S. Rosen, a logo-physical-psychological field hierarchy and associated Musès hypernumbers are ascribed to the Klein-bottle; we have shown already that logical operators are representable by hypernumbers, which are related to Pauli operators, i.e. to spin, and ultimately to torsion. In his book 'Topologies of the Flesh', Rosen further ascribes this hierarchy and the different topological and knot operations on the Klein-bottle to the mineral, vegetal, animal and human realms [116]. Thus, the Klein-bottle itself embodies an heterarchy of lower self-referentiality with respect to its integral Being, but each of these lifeworlds, is self-referential and heterarchical themselves, with respect, say, their molecules, atoms, etc.

ifested through time operators, particularly, TIME. We recall that it has been found that there exist time regularities and patterns (*chronomes*) in history (revolutions, war, peace, etc.), culture and its several manifestations (design !, architecture, philosophy, music, creativity on the most diverse fields of psycho-social action), ideas (memes, ideologies, religions, etc.) which have been ascribed to cosmo-psychological archetypes in the work of sophiologist-mathematician Emil Páles [89], which as we mentioned already, his work found antecedents in the works of Kroeber [67] and Sorokin [132]. So, again, the psychic element 'exteriorizes' the logo-physical-torsion field as well as it is the 'interiorization' of 'outer' phenomenae -cosmological chronomes-, but as we argued, it is a Klein-bottle; we shall discuss this further in the case of visual perception. The study of chronomes, a field which had physiologist Pavlov as one of its pioneers, was developed in the former USSR by several researchers grounding the ideas of 'cosmism' (Chizhevsky, Vernadsky, Kondratiev, Kozyrev, etc.). At present, it is a growing field of knowledge (which requires persistent experimentation) which may lead us to a completely new understanding in science with amazing findings which not only indicate the existence of standing time-waves acting on all systems, which manifest nonlocal cosmological correlations; see [41] and references therein and [129, 65]. These findings give support to the ideas of Musès chronotopolgy, which in our conception is the Klein-bottle, the embodiment of the Myth of the Eternal Return. We shall retake this issue on examining the self-referential Klein-bottle topographic representations of the somatosensory and auditive modes, and the self-referential gestalt given by the mammal heart. Finally, we want to mention that in the mathematical representation of the process of growth of seashells, it was found that their development, was possible due to a retrodictive character of the action of the future on their present state of growth (the Gaitlin propagator)[55]. This non-linear character of time was examined by Heidegger [45]. In physics it appeared as a branching time [100] [66]. For a study on the politics and religious beliefs behind the linear non-operator time prevailing in physics we refer to Raju [101].

To conclude these disgressions on the physic in-formational character of designed objects imbued in the design itself, by sheer consistence we are now ready to postulate a psychic in-formational character of Nature at large, of which teleology is one of its expressions, in consonance with Spinoza and several traditions standing in sharp contrast with the alienation of the Cartesian mindset. We have encountered this already in discussing the phase variable introduced by Eddington and further examined by Young. There is a time operator acting on all scale dependent systems; these operators can be associated to the TIME Operator of Matrix Logic, in a particular scale for the system under consideration; through TIME's self-referential action the system is self-determined. Furthermore, these scales are not arbitrary, but may be determined by a fractal structure discovered by H. Muller, and which now conforms the central core of the theory of Global Scaling [81].

[66] Raju's findings on the existence of a non-linear -a tilt of- time, lead him to propose a new paradigm in physics based on functional differential equations. Despite this, it is currently named as the *Atiyah Paradigm*; see http://ckraju.net/, http://www.ams.org/notices/200606/comm-walker.pdf and references therein.

7.5. Light, Torsion Fields, Quantum and Topological Non-locality, Chronomes, Visual Perception, Depth and Anticipation

Torsion as the primitive distinction introduced in the undifferentiated Plenum, thus establishing a locus and phenomenology, is the fusion of subject-with-object, as well as that of form and function, which biology and most notably, some medical practices tend to separate. This fusion unfolds in several ways, one of which is its association to spin which is a field not only notable in physics, but as well as all pervasive in biology [28]. Thus, as much as the primevally undifferentiated Plenum has the potentiality of all forms and functions, it is by torsion cleaving this plenum, that they come to manifestation. The Cartesian Cut attempts precisely to dissociate between form and function, being the case that their gestaltic fusion is due to the simple fact that function -physis- is established due to inhomogeneities that give rise to processes, which produce themselves forms and structures through symmetry, which due to the impossibility of being isolated from the environment of other forms-functions produced by other cleavages, tend to loose their symmetry and thus become processes, i.e. physis. This is the essential dynamics in semiosis [138]. Torsion is thus associated with action, giving rise to the quantization of Plenum, and thus the corresponding physical parameter is Planck's constant, which as remarked already, is not singular but multivalued as shown by Pitkanen [93] and Nottale's Scale Relativity [87]; the latter theory can be derived from torsion without recourse to forward and backward derivatives which are basic to Nottale's work [105, 107]. Thus, the Klein-bottle stands for quantized *re-entering* limited domains (QRLD), with different quantization magnitudes which correspond to different scales. We shall later return to this through stereoscopic vision and the Kozyrev phenomenae.

If thus time is subjective as in Hegelian dialectics, in the concrete realization of this stands the process of photon absortion by the subject. But the photon is no 'external' particle to the subject, as the classical formula would like us believe . Rather, the absorbed physical particle is the core of the self-referential process of fusion of object with subject (inasmuch as the geometrical fusion is the torsion). Indeed, when we visualize a photon, we are actually visualizing our seeing of the photon, the absortion process by which we complete the objectification of the photon as an independent emmiter, object-in-space-before-subject which now when absorbed becomes the fusion of object-with-subject. Self-referentially, the photon (as a lifeworld) is the observation of the link between the photon (as an element of objective reality) and the perception of the photon (i.e. the Fibonacci type reentrance of a form into itself in the calculus of indications; the 'atom of thought' as in Johansen's conception [58]); see [109]. [67]

In this context, quantum jumps play an essential role since they represent what is present

[67]This process is a lifeworld in which the body in its integrity is compromised. In a meditative state a photon may come to settle in the forefront, in the locus of the so-called third eye. It appears as an extended radiating (and per se, not in complete equilibrium) two-dimensional flat structure, which subsequently is perceived inside the brain, as a very quickly fluctuating plasma structure with philaments of great complexity and beauty, by the extended watching Self living this awesome experience with and through intensity, self-extension and detachment. Thus, there is an actual seeing of the seeing; second order. The full body of subjectivity has in the photon a first stage of self-reference which is not the last one. Finally, it exits the body in a (silvery-like shortly lived perception) helicoidal motion. We have found no account of this lifeworld in the discussions of light in the Tibetan Buddhist tradition [38] nor elsewhere.

to cognition and perception, i.e. differences which produce differences, as we also know well from physics and visual perception; these differences are crucial to the neurocortex organization and physiology under the complex logarithmic map topographic representation of vision, as explicited in the pioneering works of Schwartz [126]. We have characterized quantum jumps in terms of singularities of the torsion potential described by the differential of the logarithm of these scalar fields produced by the node set of them. Hence, concerning spacetime torsion fields, it is in the two-dimensional Riemann sphere representing the complex plane, coinciding with the cognitive plane of Matrix Logic, the plenum on which the primeval distinction creates the holonomic structure-process which fuses 'outer' and 'inner' realms, that quantum jumps, cognition, visual perception and logic are grounded on a plane, from which through holography the full structure can be retrieved [124]. This is linked to the nondual character of interior and exterior as in the topology of the Klein-bottle.

Several cues are used for the formation of the perception of depth, such as occlusion, rotation of objects (so a perceptual spin is relevant to the formation of depth perception which has already appeared as SPIN), the most important is believed possibly to be stereoscopic vision, i.e. the image formed by the joint use of two eyes. It appears that stereoscopic vision only leads to the formation of three-dimensional images if the two eyes actually sense asymmetric images for each of them, in the contrary there is no distinctive image but a blank homogeneous state [54]. This indicates that the actual concrete perception of a geometry requires an inhomogeneity at its basis, i.e. torsion, self-reference. [68] This is most remarkable since stereoscospic vision is the basis for the conceptual emergence of symmetry with which physics is constructed. These findings points out that this is only possible from actual asymmetries (which conceptually are based on the manifestation of differences); if asymmetries are lacking, only an homogeneous perception is formed, i.e. no structured perception of inhomogenities, only the triviality of sameness.

This perceptual homogeneous plane where no distinctions are present is the one that is associated with the physical symmetrical vacuum, the plenum which we have already presented. So here we have the appearance that depth to be perceived as an original dimension, a difference that makes a differences is necessary, and this is the basic asymmetry between the images of each eye [54]. We shall discuss later the complex logarithmic map that appears in the expression of the quantum jumps that originates in the differential of the complex logarithm of the photon wave function, is also the key to the multisensorial representation of the body on the neurocortex and is further related to the codification both of depth and of stereocospic vision in the neurocortex. [69] Furthermore, time dilation and space contraction are related to the baud rate of in-formation processing (which we shall elaborate in terms of ATP's metabolism) Hence, there is no 'pure objective' cognition of

[68] Further below we shall elaborate further on introducing the complex logarithm representation of the retinotopic visual and somatosensory perceptions which transform the photons impinging either in the retina or the skin in a mapping at the plane two-dimensional neurocortex. We shall see that this logarithmic map that appears explicitly in the formula for the quantum jumps, plays a essential role in providing these maps, both locally and globally, and that this map provides also for the basis of the stereoscopic vision formation in terms of differences as we have already repeatedly alluded in this work, and in terms of the vortical structures that appear in the neurocortex' hypercolumns.

[69] In the field of mathematical psychology, it has been theoretically and experimentally verified, that visual perception follows the organization of visual 'internal' geometrical representations [70, 56] that that they are described by a psychometric function dependent on the observer.

an object: Visual perception and cognition depend also on contextual *interpretations* by the subject. (As we said before, perception, cognition, truth values, action and interpretations are unseparable.) Furthermore and most surprisingly, they depend on cultural and theoretical constructs, to some extent [115]. Thus, the classical Cartesian formula is untenable and perception is not secondary to cognition, the subject is a full participant in the construction of the proper visual model. This 'interiorization' of the geometry by the subject with its dual operation of projection for the construction of the 'exterior' geometry of space and time, can still be linked with the metabolic rate of the production of ATP (adenosine triphosphate) in the brain's visual area, which in turn is linked with the quantity of light absorbed in the retina (which is related to quantum jumps and torsion, and thus to the quantization of 'outer' spacetime).

When general human metabolism (and the V5 area of the neurocortex which is believed to be associated with motion detection) is fast, time runs slow; conversely, when the metabolic rate is relatively slow, time runs relatively faster; see Harms [43] and references therein. This has support as we said already in the fact that there is a related 'inner' geometry of vision in which there is a limiting velocity of percepts processing. This is widely known to occur in extraordinary situations of stress such as high velocity drivers perceiving the 'outer' world through a time dilation and space contraction, which may be accompanied with their relative visual V5 metabolic states speed-up, compared to observers in other frames at lower speeds relative to this visual limiting perception velocity. In the Klein-bottle function-form of the neurocortex, quantized 'outer' spacetime is thus transformed into 'inner' quantized perception and thus a minimal instant is associated to it, through which motion and all differences come to be. [70] Then, at that minimal instant there is no motion. Motion, then must be the perceptual differences that exist from one instant to the next. Thus, motion may only take place between the instants in time. Here the instants are given to us by the processing speed of these instants by the V5 area, which might be related -though we are in a state of ignorance with respect to this issue- by the frequency of light entering the eye [43]. We would like to propose that TIME, Cognition Operator M and SPIN, operate at the limiting velocity of ATP production and visual geometry hyperbolic space [56], in which cognition appears due to the quantization of time, with associated limiting values of acceleration and control. This should establish the Lorentz-Fitzgerald perceptual phenomena in which the previous fundamental operators manifest. We recall that ATP production and destruction is a fundamental organizing process for the cell's operations, and is further related to the zipping and unzipping of the microtubules, in terms of which Penrose proposed his quantum non-computability source for will. So again, it is the action of time that has to do with will, and this may have a biophysical operational substrate in the creation and destruction of ATP, which may actually create at their limiting velocity, the quantum action of TIME, as the difference between creation and annihilation as suggested by its very definition; see eq. (66). The background medium which is essential to the ATP production and destruction is water, which itself has the remarkable property of having memory-like behaviour [152]. We recall that anholonomity, and particularly torsion, allows to encode memory as argued by Pattee [91]. It is important at this point to remark that the Aharonov-

[70] Experimental verifications of this quantization of time have been obtained by E. Ruhnau and elaborated in her theory of hierarchical time windows [121]; in this regard the theory of fractal time due to Vrobel can be of great relevance [149].

Bohm effect we found to be associated to the logarithmic torsion potential in our discussion of quantum jumps in Section 3.1, plays a central role producing non-local correlations in biological systems [145], especially those that are sustained by water.

Let us return to the problem of depth, its identification with the operation of time as can be perceived in the Necker cube [85], and following the phenomenological philosophies of Merleau-Ponty and S. Rosen. Stereoscopic vision which is important in establishing depth perception, brings to the fore the problem that in order to obtain a synchronous visual flow in both left and right brain hemispheres, which is needed to account for its coherence, the temporally delayed signals of both eyes' left and right visual fields (which are processed in different hemispheres), should be integrated with anticipated versions of their complements in order to close the time gap existing with the firing of neurons in the separated hemispheres and recurrent visual control [64]. As we have repeatedly elaborated in this article, control is linked to purpose and thus to anticipation and to the TIME Operator. Remarkably, Kampf's proposal for a solution of this problem (which we recall that is fundamental to the establishment of a physiological process that sustains multivalued logic as the perception of the two possible instances of the Necker cube requires as both instances unfold in time through depth perception) , is to invoke anticipation, as conceived by Dubois [24]. Kampf: "representational simultaneity, as a brain process spread over spatially distant loci, is achieved by temporally bidirectional interactions of neuronal processing on a quantum scale. Absorber effects between the presumed 'advanced' and 'retarded' signal components are proposed to generate standing time-waves pattern which might be speculatively assigned to the carrier process of an internal psychophysics of the representation of visual space." Furthermore "...absorber effects appear, on the operational side, as anticipations of future states of the system". Kampf's proposal then stems from the transactional interpretation of quantum mechanics (TIQM) due to Cramer [20]. In fact, the absorber theory which was the basis for TIQM, attempted to explain the EPR paradox in a straightforward way. It rests on the idea that a 'handshake' -we prefer the logical term 'identity transparency', or still the physical interpretation as a 'resonance'- between the ordinarily transmitted signal and an anticipative effect deeply rooted in the quantum world is *fedback* from the 'absorber' to the 'emitter'; so, in our terms, this requires control, or still, the action of the TIME Operator. [71] Coupled in-between the 'retarded' and 'advanced' components in the collapsing wave function of the quantum event under measurement, this process appears to an external observer as a seemingly time-reversed transaction. In our account, this is the action of the TIME Operator. Kampf presents arguments for the significance of advanced signals as an anticipative feedback for the synchronization of spatially distant retarded processes which he derives from an analogy between the synchronization of neuronal activity and that of chains of coupled oscillators on different scales (including a cosmological one). Thus Kampf argues for the existence of a time-loop that accounts for this synchronization and control, a proposal which fits well with our previous disgressions. He further notices that the mathematical roots of this lies in the standard quantum-mechanical procedure for calculating the 'collapse' of the wave function, by computing the square of the probability amplitude which is done by multiplying a complex number $\cos t + i\sin t$ by its conjugate $\cos t - i\sin t$, where t stands for the time angular variable i.e. by phase (as in Eddington-

[71] Further elaborations of the causality operators that transcends the mere physical causation, were applied to the Kozyrev phenomenae, in terms of his epistemology, by Johansen [58].

Young space) conjugation, which is the basis of holography as elaborated by P. Marcer and W.Schempp [124] and which we note that in matrix logic is the transformation of a cognitive state of the form $< q| =< (\cos^2 t \ \sin^2 t)|$ into $< q| =< (\cos^2 t \ \ -\sin^2 t)|$. In the case of $t = \frac{\pi}{4}$ this is the TIME Operator transformation between the superposed state $< S_+|$ into $< S_-|$. These topological entangled states form an orthogonal basis which is the transform by the Klein-bottle in-formation matrix H of the Boolean orthogonal vectors $< 0|$ and $< 1|$, and conversely, $\frac{1}{2}\mathcal{H}$ transforms $< 0|$ and $< 1|$ into $< S_+|$ and $< S_-|$; these four states generate all Matrix Logic [103]. Thus, for these states, we can 'interiorize' the transaction as an action of TIME which generates the laws of thought. [72], or conversely. [73] We already argued that the neurocortex cell visual representation is both supported by holography and the Klein-bottle [109], and that the latter is a classical-quantum-classical transformer which sustains multistate logic which is related to quantum fields, transforming the average of the cognitive operator M on cognitive states to the average of the logic spin operator SPIN (related to the perception of depth) on two-state wave functions; furthermore, we can substitute M by SPIN acting on cognitive states and M (cognition operator) acting on quantum states and this identity is still valid, so we can both have $< q|M|q >=< \psi|\text{SPIN}|\psi >$ and $< q|\text{SPIN}|q >=< \psi|M|\psi >$ where $< \psi =< (\psi(\uparrow) \ \psi(\downarrow)|$ is the spin-up spin-down quantum state; this identity stems from the decomposition $M = \text{TIME} + \text{SPIN}$, and the fact that TIME is a distinction *cognitive* operator, i.e. $< q|\text{TIME}|p >= p - q$, so that the average $< q|\text{TIME}|p >= p - p = 0$; TIME appears to be unchanged for unaltered states of consciousness. So, indeed, quantum effects, time loops as in the Klein-bottle lift of the calculus of distinctions giving thus the periodic reentering of a quantized limited space domain through time waves, the action of the TIME Operator and anticipation, which are the very embodiment of the holographic structure of the Klein-bottle, are present in the mind-brain (we use this term because of the exchangeability of cognition -mindlike- and SPIN -brainlike- observables described above). Entanglement is due to the action of the non-orientable topology of the Moebius surface, or still by the torsion introduced by the cognition operator, which also represents the non-duality of TRUE and FALSE. To resume: entanglement is topo-logical-physical. One can enquire still if this quantum entanglement is related to the 'interiorization' of quantum entanglement at a cosmological level, a question which Kampf does not raise (perhaps due to the lack of a Klein-bottle logic), restricting the arguments to the possible parallels between cosmological (as in the Kozyrev phenomena [74]) and microscopical entanglements. The natural answer is represented through the

---

[72]We have argued elsewhere that following the phenomenological philosophy due to Merleau-Ponty, depth which is precisely related to stereocospic vision is a primitive dimension associated to time and paradox. We shall later discuss the relation of this with respect to the retinotopic and somatosensory complex logarithm representation on the neurocortex and its relation with oppositely oriented vortical structures associated to the hypercolumns in the neurocortex.

[73]For the benefit of the reader who skipped the previous parts of this chapter, we recall that cognitive states are real valued vectors which we write as Dirac bras and kets formalism of quantum mechanics $< q| = (\bar{q} \ q)|$, where $\bar{q}$ is the negation of q (the *real*-valued logical variable), i.e. $1 - q$ so that they are *linearly* normalized, related to the wave-state of two-state quantum mechanics by the relation that the former are given by the complex square root of the latter, i.e. $q = \psi\bar{\psi}$.

[74]Kozyrev and Nasonov discovered on observing through a telescope with a device, that there exists a radiative field associated with, say a star or galaxy, that cannot be shielded but with polyethylene. In pointing the telescope to the future and current position of the star this radiation was also registered [66]. These experiments were repeated decades later by Lavrentyev with the same results and the theory is currently used to compute

Klein-bottle which has no inside nor outside, but a form-function which transforms a local interior to a local exterior holographically. Thus, absorption at one hemisphere of the cortex of a photon is entangled with the anticipative emission of a photon of the other hemisphere producing synchronization. This transaction is 'interiorized' in the laws of thought of multi-valued Matrix Logic, or still in the calculus of distinctions in which we incorporate paradox through reentrance of a Klein-bottle limited domain defined by a quantized distinction, a QRLD as we called them. Yet, this is the rotational action of TIME Operator; we shall later see that analytically it is embodied in the complex logarithmic map of sensorial modes (in particular, retinotopic) on the neurocortex. So, as we said before, it all (perception, cognition, interpretation, physics) boils down to the projection on a two-dimensional plane, be that a neurocortex slice, the phase space of Eddington-Young, the plane of cognitive states in Matrix Logic, or the projection of the Riemann sphere as the complex plane which can be generated by three primeval distinctions as pointed out by James [57], the phase-space of Young-Eddington and spinors, or still for characterizing quantum jumps as primeval differences. In this transaction the torsion geometry of cognitive space is essential [75], since itself produces a superposition state, $< S_+|$ which stems from the non-orientable character of the Moebius band and shows up defining the cognition operator; from $< S_+|$ the other superposed state $< S_-|$ is produced through the TIME transform, and together they form the Cognition Operator (and the Klein-bottle in-formation matrix $\mathcal{H}$) which encodes the transformation of cognition to spin observables and conversely. This 'interiorization' process, in the Klein-bottle logic is identical to a cosmological entanglement, of which the Kozyrev astronomical observation which we have already pointed out that it is a cosmological example of a chronome, a time-structure-function which also can be interpreted through the TIQM. It has been verified to exist not only in cosmological scales, but also as entanglement between solar and geophysical phenomena, as proposed and experimentally verified by Korotaev [65]. Remarkably, the Kozyrev phenomenae can be explained also through the same geometries as the mind-brain operations associated with Matrix Logic and a QRLD, i.e. in terms of torsion through spacetime (Brownian motions) fluctuations [105] -which we observe that, unless thought is programmed, is random- and spin-torsion fields [134]. This leads to enquire on how QRLDs, say a cosmological one (with a corresponding cosmological Planck constant) [87] may be entangled with a meso or microscopic domain (with an appropiate Planck constant)- as already the human 'homunculus' representation already achieves- with a neurological domain so that the entanglement that synchronizes stereoscopic vision through logical torsion entanglement or through an emission-absorption transaction, forming a self-referential loop in which one is transformed into the other. In other words, what is the process-structure inherent to the heterarchies of Klein-bottles, or still, to the Klein-bottle itself as an embodiment of heterarchies, already discussed? A natural solution perhaps can be found in the fractal structure provided by nested QRLDs in which 'interior' domains reenter through 'exterior' ones and viceversa. It would be inter-

current positions of astronomical bodies [30]. So time-loops exist at astronomical scales.

[75] In fact, already the geometro-stochastic form of the Schroedinger equation for ψ (where the torsion field $d\ln\psi$ describes the average motion of the universal Brownian motion generated by the torsion geometry together with the diffusion tensor associated with a metric)[107], incorporates *both* boundary conditions on the past *and* the future that allow to consider the probability distribution $\psi\bar{\psi}$ which we interiorize as the logical variable q depending on the universal time variable of Stuckelberg and Horwitz [51]. This variable is the variable that parametrizes quantum jumps as we characterized in the initial sections to this chapter.

esting to consider this in regards of the fractal structure of time [149]. We shall next see that the human structure-process itself is such an heterarchy.

8. The Complex Logarithm Mapping of Visual and Somatosensory Perception and the Kleinbottle

Up to this point we have presented a theory of the joint constitution of the physical, logical, cognitive and perceptual realms, yet for the latter we have mainly discussed the study of stereocospic vision and synchronization. We shall retake this issue in a wider conception which evolves from the relation between the sensorial realms and its mathematical representation in the neurocortex, and further the physical phenomena that appear in the neurocortex. This is of course related to one of the longest standing philosophical problems raised by several philosophers and the core problem of what we contemporarily know as cognitive sciences, and any attempt of dealing with it in the needed extension is unfortunately severed by length constraints. We have already presented our conception in [109] and above, yet rather than continuing with the philosophical discussions we shall restrict our disquisitions to the mathematical representations already alluded and their relations with neurology.

In the introduction to the deep work on the topology of the body representation by Werner [150], it is pointed out there is a wealth of clinical and experimental studies that established the existence of a fixed relation between regions of a body and areas on both the sensory and motor cortex, achieved in the decades 1930-1969, which led to the notion of the existence of orderly systematic mappings of the body's peripheral events in the brain [71, 80]. The evidence in this regards was that the cortical and subcortical projections of the visual system are topographically organized such that the fiber tracts and neurons preserve the spatial arrangements in the retina. Also, the spatial orderliness of representations of the body at the cortical and subcortical relay stations follows the segmental (i.e. dermatomal) innervation pattern of the body periphery.

Prior to these discoveries, there were several theories postulating that perceived space is correlated with spatial patterns established in the nervous system [76]. Thus the notion of a mapping from the peripherical sensory bodies to the central nervous system was proposed. In particular, the mapping of the body surface, and the auditory and visual receptors.

We quote E. L. Schwartz [126]. "A universal feature of the anatomical organization of the vertebrate sensory system is that the visual [140, 2], auditory [71], somatosensory [154, 150, 151] and olfactory systems are organized in terms of orderly spatial projections of a peripheric receptor mosaic to more central processing sites. Embedded within this

[76]In the Gestalt conception, "all experienced order in space is a *true* representation of a corresponding order in the underlying dynamical context of physiological processes". Remarkable to this conception, is that there is a logical value ascribed to perception, which can still be relative to the choice of particular cues. This is the basic phenomenae which we encounter when dealing with a geometry alike the Necker cube [85], in which there are two possible choices of cues and perceived cubes, which are related to the invalidity of the principle of non-contradiction of Aristotelian logic. Any of the possible perceptions are 'true' and thus the Necker cube embodies paradox very much in its design, as we discussed above. The extension of this to non-orientable surfaces, namely the Moebius surface and still more notably the Klein-bottle, led to the foundation of multivalued logics as shown above.

'receptotopic structure' is the detailed local neurophysiological processing that results in the existence of well-defined neuronal 'trigger features'. At the striate cortex in primates, cells respond optimally to oriented, elongated stimuli with well-defined velocity, direction of movement, *binocular disparity*, ocular dominance and colour. Cortical cells that are sensitive to the *orientation* of a stimulus are grouped into columns, or slabs, of common orientation [54]. These columns are themselves arranged into a highly structured geometric pattern termed 'sequence regularity' by Hubel and Wiesel, who introduced the term 'functional architecture' (we note here the blending of function and form; author's comment) to describe the anatomical arrangement of the physiological significant cortical structure." In the work by E.L. Schwartz the term 'functional architecture' was generalized to include the global, retinotopic organization of the cortical spatial map as well as the local columnar structure described by Hubel and Wiesel, and the somatosensory representation and still the relation with stereocospic vision. From the work by Schwartz appeared that all these maps act transforming input from two-dimensional sensorial surfaces to the two-dimensional cortex (albeit with a columnar structure representing each point of the cortex), and are all given by the complex logarithmic map, which in the retinotopic representation nearby to the centre of the eye, where the tangent space can be identified locally with the retina gives an excellent agreement which matches with experimental data. Furthermore, this has allowed for predictions which have later been verified, in particular, for binocular disparity tuning (which is a subject we have already raised above). Schwartz' discoveries lead him to pledge that the term functional architecture was to be taken in a literal sense: the spatial structure of neural activity in the primary sensory system turned to be of direct significance to visual perception. The problem arises as how the neural system accomplishes the task of mapping both the surface of the body and the interior of the periphery of the three-dimensional body into a *single two*-dimensional cortex surface.

The experimental findings by Werner lead to the conclusion that the "geometrical representations of the cortical body appear quite different from those of the body itself for several reasons. First, unlike the relation in the body periphery, the projection of the skin does not form a continuous boundary in the cortical map, enclosing the projection from the 'deep sensors'. A second reason for the difference between the body and its map is that the relations of proximity and distance between points on the body do not consistently remain preserved in the cortical map. The meaning of this is that receptors fields which are closed in the periphery in the two-dimensional cortex space in which the topographical map takes their image, may not be close any longer". Also remarkable, is that what is codified by the map in the neurocortex is *not* the actual points of the skin as a sensorium, but 'dermatomal trajectories' [150, 151], alike, say, shoes with helicoidal footlaces strappings or still the helicoidal Hebrew phylacteries [77]

Werner's conclusion is that the topographic map was not related to a metric as in General Relativity but to an homeomorphism, a continuous bijective invertible map whose inverse is also continuous which further transforms open sets in the domain into open sets in the image [150]. The solution compatible with the experimental findings is that the global somatosensory map is given by the Klein-bottle [150]. Furthermore, what in the periphery

[77] These phylacteries are strapped as helicoidal trajectories on the arm with further loops producing semiotic codifications (one of the 'names of God') and additionally have self-referential sentences vinculating the subject to God, written inside the box that initiate them as well as the one placed in the forefront.

may appear as broken dermatomal trajectories, in the Klein-bottle map they are represented by continuous trajectories. Werner further suggested the implications for the haptic sense (i.e. the integration of sensory modes, such as tactile and kinesthetic) are represented by one single map. Thus the Klein-bottle also plays the role of the topological codification of the integrative somatosensory mode. Yet there is something more striking on this and especially important to the very essence of the possible existence of the Kantian apriori schemata. This is the fact that "the sensory representation in the cerebral cortex can possess properties that are not inherent in the raw data originating from the peripheral sense can organs themselves, and that the nature of new properties can be a consequence of the characteristics of the neural mapping process which links body periphery and central sensory representation". This is a very strong statement which somehow inverts completely in the Cartesian mindset the ideas of where the complexity stems from: As is usually stated in the dictum "the map is not the territory", meaning that there is more to the actual complexity of what is the 'objective world' than there is in the representation. This points out to the possible fact that the complexity lies on the map, and not in the skin and efferent neuronal pathways that sustain the map itself, which here is topological; it is the Klein-bottle. Since the map plays the role of the subjective-objective fusion of the periphery and of kinestesic recordings of the raw data, with the actual percepts, this becomes a self-referential -to the map!- linkage which is absolutely irreducible to the Cartesian Cut mindset. [78] . It clearly establishes the prominence of a Platonic realm, which is essentially semiotic- mathematical-logo-physical in which by the joint constitution of the lifeworld of both subject and the physical realm does not allow for discernment of where the complexity actually resides. Through the map we map the territory i.e. self-referentially, the map is both operator and operand, and thus we cannot get 'out' [79] of it to cognize it. In fact as we have seen in the constitution of logic, the Klein-bottle is the basic operator-operand in which, by the joint constitution of

[78]These findings on the importance of the map with respect to the territory, may perhaps assist us to understand memetic blockages the reinforcement of paradigms, Taborsky has argued that the fixation (or better, in thermodynamical sense, freezing) by the signifier on his interpretation of semiosis, a kind of fetishism, leads to the dissapearance of the structure that is being signified and the desestabilization of the signifier, with its possible demise [138]. There is a widespread phrase attributed to Planck, purportedly asserting that the establishment of a new paradigm requires the demise of the whole generation that opposed to it. A similar idea is found in the Torah and retaken by the other monotheistic religions of the same branch, in which the generation that exited the land of servitude, had to expire to give place to a generation free of the previous experience of slavery, to enter a land in which they would no longer be aliens. In more explicit terms, the fixation to any particular paradigm or belief system, after a process of degradation through the impossibility to embody the new ambient conditions -thus Taborsky's disquisitions are elaborated in terms of thermodynamics-, and develop the needed knowledge, may produce a form of alienation. We have encountered this process before when we presented the Myth of Eternal Return. In the present conception, both signifier and signified are fused, as is the case of the primeval distinction already discussed. In science and also all belief systems, paradigms are established as the territory of all reality, in a process in which the initial mapping used for the construction of the paradigm is no longer examined. This territory is a reification of a consensual ideative process, which by massive addition of more information of the same conceptual quality, and the performal of -mostly hierarchical enhancement- rituals, a reinforcement of the paradigm takes place. An ideation or an experience which belies the paradigm or questions this reification, introducing at times a simpler and more integrative *mapping* principle, may be perceived as a threat to the permanence of this territory, which now has frozen frontiers and conceptual contents. Due to the identification between the signifier and its projection in a state devoided of degrees of freedom, self-preservation will be attempted through the liberation of the energy contained in the signifying process and the sign [138], of what is already a fusion of territory and signifier.

[79]Identically, we cannot *speak* about the world in its manifold realms without language.

time and thought comes to be, arising from paradox [109]. If we still take in account the self-referential being of language as pointed out in a torsioned way by psychoanalyst J. Lacan, it turns out that the Klein-bottle is the actual integral solution to this conundrum that philosophers, mystics, scientists and common folk have been trying to search for. Of course, what we are saying about the complexity of the human and primate bodies, apply as well to the Universe at large and its lifeworld. The last pending problem concerning this is on the *reality* of these schemata. There is no other consistent answer to this historically longstanding interrogation that the one we have just explained with respect to perception. Semiosis is not established independently of the subject but lies in the joint constitution of the lifeworld that constructs jointly the subject with the physical realm embodied in the subject's lifeworld. This is the *enaction* that sustains-creates the Self jointly with the World, without pledging the 'Great Doubt' [115].

Returning to the dermatomal helicoidal trajectories on the primate and human skin, these are transformed into straight lines in the planar neurocortex space. As observed by E. Schwartz, this transformation is inherent to the complex logarithmic map. Furthermore, this is not exclusive to the representation of the lymbic system but also by the retinotopic representation discovered by Schwartz consistently with experimentations [126, 127, 139]. The representation of the retinotopic stimuli, which as we mentioned above, amounts to the stimuli provided by the class of monochromatic functions in terms of which we constructed the theory of quantum jumps and the photon in which the complex logarithm appeared to give the different surfaces of the Riemann surface on which each quantized states lies, also appears to provide the mapping for the retinotopic stimuli. In fact, as discussed exhaustively by Schwartz in the case of the retinotopic mapping into the neurocortex, both the complex logarithmic map as an analytical topographic representation, and the Kleinbottle as the topological representation, are cocreated [126]. This has very important consequences to the physiology of perception. Let us briefly discuss this issue.

To start with anatomical cortical development, the complex logarithmic map being analytic (in the sense of the theory of complex functions), represent a potential flow which is -as usual in complex functions theory- subjected to *boundary* conditions imposed by the shape of the boundaries. Moreover, the dendritic summation of the afferent input to the neurocortex is also locally complex logarithmic, so the anatomical structure stands in correlation with the topographic representation by the analytic map which further is related to developmental functioning, which we shall later associate with the diffusion of morphogens. Thus we find in the retinotopic topographic map and the underlying anatomy, a perfect gestaltic superposition of function and form [80] which is further related to intra-cortical inhibition and the hypercolumnar organization. The boundaries in the neurocortex of ocular (vis-à-vis binocular) domain organize the complex logarithmic map through the fact that intra-cortical inhibition and sequence regularity run parallel to these borders while binocular summation runs perpendicular to them. The ocular dominance columns provide the link between the

[80] Further below, on discussing the somatotopic mapping, it will appear that the complexity appears to be encoded not in the body as anatomical differences per-se but on the map, while in the retinotopic visual encoding we have in the -foveal- domain of validity of the mapping, a perfect gestalt. So, for light as the most unconstrained field, the encoding is embodied as anatomical-computational-architecture while in the somatotopic case seems that the computational mapping carries the complexity as we evolve to lesser degrees of freedom. We shall not enter here, due to space limitations, to discuss the relation of this with Kant's philosophy.

axis of the global and the local mappings of the cortex. Now, to understand how is it that the actual cell density provides for the functional architecture, it is essential to note that their density is described by the same inverse square law of general potential theory in the theory of complex functions, provides a two-dimensional functional-computational-architecture for electrostatics, Newtonian potential, etc.[35]. The mathematical representation for both is the complex logarithmic map which also plays a central role in Karman vortex street model in fluid-dynamics described by the Navier-Stokes equations (which we proved to be a most basic example of a torsion geometry related to Brownian motions [105]). So, we do have vortical structures in the neurocortex which are related to the analytical retinotopic mapping. Now, how does the Klein-bottle structure appear in relation with the complex logarithmic map in this mapping? To start with, the topology of the complex logarithm map is that of the torus, which thus is the Riemann *orientable* surface for it. Anatomically, the hemispheric representation of the cortex are joined by the corpus callosum, yet in such a way that it supports the orientable torus topology [127]. Yet, the representation in cortical domains when this orientable topology is applied is asymmetric, the image by the map of the center of the local receptive field (the foveal) is on the boundary of the map, not in the central vertical axis. To reinstate symmetry with respect to this organizing center, the non-orientable topology of the Moebius band is needed, which thus becomes a non-orientable Riemann surface for the complex logarithmic map. In principle, both topologies are possible inasmuch both the Boolean and superposed basis in Matrix Logic exist and are transformed by the Klein-bottle Hadamard operator; see eq. (86). Simply, for the non-orientable topology, the field's receptive center is transformed into a symmetric organizing axis in the neurocortex, while in the Boolean orientable case it is represented in such a way that the center's field retinotopic representation in the cortex has been shifted to the boundary. Therefore, non-orientability is closely related to the preservation under the analytic map of an organizing center; again, non-orientability is related to Self and appears to be related to the integration of both hemispheres.

Returning to the issue of binocular vision and disparity which provides for a cue to depth perception, which we recall that in Merleau-Ponty's phenomenological philosophy [81], is the primeval dimension of perception , which, as also already presented, leads to multivalued logics and to the generation of time waves of synchronization, lies in the understanding of Schwartz in the fact that the complex logarithm links a cortical shift to a visual plane size change. Thus the slightly different projections of the two eyes are normalized by the complex logarithm to the identical cortical pattern, with the shift proportional to the size difference, and ultimately, the disparity of the stimulus [126].[82] Returning to the case of haptic perception, i.e. to the integration of several sensorial modes, we have a complex logarithmic map that is the representation map of both somatosensory and visual modes [126]. Now, as far as topologically the global representation for the somatosensory mode is the Klein-bottle, it is most remarkable that the Klein-bottle appears to be the solution of representation of visual topographic maps on the neurocortex. As for the topological representation of the retinotopic mode, symmetry properties of simple cell receptive properties

[81]Depth appears also as a possible spatial interpretation of the recursion of the Klein-bottle into itself , alternatively to the time-waves interpretation [109].

[82]Thus, stereoscopic vision appears as the process of a difference producing difference; the integration of both hemispheres which we also interpreted as a time-loop through a torsion field.

lead naturally to the construction of the Klein-bottle [137] coinciding with the conclusions by Schwartz. So the geometry of visual (and furthermore, of auditory and somatosensory -limbs, skin surface, etc.) space has a representation at the visual cortex, and furthermore, at the fundamental level of cells, the topology of the Klein-bottle is naturally present. Furthermore, there is experimental evidence that supports that these maps can be represented by the Klein-bottle [137, 141, 150]. Interesting enough, the starting point is the 2-dimensional Gabor function (of importance in holography [68]) commonly used to model the receptive-field profiles of simple cells [73], which make up a substantial percentage of visual cortical neurons. This function yields a topological representation which is the Klein-bottle, already present as an holographic representation of vision by the neurocortex.

9. Epilogue

To resume [83], the primary sensory projection of each limb, the kinesthesic and the visual modes are all described by the complex logarithmic map, and they all share the same topology, the Klein-bottle. So the integrated haptic modes available at the cortex are provided by a single map and a self-referential topology, which is further related to the constitution of the fusion of participatory universe on which the subject and the object are fused through the self-referential gestalt of the photon and torsion fields. So we here find more than an integration of the psychology and the neurophysiology of the spatial senses. It is further integrated with the generation of a participatory Universe, in which the thinking, perceptual, cognitive, sensory and physical realms are all one. Topologically, it is the Klein-bottle. This establishes a Universe of resonances integrated by torsion fields, in which the quantization of these resonances are provided by sheets of the complex logarithmic map. These resonances give rise to the synchronization that leads to binocular vision, as being integrated through the complex logarithmic map.

Now, the body afferents which are the mediators between the body and its topographical map, independently of whether this is the analytical or the topological one, have a clear fractal recursive structure, and thus itself is eminently self-referential. This is common also to the actual development patterns of plants, and has been studied extensively in the mathematical work of Lindenmayer. As Schwartz with profound insight points out, for the map of the sensorium to be compatible with the end image in the neurocortex be realized through the complex logarithmic map, it is sufficient that this is the map that constructs the fractal at each and all stages of the recursion [84] What about the topological map? Is this iteration of the analytical map related to non-orientability, i.e. the Klein-bottle and Moebius band, at all stages of the recursion? For the retinotopic representation, this would imply at each stage of the recursion the preservation of the stimuli local centre field, concatenated by the complex logarithmic map. More generally, we suggest that the answer may be affirmative and furthermore, and that there is anatomical evidence that this is the case. The evidence is the mammal and specifically human heart, but not seen in the usual Cartesian sense, as an object occupying a space being defined by its boundaries (were it not that it has valves

[83]Reaching this stage of ending this chapter, and due to a conjuction of fortituous and premeditated conjuction of the lack of space and the wish of the author not to spell out conclusions but allow the reader to ponder on the text by him (her)self, we shall only write an epilogue to resume the last section.

[84]We recall the anthropologists tale: It is turtles all the way down (or up).

connecting it with the rest of the body) by it. The actual structure of the human heart (dissected with previous extraction of the valves) is that of an iterated Moebius band, as was discovered in the work of his lifetime by F. Torrent-Guasp, in which each band folds another identical band. [85] Furthermore, this development is crucial to the physiology of the heart, which thus functions as a torsioned geometry through vortex motions produced by the cyclical turns of the recursive Moebius bands [143]. Including the valves, is the heart a Klein-bottle?

The next question that this elicits is: What is the relation between this recursive structure-process and the growth of organs in the human body? According to the findings of Lindenmayer the case for plants is that they are unified [69]. Plants grow following the recursive patterns. Let us return to the complex logarithmic map. If we decompose it into its real and imaginary parts, they both satisfy the diffusion equation [35]. So at each stage of development of the recurrence, we have a diffusion process that we can think of as morphogens. But now the topology is non-orientable at each stage, and thus the growth also follows the same topology , at each stage, explaining thus the Myocardial Band Model due to Torrent-Guasp.[86]

Acknowledgements

This work and those that make up the project altogether, would not have been possible without the love, care and support of my wife Sonia, and my children Tania and Tsafrir. They kept the beacons up, the fire kindled and their open smiles when my strength was fading, and this was too very often. My gratitude to my colleagues for their encouragement, notably Prof. Stein Johansen, for encouragement, syntropic action, discussions and reviewing this article with great care (if some error has persisted, it is the author's responsability, not his) and suggesting several improvements. Furthermore, for his kind invitation to Norway and the joint organization with Dr. Anita Leirfall of the Kantian Society of Norway, of a memorable day (July 24, 2009) long talk and transdisciplinary seminar at the University of Oslo based on my work and for the pleasure of communication (much too rare nowadays). Further to Profs. Daniel Dubois, Larry Horwitz, Martin Land, Steven Rosen, Walter Schempp, Gerhard Werner, Dr. Mladen Kocica and Dr. Hellmut Lockenhoff, for their kind encouragement. To Dr. Mikhail Bazanov for discussions in developmental human anatomy-physiology and the Klein-bottle. To the memory of the tragically disappeared neuropsychologist, Prof. Jacobo Grinberg-Zylberbaum, whose seminar on consciousness at the Universidad Autónoma Nacional de México during 1985-86 was certainly the most enticing experience of my life in the academic world. Jacobo urged me repeatedly to start writing down the ideas presented at the seminar. It took me more than twenty four years to take courage, and determine myself to do so, and yet time and timing unfolded. I wish he would be here to discuss with him the fruits which he so insistingly requested. This project has been gracefully acknowledged and honoured with the Telesio Galilei Academy

[85] We can actually ponder together with many loving people that there is a connection of the heart with Intelligence.

[86] Pioneering work in the recurrent growth algorithms that lead to the formation of human or animal organs and structures, we refer to the important, but unfortunately not translated to English, work by Bazanov [9], inspired in the work by Edwards [28]. It is important to remark that this is a very remarkable example of the fusion of form and function, alike the one that the Klein-bottle embodies.

of Sciences (London) 2010 Gold Award, delivered in a ceremony that took place in June 12th at Pécs, 2010 Cultural Capital of Europe. My gratitude to the founding father of the Academy and to the president, Profs. Francesco Fucilla, and Jeremy Dunning Davies, respectively. My gratitude as well to Dr. Frank Columbus, president of Nova Science, for his kind invitation to contribute to this volume.

References

[1] Aharonov,Y., Bohm, "Significance of electromagnetic potentials in quantum theory", *Physical Review* **115** 1959, 485-491.

[2] Allman, J.M and Kaas,J.H., A representation of the visual field on the medial wall of the occipital parietal-cortex in the owl monkey, *Science* **191**, 1976, 572-575.

[3] R. Amoroso and E. Rauscher, *The Holographic Anthropic Multiverse: Formalizing the Complex Geometry of Reality* (Series on Knots and Everything), World Sc., Singapore, 2009.

[4] J. Audretsch, *Entangled Systems*, Wiley-VCH, Bonn, 2006.

[5] Antoniou, I., Relativistic internal time operator, Int. J. Theor. Physics, **31**, no. 1, 1992, 119; Antoniou,I., Prigogine, I., Sadovnichii, V., Shkarin, S.A.,"Time operator for diffusion", *Chaos, Solitons and Fractals* **11**,2000, 465

[6] Bollobàs,B. *Linear Analysis*,Cambridge University Press, Cambridge , 1990.

[7] Bateman,H. *The theory of electric and optical wave motions from Maxwell's equations*, Cambridge University Press, 1915.

[8] Bateson,B. *Steps to an Ecology of Mind*, Paladin Books, 1973; ibid. *Mind and Nature:A Necessary Unity*, Bantam, New York, 1988.

[9] Bazanov, M., *Mot Stromen: En vrien leges notater*, Centrum Norsk Forlag, Oslo, 2006.

[10] Beller, " The Sokal Hoax: At whom are we laughing" http://www.mathematik.uni-muenchen.de/ bohmmech/BohmHome/sokalhoax.html

[11] Costa de Beauregard, in *Bergson and the Evolution of Physics*, Gunter, P.A. (ed.), Tennessee Univ. Press, Knoxville, 1969.

[12] Nye, J. F., Berry,M., 'Dislocations in wave trains' ,*M. V., Proc. R. Soc. A* **336**, 165-90, 1967.

[13] Bohm, D., Hiley, B. *The Implicate Order*, Routledge-Kegan, London, 1980.

[14] Bucccheri,R., di Gesù, V., Saniga,M., *Studies on the Structure of Time: From Physics to Psycho(patho)logy*, (Springer, Berlin, 2000); R. Buccheri, M. Saniga, and W.M.

Stuckey," *The Nature of Time: Geometry, Physics and Perception* (NATO Science Series II: Mathematics, Physics and Chemistry), Springer, Berlin , 2003; Buccheri,R., Elitzur,A., Saniga,M., *Endophysics, Time, Quantum And the Subjective, Proceedings of the ZIF Interdisciplinary Research Workshop,* 17-22 January 2005, Bielefeld, Germany, Springer, Berlin ,2005.

[15] Chandler, J., "An introduction to the perplex number system", *Discrete Applied Mathematics* **157**, 2009, 2296-2309.

[16] Christianto, V., Rapoport, D., Smarandache, F., Numerical solutions of time-dependent gravitational Schroedinger equation, *Progress of Physics* **2**, 56-60, 2007; ibid. an expanded version in ref. [18].

[17] Christianto,V., Smarandache,F., "Kaluza-Klein-Carmeli metric from quaternion-Clifford space, Lorenz force and some observables", *Progress in Physics* **2**, April 2002, 144-150.

[18] Christianto, V, Smarandache, F. (eds.), *Quantization in Astrophysics, Brownian Motion and Supersymmetry*, Math Tiger, Tamil Nadu, Chennai, India, 2007.

[19] Conte,E., *On some cognitive features of Clifford algebraic quantum mechanics and the origin of indeterminism in its theory,* http://philpapers.org/profile/2371.

[20] Cramer,J. "The transactional interpretation of quantum physics" *Rev. Mod. Phys.* **58**, 647-688 (1986).

[21] Dienes,I., "Consciousness - Holomatrix Quantized Dimensional Mechanics", in R.L. Amoroso, R.L., Pribram, K.H., (eds.) *The Complementarity of Mind and Body: Realizing the Dream of Descartes, Einstein and Eccles*, p. 463-471, Nova Science, New York, 2010.

[22] P. M. Dirac, "Quantized Singularities in the Electromagnetic Field",*Proc. Royal Soc. London* **133**, 6-72 (1931)

[23] J. A. Doucet,J.A., "Language and Geometry", and "Sentences with Associated Geometry", submitted to CASYS09, Conferece on Anticipative Systems, Liège, August 2009.

[24] Dubois,D., *Proceedings of CASYS (Conference on Anticipative Systems)*, Springer AIP Proceedings Series, Berlin, 1999,2000,2005, 2007, 2009.

[25] Dubois,D., P. Julià,P., 'On Two Different Conceptions of Language, Self-Referentiality and Consciousness', presented at CASYS09, Liège, 2009.

[26] Durr, H.P., Popp, F.A., Schommers, W. (Editors), *What Is Life? Scientific Approaches and Philosophical Positions: Scientific Approaches and Philosophical Positions*, World Scientific, Singapore, 2002.

[27] Eddington, A.S., *Fundamental Theory*, Cambridge University Press, London, 1946.

[28] Edwards,L., *The Vortex of Life:Nature's Patterns in Space And Time*, Floris, 2006.

[29] Eliade, M., *The Myth of Eternal Return: Or, Cosmos and History*, Princeton Univ. Press,1974.

[30] Eganova,I.A., "The World of Events Reality: Instantaneous Action as a Connection of Events Through Time", in *Relativity, Gravitation, Cosmology*, V.Dvoeglazov (ed), pag. 149-163, Nova Sc. Publ., New York, 2004.

[31] Fock,V., *The Theory of Gravitation*, Pergamon Press, London, 1962.

[32] Frankel,T. *The Geometry of Physics*, Cambridge Univ.Press, Cambridge, 2001.

[33] Hehl,F., von der Heyde,P., Kerlick, G.M., Nester,J.M., Review in Modern Physics, **48** (1976),3; Hehl,F., Dermott McCrea,J., Mielke, E., Ne'eman,Y., *Physics Reports* vol. 258 (1995), 1-157.

[34] Berezin,A.A., Gariaev,P.P., "Is it Possible to Create Laser Based on Information Biomacromolecules?", *Laser Physics*, **6**, no. 6, 1996, pp. 1211-1213; Gariaev, P., Tertishny, G. and Leonova, K., "The Wave: Probabilistic and Linguistic Representation of Cancer and HIV", J. of Non-Locality and Remote Mental Interactions **I**, no. 2, May 2002, at http://www.emergentmind.org; Maslov, M.U. and Gariaev, P.P., "Fractal Presentation of Natural Language Texts and Genetic Code", *2nd Intl. Conference on Quantitative Linguistics*, 20-24 September 1994, 1994.

[35] Gamelin, T. *Complex Analysis*, Springer, Berlin, 2001.

[36] Gelfand, I.M., Vilenkin, *Generalized Functions IV: Applications to Harmonic Analysis*, Academic Press, New York, 1966

[37] Ghalib,M. *Bibliotheca Islamica* **24**, Steiner Verlag, Wiesbaden, 1971; Halm,H., Kosmologie und Heilslehre der Fruhen Ismailiya, Steiner Verlag, Wiesbaden, 1978, 106-205.

[38] Guenther, H.V.,*Matrix of Mystery: Scientific and Humanistic Aspects of rDzogs-chen Thought*, Shambhala, Boulder and London, 2001.

[39] Gunther,G., "Time, Timeless Logic, Self-referential Systems, *Ann. New York Ac. Sciences* **138**, (1967), 317-346.

[40] Gurwitsch,A. *Die mitogenetische Strahlung*; Fischer Verlag. Jena, 1932 and 1959.

[41] Halberg,F. et al, "Chronoastrobiology", *Biomed. & Pharmacother.* **58** (2004).

[42] Haramein,N., Rauscher,E.A., *Beyond the standard model: searching for unity in physics*,Amoroso,R., Lehnert,B. ,Vigier,J.P., (eds.), The Noetic Press, USA, 2005, 153-168

[43] Harms,J., "Time-lapsed reality visual metabolic rate and quantum time and space", *Kybernetes* **32**, No. 7/8, 2003, pp. 113-1128

[44] Heelan,P., *Space Perception and the Philosophy of Science*, University of California Press, Berkeley, 1989.

[45] Heidegger,M., *Time and Being*, pags. 1-24, Harper and Row, New York,1962.

[46] Hegel, G.F., *Logic*, Evergreen Review, Inc., 2007.

[47] Hellerstein, N. S., *Diagonal Logic: A Paradox Logic*, Series on Knots and Everything, L. Kauffman editor, World Scientific, Singapore,

[48] Hill,V., Rowlands,P., *Nature's Fundamental Symmetry-Breaking* and *The Numbers of Nature's Code*, in *CASYS09, Proceedings of the Conference on Anticipatory Systems, Univ. of Liège, 2009*, D. Dubois et al (eds.), *Springer AIP Conference Proceedings Series*, Berlin, 2010.

[49] Hirschfelder,J.O., Christoph, A.C., Pale,W.E., "Quantum Mechanical Streamlines", *J. Chem. Phys.* **65**, 470-486 (1976)

[50] Ho, Mae-Wan, *The Rainbow and the Worm: The Physics of Organisms*, World Acientific, Singapore, 1996.

[51] Horwitz,L.P., Piron,C., *Helv. Physics Acta,* **46** (1973), 316; Horwitz,L.P., Piron,C., Helv. *Physica Acta* **66** (1993), 694; Land,M.C., Shnerb,N., Horwitz,L.P., *J. Math. Phys.* **36** (1995), 3263; Horwitz,L.P., Shnerb,N., N. *Found. of Phys.* **28**(1998), 1509.

[52] Horwitz,L.P., "Hypercomplex Quantum Mechanics", *Foundations of Physics,* **26**, no.6 (1996)pp. 851-.

[53] Horwitz,L.P., "Quantum Interference in Time", *Foundations of Physics,* **37**, nos. 3/4 (2007), pp.734-.

[54] Hubel,D.H. *Eye, Brain and Vision*,(Freeman and Co. New York, 1988). Arbib,M., P. Érdi,P., Szentágothai,J., *Neural Organization: Structure, Function and Dynamics*, MIT Press, Cambridge (MA), 1998.

[55] Illert,C.T., *Il Nuovo Cimento (series D)* 9(7), 791-784 (1987); 11 (5), 761-780 (1989), 12(10), 1405-21 (1990), 12 (2), 1611-32 (1990);ibid. *Foundations of Theoretical Conchology*, first edition, Hadronic Press, Palm Harbor, Fl, (1992); Illert,C.R., Santilli,R.M., *Foundations of Theoretical Conchology*, second edition, Hadronic Press, Palm Harbor, Fl, 1995.

[56] Indow,T., *The Global Structure Of Visual Space*, World Sci., Singapore, 2004.

[57] James,J., "A Calculus of Number Based on Spatial Forms", Ms. Sc. thesis, Univ. of Washington, 1993.

[58] Johansen,S., *Outline of a Differential Epistemology*, English trans. of the Norwegian (Univ. of Trondheim, 1991), to appear; ibid., "Basic Considerations about Kozyrev's Theory of Time from Recent Advances in Specialist Biology, Mathematical Physics and Philosophical Informatics", in *Time and Stars: The Centenary of N. A. Kozyrev*,

Shikhobalov, L. (ed.), 652-703, Nestor-History, Saint Petrersburg, 2008: ibid. "Non-trivial Time Flows in Anticipation and Action Revealed by Recent Advances in Natural Science, Framed in the Causality Network of Differential Ontology", Casys, *Int.J. Comp. Ant. Sys.* **22**, 2008; ibid., "Initiation of 'Hadronic Philosophy', the Philosophy Underlying Hadronic Mechanics and Chemistry", *Hadronic Journal* **28**, 111-135, 2006.

[59] Johansen,S., Complete Exposition of Non-Primes Generated From a Geometric Revolving Approach by 8x8 Sets of Related Series, and thereby ad negativo Exposition of a Systematic Pattern for the Totality of Prime Numbers, submitted to the Bulletin of the Calcutta Mathematical Society, viXra:1003.0089 (11 Mar 2010).

[60] Johansen,S., *The concept of Labour Time Content: A Theoretical Construction (A Revisionist Contribution to the Foundations of a Form-Reflected Capital)*, doctoral thesis, University of Bergen, Norway, 1991; published by Ariadne (in Norwegian), Bergen, 1991.

[61] Kauffman,L., " De Morgan Algebras, Completeness and Recursion", Proceedings of the Eighth International Symposium in Multiple Valued Logics, in http://www.math.uic.edu/ kauffman/DeMorgan.pdf

[62] The Klein-bottle, http://www.geom.uiuc.edu/zoo/toptype/klein/standard/; wikipedia.org/wiki/Kleinbottle; http://plus.maths.org/issue26/features/mathart/Build.html .

[63] Kleinert,H.,*Gauge Fields in Condensed Matter*, World Sc., Singapore, 1990.

[64] Kampf,U., "Possible Quantum Absorber Effects in Cortical Synchronization", in Vrobel,S., Rossler, O., *Simultaneity: Temporal Structures and Observer Perspectives*, World Sci., Singapore, 2008; ibid. in *Casys, Int.J. Comp. Ant. Sys.* **22**, 2008.

[65] Korotaev,S.M., Serdyuk,V.O., Sorokin,M.O., " Experimental Verification of Kozyrev's Interaction of Natural Processes",*Galilean Electrodynamics* **11** Special Issues 2, p. 23, 2000. "Experimental estimation of macroscopic nonlocality effect in solar and geomagnetic activity", Phys. Wave Phenomena, **11** (1), 46-54 (2003); Korotaev,S. in *Casys, Int.J. Comp.Anticip. Systems* **22**, 2008.

[66] Kozyrev,N.A., "On the possibility of experimental investigation of the properties of time", in *Time in Science and Philosophy*, pp. 111132, Prahue, 1971; www.chronos.msu.ru. Dadaev,A., "Astrophysics and Causal Mechanics", *Galilean Electrodynamics,* **11**, Special Issues 1, p. 4, 2000.

[67] Kroeber, A.L., *Configurations of Cultural Growth*, University of California Press, Berkeley, 1969.

[68] Liebling,M., Blu,T., Unser,M. "Fresnelets: new multiresolution wavelet bases for digital holography", *IEEE Trans. on Image Proc.,* **12**, 1, (2003), pp. 29 - 43.

[69] Lindenmayer,A., Prusinkiewicz, P., *The Algorithmic Beauty of Plants (The Virtual Laboratory)*, Springer, Berlin, 1996.

[70] Luneburg,R., *Mathematical Analysis of Binocular Vision*, (Edward Brothers, 1948).

[71] Lorente de No, "Anatomy of the eight nerve: the central projection of the nerve endings of the internal ear", *Laryngoscope* **42**,1-38, 1933

[72] Mantegna, R., Muldyrev,S., Goldberger, S.,Havlin, S, Peng, C.,Simmons, M., Stanley,C., Linguistic features of Noncoding DNA sequences, *Phys. Rev. Letts,* **73**, no. 23, 3169-3172, 1994.

[73] Marcelja S, "Mathematical description of the responses of simple cortical cells", *J. Opt. Soc. Am.* **70**, 12971300, 1980.

[74] Marcer,P., Schempp,W.,"A Mathematically Specified Template for DNA and the Genetic Code, in Terms of the Physically Realizable Processes of Quantum Holograph", in *Proceedings of the Greenwich Symposium on Living Computers* (A. Fedorec and P. Marcer eds.), (1996), pp. 45-62; ibid."The Model of the Prokaryote Cell as an Anticipatory System Working by Quantum Holography", in *Proceedings of CASYS '97, 1115 August 1997, HEC, Liège, Belgium, Inter. J. of Comp. Ant. Sys.*, **2**, (1997) pp. 307-315.

[75] Marcer,P., Schempp,W., "The brain as a conscious system", *Int. J. of Gen. Sys.* **27**, 1/3, 231-248 (1998).

[76] Mashkevich,V., "General Relativity and Quantum Jumps: The Existence of Nondiffeomorphic Solutions to the Cauchy Problem in Nonempty Spacetime and Quantum Jumps as a Provider of a Canonical Spacetime Structure" gr-qc/0403056 (March 2004); ibid. "Cosmological Quantum Jump Dynamics II. The Retrodictive Universe", gr-qc/0303046v1 (March 2003).

[77] McFadden,J., *Quantum Evolution: How Physics' Weirdest Theory Explains Life's Biggest Mystery* , W. W. Norton & Company, New York,2002;

[78] Miller,I., Miller,R.A., "From Helix to Hologram, an Ode on the Human Genome", *Nexus*, August-September, 2003; www.nexusmagazine.com .

[79] Merleau-Ponty,M., *The Visible and the Invisible*, Northwestern Univ. Press, 1968; ibid., *Phenomenology of Perception*, Routledge and Kegan Paul, London ,1962.

[80] Mountcastle, V.B., "Modality and topographic properties of single neurons of cat's somatic sensory cortex", *J. Neurophysiol.* **20**, 408-434, 1957.

[81] H. Muller, *Raum und Zeit, special issue 1*, (Ehlers, 2003),; ibid. "Fractal Scaling Models of natural oscillations in chain systems and the mass distribution of the celestial bodies in the Solar System", *Progress in Physics,* no.1, January 2010. ibid. *Raum und Zeit* **127**, pp.76-77 (2004).

[82] Musès,C., "Applied Hypemumbers: Computational Concepts", *Applied Mathematics and Computation* **3**, pp.211-226 (1976); ibid. Hypernumbers-II, **4**, pp. 45-66 (1978).

[83] Musès,C., *Destiny and Control in Human Systems* (Frontiers in System Research); (Kluwer-Nijhoff, Boston, 1984)

[84] M. Ni,*The Yellow Emperors Classic of Medicine: A New Translation of the Neijing Suwen with Commentary*, Shambhala, Boulder, 1995; Beinfield, H., Korngold, E., *Between Heaven and Earth: A Guide to Chinese Medicine*, Ballantine, 1992.

[85] The Necker Cube, in http://en.wikipedia.org/wiki/NeckerCube.

[86] Nowosad,P., MRC-Wisconsin Report, 1982; ibid. *Comm.Pure Appl. Math.***21**, 401-65 (1968). Young,L.C., Nowosad,P., J.Opt. Th. App.**41**, 261 (1983).

[87] Nottale,L., *Scale Relativity and Fractal Space-Time*, World Sc., Singapore, 1993. ibid. *Astron. Astrophys. Lett.***315**, L9 (1996); ibid. *Chaos, Solitons & Fractals,* **7**, 877 (1996).

[88] Nottale,L., Schumacher, G., Lefevre, E.T., *Astron. Astrophys. Lett.* **361**, 379-389, 2000; ibid.322 , 1018, 1997.

[89] Páles,E., *Seven Archangels: Rythms of Inspiration in the History of Culture and Nature*, Sophia Foundation, Bratislava, 2009.

[90] Papanicolaou,A., Gunter, P. A., *Bergson and Modern Thought: towards a unified science*, Harwood Academic, Switzerland, 1987.

[91] Pattee, H.H., "The Physics of Symbols:Bridging the Epistemic Gap", *BioSystems* **60**(2001) 5-21.

[92] Penrose,R., *Shadows of the Mind: A Search for the Missing Science of Consciousness* , Oxford Univ.Press (1996); Penrose,R.,M. Gardner,M. *The Emperor's new Mind: Concerning Computers, Minds, and the Laws of Physics* , Oxford University Press,(2002)

[93] Pitkanen,M.,*Topological Geometrodynamics*, Luniver,Beckington, U.K., 2006; ibid. "'Quantization of Planck constants and Dark Matter Hierarchy in Biology and Astrophysics"', 20-57, in Christianto, V, Smarandache, F. (eds.), *Quantization in Astrophysics, Brownian Motion and Supersymmetry*, Math Tiger, Tamil Nadu, Chennai, India, 2007.

[94] Pockett,S., *Susan Pockett: The Nature of Consciousness*, IUniverse, 2000.

[95] Popp,F.A. et al, "Biophoton emission: New evidence for coherence and DNA as a source", *Cell. Biophys.* **6**, p. 33-52. 1984; Popp,F.A., " On the coherence of ultraweak photoemission from living tissues" in *Disequilibrium and Self-Organization*, Kilmister,C.W., Reidel, Boston, MA, 1986 Popp,F.A., Li,K.H., Gu,K., *Recent Advances in Biophoton Research and Its Applications* (May 1992); Won-Ho,M. *The Rainbow and the Worm: The Physics of Organisms*, World Scientific, Singapore, 1998;

[96] Pribram,K., *Rethinking neural networks: quantum fields and biological data*, Hillsdale (N. J.), Erlbaum, 1993; ibid *Origins: brain and self organization*, Hillsdale (N. J.), Erlbaum, 1994; ibid. *Brain and values: is a biological science of values possible*, Mahwah (N. J.), Erlbaum, 1998.

[97] Prigogine, I., *From Being to Becoming: Time and Complexity in the Physical Sciences*, Freeman, San Francisco (CA), 1981.Prigogine, I., Stengers, I. *Order out of Chaos: Man's new dialogue with nature*, Flamingo. 1984.

[98] Purcell,M.C., *Towards a New Era (Epistemological Resolution Analysis By Through and From the Klein Bottle Wholeness)*, Ph.D. Thesis, Department of Philosophy, University of Newcastle, Australia (2006).

[99] Pylkkanen,P., *Mind, Matter and the Implicate Order*, (The Frontiers Collection), Springer, Berlin 2006.

[100] Raju,C.K. *A Consistent Theory of Time*, Fundamental Theories of Physics Series, Reidel, 1991.

[101] Raju,C.K., *The Eleven Pictures of Time*, Sage Publications, New Delhi-London, 2003.

[102] Rapoport, D. L., "The Riemann-Cartan-Weyl Quantum geometries II : The Cartan stochastic copying method, Fokker-Planck operator and Maxwell-de Rham equations", *Int. J. of Theor. Phys.* **36**, No. 10, 1997, 2115-3152; ibid., "Stochastic processes in conformal Riemann-Cartan-Weyl gravitation", *Int. J. Theor. Phy.* **30**, no. 11, (1991), 1497-1515.

[103] Rapoport,D.L. "On the unification of geometric and random structures through torsion fields: Brownian motions, viscous and magnetic fluid-dynamics", *Found. Phys.* **35**, no.7, 1205-1244 (2005).

[104] Rapoport,D.L., in *Instabilities and Nonequilibrium Structures VI*, E. Tirapegui and W. Zeller (edts.), Kluwer, Dordrecht, 2000, 359-370.

[105] Rapoport,D.L. "On the geometry of viscous fluids and a remarkable pure noise representation", *Rep. Math. Phys.* **50**, no.2, 211-250 (2002); ibid. "Random diffeomorphisms and integration of the Navier-Stokes equations", *Rep. Math. Phys.* **49**, no. 1, 1-27, (2002); ibid. "Martingale problem approach to the Navier-Stokes equations on smooth-boundary manifolds and semispace", *Rand. Oper. Stoch. Eqs.*, " Random Symplectic Geometry and the realizations of the random representations of the Navier-Stokes equations by ordinary differential equations", *Rand. Oper. Stoch. Eqs.*, **11**, no.2, 109-150, no.4, 351-382 (2003); ibid. "Viscous and magnetohydrodynamics, torsion fields, Brownian motions representations on compact manifolds and the random symplectic invariants", 276-328, in *Quantization in Astrophysics, Brownian motions and Supersymmetry*,F. Smarandache and V. Christianto (eds.), MathTiger, Tamil Nadu, India, 2007.

[106] Rapoport,D.L., "Cartan-Weyl Dirac and Laplacian operators, Brownian motions, The quantum potential and Scalar Curvature, Dirac-Hestenes equations and supersymmetric systems", *Found. Phys.* **35**, no. 8, 1383-1431 (2005).

[107] Rapoport,D.L., " Torsion Fields, CartanWeyl SpaceTime and State-Space Quantum Geometries, their Brownian Motions, and the Time Variables", *Found. of Phys.* ,**37**,nos. 4-5, 813-854 (2007); ibid., "On the state-space and spacetime geometries of geometric quantum mechanics", in *Foundations of Probability and Physics IV*, Khrennikov,A. et al (eds.), Springer,Berlin, 2007; ibid. "Torsion Fields, Cartan-Weyl Space-Time and State-Space Quantum Geometries, Brownian Motions, and their Topological Dimension", 329-385, in *Quantization in Astrophysics, Brownian motions and Supersymmetry*, F. Smarandache and V. Christianto (eds.), MathTiger, Tamil Nadu, India, 2007

[108] Rapoport,D.L., in*Hadronic Mathematics, Mechanics and Chemistry III*, R.M. Santilli, Int. Acad. Press, 2009.http://www.i-b-r.org/Hadronic-Mechanics.htm; ibid. D. Rapoport, Torsion Fields, Brownian Motions, Quantum and Hadronic Mechanics, pags 236-292 , in *Hadron Models and New Energy Issues*, F. Smarandache (ed.), Infolearnquest, Noviembre 2007

[109] Rapoport,D.L., "Surmounting the Cartesian Cut Through Philosophy, Physics, Logic, Cybernetics and Geometry: Self-reference, Torsion, the Klein Bottle, the Time Operator, Multivalued Logics and Quantum Mechanics" Found. Phys, http://www.springerlink.com/content/h168352020762932/ , in press.

[110] Rapoport,D.L., "Torsion Fields, Propagating Singularities, Nilpotence", Quantum Jumps And The Eikonal Equation, in *Proceedings of CASYS09*, Conference on Anticipative Systems 2009, D. Dubois et al (eds), Springer AIP Conference Proceedings, Berlin, 2010. Best Paper Award of CASYS09.

[111] Rapoport,D.L., "Torsion Fields, the Extended Photon, the Eikonal Equations, the Twistor Geometry of Cognitive Space and the Laws of Thought", in *Ether, Spacetime and Cosmology III*, M. Duffy and J. Levy, Apeiron, Toronto, 2010.

[112] Rapoport,D.L., Sternberg,S., "On the interactions of spin with torsion", *Annals of Physics* **158**, no.11, 447-475 (1984).

[113] Rapoport, D.L., Tilli, M. , "Scale fields as a simplicity principle", Proceedings of the Third International Workshop on Hadronic Mechanics and Nonpotential interactions", Dept. of Physics, Patras Univ., Greece, A. Jannussis (ed.), Hadronic J. Suppl. vol. 2, no.2, (1986), 682-778

[114] Rinehart,R.F., "Elements of a Theory of Intrinsic Functions on Algebras", *Duke Math.J.* **27**, 1-19, (1960).

[115] Rosch,E.,Varela,F., Thompson, E., *The Embodied Mind*, (MIT Press, Cambridge (Mass), 1991).

[116] Rosen,S.M., *Dimensions of Apeiron: A Topological Phenomenology of Space, Time and Individuation*, (Rodopi, Amsterdam-New York , 2004); ibid. *The Self-Evolving Cosmos: A Phenomenological Approach to Nature's Unity-in-Diversity* (Series on Knots and Everything), World Scientific, Singapore, 2008; ibid. *Topologies of the Flesh: A Multidimensional Exploration of the Lifeworld* (Series In Continental Thought), (Ohio University Press, 2006)

[117] S. M. Rosen, "Quantum Gravity and Phenomenological Philosophy, *Foundations of Physics* **38**, 556-582, (2008)

[118] Rossler,O., *Endophysics: The World As an Interface*, World Scientific, Singapore (1998); Vrobel,S., Rossler;O., "Simultaneity: Temporal Structures and Observer Perspectives", World Scientific, Singapore, 2008.

[119] Rowlands,P. *From Zero to Infinity*, World Scientific, Singapore, 2008.

[120] Rubcic, A., Rubcic, J., The Quantization of Solar Like Gravitational Systems, Fizika B-7, **1**, 1-13, 1998.

[121] Ruhnau, E., in *Conscious Experience , T. Metzinger* (ed.), Imprint Academic, Exeter (U.K.),1996.

[122] Santilli, R.M., *Hadronic Mathematics, Mechanics and Chemistry, vols. I-V*, International Academic Press, 2009;http://www.i-b-r.org/Hadronic-Mechanics.htm.

[123] Shakhparonov, I.M., Kozyrev-Dirac Emanation. Interaction with Matter and Methods of Detecting. http://www.rexresearch.com/kozyrev2/2-1.pdf.

[124] Schempp,W., "Rethinking VSIL Neural Networks, Cortical Linking Network Models and Quantum Holography", in Pribram,K., *Rethinking Neural Networks: Quantum Fields and Biological Data*, 233-299, Routledge, London, 1999.

[125] Schommers, W., *Space and Time, Matter and Mind: The Relationship Between Reality and Space-Time* (Series on the Foundations of Natural Science and Technology), World Scientific , Singapore, 1994.

[126] Schwartz,E.L., "Computational Anatomy and functional architecture of striate cortex: A spatial mapping approach to perceptual coding", *Vision Research* **20**, 645-669 (1980).

[127] Schwartz, E. L., Afferent Geometry in the Primate Visual Cortex and the Generation of Neuronal Trigger Features, Biol. Cybern. **28**, 1-14, 1977.

[128] Simeonov, P. L., "Integral Biomathics: A Post-Newtonian View into the Logos of Bios", J. Progress in Biophysics and Molecular Biology, Vol. 102, Issues 2/3, June/July 2010, 85-121, 2010.

[129] Shnoll,S., "Changes in the fine structure of stochastic distributions as a consequence of space-time fluctuations" , arXiv.org physics/0602017 (February 2006); Shnoll,S., Zenchenko,K.I., Berulis,I.I.,Udaltsova;n.V., Rubinstein,I.A.

; arxiv: physics/0412007 (December 2004); Panchelyuga,V., Kolombet,V., Panchelyuga,M., Shnoll,S.; physics/0612055 (December 2006); Kaminsky,A.V., Shnoll,S., physics/0605056 (May 2006); Snoll,S., physics/0602017; ibid. *Phys. Letts. A***359**, 4, 249-251 (2004); ibid. Uspekhi **43**, no. 2, 205-209 (2000); ibid. Progress in Physics , January 2007.

[130] Sokal, A. D., Bricmont, Jean, *Impostures Intellectuelles*. Editions Odile Jacob. Beller, M., " The Sokal Hoax: At whom are we laughing" http://www.mathematik.uni-muenchen.de/ bohmmech/BohmHome/sokalhoax.html

[131] Spencer-Brown,G., *Laws of Form*, George Allen and Unwin, London , 19

[132] Sorokin, P.A.,*Social and Cultural Dynamics*, Bedminster, New York, 1962.

[133] Spencer-Brown,G., *Laws of Form*, George Allen and Unwin, London , 1969.

[134] Stanislavovich,B., "On Torsion Fields", *Galilean Electrodynamics,* **12**, Special Issue 1, p. 5-9, (2000).

[135] Stern, A., *Matrix Logic and Mind*, Elsevier , Amsterdam, 1992; ibid. *Quantum Theoretic Machines*, Elsevier, Amsterdam, 2000.

[136] Stratonovich, R., *Nonlinear Nonequilibrium Thermodynamics I: Linear and Nonlinear Fluctuation-Dissipation Theorems*, Series on Synergetics, Springer Verlag, Berlin, 1992.

[137] Swindale,N.V., "Visual cortex: Looking into a Klein bottle" ,*Current Biology* **6**, No 7, (1996), 776779.

[138] Taborsky,E., *Architectonics of Semiotics*, Palmgrave MacMillan, New York , 1998.

[139] Tal,D., Schwartz,E.A., "Topological Singularities in Cortical Orientation Maps", *Network: Computation in Neural Systems*, **8**, (2), 229-238, (1997).

[140] Talbot, S.A., Marshall, W.H., Physiological studies of neural mechanisms of localization and discrimination, Amer. J. Ophtalmology **24**, 1225-1264, 1941.

[141] Tanaka,S., "Topological analysis of point singularities in stimulus preference maps of the primary visual cortex", *Proc. R. Soc. London (Biol)* **261**, (1995), 8188.

[142] Thapliyal, H., Srinivas, M., "An Extension to DNA Based Fredkin Gate Circuits: Design of Reversible Sequential Circuits using Fredkin Gates", http://arxiv4.library.cornell.edu/pdf/cs/0603092v1.

[143] Torrent-Guasp, F., Kocica, M., Corno, A., Komeda, M., Carrera-Costa, F., Flotats, A., Cosin-Aguillar, J., Wen Han, "Towards new understanding of the human heart", Eur. J. Cardiothorac. Surg. **27**, 191-201, 2005 and references therein; Nasiraei-Moghaddam, A., Gharib, M., "Evidence for the existence of a functional helical myocardial band",*Am J Physiol Heart Circ Physiol* **296**: H127H131, 2009; Grosberg, A., " A Bio-Inspired Computational Model of Cardiac Mechanics: Pathology and Development", Ph.D. thesis, Caltech, 2008.

[144] Trifonov,V., "Natural geometry of nonzero quaternions",*Int. J.Theor. Phys.* **46** (2) (2007) 251-257; ibid. *Europhys. Lett.* **32**(8) (1995) 621-626.

[145] Valenzi, V., private communication. Particularly, we have been informed that the work of academician Eduard Trukhan, Russian Academy of Sciences and Telesio Galilei Academy of Sciences, is of importance; we have no references to provide, unfortunately.

[146] F. Varela, *Principles of Biological Autonomy*, (North-Holland, New-York ,1979).

[147] L. de Broglie, *Non-linear Wave Mechanics: A Causal Interpretation.* (Elsevier, Amsterdam, 1960); J.P. Vigier, *Physics Letters A*, **135**, Issue 2, 13 February 1989, 99-105.

[148] Vitiello,G., *My Double Unveiled; The Dissipative Quantum Model of Brain (Advances in Consciousness Research)*, John Benjamins Publ., 2000.

[149] Vrobel,S., *Fractal Time*, The Inst. for Adv. Interdis. Res., Houston, Texas, U.S.A, 1998, http://www.if-online.org/Fractal

[150] Werner,G., "The Topology of the Body Representation in the Somatic Afferent Pathway, in *The Neurosciences Second Study Program*, pp. 605-617, Schmitt,F.O. (ed.), Rockefeller University, 1968.

[151] Werner,G., Whitsel,B.L., 'The Topology of the Body Representation in Somatosensory Area I of Primates', *J. of Neurophysiology* **311**, 856, 1968.

[152] Widom, A., Srivastava, Y., Valenzi, V., "The biophysical basis of Benveniste experiments: Entropy, structure, and information in water", *Intern. J. of Quantum Chem.*, **110** Issue 1, Pages 252 - 256, 2010; Del Giudice E, Preparata G, Vitiello G, "Water as a free electric dipole laser", *Physical Review Letters* **61**: 1085-1088. 1988.

[153] L. de Broglie, *Non-linear Wave Mechanics: A Causal Interpretation.* (Elsevier, Amsterdam, 1960); J.P. Vigier, *Physics Letters A*, **135**, Issue 2, 13 February 1989, 99-105.

[154] Woolsey C.N. , Marshall, W.H., Bard, P., "Cortical area for tactile sensibility", *Bull. Johns Hopkins Hospital* **70**, 339-441, 1942

[155] Wu, Y., Lin,Y. ,*Beyond Nonstructural Quantitative Analysis: Blown-ups, Spinning Currents and Modern Science*, (World Scientific, Singapore, 2002).Yi Lin, *Systemic Yoyos: Some Impacts of the Second Dimension*, (Auerbach Publications, An Imprint of Taylor and Francis, New York, 2008).

[156] Young,A.M. *The Reflexive Universe:The Evolution of Consciousness*, Robert Briggs Assoc., Lake Oswego (Oregon), 1976; ibid., "Consciousness and Cosmology", in *Consciousness and Reality: The Human Pivot Point*, Musès, C., Young, A. (eds.), pags. 153-164, Avon Books, New York, 1972.

In: Focus on Quantum Mechanics
Editor: David E. Hathaway et al. pp. 295-338
ISBN: 978-1-62100-680-0

Chapter 10

IN SITU REMARKS ON NOVEL EXACT SOLUTIONS OF QUANTUM DYNAMICAL SYSTEMS: HEAVISIDE OPERATIONAL ANSATZ IN THE QUANTUM PHASE SPACE REPRESENTATION AT THE GENERALISED HAMILTONIAN-LAGRANGIAN NEXUS

Valentino A. Simpao*
Mathematical Consultant Services, Greenville, KY, USA

A viable methodology for the exact analytical solution of the multiparticle Schrodinger and Dirac equations has long been considered a holy grail of theoretical chemistry. Since a benchmark work by Torres-Vega and Frederick in the1990's[1], the Quantum Phase Space Representation (QPSR) has been explored as an alternate method for solving various physical systems, including the harmonic oscillator[2], Morse oscillator[3], one-dimensional hydrogen atom[4], and classical Liouville dynamics under the Wigner function[5]. QPSR approaches are particularly challenging because of the complexity of phase space wave functions and the fact that the number of coordinates doubles in the phase space representation.These challenges have heretofore prevented the exact solution of the multiparticle equation in phase space. Recently, Simpao[6] has developed an exact analytical symbolic solution scheme for broad classes of differential equations utilizing the Heaviside Operational Ansatz (HOA). It is proposed to apply this novel methodology to QPSR problems to obtain exact solutions for real chemical systems and their dynamics. In his preliminary work, Simpao[6] has already applied this method to a number of simple systems, including the harmonic oscillator, with solutions in agreement to those obtained by Li [refs.2,3,4,5]. He has also demonstrated the exact solution to the radial Schrodinger Equation for an N-particle system with pairwise Coulomb interaction[7]. In addition to the Schrodinger Equation, the HOA method is capable of treating the Dirac equation[8] as well as differential systems governing both relativistic and

* E-mail address: mcs007@muhlon.com, Tel: 1-270-977-0575. 9 (Corresponding author)

non-relativistic particle dynamics. Applying these methods would allow us to pursue further exploration of this methodology, starting with the exact solution of multielectron atoms and moving toward complex molecules and reaction dynamics. It is believed that the coupling of HOA with QPSR represents not only a fundamental breakthrough in theoretical physical chemistry, but it is promising as a basis for exact solution algorithms that would have tremendous impact on the capabilities of computational chemistry/physics. As the theoretical foundation for spectroscopy is the Schrodinger equation, the significance of this discovery to the enhanced analysis of spectroscopic data is obvious. For example, the analysis of the Compton line in momentum spectroscopy necessitates the consideration of the momentum wavefunction for the molecular system under study. The novel methods refs. [6,7,8] allow the exact determination of the momentum[and configuration] space wavefunction from the QPSR wavefunction by way of a Fourier Transform. For example, the primary focus of [7] is the pairwise 1/rij interaction in context of the radial equation in the nonrelativistic Schrodinger case. This application of the exact solution ansatz developed above corresponds to the problem of n-particles with pairwise Coulomb interaction;scaling the parameters and variables of the problem yields the exact solution of the QPSR Schrodinger equation for the first principles general polyatomic molecular Hamiltonian. Upon a straightforward slight adaptation of this non-relativistic Schrodinger result, the QPSR Dirac equation addressed in [8] immediately yields the relativistic counterpart for the first –principles general polyatomic molecular Hamiltonian solution. These results form the cornerstone of the exact solution to the quantum dynamics of particular chemical systems, which shall appear elsewhere. Although all of the examples of dynamical systems mentioned above are at once solved as a special case of the general integral method already published in HOA, it is illuminating for applications to write out the solutions as the integrals are evaluated explicitly for the same. The HOA result is currently being used as the primary algorithm in the development of computer programs known as 'solver engines' for quantum chemistry/physics and plethora applications: to be reported elsewhere. In this note some remarks, examples and further directions, concerning HOA as a tool to solve and provide analytical insight into solutions of dynamical systems occurring in, but not limited to Mathematical Chemistry, are posited. Among these more general considerations are the developement of Novel Exact Analytical solutions of Generalized Hamiltonian/Lagrangian Dynamical systems in the Quantum Phase Space Representation and Classical Connections; also the exact analytical solutions of attendant differential/difference equations. To accomplish this, HOA[Heaviside Operational Ansatz] methods and Ehrenfest's Theorem are applied to general[i.e., Hamilton Extended Principle] quantum canonical and Lagrangian dynamical systems in QPSR [Quantum Phase Space Representation], yielding quadrature exact analytical solutions thereof. This novel result, arising at the nexus of; the Quantum Phase Space Representation [QPSR]; a non-traditional use of the Extended Hamilton Principle as manifest in the Generalised Lagrangian/Hamiltonian Formalism; the Heaviside Operational Ansatz [HOA] and Ehrenfest's Theorem, yields a prescription for exact analytical quadrature solutions to not only broad classes of closed/open quantum dynamical systems, but associated classical dynamical systems as well; what's more, to numerous and sundry attendant Functional Differential Equations [FDEs], For convenience, we begin with a recap of the HOA construction following from [ref. 6].

Recap of HOA

Here are the basic relations: x, p, t are respectively the configuration space position, momentum and time variables. The ^ denotes the operators, with H and Ψ denoting the Hamiltonian and wavefunction of the phase space representation[1], respectively. Also, the α, γ, are otherwise free parameters as specified in reference [1].

$$
\begin{aligned}
&H(x,p,t) \rightarrow \hat{H}(\hat{x},\hat{p},t) = \hat{H}\left(i\hbar\partial_p + \alpha x, -i\hbar\partial_x + \gamma p, t\right) \ni \alpha + \gamma = 1 \\
&x \rightarrow \hat{x} \equiv i\hbar\partial_p + \alpha x, p \rightarrow \hat{p} \equiv -i\hbar\partial_x + \gamma p, t \rightarrow t = t \\
&(x_1,\ldots,x_n) \rightarrow (\hat{x}_1,\ldots,\hat{x}_n) = \left(i\hbar\partial_{p_1} + \alpha_1 x_1,\ldots, i\hbar\partial_{p_n} + \alpha_n x_n\right) \ni \alpha_j + \gamma_j = 1, j = 1,\ldots,n \\
&(p_1,\ldots,p_n) \rightarrow (\hat{p}_1,\ldots,\hat{p}_n) = \left(-i\hbar\partial_{x_1} + \gamma_1 p_1,\ldots,-i\hbar\partial_{x_n} + \gamma_n p_n\right) \\
&H(x_1,\ldots,x_n;p_1,\ldots,p_n;t) \rightarrow \hat{H}(\hat{x}_1,\ldots,\hat{x}_n;\hat{p}_1,\ldots,\hat{p}_n;t) \\
&\equiv \hat{H}\left(i\hbar\partial_{p_1} + \alpha_1 x_1,\ldots, i\hbar\partial_{p_n} + \alpha_n x_n; -i\hbar\partial_{x_1} + \gamma_1 p_1,\ldots,-i\hbar\partial_{x_n} + \gamma_n p_n; t\right)
\end{aligned}
\tag{1}
$$

Now with the following properties of Heaviside operational methods via Laplace transforms refs. [6]

$$
\begin{aligned}
&L_{y\to z}[f(y)] = \int_{y_0}^{\infty} f(y)e^{-yz}dy = \bar{f}(z) \\
&L_{z\to y}^{-1}\left[\bar{f}(z)\right] = \frac{1}{2\pi i}\oint_{\partial} \bar{f}(z)e^{yz}dz = f(y) \\
&L_{z\to y}^{-1}\left[\bar{f}(z)\right] = \frac{1}{2\pi i}\oint_{\partial} \bar{f}(z)e^{yz}dz = f(y) = \bar{f}(D_y)U(y) \\
&L_{z\to y}^{-1}\left[\bar{f}_1(z)\bar{f}_2(z)\right] = f_1(y) * f_2(y) = \int_{y_0}^{y} f_1(y-u)f_2(u)du = \bar{f}_1(D_y)\bar{f}_2(D_y)U(y) = \bar{f}_1(D_y)f_2(y)
\end{aligned}
$$

where $U(y)$ is the Heaviside Unit Step function (2)

$$
\begin{aligned}
&L_{(y_1,\ldots,y_n)\to(z_1,\ldots,z_n)}[f(y_1,\ldots,y_n)] = \underbrace{\int_{y_{0n}}^{\infty}\cdots\int_{y_{01}}^{\infty}}_{n} f(y_1,\ldots,y_n)e^{-\sum_{j=1}^{n} y_j z_j} dy_1\ldots dy_n = \bar{f}(z_1,\ldots,z_n) \\
&L_{(z_1,\ldots,z_n)\to(y_1,\ldots,y_n)}^{-1}\left[\bar{f}(z_1,\ldots,z_n)\right] = \left(\frac{1}{2\pi i}\right)^n \oint_{\partial^n} \bar{f}(z_1,\ldots,z_n)e^{\sum_{j=1}^{n} y_j z_j} dz_1\ldots dz_n = f(y_1,\ldots,y_n) \\
&L_{(z_1,\ldots,z_n)\to(y_1,\ldots,y_n)}^{-1}\left[\bar{f}_1(z_1,\ldots,z_n)\bar{f}_2(z_1,\ldots,z_n)\right] = f_1(y_1,\ldots,y_n) \underset{(y_1,\ldots,y_n)}{\underbrace{*}} f_2(y_1,\ldots,y_n) \\
&= \underbrace{\int_{y_{0n}}^{y_n}\cdots\int_{y_{01}}^{y_1}}_{n} f_1(y_1 - y_1',\ldots,y_n - y_n')f_2(y_1',\ldots,y_n')dy_1'\ldots dy_n' \\
&= \bar{f}_1\left(\partial_{y_1},\ldots,\partial_{y_n}\right)\bar{f}_2\left(\partial_{y_1},\ldots,\partial_{y_n}\right)U(y_1,\ldots,y_n) = \bar{f}_1\left(\partial_{y_1},\ldots,\partial_{y_n}\right)f_2(y_1,\ldots,y_n)
\end{aligned}
$$

where the zero-subscripted variables (e.g., y_0) are the arbitrarily specified lower limits of integration.

With the phase-space convolution

$$f_1(x_1,....,x_n;p_1,....,p_n) \underbrace{*}_{(x_1,....,x_n;p_1,....,p_n)} f_2(x_1,....,x_n;p_1,....,p_n)$$

$$= \underbrace{\int_{x_{0n}}^{x_n} \cdots \int_{x_{01}}^{x_1}}_{n} \underbrace{\int_{p_{0n}}^{p_n} \cdots \int_{p_{01}}^{p_1}}_{n} f_1(x_1 - x_1',...x_n - x_n';p_1 - p_1',...p_n - p_n') f_2(x_1',..,x_n';p_1',..,p_n') dx_1'..dx_n'dp_1'..dp_n' \quad (3)$$

Lower bounds of respective phase space co - ordinates : $(x_{10},....,x_{n0};p_{10},....,p_{n0})$

Also the transform relation

$$L_{z\to y}\left[\tilde{f}(az-b)\right] = \frac{1}{a} e^{\frac{by}{a}} f\left(\frac{y}{a}\right)$$

$$L^{-1}_{(z_1,....,z_n)\to(y_1,....,y_n)}\left[\tilde{f}(a_1 z_1 - b_1,....,a_n z_n - b_n)\right] = \prod_{j=1}^{n} \frac{1}{a_j} e^{\frac{b_j}{a_j} y_j} f\left(\frac{y_1}{a_1},....,\frac{y_n}{a_n}\right) \quad (4)$$

From (1) the wave equation becomes

$$\hat{H}\begin{pmatrix} i\hbar\partial_{p_1} + \alpha_1 x_1,....,i\hbar\partial_{p_n} + \alpha_n x_n; \\ -i\hbar\partial_{x_1} + \gamma_1 p_1,....,-i\hbar\partial_{x_n} + \gamma_n p_n;t \end{pmatrix}\Psi(x_1,....,x_n;p_1,....,p_1;t)$$
$$= i\hbar\partial_t \Psi(x_1,....,x_n;p_1,....,p_1;t) \quad (5)$$

Applying the convolution identity and multivariable inverse transform of (2), the phase space convolution of (3) and relation (4) yields

$$\left[\begin{array}{l} L^{-1}_{\left(\substack{(\partial_{x_1},....,\partial_{x_n}) \\ \to(x_1,....,x_n)}\right)}\left[L^{-1}_{\left(\substack{(\partial_{p_1},....,\partial_{p_n}) \\ \to(p_1,....,p_n)}\right)}\left[\hat{H}\begin{pmatrix} i\hbar\partial_{p_1} + \alpha_1 x_1,....,i\hbar\partial_{p_n} + \alpha_n x_n; \\ -i\hbar\partial_{x_1} + \gamma_1 p_1,....,-i\hbar\partial_{x_n} + \gamma_n p_n;t \end{pmatrix}\right]\right] \\ \underbrace{*}_{(x_1,....,x_n;p_1,....,p_n)} \Psi(x_1,....,x_n;p_1,....,p_n;t) \end{array}\right]$$
$$\equiv \quad (6)$$
$$\hat{H}\begin{pmatrix} i\hbar\partial_{p_1} + \alpha_1 x_1,....,i\hbar\partial_{p_n} + \alpha_n x_n; \\ -i\hbar\partial_{x_1} + \gamma_1 p_1,....,-i\hbar\partial_{x_n} + \gamma_n p_n;t \end{pmatrix}\Psi(x_1,....,x_n;p_1,....,p_n;t)$$

Applying (6) to (5) with the convolution identities in (2) and transforming

$$L_{\left(\begin{smallmatrix}(p_1,\dots,p_n)\\ \to(\bar{p}_1,\dots,\bar{p}_n)\end{smallmatrix}\right)}\left[L_{\left(\begin{smallmatrix}(x_1,\dots,x_n)\\ \to(\bar{x}_1,\dots,\bar{x}_n)\end{smallmatrix}\right)}\left[\begin{array}{l}\hat{H}\begin{pmatrix}i\hbar\partial_{p_1}+\alpha_1 x_1,\dots,i\hbar\partial_{p_n}+\alpha_n x_n;\\ -i\hbar\partial_{x_1}+\gamma_1 p_1,\dots,-i\hbar\partial_{x_n}+\gamma_n p_n;t\end{pmatrix}\Psi(x_1,\dots,x_n;p_1,\dots,p_n;t)\\ =i\hbar\partial_t\Psi(x_1,\dots,x_n;p_1,\dots,p_1;t)\end{array}\right]\right]$$

$\equiv$

$$L_{\left(\begin{smallmatrix}(p_1,\dots,p_n)\\ \to(\bar{p}_1,\dots,\bar{p}_n)\end{smallmatrix}\right)}\left[L_{\left(\begin{smallmatrix}(x_1,\dots,x_n)\\ \to(\bar{x}_1,\dots,\bar{x}_n)\end{smallmatrix}\right)}\left[\begin{array}{l}L^{-1}_{\left(\begin{smallmatrix}(\partial_{x_1},\dots,\partial_{x_n})\\ \to(x_1,\dots,x_n)\end{smallmatrix}\right)}\left[L^{-1}_{\left(\begin{smallmatrix}(\partial_{p_1},\dots,\partial_{p_n})\\ \to(p_1,\dots,p_n)\end{smallmatrix}\right)}\left[\hat{H}\begin{pmatrix}i\hbar\partial_{p_1}+\alpha_1 x_1,\dots,i\hbar\partial_{p_n}+\alpha_n x_n;\\ -i\hbar\partial_{x_1}+\gamma_1 p_1,\dots,-i\hbar\partial_{x_n}+\gamma_n p_n;t\end{pmatrix}\right]\right]\\ \underset{(x_1,\dots,x_n;p_1,\dots,p_n)}{*}\Psi(x_1,\dots,x_n;p_1,\dots,p_n;t)\\ =i\hbar\partial_t\Psi(x_1,\dots,x_n;p_1,\dots,p_n;t)\end{array}\right]\right] \tag{7}$$

$\equiv$

$$\hat{H}\begin{pmatrix}i\hbar\partial_{p_1}+\alpha_1 x_1,\dots,i\hbar\partial_{p_n}+\alpha_n x_n;\\ -i\hbar\partial_{x_1}+\gamma_1 p_1,\dots,-i\hbar\partial_{x_n}+\gamma_n p_n;t\end{pmatrix}_{\left(\begin{smallmatrix}\partial_{p_1}\mapsto\bar{p}_1,\\ \dots,\partial_{p_n}\mapsto\bar{p}_n;\\ \partial_{x_1}\mapsto\bar{x}_1,\\ \dots,\partial_{x_n}\mapsto\bar{x}_n\end{smallmatrix}\right)}\breve{\Psi}(\bar{x}_1,\dots,\bar{x}_n;\bar{p}_1,\dots,\bar{p}_n;t)=i\hbar\partial_t\breve{\Psi}(\bar{x}_1,\dots,\bar{x}_n;\bar{p}_1,\dots,\bar{p}_n;t)$$

$\equiv$

$$\hat{H}\begin{pmatrix}i\hbar\bar{p}_1+\alpha_1 x_1,\dots,i\hbar\bar{p}_n+\alpha_n x_n;\\ -i\hbar\bar{x}_1+\gamma_1 p_1,\dots,-i\hbar\bar{x}_n+\gamma_n p_n;t\end{pmatrix}\breve{\Psi}(\bar{x}_1,\dots,\bar{x}_n;\bar{p}_1,\dots,\bar{p}_n;t)=i\hbar\partial_t\breve{\Psi}(\bar{x}_1,\dots,\bar{x}_n;\bar{p}_1,\dots,\bar{p}_n;t)$$

Hence the wavefunction in phase space may be analytically expressed in exact quadratures, by inverse transforming the above solution $\breve{\Psi}(\bar{x}_1,\dots,\bar{x}_n;\bar{p}_1,\dots,\bar{p}_n;t)$ of (7) as

$$\hat{H}\begin{pmatrix}i\hbar\bar{p}_1+\alpha_1 x_1,\dots,i\hbar\bar{p}_n+\alpha_n x_n;\\ -i\hbar\bar{x}_1+\gamma_1 p_1,\dots,-i\hbar\bar{x}_n+\gamma_n p_n;t\end{pmatrix}\breve{\Psi}(\bar{x}_1,\dots,\bar{x}_n;\bar{p}_1,\dots,\bar{p}_n;t)=i\hbar\partial_t\breve{\Psi}(\bar{x}_1,\dots,\bar{x}_n;\bar{p}_1,\dots,\bar{p}_n;t)$$

$$\Psi(x_1,\dots,x_n;p_1,\dots,p_n;t)=L^{-1}_{\left(\begin{smallmatrix}(\bar{x}_1,\dots,\bar{x}_n)\\ \to(x_1,\dots,x_n)\end{smallmatrix}\right)}\left[L^{-1}_{\left(\begin{smallmatrix}(\bar{p}_1,\dots,\bar{p}_n)\\ \to(p_1,\dots,p_n)\end{smallmatrix}\right)}\left[\begin{array}{l}e^{\frac{-i}{\hbar}\int_0^t \hat{H}\left(\begin{smallmatrix}i\hbar\bar{p}_1+\alpha_1 x_1,\dots,i\hbar\bar{p}_n+\alpha_n x_n;\\ -i\hbar\bar{x}_1+\gamma_1 p_1,\dots,-i\hbar\bar{x}_n+\gamma_n p_n;u\end{smallmatrix}\right)du}\\ \times\\ \breve{\Psi}_0(\bar{x}_1,\dots,\bar{x}_n;\bar{p}_1,\dots,\bar{p}_n;t=0)\end{array}\right]\right] \tag{8}$$

By definition, solution in one representation implies simultaneous solution in all representations. Along these lines, an interesting consequence arises: By equations (4.18) and (4.19) of reference [1], it is shown that the Fourier projection onto configuration x space of the phase space wavefunction, yields the configuration space wavefunction in the standard Schroedinger configuration space representation.

Symbolically

$$\hat{H}_{\text{configuration space}}\left(x,-i\hbar\partial_x,t\right)\Psi_{\text{configuration space}}(x,t)=i\hbar\partial_t\Psi_{\text{configuration space}}(x,t)$$

$$\Psi_{\text{configuration space}}(x,t)=\int_{-\infty}^{\infty}\frac{e^{\frac{ixp}{2\hbar}}}{\sqrt{4\pi\hbar}}\Psi(x,p,t)dp$$

$$\hat{H}_{\text{configuration space}}\left(x_1,\ldots,x_n,-i\hbar\partial_{x_1},\ldots,-i\hbar\partial_{x_n},t\right)\Psi_{\text{configuration space}}(x_1,\ldots,x_n,t) \tag{9}$$
$$=i\hbar\partial_t\Psi_{\text{configuration space}}(x_1,\ldots,x_n,t)$$

$$\Psi_{\text{configuration space}}(x_1,\ldots,x_n,t)=\int_{-\infty}^{\infty}\frac{e^{\frac{ix_1p_1}{2\hbar}}}{\sqrt{4\pi\hbar}}\cdots\int_{-\infty}^{\infty}\frac{e^{\frac{ix_np_n}{2\hbar}}}{\sqrt{4\pi\hbar}}\Psi(x_1,\ldots,x_n;p_1,\ldots,p_n;t)dp_1..dp_n$$

In other words, the configuration space wavefunction may be expressed in terms of exact quadratures containing the phase space wavefunction determined herein, Equation (8). Hence, Quantum Dynamics is now reduced to exact quadratures, as are all the associated purely mathematical problems that are abstracted from the physical formalism.

To wit, via HOA the configuration space solution becomes

$$\Psi_{\text{configuration space}}(x_1,\ldots,x_n,t)$$
$$=\int_{-\infty}^{\infty}\frac{e^{\frac{ix_1p_1}{2\hbar}}}{\sqrt{4\pi\hbar}}\cdots\int_{-\infty}^{\infty}\frac{e^{\frac{ix_np_n}{2\hbar}}}{\sqrt{4\pi\hbar}}L^{-1}_{\substack{(\bar{x}_1,\ldots,\bar{x}_n;\bar{p}_1,\ldots,\bar{p}_n)\\ \to(x_1,\ldots,x_n;p_1,\ldots,p_n)}}\left[\begin{array}{l} e^{\frac{-i}{\hbar}\int_0^t \hat{H}_{\text{configuration space}}\left(\substack{x_1,\ldots,x_n,\\ -i\hbar\partial_{x_1},\ldots,-i\hbar\partial_{x_n},u}\right)_{\substack{(x_1,\ldots,x_n)\to(i\hbar\bar{p}_1+\alpha_1x_1,\ldots,i\hbar\bar{p}_n+\alpha_nx_n)\\ (-i\hbar\partial_{x_1},\ldots,-i\hbar\partial_{x_n})\to(-i\hbar\bar{x}_1+\gamma_1p_1,\ldots,-i\hbar\bar{x}_n+\gamma_np_n)}}du} \\ \times \\ \Psi_{0\,\text{configuration space}}(\bar{x}_1,\ldots,\bar{x}_n;\bar{p}_1,\ldots,\bar{p}_n;t=0)\end{array}\right]dp_1..dp_n \tag{10}$$

With that said, a relatively simplistic prescription results for actually using the Ansatz to solve the problem,

Given the function $\hat{H}(\hat{x}_1,\ldots,\hat{x}_n;\hat{p}_1,\ldots,\hat{p}_n,t)$

[respectively $\hat{H}(x_1,\ldots,x_n;-i\hbar\partial_{x_1},\ldots,-i\hbar\partial_{x_n},t)$] replace $(\hat{x}_1,\ldots,\hat{x}_n;\hat{p}_1,\ldots,\hat{p}_n,t)$[respectively $(x_1,\ldots,x_n;-i\hbar\partial_{x_1},\ldots,-i\hbar\partial_{x_n},t)$] with $(i\hbar\bar{p}_1+\alpha_1x_1,\ldots,i\hbar\bar{p}_n+\alpha_nx_n;-i\hbar\bar{x}_1+\gamma_1p_1,\ldots,-i\hbar\bar{x}_n+\gamma_1p_n,t)$ in equation (10)

The result of course is the quantum phase space[respectively configuration space] wavefunction for the quantum dynamics wave equation. Just a comment on the α and γ parameters in the above formulae. From the HOA [ref.1], they are otherwise arbitrary except for the condition $\alpha+\gamma=1$. This is explained therein as a consequence of the arbitrary phase shift associated with the quantum phase space wavefunction. Further, any choice of the parameters thus constrained yields a Hamiltonian, which is dynamically equivalent[describes the same physics] as any other choice. However, it is shown therein that the Hamiltonian operator $\hat{H}\left(i\hbar\partial_p + \alpha x, -i\hbar\partial_x + \gamma p, t\right) \ni \alpha + \gamma = 1$ takes on the symmetric canonical form when $\alpha = \gamma = \frac{1}{2}$ thusly $\hat{H}\left(i\hbar\partial_p + \frac{x}{2}, -i\hbar\partial_x + \frac{p}{2}, t\right), \ni \alpha + \gamma = 1$. Notwithstanding this and with an eye towards computational simplifications for particular classes of applications, it has been found that other choices than $\alpha = \gamma = \frac{1}{2}$ sometime facilitates evaluation of the integral transforms. Unless otherwise directed, the convention for α and γ shall be specified for particular cases, presently and elsewhere. For the problems herein the $\alpha = \gamma = \frac{1}{2}$ sufficeth.

To recall the full details of HOA results, see the original work[EJTP 1 (2004), 10-16]. As pointed out therein,

> 'Notwithstanding its quantum mechanical origins, the HOA scheme takes on a life of its own and transcends the limits of quantum applications to address a wide variety of purely formal mathematical problems as well. Among other things, the result provides a formula for obtaining an exact solution to a wide variety of variable-coefficient integro-differential equations. Since the functional dependence of the Hamiltonian operator as considered is in general arbitrary upon its arguments(i.e., independent variables, derivative operator symbols[including negative powers thereof, thus the possible integral character]), then its multivariable extension can be interpreted as the most general variable coefficient partial differential operator. Moreover, it is not confined to being a scalar or even vector operator, but may be generally construed an arbitrary rank matrix operator. In all cases of course, its rank dictates the matrix rank of the wavefunction solution.'

In the present case of the Schrodinger equation, we shall be dealing with a scalar Hamiltonian structure and the solution wavefunction will be of a scalar character. (elsewhere, e.g. ref[9], the relativistic treatment demands the Dirac equation with such a 4x4 matrix Hamiltonian structure and the solution wavefunction will then be of a 4-dimensional column vector character).

At this point, we recognize a perhaps subtle aspect of the connection between Configuration and Phase Space wavefunction formulations in Equations (9) and (10):

Theory of Shifted Taylor Series yields

$$e^{-i\hbar f(x_1,\ldots,x_n,t)\partial_{x_j}}\Psi_{\text{Configuration Space}}(x_1,\ldots,x_n,t)$$
$$= \Psi_{\text{Configuration Space}}\left(x_1,\ldots,\left(x_j - i\hbar f(x_1,\ldots,x_n,t)\right)_{,\ldots} x_n,t\right)$$

Hence the Hamiltonian in the Configuration Space formalism may be used to generate "Shifted" systems of Functional Differential Equations[FDE], simultaneously containing discrete difference and continuous differential sectors in the same hybrid dynamics. This will be pursued in other work, but warrants mention here.

Some Selected HOA Example Calculations

Example 1. 1 Dim Simple Harmonic Oscillator

Having completed this brief recap of the HOA from ref.[6], now consider the first example case of the 1 dim. Simple Harmonic Oscillator dynamics in quantum phase space. Here the Hamiltonian has the form

$$\hat{H}(\hat{x}_1,\ldots,\hat{x}_n;\hat{p}_1,\ldots,\hat{p}_n,t)= \frac{\hat{p}^2}{2m}+a\hat{x}^2 \tag{11a}$$

Hence the solution of the phase space quantum dynamics is given by equation (9) as

$$\left(\frac{\left(-i\hbar\partial_x+\frac{p}{2}\right)^2}{2m}+a\left(i\hbar\partial_p+\frac{x}{2}\right)^2\right)\Psi(x;p;t)=i\hbar\partial_t\Psi(x;p;t)$$

$$\left(\frac{\left(-i\hbar\overline{x}+\frac{p}{2}\right)^2}{2m}+a\left(i\hbar\overline{p}+\frac{x}{2}\right)^2\right)\breve{\Psi}(\overline{x};\overline{p};t)=i\hbar\partial_t\breve{\Psi}(\overline{x};\overline{p};t) \tag{11b}$$

$$\Psi(x;p;t)=L^{-1}_{((\overline{x})\to(x))}\left[L^{-1}_{((\overline{p})\to(p))}\left[e^{\frac{-it\left(\frac{\left(-i\hbar\overline{x}+\frac{p}{2}\right)^2}{2m}+a\left(i\hbar\overline{p}+\frac{x}{2}\right)^2\right)}{\hbar}}\times\breve{\Psi}_0(\overline{x};\overline{p};t=0)\right]\right]$$

This is the computational construction of the Simple Harmonic Oscillator quantum phase space Schroedinger wavefunction. For the case of a Dirac delta function initial condition in quantum phase space of form $\delta(x,p)$ since the Laplace Transform of this initial condition is $\breve{\Psi}(\overline{x},\overline{p},t=0)=1,\Psi(x,p,t=0)=\delta(x,p)$.

Upon evaluation of (11b) with the above $\delta(x,p)$ initial condition, the quantum phase space wavefunction is

$$\Psi(x,p,t)=\sqrt{\frac{-m}{2a}}\,\frac{e^{i\left(\frac{p^2+2amx^2}{4a\hbar t}\right)}}{2\pi\hbar t} \tag{11c}$$

For sake of completeness, the case of a generalized to $a(t)$, an arbitrary function, yields

$$\left(\frac{\left(-i\hbar\partial_x+\frac{p}{2}\right)^2}{2m}+a(t)\left(i\hbar\partial_p+\frac{x}{2}\right)^2\right)\Psi(x;p;t)=i\hbar\partial_t\Psi(x;p;t)$$

$$\left(\frac{\left(-i\hbar\overline{x}+\frac{p}{2}\right)^2}{2m}+a(t)\left(i\hbar\overline{p}+\frac{x}{2}\right)^2\right)\breve{\Psi}(\overline{x};\overline{p};t)=i\hbar\partial_t\breve{\Psi}(\overline{x};\overline{p};t) \tag{11d}$$

$$\Psi(x;p;t)=L^{-1}_{((\overline{x})\to(x))}\left[L^{-1}_{((\overline{p})\to(p))}\left[e^{\frac{-i\left(t\frac{\left(-i\hbar\overline{x}+\frac{p}{2}\right)^2}{2m}+\int_0^t a(u)du\left(i\hbar\overline{p}+\frac{x}{2}\right)^2\right)}{\hbar}}\times\breve{\Psi}_0(\overline{x};\overline{p};t=0)\right]\right]$$

$$\Psi(x,p,t)=\Psi_0(x,p,t=0)\underbrace{*}_{x,p}\sqrt{\frac{-m}{2\int_0^t a(u)du}}\,\frac{e^{i\left(\frac{p^2+2mx^2\int_0^t a(u)du}{4\hbar t\int_0^t a(u)du}\right)}}{2\pi\hbar t} \tag{11e}$$

where $\underbrace{*}_{x,p}$ is the x,p convolutior as described in the Recap earlier.

Example 2. Schrödinger Hamiltonian with SHOs Having N-Arbitrary Masses in Pairwise Anisotropic Interaction

Now consider the example case of the nonrelativistic Schrödinger Hamiltonian with multiple pairwise interacting anisotropic SHOs of arbitrary masses in quantum phase space. Like the 1-dim example above, the oscillator strengths below may be generalized to arbitrary functions of time $a_{x_{ij}}(t), a_{y_{ij}}(t), a_{z_{ij}}(t)$ in a straightforward fashion, but the calculation is here omitted. Here the Hamiltonian has the form

$$\hat{H}_{\substack{\text{Schrodinger}\\ \text{Anisotropic}\\ \text{SHOs}}} = \sum_i \frac{1}{2m_i}\left(\left(-i\hbar\partial_{x_i} + \frac{p_{x_i}}{2}\right)^2 + \left(-i\hbar\partial_{y_i} + \frac{p_{y_i}}{2}\right)^2 + \left(-i\hbar\partial_{z_i} + \frac{p_{z_i}}{2}\right)^2\right)$$
$$+\sum_{i<j}\begin{pmatrix} a_{x_{ij}}\left(\left(i\hbar\partial_{p_{x_i}} + \frac{x_i}{2}\right) - \left(i\hbar\partial_{p_{x_j}} + \frac{x_j}{2}\right)\right)^2 \\ +a_{y_{ij}}\left(\left(i\hbar\partial_{p_{y_i}} + \frac{y_i}{2}\right) - \left(i\hbar\partial_{p_{y_j}} + \frac{y_j}{2}\right)\right)^2 \\ +a_{z_{ij}}\left(\left(i\hbar\partial_{p_{z_i}} + \frac{z_i}{2}\right) - \left(i\hbar\partial_{p_{z_j}} + \frac{z_j}{2}\right)\right)^2 \end{pmatrix} \tag{12a}$$

$$\hat{H}_{\substack{\text{Schrodinger}\\ \text{Anisotropic}\\ \text{SHOs}}}\begin{pmatrix} -i\hbar\partial_{x_i} + \frac{p_{x_i}}{2}, -i\hbar\partial_{y_i} + \frac{p_{y_i}}{2}, -i\hbar\partial_{z_i} + \frac{p_{z_i}}{2}; \\ i\hbar\partial_{p_{x_i}} + \frac{x_i}{2}, i\hbar\partial_{p_{y_i}} + \frac{y_i}{2}, i\hbar\partial_{p_{z_i}} + \frac{z_i}{2} \end{pmatrix}\Psi(x_i, y_i, z_i; p_{x_i}, p_{y_i}, p_{z_i}; t)$$
$$= i\hbar\partial_t\Psi(x_i, y_i, z_i; p_{x_i}, p_{y_i}, p_{z_i}; t)$$

$$\hat{H}_{\substack{\text{Schrodinger}\\ \text{Anisotropic}\\ \text{SHOs}}} = \sum_i \frac{1}{2m_i}\left(\left(-i\hbar\overline{x}_i + \frac{p_{x_i}}{2}\right)^2 + \left(-i\hbar\overline{y}_i + \frac{p_{y_i}}{2}\right)^2 + \left(-i\hbar\overline{z}_i + \frac{p_{z_i}}{2}\right)^2\right)$$
$$+\sum_{i<j}\begin{pmatrix} a_{x_{ij}}\left(\left(i\hbar\overline{p}_{x_i} + \frac{x_i}{2}\right) - \left(i\hbar\overline{p}_{x_j} + \frac{x_j}{2}\right)\right)^2 \\ +a_{y_{ij}}\left(\left(i\hbar\overline{p}_{y_i} + \frac{y_i}{2}\right) - \left(i\hbar\overline{p}_{y_j} + \frac{y_j}{2}\right)\right)^2 \\ +a_{z_{ij}}\left(\left(i\hbar\overline{p}_{z_i} + \frac{z_i}{2}\right) - \left(i\hbar\overline{p}_{z_j} + \frac{z_j}{2}\right)\right)^2 \end{pmatrix} \tag{12b}$$

$$\hat{H}_{\substack{\text{Schrodinger}\\ \text{Anisotropic}\\ \text{SHOs}}}\begin{pmatrix} -i\hbar\overline{x}_i+\frac{p_{x_i}}{2},-i\hbar\overline{y}_i+\frac{p_{y_i}}{2},-i\hbar\overline{z}_i+\frac{p_{z_i}}{2}; \\ i\hbar\overline{p}_{x_i}+\frac{x_i}{2},i\hbar\overline{p}_{y_i}+\frac{y_i}{2},i\hbar\overline{p}_{z_i}+\frac{z_i}{2} \end{pmatrix}\breve{\Psi}\left(\overline{x}_i,\overline{y}_i,\overline{z}_i;\overline{p}_{x_i},\overline{p}_{y_i},\overline{p}_{z_i};t\right)$$

$$=i\hbar\partial_t\breve{\Psi}\left(\overline{x}_i,\overline{y}_i,\overline{z}_i;\overline{p}_{x_i},\overline{p}_{y_i},\overline{p}_{z_i};t\right)$$

$$\Psi\left(x_i,y_i,z_i;p_{x_i},p_{y_i},p_{z_i};t\right)=$$

$$L^{-1}_{\substack{(\overline{x}_i,\overline{y}_i,\overline{z}_i;\overline{p}_{x_i},\overline{p}_{y_i},\overline{p}_{z_i})\\ \to(x_i,y_i,z_i;p_{x_i},p_{y_i},p_{z_i})}}\left[\begin{array}{l} e^{\frac{-i\int_0^t \hat{H}_{\substack{\text{Schrodinger}\\ \text{Anisotropic}\\ \text{SHOs}}}\begin{pmatrix} -i\hbar\overline{x}_i+\frac{p_{x_i}}{2},-i\hbar\overline{y}_i+\frac{p_{y_i}}{2},-i\hbar\overline{z}_i+\frac{p_{z_i}}{2}; \\ i\hbar\overline{p}_{x_i}+\frac{x_i}{2},i\hbar\overline{p}_{y_i}+\frac{y_i}{2},i\hbar\overline{p}_{z_i}+\frac{z_i}{2}\end{pmatrix}du}{\hbar}} \\ \times \\ \Psi_0\left(\overline{x}_i,\overline{y}_i,\overline{z}_i;\overline{p}_{x_i},\overline{p}_{y_i},\overline{p}_{z_i};t=0\right) \end{array}\right] \tag{12c}$$

Example 3. Schrödinger Molecular Hamiltonian with Pairwise Coulomb Interaction

Now consider the example case of the nonrelativistic Schrödinger Molecular Hamiltonian with Pairwise Coulomb interaction dynamics in quantum phase space [ref.7]. Here the Hamiltonian has the form

$$\hat{H}=\sum_A\frac{1}{2M_A}\hat{\mathbf{P}}_A^2+\sum_i\frac{1}{2m_e}\hat{\mathbf{P}}_i^2+\sum_{A<B}\frac{e^2}{8\pi\varepsilon_0}\left(\frac{Z_AZ_B}{|\hat{r}_A-\hat{r}_B|}\right)+\sum_{i<j}\frac{e^2}{8\pi\varepsilon_0}\left(\frac{1}{|\hat{r}_i-\hat{r}_j|}\right)+\sum_{A,i}\left(\frac{-Z_Ae^2}{|\hat{r}_A-\hat{r}_i|}\right)$$

$$\hat{H}_{\substack{\text{Schrodinger}\\ \text{Molecular}}}\begin{pmatrix} -i\hbar\partial_{x_A}+\frac{p_{x_A}}{2},-i\hbar\partial_{y_A}+\frac{p_{y_A}}{2},-i\hbar\partial_{z_A}+\frac{p_{z_A}}{2},-i\hbar\partial_{x_B}+\frac{p_{x_B}}{2},-i\hbar\partial_{y_B}+\frac{p_{y_B}}{2},-i\hbar\partial_{z_B}+\frac{p_{z_B}}{2}, \\ -i\hbar\partial_{x_i}+\frac{p_{x_i}}{2},-i\hbar\partial_{y_i}+\frac{p_{y_i}}{2},-i\hbar\partial_{z_i}+\frac{p_{z_i}}{2},-i\hbar\partial_{x_j}+\frac{p_{x_j}}{2},-i\hbar\partial_{y_j}+\frac{p_{y_j}}{2},-i\hbar\partial_{z_j}+\frac{p_{z_j}}{2}; \\ i\hbar\partial_{p_{xA}}+\frac{x_A}{2},i\hbar\partial_{p_{yA}}+\frac{y_A}{2},i\hbar\partial_{p_{zA}}+\frac{z_A}{2},i\hbar\partial_{p_{xB}}+\frac{x_B}{2},i\hbar\partial_{p_{yB}}+\frac{y_B}{2},i\hbar\partial_{p_{zB}}+\frac{z_B}{2}, \\ i\hbar\partial_{p_{x_i}}+\frac{x_i}{2},i\hbar\partial_{p_{y_i}}+\frac{y_i}{2},i\hbar\partial_{p_{z_i}}+\frac{z_i}{2},i\hbar\partial_{p_{x_j}}+\frac{x_j}{2},i\hbar\partial_{p_{y_j}}+\frac{y_j}{2},i\hbar\partial_{p_{z_j}}+\frac{z_j}{2} \end{pmatrix} \tag{13a}$$

$$
\begin{aligned}
\hat{H}_{\substack{\text{Schrodinger}\\ \text{Molecular}}} &= \sum_{A} \frac{1}{2M_A}\left(\left(-i\hbar\bar{x}_A + \frac{p_{x_A}}{2}\right)^2 + \left(-i\hbar\bar{y}_A + \frac{p_{y_A}}{2}\right)^2 + \left(-i\hbar\bar{z}_A + \frac{p_{z_A}}{2}\right)^2\right) \\
&+ \sum_{i} \frac{1}{2m_e}\left(\left(-i\hbar\bar{x}_i + \frac{p_{x_i}}{2}\right)^2 + \left(-i\hbar\bar{y}_i + \frac{p_{y_i}}{2}\right)^2 + \left(-i\hbar\bar{z}_i + \frac{p_{z_i}}{2}\right)^2\right) \\
&+ \sum_{A<B} \frac{e^2}{8\pi\varepsilon_0} \frac{Z_A Z_B}{\left|\begin{matrix} \left(\left(i\hbar\bar{p}_{x_A} + \frac{x_A}{2}\right) - \left(i\hbar\bar{p}_{x_B} + \frac{x_B}{2}\right)\right)^2 \\ + \left(\left(i\hbar\bar{p}_{y_A} + \frac{y_A}{2}\right) - \left(i\hbar\bar{p}_{y_B} + \frac{y_B}{2}\right)\right)^2 \\ + \left(\left(i\hbar\bar{p}_{z_A} + \frac{z_A}{2}\right) - \left(i\hbar\bar{p}_{z_B} + \frac{z_B}{2}\right)\right)^2 \end{matrix}\right|} + \sum_{i<j} \frac{e^2}{8\pi\varepsilon_0} \frac{1}{\left|\begin{matrix} \left(\left(i\hbar\bar{p}_{x_i} + \frac{x_i}{2}\right) - \left(i\hbar\bar{p}_{x_j} + \frac{x_j}{2}\right)\right)^2 \\ + \left(\left(i\hbar\bar{p}_{y_i} + \frac{y_i}{2}\right) - \left(i\hbar\bar{p}_{y_j} + \frac{y_j}{2}\right)\right)^2 \\ + \left(\left(i\hbar\bar{p}_{z_i} + \frac{z_i}{2}\right) - \left(i\hbar\bar{p}_{z_j} + \frac{z_j}{2}\right)\right)^2 \end{matrix}\right|} \\
&+ \sum_{A,i} \frac{-Z_A e^2}{\left|\begin{matrix} \left(\left(i\hbar\bar{p}_{x_A} + \frac{x_A}{2}\right) - \left(i\hbar\bar{p}_{x_i} + \frac{x_i}{2}\right)\right)^2 \\ + \left(\left(i\hbar\bar{p}_{y_A} + \frac{y_A}{2}\right) - \left(i\hbar\bar{p}_{y_i} + \frac{y_i}{2}\right)\right)^2 \\ + \left(\left(i\hbar\bar{p}_{z_A} + \frac{z_A}{2}\right) - \left(i\hbar\bar{p}_{z_i} + \frac{z_i}{2}\right)\right)^2 \end{matrix}\right|}
\end{aligned}
\tag{13b}
$$

Hence the quantum phase space Hamiltonian is specified by the Ansatz and results in the quantum phase space wave function for a given initial condition via (10) as

$$
\boldsymbol{\Psi}\begin{pmatrix} x_A, y_A, z_A, x_B, y_B, z_B, \\ x_i, y_i, z_i, x_j, y_j, z_j; \\ p_{x_A}, p_{y_A}, p_{z_A}, p_{x_B}, p_{y_B}, p_{z_B}, \\ p_{x_i}, p_{y_i}, p_{z_i}, p_{x_j}, p_{y_j}, p_{z_j}; t \end{pmatrix} =
$$

$$
L^{-1}_{\left(\begin{pmatrix} \bar{x}_A, \bar{y}_A, \bar{z}_A, \bar{x}_B, \bar{y}_B, \bar{z}_B, \\ \bar{x}_i, \bar{y}_i, \bar{z}_i, \bar{x}_j, \bar{y}_j, \bar{z}_j; \\ p_{x_A}, p_{y_A}, p_{z_A}, p_{x_B}, p_{y_B}, p_{z_B}, \\ p_{x_i}, p_{y_i}, p_{z_i}, p_{x_j}, p_{y_j}, p_{z_j} \end{pmatrix} \to \begin{pmatrix} x_A, y_A, z_A, x_B, y_B, z_B, \\ x_i, y_i, z_i, x_j, y_j, z_j; \\ p_{x_A}, p_{y_A}, p_{z_A}, p_{x_B}, p_{y_B}, p_{z_B}, \\ p_{x_i}, p_{y_i}, p_{z_i}, p_{x_j}, p_{y_j}, p_{z_j} \end{pmatrix}\right)}
\left[\begin{matrix} e^{\frac{-i}{\hbar}\int_0^t \hat{H}_{\substack{\text{Schrodinger}\\ \text{Molecular}}}\left(\begin{matrix} -i\hbar\bar{x}_A + \frac{p_{x_A}}{2}, -i\hbar\bar{y}_A + \frac{p_{y_A}}{2}, -i\hbar\bar{z}_A + \frac{p_{z_A}}{2}, -i\hbar\bar{x}_i + \frac{p_{x_i}}{2}, -i\hbar\bar{y}_i + \frac{p_{y_i}}{2}, -i\hbar\bar{z}_i + \frac{p_{z_i}}{2}; \\ i\hbar\bar{p}_{x_A} + \frac{x_A}{2}, i\hbar\bar{p}_{y_A} + \frac{y_A}{2}, i\hbar\bar{p}_{z_A} + \frac{z_A}{2}, i\hbar\bar{p}_{x_B} + \frac{x_B}{2}, i\hbar\bar{p}_{y_B} + \frac{y_B}{2}, i\hbar\bar{p}_{z_B} + \frac{z_B}{2}, \\ i\hbar\bar{p}_{x_i} + \frac{x_i}{2}, i\hbar\bar{p}_{y_i} + \frac{y_i}{2}, i\hbar\bar{p}_{z_i} + \frac{z_i}{2}, i\hbar\bar{p}_{x_j} + \frac{x_j}{2}, i\hbar\bar{p}_{y_j} + \frac{y_j}{2}, i\hbar\bar{p}_{z_j} + \frac{z_j}{2} \end{matrix}\right) du} \\ \times \\ \breve{\boldsymbol{\Psi}}_0(\bar{x}_1, \ldots, \bar{x}_n; \bar{p}_1, \ldots, \bar{p}_n; t = 0) \end{matrix}\right]
\tag{13c}
$$

Where the Schrödinger Molecular Hamiltonian is expressed as

$$\hat{H}_{\substack{\text{Schrodinger}\\ \text{Molecular}}}\begin{pmatrix} -i\hbar\bar{x}_A+\frac{p_{x_A}}{2},-i\hbar\bar{y}_A+\frac{p_{y_A}}{2},-i\hbar\bar{z}_A+\frac{p_{z_A}}{2},-i\hbar\bar{x}_B+\frac{p_{x_B}}{2},-i\hbar\bar{y}_B+\frac{p_{y_B}}{2},-i\hbar\bar{z}_B+\frac{p_{z_B}}{2}, \\ -i\hbar\bar{x}_i+\frac{p_{x_i}}{2},-i\hbar\bar{y}_i+\frac{p_{y_i}}{2},-i\hbar\bar{z}_i+\frac{p_{z_i}}{2},-i\hbar\bar{x}_j+\frac{p_{x_j}}{2},-i\hbar\bar{y}_j+\frac{p_{y_j}}{2},-i\hbar\bar{z}_j+\frac{p_{z_j}}{2}; \\ i\hbar\bar{p}_{x_A}+\frac{x_A}{2},i\hbar\bar{p}_{y_A}+\frac{y_A}{2},i\hbar\bar{p}_{z_A}+\frac{z_A}{2},i\hbar\bar{p}_{x_B}+\frac{x_B}{2},i\hbar\bar{p}_{y_B}+\frac{y_B}{2},,i\hbar\bar{p}_{z_B}+\frac{z_B}{2}, \\ i\hbar\bar{p}_{x_i}+\frac{x_i}{2},i\hbar\bar{p}_{y_i}+\frac{y_i}{2},i\hbar\bar{p}_{z_i}+\frac{z_i}{2},i\hbar\bar{p}_{x_j}+\frac{x_j}{2},i\hbar\bar{p}_{y_j}+\frac{y_j}{2},i\hbar\bar{p}_{z_j}+\frac{z_j}{2} \end{pmatrix}$$

Following from [ref.7] for the radial component of the Molecular Schrodinger equation in Phase Space, the HOA Solution of the Radial Schrödinger Equation for Nonrelativistiic N-particle system with Pairwise $\frac{1}{r_{ij}}$ Radial Potential Interaction where $\frac{1}{r_{ij}} \equiv \frac{1}{|r_i - r_j|}$ in the QPSR. Moreover, the analysis for this multiparticle system is in the Laboratory Reference Frame for the Coordinates[ref.7]. Hence no particle is fixed as a center of motion. As stated earlier the angular components separate in the usual way by [refs.7]. Here a_i, b_i, c_i are scaling constants, $L_i(L_i+1)$ are the angular coupling terms and

$$\mathbf{r} = (r_1,\ldots,r_N) = (r_{i=1,\ldots,N}),\ \mathbf{p_r} = (p_{r_1},\ldots,p_{r_N}) = (p_{r_{i=1,\ldots,N}}).$$

The resulting radial Hamiltonian yields a radial Schrodinger equation of form

$$\begin{pmatrix} \hat{H}=\sum_i \left(\frac{\left(-i\hbar\partial_{r_i}+\frac{p_{r_i}}{2}\right)^2}{2m_i} + \frac{2b_i}{\left(i\hbar\partial_{p_{r_i}}+\frac{r_i}{2}\right)}\left(-i\hbar\partial_{r_i}+\frac{p_{r_i}}{2}\right) + \frac{c_i L_i(L_i+1)}{\left(i\hbar\partial_{p_{r_i}}+\frac{r_i}{2}\right)^2} \right) \\ +\sum_{i<j} a_{ij}\left(\frac{1}{\left|\left(i\hbar\partial_{p_{r_i}}+\frac{r_i}{2}\right)-\left(i\hbar\partial_{p_{r_j}}+\frac{r_j}{2}\right)\right|} \right) \end{pmatrix} \Psi_{\mathbf{r}\,\text{phasespace}}(\mathbf{r};\mathbf{p_r};t) \tag{13d}$$

$$= i\hbar\partial_t \Psi_{\mathbf{r}\,\text{phasespace}}(\mathbf{r};\mathbf{p_r};t)$$

Hence via HOA from equation (10),

$$\left(\begin{array}{l}\sum_i\left(\frac{\left(-i\hbar\bar{r}_i+\frac{p_{r_i}}{2}\right)^2}{2m_i}+\frac{2b_i}{\left(i\hbar\bar{p}_{r_i}+\frac{r_i}{2}\right)}\left(-i\hbar\bar{r}_i+\frac{p_{r_i}}{2}\right)+\frac{c_iL_i(L_i+1)}{\left(i\hbar\bar{p}_{r_i}+\frac{r_i}{2}\right)^2}\right)\\+\sum_{i<j}a_{ij}\left(\frac{1}{\left|\left(i\hbar\bar{p}_{r_i}+\frac{r_i}{2}\right)-\left(i\hbar\bar{p}_{r_j}+\frac{r_j}{2}\right)\right|}\right)\end{array}\right)\breve{\Psi}_{\mathbf{r}\,\text{phasespace}}(\bar{\mathbf{r}};\bar{\mathbf{p}}_{\mathbf{r}};t) \tag{13e}$$

$$= i\hbar\partial_t\breve{\Psi}_{\mathbf{r}\,\text{phasespace}}(\bar{\mathbf{r}};\bar{\mathbf{p}}_{\mathbf{r}};t)$$

Yielding the exact analytical solution given in quadratures as

$$\Psi_{\mathbf{r}\,\text{phasespace}}(\mathbf{r};\mathbf{p}_{\mathbf{r}};t)=L^{-1}_{\left(\begin{array}{l}\bar{\mathbf{r}}\to\mathbf{r};\\\bar{\mathbf{p}}_{\mathbf{r}}\to\mathbf{p}_{\mathbf{r}}\end{array}\right)}\left(\begin{array}{l}e^{\frac{-it}{\hbar}\left(\begin{array}{l}\sum_i\left(\frac{\left(-i\hbar\bar{r}_i+\frac{p_{r_i}}{2}\right)^2}{2m_i}+\frac{2b_i}{\left(i\hbar\bar{p}_{r_i}+\frac{r_i}{2}\right)}\left(-i\hbar\bar{r}_i+\frac{p_{r_i}}{2}\right)+\frac{c_iL_i(L_i+1)}{\left(i\hbar\bar{p}_{r_i}+\frac{r_i}{2}\right)^2}\right)\\+\sum_{i<j}a_{ij}\left(\frac{1}{\left|\left(i\hbar\bar{p}_{r_i}+\frac{r_i}{2}\right)-\left(i\hbar\bar{p}_{r_j}+\frac{r_j}{2}\right)\right|}\right)\end{array}\right)}\\\times\breve{\Psi}_{0\mathbf{r}\,\text{phasespace}}(\bar{\mathbf{r}};\bar{\mathbf{p}}_{\mathbf{r}};t=0)\end{array}\right) \tag{13f}$$

To evaluate (13f), recall that

$$\breve{\Psi}_{\mathbf{r}\,\text{phasespace}}(\bar{\mathbf{r}};\bar{\mathbf{p}}_{\mathbf{r}};t)=$$

$$e^{\frac{-it}{\hbar}\left(\sum_i\left(\frac{\left(-i\hbar\bar{r}_i+\frac{p_{r_i}}{2}\right)^2}{2m_i}\right)\right)}\times e^{\frac{-it}{\hbar}\left(\sum_i\left(\frac{2b_i}{\left(i\hbar\bar{p}_{r_i}+\frac{r_i}{2}\right)}\left(-i\hbar\bar{r}_i+\frac{p_{r_i}}{2}\right)\right)\right)}$$

$$\times e^{\frac{-it}{\hbar}\left(\sum_i\left(\frac{c_iL_i(L_i+1)}{\left(i\hbar\bar{p}_{r_i}+\frac{r_i}{2}\right)^2}\right)\right)}\times e^{\frac{-it}{\hbar}\left(\sum_{i<j}a_{ij}\left(\frac{1}{\left|\left(i\hbar\bar{p}_{r_i}+\frac{r_i}{2}\right)-\left(i\hbar\bar{p}_{r_j}+\frac{r_j}{2}\right)\right|}\right)\right)}\times\breve{\Psi}_{0\mathbf{r}\,\text{phasespace}}(\bar{\mathbf{r}};\bar{\mathbf{p}}_{\mathbf{r}};t=0) \tag{13g}$$

Upon evaluating the inverse Laplace transforms and utilizing convolution products herein symbolized as $\underbrace{*}_{\mathbf{r}}$ or $\underbrace{*}_{\mathbf{p_r}}$ or $\underbrace{*}_{\mathbf{r},\mathbf{p_r}}$ [ref 6].

$$L^{-1}_{\left(\substack{\bar{\mathbf{r}}\to\mathbf{r};\\ \bar{\mathbf{p}}_{\mathbf{r}}\to\mathbf{p}_{\mathbf{r}}}\right)}\breve{\Psi}_{\mathbf{r}\,\text{phasespace}}\left(\bar{\mathbf{r}};\bar{\mathbf{p}}_{\mathbf{r}};t\right)=$$

$$L^{-1}_{(\bar{\mathbf{r}}\to\mathbf{r})}\left(e^{\frac{-it}{\hbar}\left(\sum_i\left(\frac{\left(-i\hbar\bar{r}_i+\frac{p_{r_i}}{2}\right)^2}{2m_i}\right)\right)}\right)\underbrace{*}_{\mathbf{r}}L^{-1}_{\left(\substack{\bar{\mathbf{r}}\to\mathbf{r};\\ \bar{\mathbf{p}}_{\mathbf{r}}\to\mathbf{p}_{\mathbf{r}}}\right)}\left(e^{\frac{-it}{\hbar}\left(\sum_i\left(\frac{2b_i}{\left(i\hbar\bar{p}_{r_i}+\frac{r_i}{2}\right)}\left(-i\hbar\bar{r}_i+\frac{p_{r_i}}{2}\right)\right)\right)}\right) \tag{13h}$$

$$\underbrace{*}_{\mathbf{p_r}}L^{-1}_{(\bar{\mathbf{p}}_{\mathbf{r}}\to\mathbf{p}_{\mathbf{r}})}\left(e^{\frac{-it}{\hbar}\left(\sum_i\left(\frac{c_iL_i(L_i+1)}{\left(i\hbar\bar{p}_{r_i}+\frac{r_i}{2}\right)^2}\right)\right)}\right)\underbrace{*}_{\mathbf{p_r}}L^{-1}_{(\bar{\mathbf{p}}_{\mathbf{r}}\to\mathbf{p}_{\mathbf{r}})}\left(e^{\frac{-it}{\hbar}\left(\sum_{i<j}a_{ij}\left(\frac{1}{\left(i\hbar\bar{p}_{r_i}+\frac{r_i}{2}\right)-\left(i\hbar\bar{p}_{r_j}+\frac{r_j}{2}\right)}\right)\right)}\right)$$

$$\underbrace{*}_{\mathbf{r},\mathbf{p_r}}L^{-1}_{\left(\substack{\bar{\mathbf{r}}\to\mathbf{r};\\ \bar{\mathbf{p}}_{\mathbf{r}}\to\mathbf{p}_{\mathbf{r}}}\right)}\left(\breve{\Psi}_{0\mathbf{r}\,\text{phasespace}}\left(\bar{\mathbf{r}};\bar{\mathbf{p}}_{\mathbf{r}};t=0\right)\right)$$

Evaluating (13h) explicitly yields

$$L^{-1}_{(\bar{\mathbf{p}}_{\mathbf{r}}\to\mathbf{p}_{\mathbf{r}})}\left(e^{\frac{-it}{\hbar}\left(\sum_{i<j}a_{ij}\left(\frac{1}{\left(i\hbar\bar{p}_{r_i}+\frac{r_i}{2}\right)-\left(i\hbar\bar{p}_{r_j}+\frac{r_j}{2}\right)}\right)\right)}\right)$$

$$=\underbrace{*}_{i<j}\left(\delta\left[p_{r_i}+p_{r_j}\right]\left(\frac{1}{\pi p_{r_i}\sqrt{\frac{-i\hbar}{a_{ij}t}}}e^{\frac{ip_{r_i}(r_i-r_j)}{2\hbar}}\left(\sqrt{\frac{p_{r_i}}{\hbar}}K_1\left[\frac{2\sqrt{\frac{p_{r_i}}{\hbar}}}{\sqrt{\frac{-i\hbar}{a_{ij}t}}}\right]-\sqrt{\frac{-p_{r_i}}{\hbar}}K_1\left[\frac{2\sqrt{\frac{-p_{r_i}}{\hbar}}}{\sqrt{\frac{-i\hbar}{a_{ij}t}}}\right]\right)\right)\right) \tag{13i}$$

example N = 3, $f_{1,2}\underbrace{*}_{p_{r_1}}f_{1,3}\underbrace{*}_{p_{r_2},p_{r_3}}f_{2,3}$

$$,\ f_{ij}=\delta\left[p_{r_i}+p_{r_j}\right]\left(\frac{1}{\pi p_{r_i}\sqrt{\frac{-i\hbar}{a_{ij}t}}}e^{\frac{ip_{r_i}(r_i-r_j)}{2\hbar}}\left(\sqrt{\frac{p_{r_i}}{\hbar}}K_1\left[\frac{2\sqrt{\frac{p_{r_i}}{\hbar}}}{\sqrt{\frac{-i\hbar}{a_{ij}t}}}\right]-\sqrt{\frac{-p_{r_i}}{\hbar}}K_1\left[\frac{2\sqrt{\frac{-p_{r_i}}{\hbar}}}{\sqrt{\frac{-i\hbar}{a_{ij}t}}}\right]\right)\right)$$

Where $\underset{i<j}{\ast}$ represents the convolution between the inverse transforms of the indicated terms as illustrated above in the case of N=3.

$$L^{-1}_{\left(\substack{\bar{\mathbf{r}}\to\mathbf{r}; \\ \bar{\mathbf{p}}_{\mathbf{r}}\to\mathbf{p}_{\mathbf{r}}}\right)}\left(e^{\frac{-it}{\hbar}\left(\sum_i\left(\frac{2b_i}{\left(i\hbar\bar{p}_{r_i}+\frac{r_i}{2}\right)}\left(-i\hbar\bar{r}_i+\frac{p_{r_i}}{2}\right)\right)\right)}\right)$$
$$=\prod_i\left(\delta\left[p_{r_i}\right]\delta\left[r_i\right]+\frac{ib_i e^{\frac{2ib_i p_{r_i}t}{\hbar r_i}}\left(\sqrt{-ip_{r_i}}-\sqrt{ip_{r_i}}\right)}{\hbar\pi\ r_i^2\sqrt{\left|p_{r_i}\right|}}\right) \tag{13j}$$

$$L^{-1}_{(\bar{\mathbf{p}}_{\mathbf{r}}\to\mathbf{p}_{\mathbf{r}})}\left(e^{\frac{-it}{\hbar}\left(\sum_i\left(\frac{c_iL_i(L_i+1)}{\left(i\hbar\bar{p}_{r_i}+\frac{r_i}{2}\right)^2}\right)\right)}\right)$$
$$=\prod_i\left(\frac{-1}{24\pi\hbar^6}\left(c_i^2L_i^2\left(L_i+1\right)^2p_{r_i}^3t^2\left(\begin{array}{l}\left(e^{\frac{ip_{r_i}r_i}{2\hbar}}-1\right)Log\left[\frac{-p_{r_i}}{\hbar}\right] \\ +Log\left[\frac{p_{r_i}}{\hbar}\right]-Log\left[\frac{p_{r_i}}{\hbar}\right]HeavisideStep\left[p_{r_i}\right]\end{array}\right)\right)\right) \tag{13k}$$

$$L^{-1}_{(\bar{\mathbf{r}}\to\mathbf{r})}\left(e^{\frac{-it}{\hbar}\left(\sum_i\left(\frac{\left(-i\hbar\bar{r}_i+\frac{p_{r_i}}{2}\right)^2}{2m_i}\right)\right)}\right)=\prod_i\frac{e^{\frac{i\left(r_i^2m_i-tr_ip_{r_i}\right)}{2\hbar t}}}{\hbar\sqrt{\frac{2\pi it}{\hbar m_i}}} \tag{13l}$$

Since $L^{-1}_{\left(\substack{\bar{\mathbf{r}}\to\mathbf{r};\\ \bar{\mathbf{p}}_{\mathbf{r}}\to\mathbf{p}_{\mathbf{r}}}\right)}\left(\bar{\Psi}_{0\mathrm{r\ phasespace}}\left(\bar{\mathbf{r}};\bar{\mathbf{p}}_{\mathbf{r}};t=0\right)\right)=\Psi_{0\mathrm{r\ phasespace}}\left(\mathbf{r};\mathbf{p}_{\mathbf{r}};t=0\right)$, the evaluated inverse transforms yield the explicit solution in the QPSR

$$\Psi_{\mathbf{r}\ \mathrm{phasespace}}\left(\mathbf{r};\mathbf{p}_{\mathbf{r}};t\right)=$$

$$\Psi_{0\mathrm{r\ phasespace}}\left(\mathbf{r};\mathbf{p}_{\mathbf{r}};t=0\right)\underbrace{*}_{\mathbf{r},\mathbf{p}_{\mathbf{r}}}\prod_i\frac{e^{\frac{i\left(r_i^2 m_i-t r_i p_{r_i}\right)}{2\hbar t}}}{\hbar\sqrt{\frac{2\pi i t}{\hbar m_i}}}\underbrace{*}_{\mathbf{r}}\prod_i\left(\delta\left[p_{r_i}\right]\delta\left[r_i\right]+\frac{i b_i e^{\frac{2 i b_i p_{r_i} t}{\hbar r_i}}\left(\sqrt{-i p_{r_i}}-\sqrt{i p_{r_i}}\right)}{\hbar\pi\, r_i^2\sqrt{\left|p_{r_i}\right|}}\right)$$

$$\underbrace{*}_{\mathbf{p}_{\mathbf{r}}}\prod_i\left(\frac{-1}{24\pi\hbar^6}\left(c_i^2 L_i^2\left(L_i+1\right)^2 p_{r_i}^3 t^2\left(\begin{array}{l}\left(e^{\frac{i p_{r_i} r_i}{2\hbar}}-1\right)Log\left[\frac{-p_{r_i}}{\hbar}\right]\\+Log\left[\frac{p_{r_i}}{\hbar}\right]-Log\left[\frac{p_{r_i}}{\hbar}\right]HeavisideStep\left[p_{r_i}\right]\end{array}\right)\right)\right)$$

$$\underbrace{*}_{\mathbf{p}_{\mathbf{r}}}\left(\underbrace{*}_{i<j}\left(\delta\left[p_{r_i}+p_{r_j}\right]\frac{1}{\pi p_{r_i}\sqrt{\frac{-i\hbar}{a_{ij}t}}}e^{\frac{i p_{r_i}\left(r_i-r_j\right)}{2\hbar}}\left(\sqrt{\frac{p_{r_i}}{\hbar}}K_1\left[\frac{2\sqrt{\frac{p_{r_i}}{\hbar}}}{\sqrt{\frac{-i\hbar}{a_{ij}t}}}\right]-\sqrt{\frac{-p_{r_i}}{\hbar}}K_1\left[\frac{2\sqrt{\frac{-p_{r_i}}{\hbar}}}{\sqrt{\frac{-i\hbar}{a_{ij}t}}}\right]\right)\right)\right)\tag{13m}$$

Although the above (13m) is explicitly given in the QPSR[down to the level of convolutions of explicitly calculated quantum phase space factors], an illustrative example will now be provided to help illuminate the actual application of the result for the case of three particles: N=3

Now for the solution of the radial Schrodinger equation for the non-relativistiic 3-particle system with pairwise $\frac{1}{r_{ij}}$ radial potential interaction where $\frac{1}{r_{ij}}\equiv\frac{1}{\left|r_i-r_j\right|}$ in the QPSR. Moreover, the analysis for this multiparticle system is in the Laboratory Reference Frame for the Coordinates[ref.7]. Hence no particle is fixed as a center of motion. As stated earlier the angular components separate in the usual way by [refs.7]. Here a_i, b_i, c_i are scaling constants, $L_i(L_i+1)$ are the angular coupling terms and

$$\mathbf{r}=\left(r_1,r_2,r_3\right)=\left(r_{i=1,2,3}\right),\ \mathbf{p}_{\mathbf{r}}=\left(p_{r_1},p_{r_2},p_{r_3}\right)=\left(p_{r_{i=1,2,3}}\right).$$

From eq(13d), the resulting radial Hamiltonian yields a radial Schrodinger equation of form

$$\left(\begin{array}{l}\dfrac{\left(-i\hbar\partial_{r_1}+\dfrac{p_{r_1}}{2}\right)^2}{2m_1}+\dfrac{\left(-i\hbar\partial_{r_2}+\dfrac{p_{r_2}}{2}\right)^2}{2m_2}+\dfrac{\left(-i\hbar\partial_{r_3}+\dfrac{p_{r_3}}{2}\right)^2}{2m_3}\\ +\dfrac{2b_1}{\left(i\hbar\partial_{p_{r_1}}+\dfrac{r_1}{2}\right)}\left(-i\hbar\partial_{r_1}+\dfrac{p_{r_1}}{2}\right)+\dfrac{2b_2}{\left(i\hbar\partial_{p_{r_2}}+\dfrac{r_2}{2}\right)}\left(-i\hbar\partial_{r_2}+\dfrac{p_{r_2}}{2}\right)\\ +\dfrac{2b_3}{\left(i\hbar\partial_{p_{r_3}}-\dfrac{r_3}{2}\right)}\left(-i\hbar\partial_{r_3}+\dfrac{p_{r_3}}{2}\right)+\dfrac{c_1L_1(L_1+1)}{\left(i\hbar\partial_{p_{r_1}}+\dfrac{r_1}{2}\right)^2}+\dfrac{c_2L_2(L_2+1)}{\left(i\hbar\partial_{p_{r_2}}+\dfrac{r_2}{2}\right)^2}\\ +\dfrac{c_3L_3(L_3+1)}{\left(i\hbar\partial_{p_{r_3}}+\dfrac{r_3}{2}\right)^2}+a_{12}\left(\dfrac{1}{\left|\left(i\hbar\partial_{p_{r_1}}+\dfrac{r_1}{2}\right)-\left(i\hbar\partial_{p_{r_2}}+\dfrac{r_2}{2}\right)\right|}\right)\\ +a_{13}\left(\dfrac{1}{\left|\left(i\hbar\partial_{p_{r_1}}+\dfrac{r_1}{2}\right)-\left(i\hbar\partial_{p_{r_3}}+\dfrac{r_3}{2}\right)\right|}\right)+a_{23}\left(\dfrac{1}{\left|\left(i\hbar\partial_{p_{r_2}}+\dfrac{r_2}{2}\right)-\left(i\hbar\partial_{p_{r_3}}+\dfrac{r_3}{2}\right)\right|}\right)\end{array}\right)\Psi_{\text{r phasespace}}\left(r_1,r_2,r_3;p_{r_1},p_{r_2},p_{r_3};t\right)$$
$$=i\hbar\partial_t\Psi_{\text{r phasespace}}\left(r_1,r_2,r_3;p_{r_1},p_{r_2},p_{r_3};t\right) \tag{13n}$$

Following eq(13e)

$$\left(\begin{array}{l}\dfrac{\left(-i\hbar\bar{r}_1+\dfrac{p_{r_1}}{2}\right)^2}{2m_1}+\dfrac{\left(-i\hbar\bar{r}_2+\dfrac{p_{r_2}}{2}\right)^2}{2m_2}+\dfrac{\left(-i\hbar\bar{r}_3+\dfrac{p_{r_3}}{2}\right)^2}{2m_3}\\ +\dfrac{2b_1}{\left(i\hbar\bar{p}_{r_1}+\dfrac{r_1}{2}\right)}\left(-i\hbar\bar{r}_1+\dfrac{p_{r_1}}{2}\right)+\dfrac{2b_2}{\left(i\hbar\bar{p}_{r_2}+\dfrac{r_2}{2}\right)}\left(-i\hbar\bar{r}_2+\dfrac{p_{r_2}}{2}\right)\\ +\dfrac{2b_3}{\left(i\hbar\bar{p}_{r_3}+\dfrac{r_3}{2}\right)}\left(-i\hbar\bar{r}_3+\dfrac{p_{r_3}}{2}\right)+\dfrac{c_1L_1(L_1+1)}{\left(i\hbar\bar{p}_{r_1}+\dfrac{r_1}{2}\right)^2}+\dfrac{c_2L_2(L_2+1)}{\left(i\hbar\bar{p}_{r_2}+\dfrac{r_2}{2}\right)^2}\\ +\dfrac{c_3L_3(L_3+1)}{\left(i\hbar\bar{p}_{r_3}+\dfrac{r_3}{2}\right)^2}+a_{12}\left(\dfrac{1}{\left|\left(i\hbar\bar{p}_{r_1}+\dfrac{r_1}{2}\right)-\left(i\hbar\bar{p}_{r_2}+\dfrac{r_2}{2}\right)\right|}\right)\\ +a_{13}\left(\dfrac{1}{\left|\left(i\hbar\bar{p}_{r_1}+\dfrac{r_1}{2}\right)-\left(i\hbar\bar{p}_{r_3}+\dfrac{r_3}{2}\right)\right|}\right)+a_{23}\left(\dfrac{1}{\left|\left(i\hbar\bar{p}_{r_2}+\dfrac{r_2}{2}\right)-\left(i\hbar\bar{p}_{r_3}+\dfrac{r_3}{2}\right)\right|}\right)\end{array}\right)\breve{\Psi}_{\text{r phasespace}}\left(\bar{r}_1,\bar{r}_2,\bar{r}_3;\bar{p}_{r_1},\bar{p}_{r_2},\bar{p}_{r_3};t\right)$$
$$=i\hbar\partial_t\breve{\Psi}_{\text{r phasespace}}\left(\bar{r}_1,\bar{r}_2,\bar{r}_3;\bar{p}_{r_1},\bar{p}_{r_2},\bar{p}_{r_3};t\right) \tag{13o}$$

Proceeding as in eq(13f) through eq(13h); for N=3 eqs (13a), (13b), (13c) and (13d) become respectively

$$L^{-1}_{(\bar{\mathbf{p}}_{\mathbf{r}} \to \mathbf{p}_{\mathbf{r}})}\left(e^{\frac{-it}{\hbar}\left(\sum_{i<j} a_{ij}\left(\frac{1}{\left|\left(i\hbar\bar{p}_{r_i}+\frac{r_i}{2}\right)-\left(i\hbar\bar{p}_{r_j}+\frac{r_j}{2}\right)\right|}\right)\right)} \right)$$

$$= \mathop{*}_{i<j}\left(\delta\left[p_{r_i}+p_{r_j}\right]\left(\frac{1}{\pi p_{r_i}\sqrt{\frac{-i\hbar}{a_{ij}t}}} e^{\frac{ip_{r_i}(r_i-r_j)}{2\hbar}}\left(\sqrt{\frac{p_{r_i}}{\hbar}}K_1\left[\frac{2\sqrt{\frac{p_{r_i}}{\hbar}}}{\sqrt{\frac{-i\hbar}{a_{ij}t}}}\right]-\sqrt{\frac{-p_{r_i}}{\hbar}}K_1\left[\frac{2\sqrt{\frac{-p_{r_i}}{\hbar}}}{\sqrt{\frac{-i\hbar}{a_{ij}t}}}\right]\right)\right)\right) \quad (13p)$$

example N = 3, $f_{1,2} \underbrace{*}_{p_{r_1}} f_{1,3} \underbrace{*}_{p_{r_2},p_{r_3}} f_{2,3}$

$$, \ f_{ij} = \delta\left[p_{r_i}+p_{r_j}\right]\left(\frac{1}{\pi p_{r_i}\sqrt{\frac{-i\hbar}{a_{ij}t}}} e^{\frac{ip_{r_i}(r_i-r_j)}{2\hbar}}\left(\sqrt{\frac{p_{r_i}}{\hbar}}K_1\left[\frac{2\sqrt{\frac{p_{r_i}}{\hbar}}}{\sqrt{\frac{-i\hbar}{a_{ij}t}}}\right]-\sqrt{\frac{-p_{r_i}}{\hbar}}K_1\left[\frac{2\sqrt{\frac{-p_{r_i}}{\hbar}}}{\sqrt{\frac{-i\hbar}{a_{ij}t}}}\right]\right)\right)$$

$$\Psi_{\mathbf{r}\,\text{phasespace}}\left(r_1,r_2,r_3;p_{r_1},p_{r_2},p_{r_3};t\right)=$$

$$\Psi_{\mathbf{r}\,\text{phasespace}}\left(r_1,r_2,r_3;p_{r_1},p_{r_2},p_{r_3};t=0\right)\underbrace{*}_{\mathbf{r},\mathbf{p}_{\mathbf{r}}}\prod_{i=1}^{3}\left(\frac{e^{\frac{i\left(r_1^2 m_1 - m_1 p_{r_1}\right)}{2\hbar t}}}{\hbar\sqrt{\frac{2\pi i t}{\hbar m_1}}}\right)\underbrace{*}_{\mathbf{r}}\prod_{i=1}^{3}\left(\delta\left[p_{r_i}\right]\delta\left[r_i\right]+\frac{ib_i e^{\frac{2ib_1 p_{r_i} t}{\hbar r_i}}\left(\sqrt{-ip_{r_i}}-\sqrt{ip_{r_i}}\right)}{\hbar\pi\ r_i^2\sqrt{\left|p_{r_i}\right|}}\right)$$

$$\underbrace{*}_{\mathbf{p}_{\mathbf{r}}}\prod_{i=1}^{3}\left(\frac{-1}{24\pi\hbar^6}\left(c_i^2 L_i^2\left(L_i+1\right)^2 p_{r_i}^3 t^2\left(\left(e^{\frac{ip_{r_i}r_i}{2\hbar}}-1\right)Log\left[\frac{-p_{r_i}}{\hbar}\right]+Log\left[\frac{p_{r_i}}{\hbar}\right]-Log\left[\frac{p_{r_i}}{\hbar}\right]HeavisideStep\left[p_{r_i}\right]\right)\right)\right)$$

$$\underbrace{*}_{\mathbf{p}_{\mathbf{r}}}\left(\mathop{*}_{i<j}^{3}\left(\delta\left[p_{r_i}+p_{r_j}\right]\left(\frac{1}{\pi p_{r_i}\sqrt{\frac{-i\hbar}{a_{ij}t}}} e^{\frac{ip_{r_i}(r_i-r_j)}{2\hbar}}\left(\sqrt{\frac{p_{r_i}}{\hbar}}K_1\left[\frac{2\sqrt{\frac{p_{r_i}}{\hbar}}}{\sqrt{\frac{-i\hbar}{a_{ij}t}}}\right]-\sqrt{\frac{-p_{r_i}}{\hbar}}K_1\left[\frac{2\sqrt{\frac{-p_{r_i}}{\hbar}}}{\sqrt{\frac{-i\hbar}{a_{ij}t}}}\right]\right)\right)\right)\right)$$

(13q)

Now consider the earlier mentioned particular case of the general nonrelativistic Schrödinger Molecular Hamiltonian with Pairwise Coulomb interaction dynamics

$$\hat{H} = \sum_A \frac{1}{2M_A}\hat{\mathbf{P}}_A^2 + \sum_i \frac{1}{2m_e}\hat{\mathbf{P}}_i^2 + \sum_{A<B}\frac{e^2}{8\pi\varepsilon_0}\left(\frac{Z_A Z_B}{|\hat{r}_A - \hat{r}_B|}\right) + \sum_{i<j}\frac{e^2}{8\pi\varepsilon_0}\left(\frac{1}{|\hat{r}_i - \hat{r}_j|}\right) + \sum_{A,i}\left(\frac{-Z_A e^2}{|\hat{r}_A - \hat{r}_i|}\right) \tag{13a}$$

Though we shall not pursue the details in the present note [viz ref. 7], clearly one can see that, after the angular components are separated in the usual way and with a straightforward scaling of the variables, the present N-particle result yields the solution of the radial Schrodinger equation for the general nonrelativistic molecular Hamiltonian in the QPSR which has the form

$$\begin{aligned}
\hat{H}_{\substack{\text{Schrodinger}\\ \text{Molecular Radial}}} &= \sum_A \frac{1}{2M_A}\left(\left(-i\hbar\partial_{r_A} + \frac{p_{r_A}}{2}\right)^2 + \frac{2}{\left(i\hbar\partial_{p_{r_A}} + \frac{r_A}{2}\right)}\left(-i\hbar\partial_{r_A} + \frac{p_{r_A}}{2}\right) + \frac{L_A(L_A+1)}{\left(i\hbar\partial_{p_{r_A}} + \frac{r_A}{2}\right)^2}\right) \\
&+ \sum_i \frac{1}{2m_e}\left(\left(-i\hbar\partial_{r_i} + \frac{p_{r_i}}{2}\right)^2 + \frac{2}{\left(i\hbar\partial_{p_{r_i}} + \frac{r_i}{2}\right)}\left(-i\hbar\partial_{r_i} + \frac{p_{r_i}}{2}\right) + \frac{L_i(L_i+1)}{\left(i\hbar\partial_{p_{r_i}} + \frac{r_i}{2}\right)^2}\right) \\
&+ \sum_{A<B}\frac{e^2}{8\pi\varepsilon_0}\frac{Z_A Z_B}{\left|\left(i\hbar\partial_{p_{r_A}} + \frac{r_A}{2}\right) - \left(i\hbar\partial_{p_{r_B}} + \frac{r_B}{2}\right)\right|} + \sum_{i<j}\frac{e^2}{8\pi\varepsilon_0}\frac{1}{\left|\left(i\hbar\partial_{p_{r_i}} + \frac{r_i}{2}\right) - \left(i\hbar\partial_{p_{r_j}} + \frac{r_j}{2}\right)\right|} \\
&+ \sum_{A,i}\frac{-Z_A e^2}{\left|\left(i\hbar\partial_{p_{r_A}} + \frac{r_A}{2}\right) - \left(i\hbar\partial_{p_{r_i}} + \frac{r_i}{2}\right)\right|}
\end{aligned} \tag{13r}$$

Hence considering the radial molecular Hamiltonian (13r) as was done in (13a) and applying HOA to (13r) as was done in (13b), the exact analytical solution of the Radial Schrodinger equation for the general nonrelativistic molecular Hamiltonian in QPSR may be expressed in quadratures via the above.

Example 4. Dirac and Majorana Equations with Minimum-Coupled Electromagnetic Gauge Field

Following [ref. 8], first consider the related Dirac equation with minimum-coupled electromagnetic gauge field interaction

$$\mathbf{A}(x_1,x_2,x_3,t)=A_1(x_1,x_2,x_3,t)\mathbf{e}_{x_1}+A_2(x_1,x_2,x_3,t)\mathbf{e}_{x_2}+A_3(x_1,x_2,x_3,t)\mathbf{e}_{x_3},\ A_0(x_1,x_2,x_3,t) \tag{14a}$$

$$\mathbf{H}_{\mathrm{Dirac}_{4\cdot 4}}\Psi_{\mathrm{D}}=\left(m\,c^2 a_0+\sum_{j=1}^{3}\left(a_j\left(p_j-eA_j\right)c+eA_0\right)\right)\Psi_{\mathrm{D}}=i\hbar\partial_t\Psi_{\mathrm{D}}$$

$$\mathbf{A}(x_1,x_2,x_3,t)=A_1(x_1,x_2,x_3,t)\mathbf{e}_{x_1}+A_2(x_1,x_2,x_3,t)\mathbf{e}_{x_2}+A_3(x_1,x_2,x_3,t)\mathbf{e}_{x_3},\ A_0(x_1,x_2,x_3,t)$$

$$\Psi_{\mathrm{D}}\equiv\begin{pmatrix}\Psi_{D_1}\\ \Psi_{D_2}\\ \Psi_{D_3}\\ \Psi_{D_4}\end{pmatrix}:\ \text{4 - component Dirac wavefunction}$$

$$\Psi_{D_0}\equiv\begin{pmatrix}\Psi_{D_{10}}\\ \Psi_{D_{20}}\\ \Psi_{D_{30}}\\ \Psi_{D_{40}}\end{pmatrix}:\ \text{4 – component Dirac Initial-State}$$

$$a_0=\begin{bmatrix}1&0&0&0\\0&1&0&0\\0&0&-1&0\\0&0&0&-1\end{bmatrix},a_1=\begin{bmatrix}0&0&0&1\\0&0&1&0\\0&1&0&0\\1&0&0&0\end{bmatrix}$$

$$a_2=\begin{bmatrix}0&0&0&-i\\0&0&i&0\\0&-i&0&0\\i&0&0&0\end{bmatrix},a_3=\begin{bmatrix}0&0&1&0\\0&0&0&-1\\1&0&0&0\\0&-1&0&0\end{bmatrix}$$

Similarly for the Majorana equation minimal-coupled to the EM gauge field

$$
\begin{aligned}
&-im\,c\sigma^2\rho_M^* + \left(i\hbar\sigma^\mu\partial_\mu - eA_\mu\right)\rho_M = 0 \\
&A_\mu \equiv \begin{pmatrix} \mathbf{A}(x_1,x_2,x_3,t) = A_1(x_1,x_2,x_3,t)\mathbf{e}_{x_1} + A_2(x_1,x_2,x_3,t)\mathbf{e}_{x_2} + A_3(x_1,x_2,x_3,t)\mathbf{e}_{x_3}, \\ A_0(x_1,x_2,x_3,t) \end{pmatrix} \\
&\rho_M = \begin{pmatrix} \rho_{M_1} \\ \rho_{M_2} \end{pmatrix} : 2-\text{component Majorana wavefunction,} \\
&\rho_{M_0} = \begin{pmatrix} \rho_{M_{10}} \\ \rho_{M_{20}} \end{pmatrix} : 2-\text{component Majorana Initial-State} \\
&\sigma^\mu : \text{usual } 2\times 2 \text{ Pauli spin matrices } \sigma^{1,2,3}, \sigma^0 = -i\mathbf{I}_{2\times 2}
\end{aligned}
\tag{14b}
$$

Now the connection between the Majorana (14b) ρ_M and Dirac (14a) $\mathbf{\Psi}_{\mathrm{D}}$ wavefunctions subject to the Majorana self-conjugacy condition $\mathbf{\Psi}_{\mathrm{D}} = \mathbf{\Psi}_{\mathrm{D}}^c$ is thoroughly discussed in the bibliography of [ref 8]; only some key relationships between them are reproduced here for convenience

$$
\begin{aligned}
&\Psi_{\mathrm{D}} \equiv \begin{pmatrix} \Psi_{\mathrm{D1}} \\ \Psi_{\mathrm{D2}} \\ \Psi_{\mathrm{D3}} \\ \Psi_{\mathrm{D4}} \end{pmatrix} = \begin{pmatrix} \chi_D \\ \sigma_2\phi_D^* \end{pmatrix}, \quad \Psi_{\mathrm{D}}^c = \begin{pmatrix} \sigma_2 & 0 \\ 0 & \sigma_2 \end{pmatrix} \Psi_{\mathrm{D}}^{*Transpose} = \begin{pmatrix} \phi_D \\ \sigma_2\chi_D^* \end{pmatrix}, \\
&\chi_D = \begin{pmatrix} \Psi_{\mathrm{D1}} \\ \Psi_{\mathrm{D2}} \end{pmatrix}, \quad \sigma_2\phi_D^* = \begin{pmatrix} \Psi_{\mathrm{D3}} \\ \Psi_{\mathrm{D4}} \end{pmatrix}, \quad \phi_{\mathrm{D}} = \begin{pmatrix} i\Psi_{\mathrm{D4}}^* \\ -i\Psi_{\mathrm{D3}}^* \end{pmatrix}, \\
&\text{Majorana Self-Conjugacy condition } \Psi_{\mathrm{D}} = \Psi_{\mathrm{D}}^c \\
&\chi_D = \frac{1}{\sqrt{2}}\left(\rho_{M_2} + i\rho_{M_1}\right), \ \rho_{M_2} = \frac{1}{\sqrt{2}}\left(\chi_D + \phi_D\right) = \frac{1}{\sqrt{2}}\begin{pmatrix} \Psi_{\mathrm{D1}} + i\Psi_{\mathrm{D4}}^* \\ \Psi_{\mathrm{D2}} - i\Psi_{\mathrm{D3}}^* \end{pmatrix} \\
&\phi_D = \frac{1}{\sqrt{2}}\left(\rho_{M_2} - i\rho_{M_1}\right), \ \rho_{M_1} = \frac{i}{\sqrt{2}}\left(\chi_D - \phi_D\right) = \frac{i}{\sqrt{2}}\begin{pmatrix} \Psi_{\mathrm{D1}} - i\Psi_{\mathrm{D4}}^* \\ \Psi_{\mathrm{D2}} + i\Psi_{\mathrm{D3}}^* \end{pmatrix} \\
&\rho_M = \begin{pmatrix} \rho_{M_1} \\ \rho_{M_2} \end{pmatrix}
\end{aligned}
\tag{14c}
$$

So by way of (14c), given the related Dirac wavefunction and subject to the Majorana self-conjugacy condition $\mathbf{\Psi}_{\mathrm{D}} = \mathbf{\Psi}_{\mathrm{D}}^c$, the Majorana wavefunction ascends naturally. Moreover, by way of the HOA method, substituting the Dirac Hamiltonian of (14a) gives the quantum phase space dynamics of the Dirac system for initial conditions and EM gauge fields of general form.

$$\hat{\mathbf{H}}_{\mathrm{Dirac}_{4\cdot 4}}\begin{pmatrix} i\hbar\partial_{p_1}+\alpha_1 x_1, i\hbar\partial_{p_2}+\alpha_2 x_2, i\hbar\partial_{p_3}+\alpha_3 x_3; \\ -i\hbar\partial_{x_1}+\gamma_1 p_1, -i\hbar\partial_{x_2}+\gamma_2 p_2, -i\hbar\partial_{x_3}+\gamma_3 p_3; t \end{pmatrix}\mathbf{\Psi}_{\mathrm{D}}(x_1,x_2,x_3;p_1,p_2,p_3;t)$$

$$= i\hbar\partial_t\mathbf{\Psi}_{\mathrm{D}}(x_1,x_2,x_3;p_1,p_2,p_3;t)$$

$$\left(m\, c^2 a_0+\sum_{j=1}^{3}\begin{pmatrix} a_j\left(-i\hbar\partial_{x_j}+\gamma_j p_j-eA_j\left(i\hbar\partial_{p_1}+\alpha_1 x_1, i\hbar\partial_{p_2}+\alpha_2 x_2, i\hbar\partial_{p_3}+\alpha_3 x_3, t\right)\right)c \\ +eA_0\left(i\hbar\partial_{p_1}+\alpha_1 x_1, i\hbar\partial_{p_2}+\alpha_2 x_2, i\hbar\partial_{p_3}+\alpha_3 x_3; t\right) \end{pmatrix}\right)\mathbf{\Psi}_{\mathrm{D}} = i\hbar\partial_t\mathbf{\Psi}_{\mathrm{D}}$$

$$\mathbf{A}\left(i\hbar\partial_{p_1}+\alpha_1 x_1, i\hbar\partial_{p_2}+\alpha_2 x_2, i\hbar\partial_{p_3}+\alpha_3 x_3, t\right)=$$

$$A_1\left(i\hbar\partial_{p_1}+\alpha_1 x_1, i\hbar\partial_{p_2}+\alpha_2 x_2, i\hbar\partial_{p_3}+\alpha_3 x_3, t\right)\mathbf{e}_{x_1}$$

$$+A_2\left(i\hbar\partial_{p_1}+\alpha_1 x_1, i\hbar\partial_{p_2}+\alpha_2 x_2, i\hbar\partial_{p_3}+\alpha_3 x_3, t\right)\mathbf{e}_{x_2}$$

$$+A_3\left(i\hbar\partial_{p_1}+\alpha_1 x_1, i\hbar\partial_{p_2}+\alpha_2 x_2, i\hbar\partial_{p_3}+\alpha_3 x_3, t\right)\mathbf{e}_{x_3}$$

$$, A_0\left(i\hbar\partial_{p_1}+\alpha_1 x_1, i\hbar\partial_{p_2}+\alpha_2 x_2, i\hbar\partial_{p_3}+\alpha_3 x_3, t\right)$$

$$\mathbf{\Psi}_{\mathrm{D}} \equiv \begin{pmatrix} \mathbf{\Psi}_{\mathrm{D1}} \\ \mathbf{\Psi}_{\mathrm{D2}} \\ \mathbf{\Psi}_{\mathrm{D3}} \\ \mathbf{\Psi}_{\mathrm{D4}} \end{pmatrix} : \text{4 - component Dirac wavefunction} \tag{14d}$$

Hence the configuration space dynamics for the related minimal-coupled Dirac system

$$\mathbf{\Psi}_{\mathrm{D\ configuration\ space}}(x_1,x_2,x_3,t)=$$

$$\begin{pmatrix} \mathbf{\Psi}_{\mathrm{D1}}(x_1,x_2,x_3,t) \\ \mathbf{\Psi}_{\mathrm{D2}}(x_1,x_2,x_3,t) \\ \mathbf{\Psi}_{\mathrm{D3}}(x_1,x_2,x_3,t) \\ \mathbf{\Psi}_{\mathrm{D4}}(x_1,x_2,x_3,t) \end{pmatrix}_{\text{configuration space}} =$$

$$= \int_{-\infty}^{\infty}\frac{e^{\frac{ix_1p_1}{2\hbar}}}{\sqrt{4\pi\hbar}}\int_{-\infty}^{\infty}\frac{e^{\frac{ix_2p_2}{2\hbar}}}{\sqrt{4\pi\hbar}}\int_{-\infty}^{\infty}\frac{e^{\frac{ix_3p_3}{2\hbar}}}{\sqrt{4\pi\hbar}} L^{-1}_{\left(\begin{pmatrix}\bar{x}_1,\ldots,\bar{x}_n; \\ \bar{p}_1,\ldots,\bar{p}_n\end{pmatrix}\to\begin{pmatrix}x_1,\ldots,x_n; \\ p_1,\ldots,p_n\end{pmatrix}\right)} e^{\frac{-i}{\hbar}\int_0^t \left(m\, c^2 a_0 + \sum_{j=1}^{3}\begin{pmatrix} a_j\begin{pmatrix}-i\hbar\bar{x}_j+\gamma_j p_j \\ -eA_j\begin{pmatrix} i\hbar\bar{p}_1+\alpha_1 x_1, \\ i\hbar\bar{p}_2+\alpha_2 x_2, \\ i\hbar\bar{p}_3+\alpha_3 x_3, u\end{pmatrix}\end{pmatrix}c \\ +eA_0\begin{pmatrix} i\hbar\bar{p}_1+\alpha_1 x_1, \\ i\hbar\bar{p}_2+\alpha_2 x_2, \\ i\hbar\bar{p}_3+\alpha_3 x_3, u\end{pmatrix}\end{pmatrix}\right)du} \begin{pmatrix} \breve{\mathbf{\Psi}}_{D0_1} \\ \breve{\mathbf{\Psi}}_{D0_2} \\ \breve{\mathbf{\Psi}}_{D0_3} \\ \breve{\mathbf{\Psi}}_{D0_4} \end{pmatrix} dp_1 dp_2 dp_3$$

$$\breve{\mathbf{\Psi}}_{D0}\begin{pmatrix} \bar{x}_1,\bar{x}_2,\bar{x}_3; \\ \bar{p}_1,\bar{p}_2,\bar{p}_3; t=0 \end{pmatrix} = \begin{pmatrix} \breve{\mathbf{\Psi}}_{D0_1} \\ \breve{\mathbf{\Psi}}_{D0_2} \\ \breve{\mathbf{\Psi}}_{D0_3} \\ \breve{\mathbf{\Psi}}_{D0_4} \end{pmatrix} : \text{Transformed Initial-condition vector} \tag{14e}$$

where the explicit form of the $\hat{\mathbf{H}}_{\mathrm{Dirac}_{4\,4}}$ in the exponent 4x4 matrix integral is supplied as

$$\begin{pmatrix}
\begin{pmatrix} mc^2 \\ +ecA_0\begin{pmatrix} i\hbar\bar{p}_1+\alpha_1x_1, \\ i\hbar\bar{p}_2+\alpha_2x_2, \\ i\hbar\bar{p}_3+\alpha_3x_3,t \end{pmatrix} \end{pmatrix} & 0 & \begin{pmatrix} c\left(-i\hbar\bar{x}_3+\gamma_3p_3\right) \\ -eA_3\begin{pmatrix} i\hbar\bar{p}_1+\alpha_1x_1, \\ i\hbar\bar{p}_2+\alpha_2x_2, \\ i\hbar\bar{p}_3+\alpha_3x_3,t \end{pmatrix} \end{pmatrix} & \begin{pmatrix} c\left(-i\hbar\bar{x}_1+\gamma_1p_1\right) \\ -eA_1\begin{pmatrix} i\hbar\bar{p}_1+\alpha_1x_1, \\ i\hbar\bar{p}_2+\alpha_2x_2, \\ i\hbar\bar{p}_3+\alpha_3x_3,t \end{pmatrix} \\ -i(c\left(-i\hbar\bar{x}_2+\gamma_2p_2\right) \\ -eA_2\begin{pmatrix} i\hbar\bar{p}_1+\alpha_1x_1, \\ i\hbar\bar{p}_2+\alpha_2x_2, \\ i\hbar\bar{p}_3+\alpha_3x_3,t \end{pmatrix}) \end{pmatrix} \\
0 & \begin{pmatrix} mc^2 \\ +ecA_0\begin{pmatrix} i\hbar\bar{p}_1+\alpha_1x_1, \\ i\hbar\bar{p}_2+\alpha_2x_2, \\ i\hbar\bar{p}_3+\alpha_3x_3,t \end{pmatrix} \end{pmatrix} & \begin{pmatrix} c\left(-i\hbar\bar{x}_1+\gamma_1p_1\right) \\ -eA_1\begin{pmatrix} i\hbar\bar{p}_1+\alpha_1x_1, \\ i\hbar\bar{p}_2+\alpha_2x_2, \\ i\hbar\bar{p}_3+\alpha_3x_3,t \end{pmatrix} \\ +i(c\left(-i\hbar\bar{x}_2+\gamma_2p_2\right) \\ -eA_2\begin{pmatrix} i\hbar\bar{p}_1+\alpha_1x_1, \\ i\hbar\bar{p}_2+\alpha_2x_2, \\ i\hbar\bar{p}_3+\alpha_3x_3,t \end{pmatrix}) \end{pmatrix} & -\begin{pmatrix} c\left(-i\hbar\bar{x}_3+\gamma_3p_3\right) \\ -eA_3\begin{pmatrix} i\hbar\bar{p}_1+\alpha_1x_1, \\ i\hbar\bar{p}_2+\alpha_2x_2, \\ i\hbar\bar{p}_3+\alpha_3x_3,t \end{pmatrix} \end{pmatrix} \\
\begin{pmatrix} c\left(-i\hbar\bar{x}_3+\gamma_3p_3\right) \\ -eA_3\begin{pmatrix} i\hbar\bar{p}_1+\alpha_1x_1, \\ i\hbar\bar{p}_2+\alpha_2x_2, \\ i\hbar\bar{p}_3+\alpha_3x_3,t \end{pmatrix} \end{pmatrix} & \begin{pmatrix} c\left(-i\hbar\bar{x}_1-\gamma_1p_1\right) \\ -eA_1\begin{pmatrix} i\hbar\bar{p}_1+\alpha_1x_1, \\ i\hbar\bar{p}_2+\alpha_2x_2, \\ i\hbar\bar{p}_3+\alpha_3x_3,t \end{pmatrix} \\ -i(c\left(-i\hbar\bar{x}_2+\gamma_2p_2\right) \\ -eA_2\begin{pmatrix} i\hbar\bar{p}_1+\alpha_1x_1, \\ i\hbar\bar{p}_2+\alpha_2x_2, \\ i\hbar\bar{p}_3+\alpha_3x_3,t \end{pmatrix}) \end{pmatrix} & \begin{pmatrix} -mc^2 \\ +ecA_0\begin{pmatrix} i\hbar\bar{p}_1+\alpha_1x_1, \\ i\hbar\bar{p}_2+\alpha_2x_2, \\ i\hbar\bar{p}_3+\alpha_3x_3,t \end{pmatrix} \end{pmatrix} & 0 \\
\begin{pmatrix} c\left(-i\hbar\bar{x}_1+\gamma_1p_1\right) \\ -eA_1\begin{pmatrix} i\hbar\bar{p}_1+\alpha_1x_1, \\ i\hbar\bar{p}_2+\alpha_2x_2, \\ i\hbar\bar{p}_3+\alpha_3x_3,t \end{pmatrix} \\ +i(c\left(-i\hbar\bar{x}_2+\gamma_2p_2\right) \\ -eA_2\begin{pmatrix} i\hbar\bar{p}_1+\alpha_1x_1, \\ i\hbar\bar{p}_2+\alpha_2x_2, \\ i\hbar\bar{p}_3+\alpha_3x_3,t \end{pmatrix}) \end{pmatrix} & -\begin{pmatrix} c\left(-i\hbar\bar{x}_3+\gamma_3p_3\right) \\ -eA_3\begin{pmatrix} i\hbar\bar{p}_1+\alpha_1x_1, \\ i\hbar\bar{p}_2+\alpha_2x_2, \\ i\hbar\bar{p}_3+\alpha_3x_3,t \end{pmatrix} \end{pmatrix} & 0 & \begin{pmatrix} -mc^2 \\ +ecA_0\begin{pmatrix} i\hbar\bar{p}_1+\alpha_1x_1, \\ i\hbar\bar{p}_2+\alpha_2x_2, \\ i\hbar\bar{p}_3+\alpha_3x_3,t \end{pmatrix} \end{pmatrix}
\end{pmatrix} \tag{14f}$$

yields by way of (14c), the associated Majorana wavefunction for the dynamics of the minimal-coupled system with arbitrary profile EM interaction and initial-conditions, in terms of the quadrature solutions for the related Dirac system just calculated above

$$\rho_{M_2} = \frac{1}{\sqrt{2}}\begin{pmatrix} \Psi_{D1}(x_1,x_2,x_3,t) + i\Psi^*_{D4}(x_1,x_2,x_3,t) \\ \Psi_{D2}(x_1,x_2,x_3,t) - i\Psi^*_{D3}(x_1,x_2,x_3,t) \end{pmatrix}$$

$$\rho_{M_1} = \frac{i}{\sqrt{2}}\begin{pmatrix} \Psi_{D1}(x_1,x_2,x_3,t) - i\Psi^*_{D4}(x_1,x_2,x_3,t) \\ \Psi_{D2}(x_1,x_2,x_3,t) + i\Psi^*_{D3}(x_1,x_2,x_3,t) \end{pmatrix}$$

$$\rho_{M\,\text{configuration space}} = \begin{pmatrix} \rho_{M_1} \\ \rho_{M_2} \end{pmatrix}_{\text{configuration space}}$$

$$\rho_{M_0\,\text{configuration space}} = \begin{pmatrix} \rho_{M_{10}} \\ \rho_{M_{20}} \end{pmatrix}_{\text{configuration space}} \quad \text{:Majorana Initial-State}$$

$$\rho_{M_{20}} = \frac{1}{\sqrt{2}}\begin{pmatrix} \Psi_{D1_0}(x_1,x_2,x_3,t) + i\Psi^*_{D4_0}(x_1,x_2,x_3,t) \\ \Psi_{D2_0}(x_1,x_2,x_3,t) - i\Psi^*_{D3_0}(x_1,x_2,x_3,t) \end{pmatrix}$$

$$\rho_{M_{10}} = \frac{i}{\sqrt{2}}\begin{pmatrix} \Psi_{D1_0}(x_1,x_2,x_3,t) - i\Psi^*_{D4_0}(x_1,x_2,x_3,t) \\ \Psi_{D2_0}(x_1,x_2,x_3,t) + i\Psi^*_{D3_0}(x_1,x_2,x_3,t) \end{pmatrix} \tag{14g}$$

Example 5. Dirac Molecular Hamiltonian with Pairwise Coulomb-Breit Interaction

Now consider the example case of the relativistic Molecular Hamiltonian with Pairwise Dirac-Coulomb-Briet interaction dynamics in quantum phase space. Here the Hamiltonian has the 4x4 matrix form

$$\begin{aligned}
\hat{\mathbf{H}}_{\substack{4\times 4\\ \text{DiracCoulombBreit}\\ \text{Molecular}}} = & \sum_A \left(a_{0_A} M_A c^2 + a_{x_A}\left(-i\hbar\partial_{x_A} + \frac{p_{x_A}}{2}\right)c + a_{y_A}\left(-i\hbar\partial_{y_A} + \frac{p_{y_A}}{2}\right)c + a_{z_A}\left(-i\hbar\partial_{z_A} + \frac{p_{zA}}{2}\right)\right) \\
& + \sum_i \left(a_{0_i} m_e c^2 + a_{x_i}\left(-i\hbar\partial_{x_i} + \frac{p_{x_i}}{2}\right)c + a_{y_i}\left(-i\hbar\partial_{y_i} + \frac{p_{y_i}}{2}\right)c + a_{z_i}\left(-i\hbar\partial_{z_i} + \frac{p_{zi}}{2}\right)\right) \\
& + \sum_{A<B} \frac{e^2}{8\pi\varepsilon_0} \frac{Z_A Z_B \left(1 + \frac{1}{2}\left((\mathbf{a}_A\cdot\mathbf{a}_B) + \frac{(\mathbf{a}_A\cdot\mathbf{r}_{AB})(\mathbf{a}_B\cdot\mathbf{r}_{AB})}{r_{AB}^2}\right)\right)}{\sqrt{\begin{aligned}&\left(\left(i\hbar\partial_{p_{xA}} + \frac{x_A}{2}\right) - \left(i\hbar\partial_{p_{xB}} + \frac{x_B}{2}\right)\right)^2 \\ &+ \left(\left(i\hbar\partial_{p_{yA}} + \frac{y_A}{2}\right) - \left(i\hbar\partial_{p_{yA}} + \frac{y_B}{2}\right)\right)^2 \\ &+ \left(\left(i\hbar\partial_{p_{zA}} + \frac{z_A}{2}\right) - \left(i\hbar\partial_{p_{zA}} + \frac{z_B}{2}\right)\right)^2\end{aligned}}} + \sum_{i<j} \frac{e^2}{8\pi\varepsilon_0} \frac{\left(1 + \frac{1}{2}\left((\mathbf{a}_i\cdot\mathbf{a}_j) + \frac{(\mathbf{a}_i\cdot\mathbf{r}_{ij})(\mathbf{a}_j\cdot\mathbf{r}_{ij})}{r_{ij}^2}\right)\right)}{\sqrt{\begin{aligned}&\left(\left(i\hbar\partial_{p_{x_i}} + \frac{x_i}{2}\right) - \left(i\hbar\partial_{p_{x_j}} + \frac{x_j}{2}\right)\right)^2 \\ &+ \left(\left(i\hbar\partial_{p_{y_i}} + \frac{y_i}{2}\right) - \left(i\hbar\partial_{p_{y_j}} + \frac{y_j}{2}\right)\right)^2 \\ &+ \left(\left(i\hbar\partial_{p_{z_i}} + \frac{z_i}{2}\right) - \left(i\hbar\partial_{p_{z_j}} + \frac{z_j}{2}\right)\right)^2\end{aligned}}} \\
& + \sum_{A,i} \frac{-Z_A e^2 \left(1 + \frac{1}{2}\left((\mathbf{a}_A\cdot\mathbf{a}_i) + \frac{(\mathbf{a}_A\cdot\mathbf{r}_{Ai})(\mathbf{a}_i\cdot\mathbf{r}_{Ai})}{r_{Ai}^2}\right)\right)}{\sqrt{\begin{aligned}&\left(\left(i\hbar\partial_{p_{xA}} + \frac{x_A}{2}\right) - \left(i\hbar\partial_{p_{x_i}} + \frac{x_i}{2}\right)\right)^2 \\ &+ \left(\left(i\hbar\partial_{p_{yA}} + \frac{y_A}{2}\right) - \left(i\hbar\partial_{p_{y_i}} + \frac{y_i}{2}\right)\right)^2 \\ &+ \left(\left(i\hbar\partial_{p_{zA}} + \frac{z_A}{2}\right) - \left(i\hbar\partial_{p_{z_i}} + \frac{z_i}{2}\right)\right)^2\end{aligned}}}
\end{aligned} \tag{15a}$$

$$a_0 = \begin{bmatrix} 1 & 0 & 0 & 0 \\ 0 & 1 & 0 & 0 \\ 0 & 0 & -1 & 0 \\ 0 & 0 & 0 & -1 \end{bmatrix} a_x = \begin{bmatrix} 0 & 0 & 0 & 1 \\ 0 & 0 & 1 & 0 \\ 0 & 1 & 0 & 0 \\ 1 & 0 & 0 & 0 \end{bmatrix} , a_y = \begin{bmatrix} 0 & 0 & 0 & -i \\ 0 & 0 & i & 0 \\ 0 & -i & 0 & 0 \\ i & 0 & 0 & 0 \end{bmatrix} , a_z = \begin{bmatrix} 0 & 0 & 1 & 0 \\ 0 & 0 & 0 & -1 \\ 1 & 0 & 0 & 0 \\ 0 & -1 & 0 & 0 \end{bmatrix}$$

$$\mathbf{a} = a_x \hat{\mathbf{e}}_{\mathbf{x}} + a_y \hat{\mathbf{e}}_{\mathbf{y}} + a_z \hat{\mathbf{e}}_{\mathbf{z}}$$

$$\sum_{i<j} \frac{1}{r_{ij}} \left(1 + \frac{1}{2} \left((\mathbf{a}_i \cdot \mathbf{a}_j) + \frac{(\mathbf{a}_i \cdot \mathbf{r}_{ij})(\mathbf{a}_j \cdot \mathbf{r}_{ij})}{r_{ij}^2} \right) \right),$$

$$\mathbf{r}_{ij} = \left(\left(i\hbar \partial_{p_{x_i}} + \frac{x_i}{2} \right) - \left(i\hbar \partial_{p_{x_j}} + \frac{x_j}{2} \right) \right) \hat{\mathbf{e}}_{\mathbf{x}} + \left(\left(i\hbar \partial_{p_{y_i}} + \frac{y_i}{2} \right) - \left(i\hbar \partial_{p_{y_j}} + \frac{y_j}{2} \right) \right) \hat{\mathbf{e}}_{\mathbf{y}} + \left(\left(i\hbar \partial_{p_{z_i}} + \frac{z_i}{2} \right) - \left(i\hbar \partial_{p_{z_j}} + \frac{z_j}{2} \right) \right) \hat{\mathbf{e}}_{\mathbf{z}}$$

$$r_{ij} = \left| \left(\left(i\hbar \partial_{p_{x_i}} + \frac{x_i}{2} \right) - \left(i\hbar \partial_{p_{x_j}} + \frac{x_j}{2} \right) \right) \hat{\mathbf{e}}_{\mathbf{x}} + \left(\left(i\hbar \partial_{p_{y_i}} + \frac{y_i}{2} \right) - \left(i\hbar \partial_{p_{y_j}} + \frac{y_j}{2} \right) \right) \hat{\mathbf{e}}_{\mathbf{y}} + \left(\left(i\hbar \partial_{p_{z_i}} + \frac{z_i}{2} \right) - \left(i\hbar \partial_{p_{z_j}} + \frac{z_j}{2} \right) \right) \hat{\mathbf{e}}_{\mathbf{z}} \right| \quad (15b)$$

Since the Hamiltonian of this system is matrix-valued $\underset{4\times 4}{\hat{\mathbf{H}}}{}_{\substack{\text{DiracCoulombBreit} \\ \text{Molecular}}}$, the quantum phase space wavefunction.is a column vector $\underset{4\times 1 \text{ column vector}}{\mathbf{\Psi}}$

$$\underset{4\times 1 \text{ column vector}}{\mathbf{\Psi}} \begin{pmatrix} x_A, y_A, z_A, x_B, y_B, z_B, \\ x_i, y_i, z_i, x_j, y_j, z_j; \\ p_{x_A}, p_{y_A}, p_{z_A}, p_{x_B}, p_{y_B}, p_{z_B}, \\ p_{x_i}, p_{y_i}, p_{z_i}, p_{x_j}, p_{y_j}, p_{z_j}; t \end{pmatrix} =$$

$$L^{-1}_{\left(\begin{pmatrix} \bar{x}_A, \bar{y}_A, \bar{z}_A, \bar{x}_B, \bar{y}_B, \bar{z}_B, \\ \bar{x}_i, \bar{y}_i, \bar{z}_i, \bar{x}_j, \bar{y}_j, \bar{z}_j \end{pmatrix} \to \begin{pmatrix} x_A, y_A, z_A, x_B, y_B, z_B, \\ x_i, y_i, z_i, x_j, y_j, z_j \end{pmatrix} \right)} \left[L^{-1}_{\left(\begin{matrix} p_{x_A}, p_{y_A}, p_{z_A}, p_{x_B}, p_{y_B}, p_{z_B}, \\ p_{x_i}, p_{y_i}, p_{z_i}, p_{x_j}, p_{y_j}, p_{z_j} \end{matrix} \to \begin{pmatrix} p_{x_A}, p_{y_A}, p_{z_A}, p_{x_B}, p_{y_B}, p_{z_B}, \\ p_{x_i}, p_{y_i}, p_{z_i}, p_{x_j}, p_{y_j}, p_{z_j} \end{pmatrix} \right)} \left[\begin{matrix} \frac{-i}{\hbar} \int_0^t \underset{4\times 4}{\hat{\mathbf{H}}}{}_{\substack{\text{DiracCoulombBreit} \\ \text{Molecular}}} \begin{pmatrix} -i\hbar \bar{x}_A + \frac{p_{x_A}}{2}, -i\hbar \bar{y}_A + \frac{p_{y_A}}{2}, -i\hbar \bar{z}_A + \frac{p_{z_A}}{2}, -i\hbar \bar{x}_i + \frac{p_{x_i}}{2}, -i\hbar \bar{y}_i + \frac{p_{y_i}}{2}, -i\hbar \bar{z}_i + \frac{p_{z_i}}{2}; \\ i\hbar \bar{p}_{x_A} + \frac{x_A}{2}, i\hbar \bar{p}_{y_A} + \frac{y_A}{2}, i\hbar \bar{p}_{z_A} + \frac{z_A}{2}, i\hbar \bar{p}_{x_B} + \frac{x_B}{2}, i\hbar \bar{p}_{y_B} + \frac{y_B}{2}, i\hbar \bar{p}_{z_B} + \frac{z_B}{2}, \\ i\hbar \bar{p}_{x_i} + \frac{x_i}{2}, i\hbar \bar{p}_{y_i} + \frac{y_i}{2}, i\hbar \bar{p}_{z_i} + \frac{z_i}{2}, i\hbar \bar{p}_{x_j} + \frac{x_j}{2}, i\hbar \bar{p}_{y_j} + \frac{y_j}{2}, i\hbar \bar{p}_{z_j} + \frac{z_j}{2} \end{pmatrix} du \\ e \\ \times \\ \underset{4\times 1 \text{ column vector}}{\bar{\mathbf{\Psi}}_0} (\bar{x}_1, \ldots, \bar{x}_n; \bar{p}_1, \ldots, \bar{p}_n; t = 0) \end{matrix} \right] \right]$$

$$(15c)$$

$$\underset{4\times 4}{\hat{\mathbf{H}}}{}_{\substack{\text{DiracCoulombBreit} \\ \text{Molecular}}} = \sum_A \left(a_{0_A} M_A c^2 + a_{x_A} \left(-i\hbar \bar{x}_A + \frac{p_{x_A}}{2} \right) c + a_{y_A} \left(-i\hbar \bar{y}_A + \frac{p_{y_A}}{2} \right) c + a_{z_A} \left(-i\hbar \bar{z}_A + \frac{p_{z_A}}{2} \right) \right)$$

$$+ \sum_i \left(a_{0_i} m_e c^2 + a_{x_i} \left(-i\hbar \bar{x}_i + \frac{p_{x_i}}{2} \right) c + a_{y_i} \left(-i\hbar \bar{y}_i + \frac{p_{y_i}}{2} \right) c + a_{z_i} \left(-i\hbar \bar{z}_i + \frac{p_{zi}}{2} \right) \right)$$

$$
+\sum_{A<B}\frac{e^{2}}{8\pi\varepsilon_{0}}\frac{Z_{A}Z_{B}\left(1+\frac{1}{2}\left(\left(\mathbf{a}_{A}\cdot\mathbf{a}_{B}\right)+\frac{\left(\mathbf{a}_{A}\cdot\mathbf{r}_{AB}\right)\left(\mathbf{a}_{B}\cdot\mathbf{r}_{AB}\right)}{r_{AB}^{2}}\right)\right)}{\left|\begin{array}{l}\left(\left(i\hbar\overline{p}_{x_{A}}+\frac{x_{A}}{2}\right)-\left(i\hbar\overline{p}_{x_{B}}+\frac{x_{B}}{2}\right)\right)^{2}\\+\left(\left(i\hbar\overline{p}_{y_{A}}+\frac{y_{A}}{2}\right)-\left(i\hbar\overline{p}_{y_{B}}+\frac{y_{B}}{2}\right)\right)^{2}\\+\left(\left(i\hbar\overline{p}_{z_{A}}+\frac{z_{A}}{2}\right)-\left(i\hbar\overline{p}_{z_{B}}+\frac{z_{B}}{2}\right)\right)^{2}\end{array}\right|}+\sum_{i<j}\frac{e^{2}}{8\pi\varepsilon_{0}}\frac{\left(1+\frac{1}{2}\left(\left(\mathbf{a}_{i}\cdot\mathbf{a}_{j}\right)+\frac{\left(\mathbf{a}_{i}\cdot\mathbf{r}_{ij}\right)\left(\mathbf{a}_{j}\cdot\mathbf{r}_{ij}\right)}{r_{ij}^{2}}\right)\right)}{\left|\begin{array}{l}\left(\left(i\hbar\overline{p}_{x_{i}}+\frac{x_{i}}{2}\right)-\left(i\hbar\overline{p}_{x_{j}}+\frac{x_{j}}{2}\right)\right)^{2}\\+\left(\left(i\hbar\overline{p}_{y_{i}}+\frac{y_{i}}{2}\right)-\left(i\hbar\overline{p}_{y_{j}}+\frac{y_{j}}{2}\right)\right)^{2}\\+\left(\left(i\hbar\overline{p}_{z_{i}}+\frac{z_{i}}{2}\right)-\left(i\hbar\overline{p}_{x_{j}}+\frac{z_{j}}{2}\right)\right)^{2}\end{array}\right|}
$$

$$
+\sum_{A,i}\frac{-Z_{A}e^{2}\left(1+\frac{1}{2}\left(\left(\mathbf{a}_{A}\cdot\mathbf{a}_{i}\right)+\frac{\left(\mathbf{a}_{A}\cdot\mathbf{r}_{Ai}\right)\left(\mathbf{a}_{i}\cdot\mathbf{r}_{Ai}\right)}{r_{Ai}^{2}}\right)\right)}{\left|\begin{array}{l}\left(\left(i\hbar\overline{p}_{x_{A}}+\frac{x_{A}}{2}\right)-\left(i\hbar\overline{p}_{x_{i}}+\frac{x_{i}}{2}\right)\right)^{2}\\+\left(\left(i\hbar\overline{p}_{y_{A}}+\frac{y_{A}}{2}\right)-\left(i\hbar\overline{p}_{y_{i}}+\frac{y_{i}}{2}\right)\right)^{2}\\+\left(\left(i\hbar\overline{p}_{z_{A}}+\frac{z_{A}}{2}\right)-\left(i\hbar\overline{p}_{z_{i}}+\frac{z_{i}}{2}\right)\right)^{2}\end{array}\right|}
\tag{15d}
$$

Comment

Though Examples 1-5 are useful applications of HOA outright for various chemical applications, so also are various combinations of these Examples . For instance, mixing the results of Example 2 (Schrödinger Hamiltonian with SHOs having N-Arbitrary Masses in Pairwise Anisotropic Interaction) with Example 3 (Schrödinger Molecular Hamiltonian with Pairwise Coulomb Interaction) utilizing the extension to time-dependent parameters in Example 1 (1dim Simple Harmonic Oscillator) , one arrives at a Quantum Phase Space description of non-relativistic molecular systems interacting with an external time-varying bath environment with anisotropic harmonic constitutive properties. Further, Example 4 (Dirac and Majorana Equations with Minimum-Coupled Electromagnetic Gauge Field) and Example 5 (Dirac Molecular Hamiltonian with Pairwise Coulomb-Breit Interaction) may be mixed to yield a Quantum Phase Space description of relativistic molecular systems in external electromagnetic fields. [ref. 9,10]

Thus far, the Examples of the HOA results with usefulness in Chemical Dynamics have been based on a direct use of particular Hamiltonians to model the particular scenarios in question. We now will consider some additional Examples where the HOA results are used as a formal tool to achieve exact analytical solutions of more general formal mathematical problems of interest in Mathematical Chemistry and Beyond.

Example 6. Exact Quadrature Solution of Linear Eigenvalue Problem for General Class of Variable Coefficient Differential Operators

Recall from the HOA Recap in this note, that

'Notwithstanding its quantum mechanical origins, the HOA scheme takes on a life of its own and transcends the limits of quantum applications to address a wide variety of purely formal mathematical problems as well. Among other things, the result provides a formula for obtaining an exact solution to a wide variety of variable-coefficient integro-differential equations. Since the functional dependence of the Hamiltonian operator as considered is in general arbitrary upon its arguments(i.e., independent variables, derivative operator symbols[including negative powers thereof, thus the possible integral character]), then its multivariable extension can be interpreted as the most general variable coefficient partial differential operator. Moreover, it is not confined to being a scalar or even vector operator, but may be generally construed an arbitrary rank matrix operator. In all cases of course, its rank dictates the matrix rank of the wavefunction solution.'

Recall Eq. (10) of the Recap

$$\hat{H}_{\text{configuration space}}\left(x_1,\ldots,x_n,-i\hbar\partial_{x_1},\ldots,-i\hbar\partial_{x_n},t\right)\Psi_{\text{configuration space}}\left(x_1,\ldots,x_n,t\right)$$
$$=i\hbar\partial_t\Psi_{\text{configuration space}}\left(x_1,\ldots,x_n,t\right)$$

$$\Psi_{\text{configuration space}}\left(x_1,\ldots,x_n,t\right)=\int_{-\infty}^{\infty}\frac{e^{\frac{ix_1p_1}{2\hbar}}}{\sqrt{4\pi\hbar}}\cdots\int_{-\infty}^{\infty}\frac{e^{\frac{ix_np_n}{2\hbar}}}{\sqrt{4\pi\hbar}}\Psi\left(x_1,\ldots,x_n;p_1,\ldots,p_n;t\right)dp_1..dp_n$$

$$\Psi_{\text{configuration space}}\left(x_1,\ldots,x_n,t\right)$$
$$=\int_{-\infty}^{\infty}\frac{e^{\frac{ix_1p_1}{2\hbar}}}{\sqrt{4\pi\hbar}}\cdots\int_{-\infty}^{\infty}\frac{e^{\frac{ix_np_n}{2\hbar}}}{\sqrt{4\pi\hbar}}L^{-1}_{\left(\bar{x}_1,\ldots,\bar{x}_n;\bar{p}_1,\ldots,\bar{p}_n\right)\to\left(x_1,\ldots,x_n;p_1,\ldots,p_n\right)}\left[\begin{array}{l} e^{\frac{-i}{\hbar}\int_0^t \hat{H}_{\text{configuration space}}\left(\begin{array}{l}x_1,\ldots,x_n,\\ -i\hbar\partial_{x_1},\ldots,-i\hbar\partial_{x_n},u\end{array}\right)_{\substack{(x_1,\ldots,x_n)\mapsto(i\hbar\bar{p}_1+\alpha_1x_1,\ldots,i\hbar\bar{p}_n+\alpha_nx_n)\\ (-i\hbar\partial_{x_1},\ldots,-i\hbar\partial_{x_n})\mapsto(-i\hbar\bar{x}_1+\gamma_1p_1,\ldots,-i\hbar\bar{x}_n+\gamma_np_n)}}du} \\ \times \\ \Psi_{0\,\text{configuration space}}\left(\bar{x}_1,\ldots,\bar{x}_n;\bar{p}_1,\ldots,\bar{p}_n;t=0\right)\end{array}\right]dp_1..dp_n$$

Now suppose that

$$\hat{H}_{\text{configuration space}}\left(x_1,\ldots,x_n,-i\hbar\partial_{x_1},\ldots,-i\hbar\partial_{x_n},t\right)=\hat{H}_{\substack{\text{configuration space}\\ \text{scleronomic}}}\left(x_1,\ldots,x_n,-i\hbar\partial_{x_1},\ldots,-i\hbar\partial_{x_n}\right)$$

$$\hat{H}_{\substack{\text{configuration space}\\ \text{scleronomic}}}\left(x_1,\ldots,x_n,-i\hbar\partial_{x_1},\ldots,-i\hbar\partial_{x_n}\right)\Psi_{\text{configuration space}}\left(x_1,\ldots,x_n,t\right)$$
$$=i\hbar\partial_t\Psi_{\text{configuration space}}\left(x_1,\ldots,x_n,t\right) \tag{16a}$$

Hence Fourier transforming (16a) with respect to t

$$F_{t\to\omega}\left[f(t)\right]=\int_{t_0}^{\infty} f(t)e^{-it\omega}dt$$

$$F_{t\to\omega}\left[\begin{array}{l}\hat{H}_{\substack{\text{configuration space}\\ \text{scleronomic}}}\left(x_1,\ldots,x_n,-i\hbar\partial_{x_1},\ldots,-i\hbar\partial_{x_n}\right)\Psi_{\text{configuration space}}\left(x_1,\ldots,x_n,t\right)\\ =i\hbar\partial_t\Psi_{\text{configuration space}}\left(x_1,\ldots,x_n,t\right)\end{array}\right]=$$

$$\hat{H}_{\substack{\text{configuration space}\\ \text{scleronomic}}}\left(x_1,\ldots,x_n,-i\hbar\partial_{x_1},\ldots,-i\hbar\partial_{x_n}\right)\Psi_{\substack{\text{configuration space}\\ \omega\text{ eigenvalue}}}\left(x_1,\ldots,x_n,\omega\right)$$

$$=\hbar\omega\Psi_{\substack{\text{configuration space}\\ \omega\text{ eigenvalue}}}\left(x_1,\ldots,x_n,\omega\right) \tag{16b}$$

So

$$\hat{H}_{\substack{\text{configuration space}\\ \text{scleronomic}}}\left(x_1,\ldots,x_n,-i\hbar\partial_{x_1},\ldots,-i\hbar\partial_{x_n}\right)\Psi_{\substack{\text{configuration space}\\ \omega\text{ eigenvalue}}}\left(x_1,\ldots,x_n,\omega\right)$$

$$=\hbar\omega\Psi_{\substack{\text{configuration space}\\ \omega\text{ eigenvalue}}}\left(x_1,\ldots,x_n,\omega\right)$$

$\ni$

$$F_{t\to\omega}\left(\breve{\Psi}_{0\,\text{configuration space}}\begin{pmatrix}\bar{x}_1,\ldots,\bar{x}_n;\\ \bar{p}_1,\ldots,\bar{p}_n;t=0\end{pmatrix}e^{\frac{-i}{\hbar}t\hat{H}_{\text{configuration space}}\begin{pmatrix}x_1,\ldots,x_n,\\ -i\hbar\partial_{x_1},\ldots,-i\hbar\partial_{x_n}\end{pmatrix}_{\begin{pmatrix}(x_1,\ldots,x_n)\\ \mapsto(i\hbar\bar{p}_1+\alpha_1x_1,\ldots,i\hbar\bar{p}_n+\alpha_nx_n)\end{pmatrix}\begin{pmatrix}(-i\hbar\partial_{x_1},\ldots,-i\hbar\partial_{x_n})\\ \mapsto(-i\hbar\bar{x}_1+\gamma_1p_1,\ldots,-i\hbar\bar{x}_n+\gamma_np_n)\end{pmatrix}}}\right)$$

$$=\breve{\Psi}_{0\,\text{configuration space}}\begin{pmatrix}\bar{x}_1,\ldots,\bar{x}_n;\\ \bar{p}_1,\ldots,\bar{p}_n\bar{p}_1,\ldots,\bar{p}_n;t=0\end{pmatrix}\delta\left(\omega-\frac{1}{\hbar}\hat{H}_{\text{configuration space}}\begin{pmatrix}x_1,\ldots,x_n,\\ -i\hbar\partial_{x_1},\ldots,-i\hbar\partial_{x_n}\end{pmatrix}_{\begin{pmatrix}(x_1,\ldots,x_n)\\ \mapsto(i\hbar\bar{p}_1+\alpha_1x_1,\ldots,i\hbar\bar{p}_n+\alpha_nx_n)\end{pmatrix}\begin{pmatrix}(-i\hbar\partial_{x_1},\ldots,-i\hbar\partial_{x_n})\\ \mapsto(-i\hbar\bar{x}_1+\gamma_1p_1,\ldots,-i\hbar\bar{x}_n+\gamma_np_n)\end{pmatrix}}\right) \tag{16c}$$

Hence

$$F_{t\to\omega}\left(\begin{array}{l}\Psi_{\text{configuration space}}\left(x_1,\ldots,x_n,t\right)\\ =\int_{-\infty}^{\infty}\frac{e^{\frac{ix_1p_1}{2\hbar}}}{\sqrt{4\pi\hbar}}\cdots\int_{-\infty}^{\infty}\frac{e^{\frac{ix_np_n}{2\hbar}}}{\sqrt{4\pi\hbar}}L^{-1}_{\begin{pmatrix}(\bar{x}_1,\ldots,\bar{x}_n;\bar{p}_1,\ldots,\bar{p}_n)\\ \to(x_1,\ldots,x_n;p_1,\ldots,p_n)\end{pmatrix}}\left[\begin{array}{l}e^{\frac{-i}{\hbar}t\hat{H}_{\text{configuration space}}\begin{pmatrix}x_1,\ldots,x_n,\\ -i\hbar\partial_{x_1},\ldots,-i\hbar\partial_{x_n}\end{pmatrix}_{\begin{pmatrix}(x_1,\ldots,x_n)\\ \mapsto(i\hbar\bar{p}_1+\alpha_1x_1,\ldots,i\hbar\bar{p}_n+\alpha_nx_n)\end{pmatrix}\begin{pmatrix}(-i\hbar\partial_{x_1},\ldots,-i\hbar\partial_{x_n})\\ \mapsto(-i\hbar\bar{x}_1+\gamma_1p_1,\ldots,-i\hbar\bar{x}_n+\gamma_np_n)\end{pmatrix}}}\\ \times\breve{\Psi}_{0\text{configuration space}}\begin{pmatrix}\bar{x}_1,\ldots,\bar{x}_n;\\ \bar{p}_1,\ldots,\bar{p}_n;t=0\end{pmatrix}\end{array}\right]dp_1..dp_n\end{array}\right)$$

$$\Psi_{\substack{\text{configuration space}\\ \omega\,\text{eigenvalue}}}(x_1,\ldots,x_n,\omega)$$

$$= \int_{-\infty}^{\infty}\frac{e^{\frac{ix_1p_1}{2\hbar}}}{\sqrt{4\pi\hbar}}\cdots\int_{-\infty}^{\infty}\frac{e^{\frac{ix_np_n}{2\hbar}}}{\sqrt{4\pi\hbar}}L^{-1}_{\left(\begin{pmatrix}\bar{x}_1,\ldots,\bar{x}_n\\ ;\bar{p}_1,\ldots,\bar{p}_n\end{pmatrix}\to\begin{pmatrix}x_1,\ldots,x_n\\ ;p_1,\ldots,p_n\end{pmatrix}\right)}\left[\begin{array}{l}\sqrt{2\pi}\delta\left(\omega-\frac{1}{\hbar}\hat{H}_{\text{configuration space}}\begin{pmatrix}x_1,\ldots,x_n,\\ -i\hbar\partial_{x_1},\ldots,-i\hbar\partial_{x_n}\end{pmatrix}_{\substack{\left((x_1,\ldots,x_n)\mapsto\begin{pmatrix}i\hbar\bar{p}_1+\alpha_1x_1,\\ \ldots i\hbar\bar{p}_n+\alpha_nx_n\end{pmatrix}\right)\\ \left((-i\hbar\partial_{x_1},\ldots,-i\hbar\partial_{x_n})\mapsto\begin{pmatrix}-i\hbar\bar{x}_1+\gamma_1p_1,\\ \ldots,-i\hbar\bar{x}_n+\gamma_np_n\end{pmatrix}\right)}}\right)\\ \times\bar{\Psi}_{0\,\text{configuration space}}\begin{pmatrix}\bar{x}_1,\ldots,\bar{x}_n;\\ \bar{p}_1,\ldots,\bar{p}_n;t=0\end{pmatrix}\end{array}\right]dp_1..dp_n \tag{16d}$$

To facilitate the evaluation of (16e) the Identity for Dirac Delta Functions is useful

$$\delta(g(u))=\sum_{j=1}^{n}\frac{\delta(u-u_j)}{\left|(D_ug(u))\right|_{u=u_j}\right|}\quad,\ni g(u_j)=0\ ,\ (D_ug(u))\big|_{u=u_j}\neq 0$$

Which completes the construction of the result.

To illustrate this, consider the Example 1 (1-dim Simple Harmonic Oscillator) with constant strength a. Applying the formalism above in (16d) to the term from (11d) yields

$$\left(\frac{(-i\hbar\partial_x)^2}{2m}+ax^2\right)\Psi_{\text{configuration space}}(x,t)=i\hbar\partial_t\Psi_{\text{configuration space}}(x,t)$$

$$F_{t\to\omega}\left(\Psi_{\text{configuration space}}(x,t)=\int_{-\infty}^{\infty}\frac{e^{\frac{ixp}{2\hbar}}}{\sqrt{4\pi\hbar}}L^{-1}_{\left((\bar{x};\bar{p})\to(x;p)\right)}\left[\bar{\Psi}_0(\bar{x},\bar{p},t=0)e^{\frac{-i}{\hbar}t\left(\frac{(-i\hbar\partial_x)^2}{2m}+ax^2\right)_{\substack{\left((x)\mapsto(i\hbar\bar{p}+\alpha x)\right)\\ \left((-i\hbar\partial_x)\mapsto(-i\hbar\bar{x}+\gamma p)\right)}}}\right]dp\right) \tag{16e}$$

$$\Psi_{\substack{\text{configuration space}\\ \omega\,\text{eigenvalue}}}(x,\omega)=\int_{-\infty}^{\infty}\frac{e^{\frac{ixp}{2\hbar}}}{\sqrt{4\pi\hbar}}L^{-1}_{\left((\bar{x};\bar{p})\to(x;p)\right)}\left[\sqrt{2\pi}\bar{\Psi}_0(\bar{x},\bar{p},t=0)\delta\left(\omega-\frac{1}{\hbar}\left(\frac{\left(-i\hbar\bar{x}+\frac{p}{2}\right)^2}{2m}+a\left(i\hbar\bar{p}+\frac{x}{2}\right)^2\right)\right)\right]dp$$

$$\left(\frac{(-i\hbar\partial_x)^2}{2m}+ax^2-\hbar\omega\right)\Psi_{\substack{\text{configuration space}\\ \omega\,\text{eigenvalue}}}(x,\omega)=0$$

Just to point out, if the potential term of the SHO above ax^2 is replaced with a potential term of arbitrary functional profile as $V(x)$ then (16e) generalizes to

$$\left(\frac{(-i\hbar\partial_x)^2}{2m}+V(x)\right)\Psi_{\text{configuration space}}(x,t)=i\hbar\partial_t\Psi_{\text{configuration space}}(x,t)$$

$$F_{t\to\omega}\left(\Psi_{\text{configuration space}}(x,t)=\int_{-\infty}^{\infty}\frac{e^{\frac{ixp}{2\hbar}}}{\sqrt{4\pi\hbar}}L^{-1}_{\left((\bar{x};\bar{p})\to(x;p)\right)}\left[\bar{\Psi}_0(\bar{x},\bar{p},t=0)e^{\frac{-i}{\hbar}t\left(\frac{(-i\hbar\partial_x)^2}{2m}+V(x)\right)_{\substack{\left((x)\mapsto(i\hbar\bar{p}+\alpha x)\right)\\ \left((-i\hbar\partial_x)\mapsto(-i\hbar\bar{x}+\gamma p)\right)}}}\right]dp\right) \tag{16f}$$

$$\Psi_{\substack{\text{configuration space}\\ \omega\,\text{eigenvalue}}}(x,\omega)=\int_{-\infty}^{\infty}\frac{e^{\frac{ixp}{2\hbar}}}{\sqrt{4\pi\hbar}}L^{-1}_{\left((\bar{x};\bar{p})\to(x;p)\right)}\left[\sqrt{2\pi}\bar{\Psi}_0(\bar{x},\bar{p},t=0)\delta\left(\omega-\frac{1}{\hbar}\left(\frac{\left(-i\hbar\bar{x}+\frac{p}{2}\right)^2}{2m}+V\left(i\hbar\bar{p}+\frac{x}{2}\right)\right)\right)\right]dp$$

$$\left(\frac{(-i\hbar\partial_x)^2}{2m}+V(x)-\hbar\omega\right)\Psi_{\substack{\text{configuration space}\\ \omega\,\text{eigenvalue}}}(x,\omega)=0$$

It is worth re-emphasizing that the result (16d) of course is immediately applicable to non-seperable Hamiltonians in general, such as Example 3 (Schrödinger Molecular Hamiltonian with Pairwise Coulomb Interaction), yielding the solufion of the radial Molecular Schrodinger Eigenvalue Equation and many others.

Example 7. Exact Quadrature Solution of General Class of Variable-Coefficient Differential Equations

In view of the arbitrary functional form of the Hamiltonian and its connection with a Lagrangian [ref. 11, 12, 13] for the given problem,

$$\begin{aligned} H(x_1,\dots,x_n;p_1,\dots,p_n;t) &= \sum_{j=1}^{n} p_j\dot{x}_j - L(x_1,\dots,x_n;\dot{x}_1,\dots,\dot{x}_n;t) \\ p_j &= \partial_{\dot{x}_j} L(x_1,\dots,x_n;\dot{x}_1,\dots,\dot{x}_n;t) \end{aligned} \tag{17a}$$

Consider now the equations of motion for the case of an external force $F(x_1,\dots,x_n;\dot{x}_1,\dots,\dot{x}_n;t)$(viz the Extended Hamilton Principle) given as

$$(-\partial_t)\partial_{\dot{x}_j} L(x_1,\dots,x_n;\dot{x}_1,\dots,\dot{x}_n;t) + \partial_{x_j} L(x_1,\dots,x_n;\dot{x}_1,\dots,\dot{x}_n;t) = F(x_1,\dots,x_n;\dot{x}_1,\dots,\dot{x}_n;t) \tag{17b}$$

It follows that (17b) extends naturally to accommodate higher-order derivatives for several variables, though the notation becomes cumbersome [refs 12,13]. As such we will below only construct the results for the case of one dependent variable x as

$$\sum_{j=1}^{n} (-D_t)^j \partial_{(D_t)^j x} L(x;\dot{x};\ddot{x};\dddot{x};\dots;D_t^n x;t) + \partial_{x_j} L(x;\dot{x};\ddot{x};\dddot{x};\dots;D_t^n x;t) = F(x;\dot{x};\ddot{x};\dddot{x};\dots;D_t^n x;t)$$

for example $n = 4$ yields

$$\left(D_t^4\partial_{D_t^4 x} - D_t^3\partial_{D_t^3 x} + D_t^2\partial_{D_t^2 x} - (D_t)\partial_{D_t x} + \partial_x\right)L\left(x;\dot{x};\ddot{x};\dddot{x};\left(\tfrac{dx}{dt}\right)^4;t\right) = F\left(x;\dot{x};\ddot{x};\dddot{x};\left(\tfrac{dx}{dt}\right)^4;t\right) \tag{17c}$$

In turn, these order-n derivative Lagrangians are reducible to first-order derivative systems by way of variable substitutions[ref. 12,13]. In this note, the properties in (17b) and (17c) are pointed out explicitly, as they form the essential basis for applying the HOA scheme to any system that can be put in the Hamiltonian form: from higher-order Lagrangian systems of discrete particles to continuous fields[refs. 11,12,13,14.15]: they all admit to HOA solution as they are all ultimately transformable to the Hamiltonian formulation of dynamics. So, without loss of generality, we consider the case of (17b) in one dependent variable x

$$(-D_t)\partial_{\dot{x}} L(x;\dot{x};t) + \partial_x L(x;\dot{x};t) = F(x;\dot{x};t) \tag{17d}$$

Now to illustrate HOA in the context of this Hamiltonian-Lagrangian connection, take the Lagrangian $L(x;\dot{x};t)=\frac{m(\dot{x}^2)}{2}-a(t)x^2$ and no external term $F(x;\dot{x};t)=0$ from the generalized Hamilton Principle, yielding from (17d)

$$\begin{aligned}&(-D_t)\partial_{\dot{x}}\left(\frac{m(\dot{x}^2)}{2}-a(t)x^2\right)+\partial_x\left(\frac{m(\dot{x}^2)}{2}-a(t)x^2\right)=0\\&\frac{m\ddot{x}}{2}+a(t)x=0\end{aligned}\tag{17e}$$

which is the equation of motion for the classical Simple Harmonic Oscillator Lagrangian with arbitrary time-dependent strength $a(t)$ therein. By way of (17a), the Hamiltonian that corresponds to the system in (17e) is $H=\frac{p^2}{2m}+a(t)x^2$: this is clearly the Hamiltonian for Example 1 herein: the quantum version of this system. Following Example 1 , the QPSR of this system is given by

$$\left(\frac{\left(-i\hbar\partial_x+\frac{p}{2}\right)^2}{2m}+a(t)\left(i\hbar\partial_p+\frac{x}{2}\right)^2\right)\Psi(x;p;t)=i\hbar\partial_t\Psi(x;p;t)$$

$$\left(\frac{\left(-i\hbar\overline{x}+\frac{p}{2}\right)^2}{2m}+a(t)\left(i\hbar\overline{p}+\frac{x}{2}\right)^2\right)\breve{\Psi}(\overline{x};\overline{p};t)=i\hbar\partial_t\breve{\Psi}(\overline{x};\overline{p};t)$$

$$\Psi(x;p;t)=L^{-1}_{((\overline{x})\mapsto(x))}\left[L^{-1}_{((\overline{p})\mapsto(p))}\left[\begin{array}{l}e^{\frac{-i\left(t\frac{\left(-i\hbar\overline{x}+\frac{p}{2}\right)^2}{2m}+\int_0^t a(u)du\left(i\hbar\overline{p}+\frac{x}{2}\right)^2\right)}{\hbar}}\\\times\\\breve{\Psi}_0(\overline{x};\overline{p};t=0)\end{array}\right]\right]\tag{11d}$$

With the configuration space wavefunction via QPSR wavefunction tor this system given as

$$\Psi_{\text{configuration space}}(x,t)=\int_{-\infty}^{\infty}\frac{e^{\frac{ixp}{2\hbar}}}{\sqrt{4\pi\hbar}}\left[\Psi_0(x,p,t=0)\underbrace{*}_{x,p}\sqrt{\frac{-m}{2\int_0^t a(u)du}}\frac{e^{i\left(\frac{p^2+2mx^2\int_0^t a(u)du}{4\hbar t\int_0^t a(u)du}\right)}}{2\pi\hbar t}\right]dp \quad (11e)$$

Now it follows by way of Ehrenfest's Theorem that the expectation value of the position[also momentum] operator follows its classical counterpart's equations of motion in any framework(e.g., Hamiltonian,Lagrangian, etc]. Following the Recap herein, applying the configuration position operator $\hat{x}\equiv x$ to the system wavefunction (11e) to yield the expectation value of the position operator as

$$\langle x(t)\rangle=\int_{\text{Configuration Space}}\left(\Psi^*_{\text{configuration space}}(x,t)(x)\Psi_{\text{configuration space}}(x,t)\right)dx$$

$$\Psi_{\text{configuration space}}(x,t)=\int_{-\infty}^{\infty}\frac{e^{\frac{ixp}{2\hbar}}}{\sqrt{4\pi\hbar}}\left[\Psi_0(x,p,t=0)\underbrace{*}_{x,p}\sqrt{\frac{-m}{2\int_0^t a(u)du}}\frac{e^{i\left(\frac{p^2+2mx^2\int_0^t a(u)du}{4\hbar t\int_0^t a(u)du}\right)}}{2\pi\hbar t}\right]dp \quad (17f)$$

So

$$\langle x(t)\rangle=\int_{\text{Configuration Space}}\left|\Psi_{\text{configuration space}}(x,t)\right|^2 xdx$$

$$\left|\Psi_{\text{configuration space}}(x,t)\right|^2=\left|\int_{-\infty}^{\infty}\frac{e^{\frac{ixp}{2\hbar}}}{\sqrt{4\pi\hbar}}\left[\Psi_0(x,p,t=0)\underbrace{*}_{x,p}\sqrt{\frac{-m}{2\int_0^t a(u)du}}\frac{e^{i\left(\frac{p^2+2mx^2\int_0^t a(u)du}{4\hbar t\int_0^t a(u)du}\right)}}{2\pi\hbar t}\right]dp\right|^2 \quad (17g)$$

The upshot of equation (17g) is that $\langle x(t)\rangle$ follows the same equation of motion (17e).

Therefore

$$\frac{m\langle \ddot{x}(t)\rangle}{2}+a(t)\langle x(t)\rangle=0$$

$$\langle x(t)\rangle=\int_{\text{Configuration Space}}\left|\int_{-\infty}^{\infty}\frac{e^{\frac{ixp}{2\hbar}}}{\sqrt{4\pi\hbar}}\left[\Psi_0(x,p,t=0)\underbrace{*}_{x,p}\sqrt{\frac{-m}{2\int_0^t a(u)du}}\,\frac{e^{i\left(\frac{p^2+2mx^2\int_0^t a(u)du}{4\hbar t\int_0^t a(u)du}\right)}}{2\pi\hbar t}\right]dp\right|^2 x\,dx \tag{17h}$$

Thus Example 7 above provides an exact analytical quadrature solution to the given related classical equations of motion as well. By way of the Hamiltonian-Lagrangian correspondence, though the variables are different, (17h) provides an exact analytical quadrature solution of the purely mathematical problem of the exact analytical quadrature solution to the second-order linear Ordinary Differential Equation with arbitrary variable coefficient in it`s canonical homogeneous form

$$\ddot{y}(t)+b(t)y(t)=0$$

a problem which should require no introduction.

Finally, consider now the Lagrangian $L(x;\dot{x};t)=\frac{5(\dot{x}^2)}{4}+\frac{x^{\frac{5}{2}}}{t^{\frac{1}{2}}}$ and no external term $F(x;\dot{x};t)=0$ from the generalized Hamilton Principle, yielding from (17d)

$$(-D_t)\partial_{\dot{x}}\left(\frac{5(\dot{x}^2)}{4}+\frac{x^{\frac{5}{2}}}{t^{\frac{1}{2}}}\right)+\partial_x\left(\frac{5(\dot{x}^2)}{4}+\frac{x^{\frac{5}{2}}}{t^{\frac{1}{2}}}\right)=0$$

$$\ddot{x}=\frac{x^{\frac{3}{2}}}{t^{\frac{1}{2}}} \tag{17i}$$

following HOA yields

$$\left(\frac{\left(-i\hbar\partial_x+\frac{p}{2}\right)^2}{2(5/2)}-\frac{1}{t^{\frac{1}{2}}}\left(i\hbar\partial_p+\frac{x}{2}\right)^{\frac{5}{2}}\right)\Psi(x;p;t)=i\hbar\partial_t\Psi(x;p;t)\ ,\ni m=\frac{5}{2}$$

$$\left(\frac{\left(-i\hbar\bar{x}+\frac{p}{2}\right)^2}{5}-\frac{1}{t^{\frac{1}{2}}}\left(i\hbar\bar{p}+\frac{x}{2}\right)^{\frac{5}{2}}\right)\breve{\Psi}(\bar{x};\bar{p};t)=i\hbar\partial_t\breve{\Psi}(\bar{x};\bar{p};t)$$

$$\Psi(x;p;t)=L^{-1}_{((\bar{x})\mapsto(x))}\left[L^{-1}_{((\bar{p})\mapsto(p))}\left[\breve{\Psi}_0(\bar{x};\bar{p};t=0)e^{\frac{-i\left(t\frac{\left(-i\hbar\bar{x}+\frac{p}{2}\right)^2}{5}-2t^{\frac{1}{2}}\left(i\hbar\bar{p}+\frac{x}{2}\right)^{\frac{5}{2}}\right)}{\hbar}}\right]\right] \tag{17j}$$

By way of the prescription of (17f) through (17h), we arrive at

$$\langle x(t)\rangle=\int_{\text{Configuration Space}}\left|\Psi_{\text{configuration space}}(x,t)\right|^2 x\,dx$$

$$\left|\Psi_{\text{configuration space}}(x,t)\right|^2=\left|\int_{-\infty}^{\infty}\frac{e^{\frac{ixp}{2\hbar}}}{\sqrt{4\pi\hbar}}L^{-1}_{\substack{(\bar{x};\bar{p})\\ \to(x;p)}}\left[\breve{\Psi}_0(\bar{x},\bar{p},t=0)e^{\frac{-i\left(t\frac{\left(-i\hbar\bar{x}+\frac{p}{2}\right)^2}{5}-2t^{\frac{1}{2}}\left(i\hbar\bar{p}+\frac{x}{2}\right)^{\frac{5}{2}}\right)}{\hbar}}\right]dp\right|^2 \tag{17k}$$

$$\langle \ddot{x}(t)\rangle=\frac{\langle x(t)\rangle^{\frac{3}{2}}}{t^{\frac{1}{2}}}$$

Hence (17k) is an exact quadrature solution of the Thomas-Fermi Equation.

Returning now to the case of a generalised Lagrange equation with the inhomogeneous term

$$(-D_t)\partial_{\dot{x}}L(x;\dot{x};t)+\partial_x L(x;\dot{x};t)=F(x;\dot{x};t) \tag{17d}$$

let us suppose that $F(x;\dot{X}t)$ is specified from the extended Hamilton Principle and

Lagrange equation with inhomogeneuous term $F(x(t);\dot{X}(t);t)$ where we recall that the arguments of this inhomogeneous term are explicitly time-dependent[i.e., rheonomic] ....

$$(-D_t)\partial_{\dot{x}(t)}L(x(t);\dot{x}(t);t)+\partial_{x(t)}L(x(t);\dot{x}(t);t)=F(x(t);\dot{x}(t);t)$$

Letting $L(x(t);\dot{x}(t);t)$ and $F(x(t);\dot{x}(t);t)$ be scalar functions of their arguements $(x(t);\dot{x}(t);t)$,

Utilising a Laplace Transform

$$\mathsf{L}_{\substack{x(t)->\bar{x}(t)\\ \dot{x}(t)->\bar{\dot{x}}(t)}}(f(x(t);\dot{x}(t);t))\equiv\int_{\dot{x}(t)_0}^{\infty}\int_{x(t)_0}^{\infty}(f(x(t);\dot{x}(t);t))e^{-\bar{x}(t)x(t)-\bar{\dot{x}}(t)\dot{x}(t)}d\dot{x}(t)dx(t)$$
$$\mathsf{L}^{-1}_{\substack{x(t)->\bar{x}(t)\\ \dot{x}(t)->\bar{\dot{x}}(t)}}(\bar{f}(\bar{x}(t);\bar{\dot{x}}(t);t))\equiv\left(\tfrac{1}{2\pi i}\right)^2\oint_{\partial_{\bar{x}(t)}}\oint_{\partial_{\bar{\dot{x}}(t)}}(\bar{f}(\bar{x}(t);\bar{\dot{x}}(t);t))e^{\bar{x}(t)x(t)+\bar{\dot{x}}(t)\dot{x}(t)}d\bar{\dot{x}}(t)d\bar{x}(t)$$

on the Lagrange equation, thus

$$(-D_t)\partial_{\dot{x}(t)}L(x(t);\dot{x}(t);t)+\partial_{x(t)}L(x(t);\dot{x}(t);t)=F(x(t);\dot{x}(t);t)$$
$$\mathsf{L}_{\substack{x(t)->\bar{x}(t)\\ \dot{x}(t)->\bar{\dot{x}}(t)}}((-D_t)\partial_{\dot{x}(t)}L(x(t);\dot{x}(t);t)+\partial_{x(t)}L(x(t);\dot{x}(t);t)=F(x(t);\dot{x}(t);t))$$
$$-D_t(\bar{\dot{x}}(t)\bar{L}(\bar{x}(t);\bar{\dot{x}}(t);t))+\bar{x}(t)\bar{L}(\bar{x}(t);\bar{\dot{x}}(t);t)=\bar{F}(\bar{x}(t);\bar{\dot{x}}(t);t)$$
$$(-D_t\bar{\dot{x}}(t))\bar{L}(\bar{x}(t);\bar{\dot{x}}(t);t)+\bar{\dot{x}}(t)(-D_t\bar{L}(\bar{x}(t);\bar{\dot{x}}(t);t))+\bar{x}(t)\bar{L}(\bar{x}(t);\bar{\dot{x}}(t);t)=\bar{F}(\bar{x}(t);\bar{\dot{x}}(t);t)$$
$$(-\bar{\dot{x}}(t)D_t+\bar{x}(t)-D_t\bar{\dot{x}}(t))\bar{L}(\bar{x}(t);\bar{\dot{x}}(t);t)=\bar{F}(\bar{x}(t);\bar{\dot{x}}(t);t)$$
$$\left(D_t+\left(\frac{D_t\bar{\dot{x}}(t)}{\bar{\dot{x}}(t)}-\frac{\bar{x}(t)}{\bar{\dot{x}}(t)}\right)\right)\bar{L}(\bar{x}(t);\bar{\dot{x}}(t);t)=\frac{\bar{F}(\bar{x}(t);\bar{\dot{x}}(t);t)}{\bar{\dot{x}}(t)}$$
$$\bar{L}(\bar{x}(t);\bar{\dot{x}}(t);t)=\frac{1}{\bar{\dot{x}}(t)}e^{\int\frac{\bar{x}(t)}{\bar{\dot{x}}(t)}dt}\int e^{-\int\frac{\bar{x}(t)}{\bar{\dot{x}}(t)}dt}\bar{F}(\bar{x}(t);\bar{\dot{x}}(t);t)dt$$
$$\mathsf{L}^{-1}_{\substack{\bar{x}(t)->x(t)\\ \bar{\dot{x}}(t)->\dot{x}(t)}}\left(\bar{L}(\bar{x}(t);\bar{\dot{x}}(t);t)=\frac{1}{\bar{\dot{x}}(t)}e^{\int\frac{\bar{x}(t)}{\bar{\dot{x}}(t)}dt}\int e^{-\int\frac{\bar{x}(t)}{\bar{\dot{x}}(t)}dt}\bar{F}(\bar{x}(t);\bar{\dot{x}}(t);t)dt\right)\tag{171}$$
$$L(x(t);\dot{x}(t);t)=\mathsf{L}^{-1}_{\substack{\bar{x}(t)->x(t)\\ \bar{\dot{x}}(t)->\dot{x}(t)}}\left(\frac{1}{\bar{\dot{x}}(t)}e^{\int\frac{\bar{x}(t)}{\bar{\dot{x}}(t)}dt}\int e^{-\int\frac{\bar{x}(t)}{\bar{\dot{x}}(t)}dt}\bar{F}(\bar{x}(t);\bar{\dot{x}}(t);t)dt\right)$$

As a result, all of the above construction for $L(x(t);\dot{x}(t);t)$ and $F(x(t);\dot{x}(t);t)$ are generalised to $L(\mathbf{x}(t);\dot{\mathbf{x}}(t);t)$ and $F(\mathbf{x}(t);\dot{\mathbf{x}}(t);t)$ thus

$$
\begin{aligned}
&(-D_t)\partial_{\dot{\mathbf{x}}(t)}L(\mathbf{x}(t);\dot{\mathbf{x}}(t);t)+\partial_{\mathbf{x}(t)}L(\mathbf{x}(t);\dot{\mathbf{x}}(t);t)=F(\mathbf{x}(t);\dot{\mathbf{x}}(t);t)\\
&\mathsf{L}_{\substack{\mathbf{x}(t)->\overline{\mathbf{x}}(t)\\ \dot{\mathbf{x}}(t)->\dot{\overline{\mathbf{x}}}(t)}}\left((-D_t)\partial_{\dot{\mathbf{x}}(t)}L(\mathbf{x}(t);\dot{\mathbf{x}}(t);t)+\partial_{\mathbf{x}(t)}L(\mathbf{x}(t);\dot{\mathbf{x}}(t);t)=F(\mathbf{x}(t);\dot{\mathbf{x}}(t);t)\right)\\
&-D_t\left(\dot{\overline{\mathbf{x}}}(t)\overline{L}(\overline{\mathbf{x}}(t);\dot{\overline{\mathbf{x}}}(t);t)\right)+\overline{\mathbf{x}}(t)\overline{L}(\overline{\mathbf{x}}(t);\dot{\overline{\mathbf{x}}}(t);t)=\overline{F}(\overline{\mathbf{x}}(t);\dot{\overline{\mathbf{x}}}(t);t)\\
&\left(-D_t\dot{\overline{\mathbf{x}}}(t)\right)\overline{L}(\overline{\mathbf{x}}(t);\dot{\overline{\mathbf{x}}}(t);t)+\dot{\overline{\mathbf{x}}}(t)\left(-D_t\overline{L}(\overline{\mathbf{x}}(t);\dot{\overline{\mathbf{x}}}(t);t)\right)+\overline{\mathbf{x}}(t)\overline{L}(\overline{\mathbf{x}}(t);\dot{\overline{\mathbf{x}}}(t);t)=\overline{F}(\overline{\mathbf{x}}(t);\dot{\overline{\mathbf{x}}}(t);t)\\
&\left(-\dot{\overline{\mathbf{x}}}(t)D_t+\overline{\mathbf{x}}(t)-D_t\dot{\overline{\mathbf{x}}}(t)\right)\overline{L}(\overline{\mathbf{x}}(t);\dot{\overline{\mathbf{x}}}(t);t)=\overline{F}(\overline{\mathbf{x}}(t);\dot{\overline{\mathbf{x}}}(t);t)\\
&\left(D_t+\left(\frac{D_t\dot{\overline{\mathbf{x}}}(t)}{\dot{\overline{\mathbf{x}}}(t)}-\frac{\overline{\mathbf{x}}(t)}{\dot{\overline{\mathbf{x}}}(t)}\right)\right)\overline{L}(\overline{\mathbf{x}}(t);\dot{\overline{\mathbf{x}}}(t);t)=\frac{\overline{F}(\overline{\mathbf{x}}(t);\dot{\overline{\mathbf{x}}}(t);t)}{\dot{\overline{\mathbf{x}}}(t)}\\
&\overline{L}(\overline{\mathbf{x}}(t);\dot{\overline{\mathbf{x}}}(t);t)=\frac{1}{\dot{\overline{\mathbf{x}}}(t)}e^{\int\frac{\overline{\mathbf{x}}(t)}{\dot{\overline{\mathbf{x}}}(t)}dt}\int e^{-\int\frac{\overline{\mathbf{x}}(t)}{\dot{\overline{\mathbf{x}}}(t)}dt}\,\overline{F}(\overline{\mathbf{x}}(t);\dot{\overline{\mathbf{x}}}(t);t)dt\\
&\mathsf{L}^{-1}_{\substack{\overline{\mathbf{x}}(t)->\mathbf{x}(t)\\ \dot{\overline{\mathbf{x}}}(t)->\dot{\mathbf{x}}(t)}}\left(\overline{L}(\overline{\mathbf{x}}(t);\dot{\overline{\mathbf{x}}}(t);t)=\frac{1}{\dot{\overline{\mathbf{x}}}(t)}e^{\int\frac{\overline{\mathbf{x}}(t)}{\dot{\overline{\mathbf{x}}}(t)}dt}\int e^{-\int\frac{\overline{\mathbf{x}}(t)}{\dot{\overline{\mathbf{x}}}(t)}dt}\,\overline{F}(\overline{\mathbf{x}}(t);\dot{\overline{\mathbf{x}}}(t);t)dt\right)\\
&L(\mathbf{x}(t);\dot{\mathbf{x}}(t);t)=\mathsf{L}^{-1}_{\substack{\overline{\mathbf{x}}(t)->\mathbf{x}(t)\\ \dot{\overline{\mathbf{x}}}(t)->\dot{\mathbf{x}}(t)}}\left(\frac{1}{\dot{\overline{\mathbf{x}}}(t)}e^{\int\frac{\overline{\mathbf{x}}(t)}{\dot{\overline{\mathbf{x}}}(t)}dt}\int e^{-\int\frac{\overline{\mathbf{x}}(t)}{\dot{\overline{\mathbf{x}}}(t)}dt}\,\overline{F}(\overline{\mathbf{x}}(t);\dot{\overline{\mathbf{x}}}(t);t)dt\right)
\end{aligned}
\tag{17m}
$$

Along these lines, the component scalar variables of the vector variables , may naturally be expressed as projections of the vector variables via suitable matrix coefficients thus

$$
\begin{aligned}
&(-D_t)\partial_{\dot{\mathbf{x}}(t)}L(\mathbf{x}(t);\dot{\mathbf{x}}(t);t)+\partial_{\mathbf{x}(t)}L(\mathbf{x}(t);\dot{\mathbf{x}}(t);t)=F(\mathbf{x}(t);\dot{\mathbf{x}}(t);t)\\
&(\mathbf{x}(t);\dot{\mathbf{x}}(t);t),\ni\\
&\mathbf{x}(t)=\begin{bmatrix}x_1(t)\\x_2(t)\\ \cdot\\ \cdot\\x_n(t)\end{bmatrix},\begin{bmatrix}1&&&\\&0&&\\&&\cdot&\\&&&0\end{bmatrix}\begin{bmatrix}x_1(t)\\x_2(t)\\ \cdot\\ \cdot\\x_n(t)\end{bmatrix}=x_1(t),\begin{bmatrix}0&&&\\&1&&\\&&\cdot&\\&&&0\end{bmatrix}\begin{bmatrix}x_1(t)\\x_2(t)\\ \cdot\\ \cdot\\x_n(t)\end{bmatrix}=x_2(t),\begin{bmatrix}0&&&\\&0&&\\&&\cdot&\\&&&1\end{bmatrix}\begin{bmatrix}x_1(t)\\x_2(t)\\ \cdot\\ \cdot\\x_n(t)\end{bmatrix}=x_n(t)\\
&\dot{\mathbf{x}}(t)=\begin{bmatrix}\dot{x}_1(t)\\\dot{x}_2(t)\\ \cdot\\ \cdot\\\dot{x}_n(t)\end{bmatrix},\begin{bmatrix}1&&&\\&0&&\\&&\cdot&\\&&&0\end{bmatrix}\begin{bmatrix}\dot{x}_1(t)\\\dot{x}_2(t)\\ \cdot\\ \cdot\\\dot{x}_n(t)\end{bmatrix}=\dot{x}_1(t),\begin{bmatrix}0&&&\\&1&&\\&&\cdot&\\&&&0\end{bmatrix}\begin{bmatrix}\dot{x}_1(t)\\\dot{x}_2(t)\\ \cdot\\ \cdot\\\dot{x}_n(t)\end{bmatrix}=\dot{x}_2(t),\begin{bmatrix}0&&&\\&0&&\\&&\cdot&\\&&&1\end{bmatrix}\begin{bmatrix}\dot{x}_1(t)\\\dot{x}_2(t)\\ \cdot\\ \cdot\\\dot{x}_n(t)\end{bmatrix}=\dot{x}_n(t)
\end{aligned}
\tag{17n}
$$

So the usual system of equations in several dependent variables in (17b) may be re-cast asa singal vector variable equation as above since

$$(\mathbf{x}(t);\dot{\mathbf{x}}(t);t)=\left(\begin{bmatrix}x_1(t)\\x_2(t)\\ \cdot\\ \cdot\\x_n(t)\end{bmatrix};\begin{bmatrix}\dot{x}_1(t)\\\dot{x}_2(t)\\ \cdot\\ \cdot\\\dot{x}_n(t)\end{bmatrix};t\right)$$

While these vector variable extensions in the Lagrangian are useful in various multi-dimensional scenarios, there is another that will be mentioned here as well. Suppose we consider the Lagrange system for explicit components of an n-dimensional vector

$$\begin{pmatrix}\left((-D_t)\partial_{\dot{x}_j(t)}+\partial_{x_j(t)}\right)L(x_1(t),x_2(t),....,x_n(t);\dot{x}_1(t),\dot{x}_2(t),....,\dot{x}_n(t);t)\\-F_j(x_1(t),x_2(t),....,x_n(t);\dot{x}_1(t),\dot{x}_2(t),....,\dot{x}_n(t);t)\end{pmatrix}\mathbf{e}_j,\ni j=1,....,\text{n}$$

$$\text{Letting }\mathbf{e}_j,\ni \text{j}=1,....,\text{n be the n - dimensional unit vector}$$

Upon taking the interior product of the Lagrangian components vector with the unit vector yields the n-dimensional equation for the system from which an L satisfying these relations to the F_j 's may be determined thus

$$\sum_j\begin{pmatrix}\left((-D_t)\partial_{\dot{x}_j(t)}+\partial_{x_j(t)}\right)L(x_1(t),x_2(t),....,x_n(t);\dot{x}_1(t),\dot{x}_2(t),....,\dot{x}_n(t);t)\\-F_j(x_1(t),x_2(t),....,x_n(t);\dot{x}_1(t),\dot{x}_2(t),....,\dot{x}_n(t);t)\end{pmatrix}=0$$

$$\mathsf{L}_{\substack{x_j(t)->\bar{x}_j(t)\\\dot{x}_j(t)->\bar{\dot{x}}_j(t)}}\left(\sum_j\begin{pmatrix}\left((-D_t)\partial_{\dot{x}_j(t)}+\partial_{x_j(t)}\right)L(x_1(t),x_2(t),....,x_n(t);\dot{x}_1(t),\dot{x}_2(t),....,\dot{x}_n(t);t)\\-F_j(x_1(t),x_2(t),....,x_n(t);\dot{x}_1(t),\dot{x}_2(t),....,\dot{x}_n(t);t)\end{pmatrix}\right)=0$$

$$\sum_j\begin{pmatrix}-D_t\left(\bar{\dot{x}}_j(t)\bar{L}(\bar{x}_1(t),\bar{x}_2(t),....,\bar{x}_n(t);\bar{\dot{x}}_1(t),\bar{\dot{x}}_2(t),....,\bar{\dot{x}}_n(t);t)\right)\\+\bar{x}_j(t)\bar{L}(\bar{x}_1(t),\bar{x}_2(t),....,\bar{x}_n(t);\bar{\dot{x}}_1(t),\bar{\dot{x}}_2(t),....,\bar{\dot{x}}_n(t);t)\\-\bar{F}_j(\bar{x}_1(t),\bar{x}_2(t),....,\bar{x}_n(t);\bar{\dot{x}}_1(t),\bar{\dot{x}}_2(t),....,\bar{\dot{x}}_n(t);t)\end{pmatrix}=0$$

$$\sum_j\begin{pmatrix}\left(-\bar{\dot{x}}_j(t)D_t+\bar{x}_j(t)-D_t\bar{\dot{x}}_j(t)\right)\bar{L}(\bar{x}_1(t),\bar{x}_2(t),....,\bar{x}_n(t);\bar{\dot{x}}_1(t),\bar{\dot{x}}_2(t),....,\bar{\dot{x}}_n(t);t)\\-\bar{F}_j(\bar{x}_1(t),\bar{x}_2(t),....,\bar{x}_n(t);\bar{\dot{x}}_1(t),\bar{\dot{x}}_2(t),....,\bar{\dot{x}}_n(t);t)\end{pmatrix}=0$$

$$\bar{L}(\bar{x}_1(t),\bar{x}_2(t),....,\bar{x}_n(t);\bar{\dot{x}}_1(t),\bar{\dot{x}}_2(t),....,\bar{\dot{x}}_n(t);t)$$
$$=e^{-\int\frac{\sum_j\bar{x}_j(t)-D_t\bar{\dot{x}}_j(t)}{\sum_j-\bar{\dot{x}}_j(t)}dt}\int\left(e^{\int\frac{\sum_j\bar{x}_j(t)-D_t\bar{\dot{x}}_j(t)}{\sum_j-\bar{\dot{x}}_j(t)}dt}\sum_j\bar{F}_j(\bar{x}_1(t),\bar{x}_2(t),....,\bar{x}_n(t);\bar{\dot{x}}_1(t),\bar{\dot{x}}_2(t),....,\bar{\dot{x}}_n(t);t)\right)dt$$

$$L(x_1(t),x_2(t),....,x_n(t);\dot{x}_1(t),\dot{x}_2(t),....,\dot{x}_n(t);t) \tag{17o}$$
$$=\mathsf{L}^{-1}_{\substack{x_j(t)->\bar{x}_j(t)\\\dot{x}_j(t)->\bar{\dot{x}}_j(t)}}\left(e^{-\int\frac{\sum_j\bar{x}_j(t)-D_t\bar{\dot{x}}_j(t)}{\sum_j-\bar{\dot{x}}_j(t)}dt}\int\left(e^{\int\frac{\sum_j\bar{x}_j(t)-D_t\bar{\dot{x}}_j(t)}{\sum_j-\bar{\dot{x}}_j(t)}dt}\sum_j\bar{F}_j(\bar{x}_1(t),\bar{x}_2(t),....,\bar{x}_n(t);\bar{\dot{x}}_1(t),\bar{\dot{x}}_2(t),....,\bar{\dot{x}}_n(t);t)\right)dt\right)$$

So the point of (17l-o) is to determine the Lagrangian L for the given inhomogeneous term F; and using the results developed herein earlier, from (17a) on, to determine the corresponding Hamiltonian of the system and via HOA and Ehrenfest's Theorem, the solution of the equations defined by $F=0$:said equations are arbitrary functions of the arguments in F.

....And a New Twist

It is at this point that the analysis of HOA has been developed up to the time of this writing. A "New Twist" on this was recently[13MAY 2010], discovered by Simpao: namely an alternative exact analytical quadrature expression for the quantum phase space wave function. By observing from (8) that

$$\Psi(x_1,\ldots,x_n;p_1,\ldots,p_n;t)=L^{-1}_{\substack{(\bar{x}_1,\ldots,\bar{x}_n)\to(x_1,\ldots,x_n),\\(\bar{p}_1,\ldots,\bar{p}_n)\to(p_1,\ldots,p_n)}}\left(\begin{array}{l}\bar{\Psi}(\bar{x}_1,\ldots,\bar{x}_n;\bar{p}_1,\ldots,\bar{p}_n;t)\\ =\bar{\Psi}_0(\bar{x}_1,\ldots,\bar{x}_n;\bar{p}_1,\ldots,\bar{p}_n;t=0)e^{\frac{-i}{\hbar}\int_0^t \hat{H}\left(\begin{array}{l}i\hbar\bar{p}_1+\frac{1}{2}x_1,\ldots,i\hbar\bar{p}_n+\frac{1}{2}x_n;\\-i\hbar\bar{x}_1+\frac{1}{2}p_1,\ldots,-i\hbar\bar{x}_n+\frac{1}{2}p_n;u\end{array}\right)du}\end{array}\right)$$

(18a)

Theory of Shifted Taylor Series yields

$$e^{\frac{-i}{\hbar}\int_0^t \hat{H}\left(\begin{array}{l}i\hbar\bar{p}_1+\frac{1}{2}x_1,\ldots,i\hbar\bar{p}_n+\frac{1}{2}x_n;\\-i\hbar\bar{x}_1+\frac{1}{2}p_1,\ldots,-i\hbar\bar{x}_n+\frac{1}{2}p_n;u\end{array}\right)du}=e^{i\hbar\bar{p}_1\partial_{\frac{1}{2}x_1}\cdots i\hbar\bar{p}_n\partial_{\frac{1}{2}x_n}}e^{-i\hbar\bar{x}_1\partial_{\frac{1}{2}p_1}\cdots -i\hbar\bar{x}_n\partial_{\frac{1}{2}p_n}}\left(e^{\frac{-i}{\hbar}\int_0^t \hat{H}\left(\begin{array}{l}\frac{1}{2}x_1,\ldots,\frac{1}{2}x_n;\\\frac{1}{2}p_1,\ldots,\frac{1}{2}p_n;u\end{array}\right)du}\right) \quad (18b)$$

and Heaviside's Alternative to the Convolution developed above yields

$$e^{i\hbar\bar{p}_1\partial_{\frac{1}{2}x_1}\cdots i\hbar\bar{p}_n\partial_{\frac{1}{2}x_n}}e^{-i\hbar\bar{x}_1\partial_{\frac{1}{2}p_1}\cdots -i\hbar\bar{x}_n\partial_{\frac{1}{2}p_n}}\left(e^{\frac{-i}{\hbar}\int_0^t \hat{H}\left(\begin{array}{l}\frac{1}{2}x_1,\ldots,\frac{1}{2}x_n;\\\frac{1}{2}p_1,\ldots,\frac{1}{2}p_n;u\end{array}\right)du}\right)$$

$$\equiv\delta\left(\tfrac{1}{2}x_1+i\hbar\bar{p}_1,\ldots,\tfrac{1}{2}x_n+i\hbar\bar{p}_n\right)\delta\left(\tfrac{1}{2}p_1-i\hbar\bar{x}_1,\ldots,\tfrac{1}{2}p_n-i\hbar\bar{x}_n\right)\underbrace{*}_{\left(\begin{array}{l}\frac{1}{2}x_1,\ldots,\frac{1}{2}x_n;\\\frac{1}{2}p_1,\ldots,\frac{1}{2}p_n\end{array}\right)}e^{\frac{-i}{\hbar}\int_0^t \hat{H}\left(\begin{array}{l}\frac{1}{2}x_1,\ldots,\frac{1}{2}x_n;\\\frac{1}{2}p_1,\ldots,\frac{1}{2}p_n;u\end{array}\right)du}$$

(18c)

Converting from the Laplace transform image variables to their Fourier counterparts

$$(\bar{p}_1,\ldots,\bar{p}_n)=(ip_{\omega 1},\ldots,ip_{\omega n}),\ (\bar{x}_1,\ldots,\bar{x}_n)=(ix_{\omega 1},\ldots,ix_{\omega n})$$

$$\breve{\Psi}(\bar{x}_1,\ldots,\bar{x}_n;\bar{p}_1,\ldots,\bar{p}_n;t)=\breve{\Psi}(ix_{\omega 1},\ldots,ix_{\omega n};ip_{\omega 1},\ldots,ip_{\omega n};t)$$

$$\delta(\tfrac{1}{2}x_1+i\hbar\bar{p}_1,\ldots,\tfrac{1}{2}x_n+i\hbar\bar{p}_n)\delta(\tfrac{1}{2}p_1-i\hbar\bar{x}_1,\ldots,\tfrac{1}{2}p_n-i\hbar\bar{x}_n)$$

$$\equiv\delta(\tfrac{1}{2}x_1-\hbar p_{\omega 1},\ldots,\tfrac{1}{2}x_n-\hbar p_{\omega n})\delta(\tfrac{1}{2}p_1+\hbar x_{\omega 1},\ldots,\tfrac{1}{2}p_n+\hbar x_{\omega n})$$

$$\delta(\tfrac{1}{2}x_1-\hbar p_{\omega 1},\ldots,\tfrac{1}{2}x_n-\hbar p_{\omega n})\delta(\tfrac{1}{2}p_1+\hbar x_{\omega 1},\ldots,\tfrac{1}{2}p_n+\hbar x_{\omega n})\underbrace{*}_{\left(\substack{\frac{1}{2}x_1,\ldots,\frac{1}{2}x_n;\\ \frac{1}{2}p_1,\ldots,\frac{1}{2}p_n}\right)}e^{\frac{-i}{\hbar}\int_0^t\hat{H}\left(\substack{\frac{1}{2}x_1,\ldots,\frac{1}{2}x_n;\\ \frac{1}{2}p_1,\ldots,\frac{1}{2}p_n;u}\right)du} \tag{18d}$$

Now the Inverse Fourier Transform of this is

$$F^{-1}_{\substack{(x_{\omega 1},\ldots,x_{\omega n})\\ \mapsto(x_1,\ldots,x_n)\\ (p_{\omega 1},\ldots,p_{\omega n})\\ \mapsto(p_1,\ldots,p_n)}}\left(\delta(\tfrac{1}{2}x_1-\hbar p_{\omega 1},\ldots,\tfrac{1}{2}x_n-\hbar p_{\omega n})\delta(\tfrac{1}{2}p_1+\hbar x_{\omega 1},\ldots,\tfrac{1}{2}p_n+\hbar x_{\omega n})\underbrace{*}_{\left(\substack{\frac{1}{2}x_1,\ldots,\frac{1}{2}x_n;\\ \frac{1}{2}p_1,\ldots,\frac{1}{2}p_n}\right)}e^{\frac{-i}{\hbar}\int_0^t\hat{H}\left(\substack{\frac{1}{2}x_1,\ldots,\frac{1}{2}x_n;\\ \frac{1}{2}p_1,\ldots,\frac{1}{2}p_n;u}\right)du}\right)$$

$$=\left(\frac{1}{2\pi\hbar^2}\right)^n\underbrace{*}_{\left(\substack{\frac{1}{2}x_1,\ldots,\frac{1}{2}x_n;\\ \frac{1}{2}p_1,\ldots,\frac{1}{2}p_n}\right)}e^{\frac{-i}{\hbar}\int_0^t\hat{H}\left(\substack{\frac{1}{2}x_1,\ldots,\frac{1}{2}x_n;\\ \frac{1}{2}p_1,\ldots,\frac{1}{2}p_n;u}\right)du}$$

$$=\left(\frac{1}{2\pi\hbar^2}\right)^n\underbrace{\int_{-\infty}^{\infty}\cdots\int_{-\infty}^{\infty}}_{n}e^{\frac{-i}{\hbar}\int_0^t\hat{H}\left(\substack{\frac{1}{2}x_1-v_{\frac{1}{2}x_1},\ldots,\frac{1}{2}x_n-v_{\frac{1}{2}x_n};\\ \frac{1}{2}p_1-v_{\frac{1}{2}p_1},\ldots,\frac{1}{2}p_n-v_{\frac{1}{2}p_n};u}\right)du}dv_{\frac{1}{2}x_1}dv_{\frac{1}{2}p_1}\cdot\cdot dv_{\frac{1}{2}x_n}dv_{\frac{1}{2}p_n} \tag{18e}$$

Hence

$$L^{-1}_{\left(\substack{(\bar{x}_1,\ldots,\bar{x}_n)\\ \to(x_1,\ldots,x_n);\\ (\bar{p}_1,\ldots,\bar{p}_n)\\ \to(p_1,\ldots,p_n)}\right)}\left(e^{\frac{-i}{\hbar}\int_0^t\hat{H}\left(\substack{i\hbar\bar{p}_1+\frac{1}{2}x_1,\ldots,i\hbar\bar{p}_n+\frac{1}{2}x_n;\\ -i\hbar\bar{x}_1+\frac{1}{2}p_1,\ldots,-i\hbar\bar{x}_n+\frac{1}{2}p_n;u}\right)du}\right)$$

$$=\left(\frac{1}{2\pi\hbar^2}\right)^n\underbrace{\int_{-\infty}^{\infty}\cdots\int_{-\infty}^{\infty}}_{n}e^{\frac{-i}{\hbar}\int_0^t\hat{H}\left(\substack{\frac{1}{2}x_1-v_{\frac{1}{2}x_1},\ldots,\frac{1}{2}x_n-v_{\frac{1}{2}x_n};\\ \frac{1}{2}p_1-v_{\frac{1}{2}p_1},\ldots,\frac{1}{2}p_n-v_{\frac{1}{2}p_n};u}\right)du}dv_{\frac{1}{2}x_1}dv_{\frac{1}{2}p_1}\cdot\cdot dv_{\frac{1}{2}x_n}dv_{\frac{1}{2}p_n} \tag{18f}$$

Thus the quantum phase space wavefunction with initial condition for the dynamics may be expressed in the direct real $(x_1,\ldots,x_n;p_1,\ldots,p_n)$ generalised coordinates and momenta.

$$\begin{aligned}&\Psi(x_1,\ldots,x_n;p_1,\ldots,p_n;t)\\&=\Psi_0(x_1,\ldots,x_n;p_1,\ldots,p_n;t=0)\underset{\binom{x_1,\ldots,x_n;}{p_1,\ldots,p_n}}{\underbrace{*}}\left(\frac{1}{2\pi\hbar^2}\right)^n\underbrace{\int_{-\infty}^{\infty}\cdots\int_{-\infty}^{\infty}}_{n}e^{\frac{-i}{\hbar}\int_0^t\hat{H}\left(\begin{matrix}\frac{1}{2}x_1-v_{\frac{1}{2}x_1},\ldots,\frac{1}{2}x_n-v_{\frac{1}{2}x_n};\\\frac{1}{2}p_1-v_{\frac{1}{2}p_1},\ldots,\frac{1}{2}p_n-v_{\frac{1}{2}p_n};u\end{matrix}\right)du}dv_{\frac{1}{2}x_1}dv_{\frac{1}{2}p_1}\cdots dv_{\frac{1}{2}x_n}dv_{\frac{1}{2}p_n}\end{aligned}\tag{18g}$$

It is interesting to note that the canonical choice of $\alpha=\gamma=\frac{1}{2}$ thusly $\hat{H}\left(i\hbar\partial_p+\frac{x}{2},-i\hbar\partial_x+\frac{p}{2},t\right),\ni\alpha+\gamma=1$, yields the result above[which shall be used throughout the present work]; the same physics is described when $\alpha\neq\gamma$ thusly $\hat{H}\left(i\hbar\partial_p+\alpha x,-i\hbar\partial_x+\gamma p,t\right)\ni\alpha+\gamma=1$

$$\begin{aligned}&\Psi(x_1,\ldots,x_n;p_1,\ldots,p_n;t)\\&=\Psi_0(x_1,\ldots,x_n;p_1,\ldots,p_n;t=0)\underset{\binom{x_1,\ldots,x_n;}{p_1,\ldots,p_n}}{\underbrace{*}}\left(\frac{1}{2\pi\hbar^2}\right)^n\underbrace{\int_{-\infty}^{\infty}\cdots\int_{-\infty}^{\infty}}_{n}e^{\frac{-i}{\hbar}\sum_{j=1}^{n}\left(\begin{matrix}(\gamma_j-\alpha_j)x_jp_j\\+\int_0^t\hat{H}\left(\begin{matrix}\alpha_1x_1-v_{\alpha_1x_1},\\\ldots,\alpha_nx_n-v_{\alpha_nx_n};\\\gamma_1p_1-v_{\gamma_1p_1},\\\ldots,\gamma_np_n-v_{\gamma_np_n};u\end{matrix}\right)du\end{matrix}\right)}dv_{\alpha_1x_1}dv_{\gamma_1p_1}\cdots dv_{\alpha_nx_n}dv_{\gamma_np_n}\end{aligned}\tag{18h}$$

Moreover, if the identification

$$\begin{aligned}&\Psi(x_1,\ldots,x_n;p_1,\ldots,p_n;t)\\&=\Psi_0(x_1,\ldots,x_n;p_1,\ldots,p_n;t=0)\underset{\binom{x_1,\ldots,x_n;}{p_1,\ldots,p_n}}{\underbrace{*}}\Psi_{\text{complimentary}}(x_1,\ldots,x_n;p_1,\ldots,p_n;t)\\&\Psi_{\text{complimentary}}(x_1,\ldots,x_n;p_1,\ldots,p_n;t)\\&=\left(\frac{1}{2\pi\hbar^2}\right)^n\underbrace{\int_{-\infty}^{\infty}\cdots\int_{-\infty}^{\infty}}_{n}e^{\frac{-i}{\hbar}\int_0^t\hat{H}\left(\begin{matrix}\frac{1}{2}x_1-v_{\frac{1}{2}x_1},\ldots,\frac{1}{2}x_n-v_{\frac{1}{2}x_n};\\\frac{1}{2}p_1-v_{\frac{1}{2}p_1},\ldots,\frac{1}{2}p_n-v_{\frac{1}{2}p_n};u\end{matrix}\right)du}dv_{\frac{1}{2}x_1}dv_{\frac{1}{2}p_1}\cdots dv_{\frac{1}{2}x_n}dv_{\frac{1}{2}p_n}\end{aligned}\tag{18i}$$

is made, then upon considering the form of a Continuous Wavelet Transform[16] as

$$X_{\text{wavelet}}(a,b)=\frac{1}{\sqrt{a}}\int_{-\infty}^{\infty}x(z)\psi^*\left(\frac{z}{a}-\frac{b}{a}\right)dz\tag{18j}$$

and applying this to the n-dimensional phase space integrals yields

$$\Psi_{\text{complimentary}}(x_1,\ldots,x_n;p_1,\ldots,p_n;t)$$

$$= \left(\frac{1}{2\pi\hbar^2}\right)^n \underbrace{\int_{-\infty}^{\infty}\cdots\int_{-\infty}^{\infty}}_{n} e^{\frac{-i}{\hbar}\int_0^t \hat{H}\left(\substack{\frac{1}{2}x_1 - v_{\frac{1}{2}x_1},\ldots,\frac{1}{2}x_n - v_{\frac{1}{2}x_n};\\ \frac{1}{2}p_1 - v_{\frac{1}{2}p_1},\ldots,\frac{1}{2}p_n - v_{\frac{1}{2}p_n};u}\right)du} dv_{\frac{1}{2}x_1}\, dv_{\frac{1}{2}p_1} \cdots dv_{\frac{1}{2}x_n}\, dv_{\frac{1}{2}p_n}$$

$$e^{\frac{-i}{\hbar}\int_0^t \hat{H}\left(\substack{\frac{1}{2}x_1 - v_{\frac{1}{2}x_1},\ldots,\frac{1}{2}x_n - v_{\frac{1}{2}x_n};\\ \frac{1}{2}p_1 - v_{\frac{1}{2}p_1},\ldots,\frac{1}{2}p_n - v_{\frac{1}{2}p_n};u}\right)du} = \psi^*\begin{pmatrix}\left(\frac{-1}{2}x_1 + v_{\frac{1}{2}x_1}\right),\ldots,\left(\frac{-1}{2}x_n + v_{\frac{1}{2}x_n}\right)\\ \left(\frac{-1}{2}p_1 + v_{\frac{1}{2}p_1}\right),\ldots,\left(\frac{-1}{2}p_n + v_{\frac{1}{2}p_n}\right)\end{pmatrix}$$

$X_{\text{wavelet}}(a,b) = \frac{1}{\sqrt{a}}\int_{-\infty}^{\infty} x(z)\psi^*\left(\frac{z}{a} - \frac{b}{a}\right)dz$, thus the n - dimensional phase space extension

$(\mathbf{a},\mathbf{b}) = \left(\mathbf{1};\left(\frac{\mathbf{x}}{2},\frac{\mathbf{p}}{2}\right)\right)$ $x(z)$ becomes a constant **1** , ψ^* the "Mother Wavelet"

and t is a free parameter with respect to the Continuous Wavelet Transform

$$X_{\text{wavelet}}\left(\mathbf{1};\left(\frac{x_1}{2},\ldots,\frac{x_n}{2};\frac{p_1}{2},\ldots,\frac{p_n}{2}\right)\right) = \underbrace{\int_{-\infty}^{\infty}\cdots\int_{-\infty}^{\infty}}_{n} \mathbf{1}e^{\frac{-i}{\hbar}\int_0^t \hat{H}\left(\substack{\frac{1}{2}x_1 - v_{\frac{1}{2}x_1},\ldots,\frac{1}{2}x_n - v_{\frac{1}{2}x_n};\\ \frac{1}{2}p_1 - v_{\frac{1}{2}p_1},\ldots,\frac{1}{2}p_n - v_{\frac{1}{2}p_n};u}\right)du} dv_{\frac{1}{2}x_1}\, dv_{\frac{1}{2}p_1} \cdots dv_{\frac{1}{2}x_n}\, dv_{\frac{1}{2}p_n}$$

$$\Psi_{\text{complimentary}}(x_1,\ldots,x_n;p_1,\ldots,p_n;t)$$

$$= \left(\frac{1}{2\pi\hbar^2}\right)^n X_{\text{wavelet}}\left(\mathbf{1};\left(\frac{x_1}{2},\ldots,\frac{x_n}{2};\frac{p_1}{2},\ldots,\frac{p_n}{2}\right)\right)$$

(18k)

Similiarly, if the Cepstrum[17] is generalised from the traditional Z-transform(discrete) case to the continuous case via the Laplace transform

$$\text{Cepstrum}(f(y)) = L^{-1}_{s\to y}\left(\log\left(L_{y\to s}(f(y))\right)\right) \tag{18l}$$

then

$$\text{Cepstrum}\left(\Psi_{\text{complimentary}}(x_1,\ldots,x_n;p_1,\ldots,p_n;t)\right)$$

$$= L^{-1}_{\begin{pmatrix}(\bar{x}_1,\ldots,\bar{x}_n)\\ \to(x_1,\ldots,x_n),\\ (\bar{p}_1,\ldots,\bar{p}_n)\\ \to(p_1,\ldots,p_n)\end{pmatrix}}\left(Log\left(L_{\begin{pmatrix}(x_1,\ldots,x_n)\\ \to(\bar{x}_1,\ldots,\bar{x}_n),\\ (p_1,\ldots,p_n)\\ \to(\bar{p}_1,\ldots,\bar{p}_n)\end{pmatrix}}\left(\Psi_{\text{complimentary}}(x_1,\ldots,x_n;p_1,\ldots,p_n;t)\right)\right)\right)$$

$$L_{\begin{pmatrix}(x_1,\ldots,x_n)\\ \to(\bar{x}_1,\ldots,\bar{x}_n),\\ (p_1,\ldots,p_n)\\ \to(\bar{p}_1,\ldots,\bar{p}_n)\end{pmatrix}}\left(\Psi_{\text{complimentary}}(x_1,\ldots,x_n;p_1,\ldots,p_n;t)\right)=\breve{\Psi}_{\text{complimentary}}(\bar{x}_1,\ldots,\bar{x}_n;\bar{p}_1,\ldots,\bar{p}_n;t)$$

$$\breve{\Psi}_{\text{complimentary}}(\bar{x}_1,\ldots,\bar{x}_n;\bar{p}_1,\ldots,\bar{p}_n;t)=e^{\frac{-i}{\hbar}\int_0^t \hat{H}\begin{pmatrix}i\hbar\bar{p}_1+\frac{1}{2}x_1,\ldots,i\hbar\bar{p}_n+\frac{1}{2}x_n;\\ -i\hbar\bar{x}_1+\frac{1}{2}p_1,\ldots,-i\hbar\bar{x}_n+\frac{1}{2}p_n;u\end{pmatrix}du}$$

$$\frac{-i}{\hbar}\int_0^t \hat{H}\begin{pmatrix}\frac{1}{2}x_1-\nu_{\frac{1}{2}x_1},\ldots,\frac{1}{2}x_n-\nu_{\frac{1}{2}x_n};\\ \frac{1}{2}p_1-\nu_{\frac{1}{2}p_1},\ldots,\frac{1}{2}p_n-\nu_{\frac{1}{2}p_n};u\end{pmatrix}du=\psi^*\begin{pmatrix}\left(\frac{-1}{2}x_1+\nu_{\frac{1}{2}x_1}\right),\ldots,\left(\frac{-1}{2}x_n+\nu_{\frac{1}{2}x_n}\right)\\ \left(\frac{-1}{2}p_1+\nu_{\frac{1}{2}p_1}\right),\ldots,\left(\frac{-1}{2}p_n+\nu_{\frac{1}{2}p_n}\right)\end{pmatrix}$$

$X_{\text{wavelet}}(a,b)=\frac{1}{\sqrt{a}}\int_{-\infty}^{\infty}x(z)\psi^*\left(\frac{z}{a}-\frac{b}{a}\right)dz$, thus the n - dimensional phase space extension

$(\mathbf{a},\mathbf{b})=\left(1;\left(\frac{x}{2},\frac{p}{2}\right)\right)$ $x(z)$ becomes a constant **1** , ψ^* the "Mother Wavelet"

and t is a free parameter with respect to the Continuous Wavelet Transform

$$\text{Cepstrum}\left(\Psi_{\text{complimentary}}(x_1,\ldots,x_n;p_1,\ldots,p_n;t)\right)$$

$$=\left(\frac{1}{2\pi\hbar^2}\right)^n\underbrace{\int_{-\infty}^{\infty}\cdots\int_{-\infty}^{\infty}}_{n}\mathbf{1}\left(\frac{-i}{\hbar}\int_0^t \hat{H}\begin{pmatrix}\frac{1}{2}x_1-\nu_{\frac{1}{2}x_1},\ldots,\frac{1}{2}x_n-\nu_{\frac{1}{2}x_n};\\ \frac{1}{2}p_1-\nu_{\frac{1}{2}p_1},\ldots,\frac{1}{2}p_n-\nu_{\frac{1}{2}p_n};u\end{pmatrix}du\right)d\nu_{\frac{1}{2}x_1}d\nu_{\frac{1}{2}p_1}\cdots d\nu_{\frac{1}{2}x_n}d\nu_{\frac{1}{2}p_n}$$

Notwithstanding their intrinsic interest, the few Examples 1-7 herein these Remarks do not even begin to touch the immense vista of problems that may be solved exactly via HOA methods. Indeed, the HOA may be modified to accommodate auxiliary conditions besides Initial-Value Problems. Moreover, the Integral and Differential Operators in the HOA may be generalized to Discrete Sum/Finite Difference Operators and sundry other extensions. Since classical and quantum dynamical systems admit to Hamiltonian/Lagrangian formulation, then indeed the various other formulations[e.g., Density Functional Theory, Path Integral Formalism, Stochastic Formalisms,etc], as well as derived abstracted mathematical problems[e.g., ref 14,15,19] may all benefit by HOA treatments. Quoting from [ref. 6,18] , "…..By definition, solution in one representation implies simultaneous solution in all representations[the Physics is the same regardless of 'pictures']. ………Hence, Quantum Dynamics is now reduced to exact quadratures, as are all the associated purely mathematical problems that are abstracted from the physical formalism." MT21:42.

References

[1] Torres-Vega, G. & Frederick, J. H. (1993). A quantum-mechanical representation in phase space. *Journal of Chemical Physics*, **98**(4), 3103-20.

[2] Li, Q. S. & Lu, J. (2000). Rigorous solutions of diatomic molecule oscillator with empirical potential function in phase space. *Journal of Chemical Physics*, **113**(11), 4565-4571.

[3] Hu, X. G. & Li, Q. S. (1999). Morse oscillator in a quantum phase-space representation: rigorous solutions. Journal of Physics A: *Mathematical and General*, **32**(1), 139-146.

[4] Li, Q. S. & Lu, J. (2001). One-dimensional hydrogen atom in quantum phase-space representation: rigorous solutions. *Chemical Physics Letters*, **336**(1,2), 118-122.

[5] Li, Q. S., Wei, G. M. & Lu, L. Q. (2004). Relationship between the Wigner function and the probability density function in quantum phase space representation. Physical Review A: *Atomic, Molecular, and Optical Physics*, **70**(2), 022105/1-022105/5.

[6] Simpao, V. A. (2004). *Electronic Journal of Theoretical Physics*, 1, 10-16.

[7] Simpao, V. A. (2008). Toward Chemical Applications of Heaviside Operational Ansatz: Exact Solution of Radial Schrodinger Equation for Nonrelativistic N-particle System with Pairwise 1/rij Radial Potential in Quantum Phase Space [published MAY *Journal of Mathematical Chemistry*.

[8] Simpao, V. A. (2006). Electronic *Journal of Theoretical Physics*, **3**, No. 10, 239-247

[9] Moss, R. E. (1973). Advanced Molecular *Quantum Mechanics*, (Wiley and Sons, NY).

[10] Lindgren, I. (1989). in Lecture Notes in Chemistry, vol 52: Many-Body Methods in Quantum Chemistry, ed. U. Kaldor (*Springer, Berlin*), 293-306.

[11] Goldstein, H., Poole, C. P. & Jr., Safko, J. L. (2000). Classical Mechanics 3rd Edition, *Pearson*.

[12] Whittaker, E. T. (1944). *Analytical Dynamics 4th Edition, Dover New York*.

[13] Weinstock, R. (1974). *Calculus of Variations, Dover*, New York.

[14] de Gossen, M. & Luef, F. (2008). *Lett Math Phys*, **85**, 173-183.

[15] Cosmos. Zachos, David B. Fairlie, & Thomas L. (2005). Curtright, *Quantum Mechanics in Phase Space*, (World Scientific, Singapore.

[16] P.S. *Addison, The illustrated wavelet transform handbook: introductory theory and applications in science, engineering, medicine and finance*. CRC2002

[17] B. P. Bogert, M. J. R. Healy, and J. W. Tukey: "The Quefrency Alanysis of Time Series for Echoes: Cepstrum, Pseudo Autocovariance, Cross-Cepstrum and Saphe Cracking". *Proceedings of the Symposium on Time Series Analysis* (M. Rosenblatt, Ed) Chapter 15, 209-243. New York: Wiley, 1963.

[18] Chapter 6 , Mathematical Chemistry, NovaScience Publishers 2010 cross-published in *International Journal of Theoretical Physics Group Theory and NonLinear Optics*. vol 14, Issue 2 Nova Science

[19] Flannery, M.R. *Am. J. Phys.*, Vol. 73, No. 3, March 2005

In: Focus on Quantum Mechanics
Editors: David E. Hathaway et al. pp. 339-343
ISBN: 978-1-62100-680-0

Chapter 11

THE QUANTUM MECHANICAL FREE PARTICLE PARADOX

***F. Gustavo Criscuolo*[1] *and M.E. Burgos*[2]**
[1]Universidad Pedagógica Experimental Libertador, Maturín, Venezuela;
[2]Universidad de Los Andes, Mérida, Venezuela

Abstract

A careless analysis of the meaning of conservation laws in quantum mechanics can lead to paradoxes. In order to illustrate this assertion, in this chapter we analyze the problem of energy conservation of a free particle. It is shown that paradoxes of this type can be solved by the adoption of a criterion referred to conservation laws which is compatible with the axiomatic of quantum mechanics and has not been previously formulated in an explicit way.

PACS: 03

Keywords: Conservation Laws, Orthodox Quantum Mechanics, Paradoxes.

I. Introduction

In classical mechanics, the free particle Hamiltonian has just one term, the kinetic energy, which is invariant under spatial translations, rotations, and time translations. Each one of these symmetries is respectively related to the conservation of a physical quantity: linear momentum, angular momentum and energy.

In quantum mechanics the free particle Hamiltonian has the same symmetry properties that it has in the classical case. Dirac points out that "the laws of conservation of energy, momentum and angular momentum hold for an isolated system... in quantum mechanics as they hold in classical mechanics."[1] Hence, conservation of linear momentum, angular momentum and energy should be expected *for every state* of the free particle.

Nevertheless, it is worth noticing that, in quantum mechanics, physical quantities do not always have sharp (definite) values. Also in the words of Dirac, "the expression that an observable 'has a particular value' for a particular state is permissible in quantum mechanics in the special case when a measurement of the observable *is certain* to lead to the particular value, so that the state is an eigenstate of the observable... In the general case we cannot speak of an observable having a value for a particular state... (emphases added)."[1]

The meaning of conservation laws concerning observables which do not have a particular value has been recently analyzed. It has been shown that the meaning of these laws is not so clear in the quantum realm as it is in the classical framework.[2,3] This is why a careless analysis can lead to paradoxes. In this article we present an example concerning the energy conservation of the free particle.

II. A Quantum Mechanical Paradox

According to Feynman, "there is a fact, or if you wish, a law, governing all natural phenomena that are known to date. There is no exception to this law—it is exact so far as we know. The law is called the conservation of energy. It states that there is a certain quantity, which we call energy, that does not change in the manifold changes which nature undergoes... it says that there is *a numerical quantity* which does not change when something happens (emphasis added)."[4]

On their hand, referring to systems which are not in a varying external field, Landau and Lifshitz say: "a Hamilton's function which is conserved is called the energy... the law of conservation of energy in quantum mechanics... signifies that, if in a given state the energy has a *definite value*, this value remains constant in time (emphases added)."[5]

We should stress that, in the two assertions previously quoted, nothing is said about the conservation of quantities which do not have definite (sharp) values.

Let us start analyzing the case of the free particle in the state

$$\varphi(\mathbf{r},t) = \exp\,[(i/\hbar)\,(\mathbf{p}.\mathbf{r} - E\,t)] \tag{1}$$

where m is the mass of the particle, $\mathbf{p}$ its linear momentum and $E = \mathbf{p}^2/2m$ its energy. So, it can be said that for the particle in this state, Feynman's and Landau's and Lifshitz's, assertions are meaningful and right: in this state the energy has the definite value E which remains a constant in time.

Nevertheless, it is important to stress that this state cannot be normalized. As a consequence, it does not belong to the Hilbert space and it cannot represent a possible state of the free particle.[6,7] This is why it has to be assumed that the possible states of a free particle are not given by (1), but by

$$\phi(\mathbf{r},t) = (2\pi)^{-3/2} \int_{-\infty}^{\infty} d^3p\, \phi^T(\mathbf{p}) \exp\,[(i/\hbar)\,(\mathbf{p}.\mathbf{r} - E\,t)] \tag{2}$$

where $\phi^T(\mathbf{p})$ is the Fourier transform of $\phi(\mathbf{r},0)$. The state given by (2), which belongs to the Hilbert space, is the most general expression of the state of a free particle; but in this state the particle's energy does not have a sharp value: As we have already pointed out, according to Dirac it is not permissible to assert that the free particle has a particular energy.[1] And, in the words of Reichenbach, "in a physical state not preceded by a measurement of the quantity *u* any statement about a value of *u* is meaningless."[8]

Since there is not a numerical quantity that can be called the energy of the free particle, we are led to conclude that the law of conservation of energy, as stated by Feynman or by Landau and Lifshitz, does not apply.

We thus arrive at the following paradox: On one hand, following Dirac and Reichenbach, it should be said that the statement "the free particle's energy is conserved" (as any other statement about its energy) is *meaningless*; so it cannot be decided whether it is *right or wrong*. On the other hand, taking into account symmetry arguments it should be said that this statement is *meaningful and right*.

To conclude this section, let us stress that this paradox is not solved by including as a possible state of the free particle that given by (1), which does not belong to the Hilbert space. If we include this state, we can say that the assertion "the energy of the free particle is conserved" is meaningful for this state, but only for this state, and *not for every state,* as required by symmetry arguments.

III. The Source of the Paradox

In quantum mechanics the physical quantity A is not represented by a function (as physical quantities in classical physics), but by an operator $\mathbf{A}$ acting in the Hilbert space. For every system's state $|\Psi\rangle$, a distribution function is associated to A. This distribution function has a characteristic function

$$f(\mu) = \langle\Psi|\exp(i\,\mu\,\mathbf{A})\,|\Psi\rangle \tag{3}$$

where μ is a parameter. The distribution function of A can be retrieved from $f(\mu)$ by means of a Fourier Transformation.[7,9] In order for $f(\mu)$ to be singular (sharp), the state of the system $|\Psi\rangle$ must be an eigenstate of $\mathbf{A}$, and only in this case it can be considered admissible to say that A has a value, as in the classical case.

Bunge asserts that the idea that all physical properties are "sharp" or "well defined" at all times is a tacit principle of classical physics which quantum theory does not obey.[10,11] In general, the usual metaphysical commitment that every observable *has* a value cannot be transferred to quantum mechanics for it collides with its axiomatic. This is why, in general, the question "Which is the numerical quantity called energy that does not change as long as the particle remains free?" *does not have an answer*.

IV. SOLVING THE PARADOX

Let $\mathbf{H}$ be the Hamiltonian of a system. In processes ruled by the Schrödinger equation, the validity of conditions

$$\partial\mathbf{A} / \partial t = 0 \tag{4a}$$

$$[\,\mathbf{H}, \mathbf{A}\,] = 0 \tag{4b}$$

ensures that the different momenta of the distribution function $\langle \mathbf{A}^n \rangle = \langle \Psi | \mathbf{A}^n | \Psi \rangle$ (n= 1, 2, 3, ...) remain constant for every state $|\Psi\rangle$. So the distribution function of *A* does not vary with time. The physical quantity *A* is termed *a constant of the motion* and it is said to be *conserved*.[7,12]

Taking into account these arguments, and in order to decide whether a physical quantity is conserved or not in the framework of quantum mechanics, we propose the following criterion:

Criterion: A physical quantity is conserved iff its distribution function does not vary with time.

This criterion, which is compatible with the axiomatic of quantum mechanics, has the following advantages:

i. Since in quantum mechanics every physical quantity has a distribution function, it allows to decide, in every case, whether this quantity is conserved or not.
ii. It can be applied to every state belonging to the Hilbert space. It does not require, in particular, that the system be in a state with a sharp energy, which is the only case considered by Landau and Lifshitz.
iii. It does not evoke the metaphysical commitment that every observable has a numerical value.
iv. If conditions (4) are fulfilled, it says that *A* is conserved in every process ruled by the Schrödinger equation. Thus, in these processes *it leads to the same conclusions resulting from symmetry arguments*.
v. This criterion can also be applied in processes not ruled by the Schrödinger equation. In these last cases, however, the validity of conditions (4) does not ensure that *A* is conserved.[2-3,13-17]

It is important to stress the difference between our criterion and that expressed by Feynman. According to this author, in cases the energy is conserved there is just a numerical quantity (the energy), which does not change. Our criterion implies that, if the energy is conserved there is an infinite set of numerical quantities which do not change. They are the momenta of the *H* distribution function: $\langle \mathbf{H} \rangle$, $\langle \mathbf{H}^2 \rangle$, $\langle \mathbf{H}^3 \rangle$,...

Now, since the evolution of the free particle is ruled by the Schrödinger equation and its Hamiltonian fulfills conditions (4), our criterion allows us to conclude that the physical quantity energy is conserved for every state, even if it is not possible to assign a numerical value to this energy. Hence the assertion "the energy of the free particle is conserved" is meaningful and right, and *the paradox is solved.*

A similar analysis can be applied to the angular momentum of the free particle. Although the different component of this momentum cannot have simultaneously definite values, our criterion leads to the conclusion that every one of them is conserved for every state of the particle.

V. Conclusions

In quantum mechanics, the most popular conception of the validity of conservation laws requires that physical quantities have sharp values. We have shown that this idea is at the origin of the free particle paradox: on one hand, since the energy of a free particle in a state that belongs to the Hilbert space does not have a definite value, it can be said that any statement about this energy is meaningless; on the other hand, taking into account symmetry arguments it can be concluded that the assertion "the energy of the free particle is conserved" is meaningful and right for every state.

In the classical realm, a physical quantity has a numerical value, and the condition for this quantity to be conserved is that the corresponding numerical value does not change. The situation is very different in quantum mechanics where, in order to decide whether this quantity is conserved or not and according to the criterion stated in the present article, the whole distribution function associated with the physical quantity has to be considered.

By adopting this criterion, the free particle paradox is solved.

Acknowledgments

We thank *Nova Science Publishers* for inviting us to publish a contribution. This work was supported by the CDCHT-ULA.

References

[1] Dirac, P. *Quantum Mechanics*; Oxford at the Clarendon Press: Oxford, 1947, pp 46-47, 115.
[2] Burgos, M. E. *Phys. Essays* 1994, 7, 1-4.
[3] Burgos, M. E. *Speculations in Science and Technology* 1997, 20, 183-187.
[4] Feynman, R.; Leighton, R.; Sands, M. *The Feynman Lectures on Physics*; Reading: Massachusetts, 1966; Chap. 4, pp 1.
[5] Landau, L.; Lifshitz, E. *Quantum Mechanics, Non-Relativistic Theory*; Pergamon Press: London, 1965; pp 28.
[6] Cohen-Tanoudji, C.; Diu, B.; Laloë, F. *Quantum Mechanics*; John Wiley & Sons: New York, 1977; pp 23, 94-95, 151, 215.
[7] Messiah, A. *Mécanique Quantique*; Dunod: Paris, 1965; pp 138, 149, 178.
[8] Reichenbach, H. *Philosophic Foundation of Quantum Mechanics*; University of California Press: Los Angeles, 1944; pp 127.
[9] Cramer, H. *Mathematical Methods of Statistics;* Princeton University Press: Princeton, 1945; pp 89.
[10] Bunge, M. *Treatise on Basic Philosophy;* Reidel: Dordrecht, 1974; Vol 6, pp 260.
[11] Bunge, M. *Treatise on Basic Philosophy*; Reidel: Dordrecht, 1974; Vol 7, pp 210.
[12] Merzbacher, E. *Quantum Mechanics*; John Wiley & Sons Inc.: New York, 1970; pp 337.
[13] Burgos, M. E.; Criscuolo, F. G.; Etter, T. L. *Speculations in Science and Technology* 1999, 21, 227-233.
[14] Criscuolo, F. G.; Burgos, M. E. Phys. Essays 2000, 13, 1-5.
[15] Burgos, M. E. *Found. Phys.* 1998, 28, 1323-1346.
[16] Burgos, M. E. *Found. Phys.* 2008, 38, 883-907.
[17] Burgos, M. E. *Journal of Modern Physics*. In press.

In: Focus on Quantum Mechanics
Editors: David E. Hathaway et al. pp. 345-403
ISBN 978-1-62100-680-0

Chapter 12

NONPERTURBATIVE ANALYSES BEYOND INSTANTONS

Toshiaki Tanaka*
Department of Physics, National Cheng Kung University,
Tainan 701, Taiwan, R.O.C.
National Center for Theoretical Sciences, Taiwan, R.O.C.

Abstract

In this chapter, we review a recent development on nonperturbative analyses in imaginary-time path integral formalism. The key ingredient of our analysis resides in considering valley configurations in the phase space of path integral. It enables us in particular to treat properly negative modes around classical configurations which often break the applicability of conventional semi-classical approximations. To see how our method works, we apply it to quantum mechanical problems of an asymmetric double-well and a symmetric triple-well potential. We show that valley configurations connect perturbative and nonperturbative regions in the phase space, from which we have a manipulation to extract purely nonperturbative contributions to physical quantities. As a by-product, we derive a dispersion relation which relates a perturbative contribution with nonperturbative one. With our method, we calculate the nonperturbative corrections to the spectrum not only for the ground state but also for all the excited states beyond the dilute-gas approximation. Using the dispersion relation, we also derive large-order behavior of the perturbation series for the corresponding energy eigenvalues. We then check our results by employing perturbation theory, the WKB approximation, and consequences of $\mathcal{N}$-fold supersymmetry to confirm their accuracy qualitatively and quantitatively.

1. Introduction

Analytical understanding of nonperturbative structure of quantum theories has been one of the most challenging issues in theoretical and mathematical physics. Among the tremendous attempts toward it, instanton calculations based on a semi-classical treatment in Euclidean path integral formalism has been one of the most successful approaches both for

*E-mail address: ttanaka@mail.ncku.edu.tw

quantum mechanical and field theoretical models since the pioneer works in Refs. [1, 2, 3]. One of the remarkable achievements of the method is the nonperturbative calculation of the spectral splitting of the lowest two energy levels of a quantum mechanical potential with two degenerate minima, namely, a double-well potential by summing up multi-instanton contributions in the dilute-gas approximation. For a review, see, e.g., Ref. [4].

However, the applicability of the semi-classical method for a quantum mechanical potential with non-degenerate minima has been questioned and controversial since the application of a multi-bounce calculation in the dilute-gas approximation to a quantum mechanically unstable system [5]. One of the most serious problems on the use of a bounce solution is that it cannot detect the difference between a stable potential with two or more local minima and an unstable potential from which a quantum particle can be escaped via quantum tunneling, see, e.g., Ref. [6] and references cited therein. One of the typical examples of the former is an asymmetric double-well potential for which there is a bounce solution but its spectrum is totally real. It was in Ref. [7] that a clear-cut resolution to the latter problem was finally given. The key ingredient there is to consider valley configurations which make dominant contributions to the path integral. They includes the bounce solution as a special case and constitutes a collective coordinate corresponding to the negative mode around the classical bounce configuration. By employing the prescription called the *valley method*, one can extract and separate out the collective coordinate from the other dynamical degrees of freedom so that one can carry out the Gaussian approximation without suffering from the negative mode. For the details, see Ref. [8].

On the other hand, the applicability of the dilute-gas approximation has been questioned relatively rarely probably because of its success on, e.g., the calculation of the spectral splitting of the lowest two states of a symmetric double-well potential. In this respect, however, we refer to Ref. [9] where it was reported that there is the discrepancy between the dilute-gas approximation and the WKB approximation for the wave functions even in the case of a symmetric double-well potential. However, it was recently found that there is at least one concrete example for which the dilute-gas approximation does not work well. It is a triple-well potential.

It is curious that a quantum mechanical triple-well potential problem had attracted little attention until around 2000 in spite of the fact that there had been hundreds of papers on double-well potentials. In Refs. [10, 11], an essentially identical triple-well potential with three degenerate potential minima was investigated independently. Both of them employed the semi-classical method in Euclidean path integral formalism and calculated multi-instanton contributions in the dilute-gas approximation to obtain the nonperturbative corrections to the lowest three energy levels. However, both of the results apparently look strange in the sense that they do not reduce to the corresponding harmonic oscillator levels when the coupling constant turns off, although both of the authors did not discuss that aspect. Later, the author and his collaborator investigated in Ref. [12] a symmetric triple-well potential, which includes the one investigated in Refs. [10, 11] as a special case, by the valley method beyond the dilute-gas approximation. Although the semi-classical approach in the dilute-gas approximation was reexamined carefully to obtain plausible results in appearance in Ref. [13], the detailed and cautious analyses entirely supported the results obtained by the valley method [14].

The main purpose of this chapter is twofold. The one is to clarify what kinds of elements can spoil the applicability of the conventional semi-classical method in the dilute-gas approximation. In this respect, it is important to recognize that there are mainly two different factors which limit the validity of the latter method. More concretely, some elements are responsible for the breakdown of the Gaussian approximation while the others are for the breakdown of the dilute-gas approximation. The other purpose is to review the valley method by putting emphasis on how the two troublesome factors which can break the applicability of the semi-classical method are tamed in it. In addition, we explain in detail how the valley method can provide a unified framework to understand the intimate relation between nonperturbative corrections and Borel summability of perturbation series.

The chapter is organized as follows. In Section 2., we first briefly review the conventional semi-classical method in Euclidean path integral formalism and then clarify under what kinds of situations it would work or fail. In particular, we show that the existence of negative modes in a certain operator would play a key role in determining its applicability. In Section 3., we introduce our mathematical definition of valley configurations in the phase space of path integral. We then show how to proceed a valid approximation for a system with negative modes based on the valley configurations.

In Section 4., we investigate a solution to the valley equation, called a valley-instanton, which is a kind of deformed instanton and constitutes a building block of various valley configurations. In Section 5., we further study and discuss relevant and significant aspects of valley configurations by applying our approach to an asymmetric double-well potential. An asymptotic analysis on the valley equation enables us to construct a valley solution which constitutes a well-separated pair of one valley and one anti-valley instantons. The asymptotic form of its action automatically contains an interaction term which is attractive. We then deform the integral path with respect to the collective coordinate of the valley configuration to extract the purely nonperturbative contribution with the knowledge of Borel summation. The analytic structure of the latter contribution naturally leads to the Bogomolny's prescription for making an instanton calculation sensible in the case where they interact attractively. It further leads to a dispersion relation which connects the nonperturbative contribution with the perturbative one. Summarizing all the obtained results, we carry out a multi-valley-instanton calculation beyond the dilute-gas approximation to obtain the nonperturbative corrections to not only the ground-state energy but also all the energy levels. We then apply the dispersion relation to obtain a prediction on the large-order behavior of the perturbation series for the corresponding spectrum. As is well-known, perturbation series are in general divergent and are at most asymptotic. However, we there observe an intriguing phenomena that the divergent behavior can disappear at infinitely many values of the parameter in the potential.

In Section 6., we apply our method to a symmetric triple-well potential. In comparison with the asymmetric double-well potential in Section 5., we find several novel features which can emerge only in a potential with more than two local minima. For example, there are inequivalent valley configurations consisting of one valley-instanton and one anti-valley-instanton which are different in interaction terms due to the different curvature of each potential well. Another significant aspect is the existence of valley configurations which consist of two successive valley-instantons or two successive anti-valley-instantons for which their interactions get repulsive. As a consequence of the different nature of interaction, the analytic structure of the integral with respect to the collective coordinate of valley is entirely different from the one in the case of the configuration with one valley-instanton and one anti-valley-instanton discussed previously. The existence of different interactions depending on different valley configurations naturally explain the breakdown of the dilute-gas approximation where all the interactions are neglected, regardless of whether the system is suffered from the existence of negative modes or not. As in the double-well potential case, we also carry out a multi-valley-instanton calculation beyond the dilute-gas approximation to obtain the nonperturbative corrections to all the energy levels and the large-order behavior of the perturbation series for them by applying the dispersion relations. Interestingly, we observe not only the phenomenon of the disappearance of the leading divergence in the large-order behavior of the perturbation series, but also the disappearance of the nonperturbative corrections at a part of the infinitely many values of the parameter in the potential where the former phenomenon takes place.

In the subsequent three sections, we examine and check the obtained results calculated by our procedure for the double-well potential in Section 5. and for the triple-well potential in Section 6. from totally different points of view. In Section 7., we calculate by the direct application of perturbation theory the perturbative coefficients for the lowest energy level of each potential well up to the 500th order and compare them with our predictions on their large-order behaviors obtained by the dispersion relation. We confirm that the exactly calculated perturbative coefficients indeed tend to our predicted asymptotic values in all the cases. In Section 8., we employ the WKB approximation for Schrödinger equation to calculate the same physical quantities as we have obtained by our method in Euclidean path integral formalism. We confirm that the semi-classical quantization conditions for the nonperturbative corrections calculated by the WKB approximation exactly coincide in analytic expression with the ones obtained by the multi-valley-instanton calculation both for the double-well and the triple-well potentials. In Section 9., we consider our results in view of hidden symmetry which underlies the potentials. Both of the phenomena of the disappearances of the leading divergence in the perturbation series and of the nonperturbative corrections observed in our results are reminiscent of the non-renormalization theorem well-known in supersymmetric quantum theories [15, 16]. In fact, we unveil hidden generalized supersymmetry, called $\mathcal{N}$-fold supersymmetry, possessed by the double-well and the triple-well potentials at the particular values of the parameters where those remarkable phenomena take place. Finally in Section 10., we summarize this chapter and make some remarks on our results.

2. The Semi-classical Method and Its Limitations

In quantum theory and statistical mechanics, one is mostly concerned with calculating a partition function which is expressed as a formal path integral

$$Z = N \int \mathcal{D}q\, \mathrm{e}^{-S_E[q]}, \tag{1}$$

with a certain boundary condition and an irrelevant normalization constant N. In the above, the quantity $S_E[q]$ which is a functional of $q(\tau)$ is called *Euclidean action*. Needless to say, however, it cannot be calculated exactly in general, and thus we must resort to an approximation. One of the most commonly employed methods is perturbation theory. It works whenever the action under consideration admits a decomposition $S_E[q] = S_E^{(0)}[q] + gS_E^{(I)}[q]$ with a parameter g called *coupling constant* such that the partition function for the free action $S_E^{(0)}[q]$ can be calculated exactly. Then, the partition function Z is approximated by the power series expansion in the coupling constant g

$$Z \approx Z^{(0)} + gZ^{(1)} + g^2 Z^{(2)} + \cdots . \tag{2}$$

Hence, a main concern in perturbation theory is to develop a manipulation to calculate systematically each expansion coefficient $Z^{(k)}$ $(k = 1, 2, \ldots)$. Unfortunately, however, perturbation theory usually suffers from mainly two defects. The one is that the perturbation series is in general divergent and at most asymptotic [17]. As a result, it can provide a good approximation only for a weak coupling regime $g \ll 1$ by retaining first several terms. The stronger the coupling constant is, the more meaningless a higher-order calculation is. The other defect is that perturbation theory cannot detect any effect which cannot be expressed in terms of a power series in the coupling constant. We call such an effect *nonperturbative*. The so-called instanton effect which gives rise to a contribution of the form $\sim \mathrm{e}^{-1/g^2}$ is a typical example.

The instanton effect is usually calculated by the semi-classical method in Euclidean path integral formalism [4]. It is actually a Gaussian approximation applied to path integral. The expression (1) tells us that the path integral is dominated by configurations which result in less values of the action. Thus, the idea of the method is to first look for the configurations $q_c(\tau)$ for which the action gets minimum. Obviously, any such configuration satisfies the equation of motion

$$\frac{\delta S_E[q_c]}{\delta q(\tau)} = 0. \tag{3}$$

We then expand the action around the classical configurations $q_c(\tau)$ in powers of the fluctuation $\tilde{q}(\tau) = q(\tau) - q_c(\tau)$ as

$$S_E[q] = S_E[q_c] + \int \mathrm{d}\tau\, \tilde{q}(\tau) \frac{\delta S_E[q_c]}{\delta q(\tau)} + \int \mathrm{d}\tau \mathrm{d}\tau' \tilde{q}(\tau) \frac{\delta^2 S_E[q_c]}{\delta q(\tau) \delta q(\tau')} \tilde{q}(\tau') + O(\tilde{q}^3), \tag{4}$$

where the second term vanishes due to the equation of motion (3). Retaining the terms up to second-order in $\tilde{q}$ and performing the Gaussian integral with respect to $\tilde{q}$, we obtain an approximation to Z as

$$Z \simeq N \frac{\mathrm{e}^{-S_E[q_c]}}{\sqrt{2\pi \det D_c}}, \tag{5}$$

where D_c is the operator $D(\tau, \tau')$ defined by

$$D(\tau, \tau') = \frac{\delta^2 S_E[q]}{\delta q(\tau) \delta q(\tau')}, \tag{6}$$

at the classical configuration $q(\tau) = q_c(\tau)$. This approximation scheme would work well if the solution $q_c(\tau)$ to the equation of motion is not only a stationary point but also a local minimum as depicted in Fig. 1. In Fig. 1, the vertical axis represents the action while the

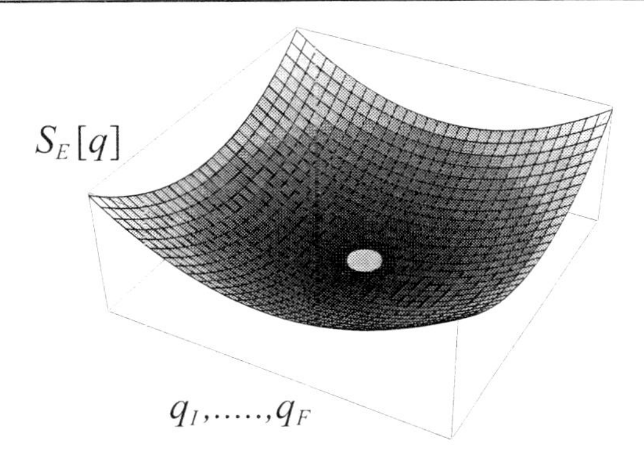

Figure 1. The geometrical structure around the stationary point with only positive modes.

horizontal plane does the configuration space of path integral which is actually an infinite dimensional though we can show at most two dimensions.

It is indeed the case when the operator D_c contains only positive modes. To see the situation more concretely, let us consider a one-dimensional quantum mechanical system for which the Euclidean action is given by

$$S_E[q] = \int_{\tau_I}^{\tau_F} \mathrm{d}\tau \left[\left(\frac{\mathrm{d}q}{\mathrm{d}\tau} \right)^2 + V(q) \right]. \tag{7}$$

If the potential $V(q)$ is bounded below and has a unique minimum at $q = q_0$, then the solution to the equation of motion

$$\frac{\mathrm{d}^2 q}{\mathrm{d}\tau^2} = V'(q), \tag{8}$$

is obviously given by $q_c(\tau) = q_0$, that is, the configuration where the classical particle stays eternally at the bottom of the potential. It is evident that the latter configuration makes the action minimum since either any deviation from the point q_0 or any τ dependence increases the value of the action. In such a situation, the operator D_c does not contain any non-positive mode and the formula (5) would provide a good approximation to the exact Z. In fact, it gives the exact value when the potential is a harmonic oscillator [18].

On the other hand, if the equation of motion (3) admits a τ-dependent solution $q_c(\tau)$, then the operator D_c always contains (at least) one zero mode. Indeed, differentiating the equation of motion with respect to τ, we immediately obtain

$$\int \mathrm{d}\tau' D_c(\tau, \tau') \dot{q}_c(\tau') = 0, \tag{9}$$

where and hereafter dot denotes the derivative with respect to the imaginary time τ. The above equation means that $\dot{q}_c(\tau)$ which is non-trivial for a τ-dependent classical configuration always corresponds to a zero mode of the operator D_c. In such a situation, there is a direction from the classical configuration to which the action is constant as depicted in Fig. 2. Hence, we cannot perform the Gaussian integration with respect to the configuration

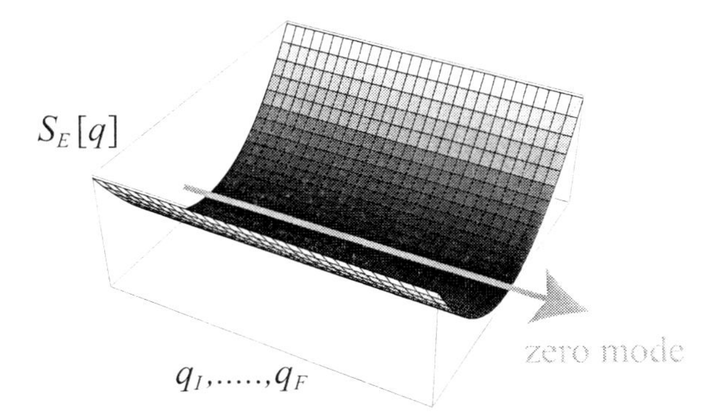

Figure 2. The geometrical structure around the stationary point with one zero mode.

proportional to the zero mode $\dot{q}_c(\tau)$. Instead, we must first separate the zero mode from the other configurations for which the Gaussian integration shall be performed. With the latter prescription, we obtain

$$\begin{aligned} Z &\simeq N \int_{\tau_I}^{\tau_F} \mathrm{d}\tau_0 \int \mathcal{D}'\tilde{q}\, \mathrm{e}^{-S_E[q_c] - \frac{1}{2}\tilde{q}\cdot D_c \cdot \tilde{q}} \\ &= (\tau_F - \tau_I) N \frac{\mathrm{e}^{-S_E[q_c]}}{\sqrt{2\pi \ \det' D_c}}, \end{aligned} \tag{10}$$

where $\mathcal{D}'\tilde{q}$ represents the path-integral measure in the space perpendicular to the zero mode and $\det' D_c$ does the determinant of D_c without the zero eigenvalue. This prescription is usually called the *collective coordinate method*, originally formulated in Ref. [19] in path-integral formalism and in Ref. [20] in canonical formalism. A typical example where the collective coordinate method works well is a symmetric double-well potential problem. In this example, there exists a τ-dependent classical solution $q_c(\tau)$ called *instanton* or *kink* solution as is shown in Fig. 3. In Fig. 3, we show three graphs. The upper-right graph

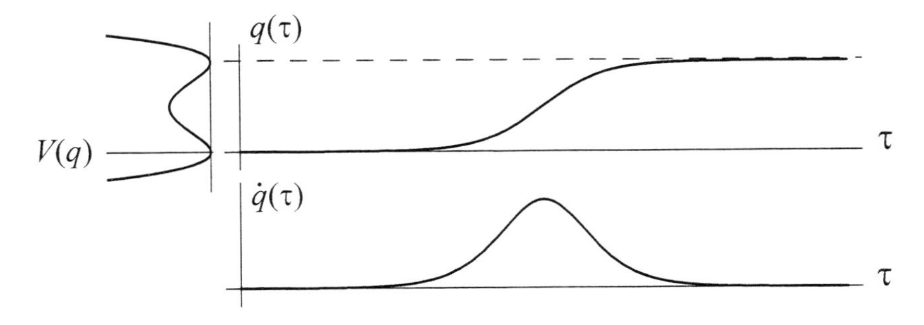

Figure 3. An instanton (kink) solution in a symmetric double-well potential.

shows the classical motion $q_c(\tau)$ (the vertical axis) as a function of the imaginary time τ (the horizontal line) while the lower-right graph does its τ-derivative $\dot{q}_c(\tau)$. The upper-left graph is the potential rotated counterclockwise by 90 degrees so that one can easily imagine how the classical particle moves in the potential.

It is quite important to note that in this example the zero mode $\dot{q}_c(\tau)$ has no nodes. In one-dimensional quantum mechanics, the zero-mode equation (9) reads as

$$\left[-\frac{\mathrm{d}^2}{\mathrm{d}\tau^2} + V''(q_c)\right] \dot{q}_c(\tau) = 0, \tag{11}$$

and actually is a Schrödinger equation. Hence, the fact that the zero mode $\dot{q}_c(\tau)$ has no nodes indicates that it corresponds to the ground-state wave function of the Schrödinger equation with the zero eigenvalue and, as a consequence, the operator D_c has no negative modes. That is exactly the reason why the formula (10) used in the instanton calculation for a symmetric double-well potential works well without suffering from negative modes.

The situation is drastically changed, however, when we deal with an asymmetric double-well potential problem. In the latter case, the classical solution $q_c(\tau)$ now becomes a so-called *bounce* solution and the zero mode $\dot{q}_c(\tau)$ associated with it has one node, see Fig. 4. The existence of one node in the zero mode means that it corresponds to the first

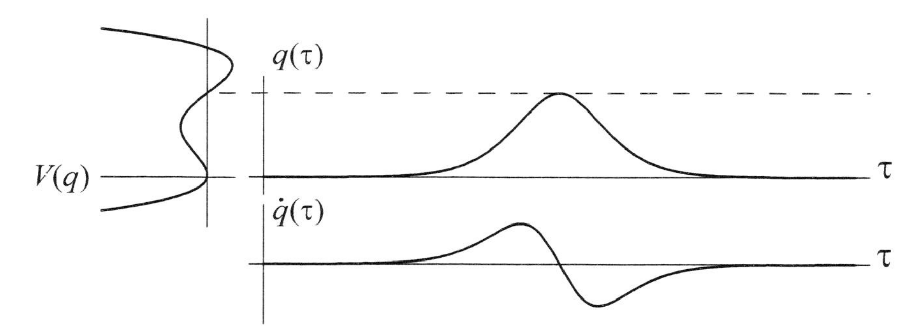

Figure 4. A bounce solution in an asymmetric double-well potential.

excited state and thus that there is the ground state of the operator D_c with a negative eigenvalue. In such a situation where D_c contains negative modes, we are still able to perform Gaussian integrations with respect to them by analytic continuation, but it results in a factor of a power of the imaginary unit i. In general, if D_c contains n negative modes $\lambda_{-n} < \cdots < \lambda_{-1} < 0$, we have the following factor in the formula (10):

$$\sqrt{\det{}' D_c} \propto \prod_{k=1}^{n} \left(\mathrm{i}\sqrt{|\lambda_{-k}|}\right), \tag{12}$$

irrespective of whether the quantum system under consideration is stable and is free from a complex spectrum. Originally, bounce solutions were employed to calculate decay rates in quantum-mechanically unstable systems [5]. However, the applicability of the semi-classical calculation using bounce solutions has been controversial, see, e.g., Ref. [6] and references cited therein.

The breakdown of the approximation in the present case is rather obvious when we consider the geometric structure of $S_E[q]$ in the configuration space. The existence of a negative mode in the operator D_c indicates that the classical configuration $q_c(\tau)$ is no longer a local extremum of $S_E[q]$ but a saddle point, as depicted in Fig. 5. To the direction of the

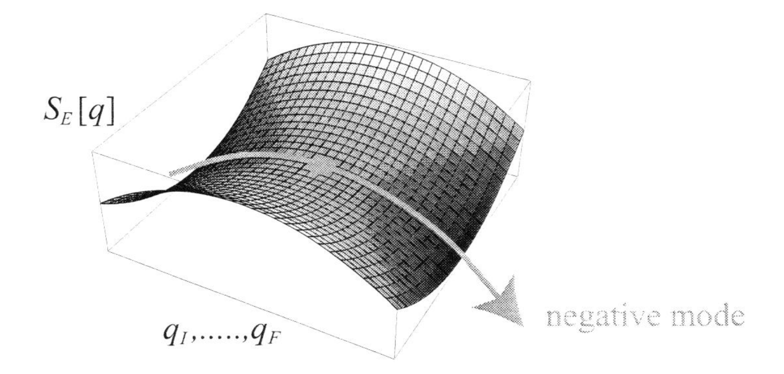

Figure 5. The geometrical structure around the stationary point with one negative mode.

negative mode, the value of the action $S_E[q]$ decreases as a configuration deviates from the classical stationary point. Hence, the classical configuration $q_c(\tau)$ gives rise to the least contribution to the path integral along the line directed to the negative mode. The validity of the semi-classical approximation, on the other hand, relies on the degree of satisfying the assumption that the classical configurations make dominant contributions. Therefore, in the case where the operator D_c has negative modes, the prerequisite of the semi-classical method is already violated.

By observing the geometrical structure in Fig. 5, it would be intuitively plausible that, we can still obtain a good approximation if we first separate as a collective coordinate a set of configurations along the line from the classical configuration to those directed to the negative mode, which looks like a *valley* in the configuration space and thus shall be hereafter called *valley configurations*, from the others and then make a Gaussian approximation at each valley configuration with respect to the other configurations. Actually, a similar idea was proposed in Ref. [21] to treat quasi-zero modes adequately. However, it was later pointed out that the stream-line equation employed there has some drawbacks and a new equation which determine valley configurations was also proposed in Ref. [22]. In the subsequent sections, we will review the method in the latter reference which can properly formulate the above intuitive consideration.

3. Definition of Valley Configurations

To realize the intuitive idea discussed at the end of the previous section, we must first define mathematically valley configurations in the configuration space where a Euclidean path integral shall be carried out. Although there have been several such attempts, the one we employ is in Ref. [22] since, as we will show shortly, there are several advantages in our purpose.

In our definition, a valley is composed of the point where the norm of the gradient vector gets minimum on each contour line, along which the 'hight' of the Euclidean action

S_E is constant. Intuitively speaking, our valley constitutes such a line (more generally a hypersurface) along which slope gets gentlest at each height. It can be translated into mathematical language as a variational problem with a constraint. Explicitly, it is formulated as the following variational equation

$$\frac{\delta}{\delta q(\tau)}\left[\frac{1}{2}\int \mathrm{d}\tau' \left(\frac{\delta S_E[q]}{\delta q(\tau')}\right)^2 - \lambda S_E[q]\right] = 0, \tag{13}$$

where λ is a Lagrange multiplier. We call it the valley equation. To make it more tractable, we first note that it is equivalent to

$$\int \mathrm{d}\tau' \frac{\delta^2 S_E[q]}{\delta q(\tau)\delta q(\tau')}\frac{\delta S_E[q]}{\delta q(\tau')} = \lambda \frac{\delta S_E[q]}{\delta q(\tau)}. \tag{14}$$

Then, by introducing an auxiliary field $F(\tau)$, the valley equation is decomposed into a set of the simultaneous equations

$$\frac{\delta S_E[q]}{\delta q(\tau)} = F(\tau), \qquad \int \mathrm{d}\tau' D(\tau,\tau')F(\tau') = \lambda F(\tau), \tag{15}$$

where D is defined by (6). We immediately notice from (15) that any solution $q(\tau) = q_c(\tau)$ to the equation of motion

$$F_c(\tau) = \frac{\delta S_E[q_c]}{\delta q(\tau)} = 0, \tag{16}$$

always satisfies the valley equation (15). Hence, a family of valley configurations always includes the classical configuration in the path integral configuration space.

Another remarkable aspect of the valley equation (15) is that any solution $F(\tau)$ is an eigenvector of the operator $D(\tau,\tau')$ with the eigenvalue λ. As we discussed in the previous section, an eigenvalue of $D(\tau,\tau')$ corresponding to an eigenvector directed to a valley is expected to be negative. Hence, if we perform Gaussian approximations only for the directions perpendicular to the gradient vector which is a solution of the valley equation with a negative eigenvalue λ, then the Gaussian integrals would be free from any problem caused by the existence of negative modes. The latter idea is formulated as follows. We first parametrize a continuous family of valley configurations by, e.g., R which would depend on λ, and the corresponding solutions to the valley equation are denoted by $q_R(\tau)$ and $F_R(\tau)$. We then introduce a new function as $\varphi_R(\tau) = q(\tau) - q_R(\tau)$ which measures the fluctuation from the valley configuration at R. To ensure that the fluctuation $\varphi_R(q)$ is always orthogonal to the gradient vector $F_R(\tau)$ at each point R, we shall insert the following identity into the path integrand:

$$\int \mathrm{d}R\, \delta\left(\int \mathrm{d}\tau\, \varphi_R(\tau)G_R(\tau)\right)\Delta_{\mathrm{FP}}[\varphi_R] = 1, \tag{17}$$

where $G_R(\tau)$ is the normalized gradient vector at R defined by

$$G_R(\tau) = F_R(\tau)\Big/\sqrt{\int \mathrm{d}\tau' F_R(\tau')^2}, \tag{18}$$

and $\Delta_{\rm FP}[\varphi_R]$ is the Faddeev–Popov (FP) determinant [23] at R given by

$$\Delta_{\rm FP}[\varphi_R] = \left| \int \mathrm{d}\tau \, \frac{\partial q_R(\tau)}{\partial R} \left(G_R(\tau) - \varphi_R(\tau) \frac{\partial G_R(\tau)}{\partial q_R(\tau)} \right) \right| . \tag{19}$$

Then, the partition function (1) under consideration reads as

$$Z = N \int \mathrm{d}R \int \mathcal{D}\varphi_R \, \delta \left(\int \mathrm{d}\tau \, \varphi_R(\tau) G_R(\tau) \right) \Delta_{\rm FP}[\varphi_R] \, \mathrm{e}^{-S_E[q]}. \tag{20}$$

To perform the Gaussian approximations with respect to φ_R, we expand the action $S_E[q]$ around the valley configuration $q(\tau) = q_R(\tau)$ at each R as

$$S_E[q] = S_E[q_R] + \int \mathrm{d}\tau \, \varphi_R(\tau) F_R(\tau) + \frac{1}{2} \int \mathrm{d}\tau \mathrm{d}\tau' \varphi_R(\tau) D_R(\tau, \tau') \varphi_R(\tau') + O(\varphi_R{}^3), \tag{21}$$

where $D_R(\tau, \tau')$ is the operator $D(\tau, \tau')$ at $q(\tau) = q_R(\tau)$. The second term vanishes in the path integral due to the δ-function term. The FP determinant (19) in the leading order in φ_R reads as

$$\Delta_{\rm FP}[\varphi_R] \simeq \left| \int \mathrm{d}\tau \, \frac{\partial q_R(\tau)}{\partial R} G_R(\tau) \right| = \left| \frac{\delta S_E[q_R]}{\delta q(\tau)} \right| \Big/ \sqrt{\int \mathrm{d}\tau' F_R(\tau')^2} := \Delta_R. \tag{22}$$

Hence, if we truncate the expansion (21) at the second order in φ_R and then perform the Gaussian integral with respect to φ_R, we obtain

$$Z \simeq N \int \frac{\mathrm{d}R}{\sqrt{2\pi \det' D_R}} \Delta_R \, \mathrm{e}^{-S[q_R]}, \tag{23}$$

where $\det' D_R$ is the determinant of the operator $D_R(\tau, \tau')$ in the configuration space perpendicular to $G_R(\tau)$. Therefore, $\det' D_R$ does not contain the eigenvalue of $G_R(\tau)$ which can be negative, and the approximation (23) would no longer suffer from the existence of a negative mode.

A generalization of the prescription which is applicable to a case where the operator D contains more than one non-positive mode would be straightforward. In such a situation, one can insert the identity (19) as many as the number of non-positive modes

$$\left[\prod_{i=1}^{n} \int \mathrm{d}R_i \, \delta \left(\int \mathrm{d}\tau_i \, \varphi_{R_i}(\tau_i) G_{R_i}(\tau_i) \right) \right] \Delta_{\rm FP}[\varphi_{\{R_i\}}] = 1, \tag{24}$$

to extract and separate out the degree of freedom corresponding to the non-positive modes. We then easily generalize with the use of (24) the procedure which leads to the formula (23) in the one negative-mode case to obtain

$$Z \simeq N \int \left(\prod_{i=1}^{n} \mathrm{d}R_i \right) \frac{\Delta_{\{R_i\}}}{\sqrt{(2\pi)^n \det{}^{(n)} D_{\{R_i\}}}} \mathrm{e}^{-S_E[q_{\{R_i\}}]}, \tag{25}$$

where $D_{\{R_i\}}$ is the operator at the configuration $q_{\{R_i\}}$ having n non-positive modes, $\Delta_{\{R_i\}}$ is the FP determinant $\Delta_{\rm FP}[\varphi_{\{R_i\}}]$ in the Gaussian approximation, and $\det{}^{(n)} D_{\{R_i\}}$ is the determinant of $D_{\{R_i\}}$ but without the n non-positive eigenvalues.

To investigate and discuss further general aspects of valley configurations, we restrict our subjects to one-body quantum mechanical systems for which a Euclidean action is given by (7). In this case, the operator $D(\tau, \tau')$ defined in (6) reads as

$$D(\tau, \tau') = \delta(\tau - \tau') \left[-\frac{\mathrm{d}^2}{\mathrm{d}\tau^2} + V''(q) \right], \tag{26}$$

and the valley equation (15) becomes

$$-\frac{\mathrm{d}^2 q(\tau)}{\mathrm{d}\tau^2} + V'(q) = F(\tau), \qquad \left[-\frac{\mathrm{d}^2}{\mathrm{d}\tau^2} + V''(q) \right] F(\tau) = \lambda F(\tau). \tag{27}$$

The first equation in (27) indicates that the auxiliary field $F(\tau)$ acts as an external force added to the usual equation of motion for $q(\tau)$. The path-integral measure is defined by

$$\mathcal{D}q = \prod_{n=0}^{\infty} \frac{\mathrm{d}c_n}{\sqrt{2\pi}}, \tag{28}$$

where c_n $(n = 0, 1, \ldots)$ are the expansion coefficients of $q(\tau)$ with respect to the harmonic oscillator basis. Then, the spectral determinant of the harmonic oscillator is given by [18]

$$\frac{N}{\sqrt{\det\left(-\partial_\tau^2 + \omega^2\right)}} = \Upsilon \mathrm{e}^{-\omega T/2}, \tag{29}$$

where $T = \tau_F - \tau_I$ and Υ is a constant which depends on the definition of Z under consideration, for instance, as follows

$$\Upsilon = \begin{cases} 1 & \text{for} \quad Z = \mathrm{Tr}\,\mathrm{e}^{-HT}, \\ (\omega/\pi)^{1/2} & \text{for} \quad Z = \langle 0 | \mathrm{e}^{-HT} | 0 \rangle. \end{cases} \tag{30}$$

4. Valley-instantons

4.1. Defining Equation

As we discussed previously, the conventional semi-classical approach works well without suffering from zero and negative modes when the potential $V(q)$ under consideration has only one local minimum. Hence, we assume that $V(q)$ has at least two local minima.

Let us now consider the configuration $q(\tau)$ connecting two adjacent local minima $q_\pm$ $(q_+ < q_-)$ of the potential $V(q)$ with the boundary condition $q(\pm\infty) = q_\mp$. As was illustrated in Fig. 3, this configuration is realized by an instanton solution if the potential energies at the local minima, $V(q_+)$ and $V(q_-)$, are the same. However, in the general situation where the potential energies are different from each other, this configuration can not be realized as a classical solution. From the energy conservation law, the configuration having the different potential energies is not allowed by the equation of motion. As a consequence, the classical solution is in general drastically changed into a bounce solution as shown in Fig. 4.

On the other hand, the valley equation admits as a solution such a configuration, denoted by $q^{(I)}(\tau)$, that connects the two local minima having different energies and thus meets the boundary condition $q^{(I)}(\pm\infty) = q_{\mp}$. We call this particular valley configuration a *valley-instanton*. From the boundary condition we immediately have

$$\left.\frac{\delta S_E[q^{(I)}]}{\delta q^{(I)}(\tau)}\right|_{\tau=\pm\infty} = 0. \tag{31}$$

Hence, the corresponding auxiliary field $F^{(I)}(\tau)$ satisfies $F^{(I)}(\pm\infty) = 0$.

As we shall show in what follows, a valley-instanton $q^{(I)}(\tau)$ is characterized by the valley equation with $\lambda = 0$

$$\frac{\delta S_E[q^{(I)}]}{\delta q(\tau)} = F^{(I)}(\tau), \qquad \int \mathrm{d}\tau' D(\tau,\tau')F^{(I)}(\tau') = 0. \tag{32}$$

To derive this, take the time-derivative of the first equation of the valley equation (15)

$$\int \mathrm{d}\tau' D(\tau,\tau')\dot{q}(\tau') = \dot{F}(\tau). \tag{33}$$

Then, multiplying $F(\tau)$ on both side of the above equation and integrating over τ from $-\infty$ to ∞, we obtain

$$\int_{-\infty}^{\infty} \mathrm{d}\tau\, \dot{F}(\tau)F(\tau) = \lambda \int_{-\infty}^{\infty} \mathrm{d}\tau\, \dot{q}(\tau)F(\tau), \tag{34}$$

where we have used the second equation of (15). The l.h.s. of (34) is evidently zero from the boundary condition of $F(\tau)$, thus we have

$$\lambda \int_{-\infty}^{\infty} \mathrm{d}\tau\, \dot{q}(\tau)F(\tau) = 0. \tag{35}$$

Therefore, λ must be zero if the integral $\int_{-\infty}^{\infty} d\tau\dot{q}F$ is not zero. In Fig.6, we illustrate a valley-instanton in an asymmetric double-well potential. As is seen clearly in this example, the coordinate $q(\tau)$ monotonically changes from q_- to q_+ like an ordinary instanton solution. The corresponding auxiliary field $F(\tau)$ has no node, so $\int_{-\infty}^{\infty} d\tau\dot{q}F \neq 0$. To the best of our knowledge, all the known valley-instantons show much or less similar behavior.

Finally, we note that the second valley equation in (32) shows that the auxiliary field $F^{(I)}(\tau)$ associated with the valley-instanton solution $q^{(I)}(\tau)$ is the zero mode of the operator D and thus (cf., Eq. (9))

$$F^{(I)}(\tau) \propto \dot{q}^{(I)}(\tau). \tag{36}$$

An *anti-valley-instanton* $q^{(\bar{I})}(\tau)$ can be analogously defined as a deformed configuration of an anti-instanton satisfying the valley equation with the boundary condition $q^{(\bar{I})}(\pm\infty) = q_{\pm}$ even when $V(q_+) \neq V(q_-)$.

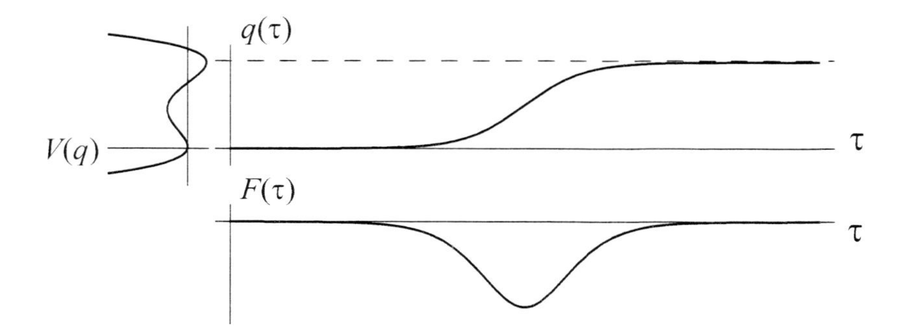

Figure 6. A valley-instanton configuration in an asymmetric double-well potential.

4.2. Single Valley-instanton Calculation

We shall now evaluate the partition function Z for a single valley-instanton. From the translation invariance of the system, a one-parameter family of the valley-instanton configurations, $q^{(I)}_{\tau_0}(\tau)$ and $F^{(I)}_{\tau_0}(\tau)$, which are generated by the translation of the valley-instanton

$$q^{(I)}_{\tau_0}(\tau) := q^{(I)}(\tau - \tau_0), \qquad F^{(I)}_{\tau_0}(\tau) := F^{(I)}(\tau - \tau_0), \tag{37}$$

also satisfy the valley equation with $\lambda = 0$ and the boundary conditions, $q^{(I)}_{\tau_0}(\pm\infty) = q_\pm$ and $F^{(I)}_{\tau_0}(\pm\infty) = 0$. All the configurations in this family have the same value of the action, and thus we should take them into account equally in the calculation of Z.

To include these configurations, it is convenient to introduce another FP determinant Δ_{FP} associated with the zero mode as

$$\int \mathrm{d}\tau_0\, \delta\left(\int \mathrm{d}\tau\, \varphi_{\tau_0}(\tau) G^{(I)}(\tau)\right) \Delta_{\mathrm{FP}}[\varphi_{\tau_0}] = 1, \tag{38}$$

where $\varphi(\tau) = q(\tau) - q^{(I)}(\tau)$ is the quantum fluctuation of the valley-instanton, φ_{τ_0} is defined by $\varphi_{\tau_0}(\tau) := \varphi(\tau + \tau_0) + q^{(I)}(\tau + \tau_0) - q^{(I)}(\tau)$, and $G^{(I)}(\tau)$ is the normalized gradient vector (18) at the valley-instanton configuration. Since we have

$$\begin{aligned}
(\varphi_{\tau_0'})_{\tau_0}(\tau) &= \varphi_{\tau_0'}(\tau + \tau_0) + q^{(I)}(\tau + \tau_0) - q^{(I)}(\tau)\\
&= \varphi(\tau + \tau_0 + \tau_0') + q^{(I)}(\tau + \tau_0 + \tau_0') - q^{(I)}(\tau + \tau_0) + q^{(I)}(\tau + \tau_0) - q^{(I)}(\tau)\\
&= \varphi(\tau + \tau_0 + \tau_0') + q^{(I)}(\tau + \tau_0 + \tau_0') - q^{(I)}(\tau)\\
&= \varphi_{\tau_0 + \tau_0'}(\tau),
\end{aligned} \tag{39}$$

the definition of $\varphi_{\tau_0}(\tau)$ leads to the relation

$$\Delta_{\mathrm{FP}}[\varphi_{\tau_0}] = \Delta_{\mathrm{FP}}[\varphi]. \tag{40}$$

Hence, inserting the FP determinant (38) into the partition function, we have

$$Z = N \int \mathrm{d}\tau_0 \int \mathcal{D}q\, \delta\left(\int \mathrm{d}\tau\, \varphi_{\tau_0}(\tau) G^{(I)}(\tau)\right) \Delta_{\mathrm{FP}}[\varphi]\, \mathrm{e}^{-S_E[q]}. \tag{41}$$

Using (40) and $S_E[q] = S_E[q^{(I)} + \varphi_{\tau 0}]$, which is a consequence of the translation invariance, we find that

$$Z = N \int \mathrm{d}\tau_0 \int \mathcal{D}\varphi_{\tau 0} \delta \left(\int \mathrm{d}\tau\, \varphi_{\tau 0}(\tau) G^{(I)}(\tau) \right) \Delta_{\mathrm{FP}}[\varphi_{\tau 0}]\, \mathrm{e}^{-S_E[q^{(I)}+\varphi_{\tau 0}]}$$
$$= N \int \mathrm{d}\tau_0 \int \mathcal{D}\varphi\, \delta \left(\int \mathrm{d}\tau\, \varphi(\tau) G^{(I)}(\tau) \right) \Delta_{\mathrm{FP}}[\varphi]\, \mathrm{e}^{-S_E[q^{(I)}+\varphi]}. \tag{42}$$

Therefore, the integration of τ_0 is manifestly factored out. Let us now expand the action $S_E[q^{(I)} + \varphi]$ around the valley-instanton

$$S_E[q^{(I)} + \varphi] = S_E[q^{(I)}] + \int \mathrm{d}\tau\, \varphi(\tau) \frac{\delta S_E[q^{(I)}]}{\delta q^{(I)}(\tau)}$$
$$+ \int \mathrm{d}\tau \mathrm{d}\tau' \varphi(\tau) \frac{\delta^2 S_E[q^{(I)}]}{\delta q^{(I)}(\tau) \delta q^{(I)}(\tau')} \varphi(\tau') + \cdots. \tag{43}$$

Due to the δ-function in the partition function, the second term of the action again vanishes. Then by the Gaussian integration, the partition function for one valley-instanton becomes

$$Z^{(I)} \simeq N \int \mathrm{d}\tau_0 \frac{\Delta_{\mathrm{FP}}^{(I)}}{\sqrt{2\pi \det{}' D^{(I)}}}\, \mathrm{e}^{-S_E[q^{(I)}]}, \tag{44}$$

where $\Delta_{\mathrm{FP}}^{(I)}$ is the leading-order FP determinant

$$\Delta_{\mathrm{FP}}^{(I)} = \int \mathrm{d}\tau\, \dot{q}^{(I)}(\tau) F^{(I)}(\tau) \Big/ \sqrt{\int \mathrm{d}\tau\, F^{(I)}(\tau)^2} = \sqrt{\int \mathrm{d}\tau\, \dot{q}^{(I)}(\tau)^2}, \tag{45}$$

where (36) has been used in the last equality, and $\det{}' D^{(I)}$ is the determinant of the operator D at the valley-instanton configuration in the configuration space perpendicular to $F^{(I)}$. Hence, $\det{}' D^{(I)}$ does not contain the contribution from the zero mode $\lambda = 0$. In addition, as was shown in Fig. 6, $F^{(I)}$ is nodeless and is a zero mode of $D^{(I)}$. Thus, the oscillator theorem tells us that the eigenvalues of $D^{(I)}$ are positive except the zero mode given by $F^{(I)}$. As a result, $\det{}' D^{(I)}$ contains only positive modes since the unique zero mode has been already removed by the definition of the modified determinant. Hence, the integral in (44) is well-defined.

For further evaluation, we need to calculate $\det{}' D^{(I)}$. Here we use the following formula (for the derivation, see, e.g., Ref. [4])

$$\frac{\det{}' D^{(I)}}{\det\left(-\partial_\tau^2 + \omega_+^2\right)} = \frac{\psi(T/2)}{\lambda \psi_0(T/2)}, \tag{46}$$

where the eigenvalues are determined with the use of the Dirichlet boundary conditions at $\pm T/2$. The eigenvalue λ is the lowest eigenvalue of $D^{(I)}$ and goes to zero in the limit $T \to \infty$. The functions $\psi(\tau)$ and $\psi_0(\tau)$ are the solutions to the following differential equations

$$\left(-\partial_\tau^2 + V''(q^{(I)})\right) \psi(\tau) = \left(-\partial_\tau^2 + \omega_+^2\right) \psi_0(\tau) = 0, \tag{47}$$

with the boundary conditions, $\psi(-T/2) = \psi_0(-T/2) = 0$ and $\dot{\psi}(-T/2) = \dot{\psi}_0(-T/2) = 1$. The function $\psi_0(\tau)$ is easily obtained as

$$\psi_0(\tau) = \frac{1}{2\omega_+}\left(\mathrm{e}^{\omega_+(\tau+T/2)} - \mathrm{e}^{-\omega_+(\tau+T/2)}\right), \tag{48}$$

and thus $\psi_0(T/2) \simeq \mathrm{e}^{\omega_+ T}/(2\omega_+)$ for $T \gg 1$. To construct the function $\psi(\tau)$, we use the zero mode $F^{(I)}$ satisfying

$$\left(-\partial_\tau^2 + V''(q^{(I)})\right) F^{(I)}(\tau) = 0, \tag{49}$$

with the boundary condition $F^{(I)}(\pm\infty) = 0$. In the asymptotic region $\tau \to \pm T/2$, the latter equation reduces to

$$\left(-\partial_\tau^2 + \omega_\mp^2\right) F^{(I)}(\tau) = 0, \tag{50}$$

where $\omega_\pm^2 = V''(q_\pm)$. Thus, the auxiliary field $F^{(I)}$ has the following asymptotic behavior at $\tau \to \pm T/2$

$$F^{(I)}(\tau) \simeq F_\pm \mathrm{e}^{\mp\omega_\mp \tau}. \tag{51}$$

We also need the other linearly independent solution $\bar{F}(\tau)$ to (49). The Wronskian of $F^{(I)}$ and $\bar{F}$

$$W = F^{(I)}(\tau)\dot{\bar{F}}(\tau) - \bar{F}(\tau)\dot{F}^{(I)}(\tau) \tag{52}$$

is τ-independent, so we can choose without any loss of generality the overall factor of $\bar{F}$ to satisfy

$$W = 2\omega_+\omega_-. \tag{53}$$

This automatically determines the asymptotic behavior of $\bar{F}(\tau)$ at $\tau \to \pm T/2$ as

$$\bar{F}(\tau) \simeq \pm\frac{\omega_\pm}{F_\pm}\mathrm{e}^{\pm\omega_\mp \tau}. \tag{54}$$

Then, the function $\psi(\tau)$ which satisfies the required boundary condition at $\tau = -T/2$ is expressed in terms of $F^{(I)}$ and $\bar{F}$ as

$$\psi(\tau) = \frac{1}{2\omega_+}\left(\frac{F^{(I)}(\tau)}{F^{(I)}(-T/2)} - \frac{\bar{F}(\tau)}{\bar{F}(-T/2)}\right). \tag{55}$$

The lowest eigenvalue λ defined by

$$D^{(I)}(\tau)\bar{\psi}(\tau) = \lambda\bar{\psi}(\tau), \tag{56}$$

with the Dirichlet boundary condition $\bar{\psi}(\pm T/2) = 0$ is calculated by using the following formal solution

$$\begin{aligned}\bar{\psi}(\tau) &= \psi(\tau) + \lambda\int_{-T/2}^{T/2} \mathrm{d}\tau' G(\tau,\tau')\bar{\psi}(\tau') \\ &= \psi(\tau) + \lambda\int_{-T/2}^{T/2} \mathrm{d}\tau' G(\tau,\tau')\psi(\tau') + O(\lambda^2),\end{aligned} \tag{57}$$

where $\psi(\tau)$ is given by (55). The Green function $G(\tau, \tau')$ in the above integral satisfying

$$D^{(I)}(\tau)G(\tau, \tau') = \delta(\tau - \tau'). \tag{58}$$

is given by

$$G(\tau, \tau') = \frac{1}{W}\left(F^{(I)}(\tau)\bar{F}(\tau') - F^{(I)}(\tau')\bar{F}(\tau)\right)\theta(\tau - \tau'). \tag{59}$$

In the above construction, the boundary condition $\bar{\psi}(-T/2) = 0$ has been already imposed, and the other boundary condition $\bar{\psi}(T/2) = 0$ determines the eigenvalue λ. Substituting (55) and (59) into the last expression in (57) and retaining the dominant term, we obtain

$$\frac{\psi(T/2)}{\lambda} \simeq \frac{\bar{F}(T/2)}{2\omega_+ W F^{(I)}(-T/2)} \int_{-T/2}^{T/2} \mathrm{d}\tau\, F^{(I)}(\tau)^2. \tag{60}$$

Finally, substituting (48) and (60) into the formula (46), we have

$$\frac{\det{}' D^{(I)}}{\det\left(-\partial_\tau^2 + \omega_+^2\right)} = \frac{\mathrm{e}^{(\omega_- - \omega_+)T/2}}{2\omega_- F_+ F_-} \int_{-T/2}^{T/2} \mathrm{d}\tau\, F^{(I)}(\tau)^2. \tag{61}$$

Then, the partition function for one valley-instanton (44) reads from (29), (45), and (61) as

$$Z^{(I)} \simeq \frac{\Upsilon}{\sqrt{2\pi\kappa^{(I)}}}\, \mathrm{e}^{-(\omega_- + \omega_+)T/4} \int_{-T/2}^{T/2} \mathrm{d}\tau_0\, \Delta_{\mathrm{FP}}^{(I)}\, \mathrm{e}^{-S_E[q^{(I)}]}, \tag{62}$$

where $\kappa^{(I)}$ is defined by

$$\kappa^{(I)} = \frac{1}{2\omega_- F_+ F_-} \int_{-T/2}^{T/2} \mathrm{d}\tau\, F^{(I)}(\tau)^2. \tag{63}$$

In a similar way, we obtain the partition function for one anti-valley-instanton as

$$Z^{(\bar{I})} \simeq \frac{\Upsilon}{\sqrt{2\pi\kappa^{(\bar{I})}}}\, \mathrm{e}^{-(\omega_- + \omega_+)T/4} \int_{-T/2}^{T/2} \mathrm{d}\tau_0\, \Delta_{\mathrm{FP}}^{(\bar{I})}\, \mathrm{e}^{-S_E[q^{(\bar{I})}]}, \tag{64}$$

where $\kappa^{(\bar{I})}$ is defined by

$$\kappa^{(\bar{I})} = \frac{1}{2\omega_+ F_- F_+} \int_{-T/2}^{T/2} \mathrm{d}\tau\, F^{(\bar{I})}(\tau)^2. \tag{65}$$

5. Analysis of an Asymmetric Double-well Potential

In this and the next sections, we shall apply our method in the previous section to some quantum mechanical systems with non-trivial vacua. The potentials we shall consider are an asymmetric double-well potential and a symmetric triple-well potential.

Let us first consider an asymmetric double-well potential given by the following functional form

$$V(q; g, \epsilon) = \frac{1}{2}q^2(1 - gq)^2 - \epsilon g q, \tag{66}$$

where g and ϵ are free parameters satisfying $|\tilde{\epsilon}| := |\epsilon g^2| < \sqrt{3}/18$. The latter restriction ensures the existence of two local minima in the potential. We also assume $\epsilon \geq 0$ without loss of generality since the reflection with respect to the axis $q = 1/2g$ results in $V(1/g - q; g, \epsilon) = V(q; g, -\epsilon) - \epsilon$. For $\epsilon = 0$, the potential gets symmetric with respect to the axis $q = 1/2g$ so that $V(1/g - q; g, 0) = V(q; g, 0)$. Thus, ϵ characterize the degree of asymmetry of the potential. More precisely, for $|\tilde{\epsilon}| \ll 1$ the potential has two local minima at $q = 0 + O(\tilde{\epsilon}) := q_+$ and $q = 1/g[1 + O(\tilde{\epsilon})] := q_-$ where $g^2 V(q_+) = 0 + O(\tilde{\epsilon})$ and $V(q_-) = -\epsilon[1 + O(\tilde{\epsilon})]$, respectively. Hence, ϵ roughly measures the energy difference between at the two potential minima. The potential barrier between the two minima is highest at $q = 1/2g[1 + O(\tilde{\epsilon})]$ with the value $V(1/2g) = 1/32g^2[1 + O(\tilde{\epsilon})]$. Hence, the inverse-square of the coupling constant g is roughly related with the height of the potential barrier. The potential form is explicitly shown in Fig. 7.

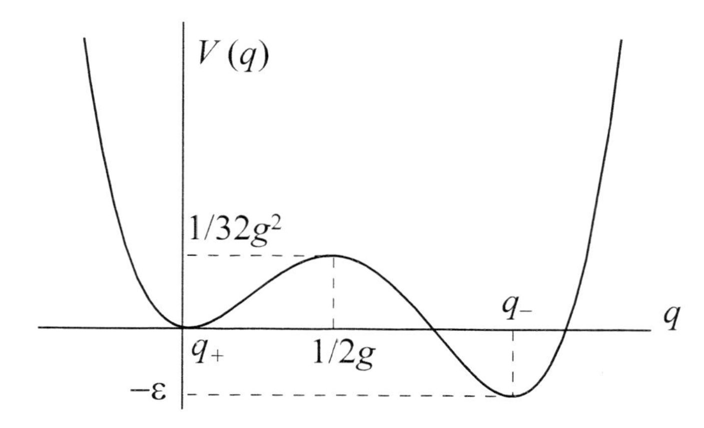

Figure 7. The form of the asymmetric double-well potential (66).

5.1. A Single Valley-instanton

For the symmetric case $\epsilon = 0$, the potential (66) admits instanton and anti-instanton solutions to the equation of motion (3), which connect the two potential minima at $q = q_+$ and $q = q_-$, as follows:

$$q_0^{(I)}(\tau - \tau_0) = \frac{1}{g}\frac{1}{1 + \mathrm{e}^{-(\tau - \tau_0)}}, \qquad q_0^{(\bar{I})}(\tau - \tau_0) = \frac{1}{g}\frac{1}{1 + \mathrm{e}^{\tau - \tau_0}}, \tag{67}$$

where τ_0 indicates the position of the (anti-)instanton. The action of one (anti-)instanton amounts to

$$S_E[q_0^{(I)}] = \frac{1}{6g^2}. \tag{68}$$

For any asymmetric case $\epsilon \neq 0$, on the other hand, there are no classical solutions which satisfy the boundary conditions $q_c(\pm\infty) = q_\mp$ or $q_\pm$ since the energies at q_+ and q_- are different. Only the possible τ-dependent classical solution is a bounce solution as has been shown in Fig. 4. Hence, the non-trivial classical solution is discontinuously changed from an (anti-)instanton solution to a bounce solution at $\epsilon = 0$. The valley equation, on the other hand, admits a (anti-)valley-instanton solution which satisfies the boundary conditions $q^{(I)}(\pm\infty) = q_{\mp}$ or $q_{\pm}$ and is a continuous deformation of a classical (anti-)instanton, as was discussed in detail in Section 4.. Although the full analytic form of the (anti-)valley-instanton solution cannot be obtained in closed form, we can still apply asymptotic analyses to the valley equation with the boundary conditions to grasp roughly the behavior of the solution. The detailed analysis show that the valley-instanton $q^{(I)}(\tau)$ and the corresponding auxiliary field $F^{(I)}(\tau)$ have the following asymptotic forms [8]:

$$q^{(I)}(\tau) = \begin{cases} \epsilon g + \dfrac{1}{g}\,\mathrm{e}^{\omega_+ \tau} + \dfrac{3\epsilon g}{\omega_+}\tau \mathrm{e}^{\omega_+ \tau} + \cdots & \text{for} \quad \tau \ll -1, \\ \dfrac{1}{g}\dfrac{1}{1+\mathrm{e}^{-\tau}} + \epsilon g + \cdots & \text{for} \quad |\tau| \ll \dfrac{1}{\epsilon g^2}, \\ \dfrac{1}{g} + \epsilon g - \dfrac{1}{g}\,\mathrm{e}^{-\omega_- \tau} - \dfrac{3\epsilon g}{\omega_-}\tau \mathrm{e}^{-\omega_- \tau} + \cdots & \text{for} \quad \tau \gg 1, \end{cases} \tag{69}$$

$$F^{(I)}(\tau) = \begin{cases} -6\epsilon g\,\mathrm{e}^{\omega_+ \tau} + \cdots & \text{for} \quad \tau \ll -1, \\ -6\epsilon g \dfrac{1 - 3\epsilon g^2 \tau}{(1+\mathrm{e}^{-\tau})(1+\mathrm{e}^{\tau})} + \cdots & \text{for} \quad |\tau| \ll \dfrac{1}{\epsilon g^2}, \\ -6\epsilon g\,\mathrm{e}^{-\omega_- \tau} + \cdots & \text{for} \quad \tau \gg 1. \end{cases} \tag{70}$$

In the present case, $\omega_{\pm}$ and $F_{\pm}$ introduced in (50) and (51), respectively, are given by

$$\omega_{\pm} = \sqrt{V''(q_{\pm})} = 1 + O(\tilde{\epsilon}), \qquad F_{\pm} = -6\epsilon g. \tag{71}$$

The above results indeed show that in the limit $\epsilon \longrightarrow 0$ they reduce to the asymptotic behavior of the classical instanton configuration $q^{(I)}(\tau) \longrightarrow q_0^{(I)}(\tau)$ and $F^{(I)}(\tau) \longrightarrow 0$.

With the asymptotic solutions (69) and (70), we can calculate its action S_E, FP determinant Δ_{FP}, and spectral determinant $\det{}' D^{(I)}$ in the leading order. For the action S_E regularized in the region $|\tau| < T/2$ ($T \gg 1$), we have

$$S_E[q^{(I)}] = S_E[q_0^{(I)}] - \epsilon T/2 + O(\tilde{\epsilon}^2), \tag{72}$$

where the first term is the action of one classical instanton given by (68) and the second term corresponds to the volume factor which comes from the configuration staying at q_- with the energy $-\epsilon$ within the period $\tau \in (0, T/2)$. For the calculation of the determinant (45), we note the fact that $F^{(I)}(\tau)$ corresponds to the zero mode associated with the valley-instanton and thus behaves like the one $\dot{q}(\tau)$ associated with the ordinary instanton, see also Fig. 6. Hence, the dominant contribution to the integral $\int \mathrm{d}\tau\, F^{(I)}(\tau)^2$ comes from the region around the position of the valley-instanton $\tau \simeq 0$. Noting this fact, we obtain from (70) that

$$\int_{-\infty}^{\infty} \mathrm{d}\tau\, F(\tau)^2 \simeq 6\epsilon^2 g^2. \tag{73}$$

The FP determinant is obtained by substituting it into (45) and reads as

$$\Delta_{\rm FP}^{(I)} = \frac{1}{\sqrt{6g^2}} \left[1 + O(\tilde{\epsilon})\right]. \tag{74}$$

The spectral determinant $\det{}' D^{(I)}$ is obtained by calculating the quantity $\kappa^{(I)}$ defined by (63). From (71) and (73) we immediately obtain

$$\kappa^{(I)} = \frac{1}{2\omega_- F_+ F_-} \int_{-\infty}^{\infty} \mathrm{d}\tau\, F^{(I)}(\tau)^2 \simeq \frac{1}{12}. \tag{75}$$

Hence, we finally obtain the partition function (44) for one valley-instanton

$$Z^{(I)} \simeq \frac{\Upsilon}{\sqrt{2\pi\kappa^{(I)}}} \int_{-T/2}^{T/2} \mathrm{d}\tau_0\, \Delta_{\rm FP}^{(I)}\, \mathrm{e}^{-T/2 - S_E[q^{(I)}]} = \alpha T \mathrm{e}^{-(1+\epsilon)T/2}\Upsilon, \tag{76}$$

where α is given by

$$\alpha = \frac{\Delta_{\rm FP}^{(I)}}{\sqrt{2\pi\kappa^{(I)}}}\, \mathrm{e}^{-S_E[q_0^{(I)}]} = \frac{\mathrm{e}^{-1/6g^2}}{\pi^{1/2} g}. \tag{77}$$

The corresponding results for one anti-valley-instanton are entirely same as for one valley-instanton since $q^{(\bar{I})}(\tau) = q^{(I)}(-\tau)$, and thus we obtain

$$S_E[q^{(\bar{I})}] = S_E[q^{(I)}], \quad \Delta_{\rm FP}^{(\bar{I})} = \Delta_{\rm FP}^{(I)}, \quad \kappa^{(\bar{I})} = \kappa^{(I)}, \quad Z^{(\bar{I})} = Z^{(I)}. \tag{78}$$

5.2. A Pair of a Valley-instanton and an Anti-valley-instanton

The valley equation with the boundary conditions $q(\pm\infty) = q_+$ (and q_-, respectively) admits solutions $q^{(I\bar{I})}(\tau)$ called an *$I\bar{I}$-valley* configuration (and $q^{(\bar{I}I)}(\tau)$ called an *$\bar{I}I$-valley* configuration, respectively) which are roughly composed of one valley-instanton and one anti-valley-instanton. As we discussed before, solutions to the valley equation includes a solution to the equation of motion. It was investigated numerically [7] that an $I\bar{I}$-valley configuration tends to and then coincides with the classical bounce configuration as the distance of the valley-instanton and the anti-valley-instanton decreases. The action, on the other hand, is monotone increasing and gets maximal at the bounce configuration. Then, it changes into being monotone decreasing until the valley configuration becomes the trivial vacuum configuration $q^{(I\bar{I})}(\tau) = q_+$. Hence, the classical bounce configuration in fact provides a local maximal of the action and thus makes the least contribution to the path integral along the valley line in the phase space. Except for the region near the vacuum configuration, the eigenvalue of the auxiliary field is negative $\lambda < 0$. Therefore, we successfully extract a collective coordinate associated with the negative mode around the bounce configuration. It also indicates that the emergence of the imaginary part in the spectrum of an asymmetric double-well potential due to a bounce solution is indeed an artifact caused by applying the semi-classical method inadequately.

The family of $I\bar{I}$-valley configurations with largely separated valley- and anti-valley-instantons corresponds to a quasi-zero mode $|\lambda| \ll 1$, and its asymptotic form is calculated by a formal series expansion in λ. For the latter purpose, let us consider an $I\bar{I}$-valley configuration which is symmetric with respect to $\tau = 0$ and where a valley-instanton is located around $\tau = -R/2$ and an anti-valley-instanton is around $\tau = R/2$. The detailed analysis shows [8] that for a large separation $R \gg 1$, the latter configuration has the eigenvalue λ which behaves like $\lambda \simeq -24\mathrm{e}^{-R}$, and its asymptotic form for $\tau > 0$ is given by

$$q^{(I\bar{I})}(\tau) = q_0^{(I)}(\tau - R/2) + 12\mathrm{e}^{-R}\int_0^\infty d\tau' G(\tau,\tau')\dot{q}_0^{(\bar{I})}(\tau' - R/2) + O(\mathrm{e}^{-2R}), \quad (79)$$

$$F^{(I\bar{I})}(\tau) = -12\mathrm{e}^{-R}\dot{q}_0^{(I)}(\tau - R/2) + O(\mathrm{e}^{-2R}), \quad (80)$$

where $G(\tau,\tau')$ is a Green function

$$G(\tau,\tau') = \frac{1}{2}\left[\psi_1(\tau)\psi_2(\tau')\theta(\tau-\tau') + \psi_1(\tau')\psi_2(\tau)\theta(\tau'-\tau)\right], \quad (81)$$

composed of two linearly independent solutions $\psi_i(\tau)$ $(i = 1,2)$ to $(-\partial_\tau^2 + V(q_0^{(I)}))\psi_i(\tau) = 0$, the one of which is given by $\psi_1(\tau) = \dot{q}_0^{(I)}(\tau - R/2)$.

With the above asymptotic solutions, we can calculate, as in the case of a single valley-instanton, the action, the FP determinant, and the spectral determinant of the $I\bar{I}$-valley configuration in the leading order. For the action, we obtain

$$S_E[q^{(I\bar{I})}] = 2S_E[q_0^{(I)}] - \epsilon R - \frac{2}{g^2}\mathrm{e}^{-R} + O(\mathrm{e}^{-2R}), \quad (82)$$

where apparently the first term corresponds to the contribution from the one instanton and the one anti-instanton at $\epsilon = 0$ while the second term to the volume factor caused by the energy difference $\epsilon \neq 0$. The third term has a dependence on the distance between a valley- and an anti-valley-instantons, and vanishes at $R \to \infty$. Hence, it can be naturally interpreted as an interaction term. We also note that the minus sign in front of the interaction term indicates that the interaction between a valley- and an anti-valley-instantons is *attractive*.

To calculate the partition function of the $I\bar{I}$ valley, we first note that we should separate out the two collective coordinates, the one is associated with the valley parameter R and the other is associated with the zero mode τ_0. Hence, we shall apply the formula (25) with $n = 2$. The FP determinant of the $I\bar{I}$-valley configuration turns out to be the product of those of one valley-instanton and one anti-valley-instanton in the leading order. Hence, we have from (74)

$$\Delta_{\mathrm{FP}}^{(I\bar{I})} \simeq \Delta_{\mathrm{FP}}^{(I)}\Delta_{\mathrm{FP}}^{(\bar{I})} = \frac{1}{6g^2}. \quad (83)$$

Similarly, the spectral determinant of the $I\bar{I}$-valley configuration in the leading order is also given by the product of those of one valley-instanton and one anti-valley-instanton as

$$\begin{aligned}\frac{N}{\sqrt{\det{}'' D^{(I\bar{I})}}} &\simeq \frac{N}{\sqrt{\det\left(-\partial_\tau^2 + \omega_+^2\right)}}\sqrt{\frac{\det\left(-\partial_\tau^2 + \omega_+^2\right)}{\det{}' D^{(I)}}}\sqrt{\frac{\det\left(-\partial_\tau^2 + \omega_-^2\right)}{\det{}' D^{(\bar{I})}}} \\ &= \frac{\Upsilon}{\sqrt{\kappa^{(I)}\kappa^{(\bar{I})}}}\mathrm{e}^{-T/2}. \end{aligned} \quad (84)$$

Hence, the partition function for an $I\bar{I}$-valley configuration is calculated as

$$Z^{(I\bar{I})} \simeq \int_0^T \mathrm{d}R \int_0^{T/2} \mathrm{d}\tau_0 \frac{\Delta_{\mathrm{FP}}^{(I\bar{I})}}{2\pi\sqrt{\kappa^{(I)}\kappa^{(\bar{I})}}} \mathrm{e}^{-T/2 - S_E[q^{(I\bar{I})}]} = \frac{T}{2}\frac{\mathrm{e}^{-T/2}}{\pi g^2} \int_0^T \mathrm{d}R\, \mathrm{e}^{-S_E[q^{(I\bar{I})}]}, \tag{85}$$

where we have employed the periodic boundary condition and thus $\Upsilon = 1$. For the region $R \gg 1$, the configuration is almost like a largely separated pair of one valley-instanton and one anti-valley-instanton, and thus has nonperturbative nature. For the region $R \ll 1$, on the other hand, the configuration is almost like the trivial vacuum, namely, $q^{(I\bar{I})}(\tau) \longrightarrow q_+$ $(R \longrightarrow 0_+)$, and thus has perturbative nature. Hence, the integration with respect to R in (85) contains both the perturbative and nonperturbative contributions from the $I\bar{I}$-valley configuration. To see the structure of the integral more transparently, we restrict ourselves to the $\epsilon = 0$ case and perform the change of variable $t = g^2 S_E[q^{(I\bar{I})}]$ (the generalization to the $\epsilon \neq 0$ case is straightforward, see Ref. [8]). Then, the expression for $Z^{(I\bar{I})}$ is rewritten as

$$Z^{(I\bar{I})} = \frac{T}{2}\frac{\mathrm{e}^{-T/2}}{\pi g^2} \int_{C_V} \mathrm{d}t\, F(t)\, \mathrm{e}^{-t/g^2}, \tag{86}$$

where the path of integration is $C_V = (0, t_0)$ with $t_0 = 1/3 - 2\mathrm{e}^{-T}$, and $F(t)$ has a singularity at $t = 1/3$. As was discussed previously, the integral in (86) contains in part the perturbative contribution as well as nonperturbative one. In this regard, we recall the fact that there is a method to express a series expansion in terms of an integral which resembles the expression (86) in form. It is called the Borel summation. So, let us first review it briefly.

For a given series in g^2

$$f(g^2) = \sum_{n=0}^{\infty} a_n g^{2n}, \tag{87}$$

the Borel function $F(g^2)$ of $f(g^2)$ is introduced as

$$F(g^2) = \sum_{n=0}^{\infty} \frac{a_n}{n!} g^{2n}. \tag{88}$$

Then, the Borel sum $f_B(g^2)$ of $f(g^2)$ is defined by

$$f_B(g^2) := \int_0^{\infty} \frac{\mathrm{d}t}{g^2} F(t)\, \mathrm{e}^{-t/g^2}. \tag{89}$$

It is evident that $f_B(g^2)$ coincides with the original series $f(g^2)$ so long as the latter series is absolutely convergent. However, the Borel sum can be well-defined even if the original series is divergent. For instance, if the expansion coefficients a_n are alternatingly and factorially divergent $a_n = (-1)^n n!$, which is a typical large-order behavior of the perturbation series for ϕ^4 theory and quantum mechanical anharmonic oscillator [17], the Borel sum reads as

$$f_B(g^2) = \int_0^\infty \frac{dt}{g^2} \frac{e^{-t/g^2}}{1+t}, \tag{90}$$

which is the well-defined integral exponential function. On the other hand, if the expansion coefficients a_n are monotone and factorially divergent $a_n = n!$, which is a typical large-order behavior of the perturbation series for quantum systems with non-trivial vacua such as a double-well potential [17], the Borel sum reads as

$$f_B(g^2) = \int_0^\infty \frac{dt}{g^2} \frac{e^{-t/g^2}}{1-t}, \tag{91}$$

which has a pole at $t = 1$ and thus is ill-defined. In such a case, the series is called *Borel non-summable*.

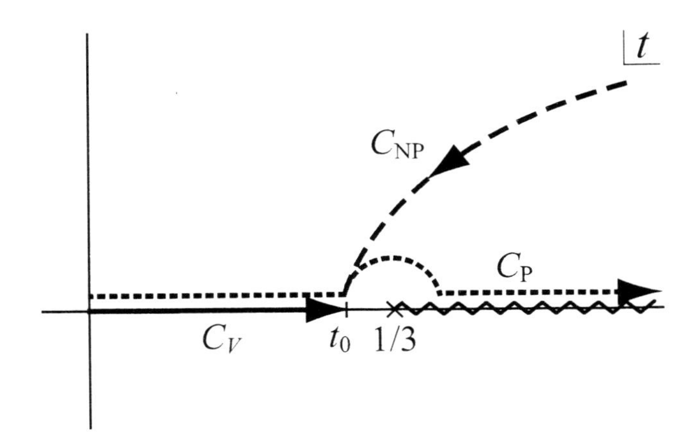

Figure 8. The deformation of the integral path for an $I\bar{I}$ valley.

Let us now come back to the integral (86). If we deform the integral path C_V as $C_V = C_{\rm P} + C_{\rm NP}$ with $C_{\rm P} = (0, \infty)$, such that it does not encircle the singularity at $t = 1/3$, see Fig. 8, and rewrite the expression (86) as

$$Z^{(I\bar{I})} = \frac{T}{2}\frac{e^{-T/2}}{\pi g^2} \int_{C_{\rm P}} dt\, F(t)\, e^{-t/g^2} + \frac{T}{2}\frac{e^{-T/2}}{\pi g^2} \int_{C_{\rm NP}} dt\, F(t)\, e^{-t/g^2}, \tag{92}$$

the first term has a structure which is quite similar to the Borel sum (91) of the divergent series which appears typically in perturbation theory for a double-well potential problem. Hence, we shall identify the first and second terms in (92) as the perturbative and nonperturbative contributions, respectively. We shall hereafter express the latter decomposition into perturbative and nonperturbative parts as $Z = Z_{\rm P} + Z_{\rm NP}$. It is worth noting that the choice of the integral path shown in Fig. 8 automatically fix the ambiguity of the Borel integral caused by the Borel singularity.

An immediate consequence of the decomposition is that

$$\operatorname{Im} Z_{\mathrm{P}} + \operatorname{Im} Z_{\mathrm{NP}} = 0, \tag{93}$$

since $Z = Z_{\mathrm{P}} + Z_{\mathrm{NP}}$ is real. The imaginary part of the perturbative contribution may originate from the existence of Borel singularities. Hence, the relation (93) indicates that the appearance of the non-zero imaginary part of the nonperturbative contribution can be regarded as a signal for Borel non-summability of the perturbation series.

On the other hand, the existence of Borel singularities indicates that $Z_{\mathrm{P}}(g^2)$, when analytically continued into the complex g^2-plane, has a cut on the real axis. Indeed, if the phase θ of $g^2 = |g^2|\mathrm{e}^{\mathrm{i}\theta}$ changes from 0 to 2π, the integral path C_{P} also rotates counterclockwise to come back to the positive real axis from below, but $Z_{\mathrm{P}}(|g^2|\mathrm{e}^{2\pi\mathrm{i}}) \neq Z_{\mathrm{P}}(|g^2|)$ due to the residues of the singularities. Hence, we have

$$\begin{aligned} Z_{\mathrm{P}}(g^2) &= \frac{1}{2\pi\mathrm{i}} \oint_{C_{g^2}} \mathrm{d}z\, \frac{Z_{\mathrm{P}}(z)}{z - g^2} = \frac{1}{\pi} \int_0^\infty \mathrm{d}z\, \frac{\operatorname{Im} Z_{\mathrm{P}}(z)}{z - g^2} + \cdots \\ &= -\frac{1}{\pi} \int_0^\infty \mathrm{d}z\, \frac{\operatorname{Im} Z_{\mathrm{NP}}(z)}{z - g^2} + \cdots, \end{aligned} \tag{94}$$

where C_{g^2} is a contour that circles the point $z = g^2$, and we have neglected any contribution from singularities other than the cut, and have applied the relation (93) to the second line. The l.h.s. of the dispersion relation (94) stands for the perturbative part which is expressed in terms of a formal series expansion in g^2

$$Z_{\mathrm{P}}(g^2) = \sum_{r=0}^{\infty} g^{2r} c^{[2r]}. \tag{95}$$

As long as it is asymptotic, it must be unique and thus must coincide with the formal series expansion of the r.h.s. of (94) in g^2. Hence, we finally have

$$c^{[2r]} \simeq -\frac{1}{\pi} \int_0^\infty \mathrm{d}g^2 \frac{\operatorname{Im} Z_{\mathrm{NP}}(g^2)}{g^{2r+2}}. \tag{96}$$

It is a quite intriguing relation since it connects the perturbative coefficients with the nonperturbative contribution. It would be almost evident that the same dispersion relation holds for other physical quantities such as energy eigenvalues. Later in Section 7., we will employ the dispersion relation (96) for energy eigenvalues to examine the validity and accuracy of our predictions.

The remaining task is now to calculate the nonperturbative contribution $Z_{\mathrm{NP}}(g^2)$. Unfortunately, however, the interaction between the valley-instantons is attractive (82), which means that a dominant contribution would come from configurations with smaller values of R where the analytic expression is not available. This observation naturally leads to the prescription suggested by Bogomolny in Ref. [24]. It was suggested that by the formal analytic continuation which changes the attractive interaction between instantons into a repulsive one, the instanton configurations get relevant and their contributions can be separated from the perturbative one. The latter manipulation was employed to obtain the spectrum for excited states in Refs. [25, 26].

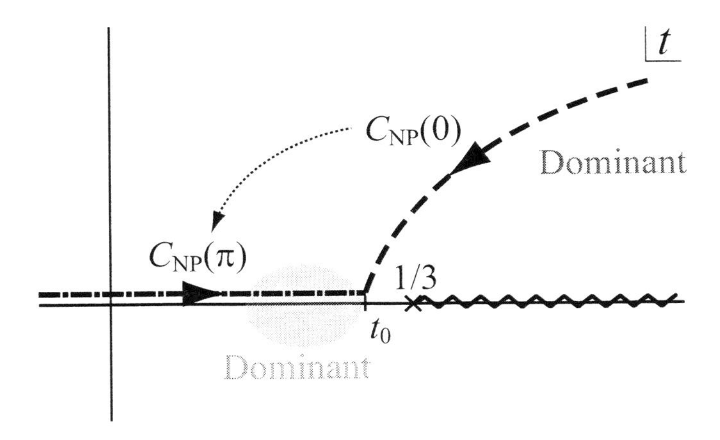

Figure 9. The analytic continuation of $Z_{\rm NP}(|g^2|{\rm e}^{{\rm i}\theta})$ from $\theta = 0$ to π.

To realize the idea in our formalism, let us consider the analytic continuation of the function $Z_{\rm NP}(g^2)$, namely, the second term in (92). To keep the integral well-defined, the integral path $C_{\rm NP}(\theta)$ changes as shown in Fig. 9 as the phase θ of the coupling constant $g^2 = |g^2|{\rm e}^{{\rm i}\theta}$ changes from $\theta = 0$ to π. For $\theta = \pi$ when the interaction between valley-instantons gets repulsive, dominant contributions to the integral would come from the configurations with $R \gg 1$ which correspond to the integral variable $t \simeq t_0$, see also Fig. 9. Hence, the integral for $Z_{\rm NP}(|g^2|{\rm e}^{{\rm i}\pi})$ with the integral path $C_{\rm NP}(\pi)$ would be well approximated by the one with the integral path C_V (compare with Fig. 8)[1], which means that

$$Z_{\rm NP}(|g^2|{\rm e}^{{\rm i}\pi}) \simeq Z(|g^2|{\rm e}^{{\rm i}\pi}). \tag{97}$$

This formula ensures that we can use the asymptotic form (82) which is valid for $R \gg 1$ in the calculation of $Z_{\rm NP}$ for $\theta = \pi$. Then, the obtained asymptotic result for $\theta = \pi$ also provides the asymptotic form of $Z_{\rm NP}$ for $\theta = 0$ unless its validity cannot be extended to the region $\theta \to 0$. We will use this prescription in the next section.

5.3. Multi-valley-instanton Calculation

In this section we shall evaluate the partition function $Z(T) = {\rm Tr}\,{\rm e}^{-HT}$ by summing over those configurations made of several (anti-)valley-instantons by utilizing the knowledge of the (anti-)valley-instantons and the interactions among them and by applying the analytic continuation. This enables us to evaluate nonperturbative contributions to excited states as well as the ground state.

We will take a valley made of n-pairs of the valley-instantons and the anti-valley-instantons with periodic boundary condition. Since we perform the calculation for $g^2 = |g^2|{\rm e}^{{\rm i}\pi} < 0$, the force between the valley-instanton and the anti-valley-instanton be-

[1]We also note that the integral path C_V for $Z(g^2)$ is not changed by the analytic continuation since it is a finite interval.

comes repulsive. Therefore, the configurations with large separations between (anti-)valley-instantons dominate. From the calculation of the action for one pair in Section 5.2., we find that the action of this n-pair configuration is approximated, if we neglect all the interactions beyond the nearest-neighbor ones, by

$$S_n \simeq 2nS_E[q_0^{(I)}] - \epsilon \sum_{i=1}^{n} R_i - \frac{2}{g^2} \sum_{i=1}^{n} \mathrm{e}^{-R_i} - \frac{2}{g^2} \sum_{i=1}^{n} \mathrm{e}^{-\bar{R}_i}, \tag{98}$$

where R_i is the distance between the ith valley-instanton and the ith anti-valley-instanton and $\bar{R}$ is the distance between the ith anti-valley-instanton and the $(i+1)$th valley-instanton $(\mathrm{mod}\ n)$, see Fig. 10. The expression (98) is valid when all the valley-instantons and the anti-valley-instantons are well-separated, that is, when $R_i, \bar{R}_i \gg 1$.

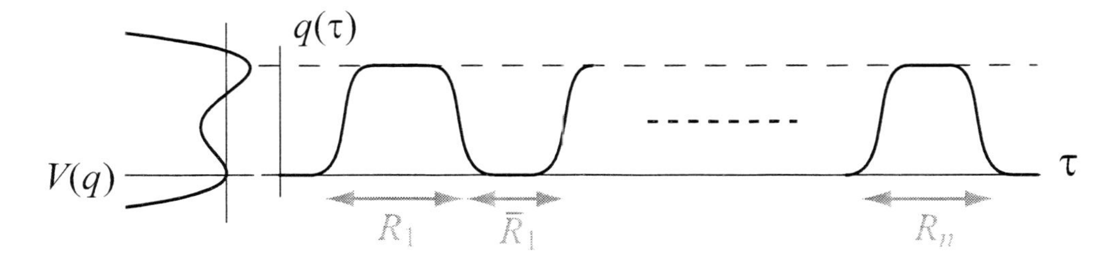

Figure 10. A multi-valley-instanton configuration in the asymmetric double-well potential.

We shall write the sum of the contributions of the n-pairs of (anti-)valley-instantons to the partition function Z for $n \geq 1$ as

$$Z_{\mathrm{NP}}(T) = \sum_{n=1}^{\infty} \alpha^{2n} J_n, \tag{99}$$

where the constant α is given by (77). It means that we have assumed the factorization properties (83) and (84) in the leading order for all $n \geq 1$. The term J_n in (99) is given by

$$J_n = \frac{T}{n} \int_0^{\infty} \left(\prod_{i=1}^{n} \mathrm{d}R_i \right) \left(\prod_{i=1}^{n} \mathrm{d}\bar{R}_i \right) \delta \left(\sum_{i=1}^{n} (R_i + \bar{R}_i) - T \right)$$
$$\times \exp \left[- \left(\frac{1}{2} - \epsilon \right) \sum_{i=1}^{n} R_i - \frac{1}{2} \sum_{i=1}^{n} \bar{R}_i + \frac{2}{g^2} \sum_{i=1}^{n} \mathrm{e}^{-R_i} + \frac{2}{g^2} \sum_{i=1}^{n} \mathrm{e}^{-\bar{R}_i} \right]. \tag{100}$$

The expression (99) can be evaluated in a manner similar to that performed by Zinn-Justin for the ordinary instanton case [25, 26]. First, we rewrite the delta function as

$$\delta \left(\sum_{i=1}^{n} (R_i + \bar{R}_i) - T \right) = \frac{1}{2\pi \mathrm{i}} \int_{-\mathrm{i}\infty - \eta}^{\mathrm{i}\infty - \eta} \mathrm{d}s \exp \left(s \sum_{i=1}^{n} (R_i + \bar{R}_i) - sT \right), \tag{101}$$

where η is a positive real number. This allows factorization of the integrals over R_i and $\bar{R}_i$ in the following manner

$$J_n = \frac{1}{2\pi\mathrm{i}} \int_{-\mathrm{i}\infty-\eta}^{\mathrm{i}\infty-\eta} \mathrm{d}s\, K_+(s)^n K_-(s)^n, \tag{102}$$

where $K_\pm(s)$ are

$$K_\pm(s) = \int_0^\infty \mathrm{d}R \, \exp\left(s_\pm R + \frac{2}{g^2}\mathrm{e}^{-R}\right) \simeq \left(-\frac{2}{g^2}\right)^{s_\pm} \Gamma(-s_\pm), \tag{103}$$

with $s_+ = s - 1/2$ and $s_- = s + \epsilon - 1/2$. This leads to the following expression:

$$\begin{aligned} Z_{\mathrm{NP}}(T) &= \frac{T}{2\pi\mathrm{i}} \int_{-\mathrm{i}\infty-\eta}^{\mathrm{i}\infty-\eta} \mathrm{d}s\, \mathrm{e}^{-Ts} \sum_{n=1}^{\infty} \frac{(\alpha^2 K_+(s) K_-(s))^n}{n} \\ &= -\frac{T}{2\pi\mathrm{i}} \int_{-\mathrm{i}\infty-\eta}^{\mathrm{i}\infty-\eta} \mathrm{d}s\, \mathrm{e}^{-Ts} \ln\left(1 - \alpha^2 K_+(s) K_-(s)\right). \end{aligned} \tag{104}$$

Integrating by parts, we then have

$$Z_{\mathrm{NP}}(T) = -\frac{1}{2\pi\mathrm{i}} \int_{-\mathrm{i}\infty-\eta}^{\mathrm{i}\infty-\eta} \mathrm{d}s\, \mathrm{e}^{-Ts} \frac{\phi'(s)}{\phi(s)}, \tag{105}$$

where

$$\phi(s) = 1 - \alpha^2 K_+(s) K_-(s). \tag{106}$$

From the formula (103), we see that $\phi(s)$ has both poles and zeros in s. Denoting the poles by $s = E_n^{(0)}$ and the zeros by $s = E_n$, we finally obtain

$$Z_{\mathrm{NP}}(T) = \sum_n \mathrm{e}^{-E_n T} - \sum_n \mathrm{e}^{-E_n^{(0)} T}. \tag{107}$$

The poles of $\phi(s)$ are given by the poles of the gamma functions in $K_\pm(s)$ and we rewrite them as $s = E_{n\pm}^{(0)}$ where

$$s = n + \frac{1}{2}, \qquad s = n + \frac{1}{2} - \epsilon, \qquad (n = 0, 1, 2, \ldots), \tag{108}$$

which correspond to the tree-level energy eigenvalues of the nth excited states of the harmonic oscillators around $q \simeq q_-$ and $q \simeq q_+$, respectively. Hence, we shall hereafter assign the quantum numbers n_- and n_+ to the former and latter states, respectively.

5.4. Nonperturbative Contributions

The subtraction of the second term in the r.h.s. of (107) indicates that the purely nonperturbative part Z_{NP} does not contain this zeroth-order contribution and in fact it vanishes for $g = 0$. Hence, the full partition function without the perturbative corrections is given by just the first term in the r.h.s. of (107). This means, in particular, that the exact energy eigenvalues without the perturbative corrections are obtained by the zeros of $\phi(E)$, or equivalently, by the solutions to the following equation:

$$\alpha^2\left(-\frac{2}{g^2}\right)^{2E-1+\epsilon}\Gamma\left(-E+\frac{1}{2}\right)\Gamma\left(-E+\frac{1}{2}-\epsilon\right)=1. \tag{109}$$

To see the asymptotic behaviors in a weak coupling regime, let us solve the quantization condition (109) by the series expansion in $\alpha \ll 1$

$$E_{n\pm}=E_{n\pm}^{(0)}+\alpha E_{n\pm}^{(1)}+\alpha^2 E_{n\pm}^{(2)}+\cdots,\quad E_{n_+}^{(0)}=n_+ +\frac{1}{2},\quad E_{n_-}^{(0)}=n_- +\frac{1}{2}-\epsilon. \tag{110}$$

In the case when $\epsilon \neq \mathcal{N} \in \mathbb{N}$, the first-order terms vanish, $E_{n\pm}^{(1)}=0$, and the second-order terms are calculated as

$$E_{n\pm}^{(2)}=\frac{(-1)^{n_\pm+1}}{n_\pm!}\left(-\frac{2}{g^2}\right)^{2n_\pm\pm\epsilon}\Gamma(-n_\pm\mp\epsilon). \tag{111}$$

The result (111) cannot be applied when $\epsilon=\mathcal{N}\in\mathbb{Z}$ since in the latter cases the gamma function in the above expression has some poles. Interestingly, the zeroth-order energy eigenvalues of the states around q_+ and q_- are degenerate when $\epsilon=\mathcal{N}\in\mathbb{Z}$, see Fig. 11 a) for $\epsilon=1$ and b) for $\epsilon=2$. These divergences in $E_{n\pm}^{(2)}$ are caused by the confluence of

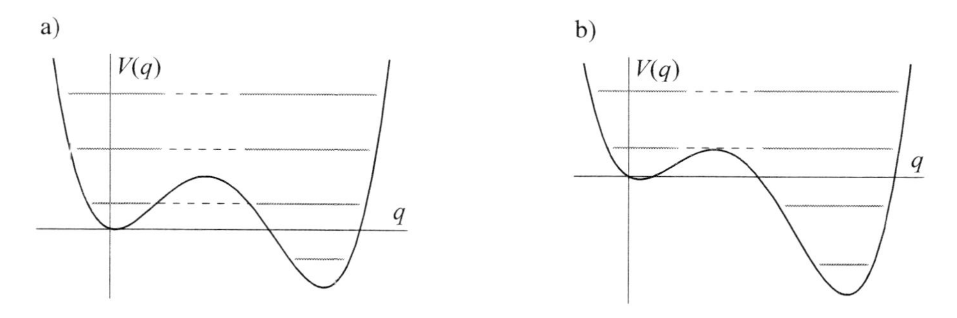

Figure 11. The degeneracies of the harmonic spectra for a) $\epsilon=1$ and b) $\epsilon=2$.

the corresponding poles of the gamma functions in K_+ and K_-. In the cases where this happens, the first-order terms do not vanish. The zeroth-order terms are identical with each other $E_{n_+}^{(0)}=E_{n_-}^{(0)}$ which means $n_-=n_++\mathcal{N}$. A straightforward calculation yields

$$E_{n\pm}^{(1)}=\pm\sqrt{\frac{1}{n_+!\,n_-!}\left(\frac{2}{g^2}\right)^{n_++n_-}}, \tag{112}$$

$$E_{n\pm}^{(2)}=\frac{(E_{n\pm}^{(1)})^2}{2}\left[2\ln\left(-\frac{2}{g^2}\right)-\psi(n_++1)-\psi(n_-+1)\right], \tag{113}$$

where ψ denotes the digamma function. The double sign in the expression (112) corresponds approximately to two linear combinations of the harmonic oscillator states of the left and right potential wells with the same energy. The situation is analogous to the lifting of the degeneracy by the instanton contribution for a symmetric double-well potential which corresponds to the $\epsilon = 0$ case. We also note that the nonperturbative corrections (111)–(113) do not disappear for all the value of ϵ.

5.5. Large Order Behavior of the Perturbation Series

The nonperturbative contributions to the energy eigenvalues obtained in the last section contain both real and imaginary parts. As discussed previously, the imaginary part is related to the large-order behavior of the perturbation series. We shall later confirm it by calculating directly the perturbative coefficients and comparing them with the results obtained here.

The expressions of the nonperturbative contributions to the energy levels (111)–(113) contain imaginary parts starting at the order α^2 and given by

$$\mathrm{Im}\, E_{n\pm} \simeq -\alpha^2 \frac{\pi}{n_\pm!\,\Gamma(n_\pm + 1 \pm \epsilon)} \left(\frac{2}{g^2}\right)^{2n_\pm \pm \epsilon}. \tag{114}$$

We note that the above formula is valid for both $\epsilon \neq \mathcal{N}$ and $\epsilon = \mathcal{N}$ and thus it is continuous in ϵ at $\epsilon = \mathcal{N}$. We recall the relation between the imaginary part of the nonperturbative contribution and the large order behavior of the perturbation series defined by

$$E_{n\pm} = E_{n\pm}^{(0)} + \sum_{r=1}^{\infty} g^{2r} c_{n\pm}^{[2r]}. \tag{115}$$

Substituting the imaginary part (114) into the dispersion relation (96) for an energy eigenvalue, we obtain the following large-order behavior of the perturbative coefficients $c_{n\pm}^{[2r]}$ for $r \gg 1$:

$$c_{n\pm}^{[2r]} \simeq A_{n\pm}(\epsilon) 3^r \Gamma(r + 2n_\pm + 1 \pm \epsilon), \tag{116}$$

where the coefficients $A_{n\pm}(\epsilon)$ are given by

$$A_{n\pm}(\epsilon) = -\frac{3}{\pi} \frac{6^{2n_\pm \pm \epsilon}}{n_\pm!\,\Gamma(n_\pm + 1 \pm \epsilon)}. \tag{117}$$

The formula (116) tells us a monotone and factorial divergence of the coefficients $c_{n\pm}^{(r)}$ as the order r increases, which indicates the Borel non-summability, provided that the factors $A_{n\pm}(\epsilon)$ do not vanish. On the other hand, the formula (117) indicates that $A_{n-}(\epsilon)$ does vanish for a number of low-lying states at certain values of $\epsilon (> 0)$. Explicitly, we have

$$A_{n-}(\epsilon) = 0 \quad \text{for} \quad n_- < \mathcal{N} \quad \text{when} \quad \epsilon = \mathcal{N}. \tag{118}$$

Later in Section 9., we will come back to this interesting phenomenon in view of an underlying symmetry hidden in the system (66).

6. Analysis of a Symmetric Triple-well Potential

Next, we shall apply our method to a triple-well potential. The form of the potential to be analyzed is given by

$$V(q;g,\epsilon)=\frac{1}{2}q^2(1-g^2q^2)^2+\frac{\epsilon}{2}(1-3g^2q^2), \tag{119}$$

where g and ϵ are free parameters. It has three local minima for all $\tilde{\epsilon}=\epsilon g^2>-1/9$. For $|\tilde{\epsilon}|\ll 1$ which we assume, the local minima are at $q=0$ and $q=\pm 1/g[1+O(\tilde{\epsilon})]:=\pm q_+$. The potential values at them are $V(0;g,\epsilon)=\epsilon/2$ and $V(\pm q_+;g,\epsilon)=-\epsilon[1+O(\tilde{\epsilon})]$. Thus, the energy difference between at the central and side potential minima is roughly given by $3\epsilon/2$. The potential barriers between the central and side potential minima are highest at $q=\pm 1/\sqrt{3}g[1+O(\tilde{\epsilon})]$ with the value $V(\pm 1/\sqrt{3}g;g,\epsilon)=2/27g^2[1+O(\tilde{\epsilon}^2)]$. Hence, the inverse-square of the coupling constant g is roughly related with the height of the potential barriers. The potential form is explicitly shown in Fig. 12.

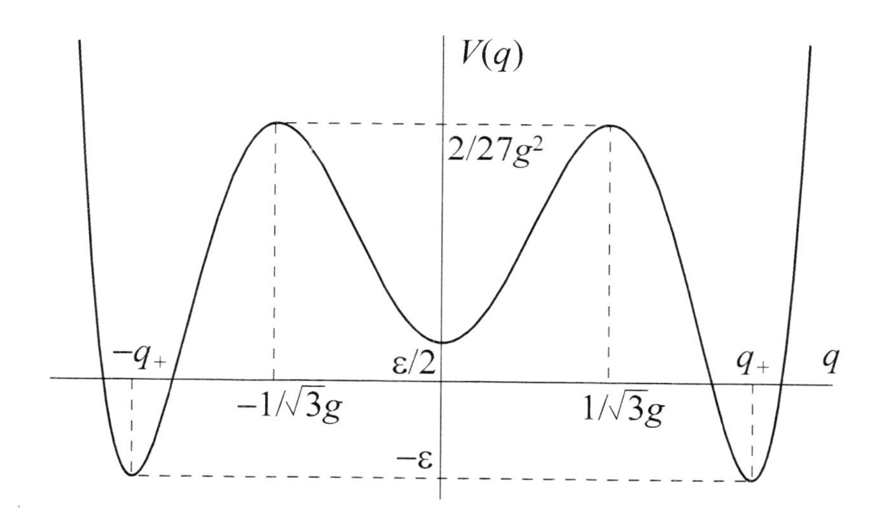

Figure 12. The form of the symmetric triple-well potential (119).

In the case of $\epsilon=0$, the three local minima of the potential have the same potential value. Thus, there are (anti-)instanton solutions of the equation of motion which describe the quantum tunneling between the neighboring vacua:

$$q_0^{(I)}(\tau-\tau_0)=\pm\frac{1}{g}\frac{1}{(1+\mathrm{e}^{\mp 2(\tau-\tau_0)})^{1/2}},\quad q_0^{(\bar{I})}(\tau-\tau_0)=\pm\frac{1}{g}\frac{1}{(1+\mathrm{e}^{\pm 2(\tau-\tau_0)})^{1/2}}. \tag{120}$$

They have the same action given by

$$S_E[q_0^{(I)}]=\frac{1}{4g^2}. \tag{121}$$

The nonperturbative corrections to the lowest three eigenvalues were calculated by the semi-classical method using these instantons with the dilute-gas approximation in Refs. [10, 11]. However, both of the results are strange in that they do not reduce to the harmonic oscillator spectrum in the limit $g \to 0$. As we will discuss later, the breakdown of the conventional semi-classical approach to the triple-well potential even in the $\epsilon = 0$ case, where no negative modes emerge, is caused by the limitations of the dilute-gas approximation.

When $\epsilon \neq 0$, the solutions of the valley equation now become the (anti-)valley-instantons. We can construct their asymptotic forms like (69) and (70) in the case of the asymmetric double-well potential. The relevant quantities for the calculation of their FP and spectral determinants are given by

$$\begin{aligned} \omega_- &= \sqrt{V''(0)} = 1 + O(\tilde{\epsilon}), & F_- &= -6\epsilon g, \\ \omega_+ &= \sqrt{V''(\pm q_+)} = 2 + O(\tilde{\epsilon}), & F_+ &= -6\epsilon g, \end{aligned} \tag{122}$$

and

$$\int_{-\infty}^{\infty} \mathrm{d}\tau\, F^{(I)}(\tau)^2 \simeq 9\epsilon^2 g^2. \tag{123}$$

Then, applying the formulas (45), (63), and (65), we obtain

$$\Delta_{\mathrm{FP}}^{(I)} = \Delta_{\mathrm{FP}}^{(\bar{I})} \simeq \frac{1}{\sqrt{4g^2}}, \qquad \kappa^{(I)} \simeq \frac{1}{16}, \qquad \kappa^{(\bar{I})} \simeq \frac{1}{8}. \tag{124}$$

6.1. Valley Configurations with Two Valley-instantons

In the case of triple-well potential, there are three kinds of the solutions of the valley equation which are asymptotically composed of two (anti-)valley-instantons. The two of them are $I\bar{I}$- and $\bar{I}I$-valley configurations, the one satisfies $q^{(I\bar{I})}(\pm\infty) = 0$ or $q^{(\bar{I}I)}(\pm\infty) = 0$, and the other satisfies $q^{(I\bar{I})}(\pm\infty) = -q_+$ or $q^{(\bar{I}I)}(\pm\infty) = q_+$. In Fig. 13, we show an $I\bar{I}$-valley of the former type while in Fig. 14, we show an $I\bar{I}$-valley of the latter type. In-

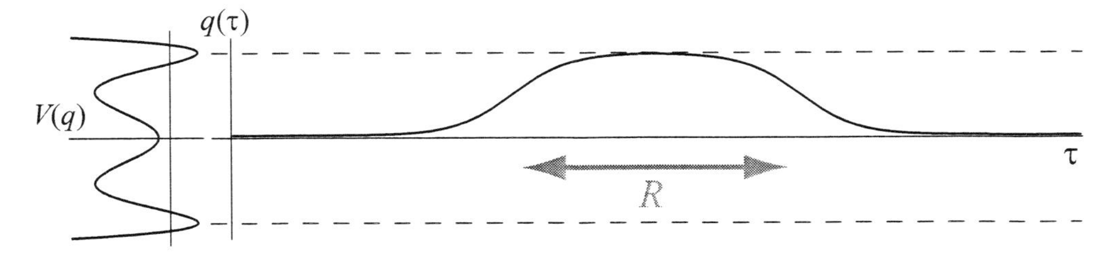

Figure 13. An $I\bar{I}$-valley configuration which satisfies $q^{(I\bar{I})}(\pm\infty) = 0$.

terestingly, the Euclidean actions of the former and the latter types are different from each other, even in the case of $\epsilon = 0$ where the apparent difference of the volume factors in the two types disappear, because of the fact that the curvature at the central potential well ω_0 is different from the one at the side potential wells $\omega_\pm$, as have been shown in (122). In fact,

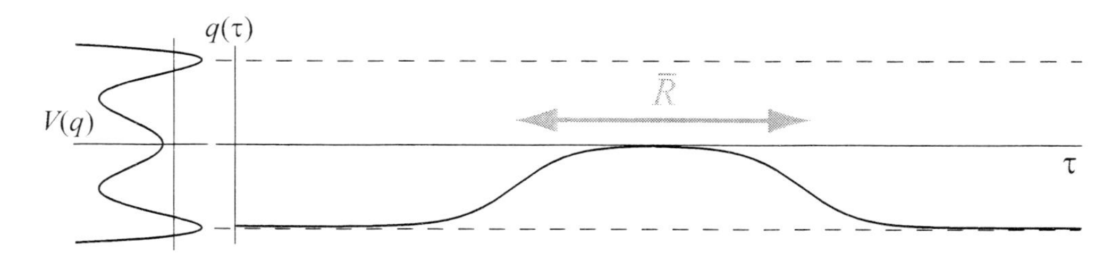

Figure 14. An $I\bar{I}$-valley configuration which satisfies $q^{(I\bar{I})}(\pm\infty) = -q_+$.

the Euclidean action of the former type with large separation R can be calculated by the perturbative expansion in $\lambda \sim O(e^{-2R})$ as

$$S_E[q^{(I\bar{I})}](R) = S_E[q^{(\bar{I}I)}](R) = 2S_E[q_0^{(I)}] - \epsilon R + \frac{\epsilon}{2}(T - R) - \frac{1}{g^2}\mathrm{e}^{-2R} + O(\mathrm{e}^{-4R}), \tag{125}$$

while the one of the latter type with large separation $\bar{R}$ can be calculated in a similar way as

$$S_E[q^{(I\bar{I})}](\bar{R}) = S_E[q^{(\bar{I}I)}](\bar{R}) = 2S_E[q_0^{(I)}] + \frac{\epsilon}{2}\bar{R} - \epsilon(T - \bar{R}) - \frac{2}{g^2}\mathrm{e}^{-\bar{R}} + O(\mathrm{e}^{-2\bar{R}}), \tag{126}$$

where $S_E[q_0^{(I)}]$ denotes the action of one classical (anti-)instanton given by (121).

In addition to the above two kinds of configurations, there emerge new types of valley configurations $q^{(II)}(\tau)$, called *II-valley*, and configurations $q^{(\bar{I}\bar{I})}(\tau)$, called *$\bar{I}\bar{I}$-valley*, which are absent in the case of double-well potential. They satisfy $q^{(II)}(\pm\infty) = \pm q_+$ and $q^{(\bar{I}\bar{I})}(\pm\infty) = \mp q_+$ and are roughly composed of two successive (anti-)valley-instantons. For an II-valley configuration, see Fig. 15. Their Euclidean actions with large separation

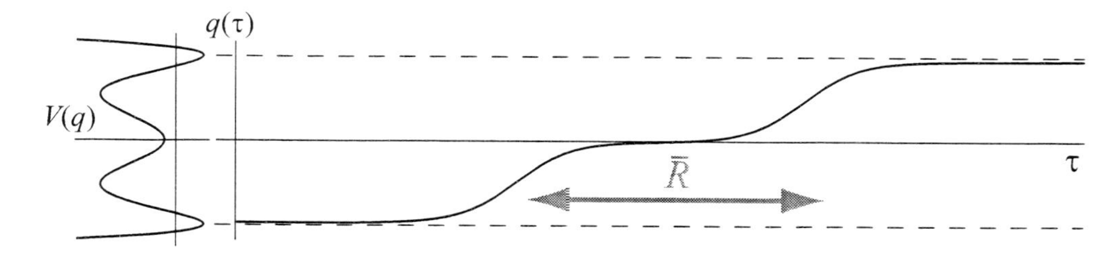

Figure 15. An II-valley configuration which satisfies $q^{(II)}(\pm\infty) = \pm q_+$.

$\bar{R}$ can be also calculated in the same way as

$$S_E[q^{(II)}](\bar{R}) = S_E[q^{(\bar{I}\bar{I})}](\bar{R}) = 2S_E[q_0^{(I)}] + \frac{\epsilon}{2}\bar{R} - \epsilon(T - \bar{R}) + \frac{2}{g^2}\mathrm{e}^{-\bar{R}} + O(\mathrm{e}^{-2\bar{R}}). \tag{127}$$

It is obvious that the first term is the action of the two instantons while the second and third terms are the volume factors. The fourth term is the interaction term between the two successive (anti-)valley-instantons. Its plus sign indicates that the interaction is *repulsive* in contrast to the attractive interactions in the $I\bar{I}$- and $\bar{I}I$-valley configurations, c.f., (82), (125), and (126). This different character of the interactions results in a different structure of the valley integration.

To see it, we make the change of integral variable $t = g^2 S_E[q^{(II)}](R)$ as in the $I\bar{I}$-valley case. The integral expression for $Z^{(II)}$ has a form which is quite similar to the one (86) for $Z^{(I\bar{I})}$. However, the integral path now becomes $C_V = (1/2, g^2 S_E[q^{(II)}](0))$ and thus is disconnected from the perturbative region $t \simeq 0$. This means, in particular, that the partition function for the II-valley configuration contains only the nonperturbative contribution

$$Z(g^2) = Z_{\rm NP}(g^2). \tag{128}$$

Hence, we need not separate the integration any more in contrast to the case of $I\bar{I}$-valley. As a consequence, any II-valley configuration does not contribute to the imaginary part and thus to the perturbative coefficients. Another relevant aspect is that the interaction in an II-valley configuration is repulsive and the integral for the partition function would be dominated by the region $t \simeq 1/2$ where the asymptotic formula (127) is valid. Hence, the prescription of analytic continuation is not needed. The situation in an $\bar{I}\bar{I}$-valley is completely the same as in the II-valley.

We now detect at least one crucial reason why the conventional semi-classical approach fails in the case of the triple-well potential even when $\epsilon = 0$. In the latter approach, we resort to the dilute-gas approximation, thus neglect all the different interaction terms and treat all the configurations of the same number of (anti-)instantons identically.

6.2. Multi-valley-instanton Calculation

The evaluation of the partition function $Z(T) = \mathrm{Tr}\, \mathrm{e}^{-HT}$ by summing over multi-valley-instanton configurations can be done in the same manner as that for the asymmetric double-well potential in Section 5.3.. One can easily see that in order to incorporate with periodic boundary condition in T, the number of the valley-instantons in a period T must be even. For a given number $2n$ of the valley-instantons, however, there are still several configurations. If we regard a configuration as n valley-instanton pairs, we have four kinds of pair, II-, $I\bar{I}$-, $\bar{I}I$-, and $\bar{I}\bar{I}$-valleys. We denote the number of the II- and $\bar{I}\bar{I}$-valley as n_{II} and $n_{\bar{I}\bar{I}}$ respectively and that of the others as $n_{I\bar{I}}$. The periodic boundary condition $q(\tau+T) = q(\tau)$ results in $n_{II} = n_{\bar{I}\bar{I}}$. As a consequence we have,

$$2n_{II} + n_{I\bar{I}} = n. \tag{129}$$

This restriction shows that for a given n the integer n_{II} can take values from 0 to $[n/2]$. For n and n_{II} fixed, however, the configuration is still not determined uniquely. There remains a freedom of permuting the location of valley-instanton pairs. The number of the cases can be calculated if one notices that the configuration is uniquely determined as far as the position of the II- and $\bar{I}\bar{I}$-valleys among n area is fixed. We denote a set of the position as

$\{i_{II}\}$. It is evident that for given n and n_{II} there are ${}_nC_{2n_{II}}$ different configurations of the multiple valley-instantons.

As in the case of the double-well potential, we shall carry out all the calculations in the sector where any nearest-neighbor interaction between two (anti-)valley-instantons is repulsive so that the asymptotic formulas (125), (126), and (127) are all valid. As we will see shortly, it is indeed possible since within our approximation none of the integrals with respect to the collective coordinates do not mix with each other and thus the analytic continuation can be done separately. Combining the results on well-separated valley-instanton pairs with the above considerations, the well-separated multi-valley-instanton action for given n, n_{II} and $\{i_{II}\}$ is approximated, if we neglect all the interactions beyond the nearest-neighbor ones, by

$$S_{n,n_{II}}^{\{i_{II}\}} \simeq 2nS_E[q_0^{(I)}] - \epsilon\sum_{i=1}^{n} R_i + \frac{\epsilon}{2}\sum_{i=1}^{n}\bar{R}_i - \frac{1}{g^2}\sum_{i=1}^{n}\mathrm{e}^{-2R_i} + \frac{2}{g^2}\sum_{i\in\{i_{II}\}}\mathrm{e}^{-\bar{R}_i} - \frac{2}{g^2}\sum_{i\notin\{i_{II}\}}\mathrm{e}^{-\bar{R}_i}, \tag{130}$$

where R_i is the distance between the $(2i-1)$th and $2i$th (anti-)valley-instanton and $\bar{R}_i$ the one between the $2i$th and the $(2i+1)$th (anti-)valley-instanton $(\mathrm{mod}\ n)$, see Fig. 16. The sum of the contributions from the configuration of $2n$ (anti-)valley-instantons can be

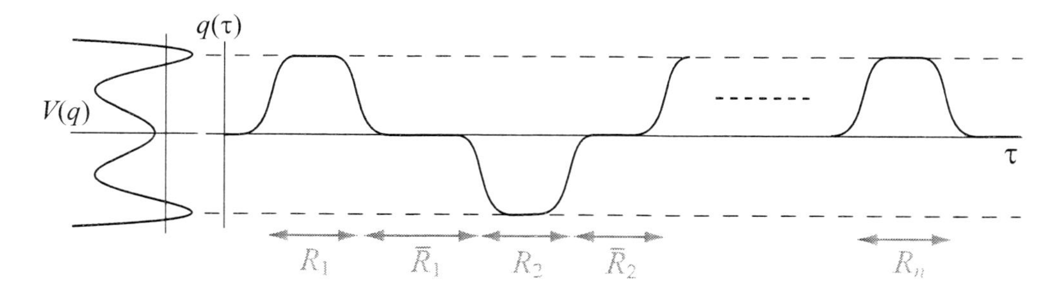

Figure 16. A multi-valley-instanton configuration in the triple-well potential.

written as

$$Z_{\mathrm{NP}}(T) = \sum_{n=1}^{\infty}\alpha^{2n}J_n, \qquad J_n = \sum_{n_{II}=0}^{[n/2]}\sum_{\{i_{II}\}}\mathcal{J}_{n,n_{II}}^{\{i_{II}\}}, \tag{131}$$

where α^2 denotes the contribution of the FP determinant and the R-independent part of the spectral determinant for a pair of one valley-instanton and one anti-valley-instanton and is calculated from (121) and (124) as

$$\alpha^2 = \frac{\Delta_{\mathrm{FP}}^{(I)}\Delta_{\mathrm{FP}}^{(\bar{I})}}{2\pi\sqrt{\kappa^{(I)}\kappa^{(\bar{I})}}}\,\mathrm{e}^{-2S_E[q_0^{(I)}]} = \frac{\sqrt{2}}{\pi g^2}\mathrm{e}^{-1/2g^2}. \tag{132}$$

Here we have again assumed the factorization properties (83) and (84) for all $n \geq 1$. The term $\mathcal{J}_{n,n_{II}}^{\{i_{II}\}}$ is given by

$$\begin{aligned}\mathcal{J}_{n,n_{II}}^{\{i_{II}\}} = \frac{T}{n}\int_0^{\infty}\left(\prod_{i=1}^{n}\mathrm{d}R_i\right)\left(\prod_{i=1}^{n}\mathrm{d}\bar{R}_i\right)\\ \delta\left(\sum_{i=1}^{n}(R_i+\bar{R}_i)-T\right)\exp\left[-(1-\epsilon)\sum_{i=1}^{n}R_i-\left(\frac{1}{2}+\frac{\epsilon}{2}\right)\sum_{i=1}^{n}\bar{R}_i\right.\\ \left.+\frac{1}{g^2}\sum_{i=1}^{n}\mathrm{e}^{-2R_i}-\frac{2}{g^2}\sum_{i\in\{i_{II}\}}\mathrm{e}^{-\bar{R}_i}+\frac{2}{g^2}\sum_{i\notin\{i_{II}\}}\mathrm{e}^{-\bar{R}_i}\right].\end{aligned} \tag{133}$$

In the above expression, we notice that for n and n_{II} fixed, the contribution $\mathcal{J}_{n,n_{II}}^{\{i_{II}\}}$ does not depend on the choice of the set $\{i_{II}\}$. This means the following equality

$$\sum_{\{i_{II}\}}\mathcal{J}_{n,n_{II}}^{\{i_{II}\}} = \binom{n}{2n_{II}}\mathcal{J}_{n,n_{II}}, \tag{134}$$

where $\mathcal{J}_{n,n_{II}}$ is the contribution $\mathcal{J}_{n,n_{II}}^{\{i_{II}\}}$ for a specific $\{i_{II}\}$ and is evaluated as

$$\begin{aligned}\mathcal{J}_{n,n_{II}} &= \frac{T}{n}\int_0^{\infty}\left(\prod_{i=1}^{n}\mathrm{d}R_i\right)\left(\prod_{i=1}^{n}\mathrm{d}\bar{R}_i\right)\\ &\quad\delta\left(\sum_{i=1}^{n}(R_i+\bar{R}_i)-T\right)\exp\left[-(1-\epsilon)\sum_{i=1}^{n}R_i-\left(\frac{1}{2}+\frac{\epsilon}{2}\right)\sum_{i=1}^{n}\bar{R}_i\right.\\ &\quad\left.+\frac{1}{g^2}\sum_{i=1}^{n}\mathrm{e}^{-2R_i}-\frac{2}{g^2}\sum_{i=1}^{2n_{II}}\mathrm{e}^{-\bar{R}_i}+\frac{2}{g^2}\sum_{i=2n_{II}+1}^{n}\mathrm{e}^{-\bar{R}_i}\right]\\ &= \frac{T}{2\pi\mathrm{i}n}\int_{-\mathrm{i}\infty-\eta}^{\mathrm{i}\infty-\eta}\mathrm{d}s\,\mathrm{e}^{-Ts}\,K_-(s)^n K_+^{(1)}(s)^{2n_{II}}K_+^{(2)}(s)^{n-2n_{II}}.\end{aligned} \tag{135}$$

In the last expression for $\mathcal{J}_{n,n_{II}}$, K_- and $K_+^{(i)}$ are given by

$$\begin{aligned}K_-(s) &= \int_0^{\infty}\mathrm{d}R\exp\left[(s-1+\epsilon)R+\frac{1}{g^2}\mathrm{e}^{-2R}\right]\\ &\simeq \frac{1}{2}\left(-\frac{1}{g^2}\right)^{\frac{s}{2}-\frac{1}{2}+\frac{\epsilon}{2}}\Gamma\left(-\frac{s}{2}+\frac{1}{2}-\frac{\epsilon}{2}\right),\end{aligned} \tag{136a}$$

$$\begin{aligned}K_+^{(1)}(s) &= \int_0^{\infty}\mathrm{d}\bar{R}\exp\left[\left(s-\frac{1}{2}-\frac{\epsilon}{2}\right)\bar{R}-\frac{2}{g^2}\mathrm{e}^{-\bar{R}}\right]\\ &\simeq \left(\frac{2}{g^2}\right)^{s-\frac{1}{2}-\frac{\epsilon}{2}}\Gamma\left(-s+\frac{1}{2}+\frac{\epsilon}{2}\right),\end{aligned} \tag{136b}$$

$$\begin{aligned}K_+^{(2)}(s) &= \int_0^{\infty}\mathrm{d}\bar{R}\exp\left[\left(s-\frac{1}{2}-\frac{\epsilon}{2}\right)\bar{R}+\frac{2}{g^2}\mathrm{e}^{-\bar{R}}\right]\\ &\simeq \left(-\frac{2}{g^2}\right)^{s-\frac{1}{2}-\frac{\epsilon}{2}}\Gamma\left(-s+\frac{1}{2}+\frac{\epsilon}{2}\right).\end{aligned} \tag{136c}$$

where the manipulation explained in Section 6. is again utilized for estimating each of the integration. Eventually, from Eqs. (131), (134) and (135) we obtain

$$Z_{\rm NP}(T) = \sum_{n=1}^{\infty} \alpha^{2n} \sum_{n_{II}=0}^{[n/2]} \binom{n}{2n_{II}} \mathcal{J}_{n,n_{II}}$$
$$= -\frac{T}{4\pi\mathrm{i}} \int_{-\mathrm{i}\infty-\eta}^{\mathrm{i}\infty-\eta} \mathrm{d}s\, \mathrm{e}^{-Ts} \ln\left(1-\alpha^2 K_-(s)K_+^{(+)}(s)\right)\left(1-\alpha^2 K_-(s)K_+^{(-)}(s)\right), \tag{137}$$

where

$$K_+^{(\pm)}(s) = K_+^{(2)}(s) \pm K_+^{(1)}(s). \tag{138}$$

Integrating (137) by parts, we have

$$Z_{\rm NP}(T) = -\frac{1}{4\pi\mathrm{i}} \int_{-\mathrm{i}\infty-\eta}^{\mathrm{i}\infty-\eta} \mathrm{d}s\, \mathrm{e}^{-Ts} \left[\frac{\phi_+'(s)}{\phi_+(s)} + \frac{\phi_-'(s)}{\phi_-(s)}\right], \tag{139}$$

where

$$\phi_\pm(s) = 1 - \alpha^2 K_-(s) K_+^{(\pm)}(s). \tag{140}$$

Hence, the partition function in this case has the same form as the one in the case of asymmetric double-well potential (99) but $s = E_n^{(0)}$ and $s = E_n$ here are respectively given by the poles and the zeros of either $\phi_+(s)$ or $\phi_-(s)$. The poles of $\phi_\pm(s)$ are identical with each other and given by the poles of the gamma functions in $K_-(s)$ and $K_+^{(\pm)}(s)$ as

$$s = n + \frac{1}{2} + \frac{\epsilon}{2}, \qquad s = 2n + 1 - \epsilon, \qquad (n = 0, 1, 2, \ldots). \tag{141}$$

The former poles correspond to the tree-level energy eigenvalues of the nth excited states of the central harmonic oscillator around the origin while the latter poles to the ones of the left or right harmonic oscillators. Hence, we shall hereafter assign n_0 to the quantum number of the former states and $n_\pm$ to the quantum numbers of the linear combination states of the latter with the definite parity $\pm$.

6.3. Nonperturbative Contributions

Following the same argument as in Section 5.4., the nonperturbative contributions to the spectra are determined by the following equation:

$$\alpha^2 \beta_\pm(E,\epsilon) \left(\frac{2}{g^2}\right)^{E-\frac{1}{2}-\frac{\epsilon}{2}} \Gamma\left(-E+\frac{1}{2}+\frac{\epsilon}{2}\right) \left(-\frac{1}{g^2}\right)^{\frac{E}{2}-\frac{1}{2}+\frac{\epsilon}{2}} \Gamma\left(-\frac{E}{2}+\frac{1}{2}-\frac{\epsilon}{2}\right) = 1, \tag{142}$$

where

$$\beta_\pm(E,\epsilon) = \frac{(-1)^{E-\frac{1}{2}-\frac{\epsilon}{2}} \pm 1}{2}. \tag{143}$$

To see the asymptotic behavior in a weak coupling region, we shall solve the above equation by the series expansion in $\alpha \ll 1$

$$E_{n_0} = E_{n_0}^{(0)} + \alpha E_{n_0}^{(1)} + \alpha^2 E_{n_0}^{(2)} + \cdots, \qquad E_{n_0}^{(0)} = n_0 + \frac{1}{2} + \frac{\epsilon}{2}, \tag{144a}$$

$$E_{n\pm} = E_{n\pm}^{(0)} + \alpha E_{n\pm}^{(1)} + \alpha^2 E_{n\pm}^{(2)} + \cdots, \qquad E_{n\pm}^{(0)} = 2n_\pm + 1 - \epsilon, \tag{144b}$$

where E_{n_0} stands for the eigenvalues corresponding to, in the limit $g \to 0$, the eigenstates of the center potential well and $E_{n\pm}$ for the ones corresponding to the parity eigenstates obtained by the linear combinations of the eigenfunctions of the each side potential well. For $\epsilon \neq \pm(2\mathcal{N}+1)/3$ ($\mathcal{N} = 0, 1, 2, \cdots$), the first-order contributions vanish, $E_{n_0}^{(1)} = E_{n\pm}^{(1)} = 0$, and the leading second-order contributions are calculated as

$$E_{n_0}^{(2)} = -\frac{1}{n_0!}\left(\frac{2}{g^2}\right)^{n_0}\left(-\frac{1}{g^2}\right)^{\frac{n_0}{2}-\frac{1}{4}+\frac{3}{4}\epsilon}\Gamma\left(-\frac{n_0}{2}+\frac{1}{4}-\frac{3}{4}\epsilon\right), \tag{145}$$

$$E_{n\pm}^{(2)} = -\left((-1)^{\frac{1-3\epsilon}{2}} \pm 1\right)\frac{1}{n_\pm!}\left(\frac{2}{g^2}\right)^{2n_\pm+\frac{1}{2}-\frac{3}{2}\epsilon}\left(\frac{1}{g^2}\right)^{n_\pm}\Gamma\left(-2n_\pm - \frac{1}{2}+\frac{3}{2}\epsilon\right). \tag{146}$$

In this case, degeneracies of the harmonic oscillator spectra for the each potential well only take place between the left- and right-side wells. The different nonperturbative contributions for the $n_\pm$th states in (146) show the splitting of the degeneracies via the quantum tunneling as in the case of a symmetric double-well potential.

When $\epsilon = (4\mathcal{N}+1)/3$ ($\mathcal{N} = 0, 1, 2, \cdots$), all the even-parity central harmonic spectra $E_{2m_0}^{(0)}$ and the higher side harmonic spectra $E_{n\pm}^{(0)}$ degenerate for $n_\pm = m_0 + \mathcal{N}$, see Fig. 17 a) for $\epsilon = 5/3$ ($\mathcal{N} = 1$).

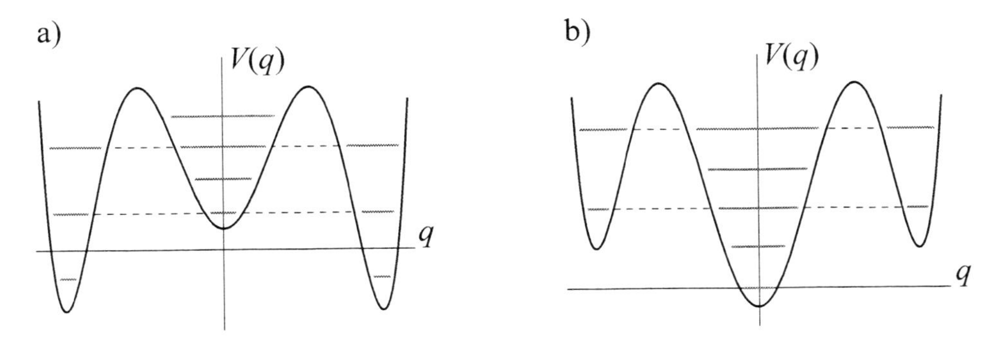

Figure 17. The degeneracies of the harmonic spectra for a) $\epsilon = 5/3$ and b) $\epsilon = -1$.

It is interesting, however, that the interference due to the quantum tunneling only occurs between the same (even-)parity states. As a consequence, E_{2m_0} and E_{n+} satisfying $n_+ =$

$m_0 + \mathcal{N}$ acquire order α^1 contributions as follows:

$$E^{(1)}_{n_+/2m_0} = \pm\sqrt{\frac{2}{n_+!\,(2m_0)!}\left(\frac{2}{g^2}\right)^{2m_0}\left(\frac{1}{g^2}\right)^{n_+}}, \tag{147}$$

$$E^{(2)}_{n_+/2m_0} = \frac{(E^{(1)}_{n_+/2m_0})^2}{4}\left[\ln\left(-\frac{2}{g^2}\right) + \ln\left(\frac{2}{g^2}\right) + \ln\left(-\frac{1}{g^2}\right) - \psi(n_+ + 1) - 2\psi(2m_0 + 1)\right]. \tag{148}$$

For the other spectra, say, E_{n_-}, E_{2m_0+1} and the lower E_{n_+} with $n_+ < \mathcal{N}$, the contributions are the same as (145) and (146). When $\epsilon = -(4\mathcal{N} - 1)/3$ ($\mathcal{N} = 1, 2, 3, \cdots$), all the side harmonic spectra $E^{(0)}_{n_\pm}$ and the higher even-parity central harmonic spectra $E^{(0)}_{2m_0}$ degenerate for $m_0 = n_\pm + \mathcal{N}$, see Fig. 17 b) for $\epsilon = -1$ ($\mathcal{N} = 1$). In this case, the interference also occurs only between the same (even-)parity states. The contributions for E_{2m_0} and E_{n_+} satisfying $m_0 = n_+ + \mathcal{N}$ are given by the same as (147) and (148). For the other spectra, say, E_{n_-}, E_{2m_0+1} and the lower E_{2m_0} with $m_0 < \mathcal{N}$, the contributions are the same as (145) and (146).

When $\epsilon = (4\mathcal{N} - 1)/3$ ($\mathcal{N} = 1, 2, 3, \cdots$), all the odd-parity central harmonic spectra $E^{(0)}_{2m_0+1}$ and the higher side harmonic spectra $E^{(0)}_{n_\pm}$ degenerate for $n_\pm = m_0 + \mathcal{N}$, see Fig. 18 a) for $\epsilon = 1$ ($\mathcal{N} = 1$).

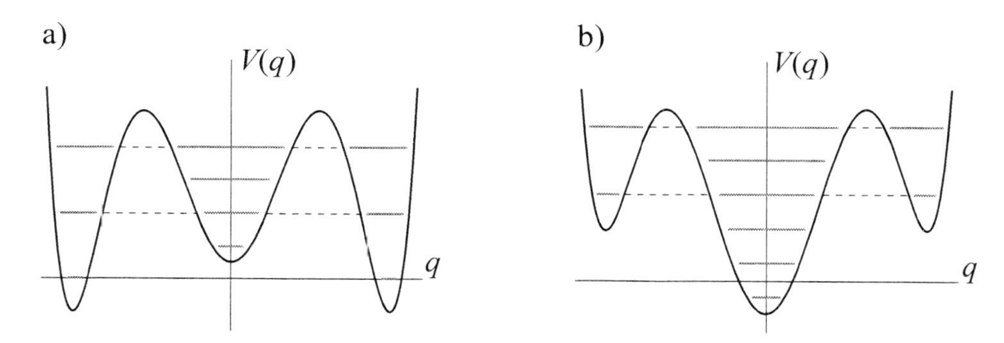

Figure 18. The degeneracies of the harmonic spectra for a) $\epsilon = 1$ and b) $\epsilon = -5/3$.

In this case, only the odd-parity states interfere and yield order α^1 contributions for $n_- = m_0 + \mathcal{N}$:

$$E^{(1)}_{n_-/2m_0+1} = \pm\sqrt{\frac{2}{n_-!\,(2m_0+1)!}\left(\frac{2}{g^2}\right)^{2m_0+1}\left(\frac{1}{g^2}\right)^{n_-}}, \tag{149}$$

$$E^{(2)}_{n_-/2m_0+1} = \frac{(E^{(1)}_{n_-/2m_0+1})^2}{4}\left[\ln\left(-\frac{2}{g^2}\right) + \ln\left(\frac{2}{g^2}\right) + \ln\left(-\frac{1}{g^2}\right) - \psi(n_- + 1) - 2\psi(2m_0 + 2)\right]. \tag{150}$$

The contributions for the other spectra, say, E_{n_+}, E_{2m_0} and the lower E_{n_-} with $n_- < \mathcal{N}$, are given by the same as (145) and (146). When $\epsilon = -(4\mathcal{N}+1)/3$ ($\mathcal{N} = 0, 1, 2, \cdots$), all the side harmonic spectra $E^{(0)}_{n\pm}$ and the higher odd-parity central harmonic spectra $E^{(0)}_{2m_0+1}$ degenerate for $m_0 = n_\pm + \mathcal{N}$, see Fig. 18 b) for $\epsilon = -5/3$ ($\mathcal{N} = 1$). Only the odd-parity states interfere in the same way and the nonperturbative contributions for E_{2m_0+1} and E_{n_-} satisfying $m_0 = n_- + \mathcal{N}$ are the same as Eqs. (149) and (150). Again, the expressions for the other spectra, say, E_{n_+}, E_{2m_0}, and the lower E_{2m_0+1} with $m_0 < \mathcal{N}$, are given by (145) and (146).

From the results obtained so far, we see that the nonperturbative corrections vanish when $\epsilon = (4\mathcal{N} \pm 1)/3$ ($\mathcal{N} = 1, 2, 3, \cdots$). More precisely, when $\epsilon = (4\mathcal{N}+1)/3$, the formula (146) is applied for the even-parity states labeled by the quantum number n_- and results in $E^{(2)}_{n_-} = 0$ for all $n_- < \mathcal{N}$. Similarly, when $\epsilon = (4\mathcal{N}-1)/3$, the formula (146) is applied for the odd-parity states labeled by the quantum number n_+ and results in $E^{(2)}_{n_+} = 0$ for all $n_+ < \mathcal{N}$. To summarize, we have

$$E^{(1)}_{n_-} = E^{(2)}_{n_-} = 0 \quad \text{for} \quad n_- < \mathcal{N} \quad \text{when} \quad \epsilon = (4\mathcal{N}+1)/3, \tag{151a}$$

$$E^{(1)}_{n_+} = E^{(2)}_{n_+} = 0 \quad \text{for} \quad n_+ < \mathcal{N} \quad \text{when} \quad \epsilon = (4\mathcal{N}-1)/3. \tag{151b}$$

6.4. Large Order Behavior of the Perturbation Series

The large order behavior of the perturbation series in g^2 for the spectra can be estimated by the same way as in the case of the double-well potential. From the nonperturbative contributions Eqs. (145)–(150), we can easily see that the imaginary parts of them are continuous, at least up to the order α^2, in ϵ and yield

$$\operatorname{Im} E_{n_0} \simeq -\alpha^2 \frac{\pi}{n_0!\,\Gamma\left(\frac{n_0}{2}+\frac{3}{4}+\frac{3}{4}\epsilon\right)} \left(\frac{2}{g^2}\right)^{n_0} \left(\frac{1}{g^2}\right)^{\frac{n_0}{2}-\frac{1}{4}+\frac{3}{4}\epsilon}, \tag{152a}$$

$$\operatorname{Im} E_{n_\pm} \simeq -\alpha^2 \frac{\pi}{n_\pm!\,\Gamma\left(2n_\pm+\frac{3}{2}-\frac{3}{2}\epsilon\right)} \left(\frac{2}{g^2}\right)^{2n_\pm+\frac{1}{2}-\frac{3}{2}\epsilon} \left(\frac{1}{g^2}\right)^{n_\pm}, \tag{152b}$$

which are valid for arbitrary ϵ. Then, if we expand the spectra in power of g^2 such that

$$E_{n_0/n_\pm} = E^{(0)}_{n_0/n_\pm} + \sum_{r=1}^{\infty} g^{2r} c^{[2r]}_{n_0/n_\pm}, \tag{153}$$

the large-order behavior of the coefficients $c^{[2r]}$ for $r \gg 1$ are calculated by substituting the imaginary parts (152) into the dispersion relation (96) for an energy eigenvalue as

$$c^{[2r]}_{n_0} \simeq A_{n_0}(\epsilon)\, 2^r\, \Gamma\left(r + \frac{3}{2}n_0 + \frac{3}{4} + \frac{3}{4}\epsilon\right), \tag{154a}$$

$$c^{[2r]}_{n_\pm} \simeq A_{n_\pm}(\epsilon)\, 2^r\, \Gamma\left(r + 3n_\pm + \frac{3}{2} - \frac{3}{2}\epsilon\right), \tag{154b}$$

where

$$A_{n_0}(\epsilon) = -\frac{\sqrt{2}}{\pi}\frac{2^{\frac{5}{2}n_0+\frac{3}{4}+\frac{3}{4}\epsilon}}{n_0!\,\Gamma\left(\frac{n_0}{2}+\frac{3}{4}+\frac{3}{4}\epsilon\right)}, \tag{155a}$$

$$A_{n_\pm}(\epsilon) = -\frac{\sqrt{2}}{\pi}\frac{2^{5n_\pm+2-3\epsilon}}{n_\pm!\,\Gamma\left(2n_\pm+\frac{3}{2}-\frac{3}{2}\epsilon\right)}. \tag{155b}$$

Equations (154) show that the perturbative coefficients diverge monotone and factorially, and thus are Borel non-summable, unless the prefactors $A(\epsilon)$ vanish. From Eq. (155), we can find that the disappearance of the leading divergence takes place when $\epsilon = \pm(2n+1)/3$ ($n = 1, 2, 3, \cdots$). More precisely, we obtain the following results:

$$A_{n_\pm}(\epsilon) = 0 \quad \text{for} \quad n_\pm < \mathcal{N} \quad \text{for} \quad \epsilon = (4\mathcal{N} \pm 1)/3, \tag{156a}$$

$$A_{2m_0+1}(\epsilon) = 0 \quad \text{for} \quad m_0 < \mathcal{N} \quad \text{for} \quad \epsilon = -(4\mathcal{N} + 1)/3, \tag{156b}$$

$$A_{2m_0}(\epsilon) = 0 \quad \text{for} \quad m_0 < \mathcal{N} \quad \text{for} \quad \epsilon = -(4\mathcal{N} - 1)/3. \tag{156c}$$

As we will later see in Section 9., both of the intriguing phenomena (151) and (156) are consequences of a hidden symmetry in the system (119).

7. Perturbation Theory

In this and subsequent sections, we shall examine, from a few different points of view, the results of our approximation scheme based on valley configurations obtained in the preceding sections. The first examination we shall make in this section is consistency with perturbative results.

In spite of the fact that perturbation series are in general divergent and at most asymptotic, they contain much information on the property of the physical quantity under consideration. An intimate relation between nonperturbative property and large order behavior of perturbation series is a typical example [17]. In fact, as we already discussed, the relation is established in our framework through the dispersion relation (94). It connect the imaginary part of the nonperturbative correction to a physical quantity with its perturbative contribution. Using it, we have predicted the large-order behavior of the perturbation series for the eigenvalues of the double-well potential, (116) and (117), and for the ones of the triple-well potential, (154) and (155). Hence, we can check our results by comparing those predictions with the exact perturbative coefficients calculated directly by perturbation theory.

7.1. The Asymmetric Double-Well Potential

Let us begin with calculating perturbative corrections to the harmonic oscillator states of each potential well in the asymmetric double-well potential. The perturbative coefficients are systematically calculated with the aid of the Bender–Wu method [27]. First, the perturbation theory around the harmonic oscillator states of the left-side potential well is set up by decomposing the Hamiltonian into the harmonic oscillator part and the remaining part:

$$H(q; g, \epsilon) = -\frac{1}{2}\frac{\mathrm{d}^2}{\mathrm{d}q^2} + \frac{1}{2}q^2 - \epsilon g q - g q^3 + \frac{1}{2}g^2 q^4. \tag{157}$$

The perturbative corrections to the eigenvalues and eigenfunctions are defined by the following formal series expansions:

$$E_{n\pm}(g,\epsilon) = -\frac{1 \mp 1}{2}\epsilon + \sum_{m=0}^{\infty} g^{2m} c_{n\pm}^{[2m]}, \quad \psi(q;g,\epsilon) = \mathrm{e}^{-q^2/2} \sum_{k=0}^{\infty} g^k \sum_{l=0}^{\infty} a_{n\pm,l}^{[k]} q^l. \tag{158}$$

Requiring that they satisfy the Schrödinger equation, we obtain a recursion relation for $c_{n+}^{[2m]}$ and $a_{n+,l}^{[k]}$:

$$\begin{aligned}(2l+1)a_{n+,l}^{[k]} &- (l+2)(l+1)a_{n+,l+2}^{[k]} \\ &- 2\epsilon a_{n+,l-1}^{[k-1]} - 2a_{n+,l-3}^{[k-1]} + a_{n+,l-4}^{[k-2]} = 2\sum_{m=0}^{[k/2]} c_{n+}^{[2m]} a_{n+,l}^{[k-2m]}.\end{aligned} \tag{159}$$

For the lowest state of the left-side potential, $c_{0_+}^{[0]} = 1/2$ and $a_{0_+,l}^{[0]} = 0$ for all $l > 0$.

Second, the perturbation theory around the harmonic oscillator states of the right-side potential well is defined by shifting the origin of the coordinate to the minimum of the right side potential $q \to q + 1/g$ and then decomposing the Hamiltonian as

$$H(q+1/g;g,\epsilon) = -\frac{1}{2}\frac{\mathrm{d}^2}{\mathrm{d}q^2} + \frac{1}{2}q^2 - \epsilon - \epsilon g q + g q^3 + \frac{1}{2}g^2 q^4. \tag{160}$$

The perturbative expansions are similarly introduced by (158), but the recursion relations are slightly modified as

$$\begin{aligned}(2l+1)a_{n-,l}^{[k]} &- (l+2)(l+1)a_{n-,l+2}^{[k]} \\ &- 2\epsilon a_{n-,l-1}^{[k-1]} + 2a_{n-,l-3}^{[k-1]} + a_{n-,l-4}^{[k-2]} = 2\sum_{m=0}^{[k/2]} c_{n-}^{[2m]} a_{n-,l}^{[k-2m]}.\end{aligned} \tag{161}$$

For the lowest state of the right-side potential, $c_{0_-}^{[0]} = 1/2$ and $a_{0_-,l}^{[0]} = 0$ for all $l > 0$.

On the other hand, the valley-instanton calculation have predicted the large-order behavior of the perturbative coefficients $c_{n\pm}^{[2m]}$ as (116) and (117). For the lowest two states we have been concerned with, they read as

$$c_{0_+}^{[2m]} \simeq -\frac{6^{\epsilon}}{\pi\,\Gamma(1+\epsilon)} 3^{m+1}\Gamma(r+1+\epsilon) := \bar{c}_{0_+}^{[2m]}, \tag{162a}$$

$$c_{0_-}^{[2m]} \simeq -\frac{6^{-\epsilon}}{\pi\,\Gamma(1-\epsilon)} 3^{m+1}\Gamma(r+1-\epsilon) := \bar{c}_{0_-}^{[2m]}. \tag{162b}$$

Hence, we can check the validity of the results (111)–(113) by comparing the predicted asymptotic forms $\bar{c}^{[2m]}$ in (162) with the exact perturbative coefficients $c^{[2m]}$ calculated using the recursion relations (159) and (161). We calculated numerically the perturbative coefficients for the lowest states of the left and right potential wells $c_{0_+}^{[2m]}$ and $c_{0_-}^{[2m]}$ up to the 500th order with $\epsilon = 8/7, 16/7, 24/7, 32/7$, and $40/7$. The results for the left and right potential wells are summarized in Tables 1 and 2, respectively. In Tables 1 and 2, we show the numerical results of the ratios $c_{0_+}^{[2m]}/\bar{c}_{0_+}^{[2m]}$ and $c_{0_-}^{[2m]}/\bar{c}_{0_-}^{[2m]}$, respectively, up to the 500th order for the five different values of ϵ. We easily see that the exact values indeed tend to the predicted asymptotic values in all the cases for the double-well potential.

Table 1. The ratios of the exact values of the perturbative coefficients $c_{0_+}^{[2m]}$ to the predicted asymptotic values $\bar{c}_{0_+}^{[2m]}$ for the lowest state of the left potential well.

m	$\epsilon = 8/7$	$\epsilon = 16/7$	$\epsilon = 24/7$	$\epsilon = 32/7$	$\epsilon = 40/7$
10	0.28691	0.09188	0.02383	0.00535	0.00108
20	0.59860	0.35078	0.17022	0.06936	0.02425
30	0.72547	0.52307	0.33635	0.19253	0.09805
50	0.83073	0.68846	0.53510	0.39002	0.26657
100	0.91349	0.83358	0.73733	0.63219	0.52543
200	0.95626	0.91395	0.86018	0.79720	0.72755
300	0.97073	0.94199	0.90480	0.86023	0.80955
400	0.97800	0.95624	0.92784	0.89341	0.85370
500	0.98238	0.96488	0.94190	0.91386	0.88125

7.2. The Symmetric Triple-Well Potential

Perturbative corrections in the symmetric triple-well potential can be calculated in the same way as in the previous asymmetric double-well potential case. First, the perturbation theory around the harmonic oscillator states of the central potential well is set up by decomposing the Hamiltonian into the harmonic oscillator part and the remaining part:

$$H(q;g,\epsilon) = -\frac{1}{2}\frac{\mathrm{d}^2}{\mathrm{d}q^2} + \frac{1}{2}q^2 + \frac{\epsilon}{2} - \frac{3}{2}\epsilon g^2 q^2 - g^2 q^4 + \frac{1}{2}g^4 q^6. \tag{163}$$

The perturbative corrections to the eigenvalues and eigenfunctions are defined by the following formal series expansions:

$$E_{n_{\mathrm{O}}}(g,\epsilon) = \frac{\epsilon}{2} + \sum_{m=0}^{\infty} g^{2m} c_{n_{\mathrm{O}}}^{[2m]}, \quad \psi_{n_{\mathrm{O}}}(q;g,\epsilon) = \mathrm{e}^{-q^2/2}\sum_{k=0}^{\infty} g^{2k}\sum_{l=0}^{\infty} a_{n_{\mathrm{O}},2l+P}^{[2k]} q^{2l+P}, \tag{164}$$

where $P = 0$ (1) for the even (odd) parity states, respectively. For the lowest state, $c_{0_{\mathrm{O}}}^{[0]} = 1/2$ and $a_{0_{\mathrm{O}},2l}^{[0]} = 0$ for all $l > 0$. Requiring that they satisfy the Schrödinger equation, we obtain a recursion relation for $c_{n_{\mathrm{O}}}^{[2m]}$ and $a_{n_{\mathrm{O}},2l+P}^{[2k]}$:

$$\begin{aligned}(4l+1+2P)a_{n_{\mathrm{O}},2l+P}^{[2k]} - (2l+2+P)(2l+1+P)a_{n_{\mathrm{O}},2(l+1)+P}^{[2k]} \\ - 3\epsilon a_{n_{\mathrm{O}},2(l-1)+P}^{[2(k-1)]} - 2a_{n_{\mathrm{O}},2(l-2)+P}^{[2(k-1)]} + a_{n_{\mathrm{O}},2(l-3)+P}^{[2(k-2)]} = 2\sum_{m=0}^{k} c_{n_{\mathrm{O}}}^{[2m]} a_{n_{\mathrm{O}},2l+P}^{[2(k-m)]}.\end{aligned} \tag{165}$$

Second, the perturbation theory around the harmonic oscillator states of the side potential wells is defined by shifting the origin of the coordinate to one of the minimum of the side

Table 2. The ratios of the exact values of the perturbative coefficients $c_{0-}^{[2m]}$ to the predicted asymptotic values $\bar{c}_{0-}^{[2m]}$ for the ground states of the right potential well.

m	$\epsilon = 8/7$	$\epsilon = 16/7$	$\epsilon = 24/7$	$\epsilon = 32/7$	$\epsilon = 40/7$
10	0.98975	0.96496	$2.9400 \cdot 10^{1}$	$-8.5898 \cdot 10^{5}$	$9.3077 \cdot 10^{8}$
20	0.99525	0.98451	0.82131	$-8.1465 \cdot 10^{2}$	$2.5867 \cdot 10^{8}$
30	0.99692	0.99005	0.88126	0.70203	$6.9129 \cdot 10^{4}$
50	0.99820	0.99420	0.92906	0.81461	0.67020
100	0.99911	0.99716	0.96467	0.90460	0.82222
200	0.99956	0.99860	0.98238	0.95164	0.90776
300	0.99971	0.99907	0.98826	0.96761	0.93775
400	0.99978	0.99930	0.99120	0.97566	0.95303
500	0.99982	0.99944	0.99296	0.98050	0.96228

potentials $q \longrightarrow q \pm 1/g$ and then decomposing the Hamiltonian as

$$\begin{aligned} H(q \pm 1/g; g, \epsilon) = & -\frac{1}{2}\frac{\mathrm{d}^2}{\mathrm{d}q^2} + \frac{4}{2}q^2 - \epsilon \mp 3\epsilon g q - \frac{3}{2}\epsilon g^2 q^2 \\ & \pm 6 g q^3 + \frac{13}{2} g^2 q^4 \pm 3 g^3 q^5 + \frac{1}{2} g^4 q^6. \end{aligned} \tag{166}$$

The perturbative corrections are similarly introduced by[2]

$$E_{n\pm}(g, \epsilon) = -\epsilon + \sum_{m=0}^{\infty} g^{2m} c_{n\pm}^{[2m]}, \quad \psi_{n\pm}(q; g, \epsilon) = \mathrm{e}^{-q^2} \sum_{k=0}^{\infty} g^k \sum_{l=0}^{\infty} a_{n\pm,l}^{[k]} q^l. \tag{167}$$

The recursion relation for $c_{n\pm}^{[m]}$ and $a_{n\pm,l}^{[k]}$ in this case is then given by

$$\begin{aligned} (4l+2)a_{n\pm,l}^{[k]} - (l+2)(l+1)a_{n\pm,l+2}^{[k]} \mp 6\epsilon a_{n\pm,l-1}^{[k-1]} - 3\epsilon a_{n\pm,l-2}^{[k-2]} \\ \pm 12 a_{n\pm,l-3}^{[k-1]} + 13 a_{n\pm,l-4}^{[k-2]} \pm 6 a_{n\pm,l-5}^{[k-3]} + a_{n\pm,l-6}^{[k-4]} = 2 \sum_{m=0}^{[k/2]} c_{n\pm}^{[2m]} a_{n\pm,l}^{[k-2m]}. \end{aligned} \tag{168}$$

For the lowest state, $c_{0\pm}^{[0]} = 1$ and $a_{0\pm,l}^{[0]} = 0$ for all $l > 0$.

On the other hand, the valley-instanton calculation have predicted the large-order behavior of the perturbative coefficients $c_{n0}^{[2m]}$ and $c_{n\pm}^{[2m]}$ as (154) and (155). For the lowest

[2]We note that all the perturbative coefficients of odd powers in g for the spectrum vanish due to the parity symmetry of the original Hamiltonian (163).

states of the central and side potential wells, they read as

$$c_{0_0}^{[2m]} \simeq -\frac{2^{5/4+3\epsilon/4}}{\pi\,\Gamma(\frac{3}{4}+\frac{3}{4}\epsilon)}\,2^m\Gamma\left(m+\frac{3}{4}+\frac{3}{4}\epsilon\right) := \bar{c}_{0_0}^{[2m]}, \tag{169a}$$

$$c_{0_\pm}^{[2m]} \simeq -\frac{2^{5/2-3\epsilon}}{\pi\,\Gamma(\frac{3}{2}-\frac{3}{2}\epsilon)}\,2^m\Gamma\left(m+\frac{3}{2}-\frac{3}{2}\epsilon\right) := \bar{c}_{0_\pm}^{[2m]}. \tag{169b}$$

Hence, we can check the validity of the results (145–150) by comparing the predicted asymptotic forms $\bar{c}^{[2m]}$ in (169) with the exact perturbative coefficients $c^{[2m]}$ calculated using the recursion relations (165) and (168). We calculated numerically the perturbative coefficients for the lowest states of the central and side potential wells $c_{0_0}^{[2m]}$ and $c_{0_\pm}^{[2m]}$ up to the 500th order with $\epsilon = -40/7, -32/7, -16/7, 0, 16/7, 32/7$, and $40/7$. The results for the central and side potential wells are summarized in Tables 3 and 4, respectively.

Table 3. The ratios of the exact values of the perturbative coefficients $c_{0_0}^{[2m]}$ to the predicted asymptotic values $\bar{c}_{0_0}^{[2m]}$ for the lowest state of the central potential well.

m	$\epsilon=-40/7$	$\epsilon=-32/7$	$\epsilon=-16/7$	$\epsilon=0$	$\epsilon=16/7$	$\epsilon=32/7$	$\epsilon=40/7$
10	0.94361	1.04368	1.12641	0.76206	0.37349	0.15429	0.09543
20	0.95634	1.03536	1.05021	0.89465	0.65222	0.41415	0.31526
30	0.97673	1.02414	1.03149	0.93167	0.76138	0.56759	0.47460
50	0.98838	1.01457	1.01806	0.95972	0.85264	0.71607	0.64310
100	0.99495	1.00728	1.00874	0.98010	0.92460	0.84754	0.80294
200	0.99765	1.00363	1.00430	0.99010	0.96185	0.92092	0.89627
300	0.99847	1.00242	1.00285	0.99341	0.97446	0.94662	0.92963
400	0.99886	1.00182	1.00214	0.99507	0.98081	0.95972	0.94676
500	0.99910	1.00145	1.00171	0.99605	0.98463	0.96765	0.95718

In Tables 3 and 4, we show the numerical results of the ratios $c_{0_0}^{[2m]}/\bar{c}_{0_0}^{[2m]}$ and $c_{0_\pm}^{[2m]}/\bar{c}_{0_\pm}^{[2m]}$, respectively, up to the 500th order for the seven different values of ϵ. We easily see that the exact values indeed tend to the predicted asymptotic values in all the cases also for the triple-well potential.

8. WKB Approximation

In the last section, we checked the accuracy of the valley-instanton calculations by comparing them with the numerical calculations of perturbation theory. In this section, we make a comparison in a different way by employing another nonperturbative approach to calculate the same physical quantities. The method we shall use here is the WKB approximation for the Schrödinger equation. In the following, we shall always consider the leading terms of the power expansion in g since we are interested in the quantization condition for the nonperturbative contribution, but not for the perturbative ones.

Table 4. The ratios of the exact values of the perturbative coefficients $c_{0\pm}^{[2m]}$ to the predicted asymptotic values $\bar{c}_{0\pm}^{[2m]}$ for the ground states of the side potential wells.

m	$\epsilon = -40/7$	$\epsilon = -32/7$	$\epsilon = -16/7$	$\epsilon = 0$	$\epsilon = 16/7$	$\epsilon = 32/7$	$\epsilon = 40/7$
10	0.00013	0.00095	0.03662	0.54770	0.87804	$-1.9005 \cdot 10^{8}$	$-1.6054 \cdot 10^{11}$
20	0.00424	0.01853	0.20077	0.77970	0.94558	$-6.3602 \cdot 10^{7}$	$-3.7131 \cdot 10^{13}$
30	0.02744	0.07687	0.36500	0.85394	0.96502	$-1.3435 \cdot 10^{4}$	$-1.2876 \cdot 10^{12}$
50	0.12329	0.22516	0.55646	0.91230	0.97960	0.67993	$-1.0437 \cdot 10^{4}$
100	0.35499	0.47901	0.74968	0.95606	0.99002	0.82945	0.70044
200	0.59684	0.69330	0.86676	0.97800	0.99506	0.91207	0.83944
300	0.70909	0.78360	0.90927	0.98533	0.99672	0.94079	0.89046
400	0.77281	0.83295	0.93123	0.98899	0.99754	0.95537	0.91690
500	0.81373	0.86401	0.94463	0.99119	0.99804	0.96419	0.93306

8.1. The Asymmetric Double-well Potential

First, we shall consider the asymmetric double-well potential problem. Around the upper minimum q_+ of the potential, the Schrödinger equation in the leading order of g is given by

$$\left(-\frac{1}{2}\frac{\mathrm{d}^2}{\mathrm{d}q^2} + \frac{1}{2}q^2\right)\psi(q) = E\psi(q), \tag{170}$$

and the local solution which vanish at $q \to -\infty$ is

$$\psi(q) = A\, D_\nu(-\sqrt{2}q), \tag{171}$$

where A is a constant, D_ν is the parabolic cylinder function with $\nu = E - 1/2$. On the other hand, around the lower minimum q_- of the potential the Schrödinger equation is approximated by

$$\left[-\frac{1}{2}\frac{\mathrm{d}^2}{\mathrm{d}q^2} + \frac{1}{2}\left(q - \frac{1}{g}\right)^2\right]\psi(q) = (E + \epsilon)\psi(q), \tag{172}$$

and the local solution which vanishes at $q \to \infty$ is

$$\psi(q) = B\, D_{\nu+\epsilon}\big(\sqrt{2}(q - 1/g)\big), \tag{173}$$

where B is another constant. We must connect these solutions with that in the classically forbidden region $q_1 \ll q \ll q_2$. In the latter region, the usual WKB formula for the wave function is available:

$$\psi(q) = \frac{C_1}{k(q)^{1/2}}\exp\left(-\int_{q_1}^{q}\mathrm{d}x\, k(x)\right) + \frac{C_2}{k(q)^{1/2}}\exp\left(\int_{q_1}^{q}\mathrm{d}x\, k(x)\right), \tag{174}$$

where C_i $(i = 1, 2)$ are constants and

$$k(x) = \sqrt{x^2(1 - gx)^2 - 2\epsilon gx - 2E}\,. \tag{175}$$

The classical turning points q_i $(i = 1, 2)$ with $q_1 < q_2$ defined by $V(q_i) = E$ are given by

$$q_1 = \sqrt{2E} + O(g), \qquad q_2 = \frac{1}{g} + \sqrt{2(E + \epsilon)} + O(g). \tag{176}$$

The integral in the wave function can be evaluated by expanding in g^2:

$$\begin{aligned}\int_{q_1}^{q} \mathrm{d}x\, k(x) &= \frac{1}{g^2}\int_{gq_1}^{gq} \mathrm{d}\omega \sqrt{\omega^2(1-\omega)^2 - 2\epsilon g^2\omega - 2g^2 E} \\ &= \frac{1}{g^2}\int_{gq_1}^{gq} \mathrm{d}\omega \left[\omega(1-\omega) - \frac{(\epsilon\omega + E)g^2}{\omega(1-\omega)} + \cdots\right] \\ &= \left[\frac{\omega^2}{2g^2} - \frac{\omega^3}{3g^2} - E\ln|\omega| + (\epsilon + E)\ln|1-\omega| + \cdots\right]_{gq_1}^{gq}.\end{aligned} \tag{177}$$

Hence, in the leading order in g it reads as

$$\int_{q_1}^{q} \mathrm{d}x\, k(x) = \frac{q^2}{2} - E\left(\ln q - \ln\sqrt{2E} + 1\right) + O(g), \tag{178}$$

where we have used (176) for the value of q_1. Substituting this result into the WKB formula (174), we obtain

$$\psi(q) \simeq \frac{C_1}{q^{1/2}}\mathrm{e}^{-q^2/2}\left(\frac{\mathrm{e}q}{\sqrt{2E}}\right)^{E} + \frac{C_2}{q^{1/2}}\mathrm{e}^{q^2/2}\left(\frac{\mathrm{e}q}{\sqrt{2E}}\right)^{-E}, \tag{179}$$

where we have used $k(q) \simeq q$. This wave function must be connected with the one given by (171). As the wave function (171) has the following asymptotic form for $q \gg 1$ (cf. Ref. [28])

$$\psi(q) \simeq A\left[\mathrm{e}^{-q^2/2}(-\sqrt{2}q)^{\nu} + \frac{\sqrt{2\pi}}{\Gamma(-\nu)}\mathrm{e}^{q^2/2}(\sqrt{2}q)^{-\nu-1}\right], \tag{180}$$

we obtain the following matching condition

$$\frac{\sqrt{2\pi}}{(-1)^{\nu}\Gamma(-\nu)}2^{-\nu-1/2} = \frac{C_2}{C_1}\left(\frac{\sqrt{2E}}{\mathrm{e}}\right)^{2E}. \tag{181}$$

It should be noted that the asymptotic expansion of the parabolic cylinder function $D_\nu(z)$ with $|\arg z| \geq 3\pi/4$ has the phase ambiguity. The latter fact has an intimate relation with the ambiguity of the Borel integral due to the existence of Borel singularities discussed in Section 5.2.. In the asymptotic formula (180), and also in those appeared afterward, we fix the ambiguity so that it is consistent with our choice of the Borel integral. For more details on the issue, see Ref. [24]. On the other hand, the integral in the exponent in the

WKB wave function (174) viewed from the point around the right side potential well, where $1/g - q \sim O(1)$, is evaluated as

$$\begin{aligned}\int_{q_1}^{q} \mathrm{d}x\, k(x) &= \frac{1}{g^2}\int_{gq_1-1}^{gq-1} \mathrm{d}\omega \sqrt{\omega^2(1+\omega)^2 - 2\epsilon g^2(1+\omega) - 2Eg^2} \\ &= -\frac{1}{g^2}\int_{gq_1-1}^{gq-1} \mathrm{d}\omega \left[\omega(1+\omega) - \frac{\epsilon g^2(1+\omega)+Eg^2}{\omega(1+\omega)} + \cdots\right] \\ &= \left[-\frac{1}{g^2}\left(\frac{\omega^2}{2}+\frac{\omega^3}{3}\right) + (\epsilon+E)\ln|\omega| - E\ln|1+\omega| + \cdots\right]_{gq_1-1}^{gq-1}. \quad (182)\end{aligned}$$

Thus, in the leading order in g it reads, with the value of q_1 given by (176), as

$$\begin{aligned}\int_{q_1}^{q} \mathrm{d}x\, k(x) &= \frac{1}{6g^2} + (2E+\epsilon)\ln g - \frac{(1/g-q)^2}{2} + (E+\epsilon)\ln(1/g-q) \\ &\quad + E\left(\ln\sqrt{2E} - 1\right) + O(g). \quad (183)\end{aligned}$$

Substituting this formula into the WKB formula (174), we have

$$\begin{aligned}\psi(q) \simeq{}& \frac{C_1 \mathrm{e}^{-1/6g^2} g^{-(2E+\epsilon)}}{(1/g-q)^{1/2}} \mathrm{e}^{(1/g-q)^2/2}(1/g-q)^{-(E+\epsilon)}\left(\frac{\mathrm{e}}{\sqrt{2E}}\right)^{E} \\ &+ \frac{C_2 \mathrm{e}^{1/6g^2} g^{2E+\epsilon}}{(1/g-q)^{1/2}} \mathrm{e}^{-(1/g-q)^2/2}(1/g-q)^{E+\epsilon}\left(\frac{\mathrm{e}}{\sqrt{2E}}\right)^{-E}, \quad (184)\end{aligned}$$

where we have used $k(q) \simeq 1/g - q$. The latter WKB wave function is to be connected with the one given by (173) near the lower minimum of the potential which has the following asymptotic form for $1/g - q \gg 1$:

$$\psi(q) \simeq B\left[\mathrm{e}^{-(q-1/g)^2/2}\left(\sqrt{2}(q-1/g)\right)^{\nu+\epsilon} - \frac{\sqrt{2\pi}(-1)^{\nu+\epsilon}}{\Gamma(-\nu-\epsilon)}\mathrm{e}^{(q-1/g)^2/2}\left(\sqrt{2}(q-1/g)\right)^{-\nu-\epsilon-1}\right]. \quad (185)$$

Hence, we obtain another matching condition

$$\frac{\sqrt{2\pi}}{(-1)^{\nu+\epsilon}\Gamma(-\nu-\epsilon)} 2^{-(\nu+1/2+\epsilon)} = \frac{C_1}{C_2}\mathrm{e}^{-1/3g^2} g^{-(4E+2\epsilon)}\left(\frac{\mathrm{e}}{\sqrt{2E}}\right)^{2E}. \quad (186)$$

Finally, if we eliminate the yet undetermined ratio of the coefficients C_1/C_2 in the matching conditions (181) and (186), we obtain the quantization condition as follows:

$$\frac{\mathrm{e}^{-1/3g^2}}{\pi g^2}\left(-\frac{2}{g^2}\right)^{2E-1+\epsilon}\Gamma\left(-E+\frac{1}{2}\right)\Gamma\left(-E+\frac{1}{2}-\epsilon\right) = 1. \quad (187)$$

It is in complete agreement with the formula (109) obtained by the valley-instanton calculation. We note that they are identical not only for the case $\epsilon = 0$ but also for all $\epsilon \neq 0$ where we cannot apply the usual instanton technique since the classical configuration in the latter case becomes a bounce solution.

8.2. The Symmetric Triple-well Potential

Let us next consider the symmetric triple-well potential (119) for all $\epsilon g^2 \ll 1$. The system has parity symmetry and thus it is sufficient to study the connection condition of the WKB wave function only on the half-line $q \in (0, \infty)$. In the vicinity of the minimum of the central potential well, the Schrödinger equation for the potential (119) in the leading order of g is given by

$$\left(-\frac{1}{2}\frac{\mathrm{d}^2}{\mathrm{d}q^2} + \frac{1}{2}q^2\right)\psi(q) = \left(E - \frac{\epsilon}{2}\right)\psi(q). \tag{188}$$

The local solutions possessing a definite parity $\pm$ are expressed as

$$\psi(q) = A_\pm\left(D_\nu(-\sqrt{2}q) \pm D_\nu(\sqrt{2}q)\right), \tag{189}$$

where $A_\pm$ are constants and D_ν is the parabolic cylinder function with $\nu = E - \epsilon/2 - 1/2$. In a similar way, around the minimum of the right side potential well, the Schrödinger equation is approximated by

$$\left[-\frac{1}{2}\frac{\mathrm{d}^2}{\mathrm{d}q^2} + 2\left(q - \frac{1}{g}\right)^2\right]\psi(q) = (E + \epsilon)\psi(q), \tag{190}$$

and the local solution which vanishes at $q \to \infty$ is given by

$$\psi(q) = BD_\lambda(2(q - 1/g)), \tag{191}$$

where B is a constant and $\lambda = E/2 + \epsilon/2 - 1/2$. The solutions (189) and (191) are to be connected with the WKB solution (174) valid in the classically forbidden region $q_1 \ll q \ll q_2$ with

$$k(x) = \sqrt{x^2(1 - g^2x^2)^2 + \epsilon(1 - 3g^2x^2) - 2E}\,. \tag{192}$$

The positive classical turning points q_i $(i = 1, 2)$ with $0 < q_1 < q_2$ defined by the solutions of $V(q_i) = E$ are

$$q_1 = \sqrt{2E - \epsilon} + O(g^2), \qquad q_2 = \frac{1}{g} - \sqrt{\frac{E + \epsilon}{2}} + O(g). \tag{193}$$

In order to connect the wave functions obtained in the each region, it is important to note that the leading term in g of the WKB solution (174) varies according to the position it is viewed from. If it is viewed from the point around the central potential well, where $q \sim O(1)$, the integral in the exponent in (174) is evaluated as

$$\begin{aligned}\int_{q_1}^{q} \mathrm{d}x\, k(x) &= \frac{1}{g^2}\int_{gq_1}^{gq} \mathrm{d}\omega \sqrt{w^2(1 - w^2)^2 + \epsilon g^2(1 - 3w^2) - 2Eg^2} \\ &= \frac{1}{g^2}\int_{gq_1}^{gq} \mathrm{d}\omega \left[w(1 - w^2) - \frac{1}{2}\frac{(2E - \epsilon)g^2}{w(1 - w^2)} - \frac{3\epsilon g^2}{2}\frac{w}{1 - w^2} + \cdots\right] \\ &= \left[\frac{\omega^2}{2g^2} - \frac{\omega^4}{4g^2} + \frac{\epsilon - 2E}{2}\ln|\omega| + \frac{\epsilon + E}{2}\ln|1 - \omega^2| + \cdots\right]_{gq_1}^{gq}. \end{aligned} \tag{194}$$

Thus, in the leading order in g it reads as

$$\int_{q_1}^{q} \mathrm{d}x\, k(x) = \frac{q^2}{2} - \frac{2E-\epsilon}{2}\left(\ln q - \ln\sqrt{2E-\epsilon} + 1\right) + O(g^2), \tag{195}$$

where we have used (193) for the value of q_1. Substituting this result into the WKB formula (174), we obtain

$$\psi(q) \simeq \frac{C_1}{q^{1/2}}\,\mathrm{e}^{-q^2/2}\left(\frac{\mathrm{e}q}{\sqrt{2E-\epsilon}}\right)^{(2E-\epsilon)/2} + \frac{C_2}{q^{1/2}}\,\mathrm{e}^{q^2/2}\left(\frac{\mathrm{e}q}{\sqrt{2E-\epsilon}}\right)^{-(2E-\epsilon)/2}, \tag{196}$$

where we have used $k(q) \simeq q$. Comparing this with the following asymptotic form for $q \gg 1$ of the wave function (189) determined in the region of the central potential well

$$\psi(q) \simeq A_{\pm}\left[((-1)^{\nu} \pm 1)\mathrm{e}^{-q^2/2}(\sqrt{2}q)^{\nu} + \frac{\sqrt{2\pi}}{\Gamma(-\nu)}\mathrm{e}^{q^2/2}(\sqrt{2}q)^{-\nu-1}\right], \tag{197}$$

we have the following connection condition

$$\frac{\sqrt{2\pi}}{((-1)^{\nu} \pm 1)\Gamma(-\nu)} = \frac{C_2}{C_1}\left(\frac{2\sqrt{E-\epsilon/2}}{\mathrm{e}}\right)^{2E-\epsilon}. \tag{198}$$

On the other hand, the integral in the exponent in the WKB wave function (174) viewed from the point around the right side potential well, where $1/g - q \sim O(1)$, is evaluated as

$$\begin{aligned}
\int_{q_1}^{q} \mathrm{d}x\, k(x) &= \frac{1}{g^2}\int_{gq_1-1}^{gq-1} \mathrm{d}\omega \sqrt{\omega^2(1+\omega)^2(2+\omega)^2 - \epsilon g^2(2+6\omega+3\omega^2) - 2Eg^2} \\
&= -\frac{1}{g^2}\int_{gq_1-1}^{gq-1} \mathrm{d}\omega \left[\omega(1+\omega)(2+\omega) - \frac{\epsilon g^2(2+6\omega+3\omega^2)+2Eg^2}{2\omega(1+\omega)(2+\omega)} + \cdots\right] \\
&= \left[-\frac{1}{g^2}\left(\omega^2+\omega^3+\frac{\omega^4}{4}\right) + \frac{\epsilon+E}{2}\ln|\omega(2+\omega)| + \frac{\epsilon-2E}{2}\ln|1+\omega| + \cdots\right]_{gq_1-1}^{gq-1}.
\end{aligned} \tag{199}$$

Thus, in the leading order in g it reads, with the value of q_1 given by (193), as

$$\begin{aligned}
\int_{q_1}^{q} \mathrm{d}x\, k(x) &= \frac{1}{4g^2} + \frac{3E}{2}\ln q - (1/g-q)^2 + \frac{E+\epsilon}{2}\ln 2(1/g-q) \\
&\quad + \frac{2E-\epsilon}{2}\left(\ln\sqrt{2E-\epsilon} - 1\right) + O(g).
\end{aligned} \tag{200}$$

Substituting this expression into the WKB formula (174), we obtain

$$\begin{aligned}
\psi(q) \simeq{}& \frac{C_1\,\mathrm{e}^{-1/4g^2}g^{-3E/2}}{(1/g-q)^{1/2}}\,\mathrm{e}^{(q-1/g)^2}\left(2(1/g-q)\right)^{-(E+\epsilon)/2}\left(\frac{\mathrm{e}}{\sqrt{2E-\epsilon}}\right)^{(2E-\epsilon)/2} \\
&+ \frac{C_2\,\mathrm{e}^{1/4g^2}g^{3E/2}}{(1/g-q)^{1/2}}\,\mathrm{e}^{-(q-1/g)^2}\left(2(1/g-q)\right)^{(E+\epsilon)/2}\left(\frac{\mathrm{e}}{\sqrt{2E-\epsilon}}\right)^{-(2E-\epsilon)/2},
\end{aligned} \tag{201}$$

where we have used $k(q) \simeq 1/g - q$. Matching this with the following asymptotic form for $1/g - q \gg 1$ of the wave function (191) determined in the region of the right side potential well:

$$\psi(q) \simeq B \left[e^{-(q-1/g)^2} \bigl(2(q-1/g)\bigr)^{\lambda} - \frac{\sqrt{2\pi}(-1)^{\lambda}}{\Gamma(-\lambda)} e^{(q-1/g)^2} \bigl(2(q-1/g)\bigr)^{-\lambda-1} \right], \quad (202)$$

we get another connection condition as follows:

$$-\frac{\sqrt{2\pi}}{(-1)^{\lambda+1}\Gamma(-\lambda)} = \frac{C_1}{C_2} e^{-1/2g^2} g^{-3E} \left(\frac{e}{\sqrt{2E-\epsilon}} \right)^{2E-\epsilon}. \quad (203)$$

Therefore, eliminating the coefficient C_1/C_2 in (198) and (203), we finally obtain the following quantization condition:

$$\frac{\sqrt{2}}{\pi g^2} e^{-1/2g^2} \frac{(-1)^{E-1/2-\epsilon/2} \pm 1}{2} \left(\frac{2}{g^2} \right)^{E-1/2-\epsilon/2} \Gamma\left(-E + \frac{1}{2} + \frac{\epsilon}{2} \right) \\ \times \left(-\frac{1}{g^2} \right)^{E/2-1/2+\epsilon/2} \Gamma\left(-\frac{E}{2} + \frac{1}{2} - \frac{\epsilon}{2} \right) = 1. \quad (204)$$

It is again in complete agreement with the formula (142) obtained by the valley-instanton calculation, not only for $\epsilon = 0$ but also for all $\epsilon \neq 0$.

9. $\mathcal{N}$-fold Supersymmetry and Quasi-solvability

In the last two sections, we have examined the accuracy of our approximations from the two different points of view, the one is perturbation theory and the other is the WKB method. In this section, we shall check our results in a totally different way. To make the point clear, let us first recall the intriguing phenomena summarized in (118), (151), (156) that the nonperturbative corrections and/or the leading divergences of the perturbation series disappear for a number of low-lying states when the parameter ϵ in each the model takes certain values. These phenomena are reminiscent of the result in supersymmetric quantum field theory known as the *non-renormalization theorem* [15, 16] which states that any perturbative correction to the vacuum amplitude vanishes in all order unless supersymmetry is explicitly broken at the tree level.

In fact, the disappearance of divergence in large-order behavior of the perturbation series for the ground-state energy of supersymmetric quantum mechanical systems are investigated in Refs. [29, 30]. In our present two cases, we can easily check that both of the double-well (66) and the triple-well potentials (119) have supersymmetry when the integer $\mathcal{N}$, which characterizes the phenomena and is introduced in (118), (151), and (156), is equal to 1. Hence, it is natural to ask whether there exists another symmetry which resembles supersymmetry in nature and which are realized in the two potentials when the parameter ϵ takes the values corresponding to the integral values of $\mathcal{N} \geq 2$.

As we will show in what follows, it is indeed the case, and all the phenomena can be explained as consequences of a class of generalized supersymmetry, called $\mathcal{N}$-fold supersymmetry (SUSY), hidden in both the double-well and the triple-well potentials.

9.1. General Aspects of $\mathcal{N}$-fold Supersymmetry

$\mathcal{N}$-fold SUSY in quantum mechanics is a natural generalization of supersymmetric quantum mechanics [31], and has a lot of significant features and rich structures. In this section, however, we only pick some of its general aspects which are most relevant for our current purpose. For a review of $\mathcal{N}$-fold SUSY, please refer to Ref. [32]. Roughly speaking, an $\mathcal{N}$-fold SUSY quantum mechanical system is composed of a pair of Hamiltonians $H^{\pm}$ and an $\mathcal{N}$th-order linear differential operator $P_{\mathcal{N}}^{-}$

$$H^{\pm} = -\frac{1}{2}\frac{\mathrm{d}^2}{\mathrm{d}q^2} + V^{\pm}(q), \qquad P_{\mathcal{N}}^{-} = \frac{\mathrm{d}^{\mathcal{N}}}{\mathrm{d}q^{\mathcal{N}}} + \sum_{k=0}^{\mathcal{N}-1} w_k(q)\frac{\mathrm{d}^k}{\mathrm{d}q^k}, \tag{205}$$

which satisfy the following intertwining relations

$$P_{\mathcal{N}}H^{-} - H^{+}P_{\mathcal{N}}^{-} = 0, \qquad P_{\mathcal{N}}^{+}H^{+} - H^{-}P_{\mathcal{N}}^{+} = 0. \tag{206}$$

In the above, $P_{\mathcal{N}}^{+}$ is defined as the transposition of $P_{\mathcal{N}}^{-}$ and thus reads as

$$P_{\mathcal{N}}^{+} = (P_{\mathcal{N}}^{-})^{\mathrm{T}} = (-1)^{\mathcal{N}}\frac{\mathrm{d}^{\mathcal{N}}}{\mathrm{d}q^{\mathcal{N}}} + \sum_{k=0}^{\mathcal{N}-1}(-1)^k\frac{\mathrm{d}^k}{\mathrm{d}q^k}w_k(q). \tag{207}$$

In the case of $\mathcal{N} = 1$, $\mathcal{N}$-fold SUSY systems reduce to ordinary SUSY ones.

One of the most significant consequences of the intertwining relations is *weak quasi-solvability*. That is, the Hamiltonians $H^{\pm}$ preserve the linear spaces $\mathcal{V}_{\mathcal{N}}^{\pm}$ spanned by the kernels of the operators $P_{\mathcal{N}}^{\pm}$

$$H^{\pm}\mathcal{V}_{\mathcal{N}}^{\pm} \subset \mathcal{V}_{\mathcal{N}}^{\pm}, \qquad \mathcal{V}_{\mathcal{N}}^{\pm} = \ker P_{\mathcal{N}}^{\pm}. \tag{208}$$

If each $\mathcal{V}_{\mathcal{N}}^{\pm}$ admits an expression in closed form,

$$\mathcal{V}_{\mathcal{N}}^{\pm} = \langle \phi_1^{\pm}(q), \ldots, \phi_{\mathcal{N}}^{\pm}(q)\rangle, \tag{209}$$

then the corresponding $H^{\pm}$ is said to be *quasi-solvable*. In the latter case, we can diagonalize algebraically $H^{\pm}$ in the $\mathcal{N}$-dimensional linear spaces $\mathcal{V}_{\mathcal{N}}^{\pm}$ since the quasi-solvability assures that for an arbitrary $\phi_k^{\pm}(q) \in \mathcal{V}_{\mathcal{N}}^{\pm}$ we have

$$H^{\pm}\phi_k^{\pm}(q) = \sum_{l=1}^{\mathcal{N}} \mathsf{H}_{k,l}^{\pm}\phi_l^{\pm}(q) = E\phi_k^{\pm}(q). \tag{210}$$

Owing to this fact, the spaces $\mathcal{V}_{\mathcal{N}}^{\pm}$ are called *solvable sectors* of $H^{\pm}$. The diagonal elements of the matrix $\mathsf{H}^{\pm}$ are obviously given by the roots of the characteristic polynomial

$$\det\left(H^{\pm}|_{\mathcal{V}_{\mathcal{N}}^{\pm}} - E\right) = 0. \tag{211}$$

However, as it sometimes happens, any element of $\mathcal{V}_{\mathcal{N}}^{\pm}$ is not necessarily normalizable on $S \subset \mathbb{R}$ where $H^{\pm}$ are defined. Hence, quasi-solvability by itself does not guarantee that a number of exact eigenvectors and eigenvalues is available. Then, if each $\mathcal{V}_{\mathcal{N}}^{\pm}$ is a subspace of the Hilbert space $L^2(S)$ on which the corresponding $H^\pm$ acts, $\mathcal{V}_\mathcal{N}^\pm(S) \subset L^2(S)$, the elements of $\mathcal{V}_\mathcal{N}^\pm$ which diagonalize $H^\pm$ in the solvable sector are the normalizable exact eigenvectors of $H^\pm$, and the corresponding diagonal elements of $H^\pm$ determined by the equation (211) are the exact eigenvalues of $H^\pm$. In the latter case, the corresponding $H^\pm$ is said to be *quasi-exactly solvable*. For a review of quasi-exact solvability, please refer to Ref. [33].

The $\mathcal{N}$th-order intertwining operators $P_\mathcal{N}^\pm$ actually correspond to components of $\mathcal{N}$-fold supercharges. Hence, the situation where neither $\mathcal{V}_\mathcal{N}^+$ nor $\mathcal{V}_\mathcal{N}^-$ has any element of the Hilbert space L^2 means that there are no physical states which have $\mathcal{N}$-fold SUSY. In other words, $\mathcal{N}$-fold SUSY is dynamically broken in the latter situation. It is thus clear, for instance, that $\mathcal{N}$-fold SUSY is not broken dynamically if either H^+ or H^- is not only quasi-solvable but also quasi-exactly solvable with the solvable sector $\mathcal{V}^+$ or $\mathcal{V}^-$, respectively.

Let us now consider an $\mathcal{N}$-fold SUSY system to which perturbation theory is applicable. It means that the pair of Hamiltonians $H^\pm$ admits a decomposition $H^\pm = H_0^\pm + gH_I^\pm$ with a parameter g such that the complete set of exact eigenvectors $\psi_n^{(0)}$ and eigenvalues $E_n^{(0)}$ of H_0 is available. In such a situation, we can also apply perturbation theory to the equation (210) in $\mathcal{V}_\mathcal{N}^\pm$ by expanding formally $\phi_k^\pm$ in terms of $\psi_n^{(0)}$, regardless of the normalizability of $\phi_k^\pm$. Perturbation series are usually asymptotic, and asymptotic series for a given function is uniquely determined. Hence, the perturbative results in each the solvable sector $\mathcal{V}_\mathcal{N}^\pm$ would provide a number of perturbative eigenfunctions and eigenvalues of the corresponding $H^\pm$ defined in the full Hilbert space $L^2(S)$.

The latter fact results in two remarkable consequences. The one is that perturbation theory would not detect the normalizability of any element of $\mathcal{V}_\mathcal{N}^\pm$, and thus would fail to observe dynamical breaking of $\mathcal{N}$-fold SUSY. The other is that each the perturbation series for the eigenvalues corresponding to the states which belong to $\mathcal{V}_\mathcal{N}^\pm$ when $g \to 0$ would have a non-zero convergence radius. This is because the eigenvalues for such states are, up to nonperturbative contributions, given by the roots of the characteristic polynomial (211), and they would be analytic around $g = 0$ as long as all the coefficients of the characteristic polynomial are so. These two consequences can be regarded as generalized non-renormalization theorems in $\mathcal{N}$-fold SUSY. Although the arguments here are not mathematically rigorous, but have been shown to be valid in several examples. In the subsequent two sections, we will in fact see their validity in both the double-well and triple-well potentials.

9.2. The Asymmetric Double-well Potential

The class of $\mathcal{N}$-fold SUSY which is responsible for the double-well potential is Case I type A $\mathcal{N}$-fold SUSY [32, 34] given by

$$V^\pm(q) = \frac{b_2^2}{2}q^4 + b_2 b_1 q^3 + \frac{b_1^2 + 2b_2 b_0}{2}q^2 + (b_1 b_0 \mp \mathcal{N} b_2)q + \frac{b_0^2}{2} \mp \frac{\mathcal{N} b_1}{2} - R, \quad (212a)$$

$$P_\mathcal{N}^- = \left(\frac{\mathrm{d}}{\mathrm{d}q} - b_2 q^2 - b_1 q - b_0\right)^\mathcal{N}, \quad (212b)$$

$$\mathcal{V}_\mathcal{N}^+ = \langle 1, q, \dots, q^{\mathcal{N}-1}\rangle \exp\left(-b_0 q - \tfrac{b_1}{2}q^2 - \tfrac{b_2}{3}q^3\right), \quad (212c)$$

$$\mathcal{V}_\mathcal{N}^- = \langle 1, q, \dots, q^{\mathcal{N}-1}\rangle \exp\left(+b_0 q + \tfrac{b_1}{2}q^2 + \tfrac{b_2}{3}q^3\right). \quad (212d)$$

In fact, if we put $b_2 = g$, $b_1 = -1$, $b_0 = 0$, and $R = \mathcal{N}/2$, then the pair of $\mathcal{N}$-fold SUSY potentials in the above reads as

$$V^+(q) = \frac{g^2}{2}q^4 - gq^3 + \frac{1}{2}q^2 - \mathcal{N}gq,$$
$$V^-(q) = \frac{g^2}{2}q^4 - gq^3 + \frac{1}{2}q^2 + \mathcal{N}gq - \mathcal{N}. \tag{213}$$

Comparing the double-well potential (66) with the above $V^{\pm}(q)$, we see that it coincides with $V^+(q)$ and thus has $\mathcal{N}$-fold SUSY when $\epsilon = \mathcal{N}$. It is exactly the underlying reason for the phenomenon (118) that the leading divergence of the perturbation series vanishes in the case of the latter values of ϵ by the generalized non-renormalization theorem. On the other hand, its solvable sector $\mathcal{V}_{\mathcal{N}}^{+}$ is given by

$$\mathcal{V}_{\mathcal{N}}^{+} = \langle 1, q, \ldots, q^{\mathcal{N}-1}\rangle \exp\left(+\tfrac{1}{2}q^2 - \tfrac{g}{3}q^3\right), \tag{214}$$

and thus is not a subspace of the Hilbert space, $\mathcal{V}_{\mathcal{N}}^{+}(\mathbb{R}) \not\subset L^2(\mathbb{R})$. Hence, the double-well potential (66) is only quasi-solvable but *not* quasi-exactly solvable when $\epsilon = \mathcal{N}$, and $\mathcal{N}$-fold SUSY in the plus-component is broken dynamically. It explains the fact that the nonperturbative corrections do not vanish for any value of ϵ in the case of the double-well potential although the system has $\mathcal{N}$-fold SUSY when $\epsilon = \mathcal{N}$.

9.3. The Symmetric Triple-well Potential

The situation in the symmetric triple-well potential (119) is more complicated as can be anticipated from the involved structures in (151) and (156). The origin of the complexity resides in the fact that it has two different $\mathcal{N}$-fold SUSYs depending on the value of ϵ, as we will show shortly. The class of $\mathcal{N}$-fold SUSY which is responsible for the triple-well potential is Case II type A $\mathcal{N}$-fold SUSY [32, 34] given by

$$V^{\pm}(q) = \frac{b_2^2}{8}q^6 + \frac{b_2 b_1}{4}q^4 + \frac{b_1^2 + 2b_2 b_0 \mp 6\mathcal{N} b_2}{8}q^2 + \frac{(\mathcal{N}+1 \pm b_0)(\mathcal{N}-1 \pm b_0)}{8q^2} + \frac{b_1(b_0 \mp \mathcal{N})}{4} - R, \tag{215a}$$

$$P_{\mathcal{N}}^{-} = \prod_{k=0}^{\mathcal{N}-1}\left(\frac{\mathrm{d}}{\mathrm{d}q} - \frac{b_2}{2}q^3 - \frac{b_1}{2}q + \frac{\mathcal{N}-1-b_0-2k}{2q}\right), \tag{215b}$$

$$\mathcal{V}_{\mathcal{N}}^{+} = \langle 1, q^2, \ldots, q^{2(\mathcal{N}-1)}\rangle\, q^{-(\mathcal{N}-1+b_0)/2} \exp\left(-\tfrac{b_1}{4}q^2 - \tfrac{b_2}{8}q^4\right), \tag{215c}$$

$$\mathcal{V}_{\mathcal{N}}^{-} = \langle 1, q^2, \ldots, q^{2(\mathcal{N}-1)}\rangle\, q^{-(\mathcal{N}-1-b_0)/2} \exp\left(+\tfrac{b_1}{4}q^2 + \tfrac{b_2}{8}q^4\right). \tag{215d}$$

Let us first put $b_2 = 2g^2$, $b_1 = -2$, $b_0 = -(\mathcal{N} \pm 1)$, and $R = (\mathcal{N} \pm 1)/3$. Then, the pair of $\mathcal{N}$-fold SUSY potentials reads as

$$V^+(q) = \frac{g^4}{2}q^6 - g^2q^4 + \frac{1-(4\mathcal{N}\pm 1)g^2}{2}q^2 + \frac{4\mathcal{N}\pm 1}{6},$$
$$V^-(q) = \frac{g^4}{2}q^6 - g^2q^4 + \frac{1+(2\mathcal{N}\mp 1)g^2}{2}q^2 + \frac{\mathcal{N}(\mathcal{N}\pm 1)}{2q^2} - \frac{2\mathcal{N}\mp 1}{6}. \tag{216}$$

Comparing the triple-well potential (119) with the above $V^{\pm}(q)$, we see that it coincides with $V^{+}(q)$ and thus has $\mathcal{N}$-fold SUSY when $\epsilon = (4\mathcal{N} \pm 1)/3$. Thus, the generalized non-renormalization theorem explains the reason why the leading divergences of the perturbation series vanish when $\epsilon = (4\mathcal{N} \pm 1)/3$ as stated in (156a). The corresponding solvable sector $\mathcal{V}_{\mathcal{N}}^{+}$ is given by

$$\mathcal{V}_{\mathcal{N}}^{+} = \left\langle 1, q^2, \dots, q^{2(\mathcal{N}-1)} \right\rangle q^{(1\pm 1)/2} \exp\left(+\tfrac{1}{2}q^2 - \tfrac{g^2}{4}q^4 \right), \tag{217}$$

and thus it is a subspace of the Hilbert space, $\mathcal{V}_{\mathcal{N}}^{+}(S) \subset L^2(S)$, irrespective of whether the system is defined on the whole real line $S = (-\infty, \infty)$ or on the half line $S = (0, \infty)$. Hence, the triple-well potential (119) is quasi-exactly solvable when $\epsilon = (4\mathcal{N} \pm 1)/3$ and $\mathcal{N}$-fold SUSY is not broken dynamically. The latter fact is indeed consistent with the fact (151) that the nonperturbative corrections vanish when $\epsilon = (4\mathcal{N} \pm 1)/3$. Furthermore, we see from (217) that all the exactly solvable states have odd (even) parity for $\epsilon = (4\mathcal{N}+1)/3$ ($\epsilon = (4\mathcal{N} - 1)/3$), respectively, which also coincide with the different patterns of the disappearances in the $\epsilon = (4\mathcal{N} + 1)/3$ and $\epsilon = (4\mathcal{N} - 1)/3$ cases. Finally, it is worth noting that the disappearances of the leading divergences of the perturbation series take place both for the states with the quantum numbers n_- and n_+ when $\epsilon = (4\mathcal{N} \pm 1)/3$, in spite of the fact that the states which subject to the non-renormalization theorem carry only n_- when $\epsilon = (4\mathcal{N} - 1)/3$ or only n_+ when $\epsilon = (4\mathcal{N} + 1)/3$. The reason is that the perturbative corrections to the eigenvalues for the n_- and n_+ states are always identical, as was already discussed in Section 7.1..

Next, let us put $b_2 = 2g^2$, $b_1 = -2$, $b_0 = \mathcal{N} \pm 1$, and $R = -(\mathcal{N} \pm 1)/3$. Then, the pair of $\mathcal{N}$-fold SUSY potentials in this case become

$$\begin{aligned} V^{+}(q) &= \frac{g^4}{2}q^6 - g^2 q^4 + \frac{1 - (2\mathcal{N} \mp 1)g^2}{2}q^2 + \frac{\mathcal{N}(\mathcal{N} \pm 1)}{2q^2} + \frac{2\mathcal{N} \mp 1}{6}, \\ V^{-}(q) &= \frac{g^4}{2}q^6 - g^2 q^4 + \frac{1 + (4\mathcal{N} \pm 1)g^2}{2}q^2 - \frac{4\mathcal{N} \pm 1}{6}. \end{aligned} \tag{218}$$

Comparing again the triple-well potential (119) with the above $V^{\pm}(q)$, we see that it coincides with $V^{-}(q)$ and thus have $\mathcal{N}$-fold SUSY when $\epsilon = -(4\mathcal{N} \pm 1)/3$. It is again consistent with the disappearances of the perturbation series (156b) and (156c). The corresponding solvable sector $\mathcal{V}_{\mathcal{N}}^{-}$ is, however, given by

$$\mathcal{V}_{\mathcal{N}}^{-} = \left\langle 1, q^2, \dots, q^{2(\mathcal{N}-1)} \right\rangle q^{(1\pm 1)/2} \exp\left(-\tfrac{1}{2}q^2 + \tfrac{g^2}{4}q^4 \right), \tag{219}$$

and thus it is *not* a subspace of the Hilbert space, $\mathcal{V}_{\mathcal{N}}^{+}(S) \not\subset L^2(S)$, irrespective of whether $S = (-\infty, \infty)$ or $S = (0, \infty)$. Hence, the triple-well potential (119) is only quasi-solvable but *not* quasi-exactly solvable when $\epsilon = -(4\mathcal{N} \pm 1)/3$, and $\mathcal{N}$-fold SUSY in the minus-component is broken dynamically. The latter fact explains the reason why the nonperturbative corrections persist when $\epsilon = -(4\mathcal{N} \pm 1)/3$ in spite of the $\mathcal{N}$-fold SUSY. We also notice from (156b) and (156c) that, in contrast to the $\epsilon = (4\mathcal{N} \pm 1)/3$ cases, the states which subject to the non-renormalization theorem carry the quantum number n_0, rather than $n_{\pm}$, with definite parity. Then, the formula (219) indicates that they must have odd (even) parity when $\epsilon = -(4\mathcal{N}+1)/3$ ($\epsilon = -(4\mathcal{N}-1)/3$), respectively, which also agree with the results (156b) and (156c).

10. Discussion and Summary

In this chapter, we have reviewed the nonperturbative method based on valley configurations in Euclidean path-integral formalism. It can be regarded as a generalization of the conventional collective coordinate method which enables one to treat zero modes properly. In fact, our method can treat not only zero modes but also negative and/or quasi-zero modes which usually make the semi-classical method invalid. We have shown that our suitable definition of valley configurations in cooperation with the prescription of Faddeev–Popov determinants enables us to extract from the configuration space as collective coordinates the dynamical degrees corresponding to non-positive and/or quasi-zero modes so that the validity of the Gaussian approximation remains intact. To demonstrate how our method works, we have applied it to the asymmetric double-well and the symmetric triple-well potentials to calculate the nonperturbative corrections to their spectra due to quantum tunneling.

In the case of the asymmetric double-well potential, it turns out that a classical instanton configuration in the symmetric case is continuously deformed into a valley-instanton as a solution to the valley equation. On the other hand, a classical bounce configuration in the asymmetric case is realized as an intermediate configuration of a family of $I\bar{I}$-valley configurations which connects the trivial vacuum configuration with the one which consists of an infinitely separated pair of a valley-instanton and an anti-valley-instanton. Among the $I\bar{I}$-valley configurations, the action gets maximal at the bounce configuration, which means that we have successfully extract a collective coordinate associated with the negative mode around the classical bounce solution. The asymptotic action of the $I\bar{I}$-valley configuration with a larger separation automatically includes the attractive interaction term between a valley-instanton and an anti-valley-instanton. The analytic structure of the partition function of an $I\bar{I}$-valley naturally leads to the separation into the perturbative and the nonperturbative parts. From this separation, we obtain the dispersion relation which connects the perturbative contribution with the imaginary part of the nonperturbative one. The analytic structure of the nonperturbative part supports the Bogomolny's prescription for its calculation.

In the case of the symmetric triple-well potential, we have found that there are two novel and salient aspects which are absent in the case of double-well potential. The one is the existence of different kinds of $I\bar{I}$-valley and $\bar{I}I$-valley configurations which differ in functional form of their attractive interactions. The other is the existence of II-valley and $\bar{I}\bar{I}$-valley configurations which have repulsive interaction terms. Hence, a proper treatment of those different interactions turns to be crucial, regardless of the existence of non-positive or quasi-zero modes.

In both the cases, we have carried out the multi-valley-instanton calculations beyond the dilute-gas approximation by including up to the nearest-neighbor interactions among (anti-)valley-instantons. We have obtained the nonperturbative corrections to the spectra not only for the ground state but also for all the excited states. We have also calculated large-order behavior of the perturbation series for the corresponding energy eigenvalues with the aid of the dispersion relation. The obtained results have been checked by comparing with the perturbative calculations, the WKB approximation, and the consequences of $\mathcal{N}$-fold SUSY. All the comparisons strongly support our nonperturbative calculations both qualitatively and quantitatively. Some final remarks are in order.

1. The WKB approximation for Schrödinger equation and the Gaussian approximation around the classical configuration in Euclidean path integral are both often called semi-classical methods since both of them can be formulated as formal $\hbar$ expansion. However, it does not mean at all that their results always coincides with each other. In fact, for a quantum system with more than one local potential minima, the latter approach would fail due to the existence of negative modes while the validity of the former would remain intact. The complete agreement between our results obtained by our method based on valley configurations and the WKB results shown in Section 8. is an indirect indication of the disagreement between the two semi-classical methods. An important lesson one should draw from this fact would be that one must be more cautious with semi-classical description of a given quantum system. A quantum system is often regarded just as the corresponding classical one perturbed by small quantum fluctuations. Our present analysis clearly indicates that the emergence of negative modes is a signal for breakdown of the latter semi-classical description in Euclidean path integral formalism.

2. We note that the agreement of our results with the WKB ones actually has more meanings than the agreement itself. In the WKB calculations, it is quite difficult to interpret what the imaginary part of the nonperturbative spectrum which results from the obtained semi-classical quantization condition means. In our approach, on the other hand, it is indispensable for canceling the imaginary part of the perturbative corrections caused by the existence of Borel singularities. Hence, we can interpret the imaginary part as a signal for Borel non-summability of the corresponding perturbation series. In this way, the agreement of the resulting formulas can provide a way to interpret from a different point of view some peculiar aspects of them for which the WKB method itself can tell us virtually nothing.

3. As we already pointed out, the emergence of the different interactions both in nature, namely, attractive and repulsive, and in functional form depending on configurations naturally explain the breakdown of the dilute-gas approximation in triple-well potentials [10, 11, 13]. Then, it would be almost evident that the number of different nearest-neighbor interactions increases as the number of potential wells does. Hence, we anticipate the breakdown of the latter approximation scheme not only for triple-well but also for n-tuple-well potentials for all $n > 3$. It in particular indicates that the success of the dilute-gas approximation in the calculation of the energy splitting between the ground and the first excited states of a symmetric double-well potential is rather exceptional.

A reason for the latter success can be explained intuitively as follows. In any symmetric double-well potential, any multi-instanton configuration consists of alternating instanton and anti-instanton which interact each other *attractively*. As a consequence, any such a configuration in the topologically trivial sector tends to collapse into the trivial vacuum configuration by pair-annihilation of an instanton and an anti-instanton to lower the total energy of the system. Hence, lower energy eigenstates would receive a dominant contributions from configurations with fewer numbers of (anti-)instantons, namely, such configurations well described by dilute-gas. That would be one of the central physical reasons why the dilute-gas approximation works well for the lowest two energy levels of a symmetric double-well potential.

The situation would be drastically changed if there exist a multi-(valley-)instanton configuration in which some pairs of (anti-)(valley-)instantons interact each other *repulsively*. In such a case, the system might lower its total energy by pair-creation of a (valley-)instanton and an anti-(valley-)instanton[3]. The final configuration which minimizes the total energy would strongly depend on the balance of both the attractive and repulsive interactions. Hence, we must take into account all the (at least) leading-order interactions properly in any quantum potential with more than two potential wells. That naturally explains the very reason why the dilute-gas approximation fails even for the lowest three energy levels of a triple-well potential.

4. The complete correspondence between the disappearance of the leading divergence in the perturbation series for the spectrum and $\mathcal{N}$-fold SUSY in the present two cases leads us to ask the following question: is $\mathcal{N}$-fold SUSY sufficient and/or necessary for such a phenomenon? If, for instance, it is not a necessary condition, it then means that such a phenomenon could take place without $\mathcal{N}$-fold SUSY. In the latter case, it might be that there is another hidden symmetry which is different from $\mathcal{N}$-fold SUSY. Therefore, observing a similar phenomena might provide us a clue to discover a new type of symmetry which has been still unknown.

Acknowledgments

The author would like to thank M. Sato for his cooperation on a part of the numerical calculations in Section 7. and his valuable suggestions on Section 4.. This work was partially supported by the National Cheng Kung University under the grant No. HUA:98-03-02-227.

References

[1] Polyakov A M 1977 Quark confinement and topology of gauge theories *Nucl. Phys. B* **120** 429–458

[2] Gildener E and Patrascioiu A 1977 Pseudoparticle contributions to the energy spectrum of a one-dimensional system *Phys. Rev. D* **16** 423–430

[3] Callan C G, Dashen R and Gross D J 1978 Toward a theory of the strong interactions *Phys. Rev. D* **17** 2717–2763

[4] Coleman S 1985 *Aspects of Symmetry* (New York: Cambridge University Press)

[5] Callan C G and Coleman S 1977 Fate of the false vacuum. II. first quantum corrections *Phys. Rev. D* **16** 1762–1768

[6] Boyanovsky D, Willey R and Holman R 1992 Quantum mechanical metastability: When and why? *Nucl. Phys. B* **376** 599–634

[3]It is reminiscent of the famous Dyson's argument on the vacuum structure of QED [35].

[7] Aoyama H, Harano T, Kikuchi H, Sato M and Wada S 1997 Fake instability in the Euclidean formalism of quantum tunneling *Phys. Rev. Lett.* **79** 4052–4055 arXiv:hep-th/9606159

[8] Aoyama H, Kikuchi H, Okouchi I, Sato M and Wada S 1999 Valley views: instantons, large order behaviors, and supersymmetry *Nucl. Phys. B* **553** 644–710 arXiv:hep-th/9808034

[9] Rossi G C and Testa M 1983 Ground state wave function from Euclidean path integral *Ann. Phys.* **148** 144–167

[10] Lee S Y, Kahng J R, Yoo S K, Park D K, Lee C H, Park C S and Yim E S 1997 Instantonic approach to triple well potential *Mod. Phys. Lett. A* **12** 1803–1813

[11] Casahorrán J 2001 Analysis of nonperturbative fluctuations in a triple-well potential *Phys. Lett. A* **283** 285–290

[12] Sato M and Tanaka T 2002 $\mathcal{N}$-fold supersymmetry in quantum mechanics–analyses of particular models *J. Math. Phys.* **43** 3484–3510 arXiv:hep-th/0109179

[13] Alhendi H A and Lashin E I 2004 Symmetric triple well with non-equivalent vacua: Instantonic approach *Mod. Phys. Lett. A* **19** 2103–2112 arXiv:quant-ph/0406200

[14] Sato M and Tanaka T 2005 On disagreement about nonperturbative corrections in triple-well potential *Mod. Phys. Lett. A* **20** 881–896 arXiv:quant-ph/0411207

[15] Zumino B 1975 Supersymmetric and the vacuum *Nucl. Phys. B* **89** 535–546

[16] West P C 1976 The supersymmetric effective potential *Nucl. Phys. B* **106** 219–227

[17] Guillou J C L and Zinn-Justin J, eds. 1990 *Large-Order Behavior of Perturbation Theory* (Amsterdam: North-Holland)

[18] Zinn-Justin J 2005 *Path Integrals in Quantum Mechanics* (New York: Oxford University Press)

[19] Gervais J L and Sakita B 1975 Extended particles in quantum field theories *Phys. Rev. D* **11** 2943–2945

[20] Tomboulis E 1975 Canonical quantization of nonlinear waves *Phys. Rev. D* **12** 1678–1683

[21] Balitsky I I and Yung A V 1986 Collective-coordinate method for quasizero modes *Phys. Lett. B* **168** 113–119

[22] Aoyama H and Kikuchi H 1992 A new valley method for instanton deformation *Nucl. Phys. B* **369** 219–234

[23] Faddeev L and Popov V 1967 Feynman diagrams for the Yang–Mills field *Phys. Lett. B* **25** 29–30

[24] Bogomolny E B 1980 Calculation of instanton–anti-instanton contributions in quantum mechanics *Phys. Lett. B* **91** 431–435

[25] Zinn-Justin J 1981 Multi-instanton contributions in quantum mechanics *Nucl. Phys. B* **192** 125–140

[26] Zinn-Justin J 1983 Multi-instanton contribution in quantum mechanics (II) *Nucl. Phys. B* **218** 333–348

[27] Bender C M and Wu T T 1969 Anharmonic oscillator *Phys. Rev.* **184** 1231–1260

[28] Gradshteyn I S and Ryzhik I M 2000 *Table of Integrals, Series, and Products* (San Diego: Academic Press) sixth edition

[29] Verbaarschot J J M, West P and Wu T T 1990 Large-order behavior of the supersymmetric anharmonic oscillator *Phys. Rev. D* **42** 1276–1284

[30] Verbaarschot J J M and West P 1991 Instantons and Borel resummability for the perturbed supersymmetric anharmonic oscillator *Phys. Rev. D* **43** 2718–2725

[31] Witten E 1981 Dynamical breaking of supersymmetry *Nucl. Phys. B* **188** 513–554

[32] Tanaka T 2009 $\mathcal{N}$-fold supersymmetry and quasi-solvability in M B Levy, ed., *Mathematical Physics Research Developments* (New York: Nova Science Publishers, Inc.) chapter 18 pp. 621–679

[33] Ushveridze A G 1994 *Quasi-exactly Solvable Models in Quantum Mechanics* (Bristol: IOP Publishing)

[34] Tanaka T 2003 Type A $\mathcal{N}$-fold supersymmetry and generalized Bender–Dunne polynomials *Nucl. Phys. B* **662** 413–446 `arXiv:hep-th/0212276`

[35] Dyson F J 1952 Divergence of perturbation theory in quantum electrodynamics *Phys. Rev.* **85** 631–632

In: Focus on Quantum Mechanics
Editors: David E. Hathaway et al. pp. 405-446
ISBN 978-1-62100-680-0

Chapter 13

REINTERPRETING QUANTUM MECHANICS INTO A NONCONTEXTUAL FRAMEWORK: A SURVEY OF THE ESR MODEL

Claudio Garola [*] ***and Sandro Sozzo*** [†]
Dipartimento di Fisica dell'Università del Salento
Via Arnesano, 73100 Lecce, Italy, EU
and
INFN-Sezione di Lecce
Via Arnesano, 73100 Lecce, Italy, EU

Abstract

Based on known theoretical results (mainly the Bell and the Bell–Kochen–Specker theorems) most quantum physicists maintain that contextuality and nonlocality are unavoidable consequences of the mathematical apparatus of quantum mechanics (QM). Yet contextuality and nonlocality entail known paradoxes and unsolved problems in quantum theory measurement. A critical analysis shows, however, that the premise in the proofs of these theorems rest on an implicit assumption of a metatheoretical classical principle (MCP) which does not fit in well with the operational philosophy of QM. If MCP is replaced by a weaker assumption such as a metatheoretical generalized principle (MGP) such proofs cannot be carried out, which opens the way to a noncontextual; hence a local, interpretation of QM. This topic has been discussed in several papers, and a *semantic realism* (*SR*) interpretation of QM has been provided which is noncontextual; and thus, avoids the aforementioned paradoxes and the problems of the quantum theory of measurement. More recently, an *extended semantic realism* (*ESR*) model has been worked out which modifies and extends the original SR interpretation but preserves its basic features. The ESR model consists of a microscopic and a macroscopic parts. The former is a noncontextual hidden variables theory for QM that provides a justification for the assumptions introduced in the macroscopic part. The latter can be considered as an autonomous theory that embodies the formalism of QM into a noncontextual (hence local) framework, reinterpreting quantum probabilities as conditional (in a nonconventional sense) rather than absolute. One can then show that

[*]E-mail address: Garola@le.infn.it
[†]E-mail address: Sozzo@le.infn.it

the ESR model implies predictions that differ in some cases from those of QM and make it *falsifiable*. In particular the ESR model predicts that, whenever idealized measurements are performed and some additional assumptions are introduced, a *modified Bell–Clauser–Horne-Shimony–Holt (BCHSH)* inequality holds which is compatible with the reinterpreted quantum probabilities. Hence, the long–standing conflict between "local realism" and QM is settled in the ESR model. By referring to the macroscopic part of the ESR model, one can also supply a Hilbert space representation of the generalized observables introduced by the model and a generalization of the projection postulate of elementary QM (*generalized projection postulate*). These results imply a new mathematical representation of mixtures that does not coincide with the standard quantum representation and avoids some deep interpretational problems that follow from the standard representation in QM. A further generalization of the projection postulate (*generalized Lüders postulate*) can then be provided, and the generalized projection postulate can be partially justified by introducing a nonlinear time evolution of the compound system made up of the measured object and the measuring apparatus.

PACS 03.65.-w, 03.65.Ca, 03.65.Ta

Keywords: quantum mechanics, generalized observables, mixtures, quantum theory of measurement, Bell inequalities

1. Introduction

In their book on the quantum theory of measurement [1] Busch, Lahti, and Mittelstaedt stated that the crucial *objectification problem* arises whenever an interpretation of the mathematical apparatus of standard (Hilbert space) quantum mechanics (QM) is adopted which is *realistic* (in the sense that it assumes that QM deals with individual objects and their properties) and *complete* (in the sense that it assumes that all elements of physical reality are described by QM).

It is expedient in our opinion to formulate the objectification problem in the terms of the notion of *objectivity*. Let us consider a physical theory $\mathcal{T}$ which is realistic in the sense as explained earlier. We say that a property E of a physical system Ω is objective for a given state S of Ω if, for every individual example x of Ω (*physical object*) in the state S, E is either possessed or not possessed by x, independently of any measurement that may be performed on x.[1] Furthermore, we say that $\mathcal{T}$ is an objective theory if all macroscopic properties are objective in it for every state S of Ω, nonobjective otherwise. Then some celebrated *no–hidden–variables* theorems (in particular, the Bell–Kochen–Specker, or Bell–KS, and the Bell theorems [2, 3, 4]) show, according to most scholars, that QM is a *contextual*, hence a nonobjective theory. In particular, if S is a pure state, a macroscopic property E is nonobjective in S if and only if one has probability different from 1 or 0 that a measurement on a physical object x in the state S shows that x possesses the property F. If one now maintains that QM is a complete theory, one cannot explain how macroscopic properties that are not objective may become objective (and conversely) when a measurement is performed.

[1]The terms "possessed" and "not possessed" introduced here are rather loose. We show in the following that our definition of objectivity acquires a more precise meaning whenever its implications are considered on the semantics of the observational language of QM, or on the features of hidden–variables theories for QM.

According to most scholars, no satisfactory solution of the objectification problem has been found in the framework of the realistic and complete interpretation of QM or its unsharp extensions [1, 5, 6, 7, 8, 9]. Therefore, many proposals have been made to avoid it (Bohmian mechanics, GRW theory, many–worlds interpretations, decoherence program, quantum histories, etc. [10]), some of which consider QM as an incomplete theory. If QM is incomplete, indeed, one could contrive a hidden–variables theory containing parameters whose unknown values determine whether a property is possessed or not by a given physical object x in the state S, hence determine the outcome of any measurement on x. The objectification problem thus disappears. But the theorems quoted here oblige one to introduce hidden variables whose values depend on the measurement context and not only on the features of the physical object that is considered (*e.g.*, Bohm's theory), hence macroscopic properties are nonobjective anyway. Moreover, the Bell theorem implies that contextuality holds also at a distance (*nonlocality*) [4, 11], which is counterintuitive because it implies that performing a measurement on a part of a physical system may make objective a property of another far away part of the physical system which was previously nonobjective. In addition, from a semantic point of view nonobjectivity implies that a statement $E(x)$ attributing a macroscopic property E to a physical object x in a state S has a truth value if and only if E is objective in the state S, that implies adopting a highly problematical verificationist theory of truth for the observational language of QM [12].

Because of the consequences of nonobjectivity as mentioned, one may wonder whether an interpretation of the mathematical apparatus of QM can be supplied that recovers objectivity for macroscopic properties. Although, this possibility seems to be excluded by the no–hidden–variables theorems, hence every attempt at vindicating objectivity for QM must begin with a preliminary criticism of these theorems. Such a criticism had been carried out more than ten years ago by one of us in collaberation with another scholar. Some papers aiming to show that a noncontextual (hence local) interpretation of the mathematical formalism of quantum mechanics (QM) is, in principle, possible, at variance with the orthodox view [13, 14]. This conclusion rests on an argument that has never been introduced before in the literature, that is, on the remark that the proofs of the no–hidden–variables theorems, though technically rigorous, rest on an implicit epistemological assumption about the unrestricted validity of the empirical physical laws of QM (*metatheoretical classical principle*, or *MCP*) which can be questioned from an epistemological point of view and is not consistent with the operational and antimetaphysical attitude of QM. If only a weaker assumption (*metatheoretical generalized principle*, or *MGP*) is accepted, the proofs of the no–hidden-variables theorems cannot be completed [13, 14, 15, 16, 17, 18, 19, 20]. Based on this criticism, a noncontextual *semantic realism*, or *SR*, *interpretation* has been provided in these quoted papers, with the aim of recovering objectivity of properties and avoiding those known quantum paradoxes that find their roots in the contextuality of the orthodox interpretation of QM.

The SR interpretation is, however, rather abstract. Therefore some models have been propounded [21, 22, 23, 24, 25, 26, 27, 28] to prove its consistency, among which an *extended semantic realism* (*ESR*) *model* that supplies a set–theoretical picture of the physical world that preserves the basic features of the SR interpretation; that is, semantic realism and the substitution of MCP with the weaker principle MGP, but modifies and in some sense extends it. The ESR model satisfies physical intuition and it can be split into a microscopic and a macroscopic part. The former is a new kind of noncontextual hidden variables theory for QM which introduces, besides hidden variables, a reinterpretation of standard quantum probabilities, providing a justification of the assumptions introduced in the macroscopic part. The latter can be considered as a self–consistent theory that embodies the formalism of QM into a broader noncontextual framework.

As most hidden variables theories, the ESR model presupposes that "something is happening" at a microscopic level which underlies the standard quantum picture of the physical world and does not reduce to it (which implies that improvements of the measurements' precision and/or technological developments within the established framework of QM can hardly help in solving the conceptual problems of this theory). One therefore needs a broader theory, and the ESR model aims to be a first step in this direction. According to this model, the macroscopic properties of a given physical system that can be measured according to QM by macroscopic devices on a physical object x bijectively correspond to the microscopic properties that can be possessed or not possessed by x, which are hidden variables of the ESR model (together with further parameters that we will not take explicitly into account here for the sake of simplicity; these parameters, however, do not appear in the *deterministic* ESR model [24]). Then, the set of all microscopic properties possessed by x is called the *microscopic state* of x. If a macroscopic property F is measured on x and x displays F, then x possesses the microscopic property f corresponding to F. But the converse implication does not hold, because it can occur that the set of microscopic properties possessed by x is such that x is not detected when F is measured, even if x possesses the property f, independently of the specific features of the apparatus measuring F. Hence, a detection probability is associated with the measurement of F which depends on the microscopic state of x, not only on f, and must not be mistook for the detection probability that occurs because of the reduced efficiencies of the real measuring apparatuses.

The introduction of a detection probability depending on the microscopic state of the physical object x is crucial and distinguishes the ESR model from other hidden variables theories in the literature (though some assumptions of the ESR model have been anticipated by several authors when studying particular cases, see, *e.g.*, [29]). But the ESR model does not say anything about the deep causes of this detection probability: rather, introduces it as a global effect of such causes. At a statistical level overall detection probabilities then occur, to be considered as unknown parameters, whose value is not predicted by any existing theory. More generally, the main features of the ESR model can be resumed as follows.

(i) A *no–registration outcome* a_0 must be added to the set of possible values of any observable A of QM, which produces a *generalized observable* A_0 in which a_0 is a possible outcome, providing physical information, whenever a measurement of A_0 is performed on a physical object x (we stress that a_0 occurs also in the case of *idealized* measurements, which correspond to the *ideal first kind* measurements of QM in the ESR model).

(ii) The mathematical formalism of standard (Hilbert space) QM can be embodied into the noncontextual (hence, local) framework provided by the ESR model. To be precise, the quantum rules for calculating the probability that a physical object x in a state S display a macroscopic property F when a measurement of F is performed are reinterpreted as referring to the subset of all physical objects for which the values of the hidden variables are such that the objects can be detected (hence, they do not refer to the set of all physical objects that are actually prepared in the state S, as in QM: in this sense we say that quantum probabilities must be interpreted as *conditional* rather than *absolute* in the ESR model[2]). This reinterpretation implies, in particular, that the validity of the empirical physical laws of QM is ruled by MGP rather than MCP.

(iii) Because of noncontextuality, the standard distinction between *actual* and *potential* properties of a physical system in a given state does not occur in the ESR model. The objectification problem of the quantum theory of measurement disappears, together with such paradoxes as "Schrödinger's cat" and "Wigner's friend", because all macroscopic properties are *objective*: the measurement (or the observer) does not actualize them, and the values of the generalized observables of the physical system can be thought of as assigned for each physical object, independently of any measurement (but, of course, it is impossible to predict all of them even if the quantum state of the object is specified).

(iv) The detection probabilities introduced by the ESR model can be hardly distinguished from the inefficiencies of real measuring apparatuses, which explains why the former are ignored in QM. But the introduction of these detection probabilities implies also predictions that substantially differ from those of QM and make the ESR model *falsifiable*.

(v) In the physical situation considered by Clauser, Horne, Shimony, and Holt to obtain their version of the Bell inequality [31] (briefly, *BCHSH inequality*) one gets a *modified BCHSH inequality*, in which trichotomic instead of dichotomic observables occur.

These conclusions can be achieved without providing an explicit mathematical representation of the new physical entities introduced by the ESR model at a macroscopic level. Whenever such a representation is supplied, one can deduce from it some relevant theoretical consequences [24, 25, 26, 27, 28, 32, 33, 34, 35] that can be synthetized as follows.

(vi) Each generalized observable is represented by a (commutative) *family* of positive operator valued (POV) measures parametrized by the set of all pure states of the physical system that is considered. It follows that every macroscopic property is represented by a family of bounded positive operators (*effects*) rather by a single projection operator.

(vii) A *generalized projection postulate* (*GPP*) rules the transformations of pure states induced by nondestructive idealized measurements.

(viii) The modified BCHSH inequality can be expressed by means of overall detection probabilities and (reinterpreted) standard quantum expectation values, and provides a specific example of falsifiable prediction in the ESR model. In particular, it implies that upper limits must exist for the detection efficiencies in the experiments on Bell's inequalities, which can be experimentally checked, at least in principle.

(ix) The BCHSH inequality holds in the ESR model at a microscopic, purely theoretical, level, the modified BCHSH inequality holds at a macroscopic level if one takes into account all physical objects that are prepared, and standard quantum formulas hold if one takes into account only those physical objects that can be detected (see (iii)). Thus all these results hold together because they refer to different parts of the picture of the physical world supplied by the ESR model (in which, in particular, *local realism* [4, 36] holds).

(x) The violation of the BCHSH inequality at a macroscopic level can be interpreted in terms of a unconventional kind of *unfair sampling*.

---

[2]The idea of considering quantum probabilities as conditional has a long history, and we refer to [30] for a short bibliography on this topic. The kinds of conditioning that are usually considered, however, are different from conditioning with respect to detection introduced in the ESR model, which, at the best of our knowledge, has no precedent in the literature.

(xi) Each mixture is represented by a *family* of density operators parametrized by the set of macroscopic properties characterizing the physical system that is considered. This representation associates different families of density operators with mixtures having different operational definitions, even if such mixtures are probabilistically equivalent, which avoids some interpretational problems that arise in QM [34, 37, 38, 39]. In particular, the distinction between *proper* and *improper* mixtures [40] does not occur because all probabilities are epistemic in the ESR model.

(xii) A *generalized Lüders postulate* (GLP) that generalizes GPP rules the general transformations of states induced by nondestructive idealized measurements.

(xiii) GPP can be (partially) justified by describing a measurement as a dynamical process in which a *nonlinear* evolution occurs of the compound system made up of the (microscopic) measured object plus the (macroscopic) measuring apparatus.

Let us briefly sketch now the outline of this paper. We preliminarily supply in Sec. 2. an example of the criticism of the standard no–go theorems that justifies the proposal of an SR interpretation and, successively, the ESR model. We then provide a general treatment of this model in Sec. 3. (see items (i) and (ii)) and introduce noncontextuality (objectivity) in Sec. 4. (see items (iii) and (iv)). Furthermore we obtain the modified BCHSH inequality in Sec. 5., the mathematical representation of generalized observables in Sec. 6. and GPP in Sec. 7. (see items (v), (vi), and (vii), respectively). Based on these results we obtain new predictions in Sec. 8. (see item (viii)), and show how the BCHSH inequality can be reconciled with local realism, justifying our conclusions in terms of an unconventional kind of unfair sampling, in Sec. 9. (see items (ix) and (x)). We then obtain a new mathematical representation of mixtures according to the ESR model in Sec. 10., and prove that it admits an ignorance interpretation in Sec. 11. (see item (xi)). Finally, we discuss GLP in Sec. 12. (see item (xii)) and give a partial dynamical justification of GPP in Sec. 13. (see item (xiii)).

Let us close this section with two remarks.

Firstly, we observe that the results expounded in this paper do not provide yet a complete theory. One should still introduce a general (possibly nonlinear) evolution law and a treatment of compound physical systems, in particular to deal with measurements in terms of interactions between macroscopic apparatuses and microscopic physical objects (we plan to concentrate future research on these topics). The need of such an extended treatment for describing the measurement procedure in a unified way explains, from the point of view of the ESR model, the failure of the attempts at providing a unified theory of measurement in QM.

Secondly, we observe that the ESR model is noncontextual (hence, local) as far as one considers only *idealized* measurements, as we do in this paper. It has been proven by other authors that, whenever actual measurements are taken into account, a statistical description of the experiments on spatially separated systems that adopts a multi–Kolmogorovian rather than a simple Kolmogorovian model implies contextuality. This kind of contextuality, however, has not a quantum basis and may appear also in classical theories. It occurs, in particular, in the Växjö interpretation of QM [41, 42, 43], it can be supported by a wave model in which *unfair sampling* occurs if the measurement apparatuses have a threshold [44], and it implies that experimental violations of Bell's inequalities may be compatible with locality. It is then interesting to note that it can be recovered within the ESR model by introducing additional hidden variables associated with actual measuring apparatuses.

2. Criticizing a No–hidden–variables Theorem

As we have seen in Sec. 1., the SR interpretation of QM and the ESR model rest on a preliminary criticism of the epistemological position that is implicit in the "no–hidden–variables" theorems which aim to prove that the mathematical apparatus of QM implies contextuality [13, 14, 15, 18, 20]. Because of its significance, the essentials of this criticism will be illustrated here referring to the specific example of the Bell–KS theorem.

The first step for carrying out a critical analysis of the Bell–KS theorem consists in specifying an epistemological position about the laws of a physical theory $\mathcal{T}$. To this end *theoretical laws* must be distinguished from *empirical laws*. The former are general mathematical schemes stated by means of the *theoretical language* of $\mathcal{T}$, that have no direct interpretation on the domain of physical facts. The latter reformulate (via *correspondence rules*) some theoretical laws by means of the *observational language* of $\mathcal{T}$, hence are interpreted on the domain of physical facts. Then, two different positions can be assumed with respect to the truth value of a sentence α expressing an empirical physical law.

(i) *Metatheoretical generalized principle* (*MGP*). α is *true* in every physical situation in which the law can be checked (*epistemically accessible physical situation*), while it may be *true* as well as *false* in physical situations in which it cannot be checked because $\mathcal{T}$ prohibits any test.

(ii) *Metatheoretical classical principle* (*MCP*). α is *true* in every physical situation that can be devised, even if this situation is such that $\mathcal{T}$ prohibits that the empirical law expressed by α be checked.

Let the theory $\mathcal{T}$ be identified with QM. Then MGP sounds consistent with the operational and antimetaphysical attitude of QM, while the stronger principle MCP does not fit in with it.

Let us come now to the Bell–KS theorem. It is easy to see that MCP is implicitly used in the proofs of this theorem, which instead cannot be carried out if MGP is adopted in place of MCP. Indeed, all proofs assume the following condition [2, 3, 11].[3]

KS. Let $A, B, \ldots$ be compatible observables and let

$$f(A, B, \ldots) = 0 \tag{1}$$

express an empirical quantum law. Then, whenever measurements of $A, B, \ldots$ are performed obtaining the outcomes $a, b, \ldots$, respectively, the following equation holds

$$f(a, b, \ldots) = 0. \tag{2}$$

Condition KS is needed if one wants to get physical predictions from empirical laws, hence it must be accepted for physical reasons. All proofs then proceed *ab absurdo*. They

[3] It is well known that many proofs of the Bell–KS theorem have been given. For the sake of brevity only the original proofs and Mermin's simple and straightforward proofs are explicitly mentioned here, but the previous arguments apply to all existing proofs of the theorem.

consider several empirical quantum laws

$$\begin{cases} f(A, B, \ldots) = 0 \\ g(A', B', \ldots) = 0 \quad , \\ \quad \ldots\ldots \end{cases} \tag{3}$$

assume that the values $a, b, \ldots; a', b', \ldots$ of the observables $A, B, \ldots; A', B', \ldots$, respectively, are defined independently of any measurement procedure for any individual example x of the physical system that is considered,[4] apply the KS condition *repeteadly*, this implies that $f(a, b, \ldots) = 0$, $g(a', b', \ldots) = 0$ must hold simultaneously, and finally show that a contradiction occurs. A seemingly unavoidable conclusion is that the values $a, b, \ldots; a', b', \ldots; \ldots$ are not defined independently of the set of measurements that are performed (contextuality), hence cannot be considered as preexisting to the measurements.

The proofs schematized here are mathematically correct. However, a careful analysis shows that their premise follows from the adoption, implicit but essential, for the epistemological position expressed by MCP. Indeed, we have seen that each proof requires a repeated application of the KS condition. But direct inspection shows that in every proof there are observables in a law (say, $f(A, B, \ldots) = 0$) that are not compatible with some observables in another law (say, $g(A', B', \ldots) = 0$). Whenever the values $a, b, \ldots; a', b', \ldots$ are simultaneously attributed to the physical object x, a nonaccessible physical situation is devised in which only one (at choice) of the empirical laws in Eqs. (3) can be checked. If one adopts the position expressed by the weaker principle MGP, the proofs of the Bell-KS theorem cannot be completed because one cannot assert that $f(a, b, \ldots) = 0$ and $g(a', b', \ldots) = 0$ hold simultaneously. Hence the repeated application of the KS condition implies postulating the unrestricted simultaneous validity of all empirical quantum laws, that is, MCP. As stated earlier, this sounds problematic just from a standard quantum point of view.

To conclude this section, let us note that the adoption of MCP can be defended by observing that physicists can choose the empirical law that they want to check among the laws listed in Eqs. (3). Since all experiments show that, for every choice, quantum predictions are fulfilled, it is difficult to understand how a breakdown of a law in Eqs. (3) may occur just when another law is experimentally proven to hold without appealing to a sort of *conspiracy of nature* [45]. But we prove in the next section that the ESR model implies that such a failure can actually occur without implying any conspiracy of nature.

3. The ESR Model

The ESR model had been first proposed by one of the authors several years ago [21, 22] and successively refined together with other authors [23]. We present here an improved version of it, mainly deepening the treatment in [24].

The primitive notions of the ESR model are standard: *physical system*, *preparing device*, *measuring apparatus*, *outcome* (on the real line $\Re$) of a measuring apparatus. Let us therefore refer to a specific physical system Ω. We then introduce the set Π of preparing

[4]It is interesting to observe that this assumption implies adopting a perspective that is realistic in the sense specified at the beginning of Sec. 1.. As Busch, Lahti, and Mittelstaedt have pointed out [1], no objectification problem occurs if only a statistical interpretation of QM is accepted.

devices and a set $\mathscr{R}$ of measuring apparatuses associated with Ω. We call *physical object* every individual example of Ω obtained by activating a preparing device $\pi \in \Pi$ and assume that, for every $\pi \in \Pi$, $r \in \mathscr{R}$, $X \in \mathbb{B}(\Re)$ (where $\mathbb{B}(\Re)$ is the σ–algebra of all Borel subsets of $\Re$), a probability $p(\pi, r, X)$ is defined that x yield an outcome a in X when r performs a measurement on x. Then, p induces an equivalence relation $\equiv$ on Π, which is defined by setting, for every $\pi_1, \pi_2 \in \Pi$, $\pi_1 \equiv \pi_2$ iff, for every $r \in \mathscr{R}$ and $X \in \mathbb{B}(\Re)$, $p(\pi_1, r, X) = p(\pi_2, r, X)$. We call $\mathcal{S} = \Pi/_{\equiv}$ the *set of all states of* Ω and, for every physical object x and state $S = [\pi]_{\equiv} \in \mathcal{S}$, we say that x *is in the state* S whenever x is prepared by a preparation that belongs to $[\pi]_{\equiv}$ [37, 46]. Furthermore, p induces an equivalence relation $\simeq$ on $\mathscr{R}$, which is defined by setting, for every $r_1, r_2 \in \mathscr{R}$, $r_1 \simeq r_2$ iff a bijective mapping ω of the set of outcomes of r_1 onto the set of outcomes of r_2 exists such that, for every $\pi \in \Pi$ and $X \in \mathbb{B}(\Re)$, $p(\pi, r_1, X) = p(\pi, r_2, \omega(X))$. We assume now that $\mathscr{R}$ consists of all measuring apparatuses that correspond to standard observables of QM modified by adding a further outcome in each $r \in \mathscr{R}$, called the *no–registration outcome* of r. Thus, the set $\mathcal{O} = \mathscr{R}/_{\simeq}$ is called the *set of all generalized observables of* Ω. Because of this definition, every generalized observable A_0 (here, this is meant as a class of measuring apparatuses, without any reference to a mathematical representation) is obtained in the ESR model by considering an observable A of QM with set of possible values Ξ on $\Re$ and adding a further no–registration outcome a_0 that does not belong to Ξ, so that the set of possible values of A_0 is $\Xi_0 = \{a_0\} \cup \Xi$.[5]

The introduction of generalized observables allows us to define the set $\mathcal{F}_0$ of all *macroscopic properties* of Ω,

$$\mathcal{F}_0 = \{(A_0, X) \mid A_0 \in \mathcal{O},\ X \in \mathbb{B}(\Re)\}, \tag{4}$$

and the set $\mathcal{F} \subset \mathcal{F}_0$ of all macroscopic properties associated with observables of QM,

$$\mathcal{F} = \{(A_0, X) \mid A_0 \in \mathcal{O},\ X \in \mathbb{B}(\Re),\ a_0 \notin X\}. \tag{5}$$

For every $A_0 \in \mathcal{O}$, different Borel sets containing the same subset of Ξ_0 define physically equivalent properties. For the sake of simplicity we convene that, whenever we mention macroscopic properties in the following, we actually understand such classes of physically equivalent macroscopic properties. Furthermore, we agree to write simply *observable* in place of *generalized observable* so that no misunderstanding is possible.

We introduce now some new theoretical entities by assuming that every physical system Ω is characterized by a set $\mathcal{E}$ of *microscopic properties*, and that, for every physical object x, the set $\mathcal{E}$ is partitioned in two classes, the class of properties that are possessed by x and the class of properties that are not possessed by x, independently of any measurement procedure (we stress that we do not assume instead that different physical objects in the same state S must possess the same microscopic properties, even if assigning the state S of x imposes some limits on the subset of microscopic properties that can be possessed by x, as we show in footnote 7). Then we establish a link between microscopic properties of $\mathcal{E}$ and macroscopic properties of $\mathcal{F}$ by means of the following assumption.

[5] If $\Xi = \Re$ the observable A can be substituted, without loss of generality, by the observable $f(A)$, with f a bijective mapping that maps $\Re$ onto a proper subset, say $\Re^+$, of $\Re$.

Ax. 1. Every microscopic property f has a macroscopic counterpart $F = \varphi(f)$, and the mapping $\varphi : \mathcal{E} \longrightarrow \mathcal{F} \subset \mathcal{F}_0$ is bijective.

Let us describe now an *idealized* measurement of a macroscopic property $F = (A_0, X)$ on a physical object x in the state S. We assume that such a measurement is performed by means of a dichotomic measuring apparatus (which may be constructed by using one of the apparatuses associated with A_0 but does not necessarily belong to $\mathscr{R}$), whose outcomes we denote by *yes* and *no*. The measurement yields outcome yes if the value of A_0 belongs to X, and we say in this case that x *displays* F.[6] Consistently, the measurement yields outcome no if the value of A_0 does not belong to X, and we say in this case that x *does not display* F. It is then important to note that x displays F if it does not display $F^c = (A_0, \Re \setminus X)$, and conversely. Now, we assume that the set of all microscopic properties possessed by x induces a probability that the apparatus react (*detection probability*). Whenever the apparatus reacts, x displays F if it possesses the microscopic property $f = \varphi^{-1}((A_0, X \setminus \{a_0\}))$ (where $(A_0, X \setminus \{a_0\})$ coincides with F iff $F \in \mathcal{F}$), otherwise it displays F^c. Whenever the apparatus does not react, x displays F if $F \in \mathcal{F}_0 \setminus \mathcal{F}$, while it displays F^c if $F \in \mathcal{F}$. Furthermore, it may occur that the detection probability is always either 0 or 1. In this case, we say that the ESR model is *deterministic*.

This description implies that the microscopic properties determine the probability of an outcome (or the outcome itself if the model is deterministic), which therefore does not depend on features of the measuring apparatus (flaws, termal noise, etc.) nor is influenced by the environment. In this sense idealized measurements are "perfectly efficient," and must be considered as a limit of concrete measurements in which the specific features of apparatuses and environment must instead be taken into account.

We must still place quantum probability into our picture. To this end, let us suppose that the preparing device $\pi \in S$ is activated repeatedly; hence, a finite set $\mathscr{S}$ of physical objects in the state S is prepared. Then, $\mathscr{S}$ can be partitioned into subsets $\mathscr{S}^1, \mathscr{S}^2, \ldots, \mathscr{S}^n$ such that each subset collects all objects possessing the same microscopic properties. We briefly say that the objects in $\mathscr{S}^i$ $(i = 1, 2, \ldots, n)$ are in some *microscopic state* S^i. This suggests for us to associate every state S with a family of microscopic states $S^1, S^2, \ldots$ and characterize S^i $(i = 1, 2, \ldots)$ by means of the set of all microscopic properties that are possessed by any physical object in S^i (also, microscopic states play the role for theoretical entities in the ESR model). Let us now consider a physical object x in S^i, and let us suppose that a measurement of a macroscopic property $F = (A_0, X) \in \mathcal{F}$ is performed on it. Our description of the measurement process then implies that, whenever x is detected, x displays F if and only if the microscopic property $f = \varphi^{-1}(F)$ is one of the microscopic properties characterizing S^i. We are thus led to introduce the following probabilities.

$p_S^{i,d}(F)$: the probability that x be detected when F is measured on it.

$p_S^i(F)$: the conditional probability that x display F when it is detected (which is 0 or 1 because x either possesses $\varphi^{-1}(F)$ or does not possess it).

$p_S^{i,t}(F)$: the joint probability that x be detected and display F.

[6]The introduction of microscopic properties implies that the terms "possessed" and "not possessed" may be misleading if referred to macroscopic properties in our framework. Hence we use them with reference to microscopic properties only and introduce a new terminology to describe the results that are obtained when macroscopic properties are measured on physical objects.

Hence, we get

$$p_S^{i,t}(F) = p_S^{i,d}(F) p_S^i(F). \tag{6}$$

Eq. (6) is purely theoretical, since one can never know whether a physical object is in the microstate S^i. Therefore, let us consider a physical object in the state S and introduce a further conditional probability, as follows.

$p(S^i|S)$: the conditional probability that x, which is in the macroscopic state S, be in the microstate S^i.

The joint probability that x be in the state S^i, be detected and display F is thus given by $p(S^i|S)p_S^{i,t}(F)$. Hence, the overall probability $p_S^t(F)$ that x be detected and display F is given by

$$p_S^t(F) = \sum_i p(S^i|S) p_S^{i,t}(F). \tag{7}$$

Moreover, the probability $p_S^d(F)$ that x be detected when F is measured is given by

$$p_S^d(F) = \sum_i p(S^i|S) p_S^{i,d}(F). \tag{8}$$

Let us define now

$$p_S(F) = \frac{\sum_i p(S^i|S) p_S^{i,t}(F)}{\sum_i p(S^i|S) p_S^{i,d}(F)}. \tag{9}$$

Then, we get

$$p_S^t(F) = p_S^d(F) p_S(F). \tag{10}$$

Eq. (10) is the fundamental equation of the ESR model. Let us therefore discuss the two factors that appear on the right.

Let us begin with the *detection probability* $p_S^d(F)$. We are dealing here with idealized measurements, where the occurrence of the outcome a_0 is attributed only to the set of microscopic properties possessed by x, and determines the probability $p_S^{i,d}(F)$. Therefore, $p_S^{i,d}(F)$ neither depends on features of the measuring apparatus nor is influenced by the environment. Furthermore, the conditional probability $p(S^i|S)$ depends only on S. It follows that Eq. (8) implies that $p_S^d(F)$ depends only on the microscopic properties of the physical objects in S.

Let us come to $p_S(F)$. Its definition in Eq. (9) is purely formal. Yet, Eqs. (6) and (9) imply $0 \leq p_S(F) \leq 1$. Moreover, the interpretations of $p_S^t(F)$ and $p_S^d(F)$ in Eq. (10) show that $p_S(F)$ can be interpreted as the conditional probability that a physical object x display F when it is detected. We thus get an interpretation of $p_S(F)$ which provides a basis for the introduction of the main assumption of the ESR model, as follows.

Ax. 2. If S is a pure state, the probability $p_S(F)$ can be evaluated by using the same rules that yield the probability of F in the state S according to QM.

Ax. 2 implies a new interpretation of the probabilities provided by standard quantum rules, which are now considered as conditional rather than absolute.[7] The old and the new interpretation of quantum probabilities coincide if $p_S^d(F) = 1$ for every state S and

[7]Note that, if F is such that $p_S(F) = 1$, then every physical object that is in the state S and is detected necessarily possesses the microscopic property $f = \varphi^{-1}(F)$. Hence, every detected physical object in the state S possesses all microscopic properties in the set $\{f \in \mathcal{F} \mid p_S(\varphi(f)) = 1\}$.

property F. If there are states and properties such that $p_S^d(F) < 1$, instead, the difference between the two interpretations is conceptually relevant. To better grasp this difference, let us interpret probabilities as large number limits of frequencies in ensembles of physical objects,[8] and let us consider ensembles of physical objects in the state S. Then, $p_S(F)$ is the limit of the ratio between the number of objects in a given ensemble that are detected and display the property F, and the number of objects that are detected. The same quantity would be interpreted in QM as the limit of the ratio between the number of objects in a given ensemble that display the property F and the number of objects in the ensemble. Hence, the ESR model introduces a non–orthodox interpretation of quantum probabilities. But Ax. 2 implies that, as far as the probability $p_S(F)$ is concerned, the physical system Ω can be associated with a Hilbert space $\mathscr{H}$ and its pure states and macroscopic properties in $\mathcal{F}$ can be represented by means of (unit) vectors on $\mathscr{H}$ and (orthogonal) projection operators on $\mathscr{H}$, respectively.

We have discussed so far only macroscopic properties in $\mathcal{F}$. To complete our treatment, let us consider a macroscopic property $G = (A_0, Y) \in \mathcal{F}_0 \setminus \mathcal{F}$. In this case we can introduce the Borel set $X = Y \setminus \{a_0\}$ and associate the macroscopic property $F = (A_0, X) \in \mathcal{F}$ with G. We get, because of our description of idealized measurements,

$$p_S^{i,t}(G) = 1 - p_S^{i,d}(F) + p_S^{i,t}(F) \tag{11}$$

where $p_S^{i,t}(G)$ is the probability that x display the property G whenever it is in the microscopic state S^i. By using Eqs. (7)–(10) we then obtain

$$p_S^t(G) = 1 - p_S^d(F) + p_S^t(F) = 1 - p_S^d(F)(1 - p_S(F)) \tag{12}$$

where $p_S^t(G)$ is the overall probability that x display G and $p_S(F)$ can be evaluated by using standard quantum rules. Therefore, Eq. (12) provides the overall probability of a property in $\mathcal{F}_0 \setminus \mathcal{F}$ in terms of the overall probability of a property in $\mathcal{F}$.

Let us introduce now a further assumption that led to a result that is fundamental in Sec. 6.. Let us associate with F a macroscopic property $F^\perp = (A_0, \Re \setminus Y) = G^c$, and assume that $p_S^d(F^\perp) = p_S^d(F)$, which is physically reasonable because $F^\perp$ can be measured by the same apparatus measuring F, suitably modifying the outcomes. Then, obviously, $p_S(F^\perp) = 1 - p_S(F)$, hence, we get from Eq. (10)

$$p_S^t(F^\perp) = p_S^d(F^\perp)p_S(F^\perp) = p_S^d(F)(1 - p_S(F)) = p_S^d(F) - p_S^t(F). \tag{13}$$

By introducing the probability $p_S^{t,F}((A_0, \{a_0\})) = 1 - p_S^d(F)$ that x be not detected when F is measured, we finally obtain

$$p_S^{t,F}((A_0, \{a_0\})) + p_S^t(F) + p_S^t(F^\perp) = 1, \tag{14}$$

which expresses the promised result.

Let us conclude this section with two remarks.

[8]We adopt this naïve interpretation of physical probabilities here for the sake of simplicity. A more sophisticated treatment would associate quantum measurements with random variables, require that distribution functions approach experimental frequencies, etc. Our conclusions, however, would not be modified by the adoption of this more general and rigorous machinery.

Firstly, we note that Ax. 2 entails that, whenever an empirical quantum law is checked, it proves to be true for all physical objects that are detected, that implies the general principle MGP (see Sec. 2.) holds in the ESR model. On the contrary, MCP may break down, because an empirical quantum law may fail to be true in a physical situation where another empirical quantum law is checked and proven to hold. Indeed, let us consider Eq. (3). If one measures, say, the observables $A, B, \ldots$ and a physical object x is detected, then $f(a, b, \ldots) = 0$ holds because of Ax. 2, hence the empirical quantum law $f(A, B, \ldots) = 0$ is true in this physical situation. But x could possess such microscopic properties that it would not be detected if instead the observables $A', B', \ldots$ were measured. Thus, one cannot assert that the equation $g(a', b', \ldots) = 0$ also holds, which implies that $g(A', B', \ldots) = 0$ could be false, proving our previous statement (but, of course, if $A', B', \ldots$ were measured instead of $A, B, \ldots$ and x were detected, then $g(a', b', \ldots) = 0$ would hold). Bearing in mind the possible objection to our criticism of the Bell–KS theorem expounded at the end of Sec. 2., we stress that this possible failure of MCP has been obtained as a consequence of the reinterpretation of quantum probabilities as conditional rather than absolute in the ESR model, without resorting to any conspiracy of nature.

Secondly, we observe that every microscopic property f can be associated with a dichotomic *hidden variable*, which takes value 1 (0) if f is possessed (not possessed) by the physical object x that is considered. Equivalently, a microscopic state can be seen as the value of a hidden variable λ specifying all microscopic properties of x. Hence, the ESR model provides a general scheme for a hidden–variables theory that recovers the formalism of QM but is different from all previous proposals because of its reinterpretation of quantum probabilities. Besides, it must be stressed that no hidden variable associated with the measuring apparatuses occurs in this theory if idealized measurements only are considered.

4. Recovering Objectivity in the ESR Model

We intend to show in this section that the ESR model provides, by introducing microscopic properties, an intuitive set–theoretical picture of the physical world in which macroscopic properties can be considered objective. To this end, let us consider hidden–variables theories and particularize the definition of objectivity introduced in Sec. 1. as follows.

Objectivity. A hidden–variables theory is objective iff for every macroscopic property F and state S the outcome of a measurement of F on a physical object x in the state S is determined only by hidden variables associated with x.

This notion of objectivity is strictly linked with the notion of *contextuality*, which is broadly used in the literature where it occurs in two different ways.

(i) *Contextuality*$_1$. A hidden–variables theory is contextual if the outcome of a measurement of a (macroscopic) property F on a physical object x may depend on hidden variables associated with the set of (compatible) observables that are simultaneously measured on x, not only on hidden variables associated with x (standard notion, see, *e.g.*, [3, 11]).

(ii) *Contextuality*$_2$. A hidden–variables theory is contextual if the outcome of the measurement of a (macroscopic) property F on a physical object x may depend on hidden variables associated with the specific registration that is used to perform the measurement, not only on hidden variables associated with x (*hidden measurement approach*, see, *e.g.*, [47] and *probabilistic opposition*, see, *e.g.*, [43]).

These kinds of contextuality may coexist and both imply nonobjectivity; hence, conversely, objectivity implies both noncontextuality$_1$ and noncontextuality$_2$. Therefore, we use the word *objectivity* as a synonym of *noncontextuality* in the following, without specifying whether contextuality$_1$ or contextuality$_2$ is understood.

Let us come to Eq. (6). Because of the definition of $p^i_S(F)$, and because $p^{i,d}_S(F)$ only depends on microscopic properties, this equation shows that the probability $p^{i,t}_S(F)$ is determined by the value S^i of the hidden variable (equivalently, by the set of all microscopic properties that are possessed by any physical object in S^i), where it is independent of the measurement context. But, of course, this *noncontextuality of probabilities* (which holds also in QM, see *e.g.*, [11]) does not imply objectivity in the sense specified earlier. Nevertheless, whenever the ESR model is deterministic, the microscopic properties of a physical object x in S actually determine the outcome of a measurement of the macroscopic property F on x (Sec. 3.); hence, the objectivity holds in this case. If the ESR model, instead, is nondeterministic, we cannot deduce objectivity but can introduce a new assumption which implies it.

Ax. 3. For every microscopic state S^i, the probability $p^{i,d}_S(F)$ admits an epistemic *(or* ignorance*) interpretation* (see, *e.g.*, [37]) *in terms of further unknown features of the physical objects in the state S^i.*

Indeed, Ax. 3 implies that a parameter μ exists which determines, together with S^i, whether a physical object x in the state S is detected when the property F is measured on it, *i.e.*, μ and S^i determine whether the outcome a_0 occurs or not (note that μ can be interpreted as denoting a subset of further microscopic properties possessed by x, selected in a new set of microscopic properties that do not correspond to macroscopic properties via φ). Since S^i determines all macroscopic properties of x whenever x is detected, all macroscopic properties are determined by the pair (μ, S^i); thus, the ESR model is objective in the sense specified before.

Bearing in mind the second remark at the end of Sec. 3. we conclude that the scheme for a hidden–variables theory provided by the ESR model is noncontextual.[9]

Let us consider the predictions that can be obtained within the ESR model. Whenever a set of idealized measurements is performed on an ensemble of physical objects in a state S the general features of this model (in particular, Ax. 2) imply that its predictions can be partitioned in two classes.

(a) Predictions concerning the subensemble of all physical objects that are detected by the measurements. They can be obtained by using the quantum formalism (see Ax. 2). This formally coincides with the predictions of QM, but QM would interpret them as referring to the whole ensemble.

(b) Predictions concerning the whole ensemble. To get this kind of prediction, we need a mathematical representation of the new theoretical entities introduced by the ESR model

[9]Note that if the ESR model is nondeterministic and one avoids introducing Ax. 3, contextuality cannot be excluded. In this case, one can assume that the outcome of a measurement of a macroscopic property is determined by a pair (ν, S^i), where ν denotes a hidden variable (or a set of hidden variables) associated with the measurement context (*contextual ESR* model). The contextual ESR model provides a scheme for a general hidden–variables theory which embodies the quantum formalism and reinterprets quantum probabilities but preserves contextuality. It is interesting to observe that such a theory may be *local* or *nonlocal*, depending on the assumptions on the parameter ν. If it is local it converges with the proposals of the probabilistic opposition [43], if it is nonlocal it converges with the perspective of the hidden measurement approach [47].

at a macroscopic level, *i.e.*, generalized observables and related macroscopic properties (though some general predictions can already be done at this stage, as we show in the next section). This representation is introduced in Sec. 6., and its consequences are discussed in the sections that follow. It must be stressed that detection probabilities of the form $p_S^d(F)$ occur in all physical predictions obtained by using the new mathematical apparatus, and that we have at present no physical theory to allow us to predict such probabilities. Nevertheless, one can consider them as unknown parameters in the ESR model and contrive experiments to determine them,[10] then inserting the obtained values into the mathematical framework of the model to obtain numerical predictions.

It follows from (a) and (b) that the predictions of the ESR model are generally different from those of QM. Hence, the ESR model is not only *falsifiable*, but it can also be compared with QM to establish which theory is correct.

5. A New Bell's Inequality

We have seen in Sec. 4. that the ESR model is objective. We intend to discuss in this section some consequences of this crucial feature that can be obtained directly from Eq. (10), without resorting to the microscopic part of the ESR model (microscopic properties and states) nor using the mathematical representation of the new macroscopic entities introduced by the model, to be provided in the following sections.

Let us first observe that objectivity, if defined as in Sec. 4., implies *local realism* in the standard sense in the literature [4, 31, 36],[11] *i.e.*, the join of the assumptions of *realism*,

R: *the values of all observables of a physical system in a given state are predetermined for any measurement context*,

and *locality*,

LOC: *if measurements are made at places remote from one another on parts of a physical system which no longer interact, the specific features of one of the measurements do not influence the results obtained with the others.*

Because of R and LOC it seems at first sight that the BCHSH inequality must hold in the ESR model. But such a conclusion is wrong, and a different inequality holds in the ESR model. To prove this statement let us first resume the argument that leads to the BCHSH inequality. One considers a compound physical system Ω made up of two far away subsystems Ω_1 and Ω_2, and two dichotomic observables $A(\mathbf{a})$ and $B(\mathbf{b})$ of Ω_1 and Ω_2, respectively, depending on the parameters $\mathbf{a}$ and $\mathbf{b}$ and taking either value -1 or $+1$. Then,

[10] A major difficulty when performing an experiment for determining $p_S^d(F)$ is counting the physical objects that are actually prepared in the state S, even if they are not detected by the measurement of F. Furthermore, if the intrinsic lack of efficiency of any measuring device is schematized by multiplying $p_S^d(F)$ by a factor k, with $0 \leq k \leq 1$, it may be difficult to distinguish empirically k from $p_S^d(F)$.

[11] The use of the term *local realism* in the context of Bell's Theorem has been recently disputed [48]. We add that the meaning of the word "realism" in this term does not match the meaning of the word "realistic" used by Busch *et al.* [1] and mentioned in Sec. 1.. Nevertheless, we will not break with the standard language here because definition R is widely used in the literature and has a precise meaning when translated in terms of hidden variables, even thought it does not fit in well with traditional notions of realism adopted by philosophers (see in particular [49] for an exhaustive review on existing approaches and experimental tests of local realism).

one defines a correlation function

$$E(\mathbf{a},\mathbf{b})=\int_{\Lambda} d\lambda \rho(\lambda) A(\lambda,\mathbf{a}) B(\lambda,\mathbf{b}), \tag{15}$$

where λ is a deterministic *hidden variable* whose value ranges over a domain Λ when measurements on different examples of Ω in a given state S are considered, $\rho(\lambda)$ is a probability distribution on Λ, $A(\lambda,\mathbf{a})$ and $B(\lambda,\mathbf{b})$ are the values of $A(\mathbf{a})$ and $B(\mathbf{b})$, respectively. By using Eq. (15), R and LOC one gets the *BCHSH inequality*

$$|E(\mathbf{a},\mathbf{b})-E(\mathbf{a},\mathbf{b}')|+|E(\mathbf{a}',\mathbf{b})+E(\mathbf{a}',\mathbf{b}')|\leq 2. \tag{16}$$

Let us come to the ESR model. Here, the no–registration outcome occurs in every generalized observable that is considered, where the dichotomic observables $A(\mathbf{a})$, $B(\mathbf{b})$, $A(\mathbf{a}')$ and $B(\mathbf{b}')$ must be substituted by the trichotomic observables $A_0(\mathbf{a})$, $B_0(\mathbf{b})$, $A_0(\mathbf{a}')$ and $B_0(\mathbf{b}')$, respectively, in each of which a no–registration outcome is added to the outcomes $+1$ and -1. Let us agree to consider trichotomic observables such that all no–registration outcomes coincide with 0 (which is not restrictive, see footnote 13). Then, let us recall that the range of values of the hidden variable λ is the subset of all microscopic states associated with the macroscopic state S (see Sec. 3.). Hence the correlation function in Eq. (15) must be substituted by the expectation value of the product of the trichotomic generalized observables $A_0(\mathbf{a})$ and $B_0(\mathbf{b})$

$$E(A_0(\mathbf{a}),B_0(\mathbf{b}))=\sum_i p(S^i|S)A_0(S^i,\mathbf{a})B_0(S^i,\mathbf{b}), \tag{17}$$

where $A_0(S^i,\mathbf{a})$ and $B_0(S^i,\mathbf{b})$ denote the values of $A_0(\mathbf{a})$ and $B_0(\mathbf{b})$, respectively, when the hidden variable takes value S^i (which implies that we must consider an ESR model which is deterministic in the sense explained in Sec. 3.). We can now follow the standard procedures leading to Eq. (16). Since $|A_0(S^i,\mathbf{a})|\leq 1$ we get

$$|E(A_0(\mathbf{a}),B_0(\mathbf{b}))-E(A_0(\mathbf{a}),B_0(\mathbf{b}'))|\leq\sum_i p(S^i|S)|B_0(S^i,\mathbf{b})-B_0(S^i,\mathbf{b}')| \tag{18}$$

and, similarly,

$$|E(A_0(\mathbf{a}'),B_0(\mathbf{b}))+E(A_0(\mathbf{a}'),B_0(\mathbf{b}'))|\leq\sum_i p(S^i|S)|B_0(S^i,\mathbf{b})+B_0(S^i,\mathbf{b}')|. \tag{19}$$

Now, we have

$$|B_0(S^i,\mathbf{b})-B_0(S^i,\mathbf{b}')|+|B_0(S^i,\mathbf{b})+B_0(S^i,\mathbf{b}')|\leq 2 \tag{20}$$

and

$$\sum_i p(S^i|S)=1, \tag{21}$$

hence we obtain the *modified BCHSH inequality*

$$|E(A_0(\mathbf{a}),B_0(\mathbf{b}))-E(A_0(\mathbf{a}),B_0(\mathbf{b}'))|+|E(A_0(\mathbf{a}'),B_0(\mathbf{b}))+E(A_0(\mathbf{a}'),B_0(\mathbf{b}'))|\leq 2, \tag{22}$$

which replaces Eq. (16) in the ESR model [24, 26, 27].

We have thus proven our statement. It remains to compare the orthodox position about the BCHSH inequality with the interpretation of the modified BCHSH inequality in the conceptual framework of the ESR model.

According to the orthodox position, all terms in the sum at the left in the BCHSH inequality are expectation values of products of compatible observables that can be calculated by using standard QM rules. If one then puts them in the BCHSH inequality, one sees that there are choices of the physical system Ω, state S and parameters $\mathbf{a}$, $\mathbf{a}'$, $\mathbf{b}$, $\mathbf{b}'$ such that the inequality is violated. The seemingly unavoidable conclusion is that the assumptions from which the inequality is deduced, that is, R and LOC, are not consistent with QM. One can then make experimental tests to check whether the BCHSH inequality holds or the predictions of QM are correct. Most physicists then maintain that the experimental data that have been obtained [50, 51, 52, 53] show that the BCHSH inequality is violated and confirm QM. Hence, local realism is rejected.

We have criticized this conclusion in previous papers [13, 14, 23], based on the same argument that has been used to criticize the proofs of the Bell–KS theorem in Sec. 2.. One can indeed observe that all proofs of the BCHSH inequality are based on the premise to implicitly assume the problematic principle MCP, while no contradiction between local realism and QM can be proven if the weaker principle MGP is adopted. It follows, in particular, that we expect no contradiction between the modified BCHSH inequality and the reinterpreted mathematical apparatus of QM which is embodied in the ESR model, where MGP holds, while MCP may break down (see the first remark at the end of Sec. 2.). In the case of the ESR model, in addition, we can describe in more details how objectivity can coexist with the mathematical apparatus of QM. This topic is discussed in Sec. 8., after introducing a mathematical representation of generalized observables and a generalization of the projection postulate of QM. We can, however, present here an intuitive anticipation of our results. To this end, observe that the terms in the sum at the left in the modified BCHSH inequality are expectation values of products of generalized observables that cannot be calculated by using standard QM rules (they refer indeed to all physical objects that are prepared in the state S, while quantum rules refer to detected physical objects only according to the ESR model, see Sec. 3.). Hence the modified BCHSH inequality provides conditions on the values of the detection probabilities rather than proving a contradiction between local realism and the mathematical formalism of QM embodied in the ESR model. Of course, these conditions could be experimentally tested, and a violation of them would falsify the ESR model.

Let us close this section with a comparison between some arguments of the minority of scholars defending local realism against seeming experimental evidence and the point of view introduced by the ESR model. To this end, let us briefly write BI in place of "BCHSH inequality" and note that this argument showing that QM conflicts with local realism can be schematized, by using standard logical symbols, as follows.

$$R \wedge LOC \Longrightarrow BI,$$
$$QM \Longrightarrow \neg BI,$$
$$QM \Longrightarrow \neg R \vee \neg LOC.$$

Scholars aiming to defend local realism do not reject this scheme and accept that QM conflicts with local realism, but uphold that experimental evidence does not allow one to make a choice between these issues. They stress indeed that the measurements that can actually be performed have low efficiencies, hence the BCHSH inequality cannot be tested directly. Empirical tests refer to derived inequalities (BI^*) that are obtained by adding additional assumptions (AA) to R and LOC. Since experimental data (ED) show that BI^* are violated, one can write, by using standard logical symbols, the following sequence of implications.

$$
\begin{aligned}
R \wedge LOC \wedge AA &\Longrightarrow BI^*, \\
ED &\Longrightarrow \neg BI^*, \\
ED &\Longrightarrow \neg(R \wedge LOC \wedge AA), \\
ED &\Longrightarrow \neg(R \wedge LOC) \vee (\neg AA).
\end{aligned}
$$

The last implication shows that the experimental data proving that the derived inequalities are violated do not disprove the BCHSH inequality, hence local realism, for they could simply disprove the additional assumptions that have been introduced to obtain such inequalities [29, 54, 55, 56, 57, 58, 59, 60].[12]

The arguments led instead to the unified perspective expressed by the ESR model follow a different logical scheme. They start from the remark that the proof of the conflict between QM and local realism implicitly assumes, besides R and LOC, also the general principle MCP (which follows from considering quantum probabilities as absolute, see Sec. 3.). Here its scheme should be rewritten as follows.

$$
\begin{aligned}
R \wedge LOC \wedge MCP &\Longrightarrow BI, \\
QM &\Longrightarrow \neg BI, \\
QM &\Longrightarrow \neg(R \wedge LOC) \vee \neg MCP.
\end{aligned}
$$

The last implication shows that QM does not necessarily contradict local realism, for it could simply imply that MCP must be rejected. This is just what occurs in the ESR model, where the reinterpretation of quantum probabilities as conditional instead of absolute implies that MCP may break down, while MGP holds (see the first remark at the end of Sec. 3.). More specifically, the ESR model predicts that the experimental data, taking into account only those physical objects that are detected, must match the same predictions made by QM with reference to the set of all physical objects that are prepared. Although these predictions do not conflict with local realism, which would instead be challenged if the modified BCHSH inequality were violated.

[12]This argument, of course, does not imply that the experimental data may be consistent with local realism in all performed experiments, nor explains why the ED match the predictions of QM. Therefore, the foregoing reasonings are usually completed by denying the additional assumptions, introducing some specific hypotheses (SH) and showing that

$$(R \wedge LOC) \wedge (\neg AA) \wedge SH \Longrightarrow ED.$$

6. A Mathematical Representation of Observables and Properties

It follows from Ax. 2 in Sec. 3. that, for every pure state S, the probability $p_S(F)$ in Eq. (10) can be evaluated by using the formalism of QM. As far as $p_S(F)$ is concerned, the physical system Ω can therefore be associated with a (separable) complex Hilbert space $\mathscr{H}$, every pure state S of Ω can be represented by a unit vector $|\psi\rangle \in \mathscr{H}$ or by a one–dimensional (orthogonal) projection operator $\rho_\psi = |\psi\rangle\langle\psi|$ on $\mathscr{H}$, and every $F \in \mathcal{F}$ can be represented by an (orthogonal) projection operator on $\mathscr{H}$. The probability $p_S^d(F)$ (hence, $p_S^t(F)$) cannot be obtained instead by using quantum rules, and we have as yet no theory which allow us to predict it, even if one can try to contrive experiments to determine it empirically (see Sec. 4.). Nevertheless, one can provide a mathematical expression for $p_S^t(F)$, hence a mathematical representation of the generalized observables and macroscopic properties introduced by the ESR model, by considering $p_S^d(F)$ as an unknown parameter [28, 32, 33]. We intend to supply a synthetic approach to this topic in the present section mainly following our treatment in [34].

Let A be an observable of QM, let $\Xi \subset \Re$ be the set of its possible outcomes and let A_0 be the generalized observable obtained from A, whose set of possible outcomes is $\{a_0\} \cup \Xi$ (see footnote 5). Furthermore, let $\widehat{A}$ be the self–adjoint operator representing A (the spectrum of which obviously coincides with Ξ) and let $P^{\widehat{A}}$ be the projection valued (PV) measure associated with $\widehat{A}$ by the spectral theorem,

$$P^{\widehat{A}} : X \in \mathbb{B}(\Re) \longmapsto P^{\widehat{A}}(X) \in \mathscr{L}(\mathscr{H}), \tag{23}$$

where $\mathscr{L}(\mathscr{H})$ is the set of all orthogonal projection operators on $\mathscr{H}$ (hence $\widehat{A} = \int_\Re \lambda \mathrm{d}P_\lambda^{\widehat{A}}$, $\int_\Re \mathrm{d}P_\lambda^{\widehat{A}} = I$, and, for every $X \in \mathbb{B}(\Re)$, $P^{\widehat{A}}(X) = \int_X \mathrm{d}P_\lambda^{\widehat{A}}$). Measuring A_0 is then equivalent to measuring all macroscopic properties of the form $F = (A_0, X)$, with $X \in \mathbb{B}(\Re)$, simultaneously. In particular, if one considers an interval $\mathrm{d}\lambda$, with $a_0 \notin \mathrm{d}\lambda$, and the infinitesimal overall probability $\mathrm{d}p_S^t$ that an idealized measurement of A_0 on a physical object x in a pure state S represented by the unit vector $|\psi\rangle \in \mathscr{H}$ yield an outcome in $\mathrm{d}\lambda$, Eq. (10) and Ax. 2 in Sec. 3. naturally led to assume that

$$\mathrm{d}p_S^t = p_\psi^d(\widehat{A}, \lambda)\langle\psi|\mathrm{d}P_\lambda^{\widehat{A}}|\psi\rangle, \tag{24}$$

where $p_\psi^d(\widehat{A}, \lambda)$ is a detection probability such that $\langle\psi|p_\psi^d(\widehat{A}, \lambda)\frac{dP_\lambda^{\widehat{A}}}{d\lambda}|\psi\rangle$ is a measurable function on $\Re$. Let $X \in \mathbb{B}(\Re)$. Whenever $a_0 \notin X$, Eq. (24) implies

$$p_S^t((A_0, X)) = \langle\psi| \int_X p_\psi^d(\widehat{A}, \lambda)\mathrm{d}P_\lambda^{\widehat{A}}|\psi\rangle. \tag{25}$$

Furthermore, putting $F = (A_0, X)$, Eq. (14) can be written as follows,

$$p_S^{t,F}((A_0, \{a_0\})) + \langle\psi| \int_X p_\psi^d(\widehat{A}, \lambda)\mathrm{d}P_\lambda^{\widehat{A}}|\psi\rangle + \langle\psi| \int_{\Re\setminus(X\cup\{a_0\})} p_\psi^d(\widehat{A}, \lambda)\mathrm{d}P_\lambda^{\widehat{A}}|\psi\rangle = 1. \tag{26}$$

Since $a_0 \notin \Xi$, we get, choosing $X = \emptyset$,

$$p_S^t((A_0, \{a_0\})) + \langle \psi | \int_{\Re} p_\psi^d(\widehat{A}, \lambda) \mathrm{d}P_\lambda^{\widehat{A}} | \psi \rangle = 1, \quad (27)$$

where $p_S^t((A_0, \{a_0\}))$ can be interpreted as the overall probability that the measurement of A_0 yield the a_0 outcome. Bearing in mind this interpretation we get, whenever $a_0 \in X$,

$$p_S^t((A_0, X)) = p_S^t((A_0, \{a_0\})) + p_S^t((A_0, X \setminus \{a_0\})), \quad (28)$$

hence, by using Eqs. (25) and (27),

$$p_S^t((A_0, X)) = \langle \psi | (I - \int_{\Re \setminus X} p_\psi^d(\widehat{A}, \lambda) \mathrm{d}P_\lambda^{\widehat{A}}) | \psi \rangle. \quad (29)$$

It follows at once from Eqs. (25) and (29) that $p_S^t((A_0, \cdot))$ is a probability measure on the σ–algebra $\mathbb{B}(\Re)$ of all Borel subsets of $\Re$.

Because of Eqs. (25) and (29) one can introduce, for every unit vector $|\psi\rangle \in \mathscr{H}$, a mapping

$$T_\psi^{\widehat{A}} : X \in \mathbb{B}(\Re) \longmapsto T_\psi^{\widehat{A}}(X) \in \mathscr{B}(\mathscr{H}), \quad (30)$$

where $\mathscr{B}(\mathscr{H})$ is the set of all bounded operators on $\mathscr{H}$, defined by setting

$$T_\psi^{\widehat{A}}(X) = \begin{cases} \int_X p_\psi^d(\widehat{A}, \lambda) \mathrm{d}P_\lambda^{\widehat{A}} & \text{if } a_0 \notin X \\ I - \int_{\Re \setminus X} p_\psi^d(\widehat{A}, \lambda) \mathrm{d}P_\lambda^{\widehat{A}} & \text{if } a_0 \in X \end{cases}. \quad (31)$$

It follows from Eq. (31) that $T_\psi^{\widehat{A}}$ is a POV measure on $\mathbb{B}(\Re)$ which is *commutative, i.e.*, for every $X, Y \in \mathbb{B}(\Re)$, $T_\psi^{\widehat{A}}(X) T_\psi^{\widehat{A}}(Y) = T_\psi^{\widehat{A}}(Y) T_\psi^{\widehat{A}}(X)$ [1]. Hence, the discrete generalized observable A_0 can be represented by the *family of commutative POV measures*

$$\left\{ T_\psi^{\widehat{A}} : X \in \mathbb{B}(\Re) \longmapsto T_\psi^{\widehat{A}}(X) \in \mathscr{B}(\mathscr{H}) \right\}_{\| |\psi\rangle \| = 1}. \quad (32)$$

Indeed, bearing in mind Eqs. (25) and (29), one gets that the probability that the outcome of a measurement of A_0 on a physical object x in the pure state S represented by the unit vector $|\psi\rangle$ lie in the Borel set X is given by

$$p_S^t((A_0, X)) = \langle \psi | T_\psi^{\widehat{A}}(X) | \psi \rangle. \quad (33)$$

Equivalently, one gets

$$p_S^t((A_0, X)) = Tr[\rho_\psi T_\psi^{\widehat{A}}(X)] \quad (34)$$

if S is represented by the one–dimensional projection operator ρ_ψ.

The representation of generalized observables provided by Eq. (32) immediately induces a representation of macroscopic properties. Indeed, every $F = (A_0, X) \in \mathcal{F}_0$ can be associated with the family $\{T_\psi^{\widehat{A}}(X)\}_{\| |\psi\rangle \| = 1}$ of bounded positive operators (*effects*) [1, 6, 7] in the ESR model.

We have thus obtained a mathematical representation of the macroscopic entities that have been introduced in Sec. 3.. To relate this representation with the theoretical framework in Sec. 3. let us consider the property $F = (A_0, X)$, with $a_0 \notin X$. We get from Eq. (10) and Ax. 2

$$p_S^t(F) = p_S^d(F)\langle\psi| \int_X \mathrm{d}P_\lambda^{\widehat{A}}|\psi\rangle, \tag{35}$$

while Eqs. (31) and (33) yield

$$p_S^t(F) = \langle\psi| \int_X p_\psi^d(\widehat{A}, \lambda)\mathrm{d}P_\lambda^{\widehat{A}}|\psi\rangle\ , \tag{36}$$

hence

$$p_S^d(F) = \frac{\langle\psi| \int_X p_\psi^d(\widehat{A}, \lambda)\mathrm{d}P_\lambda^{\widehat{A}}|\psi\rangle}{\langle\psi| \int_X \mathrm{d}P_\lambda^{\widehat{A}}|\psi\rangle}, \tag{37}$$

or, equivalently,

$$p_S^d(F) = \frac{Tr[\rho_\psi T_\psi^{\widehat{A}}(X)]}{Tr[\rho_\psi P^{\widehat{A}}(X)]}. \tag{38}$$

Whenever $p_\psi^d(\widehat{A}, \lambda)$ does not depend on λ, Eq. (37) implies that $p_S^d(F)$ depends on A_0 but not on X, which is the case studied in [28, 33]. More generally, Eq. (37) (equivalently, Eq. (38)) establishes a relation among detection probabilities which is a direct consequence of Eq. (27). We note explicitly that it entails, for every $|\psi\rangle \in \mathscr{H}$,

$$\langle\psi| \int_X (p_S^d(F) - p_\psi^d(\widehat{A}, \lambda))\mathrm{d}P_\lambda^{\widehat{A}}|\psi\rangle = 0, \tag{39}$$

which does not imply $p_S^d(F) - p_\psi^d(\widehat{A}, \lambda) = 0$, because $p_S^d(F) - p_\psi^d(\widehat{A}, \lambda)$ generally depends on $|\psi\rangle$.

It is interesting to compare the representation of generalized observables in the ESR model with the representation of observables in unsharp QM [1, 5, 6, 7, 46]. It follows from Eq. (32) that there are two basic differences.

(i) A generalized observable in the ESR model is represented by a family of POV measures parametrized by the set of all vectors representing pure states, while an observable of unsharp QM is represented by a single POV measure.

(ii) Only commutative POV measures appear in the representation of a generalized observable, while the POV measure representing an observable of unsharp QM may be noncommutative.

Difference (i) is relevant since it shows that the generalized observables introduced by the ESR model do not coincide, in general, with the observables introduced by unsharp QM. This can be intuitively explained by reminding that the occurrence of the no–registration outcome, hence the detection probabilities, is assumed to depend on intrinsic features of the physical object that is considered, while it neither depends on the measuring apparatus nor it has an unsharp source (see Sec. 3.). Of course this assumption is introduced to recover objectivity of macroscopic properties, avoiding in particular the objectification problem, which remains instead unsolved in unsharp QM [8, 9].

Difference (ii) seems less relevant, because the ESR model considers only idealized measurements, which correspond to sharp measurements in unsharp QM. One could indeed devise an unsharp extension of the ESR model by introducing unsharp generalized observables represented by families of noncommutative POV measures.

To conclude this section, let us illustrate our results by considering a special case.

Let A be a discrete observable of QM, let $\Xi = \{a_1, a_2, \ldots\}$ be the set of all its possible outcomes, and let A_0 be a generalized observable obtained from A, with set of possible outcomes $\Xi_0 = \{a_0\} \cup \{a_1, a_2, \ldots\}$. We denote by $\widehat{A}$ the self–adjoint operator representing A, and by $P_1^{\widehat{A}}$, $P_2^{\widehat{A}}$, ...the (orthogonal) projection operators associated with a_1, a_2, ..., respectively, by the spectral decomposition of $\widehat{A}$. For the sake of brevity we also put, for every $n \in \mathbb{N}$, $p_{\psi n}^d(\widehat{A}) \equiv p_\psi^d(\widehat{A}, a_n)$. We then get from Eq. (31)

$$T_\psi^{\widehat{A}}(X) = \begin{cases} \sum_{n, a_n \in X} p_{\psi n}^d(\widehat{A}) P_n^{\widehat{A}} & \text{if } a_0 \notin X \\ I - \sum_{n, a_n \in \Re \setminus X} p_{\psi n}^d(\widehat{A}) P_n^{\widehat{A}} & \text{if } a_0 \in X \end{cases} . \tag{40}$$

Thus, the probability that a measurement of A_0 on a physical object x in the pure state S represented by the unit vector $|\psi\rangle$ yield an outcome in $X \in \mathbb{B}(\Re)$ is

$$p_S^t((A_0, X)) = \begin{cases} \langle\psi| \sum_{n, a_n \in X} p_{\psi n}^d(\widehat{A}) P_n^{\widehat{A}} |\psi\rangle & \text{if } a_0 \notin X \\ \langle\psi|(I - \sum_{n, a_n \in \Re \setminus X} p_{\psi n}^d(\widehat{A}) P_n^{\widehat{A}})|\psi\rangle & \text{if } a_0 \in X \end{cases} , \tag{41}$$

consistently with Eq. (33). Let $X = \{a_n\}$, with $n \in \mathbb{N}_0$. Then Eq. (40) yields

$$T_\psi^{\widehat{A}}(\{a_n\}) = \begin{cases} p_{\psi n}^d(\widehat{A}) P_n^{\widehat{A}} & \text{if } n \neq 0 \\ \sum_{m \in \mathbb{N}} (1 - p_{\psi m}^d(\widehat{A})) P_m^{\widehat{A}} & \text{if } n = 0 \end{cases} . \tag{42}$$

Furthermore, if we put $F_n = (A_0, \{a_n\})$, Eq. (33) yields

$$p_S^t(F_n) = \begin{cases} p_{\psi n}^d(\widehat{A}) \langle\psi|P_n^{\widehat{A}}|\psi\rangle & \text{if } n \neq 0 \\ \sum_{m \in \mathbb{N}} (1 - p_{\psi m}^d(\widehat{A})) \langle\psi|P_m^{\widehat{A}}|\psi\rangle & \text{if } n = 0 \end{cases} . \tag{43}$$

7. The Generalized Projection Postulate

The mathematical representations of generalized observables and macroscopic properties provided in Sec. 6. are strictly connected with the state transformations induced by measurements of macroscopic properties. Indeed, if a *nondestructive* idealized measurement is performed on a physical object x in a pure state S, x is detected and a sharp value of a discrete observable is obtained, then Ax. 2 suggests assuming that S be modified according to standard QM rules. This assumption, together with the results obtained in Sec. 6., supports the introduction of the following *generalized projection postulate*.

GPP. Let S be a pure state represented by the unit vector $|\psi\rangle$ or, equivalently, by the density operator $\rho_\psi = |\psi\rangle\langle\psi|$, and let a nondestructive idealized measurement of a macroscopic property $F = (A_0, X) \in \mathcal{F}_0$ be performed on a physical object x in the state S.

Let the measurement yield the yes outcome. Then, the state S_F of x after the measurement is a pure state represented by the unit vector

$$|\psi_F\rangle = \frac{T_\psi^{\widehat{A}}(X)|\psi\rangle}{\sqrt{\langle\psi|T_\psi^{\widehat{A}\dagger}(X) T_\psi^{\widehat{A}}(X)|\psi\rangle}} , \tag{44}$$

or, equivalently, by the density operator

$$\rho_{\psi_F} = \frac{T_\psi^{\widehat{A}}(X)\rho_\psi T_\psi^{\widehat{A}\dagger}(X)}{Tr[T_\psi^{\widehat{A}}(X)\rho_\psi T_\psi^{\widehat{A}\dagger}(X)]} \,. \tag{45}$$

Let the measurement yield the no outcome. Then, the state S'_F of x after the measurement is a pure state represented by the unit vector

$$|\psi'_F\rangle = \frac{T_\psi^{\widehat{A}}(\Re \setminus X)|\psi\rangle}{\sqrt{\langle\psi|T_\psi^{\widehat{A}\dagger}(\Re \setminus X)T_\psi^{\widehat{A}}(\Re \setminus X)|\psi\rangle}} \,, \tag{46}$$

or, equivalently, by the density operator

$$\rho_{\psi'_F} = \frac{T_\psi^{\widehat{A}}(\Re \setminus X)\rho_\psi T_\psi^{\widehat{A}\dagger}(\Re \setminus X)}{Tr[T_\psi^{\widehat{A}}(\Re \setminus X)\rho_\psi T_\psi^{\widehat{A}\dagger}(\Re \setminus X)]} \,. \tag{47}$$

GPP replaces the projection postulate of QM [37] introducing two basic changes.

(i) The operator $T_\psi^{\widehat{A}}(X)$ that depends on $|\psi\rangle$ replaces the projection operator which appears in the projection postulate and does not depend on $|\psi\rangle$.

(ii) The terms in the denominators in Eqs. (44)–(47) do not coincide with the probabilities of the yes and no outcomes, respectively (see Eqs. (33) and (34)).

To illustrate GPP, let us consider the special case of a discrete generalized observable discussed at the end of Sec. 6.. Let the property F_n be measured. Whenever the yes outcome is obtained, Eq. (44) yields, because of Eq. (42),

$$|\psi_{F_n}\rangle = \begin{cases} \frac{P_n^{\widehat{A}}|\psi\rangle}{\sqrt{\langle\psi|P_n^{\widehat{A}}|\psi\rangle}} & \text{if } n \neq 0 \\ \frac{\sum_{m\in\mathbb{N}}(1-p_{\psi m}^d(\widehat{A}))P_m^{\widehat{A}}|\psi\rangle}{\sqrt{\sum_{m\in\mathbb{N}}(1-p_{\psi m}^d(\widehat{A}))^2\|P_m^{\widehat{A}}|\psi\rangle\|^2}} & \text{if } n = 0 \end{cases} \,. \tag{48}$$

Whenever the no outcome is obtained, Eq. (46) yields, because of Eq. (40),

$$|\psi'_{F_n}\rangle = \frac{T_\psi^{\widehat{A}}(\Re \setminus \{a_n\})|\psi\rangle}{\sqrt{\langle\psi|T_\psi^{\widehat{A}\dagger}(\Re \setminus \{a_n\})T_\psi^{\widehat{A}}(\Re \setminus \{a_n\})|\psi\rangle}} =$$

$$= \begin{cases} \frac{(I-p_{\psi n}^d(\widehat{A})P_n^{\widehat{A}})|\psi\rangle}{\sqrt{\sum_{m\in\mathbb{N}\setminus\{n\}}\|P_m^{\widehat{A}}|\psi\rangle\|^2+(1-p_{\psi n}^d(\widehat{A}))^2\|P_n^{\widehat{A}}|\psi\rangle\|^2}} & \text{if } n \neq 0 \\ \frac{\sum_{m\in\mathbb{N}}p_{\psi m}^d(\widehat{A})P_m^{\widehat{A}}|\psi\rangle}{\sqrt{\sum_{m\in\mathbb{N}}(p_{\psi m}^d(\widehat{A}))^2\|P_m^{\widehat{A}}|\psi\rangle\|^2}} & \text{if } n = 0 \end{cases} \,. \tag{49}$$

If $n \neq 0$, Eq. (48) is consistent with the assumption at the beginning of this section. If $n = 0$, Eq. (48) shows that the final state after the measurement of F_0 does not coincide in general with the initial state, though this may occur for special classes of generalized observables [28, 33]. Moreover, when a measurement of the discrete generalized observable A_0 is performed on a physical object x in the pure state S represented by the unit vector

$|\psi\rangle$, a natural extension of GPP consists in assuming that, if the outcome a_n is obtained and the measurement is idealized and nondestructive, the final state of x is given by Eq. (48). It is then natural to wonder whether such a measurement can be classified as an *ideal measurement of the first kind* according to standard definitions in QM [37, 61, 62]. Let us therefore suppose that a first measurement of A_0 is performed on x in the state S and then repeated on x in the final state. If the first measurement yields outcome $a_n \neq a_0$, the second measurement could yield a_n as well as a_0. If the first measurement yields outcome a_0, the second measurement could yield a_0 or any $a_m \neq a_0$ such that $(1 - p^d_{\psi m}(\widehat{A}))P^{\widehat{A}}_m|\psi\rangle \neq 0$. Strictly speaking, the measurement is not a first kind measurement. Since, however, if the first measurement yields outcome $a_n \neq a_0$, the second measurement can never yield outcome a_m, with $0 \neq m \neq n$, we say that our measurement is a *generalized* measurement of the first kind. Moreover, the outcome of the measurement determines the final state of the physical object x, hence we also say that the measurement is a *generalized* ideal measurement. Summarizing, we say that the nondestructive idealized measurement that we are considering is a *generalized ideal measurement of the first kind*.

8. ESR's Model versus QM's Predictions

We have already seen in Sec. 6. that, if one describes the physical situation that led to the BCHSH inequality from the point of view of the ESR model, then the BCHSH inequality must be substituted by a modified BCHSH inequality and no conflict between local realism and reinterpreted quantum predictions occurs. We intend to deepen this topic in the present section by using the mathematical apparatus presented in Secs. 6. and 7..

Let us introduce some preliminary technical remarks on joint measurements of discrete generalized observables in the ESR model.

Let A_0 be a discrete generalized observable. By using the symbols introduced at the end of Sec. 6., the *expectation value* $\langle A_0\rangle_S$ of A_0 in the pure state S represented by the unit vector $|\psi\rangle$ can be written as follows

$$\langle A_0\rangle_S = \sum_{n\in\mathbb{N}_0} a_n p^t_S(F_n) \tag{50}$$

and can be evaluated by using Eq. (43)

$$\langle A_0\rangle_S = \sum_{n\in\mathbb{N}_0} a_n\langle\psi|T^{\widehat{A}}_\psi(\{a_n\})|\psi\rangle = a_0 + \sum_{n\in\mathbb{N}}(a_n - a_0)p^d_{\psi n}(\widehat{A})\langle\psi|P^{\widehat{A}}_n|\psi\rangle\,. \tag{51}$$

Let us consider another discrete observable B of QM represented by the self–adjoint operator $\widehat{B}$, let $\{b_1, b_2, \ldots\}$ be the set of all its possible outcomes, and let B_0 be a generalized observable obtained from B, with set of possible outcomes $\{b_0\} \cup \{b_1, b_2, \ldots\}$. Let us assume that nondestructive idealized measurements of A_0 and B_0 are performed. By using GPP we can calculate the probability $p^t_S(a_n, b_p)$ (with $n, p \in \mathbb{N}_0$) of obtaining the pair of outcomes (a_n, b_p) when first measuring A_0 and then B_0 on a physical object x in the state S. We get

$$p^t_S(a_n, b_p) = \langle\psi|T^{\widehat{A}}_\psi(\{a_n\})|\psi\rangle\langle\psi_{F_n}|T^{\widehat{B}}_{\psi_{F_n}}(\{b_p\})|\psi_{F_n}\rangle, \tag{52}$$

where $|\psi_{F_n}\rangle$ is given by Eq. (48) and $T^{\widehat{B}}_{\psi_{F_n}}(\{b_p\})$ follows from Eq. (42), with obvious substitutions. Whenever $n \neq 0 \neq p$, Eq. (52) yields

$$p^t_S(a_n, b_p) = p^d_{\psi n}(\widehat{A}) p^d_{\psi_{F_n} p}(\widehat{B}) \langle \psi | P^{\widehat{A}}_n P^{\widehat{B}}_p P^{\widehat{A}}_n | \psi \rangle \,. \tag{53}$$

Let now Ω be a compound system made up of two subsystems Ω_1 and Ω_2 that are associated in QM with the Hilbert spaces $\mathscr{H}_1$ and $\mathscr{H}_2$, respectively, so that Ω is associated with the Hilbert space $\mathscr{H} = \mathscr{H}_1 \otimes \mathscr{H}_2$.

Let $A(1)$ be a discrete quantum observable of Ω_1, with set of possible outcomes $\Xi_1 = \{a_1, a_2, \ldots\}$, represented by the self–adjoint operator $\widehat{A}(1)$ on $\mathscr{H}_1$, and let $B(2)$ be a discrete quantum observable of Ω_2, with set of possible outcomes $\Xi_2 = \{b_1, b_2, \ldots\}$, represented by the self–adjoint operator $\widehat{B}(2)$ on $\mathscr{H}_2$. When considered as observables of Ω, $A(1)$ and $B(2)$ are represented in QM by the self–adjoint operators $\widehat{A}(1) \otimes I(2)$ and $I(1) \otimes \widehat{B}(2)$, respectively, where $I(2)$ and $I(1)$ are the identity operators on $\mathscr{H}_2$ and $\mathscr{H}_1$, respectively, that are understood in the following, for the sake of simplicity. Let $A_0(1)$ be a generalized observable obtained from $A(1)$ by adding the no–registration outcome a_0 to Ξ_1, and, similarly, let $B_0(2)$ be a generalized observable obtained from $B(2)$ by adding the no–registration outcome b_0 to Ξ_2. Whenever simultaneous measurements of $A_0(1)$ and $B_0(2)$ are performed on a physical object x (individual example of Ω) in a pure state S such that Ω_1 and Ω_2 are spatially separated, noncontextuality implies that the transformation of S induced by the measurement of $A_0(1)$ must not affect the detection probability associated with the measurement of $B_0(2)$. If S is represented by the unit vector $|\Psi\rangle \in \mathscr{H}$, we obtain

$$p^d_{\Psi_{F_n} p}(\widehat{B}(2)) = p^d_{\Psi p}(\widehat{B}(2)), \tag{54}$$

hence, Eq. (53) yields

$$p^t_S(a_n, b_p) = p^d_{\Psi n}(\widehat{A}(1)) p^d_{\Psi p}(\widehat{B}(2)) \langle \Psi | P^{\widehat{A}(1)}_n P^{\widehat{B}(2)}_p | \Psi \rangle \,, \tag{55}$$

because $P^{\widehat{A}(1)}_n$ and $P^{\widehat{B}(2)}_p$ commute.

Let us consider the expectation value of the product of the generalized observables $A_0(1)$ and $B_0(2)$ in the state S, defined as follows,

$$E(A_0(1), B_0(2)) = \sum_{n,p \in \mathbb{N}} a_n b_p p^t_S(a_n, b_p) +$$

$$+ \sum_{n \in \mathbb{N}} a_n b_0 p^t_S(a_n, b_0) + \sum_{p \in \mathbb{N}} a_0 b_p p^t_S(a_0, b_p) + a_0 b_0 p^t_S(a_0, b_0). \tag{56}$$

By using Eq. (55) and restricting to generalized observables such that $a_0 = 0 = b_0$[13] (hence, for every $n, p \in \mathbb{N}$, $a_n \neq 0 \neq b_p$) we get

$$E(A_0(1), B_0(2)) = \sum_{n,p \in \mathbb{N}} a_n b_p p^d_{\Psi n}(\widehat{A}(1)) p^d_{\Psi p}(\widehat{B}(2)) \langle \Psi | P^{\widehat{A}(1)}_n P^{\widehat{B}(2)}_p | \Psi \rangle \,. \tag{57}$$

[13]Note that, for every generalized observable A_0, with $a_0 \neq 0$, one can construct a new observable whose no–registration outcome is 0. Indeed, one can select a Borel function on $\Re$ which is bijective on Ξ_0 and such that $\chi(a_0) = 0$, and consider the generalized observable $\chi(A_0)$ obtained from $\chi(A)$ by adjoining the outcome 0 and putting, for every $\lambda \in \Re$, $p^d_\psi(\chi(\widehat{A}), \lambda) = p^d_\psi(\widehat{A}, \chi^{-1}(\lambda))$ (hence $p^t_S(\chi(A_0, \{0\})) = p^t_S(A_0, \{a_0\})$ because of Eq. (27)).

Coming to the modified BCHSH inequality, one can now particularize Eq. (57) to trichotomic generalized observables that can take only values $+1$, 0, and -1, and substitute it into Eq. (22). The resulting equation contains eight detection probabilities which make it uneasy to handle. Therefore, let us denote by $\mathcal{O}_R$ the set of trichotomic generalized observables such that, for every $A_0 \in \mathcal{O}_R$, the detection probability in a given state depends on A_0 but not on its specific value, let us assume that $\mathcal{O}_R$ is non–void, and let us consider only observables in $\mathcal{O}_R$. Then we get from Eq. (57)

$$\begin{aligned} E(A_0(\mathbf{a}), B_0(\mathbf{b})) = p_\Psi^d(\widehat{A}(\mathbf{a}))p_\Psi^d(\widehat{B}(\mathbf{b}))[\langle\Psi|P_1^{\widehat{A}(\mathbf{a})}P_1^{\widehat{B}(\mathbf{b})}|\Psi\rangle + \\ -\langle\Psi|P_1^{\widehat{A}(\mathbf{a})}P_{-1}^{\widehat{B}(\mathbf{b})}|\Psi\rangle - \langle\Psi|P_{-1}^{\widehat{A}(\mathbf{a})}P_1^{\widehat{B}(\mathbf{b})}|\Psi\rangle + \\ +\langle\Psi|P_{-1}^{\widehat{A}(\mathbf{a})}P_{-1}^{\widehat{B}(\mathbf{b})}|\Psi\rangle] = p_\Psi^d(\widehat{A}(\mathbf{a}))p_\Psi^d(\widehat{B}(\mathbf{b}))\langle\widehat{A}(\mathbf{a})\widehat{B}(\mathbf{b})\rangle_\Psi. \end{aligned} \quad (58)$$

According to the ESR model, the term $\langle\widehat{A}(\mathbf{a})\widehat{B}(\mathbf{b})\rangle_\Psi$ in Eq. (58) is interpreted as a *conditional* expectation value, that is, as the mean value of the product of the trichotomic generalized observables $A_0(\mathbf{a})$ and $B_0(\mathbf{b})$ whenever only detected objects are taken into account, and formally coincides with the quantum expectation value, in the state S, of the product of the quantum observables $A(\mathbf{a})$ and $B(\mathbf{b})$ from which $A_0(\mathbf{a})$ and $B_0(\mathbf{b})$, respectively, are obtained. Since similar equations hold if we consider $A_0(\mathbf{a})$ and $B_0(\mathbf{b}')$, $A_0(\mathbf{a}')$ and $B_0(\mathbf{b})$, $A_0(\mathbf{a}')$ and $B_0(\mathbf{b}')$, we obtain from Eq. (22)

$$\begin{aligned} p_\Psi^d(\widehat{A}(\mathbf{a}))|p_\Psi^d(\widehat{B}(\mathbf{b}))\langle\widehat{A}(\mathbf{a})\widehat{B}(\mathbf{b})\rangle_\Psi - p_\Psi^d(\widehat{B}(\mathbf{b}'))\langle\widehat{A}(\mathbf{a})\widehat{B}(\mathbf{b}')\rangle_\Psi| + \\ +p_\Psi^d(\widehat{A}(\mathbf{a}'))|p_\Psi^d(\widehat{B}(\mathbf{b}))\langle\widehat{A}(\mathbf{a}')\widehat{B}(\mathbf{b})\rangle_\Psi + p_\Psi^d(\widehat{B}(\mathbf{b}'))\langle\widehat{A}(\mathbf{a}')\widehat{B}(\mathbf{b}')\rangle_\Psi| \leq 2. \end{aligned} \quad (59)$$

Eq. (59) constitutes our main result in this section.[14] Let us briefly comment on it.

Eq. (59) contains four detection probabilities and four conditional expectation values. The latter can be calculated by using the rules of QM because of Ax. 2 in Sec. 3., and formally coincide with expectation values of QM. If one puts them into Eq. (59) the inequality can be interpreted as a condition that must be fulfilled by the detection probabilities in the ESR model, as we have anticipated in Sec. 5.. It is important to stress that this condition implies that the detection probability $p_S^d(F)$ in Eq. (10) cannot be equal to 1 for every state S and property F, hence the physical predictions of the ESR model are necessarily different from those of QM (see (b) at the end of Sec. 4.). Of course, we have as yet no theory allowing us to calculate precise values for $p_S^d(F)$. Nevertheless, should one be able to perform actual measurements that are close to idealized measurements, the detection probabilities could be determined experimentally and then inserted into Eq. (59). Two possibilities occur.

(i) There exist states and observables such that the conditional expectation values violate Eq. (59). In this case the ESR model (hence, R and LOC) is called into question.

(ii) For every choice of states and observables the conditional expectation values fit in with Eq. (59). In this case the ESR model is supported by the experimental data.

These alternatives show explicitly that the ESR model is, in principle, falsifiable, as we have stated at the end of Sec. 4..

[14]We observe that this result can be obtained without using the mathematical representation of generalized observables and GPP [24, 26], but requires in this case some additional assumptions that are avoided in our present treatment.

Let us illustrate the previous remarks by dealing with a special case. Let Ω be a compound system made up of two far apart spin$-\frac{1}{2}$ quantum particles 1 and 2 in the singlet spin state represented by the unit vector $|\eta\rangle = \frac{1}{\sqrt{2}}(|+,-\rangle - |-,+\rangle)$. Let $A(\mathbf{a})$ and $A(\mathbf{a}')$ be the quantum observables "spin of particle 1 along the direction $\mathbf{a}$" and "spin of particle 1 along the direction $\mathbf{a}'$," represented by the self–adjoint operator $\sigma_a(1)$ and $\sigma_{a'}(1)$, respectively. Analogously, let $B(\mathbf{b})$ and $B(\mathbf{b}')$ be the quantum observables "spin of particle 2 along the direction $\mathbf{b}$" and "spin of particle 2 along the direction $\mathbf{b}'$," represented by the self–adjoint operator $\sigma_b(2)$ and $\sigma_{b'}(2)$, respectively (for the sake of simplicity we have obviously omitted a factor $\frac{\hbar}{2}$ in the representations). By using standard quantum rules for probabilities one gets from Eq. (58), by considering the pair $(A_0(\mathbf{a}), B_0(\mathbf{b}))$,

$$E(A_0(\mathbf{a}), B_0(\mathbf{b})) = p_\eta^d(\sigma_a(1))p_\eta^d(\sigma_b(2))(-\mathbf{a}\cdot\mathbf{b}). \tag{60}$$

Of course, similar equations hold for the remaining pairs. Now, symmetry reasons suggest that all detection probabilities have the same value in the singlet spin state, say, e_η. Hence, Eq. (59) yields

$$(e_\eta)^2 \leq \frac{2}{|-\mathbf{a}\cdot\mathbf{b}+\mathbf{a}\cdot\mathbf{b}'| + |-\mathbf{a}'\cdot\mathbf{b}-\mathbf{a}'\cdot\mathbf{b}'|}. \tag{61}$$

Since e_η does not depend on $\mathbf{a}$, $\mathbf{a}'$, $\mathbf{b}$, $\mathbf{b}'$, Eq. (61) provides for e_η the upper bound [24, 26]

$$e_\eta \leq \frac{1}{\sqrt[4]{2}} \approx 0.841. \tag{62}$$

Eq. (62) implies that no spin measurement on particle 1 or 2, even if idealized, can have a detection efficiency greater than 0.841, this yields a prediction which is typical of the ESR model. Should Eq. (62) be contradicted by experimental data, the ESR model would be questioned. If not, one would get a clue that the ESR model is correct.

9. A New Integrated Perspective

We have seen in Sec. 4. that the ESR model provides a set–theoretical picture of the physical world in which macroscopic properties can be considered objective. The basic elements of this picture are microscopic properties (primitive notion) and microscopic states (derived notion). We intend to broaden such picture in the present section by introducing theoretical probabilities of microscopic properties and microscopic observables. These notions will be used to show that the BCHSH inequality, the modified BCHSH inequality and the standard quantum predictions do not conflict in the framework of the ESR model because they refer to different parts of the picture provided by the model.

Let us first recall from Sec. 3. that, whenever an ensemble Σ of physical objects is prepared in a state S, the microscopic properties possessed by each object depend on the microscopic state S^i of the object but not on the measurement context. It follows that, for every $f \in \mathscr{E}$, one can introduce a theoretical probability $p_S(f)$ that a physical object x in the state S possess f. Furthermore, let us consider the macroscopic property $F = \varphi(f)$ corresponding to f. The probability $p_S(f) = p_S(\varphi^{-1}(F))$ generally does not coincide with the probability $p_S^t(F)$ in Eq. (10) because there may be physical objects that possess f and yet are not detected, which implies that they do not display F (hence, $p^t_S(F) \leq p_S(f)$). As far as $p_S(f)$ and $p_S(F)$ are concerned, instead, two possibilities occur.

(i) The subensemble Σ^d of all physical objects that are detected is a *fair sample* of Σ, that is, the percentage of physical objects possessing f in Σ^d is identical to the percentage of physical objects possessing f in Σ. Since all detected objects possessing f turn out to display $F = \varphi(f)$ when a measurement is done, $p_S(f)$ and $p_S(F)$ coincide.

(ii) Σ^d is not a fair sample of Σ. In this case $p_S(f)$ does not coincide with $p_S(F)$.

Let us now introduce microscopic observables and their expectation values in the ESR model, as follows.

Let A_0 be a discrete generalized observable and let us use the symbols introduced at the end of Sec. 6.. Hence, A_0 is characterized by the macroscopic properties $F_0 = (A_0, \{a_0\})$, $F_1 = (A_0, \{a_1\})$, $F_2 = (A_0, \{a_2\})$, ... The macroscopic property F_0 has no microscopic counterpart, while $F_1, F_2, \ldots$ correspond to the microscopic properties $f_1 = \varphi^{-1}(F_1)$, $f_2 = \varphi^{-1}(F_2), \ldots$, respectively. Then, we define the microscopic observable $\mathbb{A}$ corresponding to A_0 by means of the family $\{f_n\}_{n \in \mathbb{N}}$. The possible values of $\mathbb{A}$ are the outcomes $a_1, a_2, \ldots$ and its expectation value $\langle \mathbb{A} \rangle_S$ in the state S is given by

$$\langle \mathbb{A} \rangle_S = \sum_n a_n p_S(f_n), \tag{63}$$

where $p_S(f_n)$ is the theoretical probability of the microscopic property f_n.

We are thus ready to discuss what is going on at a microscopic level. Indeed, by using this definition we can consider the (dichotomic) microscopic observables $\mathbb{A}(\mathbf{a})$, $\mathbb{A}(\mathbf{a}')$, $\mathbb{B}(\mathbf{b})$ and $\mathbb{B}(\mathbf{b}')$, each of which has possible values -1 and $+1$, corresponding to the (trichotomic) macroscopic observables $A_0(\mathbf{a})$, $A_0(\mathbf{a}')$, $B_0(\mathbf{b})$ and $B_0(\mathbf{b}')$ introduced in Sec. 5., respectively. Since all microscopic properties are either possessed or not possessed by a given physical object, the usual procedures leading to Eq. (16) can be applied. Therefore, one gets the BCHSH inequality, with $E(\mathbf{a}, \mathbf{b})$, $E(\mathbf{a}, \mathbf{b}')$, $E(\mathbf{a}', \mathbf{b})$ and $E(\mathbf{a}', \mathbf{b}')$ reinterpreted in terms of microscopic observables.

Bearing in mind our results in Sec. 8., we can draw the conclusion that, under suitable assumptions on the observables that are taken into account, different inequalities hold for different parts of the picture provided by the ESR model.

(a) The BCHSH inequality holds at a microscopic level (this is purely theoretical and cannot be experimentally checked).

(b) The modified BCHSH inequality holds at a macroscopic level whenever all physical objects that are prepared are considered (it can be experimentally checked, at least in principle, see Sec. 8.).

(c) The quantum predictions deduced by using QM rules hold at a macroscopic level whenever only the physical objects that are detected are considered (this can be experimentally checked). In this case, there are physical situations in which quantum inequalities hold which do not coincide with the BCHSH inequality.

This conclusion is "conciliatory" in the sense that it settles the conflict between the BCHSH inequality and quantum predictions, as anticipated at the beginning of this section. It is then interesting to observe that the ESR model allows us to explain the violation of the BCHSH inequality which occurs when quantum expectation values are substituted in this inequality in terms of a (unconventional) kind of unfair sampling. Indeed, let us suppose that A_0 is measured on each physical object in Σ. Then, several physical objects display the property F_0 (hence, the expectation value $\langle A_0 \rangle_S$ of A_0 is given by Eq. (50)). Therefore, the objects for which the outcomes $a_1, a_2, \ldots$ are obtained belong to the subset $\Sigma^d \subseteq \Sigma$. Furthermore, the probabilities $p_S(F_1) = p_S(a_1), p_S(F_2) = p_S(a_2), \ldots$ must be interpreted as the large number limits of the frequencies of $a_1, a_2, \ldots$, respectively, in Σ^d. Let us consider the conditional expectation value $\langle A \rangle_S = \sum_n a_n p_S(F_n) = \langle \psi | \widehat{A} | \psi \rangle$ in Sec. 8. and compare it with $\langle \mathbb{A} \rangle_S$. It is apparent that $\langle A \rangle_S$ and $\langle \mathbb{A} \rangle_S$ must coincide if case (i) occurs, while they generally do not coincide if case (ii) occurs. Analogous remarks hold if we consider the conditional expectation value $\langle AB \rangle_S$ which can also be evaluated by using standard QM rules. It follows that, if we substitute $P(\mathbf{a}, \mathbf{b})$, $P(\mathbf{a}, \mathbf{b}')$, $P(\mathbf{a}', \mathbf{b})$ and $P(\mathbf{a}', \mathbf{b}')$ in Eq. (16) with the conditional expectation values $\langle A(\mathbf{a})B(\mathbf{b}) \rangle_S$, $\langle A(\mathbf{a})B(\mathbf{b}') \rangle_S$, $\langle A(\mathbf{a}')B(\mathbf{b}) \rangle_S$ and $\langle A(\mathbf{a}')B(\mathbf{b}') \rangle_S$, respectively, the inequality must be fulfilled in case (i), while it can be violated in case (ii). Since the conditional expectation values formally coincide with quantum expectation values, there are physical situations in which Eq. (16) is violated; thus, we conclude that case (ii) occurs and Σ^d is not a fair sample of Σ.

Let us close with a remark. We have seen in Sec. 5. that the tests of the BCHSH inequality actually check derived inequalities, obtained by adding additional assumptions to *local realism*, where many scholars uphold that the experimental data that disprove these inequalities could actually show that the additional assumptions are false, not that local realism is untenable. We add here that some authors point out that the proof of Bell's inequality requires a *hidden Bell's postulate* (HBP) besides local realism, *i.e.*, the assumption that "an experiment involving several incompatible measurements can be written on a single probability space, independently of the measurement context" [44]. HBP implies a *fair sampling assumption* on the measuring apparatuses, which has been recently questioned by reconsidering some available experimental data [63]; moreover, a wave model has been devised in which unfair sampling occurs [44]. These results provide further support to the Växjö interpretation of QM, which rejects HBP; and, according to the authors, is contextual, statistical, and realistic [43]. It is then interesting to observe that the foregoing criticism to the fair sampling assumption rests on investigations into real measuring processes, hence unfair sampling is ascribed to features of the measuring apparatuses (*e.g.*, existence of thresholds) and does not follow from intrinsic properties of the physical objects that are considered, as in the ESR model (where therefore no hidden variable associated with measuring apparatuses occurs, see Sec. 3.). This makes the Växjö interpretation of QM basically different from the ESR model. Nevertheless, the two theories do not seem incompatible, because the former could be recovered within the latter by introducing additional hidden variables associated with actual measuring apparatuses in the ESR model, as we have already noted at the end of Sec. 1..

10. A Mathematical Representation of Mixtures

The probability $p_S(F)$ in Eq. (10) can be calculated by using standard QM rules whenever S is a pure state because of Ax. 2. But Eq. (10) has been derived without making assumptions on S, since it holds also if S is a mixture. One is thus led to wonder whether $p_S(F)$ can be calculated by means of standard quantum rules also in this case. We intend to show in the present section that the answer is negative. Moreover, we propose to supply new rules for evaluating $p_S(F)$ and $p^t_S(F)$ in the case of mixtures by using the mathematical representations introduced in Sec. 6..

Let S be a mixture of the pure states $S_1, S_2, \ldots$, represented by the density operators $\rho_{\psi_1}, \rho_{\psi_2}, \ldots$, with probabilities $p_1, p_2, \ldots$, respectively. The probability $p^t_S((A_0, X))$ that a measurement of the generalized observable A_0 on a physical object x in the state S yield an outcome in the Borel set $X \in \mathbb{B}(\Re)$, with $a_0 \notin X$, or, equivalently, the probability $p^t_S(F)$ that x display the macroscopic property $F = (A_0, X) \in \mathcal{F}$ when F is measured on it, is given by

$$p^t_S((A_0, X)) = p^t_S(F) = \sum_j p_j p^t_{S_j}(F) = \sum_j p_j p^d_{S_j}(F) p_{S_j}(F), \tag{64}$$

because of Eq. (10), with S_j in place of S. The term $p^t_{S_j}(F)$ in Eq. (64) denotes the overall probability that a physical object x in the pure state S_j display F when an (idealized) measurement of F is performed on x, $p^d_{S_j}(F)$ denotes the probability that x be detected and $p_{S_j}(F)$ denotes the conditional probability that x display F when detected. By using again Eq. (10) we get

$$p_S(F) = \sum_j p_j \frac{p^d_{S_j}(F)}{p^d_S(F)} p_{S_j}(F). \tag{65}$$

Eq. (65) is reasonable from an intuitive point of view. Indeed, bearing in mind the interpretation of the probabilities that appear in it, the term $p_j \frac{p^d_{S_j}(F)}{p^d_S(F)}$ can be interpreted, because of the Bayes theorem, as the conditional probability that x be in the state S_j whenever F is measured and x is detected.

We can rewrite Eq. (65) by using the mathematical representations of generalized observables provided in Sec. 6.. We get from Ax. 2

$$p_{S_j}(F) = Tr[\rho_{\psi_j} P^{\widehat{A}}(X)], \tag{66}$$

where $P^{\widehat{A}}$ is the (spectral) PV measure associated with the self–adjoint operator $\widehat{A}$ representing the observable A of QM from which A_0 is obtained, as in Sec. 6.. Hence

$$p_S(F) = Tr\left[(\sum_j p_j \frac{p^d_{S_j}(F)}{p^d_S(F)} \rho_{\psi_j}) P^{\widehat{A}}(X) \right]. \tag{67}$$

Furthermore, if we introduce the obvious assumption

$$p^d_S(F) = \sum_j p_j p^d_{S_j}(F) \tag{68}$$

and use Eq. (38) (holds for pure states only) with S_j in place of S, we obtain

$$\frac{p^d_{S_j}(F)}{p^d_S(F)} = \frac{\frac{Tr[\rho_{\psi_j} T^{\widehat{A}}_{\psi_j}(X)]}{Tr[\rho_{\psi_j} P^{\widehat{A}}(X)]}}{\sum_j p_j \frac{Tr[\rho_{\psi_j} T^{\widehat{A}}_{\psi_j}(X)]}{Tr[\rho_{\psi_j} P^{\widehat{A}}(X)]}}, \tag{69}$$

hence, we get from Eq. (67)

$$p_S(F) = Tr[\rho_S(F)P^{\widehat{A}}(X)] \,, \tag{70}$$

with

$$\rho_S(F) = \sum_j p_j \frac{p^d_{S_j}(F)}{p^d_S(F)} \rho_{\psi_j} = \frac{\sum_j p_j \frac{Tr[\rho_{\psi_j} T^{\widehat{A}}_{\psi_j}(X)]}{Tr[\rho_{\psi_j} P^{\widehat{A}}(X)]} \rho_{\psi_j}}{\sum_j p_j \frac{Tr[\rho_{\psi_j} T^{\widehat{A}}_{\psi_j}(X)]}{Tr[\rho_{\psi_j} P^{\widehat{A}}(X)]}} \,. \tag{71}$$

Eqs. (70) and (71) show that $p_S(F)$ does not coincide, in general, with the probability obtained by applying standard QM rules, that is, calculating $Tr[\rho_S P^{\widehat{A}}(X)]$, where $\rho_S = \sum_j p_j \rho_{\psi_j}$ is the density operator that represents the mixture S in QM. This can be intuitively explained by observing that an ensemble Σ of physical objects prepared in S can be partitioned into subensembles $\Sigma_1, \Sigma_2, \ldots$ of physical objects prepared in the states $S_1, S_2, \ldots$, respectively.[15] Whenever a macroscopic property F is measured on Σ, for every Σ_j the subensemble Σ^d_j of detected objects depends not only on F but also on S_j hence, generally, the subensemble Σ^d of all detected objects is not a fair sample of Σ. As far as $p_S(F)$ is concerned, S must then be represented by the density operator $\rho_S(F)$, which depends on F and coincides with ρ_S only in special cases. More generally, S must be associated with the family of density operators

$$\{\rho_S(F)\}_{F \in \mathcal{F}} = \{\sum_j p_j \frac{p^d_{S_j}(F)}{p^d_S(F)} \rho_{\psi_j}\}_{F \in \mathcal{F}}. \tag{72}$$

Eq. (72) provides a representation of S in the ESR model. If pure states are considered as limiting cases of mixtures, this family reduces to the constant $\{\rho_S\}_{F \in \mathcal{F}}$ whenever S is a pure state, which implies that Eq. (70) embodies Ax. 2. But Eq. (70) also shows that Ax. 2 cannot be extended to nonpure states.

Let us come to the probability $p^t_S(F)$. By using Eq. (34) we get

$$p^t_{S_j}(F) = Tr[\rho_{\psi_j} T^{\widehat{A}}_{\psi_j}(X)], \tag{73}$$

where $T^{\widehat{A}}_{\psi_j}(X) = \int_X p^d_{\psi_j}(\widehat{A}, \lambda) \mathrm{d}P^{\widehat{A}}_\lambda$ because of Eq. (31). Hence, Eq. (64) yields

$$p^t_S(F) = Tr\Big[\sum_j p_j \rho_{\psi_j} T^{\widehat{A}}_{\psi_j}(X)\Big]. \tag{74}$$

Eq. (74) shows that, generally, $p^t_S(F)$ cannot be written as the trace of the product of two operators, one of which represents S and the other represents F. However, by using Eqs. (38) and (71) we get from Eq. (74)

$$p^t_S(F) = Tr\Big[\sum_j p_j \rho_{\psi_j} p^d_{S_j}(F) P^{\widehat{A}}(X)\Big] = p^d_S(F) Tr\Big[\rho_S(F) P^{\widehat{A}}(X)\Big] \,, \tag{75}$$

[15] We use here an epistemic interpretation of the probabilities $p_1, p_2, \ldots$ which is not generally accepted in QM [37], but can be justified in the conceptual framework of the ESR model, as we show in Sec. 11..

consistently with Eq. (10).

Let us now consider the macroscopic property $G=(A_0,Y)\in\mathcal{F}_0\setminus\mathcal{F}$, hence $a_0\in Y$, and put $F=(A_0,X)$, with $X=Y\setminus\{a_0\}$, as in Sec. 3.. We get from Eqs. (12), (64) and (68)

$$p_S^t(G)=1-p_S^d(F)+p_S^t(F)=\sum_j p_j-\sum_j p_j p_{S_j}^d(F)+\sum_j p_j p_{S_j}^t(F)=$$
$$=\sum_j p_j(1-p_{S_j}^d(F)+p_{S_j}^t(F))=\sum_j p_j p_{S_j}^t(G). \tag{76}$$

By using Eq. (34) we obtain

$$p_{S_j}^t(G)=Tr[\rho_{\psi_j}T_{\psi_j}^{\widehat{A}}(Y)] \tag{77}$$

where $T_{\psi_j}^{\widehat{A}}(Y)=I-\int_{\Re\setminus Y}p_{\psi_j}^d(\widehat{A},\lambda)\mathrm{d}P_\lambda^{\widehat{A}}$ because of Eq. (31), hence

$$p_S^t(G)=Tr\Big[\sum_j p_j\rho_{\psi_j}T_{\psi_j}^{\widehat{A}}(Y)\Big]. \tag{78}$$

Putting together Eqs. (74) and (78), we conclude that, for every macroscopic property (A_0,X), with $X\in\mathbb{B}(\Re)$,

$$p_S^t((A_0,X))=Tr\Big[\sum_j p_j\rho_{\psi_j}T_{\psi_j}^{\widehat{A}}(X)\Big], \tag{79}$$

where $T_{\psi_j}^{\widehat{A}}(X)$ is given by Eq. (31) with ψ_j in place of ψ.

11. The Operational Definitions of Mixtures

Let S be the mixture introduced in Sec. 10.. A typical preparation procedure of a physical object x in the state S can be described as follows.

Select a preparing device π_j for every pure state S_j, use each π_j to prepare an ensemble $\mathscr{E}_{S_j}$ of n_j physical objects in the state S_j and choose n_j such that $n_j=Np_j$, with $N=\sum_j n_j$. Then mingle the ensembles $\mathscr{E}_{S_1},\mathscr{E}_{S_2},\ldots$ to prepare an ensemble $\mathscr{E}_S$ of N physical objects, remove any memory of the way in which $\mathscr{E}_{S_1},\mathscr{E}_{S_2},\ldots$ have been mingled and select a physical object x in $\mathscr{E}_S$.

This description implies that one assumes that frequencies converge to probabilities in the large number limit (see Sec. 3. and footnote 8) and interprets probabilities as epistemic, that is, formalizing the loss of memory about the pure state in which each physical object has been actually prepared (*ignorance interpretation*). It also implies that many preparation procedures of a physical object in a state S can be constructed, hence we call *operational definition* of S and denote by σ_S the set of preparation procedures of S that can be obtained by selecting the preparing devices in the states S_1, $S_2,\ldots$ in all possible ways.

Coming to QM, states are defined in this theory as classes of probabilistically equivalent preparing devices [37, 46], and every equivalence class is represented by a density operator (Gleason theorem). But the density operator that represents a given mixture can generally be decomposed in different ways as a convex combination of density operators representing pure states (*nonunique decomposability of quantum mixtures* [37]), which is usually considered as a distinguishing feature of QM but is a source of interpretative problems. Indeed, let us consider the mixture S as discussed earlier. The operational definition of it implies that S can be represented in QM by the density operator $\rho = \sum_j p_j \rho_{\psi_j}$, where the probabilities $p_1, p_2, \ldots$ are epistemic, as we have seen here. But it is well known that one–dimensional projection operators $\rho_{\chi_1}, \rho_{\chi_2}, \ldots$ generally exist, none of which coincides with one of the projection operators $\rho_{\psi_1}, \rho_{\psi_2}, \ldots$, which are such that $\rho_S = \sum_l q_l \rho_{\chi_l}$, with $0 \leq q_l \leq 1$ and $\sum_l q_l = 1$. If this expression of ρ_S is adopted, the coefficients q_l cannot be interpreted as probabilities bearing an ignorance interpretation. Consider now a mixture T of the pure states $T_1, T_2, \ldots$ represented by the density operators $\rho_{\chi_1}, \rho_{\chi_2}, \ldots$, with probabilities $q_1, q_2, \ldots$, respectively, prepared following the procedure described in the case of S with obvious changes, so that T has an operational definition σ_T which is different from σ_S. According to QM, T and S must be identified because they are represented by the same density operator. But the probabilities $q_1, q_2, \ldots$ now admit an ignorance interpretation, which contradicts the conclusion expounded earlier.

Because of these problems many scholars maintain that an ignorance interpretation of the probabilities that appear in the various possible expressions of ρ_S must be avoided [37], even if this position clashes with the interpretation of the probabilities p_j and q_l in the operational definitions of S and T. The origin of the problem is, however, rather clear. Indeed, different operational definitions of mixtures have different physical interpretations, hence they are mutually exclusive, but may be probabilistically equivalent. If this is the case, they correspond to the same mixture, say S, which is bijectively represented by the density operator ρ_S. The different decompositions of ρ_S correspond to the different operational definitions collected in S, but assigning ρ_S does not single out one of them, hence it does not allow one to privilege a particular operational definition of S, which causes the ambiguity in the interpretation of the probabilities p_j and q_l as illustrated.[16] We add that there is in QM another deep argument supporting the position of these scholars mentioned. Indeed the subsystems of a compound physical system in an entangled state are described by convex combinations of pure states whose coefficients do not bear an ignorance interpretation because of nonobjectivity of macroscopic properties (*improper mixtures* [38, 40]).

Synthetically, we can draw in QM the commutative diagram

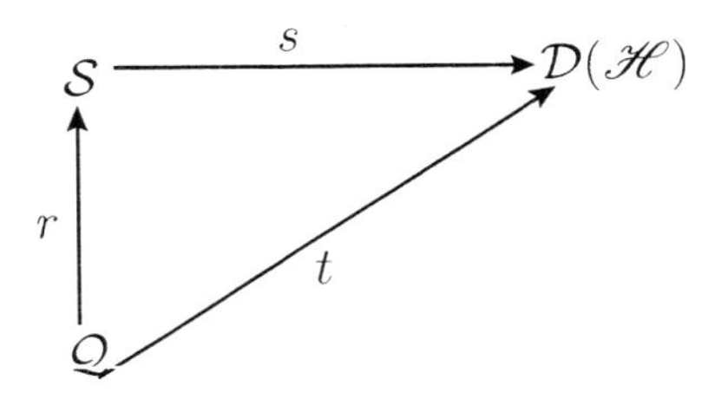

(80)

where $\mathcal{Q}$ is the set of all operational definitions of states, $\mathcal{S}$ is the set of all states and $\mathcal{D}(\mathscr{H})$ is the convex set of all density operators on $\mathscr{H}$. The mappings in diagram (80) are

[16] It seems to us that these features of the mathematical representation of mixtures in QM point out some limits of the formal description supplied by QM rather than illustrating some peculiarities of the microscopic world, as maintained by many authors [37, 39].

then defined as follows.

$r : \sigma_S \in \mathcal{Q} \longmapsto S \in \mathcal{S}$ maps each class of probabilistically equivalent operational definitions into a state.

$s : S \in \mathcal{S} \longmapsto \rho_S \in \mathcal{D}(\mathscr{H})$ is the *standard representation* of $\mathcal{S}$.

$t : \sigma_S \in \mathcal{Q} \longmapsto \rho_S \in \mathcal{D}(\mathscr{H})$ maps each class of probabilistically equivalent operational definitions into a density operator.

Because of our previous remarks, the mapping s is bijective, while r and t are surjective but not bijective (though their restriction to operational definitions of pure states only is obviously bijective).

Let us come to the ESR model. As in QM, a state is defined as a class of preparing devices that are probabilistically equivalent (see Sec. 3.) and it may occur that different operational definitions of mixtures, say σ_S and σ_T, are made up of preparing devices that are probabilistically equivalent (in this case we briefly say that σ_S and σ_T are probabilistically equivalent and write $\sigma_S \equiv \sigma_T$). Whenever $\sigma_S \equiv \sigma_T$ the mixtures S and T, operationally defined by σ_S and σ_T, respectively, must be identified. But, now, the families of density operators associated with S and T (see Eq. (72)) are built up by referring to the operational definitions of σ_S and σ_T, respectively, and do not necessarily coincide.[17] Synthetically, we can draw in the ESR model the commutative diagram

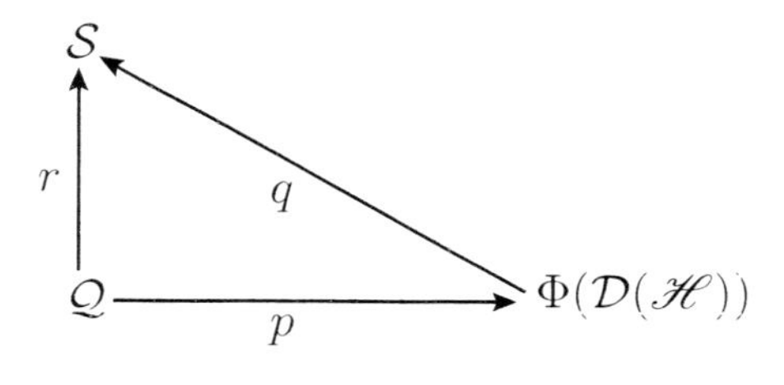

(81)

The symbols $\mathcal{Q}$ and $\mathcal{S}$ in diagram (81) are defined as in diagram (80), while $\Phi(\mathcal{D}(\mathscr{H}))$ is the set of all families of the form $\{\rho_S(F)\}_{F\in\mathcal{F}}$, with $S \in \mathcal{S}$. Since $\{\rho_S(F)\}_{F\in\mathcal{F}}$ is built up by referring to π_S, we can assume that the mapping

$p : \sigma_S \in \mathcal{Q} \longmapsto \{\rho_S(F)\}_{F\in\mathcal{F}} \in \Phi(\mathcal{D}(\mathscr{H}))$

is bijective. The mappings

$r : \sigma_S \in \mathcal{Q} \longmapsto S \in \mathcal{S}$

and

$q : \{\rho_S(F)\}_{F\in\mathcal{F}} \in \Phi(\mathcal{D}(\mathscr{H})) \longmapsto S \in \mathcal{S}$

are instead not bijective, since it may occur that a mixture S has probabilistically equivalent but physically different operational definitions (equivalently, the mathematical representation of S is not unique). These features of p, q and r have some relevant consequences. From one side, the bijectivity of p implies that it does not occur that a given mathematical representation may correspond to several different operational definitions, as in QM, which avoids the interpretational ambiguities following from the non–unique decomposition of

[17]To be precise, one has $\sigma_S \equiv \sigma_T$ if and only if, for every $F \in \mathcal{F}_0$, $p_S^t(F) = p_T^t(F)$. By using Eq. (10), with $F = (A_0, X) \in \mathcal{F}$, and Eq. (12), with $G = (A_0, X \cup \{a_0\})$, one gets at once that $\sigma_S \equiv \sigma_T$ if and only if, for every $F \in \mathcal{F}$, $p_S^d(F) = p_T^d(F)$ and $p_S(F) = p_T(F)$. Because of Eq. (70) the latter condition is satisfied if and only if, for every $F \in \mathcal{F}$, the equality $Tr[\rho_S(F)P^{\widehat{A}}(X)] = Tr[\rho_T(F)P^{\widehat{A}}(X)]$ holds, which does not imply that $\{\rho_S(F)\}_{F\in\mathcal{F}} = \{\rho_T(F)\}_{F\in\mathcal{F}}$.

density operators in QM (the ambiguities following from the nonobjectivity of macroscopic properties in QM are avoided *a priori* in the ESR model, which is an objective theory, see Secs. 1. and 4.). On the other side, the non–bijectivity of r and q shows that the class of preparing devices that characterizes a mixture S can be partitioned in subclasses, each of which corresponds to a different operational definition of S. This suggests that the standard equivalence relations based on probability should be refined to take into account some physically important operational differences (see also [38]).

To close up, let us come back to the standard representation of mixtures in QM and consider the state S introduced at the beginning of this section. Adopting the point of view of the ESR model does not prohibit one to associate, formally, a density operator $\rho_S = \sum_j p_j \rho_{\psi_j}$ with the operational definition σ_S of S, even if ρ_S cannot be used to calculate probabilities. However, one cannot guarantee that $\sigma_S \equiv \sigma_T$ implies $\rho_S = \rho_T$ in the ESR model (indeed, $\sigma_S \equiv \sigma_T$ does not imply that, for every self–adjoint operator $\widehat{A}$ and Borel set X, $Tr[\rho_S P^{\widehat{A}}(X)] = Tr[\rho_T P^{\widehat{A}}(X)]$), hence a mixture S cannot be associated in a unique way with a density operator ρ_S. But if we assume that, actually, $\sigma_S \equiv \sigma_T$ implies $\rho_S = \rho_T$,[18] diagram (81) can be completed as follows.

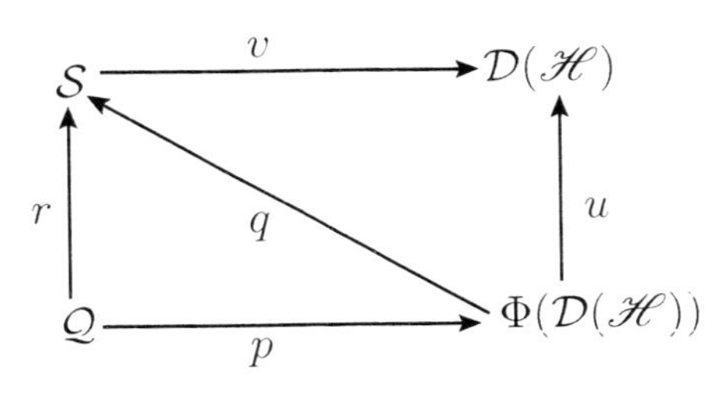

(82)

In diagram (82) the mapping $s : S \in \mathcal{S} \longmapsto \rho_S \in \mathcal{D}(\mathscr{H})$ is the *standard representation* of $\mathcal{S}$ (which is not necessarily bijective unless we also assume that $\rho_S = \rho_T$ implies $\sigma_S \equiv \sigma_T$, which yet does not seem physically justifiable in the framework of the ESR model) while the mapping $u : \{\rho_S(F)\}_{F \in \mathcal{F}} \in \Phi(\mathcal{D}(\mathscr{H})) \longmapsto \rho_S \in \mathcal{D}(\mathscr{H})$ is surjective and makes the diagram commutative.

12. The Generalized Lüders' Postulate

We have seen in Sec. 7. that GPP rules the transformation of a pure state induced by an idealized nondestructive measurement. We now intend to show that our results in Sec. 10., together with GPP, allow us to predict the state transformation induced by a measurement of the same kind if the state of the measured object is a mixture.

Let S be the mixture introduced in Sec. 10.. Let $F = (A_0, X) \in \mathcal{F}_0$ be any property of Ω. Whenever an idealized measurement of F is performed on a physical object x in the state S, the probabilities $p^t_{S_j}(F)$ and $p^t_S(F)$ can be deduced from Eq. (79). If the measurement yields the yes outcome, the final state S_F is a mixture of the pure states $S_{1F}, S_{2F}, \ldots$

[18] By considering Eq. (67) we see that this assumption requires, intuitively, that the detection probabilities $p^d_{S_1}(F), p^d_{S_2}(F), \ldots$ take a sufficiently large number of values when F varies in $\mathcal{F}$, so that for every $F \in \mathcal{F}$, $p_S(F) = p_T(F)$ may occur only if pure states and probabilities are the same in π_S and π_T, which is physically reasonable.

represented by the density operators $\rho_{\psi_{1F}}, \rho_{\psi_{2F}}, \ldots$, respectively, obtained by applying Eq. (45), with probabilities $p_{1F}, p_{2F}, \ldots$, respectively, obtained by using the Bayes theorem (indeed, p_{jF}, with $j = 1, 2, \ldots$ denotes the conditional probability that x be in the state S_{jF} whenever a measurement of F on x has yielded the yes outcome). Thus, GPP can be extended to mixtures, taking the form of a *generalized Lüders postulate*, as follows.

GLP. Let S be a mixture of the pure states $S_1, S_2, \ldots$, represented by the density operators $\rho_{\psi_1}, \rho_{\psi_2}, \ldots$, with probabilities $p_1, p_2, \ldots$, respectively, and let a nondestructive idealized measurement of a macroscopic property $F = (A_0, X) \in \mathcal{F}_0$ be performed on a physical object x in the state S.

Let the measurement yield the yes outcome. Then, the state S_F of x after the measurement is a mixture of the pure states $S_{1F}, S_{2F}, \ldots$ represented by the density operators $\rho_{\psi_{1F}}, \rho_{\psi_{2F}}, \ldots$, respectively, with

$$\rho_{\psi_{jF}} = \frac{T^{\widehat{A}}_{\psi_j}(X)\rho_{\psi_j}T^{\widehat{A}\dagger}_{\psi_j}(X)}{Tr[T^{\widehat{A}}_{\psi_j}(X)\rho_{\psi_j}T^{\widehat{A}\dagger}_{\psi_j}(X)]}, \tag{83}$$

and probabilities $p_{1F}, p_{2F}, \ldots$, respectively, with

$$p_{jF} = p_j \frac{p^t_{S_j}((A_0, X))}{p^t_S((A_0, X))} = p_j \frac{Tr[\rho_{\psi_j}T^{\widehat{A}}_{\psi_j}(X)]}{Tr\left[\sum_j p_j \rho_{\psi_j} T^{\widehat{A}}_{\psi_j}(X)\right]}, \tag{84}$$

hence S_F is represented by the family of density operators

$$\{\rho_{S_F}(H)\}_{H\in\mathcal{F}} = \{\sum_j p_{jF} \frac{p^d_{S_jF}(H)}{p^d_{S_F}(H)} \rho_{\psi_{jF}}\}_{H\in\mathcal{F}}. \tag{85}$$

Let the measurement yield the no outcome. Then, the state S'_F of x after the measurement is a mixture of the pure states $S'_{1F}, S'_{2F}, \ldots$ represented by the density operators $\rho_{\psi'_{1F}}, \rho_{\psi'_{2F}}, \ldots$, respectively, with

$$\rho_{\psi'_{jF}} = \frac{T^{\widehat{A}}_{\psi_j}(\Re \setminus X)\rho_{\psi_j}T^{\widehat{A}\dagger}_{\psi_j}(\Re \setminus X)}{Tr[T^{\widehat{A}}_{\psi_j}(\Re \setminus X)\rho_{\psi_j}T^{\widehat{A}\dagger}_{\psi_j}(\Re \setminus X)]}, \tag{86}$$

and probabilities $p'_{1F}, p'_{2F}, \ldots$, respectively, with

$$p'_{jF} = p_j \frac{p^t_{S_j}((A_0, \Re \setminus X))}{p^t_S((A_0, \Re \setminus X))} = p_j \frac{Tr[\rho_{\psi_j}T^{\widehat{A}}_{\psi_j}(\Re \setminus X)]}{Tr\left[\sum_j p_j \rho_{\psi_j} T^{\widehat{A}}_{\psi_j}(\Re \setminus X)\right]}, \tag{87}$$

hence S'_F is represented by the family of density operators

$$\{\rho_{S'_F}(H)\}_{H\in\mathcal{F}} = \{\sum_j p'_{jF} \frac{p^d_{S'_jF}(H)}{p^d_{S'_F}(H)} \rho_{\psi'_{jF}}\}_{H\in\mathcal{F}}. \tag{88}$$

It is apparent that GLP generalizes GPP, as required, because Eqs. (83) and (86) coincide with Eqs. (45) and (47), respectively, if S is a pure state. We note explicitly that GPP allows one, in particular, to calculate the overall and the conditional probability that a physical object x display a macroscopic property $H \in \mathcal{F}_0$ whenever H is measured after the measurement of another macroscopic property $F \in \mathcal{F}$ that has yielded the yes outcome. To this end one can indeed use Eqs. (70) and (75), replacing S and F with S_F and H, respectively. This procedure introduces the density operator $\rho_{S_F}(H)$ but, of course, it makes no reference to the standard representation of S_F. However, the standard representation of mixtures has been considered in Sec. 11. and enters into play in the next section. It is interesting to compare the representation of S_F supplied by Eq. (85) with its standard representation, which is provided by the density operator

$$\rho_{S_F} = \sum_j p_{jF} \rho_{\psi_j F} = \sum_j p_j \frac{Tr[\rho_{\psi_j} T_{\psi_j}^{\widehat{A}}(X)]}{Tr\left[\sum_j p_j \rho_{\psi_j} T_{\psi_j}^{\widehat{A}}(X)\right]} \frac{T_{\psi_j}^{\widehat{A}}(X) \rho_{\psi_j} T_{\psi_j}^{\widehat{A}\dagger}(X)}{Tr[T_{\psi_j}^{\widehat{A}}(X) \rho_{\psi_j} T_{\psi_j}^{\widehat{A}\dagger}(X)]} \tag{89}$$

(note that Eq. (89) reduces to the Lüders formula whenever all detection probabilities coincide with 1).

These results can be illustrated by considering the special case of a discrete generalized observable A_0. By using the symbols introduced at the end of Sec. 6., we get $p_{S_j}^d(F_n) = p_{\psi_j n}^d(\widehat{A})$, hence Eq. (68) yields

$$p_S^d(F_n) = \sum_j p_j p_{\psi_j n}^d(\widehat{A}). \tag{90}$$

Therefore, if $n \neq 0$ we get from Eq. (71)

$$\rho_S(F_n) = \frac{\sum_j p_j p_{\psi_j n}^d(\widehat{A}) \rho_{\psi_j}}{\sum_j p_j p_{\psi_j n}^d(\widehat{A})}. \tag{91}$$

By using Eq. (70) we then obtain

$$p_S(F_n) = \frac{1}{\sum_j p_j p_{\psi_j n}^d(\widehat{A})} Tr\left[\sum_j p_j \rho_{\psi_j} p_{\psi_j n}^d(\widehat{A}) P_n^{\widehat{A}}\right]. \tag{92}$$

Furthermore, Eq. (79) yields

$$p_S^t(F_n) = Tr\left[\sum_j p_j \rho_{\psi_j} p_{\psi_j n}^d(\widehat{A}) P_n^{\widehat{A}}\right] \tag{93}$$

if $n \neq 0$, and

$$p_S^t(F_0) = Tr\left[\sum_j p_j \rho_{\psi_j} \sum_{m \in \mathbb{N}} (1 - p_{\psi_j m}^d(\widehat{A})) P_m^{\widehat{A}}\right] \tag{94}$$

if $n = 0$.

Let us consider now a measurement of F_n which yields the yes outcome. If $n \neq 0$ we get from Eqs. (83), (84) and (85), by using Eq. (42),

$$\rho_{\psi_j F_n} = \frac{P_n^{\widehat{A}} \rho_{\psi_j} P_n^{\widehat{A}}}{Tr[\rho_{\psi_j} P_n^{\widehat{A}}]}, \tag{95}$$

$$p_{jF_n} = p_j p^d_{\psi_j n}(\widehat{A}) \frac{Tr[\rho_{\psi_j} P_n^{\widehat{A}}]}{Tr[\sum_j p_j p^d_{\psi_j n}(\widehat{A}) \rho_{\psi_j} P_n^{\widehat{A}}]} \tag{96}$$

and

$$\{\rho_{S_{F_n}}(H)\}_{H \in \mathcal{F}} = \{\sum_j p_{jF_n} \frac{p^d_{S_j F_n}(H)}{p^d_{S_{F_n}}(H)} \rho_{\psi_j F_n}\}_{H \in \mathcal{F}}, \tag{97}$$

respectively. If $n = 0$ one gets more complicate formulas, but their deduction is also straightforward. Moreover, similar formulas hold if the measurement of F_n yields the no outcome.

13. Nonlinear Dynamics for Quantum Measurements

As we have already observed at the end of Sec. 1., some topics have not yet been studied in the framework of the ESR model. In particular, we have not provided a general theory of compound physical systems (though we have already considered examples of systems of this kind in Secs. 5. and 8.), nor we have discussed the time evolution of quantum systems. Therefore no general quantum theory of measurement as interaction of quantum systems can be given at this stage. Nevertheless, we can provide a partial justification of GPP in the discrete case by introducing a reasonable physical assumption on the evolution of the compound system made up of the (microscopic) physical object plus the (macroscopic) measuring apparatus and an elementary measurement scheme whose simplicity allows one to better grasp the conceptual novelties introduced by our approach.

Let A_0 be a discrete generalized observable, let A_0 be measured on a physical object x in the pure state S represented by the unit vector $|\psi\rangle$, let the outcome a_n be obtained, and let the measurement be idealized and nondestructive. Then, the final state of x is given by Eq. (48) because of our assumptions at the end of Sec. 7.. Hence, if the measurement is *nonselective* (*i.e.*, the outcome of the measurement remains unknown), the final state of x is a mixture $\tilde{S}$ of the pure states S_{F_0}, S_{F_1}, S_{F_2}, ... represented by the density operators $|\psi_{F_0}\rangle\langle\psi_{F_0}|$, $|\psi_{F_1}\rangle\langle\psi_{F_1}|$, $|\psi_{F_2}\rangle\langle\psi_{F_2}|$, ..., with probabilities $p^t_S(F_0)$, $p^t_S(F_1)$, $p^t_S(F_2)$, ..., respectively, where the unit vectors $|\psi_{F_0}\rangle$, $|\psi_{F_1}\rangle$, $|\psi_{F_2}\rangle$, ... are given by Eq. (48) and the probabilities $p^t_S(F_0)$, $p^t_S(F_1)$, $p^t_S(F_2)$, ..., by Eq. (43). The representation of $\tilde{S}$ in the ESR model is therefore provided by the family

$$\{\sum_{n \in \mathbb{N}_0} p^t_S(F_n) \frac{p^d_{S_{F_n}}(F)}{p^d_{\tilde{S}}(F)} |\psi_{F_n}\rangle\langle\psi_{F_n}|\}_{F \in \mathcal{F}}, \tag{98}$$

while the standard representation of $\tilde{S}$ is provided instead by the density operator

$$\tilde{\rho} = \sum_{n \in \mathbb{N}_0} p^t_S(F_n)|\psi_{F_n}\rangle\langle\psi_{F_n}| = p^t_S(F_0)|\psi_{F_0}\rangle\langle\psi_{F_0}| + \sum_{n \in \mathbb{N}} p^d_{\psi n}(\widehat{A}) P_n^{\widehat{A}}|\psi\rangle\langle\psi|P_n^{\widehat{A}}. \tag{99}$$

Let us denote by $g_1, g_2, \ldots$ the dimensions of the subspaces $\mathscr{S}_1, \mathscr{S}_2, \ldots$ associated with the eigenvalues $a_1, a_2, \ldots$, respectively, of $\widehat{A}$. Then, for every $n \in \mathbb{N}$, $P_n^{\widehat{A}} = \sum_\mu |a_n^\mu\rangle\langle a_n^\mu|$, where $\mu = 1, \ldots, g_n$ and $\{|a_n^\mu\rangle\}_{\mu=1,\ldots,g_n}$ is an orthonormal basis in $\mathscr{S}_n$. Putting $|\psi\rangle = \sum_{n\in\mathbb{N}} \sum_\mu c_n^\mu |a_n^\mu\rangle$, Eq. (99) yields

$$\tilde{\rho} = p_S^t(F_0)|\psi_{F_0}\rangle\langle\psi_{F_0}| + \sum_{n\in\mathbb{N}} p_{\psi n}^d(\widehat{A}) \sum_{\mu,\nu} c_n^\mu c_n^{\nu *} |a_n^\mu\rangle\langle a_n^\nu|. \tag{100}$$

Let us now consider the macroscopic apparatus measuring A_0 as an individual example of a physical system Ω_M associated with the Hilbert space $\mathscr{H}_M$. By adopting a very simplified measurement scheme, let $|0\rangle, |1\rangle, |2\rangle, \ldots$ be the unit vectors of $\mathscr{H}_M$ representing the macroscopic states of Ω_M which correspond to the outcomes $a_0, a_1, a_2, \ldots$, respectively (hence $|0\rangle$ represents the macroscopic state of the apparatus when it is ready to perform a measurement or when the physical object x is not detected), and let us assume that $\{|0\rangle, |1\rangle, |2\rangle, \ldots\}$ is an orthonormal basis in $\mathscr{H}_M$. Let S_0 be the initial state of the compound system consisting of the physical object x and the macroscopic apparatus, and let S_0 be represented by the unit vector $|\Psi_0\rangle = |\psi\rangle|0\rangle$. Because of the interpretation of the a_0 outcome provided in Sec. 3., the time evolution of the compound system must be such that the term $|\psi_{F_0}\rangle|0\rangle$ occurs in the expression of the unit vector $|\Psi\rangle$ representing the final state of the system in such a way that the probability of the a_0 outcome is $p_S^t(F_0)$. This makes it reasonable to suppose that the compound system undergoes the (generally *nonlinear*, hence *nonunitary*) time evolution

$$|\Psi_0\rangle = |\psi\rangle|0\rangle = \sum_{n\in\mathbb{N}} \sum_\mu c_n^\mu |a_n^\mu\rangle|0\rangle \longrightarrow |\Psi\rangle = \sum_{n\in\mathbb{N}} \alpha_{\psi n} \sum_\mu c_n^\mu |a_n^\mu\rangle|n\rangle + \beta_{\psi 0}|\psi_{F_0}\rangle|0\rangle, \tag{101}$$

where $\alpha_{\psi n} = \sqrt{p_{\psi n}^d(\widehat{A})}e^{i\theta_{\psi n}}$ and $\beta_{\psi 0} = \sqrt{p_S^t(F_0)}e^{i\varphi_{\psi 0}}$, with $\theta_{\psi n}, \varphi_{\psi 0} \in \Re$, hence $\langle\Psi|\Psi\rangle = \sum_{n\in\mathbb{N}} |\alpha_{\psi n}|^2 \sum_\mu |c_n^\mu|^2 + |\beta_{\psi 0}|^2 = 1$ because of Eq. (43).

By performing the partial trace with respect to $\mathscr{H}_M$ of the density operator $\rho_C = |\Psi\rangle\langle\Psi|$, which also represents the final state of the compound system after the interaction, we get

$$Tr_M \rho_C = p_S^t(F_0)|\psi_{F_0}\rangle\langle\psi_{F_0}| + \sum_{n\in\mathbb{N}} p_{\psi n}^d(\widehat{A}) \sum_{\mu,\nu} c_n^\mu c_n^{\nu *} |a_n^\mu\rangle\langle a_n^\nu|, \tag{102}$$

hence, comparing Eqs. (100) and (102),

$$Tr_M \rho_C = \tilde{\rho}\,. \tag{103}$$

Eq. (103) provides a partial justification of GPP. This justification is not rigorous nor complete, in particular because Eq. (103) does not imply that the state of x is $\tilde{S}$, since the mapping v in Eq. (82) may be not bijective, and $Tr_M \rho_C$ does not provide the representation that must be used to calculate probabilities of macroscopic properties in the ESR model (Sec. 10.). On the other side, if we recall that all probabilities are epistemic according to the ESR model (Sec. 11.) we see that our justification does not introduce the problematic distinction between *proper* and *improper* mixtures that occurs in standard and unsharp QM, where the states obtained by performing partial traces are improper mixtures [1, 8, 9, 38, 40].

References

[1] Busch, P.; Lahti, P. J.; Mittelstaedt, P. *The Quantum Theory of Measurement*; Springer: Berlin, 1991.

[2] Bell, J. S. *Rev. Mod. Phys.* 1966, 38, 447–452.

[3] Kochen, S.; Specker, E.P. *J. Math. Mech.* 1967, 17, 59–87.

[4] Bell, J. S. *Physics* 1964, 1, 195–200.

[5] Davies, E. B. *Quantum Theory of Open Systems*; Academic Press: London, 1976.

[6] Busch, P.; Grabowski, M.; Lahti, P. J. *Operational Quantum Physics*; Springer: Berlin, 1996.

[7] Busch, P.; Lahti, P. J. *Found. Phys.* 1996, 26, 875–893.

[8] Busch, P.; Shimony, A. *Stud. His. Phil. Mod. Phys.* 1996 27B, 397–404.

[9] Busch, P. *Int. J. Theor. Phys.* 1998 37, 241–247.

[10] Schlosshauer, M. *Rev. Mod. Phys.* 2004, 76, 1267–1305.

[11] Mermin, N. D. *Rev. Mod. Phys.* 1993, 65, 803–815.

[12] Garola, C.; Sozzo, S. *Found. Phys.* 2004, 34, 1249–1266.

[13] Garola, C.; Solombrino, L. *Found. Phys.* 1996, 26, 1121–1164.

[14] Garola, C.; Solombrino, L. *Found. Phys.* 1996, 26, 1329–1356.

[15] Garola, C. In *Language, Quantum, Music*; Dalla Chiara, M.; et al.; Eds.; Kluwer: Dordrecht, 1999; pp 101–116.

[16] Garola, C. In *Quantum Structures and the Nature of Reality*; Aerts, D.; Pykacz, J.; Eds.; Kluwer: Dordrecht, 1999; pp 103–140.

[17] Garola, C. *Int. J. Theor. Phys.* 1999, 38, 3241–3252.

[18] Garola, C. *Found. Phys.* 2000, 30, 1539–1565.

[19] Garola, C. *Int. J. Theor. Phys.* 2005 44, 807–814.

[20] Garola, C. In *Foundations of Probability and Physics-5*; Accardi, L.; et al.; Eds.; American Institute of Physics: New York, 2009; pp 51–52.

[21] Garola, C. *Found. Phys.* 2002, 32, 1597–1615.

[22] Garola, C. *Found. Phys. Lett.* 2003, 16, 605–612.

[23] Garola, C.; Pykacz, J. *Found. Phys.* 2004, 34, 449–475.

[24] Garola, C.; Sozzo, S. *Int. J. Theor. Phys.* 2009, DOI 10.1007/s10773-009-0222-8.

[25] Garola, C. In *Quantum Theory: Reconsideration of Foundations-4*; Adenier, G.; et al.; Eds.; American Institute of Physics: New York, 2007; pp 247–252.

[26] Sozzo, S. In *Quantum Theory: Reconsideration of Foundations-4*; Adenier, G.; et al.; Eds.; American Institute of Physics: New York, 2007; pp 334–338.

[27] Garola, C. In *Foundations of Probability and Physics-5*; Accardi, L.; et al.; Eds.; American Institute of Physics: New York, 2009; pp 42–50.

[28] Sozzo, S. In *Foundations of Probability and Physics-5*; Accardi, L.; et al.; Eds.; American Institute of Physics: New York, 2009; pp 381–385.

[29] Szabo, L. E.; Fine, A. *Phys. Lett. A* 2002, 295, 229–240.

[30] Nánásiová, O.; Khrennikov, A. Y. *Int. J. Theor. Phys.* 2006, 45, 481–494.

[31] Clauser, J. F.; Horne, M. A.; Shimony, A.; Holt, R. A. *Phys. Rev. Lett.* 1969, 23, 880–884.

[32] Garola, C.; Sozzo, S. *Europhys. Lett.* 2009, 86, 20009.

[33] Sozzo, S.; Garola, C. *Int. J. Theor. Phys.* 2010, DOI 10.1007/s10773-010-0264-y.

[34] Garola, C.; Sozzo, S. *Found. Phys.* 2010, DOI 10.1007/s10771-010-9435-1.

[35] Garola, C.; Sozzo, S. In *Quantum Theory: Reconsideration of Foundations-5*; Khrennikov, A. Y.; et al.; Eds.; American Institute of Physics: New York, in print.

[36] Einstein, A.; Podolsky, B.; Rosen, N. *Phys. Rev.* 1935; 47, 777 (1935).

[37] Beltrametti, E. G.; Cassinelli, G. *The Logic of Quantum Mechanics*; Addison–Wesley: Reading, MA, 1981.

[38] Garola, C.; Sozzo, S. *Theor. Math. Phys.* 2007, 152(2), 1087–1098.

[39] Garola, C.; Sozzo, S. *Humana.Mente - Jour. Phil. Stud.* 2010, Issue 13, 81–101.

[40] d'Espagnat, B. *Conceptual Foundations of Quantum Mechanics*; Benjamin: Reading, MA, 1976.

[41] Khrennikov, A. Y. *Found. Phys.* 2005, 35, 1655–1693.

[42] Khrennikov, A. Y. *Physica E* 2005, 29, 226–236.

[43] Khrennikov, A. Y. *Contextual Approach to Quantum Formalism*; Springer: Berlin, 2009.

[44] Adenier, G. In *Foundations of Probability and Physics-5*; Accardi, L.; et al.; Eds.; American Institute of Physics: New York; 2009; pp 8–18.

[45] Laloë, F. *Am. J. Phys.* 2001, 69, 655–701.

[46] Ludwig, G. *Foundations of Quantum Mechanics I*; Springer: Berlin, 1983.

[47] Aerts, D.; Aerts, S. In *Quo Vadis Quantum Mechanics? Possible Developments in Quantum Theory in the 21st Century*; Elitzur, A. C.; Dolev, S.; Kolenda, N.; Eds.; Springer: Berlin, 2004; pp 153–207.

[48] Norsen, T. *Found. Phys.* 2007 37, 311–340.

[49] Genovese, M. *Phys. Repts.* 2005 413, 319–396.

[50] Aspect, A.; Dalibard, J.; Rogers, G. *Phys. Rev. Lett.* 1982 49, 1804–1807.

[51] Kwiat, P. G.; Steinberg, A. M.; Chiao, R. Y. *Phys. Rev. A* 1993 47, R2472–R2475.

[52] Aspect, A.; Grangier, P. In *Advances in Quantum Phenomena*; Beltrametti, E. G.; Lévy–Leblond, J. M.; Eds.; Plenum Press: New York, 1995; pp 201–214.

[53] Rowe, M. A.; Kielpinski, D.; Meyer, V.; Sackett, C. A.; Itano, W. M.; Monroe, C.; Wineland, D. J. *Nature* 2001 409, 791–794.

[54] Fine, A. *Phys. Rev. Lett.* 1982 48, 291–295.

[55] Fine, A. *Found. Phys.* 1989, 19, 453–477.

[56] Santos, E. *Found. Phys.* 2004, 34, 1643–1673.

[57] Santos, E. *Stud. Hist. Phil. Mod. Phys.* 2005, 36B, 544–565.

[58] De Caro, L.; Garuccio, A. *Phys. Rev. A* 1996, 54, 174–181.

[59] Szabo, L. E. *Found. Phys.* 2000, 30, 1891–1909.

[60] Gisin, N.; Gisin, B. *Phys. Lett. A* 1999, 260, 323–327.

[61] Jauch, J. M. *Foundations of Quantum Mechanics*; Addison–Wesley: Reading, MA, 1968.

[62] Piron, C. *Foundations of Quantum Physics*; Benjamin: Reading, MA, 1976.

[63] Adenier, G.; Khrennikov, A. Y. *J. Phys. B* 2007, 42, 131–141.

Index

A

B

C

D

E

F

G

H

I

J

K

L

O

P

Q

R

S

T

U

V

W

Y